United Nations Development Programme

THE UNITED NATIONS DEVELOPMENT PROGRAMME (UNDP) is committed to the principle that development is inseparable from the quest for peace and human security and that the United Nations must be a strong force for development as well as peace. UNDP's mission is to help countries in their efforts to achieve sustainable human development by assisting them to build their capacity to design and carry out development programmes in poverty eradication, employment creation and sustainable livelihoods, the empowerment of women, the protection and regeneration of the environment–giving first priority to poverty eradication.

UNDP, at the request of governments and in support of its areas of focus, assists in building capacity for good governance, popular participation, private and public sector development and growth with equity, stressing that national plans and priorities constitute the only viable frame of reference for the national programming of operational activities for development within the United Nations system.

UNDP strives to be an effective development partner for the United Nations relief agencies, working to sustain livelihoods while they seek to sustain lives. It acts to help countries to prepare for, avoid, and manage complex emergencies and disasters.

Visit the UNDP website
http://www.undp.org/info/discover/mission.html

United Nations Environment Programme

THE UNITED NATIONS ENVIRONMENT PROGRAMME (UNEP) was established as the environmental conscience of the United Nations. UNEP has created a basis for comprehensive consideration and coordinated action within the UN on problems of the human environment. UNEP's mission is to provide leadership and encourage partnerships in caring for the environment by inspiring, informing, and enabling nations and people to improve their quality of life without compromising that of future generations.

One of the most important functions of UNEP is the promotion of environmental science and information. UNEP has always recognized that the environment is a system of interacting relationships that extends through all sectors of activity. To manage these relationships requires an integrated approach. UNEP's uniqueness lies in its advocacy of environmental concerns within the international system. UNEP nurtures partnerships with other UN bodies possessing complementary skills and delivery capabilities and enhances the participation of the private sector, scientific community, NGOs, youth, women, and sports organizations in achieving sustainable development.

UNEP derives its strength and influence from the authority inherent in its mission–environmental management. UNEP has and will continue to play a pivotal role in caring for the environment for the future.

Visit the UNEP website
http://www.unep.org/unep/about.htm

World Bank Group

FOUNDED IN 1944, THE WORLD BANK GROUP consists of five closely associated institutions: the International Bank for Reconstruction and Development (IBRD); International Development Association (IDA), International Finance Corporation (IFC); Multilateral Investment Guarantee Agency (MIGA); and the International Centre for Settlement of Investment Disputes (ICSID).

The World Bank is the world's largest source of development assistance, providing nearly $30 billion in loans annually to its client countries. The Bank uses its financial resources, its highly trained staff, and its extensive knowledge base to individually help each developing country onto a path of stable, sustainable, and equitable growth. The main focus is on helping the poorest people and the poorest countries, but for all its clients the Bank emphasizes the need for: • Investing in people, particularly through basic health and education • Protecting the environment • Supporting and encouraging private business development • Strengthening the ability of the governments to deliver quality services, efficiently and transparently • Promoting reforms to create a stable macroeconomic environment, conducive to investment and long-term planning • Focusing on social development, inclusion, governance, and institution-building as key elements of poverty reduction.

Visit the World Bank website
http://www.worldbank.org/html/extdr/about/

World Resources Institute

THE WORLD RESOURCES INSTITUTE provides information, ideas, and solutions to global environmental problems. Our mission is to move human society to live in ways that protect Earth's environment for current and future generations.

Our program meets global challenges by using knowledge to catalyze public and private action • to reverse damage to ecosystems • to expand participation in environmental decisions • to avert dangerous climate change • to increase prosperity while improving the environment.

For hundreds of years, enterprises have expanded and national economies have grown by using more resources, burning more energy, creating more waste. That economic growth has improved human well-being dramatically by providing more goods and services, by creating more opportunities for trade and employment, and by underwriting more investments in technology and education. But the historical process of growth has also degraded biological resources, depleted energy supplies, and polluted the water, the land, and the air. WRI believes the remedy to environmental problems lies not in reducing growth, but in breaking the connection between expanded prosperity and depleted resources. We are working with governments, businesses, and civil society to find new ways to use resources more efficiently and to take advantage of new technologies and new markets.

Visit the World Resources Institute website
http://www.wri.org/

WORLD RESOURCES
2000–2001

People and Ecosystems

The Fraying Web of Life

UNITED NATIONS DEVELOPMENT PROGRAMME

UNITED NATIONS ENVIRONMENT PROGRAMME

WORLD BANK

WORLD RESOURCES INSTITUTE

2000

ELSEVIER SCIENCE

Amsterdam • Lausanne • New York • Oxford • Shannon • Singapore • Tokyo

World Resources 2000–2001: People and Ecosystems: The Fraying Web of Life

Casebound edition published by
Elsevier Science
The Boulevard, Langford Lane
Kidlington, Oxford OX5 1GB, UK

Printed in Canada on recycled paper.
First printing September 2000

The **World Resources Series** is a collaborative product of four organizations: the United Nations Development Programme, the United Nations Environment Programme, the World Bank, and the World Resources Institute. The views expressed in this volume are those of staff from each organization and do not necessarily reflect the judgments of the organizations' boards of directors or member governments.

English paperback edition:
World Resources Institute
10 G Street, NE
Washington, DC 20002 USA
http://www.wri.org

Japanese edition:
Nikkei Business Publications, Inc.
2-7-6, Hirakawacho, Chiyoda-ku
Tokyo 102-8622, Japan

French edition:
Editions Eska
12, rue du Quatre-Septembre
75002 Paris, France

Spanish edition:
Ecoespaña Editorial
Apto. 16.158
28080 Madrid, Spain

British Library of Congress Cataloging in Publication Data
A catalog record from the British Library of Congress has been applied for.
ISBN: 0-08-0437818
ISSN: 0887-0403

Photo Credits: Cover: fall harvest in the Katmandu Valley of Nepal, Sara Elder. **Chapter 1:** (p. 2) Inle Lake, Shan State, Myanmar, Kevin Morris/Stone; (p. 3 within collage) fisherman threading net, Burundi, Bruno de Hogues/Stone; (p. 4) man washing baskets, Fez, Morocco, Gerard Del Vecchio/Stone; (p. 8) woman drinking coffee and boat on Borneo river, Corbis; Achuar hut in Ecuador, Elizabeth O'Neill; (p. 10) men working in fields, West Thebes, Egypt, Thierry Borredon/Stone; (p. 17) Street in Mexico City, Corbis; (p. 18) clear-cut forest and dead fish, Corbis; (p. 31) marketplace, Indonesia, Paul Chesley/Stone; (p. 34) Safari in Kenya, Renee Lynn/Stone; (p. 36) Indonesian forest village, Frances Seymour (WRI); (p. 40) woman collecting sap from rubber trees, Vietnam, Paula Bronstein/Stone. **Chapter 2:** (p. 135) gold mine, Mineral Policy Center; (p. 143) Baltimore-Washington corridor, American Forests; (p. 144) Cuban agriculture, Institute for Food and Development Policy (Food First). **Chapter 3 Case Studies:** Machakos, Kenya, from Fred Zaal, AGIDS. Cuba, from Institute for Food and Development Policy (Food First). Florida Everglades, from SFWMD. Bolinao, Philippines from Corbis. Mankóté Mangroves, St. Lucia, VI, from Allan Smith, CANARI. Dhani Forest, Orissa, India, from Vasundhara (Dhani research organization); South Africa's Working for Water Programme, from David Richardson, University of Cape Town; Mongolian Steppe, from David Sneath, University of Cambridge. **Chapter 4:** (p. 228) weaver, Kalimpong, India, Sara Elder. **Sources Section Opener:** (pp. 358–359) roots, Dave Schiefelbein/Stone.

Reprinted with Permission: Chapter 1: (table p. 13) from S.L. Buchmann and G.P. Nabhan, *The Forgotten Pollinators* © 1996 by Island Press; (table p. 14) from F. Grifo and J. Rosenthal, eds., *Biodiversity and Human Health* © 1997 by Island Press; (line graph p. 20) from Ruesink et al. "Reducing the Risks of Nonindigenous Species Introductions," *BioScience* 45(7):465–477 © 1995; (line graph p. 26) from the Population Reference Bureau; (bar chart p. 27) from D. Hoornweg, *What a Waste! Solid Waste Management in Asia* © 1999 World Bank; (table p. 39) from "Poverty Incidence in West and Central Africa," *Rural Poverty Report 2000* (n.d.) IFAD. **Chapter 3 Case Studies:** (bar chart p. 154) from *More People, Less Erosion: Environmental Recovery in Kenya*, M. Tiffen et al. © 1994 ODI, permission from Wiley & Sons; (line graph p. 202) from *The Contribution of Plantation Forestry to the Problem of Invading Alien Trees in South Africa: A Preliminary Assessment,* J.L. Nel et al. © 1999 CSIR. Mongolia; (line graph, p. 219; table and bar chart, p. 223) from C. Humphrey and D. Sneath, *The End of Nomadism? Society, State, and the Environment in Inner Asia* © 1999 Duke University Press.

WORLD RESOURCES 2000-2001

CONTENTS

FOREWORD

REPAIRING THE FRAYING WEB

There are times when the most difficult decision of all is to acknowledge the obvious. It is obvious that the world's national economies are based on the goods and services derived from ecosystems; it is also obvious that human life itself depends on the continuing capacity of ecosystems to provide their multitude of benefits. Yet for too long in both rich and poor nations, development priorities have focused on how much humanity can take from our ecosystems, with little attention to the impact of our actions. With this report, the United Nations Development Programme, the United Nations Environment Programme, the World Bank, and the World Resources Institute reconfirm their commitment to making the viability of the world's ecosystems a critical development priority for the 21st century.

While our dependence on ecosystems may be obvious, the task of integrating considerations of ecosystem capacity into decisions about development is difficult. It requires governments and businesses to rethink some basic assumptions about how we measure and plan economic growth. Poverty forces many people to jeopardize the ecosystems on which they depend, even when they know that they are cutting timber or extracting fish at unsustainable levels. Greed or enterprise, ignorance or inattention also leads people to disregard the natural limits that sustain ecosystems. The biggest difficulty of all, however, is that people at all levels, from the farmers at the grassroots to the policy makers in the capitals, either can't make good use of the knowledge at hand or lack basic information about the condition and long-term prospects of ecosystems. This report, and the Pilot Analysis of Global Ecosystems on which it is based, is a step toward addressing this problem.

In our unique collaboration on the World Resources Report Series, our four organizations undertook this edition in a genuine partnership to develop recommendations that would safeguard the world's ecosystems. We bring together different perspectives and decades of experience working on environment and development issues. We are motivated by the urgent need for solutions that will benefit both people and ecosystems.

At this moment, in all nations–rich and poor–people are experiencing the effects of ecosystem decline in one guise or another: water shortages in the Punjab, India; soil erosion in Tuva, Russia; fish kills off the coast of North Carolina in the United States; landslides on the deforested slopes of Honduras; fires in the disturbed forests of Borneo and Sumatra in Indonesia. The poor, who often depend directly on ecosystems for their livelihoods, suffer most when ecosystems are degraded.

At the same time, people in all parts of the world are working to find solutions: community forest conservation programs in Dhani, India; collective management of grasslands in Mongolia; agricultural transformation in Machakos, Kenya; removal of invasive tree species to protect water resources in South Africa; and restoration of the Everglades in the United States. Governments and private interests are spending billions trying to rectify ecosystem degradation or, at least, stave off the consequences–and countless billions more may be needed to restore ecosystems on a global scale.

As these examples and many others in this volume demonstrate, our knowledge of ecosystems has increased dramatically, but it has simply not kept pace with our ability to alter them. Unless we use the knowledge we've gained to sustainably develop Earth's ecosystems, we risk inflicting ever greater damage on them with dire consequences for economic development and human well-being. Thus, the urgency of this issue: shortsighted, avoidable mistakes can affect the lives of millions of people, now and in the future. We can continue blindly altering Earth's ecosystems, or we can learn to use them more sustainably.

If we choose to continue our current patterns of use, we face almost certain declines in the ability of ecosystems to yield their broad spectrum of benefits–from clean water to stable climate, fuelwood to food crops, timber to wildlife habitat. We can choose another option, however. It requires reorienting how we see ecosystems, so that we learn to view their sustainability as essential to our own. Adopting this "ecosystem approach" means we evaluate our decisions on land and resource use in terms of how they affect the capacity of ecosystems to sustain life, not only human well-being but also the health and productive potential of plants, animals, and natural systems. Maintaining this capacity becomes our passkey to human and national development, our hope to end poverty, our safeguard for biodiversity, our passage to a sustainable future.

It's hard, of course, to know what will be truly sustainable in either the physical or political environments of the future. That's why the ecosystem approach emphasizes the need for both good scientific information and sound policies and institutions. On the scientific side, an ecosystem approach should:

- Recognize the "system" in ecosystems, respecting their natural boundaries and managing them holistically rather than sectorally.
- Regularly assess the condition of ecosystems and study the processes that underlie their capacity to sustain life so that we understand the consequences of our choices.

On the political side, an ecosystem approach should:

- Demonstrate that much can be done to improve ecosystem management by developing wiser policies and more effective institutions to implement them.
- Assemble the information that allows a careful weighing of the trade-offs among various ecosystem goods and services and among environmental, political, social, and economic goals.
- Include the public in the management of ecosystems, particularly local communities, whose stake in protecting ecosystems is often greatest.

The goal of this approach is to optimize the array of goods and services ecosystems produce while preserving or increasing their capacity to produce these things in the future. *World Resources 2000–2001* advocates an ecosystem approach and recommends how we can apply it.

A critical step in taking care of our ecosystems is taking stock of their condition and their capacity to continue to provide what we need. Yet, there has never been a global assessment of the state of the world's ecosystems. This report starts to address this knowledge gap by presenting results from the Pilot Analysis of Global Ecosystems, a new study undertaken to be the foundation for more comprehensive assessment efforts.

What makes the pilot analysis valuable now, before any other assessment, is that it compares information already available on a global scale about the condition of five major classes of ecosystems: agroecosystems, coastal areas, forests, freshwater systems, and grasslands. The pilot analysis examines not only the quantity and quality of outputs but also the biological basis for production, including soil and water condition, biodiversity, and changes in land use over time. And rather than looking just at marketed products, such as food and timber, the pilot analysis evaluates the condition of a broad array of ecosystem goods and services that people rely on but don't buy in the marketplace. The bottom line is a comprehensive evaluation, based on available information, of the current condition of five major ecosystems.

It's an evaluation that clearly shows the strengths and weaknesses of the information at hand. The pilot analysis identifies significant gaps in the data and what it would take to fill those gaps. Satellite imaging and remote sensing, for example, have added to information about certain features of ecosystems, such as their extent, but on-the-ground information for such indicators as freshwater quality and river discharge is less available today than in the past.

Although some data are being created in abundance, the pilot analysis shows that we have not yet succeeded in coordinating our efforts. Scales now diverge, differing measures defy integration, and different information sources may not know of each other's relevant findings.

Our partner organizations began work on this edition of the World Resources Report with a conviction that the challenge of managing Earth's ecosystems–and the consequences of failure–will increase significantly during the 21st century. We end with a keen awareness that the scientific knowledge and political will required to meet this challenge are often lacking today. To make sound ecosystem management decisions in the 21st century, dramatic changes are needed in the way we use the knowledge and experience at hand, as well as the range of information brought to bear on resource management decisions.

A truly comprehensive and integrated assessment of global ecosystems that goes well beyond our pilot analysis is needed to meet information needs and to catalyze regional and local assessments. Planning for such a Millennium Ecosystem Assessment is already under way. In 1998, representatives from a broad range of international scientific and political bodies began to explore the merits of and to recommend the structure for such an assessment. After consulting for a year and considering the preliminary findings in this report, they concluded that a global assessment of the past, present, and future of ecosystems was feasible and urgently needed. They urged local, national, and international institutions to support the effort as stakeholders, users, and sources of expertise. If concluded successfully, the Millennium Ecosystem Assessment will generate new information, integrate current knowledge, develop methodological tools, and increase public understanding. At local, national, and regional scales it will build the capacity to obtain, analyze, and act on improved information. Our institutions are united in supporting this call for the Millennium Ecosystem Assessment.

At the dawn of a new century, we have the ability to change the vital systems of this planet, for better or worse. To change them for the better, we must recognize that the well-being of people and ecosystems is interwoven and that the fabric is fraying. We need to repair it, and we have the tools at hand to do so. What better time than now?

Mark Malloch Brown
Administrator,
United Nations Development Programme

Klaus Töpfer
Executive-Director,
United Nations Environment Programme

James D. Wolfensohn
President,
World Bank

Jonathan Lash
President,
World Resources Institute

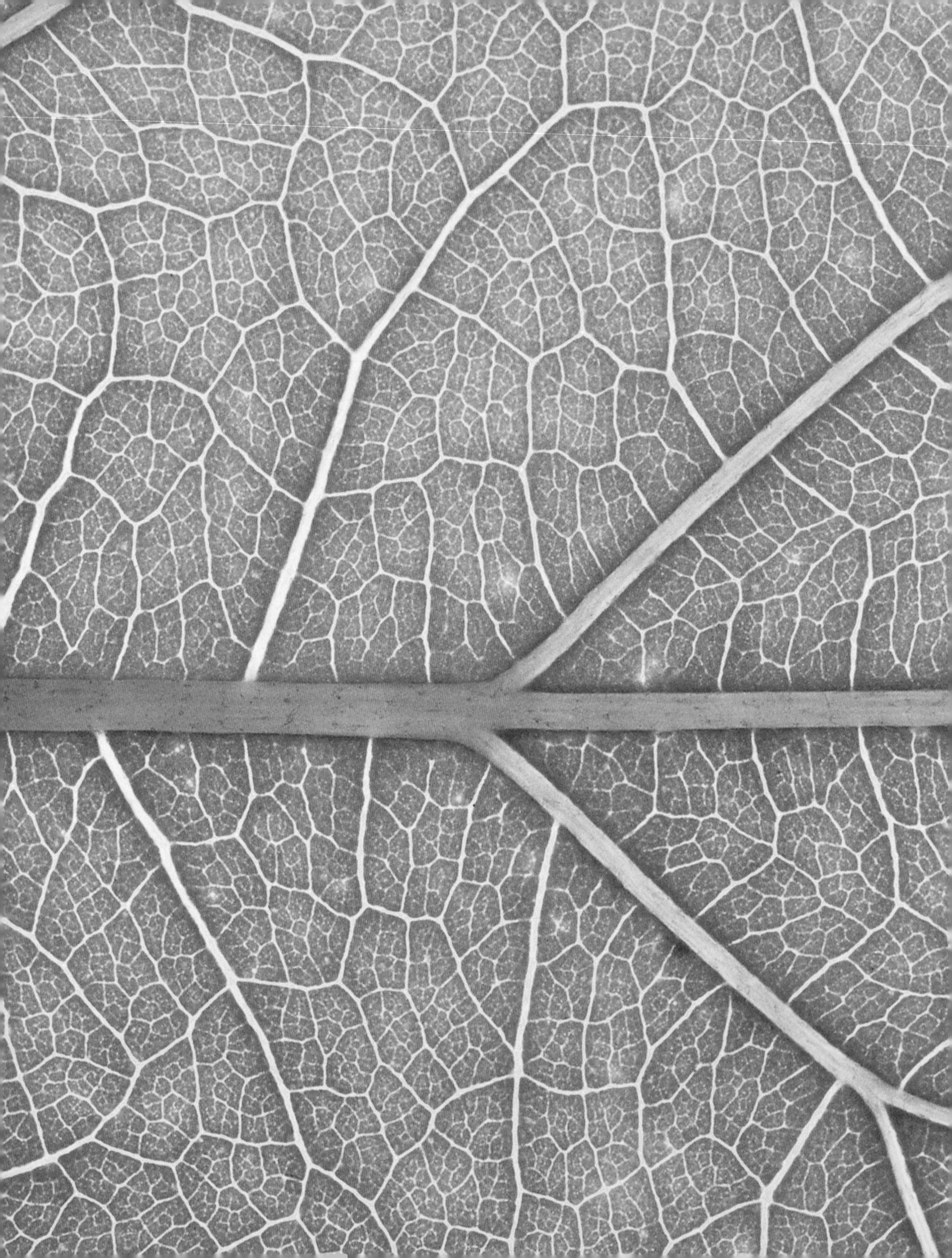

PART I

WORLD RESOURCES 2000-2001

RETHINKING THE LINK

A spring flowing out of the
ground appears new.
We call it a source
of fresh water.
Yet the water is ancient,
having circulated between
earth and sky for eons.
We rely on the land
to purify the water
as it moves
through this cycle.

CHAPTER 1

LINKING PEOPLE AND ECOSYSTEMS

Try to imagine Earth without ecosystems. Ecosystems are the productive engines of the planet–communities of species that interact with each other and with the physical setting they live in. They surround us as forests, grasslands, rivers, coastal and deep-sea waters, islands, mountains–even cities. Each ecosystem represents a solution to a particular challenge to life, worked out over millennia; each encodes the lessons of survival and efficiency as countless species scramble for sunlight, water, nutrients, and space. Stripped of its ecosystems, Earth would resemble the stark, lifeless images beamed back from Mars by NASA cameras in 1997.

That image also underscores the difficulty of recreating the natural life-support systems that ecosystems provide, should we damage them beyond their capacity to rebound. The world's fertile soils, for instance, are a gift of millions of years of organic and inorganic processes. Technology can replicate the nutrients soils provide for crops and native flora, but on a global scale the costs would be prohibitive.

The fact is, we are utterly dependent on ecosystems to sustain us. From the water we drink to the food we eat, from the sea that gives up its wealth of products, to the land on which we build our homes, ecosystems yield goods and services that we can't do without. Ecosystems make the Earth habitable: purifying air and water, maintaining biodiversity, decomposing and recycling nutrients, and providing myriad other critical functions.

Harvesting the bounty of ecosystems roots our economies and provides us employment, particularly in low- and middle-income countries. Agriculture, forestry, and fishing are responsible for one of every two jobs worldwide and seven of ten jobs in sub-Saharan Africa, East Asia, and the Pacific. In a quarter of the world's nations, crops, timber, and fish still contribute more to the economy than industrial goods (World Bank 1999b:28–31, 192–195). Global agriculture alone produces US$1.3 trillion in food and fiber each year (Wood et al. [PAGE] 2000).

Ecosystems feed our souls as well, providing places for religious expression, aesthetic enjoyment, and recreation. In every respect, human development and human security are closely linked to the productivity of ecosystems. Our future rests squarely on their continued viability.

If our life on Earth is unimaginable without ecosystems, then we need to know how to live better within them. The world is large, nature is resilient, and humans have been altering the landscape for tens of thousands of years, all of which makes it easy to ignore warning signs that human activities might be damaging the capacity of an ecosystem to continue to deliver goods and services.

In fact, many nations and societies have completely altered the landscape, converting wetlands, prairies, and forests to other uses, and continue to prosper. What was once 200 Mha of tallgrass prairie in the heartland of the United States has been converted almost entirely to cropland and urban areas. The once-extensive forests of Europe have suffered much the same fate. These conversions have brought obvious benefits, such as stable food supplies and industrial production, that have made the United States and some European nations economic powerhouses. But they also impose costs–eroded topsoil, polluted wells and waterways, reduced fish yields, and lost wildlands and scenic places–that threaten to erode the wealth and quality of life these nations enjoy.

We don't have to look far to see how high the costs of degrading ecosystems can be. The rich waters of the Black Sea

In every respect, human
development and human
security are closely
linked to the
productivity of
ecosystems.
Our future
rests squarely
on their continued
viability.

used to yield more than 700,000 tons of anchovy, sturgeon, bonito, and other valuable fish annually. But over the last 30 years, human pressures have radically altered the Black Sea ecology. Beginning in the 1970s, increasing pollution brought on frequent algal blooms. A rapid rise in fishing in the 1980s depleted key fish stocks. In 1982, the final blow came with the accidental introduction of a jellyfish-like creature, a ctenophore, that soon dominated the aquatic food web, directly competing with native fish for food. By 1992, the Black Sea fish catch had collapsed to one-third of its former volume (Prodanov et al. 1997:1–2). Now most fishers from the six nations surrounding the sea bring up nearly empty nets, and the once prominent fishing industry hemorrhages jobs and profits (Travis 1993:262–263).

Ecosystem degradation showed a different face to the Chinese living alongside the Yangtze River in 1998. In prior years, loggers had cut forests in the river's vast watershed, while farmers and urban developers drained lakes and wetlands and occupied the river's flood plains. In the meantime, little heed to soil conservation allowed 2.4 billion metric tons of earth to wash downstream each year, silting lakes and further reducing the buffers that formerly absorbed floodwaters (Koskela et al. 1999:342). When record rains fell in the Yangtze basin in the summer of 1998, these degrading practices amplified the flooding, which left 3,600 people dead, 14 million homeless, and $36 billion in economic losses (NOAA 1998; World Bank 1999a). The Chinese government is now trying to restore the ecosystem's natural flood-control services, but it could take decades and billions of dollars to reforest denuded slopes and reclaim wetlands, lakes, and flood plains.

How Viable Are Earth's Ecosystems?

In spite of the costs of degrading ecosystems and our dependence on their productivity, we know surprisingly little about the overall state of Earth's ecosystems or their capacity to provide for the future. We need to know: How viable are Earth's ecosystems today? How best can we manage ecosystems so that they remain healthy and productive in the face of increasing human demands?

This special millennial edition of the World Resources Report, *World Resources 2000–2001*, tries to answer these questions, focusing on ecosystems as the biological underpinning of the global economy and human well-being. It considers both predominantly natural ecosystems like forests and grasslands as well as human-constructed ecosystems like croplands, orchards, or other agroecosystems. Both ecosystem types are capable of producing an array of benefits, and both are crucial to human survival.

This chapter examines how people rely on ecosystems and surveys the factors that drive how people use, and often degrade, ecosystems. Chapter 2 assesses the current state of global ecosystems, presenting the results of a major new analysis of ecosystem conditions and pressures undertaken by World Resources Institute, the International Food Policy Research Institute, and many other collaborators. In Chapter 3, case studies illustrate trade-offs involved in managing ecosystems and ways that some communities responded as their local ecosystems declined. Chapter 4 considers the greater challenge of managing ecosystems in the 21st century to keep them productive and vital, even as our population and consumption grow.

All these chapters focus on the goods and services that ecosystems yield as fundamental measures of ecosystem health. This "goods and services" approach emphasizes how we depend on ecosystems on a daily basis.

Losing the Link?

It is easy to lose touch with our link to ecosystems, despite their importance. For the millions of us who depend directly on forests or fisheries for our survival, the vital importance of ecosystems is a fact of daily life. But for the millions of us who live in cities or suburbs and have transitioned from working the soil to working at computer keyboards, our link to ecosystems is less direct. We buy our food and clothing in stores and depend on technology to deliver water and energy. We take for granted that there will be food in the market, that transportation and housing will be available, and all at reasonable cost. Too often, we're only reminded of our link to natural systems when a fishery collapses, a reservoir goes dry, or air pollution begins to make us sick–when the flow of goods and services is disrupted. Then we suddenly become aware of the real value of these resources and the potential economic and biological costs of mismanagement.

Unfortunately, mismanagement of ecosystems abounds. Worldwide, human overuse and abuse of major ecosystems from rainforests to coral reefs to prairie grasslands have degraded or destroyed hectare upon hectare of once-productive habitat. This has harmed wildlife, to be sure, as the number of endangered species attests. But it has also harmed human interests by depleting the flow of the very goods and services we depend on.

Decline in the productive capacity of ecosystems can have devastating human costs. Too often, the poor are first and most directly affected by the degradation of ecosystems. Impoverished people are generally the most dependent on ecosystems for subsistence and cash, but usually exert the least control over how ecosystems are used or who reaps the benefits of that use.

In many areas, declining agricultural productivity, diminished supplies of freshwater, reduced timber yields, and declining fish harvests have already taken a significant toll on local economies.

(continues on p. 10)

Box 1.1 History of Use and Abuse

Many of the challenges we face today—deforestation, soil erosion, desertification, salinization, and loss of biodiversity—were problems even in ancient times. What is different now is the scale, speed, and long-term nature of modern civilization's challenges to Earth's ecosystems. Before the industrial revolution, environmental degradation was much more gradual—occurring over hundreds or thousands of years—and relatively localized. The cumulative actions of rapidly growing and industrializing societies, however, have given rise to more complex problems. Acid rain, greenhouse gas emissions, ozone depletion, toxic waste, and large-scale industrial accidents are examples of such problems with global or regional consequences.

7000 BC–1800 BC 	Mesopotamia/Sumer *Salinization and waterlogging of Sumer's agroecosystem*	Around 7000 BC, people in this region (now, largely, Iraq) began to modify the natural environment. Lacking adequate rainfall, land had to be irrigated for cultivation, and the demand for food increased as the population grew. The irrigated land became salinized and waterlogged. Records noting "the earth turned white" with salt date back to 2000 BC. By 1800 BC, the agricultural system—the foundation of Sumerian civilization—collapsed.
2600 BC–present 	Lebanon *Overuse and exploitation of Lebanon's cedar forest*	At one time, Mount Lebanon was covered with a forest of cedars that were famous for their beauty and strength. Solomon's temple was built of cedar from this area as were many Phoenician ships. In the third millennium BC, Byblos grew wealthy from its timber trade. The Egyptians used cedar timber for construction and used the resin for mummification. The exploitation continued through the centuries. Only four small groves remain today.
2500 BC–900 	Mayan Empire *Soil erosion, loss of agroecosystem viability, and water siltation in Central America*	Mayans lived in what are now parts of Mexico, Guatemala, Belize, and Honduras. The agriculture techniques they used were creative and intensive—clearing hillsides of jungle, terracing fields to contain soil erosion, draining swamps by digging ditches and using the soil from the ditches to form raised fields. Eventually too much was demanded of this system. Soil erosion reduced crop yields, and higher levels of silt in rivers damaged the raised fields. Decreased food production and competition for the remaining resources may have led to that civilization's demise.
800 BC–200 BC	Greece *Conversion and deforestation in the Mediterranean*	In Homeric times, Greece was still largely covered with mixed evergreen and deciduous forests. Over time the trees were cleared to provide land for agriculture, fuel for cooking and heating, and construction materials. Overgrazing prevented regeneration. The olive tree, favored for its economic value, began to flourish in ancient Greece because it grew well on the degraded land.
200 BC–present	China *Desertification along the Silk Road* 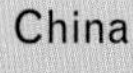	The fortification of the Great Wall during the Han dynasty gave rise to intensive cultivation of farmland in northern and western China and to the growth of a major travel and trade route that came to be known as the Silk Road. Deserts began irreversibly expanding in this area as a result of the demands of a growing population and gradual climate changes.
50 BC–450 	Roman Empire *Desertification and loss of agroecosystem viability in North Africa*	The challenge of providing food for the population of Rome and its large standing armies plagued the empire. The North African provinces, once highly productive granaries, gradually became degraded as Roman demands for grain pushed cultivation onto marginal lands, prone to erosion. Scrub vegetation spread and some intensively cultivated areas became desertified. The irrigation systems the Romans used depended on watersheds that have since been deforested, and now yield less runoff, reducing the chance of restoring productivity.

1400–1600 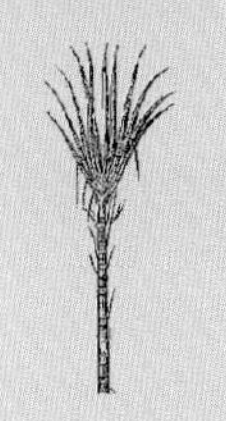	Canary Islands *Human and natural resource exploitation, degradation and extinctions in many regions*	Originally from North Africa, the Guanches were a people who inhabited the Canary Islands for more than 1,000 years before the Spanish arrived in the 1400s. The Spanish enslaved the Guanches, cleared the forests, and built sugar cane plantations. By 1600 the Guanches were dead, victims of Eurasian diseases and plantation conditions. As in the Canary Islands, regions in the Americas, Africa, and Asia where people were forced to grow and export cash crops such as sugar, tobacco, cotton, rubber, bananas, or palm oil, continue to suffer from deforestation, soil damage, biodiversity losses, and economic dependency instituted during colonization.
1800 	Australia and New Zealand *Loss of biodiversity and proliferating invasive species in island ecosystems*	There were no hoofed animals in Australia and New Zealand before Europeans arrived at the end of the 18th century and began importing them. Within 100 years there were millions of sheep and cattle. The huge increase in grazing animals killed off many of the native grasses that were not well adapted to intensive grazing. Island biodiversity worldwide suffered some of the most dramatic losses after nonnative plants and animals were introduced. Island flora and fauna had developed in isolation over millennia and thus lacked natural predators. Many island bird species, for example, were flightless and became easy prey for invaders. It is estimated that 90 percent of all bird extinctions occurred on islands.
1800 	North America *Conversion, loss of habitat, and unrestrained killing of wildlife in North America*	As land was cleared for settlement and cultivation around the world, animal habitats of almost every kind were reduced; animals were killed for food, hides, or recreation as commerce spread. In North America, herds of bison, totaling perhaps as many as 50 million, were hunted to near extinction by the end of the 19th century. Aquatic as well as terrestrial species became targets of exploitation and extincition. In the 19th century, whales were killed in large numbers to support industrializing economies in need of whale oil in great quantity, mainly for lighting and lubricants. On the northwest coast of North America, whale populations were on the verge of extinction by the 20th century.
1800–1900 	Germany and Japan *Industrial chemical poisoning of freshwater systems*	The industrial revolution had a profound impact on the waters of the world. Rivers that ran through industrial zones, like the Rhine in Germany, or rivers that ran through mining zones, like the Watarase in Japan, became heavily polluted in the 19th century. The German chemical industry poisoned the Rhine so badly that salmon, which had been plentiful as late as 1765, were rare by 1914. Japan's most important copper mine in the 1800s dumped mine tailings in the Watarase River, and sulfuric acid from smelters contaminated the water and killed thousands of hectares of forest trees and vegetation. Fish and fowl died and local residents became sick. The human birth rate dipped below the death rate in the nearby town of Ashio in the 1890s.
1900 	United States and Canada *Soil erosion and loss of biodiversity in the United States and Canada*	The Great Plains of the United States and Canada were ploughed in the late 19th and early 20th centuries and planted with new forms of drought-resistant wheat. Once the protective original grass cover was destroyed, drought in the 1930s enabled high, persistent wind storms to blow away much of the dry soil. Soil conservation methods were subsequently introduced such that when wind erosion again affected the area in the 1950s and in the 1970s, the consequences were less severe.
1928–present 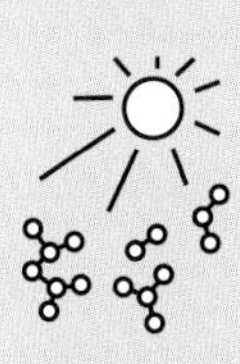	Worldwide *Industrial chemicals deplete the world's protective ozone layer*	Chlorofluorocarbons (CFCs) are a family of volatile compounds invented in 1928. Thought to be the world's first nontoxic, nonflammable refrigerants, their use grew rapidly. They also were used as industrial solvents, foaming agents, and aerosol propellants. CFC production peaked in 1974, the same year researchers noted that CFC emissions could possibily damage human health and the ozone layer. In 1985, the discovery of an "ozone hole" over the Antarctic coincided with a first-ever coordinated international effort to phase out production of CFCs and other ozone-depleting substances. Worldwide phase out of CFC production is scheduled for 2010.

Box 1.2 Linking Ecosystems and People

An urban professional in Tokyo reads a newspaper printed on pulped trees from North American forests. Her food and clothing come from plants and animals raised around the world—cotton and cashmere from Asia, fish from the Pacific and Indian oceans, beef from Australian and North American grasslands, fruits and vegetables from farmlands on four continents. The coffee she sips comes from tropical Central American plantations, but it is brewed with water from wells near the city.

In a Borneo village children get to school via river, poled in long boats handmade from local trees. In nearby paddies, families grow rice, their main dietary staple as well as a source of pepper, a cash crop, and wine.

The Shuar of Amazonian Ecuador find shelter in houses with thatched roofs made from the local palm leaves. They also use palm-leaf stems for weaving baskets and containers. They grow manioc, papaya, sweet potato, and other crops derived from the rainforest, for their own subsistence and for cash. The forest is also the source of their woodfuel and medicines, as well as fish and game.

Ecosystems sustain us. They are Earth's primary producers, solar-powered factories that yield the most basic necessities—food, fiber, water. Ecosystems also provide essential services—air and water purification, climate control, nutrient cycling, and soil production—services we can't replace at any reasonable price.

Primary Goods and Services Provided by Ecosystems

Ecosystem	Goods	Services
Agroecosystems	■ Food crops ■ Fiber crops ■ Crop genetic resources	■ Maintain limited watershed functions (infiltration, flow control, partial soil protection) ■ Provide habitat for birds, pollinators, soil organisms important to agriculture ■ Build soil organic matter ■ Sequester atmospheric carbon ■ Provide employment
Coastal Ecosystems	■ Fish and shellfish ■ Fishmeal (animal feed) ■ Seaweeds (for food and industrial use) ■ Salt ■ Genetic resources	■ Moderate storm impacts (mangroves; barrier islands) ■ Provide wildlife (marine and terrestrial) habitat ■ Maintain biodiversity ■ Dilute and treat wastes ■ Provide harbors and transportation routes ■ Provide human habitat ■ Provide employment ■ Provide for aesthetic enjoyment and recreation
Forest Ecosystems	■ Timber ■ Fuelwood ■ Drinking and irrigation water ■ Fodder ■ Nontimber products (vines, bamboos, leaves, etc.) ■ Food (honey, mushrooms, fruit, and other edible plants; game) ■ Genetic resources	■ Remove air pollutants, emit oxygen ■ Cycle nutrients ■ Maintain array of watershed functions (infiltration, purification, flow control, soil stabilization) ■ Maintain biodiversity ■ Sequester atmospheric carbon ■ Moderate weather extremes and impacts ■ Generate soil ■ Provide employment ■ Provide human and wildlife habitat ■ Provide for aesthetic enjoyment and recreation
Freshwater Systems	■ Drinking and irrigation water ■ Fish ■ Hydroelectricity ■ Genetic resources	■ Buffer water flow (control timing and volume) ■ Dilute and carry away wastes ■ Cycle nutrients ■ Maintain biodiversity ■ Provide aquatic habitat ■ Provide transportation corridor ■ Provide employment ■ Provide for aesthetic enjoyment and recreation
Grassland Ecosystems	■ Livestock (food, game, hides, fiber) ■ Drinking and irrigation water ■ Genetic resources	■ Maintain array of watershed functions (infiltration, purification, flow control, soil stabilization) ■ Cycle nutrients ■ Remove air pollutants, emit oxygen ■ Maintain biodiversity ■ Generate soil ■ Sequester atmospheric carbon ■ Provide human and wildlife habitat ■ Provide employment ■ Provide for aesthetic enjoyment and recreation

- In Canada's maritime provinces, collapse of the cod fishery in the early 1990s left 30,000 fishers dependent on government welfare payments and decimated the economies of 700 communities in Newfoundland alone (Milich 1999:628).

- Urban water shortages in China–greatly aggravated by overextraction and pollution of nearby rivers and groundwater sources–cost urban economies an estimated US$11.2 billion per year in reduced industrial output and afflict nearly half of the nation's major cities (WRI et al. 1998:120).

- Commercial cutting of India's forests and conversion of forests to agriculture have left the traditional system of village management of local forests in shambles. This has brought shortages of fuelwood and building materials to many of the 275 million rural Indians who draw on local forest resources (Gadgil and Guha 1992:113–145, 181–214; WCFSD 1999:59).

If this pattern holds, the loss of healthy ecosystems will ultimately act as a brake not just on local economies, but on national and global development as well.

Adopting a Human Perspective

All organisms have intrinsic value; grasslands, forests, rivers, and other ecosystems do not exist to serve humans alone. Nonetheless, *World Resources 2000–2001* deliberately examines ecosystems, and their management, from a human perspective because human use is the primary source of pressure on ecosystems today, far outstripping the natural processes of ecosystem change. In the modern world, virtually every human use of the products and services of ecosystems translates into an impact on those ecosystems. Thus, every use becomes either an opportunity for enlightened management or an occasion for degradation.

Responsible use of ecosystems faces fundamental obstacles, however. Typically, we don't even recognize ecosystems as cohesive units because they often extend across political and management boundaries. We look at them in pieces or concentrate on the specific products they yield. We miss their complexity, the interdependence of their organisms–the very qualities that make them productive and stable.

The challenge for the 21st century, then, is to understand the vulnerabilities and resilience of ecosystems, so that we can find ways to reconcile the demands of human develop-

An ecosystem is a
community of
interacting organisms
and the physical
environment they live
in. Every hectare of the
planet is part of an
ecosystem.

ment with the tolerances of nature. That requires learning to look at our activities through the living lens of ecosystems. In the end, it means adopting an ecosystem-oriented approach to managing the environment–an approach that respects the natural boundaries of ecosystems and takes into account their interconnections and feedbacks.

Sources of Wealth and Well-Being

Ecosystems are not just assemblages of species, they are systems combined of organic and inorganic matter and natural forces that interact and change. The energy that runs the system comes from the sun; solar energy is absorbed and turned into food by plants and other photosynthesizing organisms at the base of food chains. Water is the crucial element flowing through the system. The amount of water available, along with the temperature extremes and the sunlight the site receives, largely determine what types of plants, insects, and animals live there, and how the ecosystem is categorized.

Ecosystems are dynamic, constantly remaking themselves, reacting to natural disturbances and the competition among and between species. It is the complex, local interaction of the physical environment and the biological community that gives rise to the particular package of services and products that each ecosystem yields; it also is what makes each ecosystem unique and vulnerable.

Scale also is important. A small bog, a single sand dune, or a tiny patch of forest may be viewed as an ecosystem, unique in its mix of species and microclimate–a microenvironment. On a much larger scale, an ecosystem refers to more extensive communities–a 100 or 1,000 km^2 forest, or a major river system, each having many such microenvironments.

This edition of the World Resources Report examines ecosystems on an even larger scale. It considers five main types or categories of ecosystems: grasslands, forests, agroecosystems, freshwater systems, and coastal ecosystems. Together, these five major ecosystem types cover most of the Earth's surface and render the bulk of the goods and services people derive from ecosystems. Dividing ecosystems in this way allows us to examine them on a global scale and think in broad terms about the challenges of managing them sustainably.

Divisions between ecosystems are less important, however, than the linkages between them. Grasslands give way to savannas that segue into forests. Freshwater becomes brackish as it approaches a coastal area. Polar, island, mountain, and even urban ecosystems blend into and add to the mix. All these systems are tightly knit into a global continuum of energy and nutrients and organisms–the biosphere in which we live.

Direct and Indirect Benefits

The benefits that humans derive from ecosystems can be direct or indirect (Daily 1997:1-10; ESA 1997a:1-13). Direct benefits are harvested largely from the plants and animals in an ecosystem in the form of food and raw materials. These are the most familiar "products" an ecosystem yields–crops, livestock, fish, game, lumber, fuelwood, and fodder. Genetic resources that flow from the biodiversity of the world's ecosystems also provide direct benefits by contributing genes for improving the yield and disease resistance of crops, and for developing medicines and other products.

Indirect benefits arise from interactions and feedback among the organisms living in an ecosystem. Many of them take the form of services, like the erosion control and water purification and storage that plants and soil microorganisms provide in a watershed, or the pollination and seed dispersal that many insects, birds, and mammals provide. Other benefits are less tangible, but nonetheless highly valued: the scenic enjoyment of a sunset, for example, or the spiritual significance of a sacred mountain or forest grove (Kellert and Wilson 1993). Every year, millions of people make pilgrimages to outdoor holy places, vacation in scenic regions, or simply pause in a park or their gardens to reflect or relax. As the manifestation of nature, ecosystems are the psychological and spiritual backdrop for our lives.

Some benefits are global in nature, such as biodiversity or the storage of atmospheric carbon in plants and soils. Others are regional; watershed protection that prevents flooding far downstream is an example. But many ecosystem benefits are local, and these are often the most important, affecting people directly in many aspects of their daily lives. Homes, industries, and farms usually get their water supplies from local sources, for instance. Jobs associated with agriculture and tourism are local benefits as well. Urban and suburban parks, scenic vistas, and the enjoyments of backyard trees and wildlife are all local products that define our sense of place.

Because so many ecosystem goods and services are enjoyed locally, it follows that local inhabitants often suffer most when these benefits are lost. By the same token, it is local inhabitants who usually have the greatest incentive to preserve the ecosystems they depend on. In fact, local people hold enormous potential both for managing ecosystems sustainably and for damaging them through careless use. But local communities rarely exert full control over the ecosystems they inhabit; with the market for ecosystem goods becoming increasingly global, outside economic forces and government policies can overwhelm the best local intentions.

(continues on p. 16)

Box 1.3 Water Filtration and Purification

At every stage of its journey between earth and sky, water can pick up pollutants and wastes—as it flows from a spring into streams, rivers, and the sea; as it pools into ponds and lakes; when it returns from the atmosphere as rain; when it soaks back into the soil after use on croplands or as effluent from sewage systems.

Fortunately, ecosystems can cleanse the water for us.

- Soils are inhabited by microorganisms that consume and recycle organic material, human and animal feces, and other potential toxins and pathogens. Deeper rocky layers of an aquifer may continue the cleansing process as water seeps through.

- Plants and trees hold soil in place as the water filters through. The vegetation interacts with fungi and soil microorganisms to generate many of soil's filtering capabilities.

- Freshwater bodies dilute pollutants where large quantities of municipal, agricultural, and industrial waters are drained or released.

- Wetlands intercept surface runoff, trap sediments from floodwaters, sequester metals, and excel at removing nitrogen and minerals from the water. A hectare of cattail marsh can consume three times as many nutrients as a hectare of grassland or forest (Trust for Public Land 1997:16).

In many places, however, we are straining nature's ability to filter and purify water. Where land is stripped of vegetation or overcultivated, rainwater flows downstream—unfiltered—over compacted and crusted soils. We have drained and converted half of all wetlands worldwide (Revenga et al. [PAGE] 2000), and we add levels of pollutants to watersheds that overwhelm their natural purification and dilution capacities.

To an extent, we can replace ecosystems' natural cleaning service with wastewater treatment plants, chlorination and other disinfectant processes, and artificial wetlands. But these options typically are expensive and do not provide the many other benefits supplied by forests and natural wetlands, such as wildlife habitat, open space, and flood protection.

The Costs of Clean Water

Here are some global and local indicators of our dependence on the water filtration and purification services that ecosystems provide. The human and economic costs of trying to replace them can be high.

- **Percentage of the world's population that lacks access to clean drinking water:**
  28 percent, or as many as 1.7 billion people (UNICEF 2000)

- **Number of people who die each year because of polluted drinking water, poor sanitation, and domestic hygiene:**
  5 million. Additionally, waterborne diseases such as diarrhea, ascariasis, dracunculiasis, hookworm, schistosomiasis, and trachoma cause illness in perhaps half the population of the developing world each year (WHO 1996).

- **Percentage of urban sewage in the developing world that is discharged into rivers, lakes, and coastal waters without any treatment:**
  90 percent (WRI et al. 1996:21)

- **Amount spent on bottled water worldwide in 1997:**
  $42 billion (Beverage Industry 1999)

- **Amount U.S. consumers spent on home water filtration systems in 1996:**
  $1.4 billion (Trust for Public Land 1997:24)

- **Cost incurred by households in Jakarta that must buy kerosene to boil the city's public water before use:**
  Rp 96 billion or US$52 million a year (1987 prices) (Bhatia and Falkenmark 1993:9)

- **Replacement cost of the water that would be lost if thirteen of Venezuela's National Parks that provide critical protection for urban water supplies were deforested:**
  $103 million to $206 million (net present value) (Reid forthcoming:6)

- **Typical cost to desalinize seawater:**
  $1.00–$1.50 per cubic meter (UNEP 1999:166)

- **Amount of open space and critical recharge area paved over every day in the United States:**
  11.7 km^2 (TPL 1997:3)

- **Estimated annual value of water quality improvement provided by wetlands along a 5.5-km stretch of the Alchovy River in Georgia, USA**
  $3 million (Lerner and Poole 1999:41)

- **Cost to construct wetlands to help process and recycle sewage produced by the 15,000 residents of Arcata, California:**
  $514,600 for a 40-ha system (Marinelli 1990). The city's alternative was to build a larger wastewater treatment plant at a cost of $25 million (Neander n.d.).

Box 1.4 Pollination

To many people, bees are known simply as prodigious honey makers and bats as cohorts of vampires and darkness. Rarely do we recognize that thousands of species of plants could not reproduce without their help. Wind pollinates some plants, but 90 percent of all flowering plants—including the great majority of the world's food crops—would not exist without animals and insects transporting pollen from one plant to another. Of the world's 100 most important crops, bees alone pollinate more than 70 percent (Nabhan and Buchmann 1997:136, 138). Besides food, pollinators help produce other agricultural products that enhance our lives, including dyes, fuelwood, tropical timbers, and textile fibers such as cotton and flax. The diets of many birds and mammals also are based on seeds and fruits produced by pollination.

No wonder, then, that agricultural specialists consider the current worldwide decline in pollinators a cause for alarm. Losses of pollinators have been reported on every continent except Antarctica. Some are on the verge of extinction; pesticides, mites, invasive species, and habitat loss and fragmentation are major killers. The consequences of continued pollinator declines could include billions of dollars in reduced harvests, cascades of plant and animal extinctions, and a less stable food supply.

Few studies have calculated the economic contribution of all pollinators, globally, to agricultural production and biodiversity, but

- The FAO recently estimated the 1995 contribution from pollination to the worldwide production of just 30 of the major fruit, vegetable, and tree crops (not including pasture or animal feeds) to be in the range of $54 billion (international dollars) per year (Kenmore and Krell 1998).
- Estimates of the value of pollination just for crop systems in the United States range from US$20 to $40 billion (Kearns et al. 1998:84).

Dependence of Selected U.S. Crops on Honey Bee Pollination

Crops	1998 Quantity Produced (metric tons)	Percentage of Crop Loss Without Honey Bee Pollination*
Temperate Fruits		
Almonds	393,000	90
Apples	5,165,000	80
Cherries	190,000	60
Oranges	12,401,000	30
Pears	866,500	50
Strawberries	765,900	30
Vegetables and Seeds		
Asparagus	92,800	90
Cabbage	2,108,200	90
Carrots	2,201,000	60
Cottonseed	7,897,000	30
Sunflowers	2,392,000	80
Watermelons	1,673,000	40

*Crop losses are estimates of loss if managed honey bee populations were eliminated in the United States, with no replacement of their services by alternative pollinators.
Sources: FAO 2000; Southwick and Southwick 1992.

Pollinators for the World's Flowering Plants (Angiosperms)

Pollinators	Estimated Number of Plant Species Pollinated	Total Percentage of Plant Species Pollinated*
Wind	20,000	8.30
Water	150	0.63
Bees	40,000	16.60
Hymenoptera	43,295	18.00
Butterflies/Moths	19,310	8.00
Flies	14,126	5.90
Beetles	211,935	88.30
Thrips	500	0.21
Birds	923	0.40
Bats	165	0.07
All Mammals	298	0.10
All Vertebrates	1,221	0.51
	351,923	

*Total percentage does not equal 100, reflecting pollination by more than one pollinator.
Source: Buchmann and Nabhan 1996:274.

Box 1.5 Biological Diversity

With an estimated 13 million species on Earth (UNEP 1995:118), few people take notice of an extinction of a variety of wheat, a breed of sheep, or an insect. Yet it is the very abundance of species on Earth that helps ecosystems work at their maximum potential. Each species makes a unique contribution to life.

- Species diversity influences ecosystem stability and undergirds essential ecological services. From water purification to the cycling of carbon, a variety of plant species is essential to achieving maximum efficiency of these processes. Diversity also bolsters resilience—an ecosystem's ability to respond to pressures—offering "insurance" against climate change, drought, and other stresses.

- The genetic diversity of plants, animals, insects, and microorganisms determines agroecosystems' productivity, resistance to pests and disease, and, ultimately, food security for humans. Extractions from the genetic library are credited with annual increases in crop productivity worth about $1 billion per year (WCMC 1992:433); yet the trend in agoecosystems is toward the replacement of polycultures with monocultures and diverse plant seed varieties with uniform seed varieties (Thrupp 1998: 23–24). For example, more than 2,000 rice varieties were found in Sri Lanka in 1959, but just five major varieties in the 1980s (WCMC 1992:427).

- Genetic diversity is fundamental to human health. From high cholesterol to bacteria fighters, 42 percent of the world's 25 top-selling drugs in 1997 were derived from natural sources. The global market value of pharmaceuticals derived from genetic resources is estimated at $75–$150 billion. Botanical medicines like ginseng and echinacea represent an annual market of another $20–$40 billion, with about 440,000 tons of plant material in trade, much of it originating in the developing world. Not fully captured by this commercial data is the value of plant diversity to the 75 percent of the world's population that relies on traditional medicine for primary health care (ten Kate and Laird 1999:1–2, 34, 101, 334–335).

Origins of Top 150 Prescription Drugs in the United States of America

Origin	Total Number of Compounds	Natural Product	Semi-synthetic	Synthetic	Percent
Animal	27	6	21	—	23
Plant	34	9	25	—	18
Fungus	17	4	13	—	11
Bacteria	6	5	1	—	4
Marine	2	2	0	—	1
Synthetic	64	—	—	64	43
Totals	150	26	60	64	100

Source: Grifo et al. 1997:137.

The threat to biodiversity is growing. Among birds and mammals, rates may be 100–1,000 times what they would be without human-induced pressures—overexploitation, invasive species, pollution, global warming, habitat loss, fragmentation, and conversion (Reid and Miller 1989). Regional extinctions, particularly the loss of populations of some species in tropical forests, may be occurring 3–8 times faster than global species extinctions (Hughes et al. 1997:691).

Such localized extinctions may be just as significant as the extinction of an entire species worldwide. Most of the benefits and services provided by species working together in an ecosystem are local and regional. If a keystone species is lost in an area, a dramatic reorganization of the ecosystem can occur. For example, elephants disperse seeds, create water holes, and trample vegetation through their movements and foraging. The extinction of elephants in a piece of savanna can cause the habitat to become less diverse and open and cause water holes to silt up, which would have dramatic repercussions on other species in the region (Goudie 2000:67).

Vascular Plants Threatened on a Global Scale

Of the estimated 250,000–270,000 species of plants in the world, only 751 are known or suspected to be extinct. But an enormous number—33,047, or 12.5 percent—are threatened on a global scale. Even that grim statistic may be an underestimate because much information about plants is incomplete, particularly in the tropics.

Source: WCMC/IUCN 1998.

Box 1.6 Carbon Storage

Carbon is the basis of life, cycling through the oceans, atmosphere, vegetation, and soils. Through photosynthesis, plants take up carbon as carbon dioxide (CO_2) and convert it to sugar for energy; animals consume the plants; and when both plants and animals die, carbon is returned to the atmosphere as the organisms decay. But ever-increasing emissions of carbon from fossil fuel combustion and deforestation are unbalancing the global carbon cycle; there's less carbon in the soil and vegetation and more in the atmosphere. Because CO_2 in the atmosphere captures the sun's heat, increasing amounts destabilize the global climate.

It is estimated that prior to the 18th century, increases in atmospheric carbon were less than 0.01 billion metric tons of carbon (GtC) per year (Ciaias 1999). The Industrial Revolution and subequent global development greatly increased fossil fuel emissions, as did the clearing of forests and other land-use changes that release carbon. By 1998, there was approximately 176 GtC more carbon in the atmosphere than in 1850, an increase of nearly 30 percent (IPCC 2000:4). Today, human activities emit an estimated 7.9 GtC to the atmosphere annually (IPCC 2000:5). The oceans absorb slightly less than 30 percent of this carbon and terrestrial ecosystems absorb slightly more, but that leaves 40 percent of yearly emissions to accumulate in the atmosphere (IPCC 2000:5).

Reducing anthropogenic carbon emissions is one way to mitigate climate change. Other ways depend on maintaining the ability of ecosystems to absorb carbon. Through photosynthesis, plants provide the most effective and efficient way to recapture and store atmospheric carbon.

- Oceans are the major carbon reservoir or "sink." Through chemical and biological processes, including phytoplankton's growth and decay, oceans store roughly 50 times more carbon than is in the atmosphere, mostly as dissolved inorganic carbon (IPCC 2000:30).
- Soil and its organic layer store about 75 percent of total terrestrial carbon (Brown 1998:16). Most of the carbon released to the atmosphere in the last 2 centuries occurred as grasslands and forests were converted to agricultural uses.
- Forests are the most effective terrestrial ecosystem for recapturing carbon, but not all forests offer the same sequestration benefits. Faster-growing young trees absorb about 30 percent more carbon than mature wood, but an older forest stores more carbon overall in the soil and in above- and below-ground vegetation than a tree plantation of the same size. Latitude, climate, species mix, and other biological and ecosystem factors also affect carbon fluxes in forests (see Brown 1998:10).

Earth's Annual Carbon Budget, 1989–98

Type of emission or uptake	Gigatons of carbon per year
Human-induced emissions into the atmosphere	
Emissions from consumption and production (fossil fuel combustion and cement production)	6.3 ± 0.6
Net emissions from land use change (fires, deforestation, agriculture)	1.6 ± 0.8
Ocean and terrestrial capture from the atmosphere	
Net uptake by oceans (photosynthesis and ocean capture minus ocean release)	2.3 ± 0.8
Net uptake by terrestrial ecosystems (photosynthesis and terrestrial storage minus decay and respiration)	2.3 ± 1.3
Carbon added to the atmosphere each year	3.3 ± 0.2

Source: IPCC, 2000:5. Error limits correspond to an estimated 90 percent confidence interval. Emissions from consumption and production are calculated with high confidence. Net emissions from land use change are estimated from observed data and models. Uptake by oceans is based on models. Carbon added to the atmosphere each year is measured with high accuracy. Uptake by terrestrial ecosystems is an imputed amount (the difference between total emissions and estimated uptake by oceans and atmosphere).

Global Carbon Storage

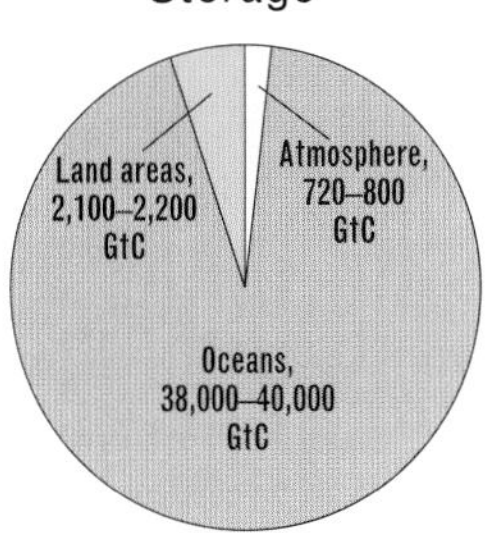

Carbon Stored in Soil versus Vegetation

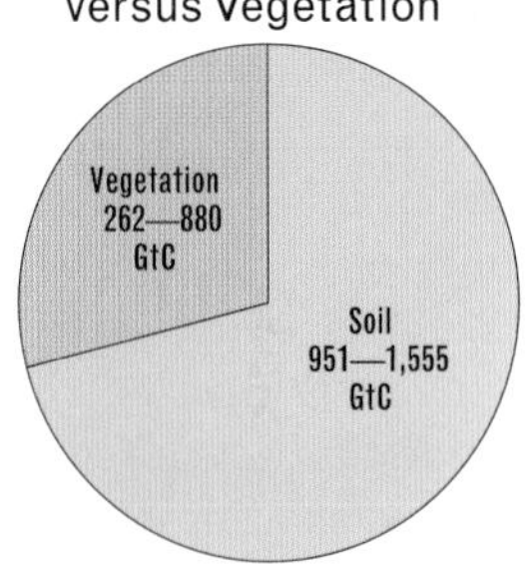

Carbon Storage in Terrestrial Ecosystems

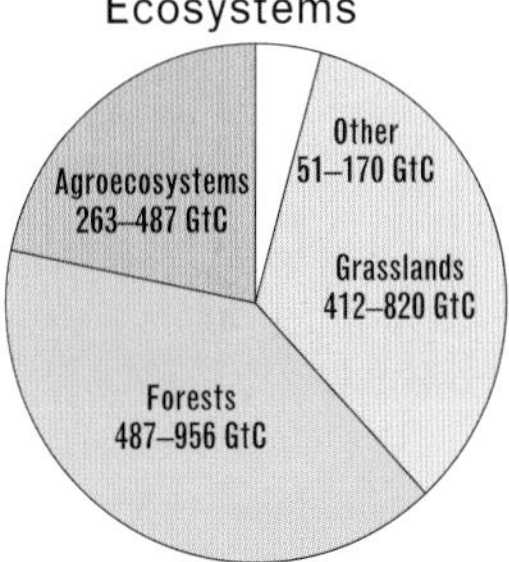

Sources: IPCC 1996:63; Matthews et al. [PAGE] 2000. Data on carbon stored in soil versus vegetation and in terrestrial ecosystems is derived from the International Geosphere-Biosphere Programme. Thus estimated share of carbon in each ecosystem varies slightly from PAGE results in Chapter 2, because PAGE definitions of ecosystems accommodate some overlap of transitional areas.

Managing Ecosystems: Trade-Offs and Costs

People often modify or manage ecosystems to enhance the production of one or more goods, such as crops or trees or water storage. The degree of modification varies widely. Some ecosystems are heavily affected, others remain relatively unaltered, and management ranges through various types of use–from nondestructive rubber tapping, to clear-cutting, and even to single-species tree plantations. Similarly, aquatic ecosystems can range from free-flowing rivers to artificial ponds for raising fish or shrimp.

Sometimes the dividing line between "natural" and "managed" ecosystems is clear. A farm is obviously a highly managed ecosystem–an *agroecosystem*. But often management is more subtle: a fence dividing a rangeland, a forest access road, a seawall protecting a private beach, a mountain stream diverted to supply a village with water. In any case, human influence, even if it is not intensive management, is pervasive among all ecosystem types.

The decision to manage or alter an ecosystem involves trade-offs. Not all benefits can be obtained at the same time, and maximizing one benefit may reduce or eliminate others. For example, converting a natural forest to a tree plantation may increase the production of marketable pulp or lumber, bringing high monetary returns per hectare, but it generally decreases biodiversity and habitat value compared with a natural forest. Likewise, damming a river may increase the water available for irrigation or hydroelectric power production and decrease the danger of floods, but it may also disrupt natural breeding cycles of fish and damage aquatic habitats downstream by diverting water or releasing it at inappropriate times.

To a certain extent, we accept these trade-offs as necessary to efficiently produce food, power, and the other things we need. Historically, we have been hugely successful at selectively increasing those ecosystem goods we value most. It is only recently that we have begun to focus on the dangers of such trade-offs.

The environmental awareness and knowledge we have gained over the last 30 years have taught us that there are limits to the amount of alteration that ecosystems can tolerate and still remain productive. The loss of a hectare of forest habitat or a single plant or insect species in a grassland may not affect the functioning of the system drastically or immediately, but it may push the system toward a threshold from which it cannot recover.

Biological thresholds remind us that it is the cumulative effects of human activities that factor most in ecosystem decline. A series of small changes, each seemingly harmless, can result in cumulative impacts that are irreversible; this is sometimes called the "tyranny of small decisions." The progressive conversion of a mangrove forest is a good example.

Mangroves serve as nurseries for many species of fish and shellfish that then leave the mangrove and are later caught in surrounding waters. The value of this seafood is often many times greater than the wood, crabs, and other fish harvested within the mangrove forest itself. But in regions where mangroves grow, raising shrimp is a profitable enterprise. Converting small sections of the mangrove to shrimp ponds may have little impact on the fish harvest in surrounding waters. But if shrimp growers gradually convert the entire mangrove to ponds, the local fishery will collapse at some point.

Determining the threshold between sustainability and collapse is no easy matter. This is one reason why it is difficult to manage ecosystems responsibly. Ecosystems are naturally resilient and can accommodate considerable disturbance. But how much? Our understanding of ecosystems, although it has increased rapidly, is still too limited to answer this crucial question. For most ecosystems, we have yet to master the details of how organisms and environment interact and connect, how changes in one element of the system reverberate through the whole, or what factors moderate the speed of change in an ecosystem. At a global level, we still lack even the most basic statistics on ecosystems–how much and where they have been modified, for example, or how their productivity has changed over time. So at both an individual ecosystem level and at a larger national or regional level, we find it nearly impossible to predict how close to the edge our management has brought us, or to determine the extent of the trade-offs we have already made.

How Are Ecosystems Degraded?

Human activities have put global ecosystems under siege:

- Some 75 percent of the major marine fish stocks are either depleted from overfishing or are being fished at their biological limit (Garcia and Deleiva In press).

- Logging and conversion have shrunk the world's forest cover by as much as half, and roads, farms, and residences are rapidly fragmenting what remains into smaller forest islands (Bryant et al. 1997:9).

- Some 58 percent of coral reefs are potentially threatened by destructive fishing practices, tourist pressures, and pollution (Bryant et al. 1998:6).

- Fully 65 percent of the roughly 1.5 billion ha of cropland worldwide have experienced some degree of soil degradation (Wood et al. [PAGE] 2000).

- Overpumping of groundwater by the world's farmers exceeds natural recharge rates by at least 160 billion m^3 per year (Postel 1999:255).

The pressures responsible for these declines continue to increase in most cases, accelerating ecosystem change (Vitousek et al. 1997:498). (See Chapter 2 for a detailed look at ecosystem conditions.)

In many instances, the principal pressure on ecosystems is simple overuse–too much fishing, logging, water diversion, or tourist traffic. Overuse not only depletes the plants and wildlife that inhabit the ecosystem, but also can fragment the system and disrupt its integrity–all factors that diminish its productive capacity.

Outright conversion of forests, grasslands, and wetlands to agriculture or other uses is a second principal pressure reshaping global ecosystems and the benefits they give. Invasive species, air and water pollution, and the threat of climate change are key ecosystem pressures as well.

AGRICULTURAL CONVERSION

When farmers convert a natural ecosystem to agriculture, they change both the composition of the ecosystem and how it functions. In agroecosystems, naturally occurring plants give way to a few nonnative crop species. Wildlife is pushed to the margins of the system. Pesticides may decimate insect populations and soil microorganisms. Soil compaction causes water to infiltrate the soil differently, and runoff and erosion may increase. The cycle of nutrients through the system shifts as fertilizers are applied and soil bacteria and vegetation change.

The result is a substantial change in benefits. Food production–clearly a boon–surges, but most other benefits suffer to some degree. Biodiversity and the benefits associated with it, such as production of a wide variety of wild plants and animals and the availability of diverse genetic material, often decline substantially. At the scale of conversion prevalent today, that can mean huge biodiversity losses in the aggregate. One study estimates that in the species-rich tropics, forest conversion commits two to five species of plants, insects, birds, or mammals to extinction each hour (Hughes et al. 1997:691).

Agriculture in converted areas may also increase pressures on surrounding ecosystems through the introduction of nonnative species that become invasive and displace indigenous species. Bioinvasions are second only to habitat loss, usually through conversion, as a threat to global biodiversity. In South Africa, nonnative tree species originally imported for forest plantations have invaded a third of the nation's mountain watersheds. The invading plants have depleted freshwater supplies, displaced thousands of native plants, and altered animal habitats, precipitating a countrywide eradication program (see Chapter 3, Working for Water).

(continues on p. 22)

Conversion represents the ultimate in human impact on an ecosystem, and the most abrupt change in the goods and services it produces.

Box 1.7 Linking People and Ecosystems: Human-Induced Pressures

Thousands of used tires are shipped into the United States from Asia for retreading and resale every year. Some have contained larvae of the Asian tiger mosquito. Already the mosquito has established itself in 25 states, feeding on mammals and birds. Some of the mosquitos carry the equine encephalitis virus, often fatal to horses and people.

Behind all the pressures impinging on ecosystems are two basic drivers: human population growth and increasing consumption.

A logging concessionaire in Gabon clear-cuts areas in its assigned tract, paying the government a sizable permit fee. Its contract with the government, which owns the tract, allows it to harvest timber at below market rates if it replants the area. The concessionaire plants seedlings but does nothing to stop the ensuing erosion of topsoil, the siltation of nearby streams, and the migration or loss of wildlife that depended on the mature forest.

Small-scale, artisanal miners from Venezuela illegally cross the unmarked border into Brazil deep in the Amazonian rainforest. Although they have no legal right to mine there for gold, they can eke out a living for their families if they keep their operation small and move frequently from place to place. To increase their chances of extracting gold, they add mercury to the sluice, although the toxic metal is technically banned. Like thousands of other independents in the area, they let the mixture run off directly into a tributary where it poisons local fish.

Primary Human-Induced Pressures on Ecosystems

Ecosystem	Pressures	Causes
Agroecosystems	■ Conversion of farmland to urban and industrial uses ■ Water pollution from nutrient runoff and siltation ■ Water scarcity from irrigation ■ Degradation of soil from erosion, shifting cultivation, or nutrient depletion ■ Changing weather patterns	■ Population growth ■ Increasing demand for food and industrial goods ■ Urbanization ■ Government policies subsidizing agricultural inputs (water, research, transport) and irrigation ■ Poverty and insecure tenure ■ Climate change
Coastal Ecosystems	■ Overexploitation of fisheries ■ Conversion of wetlands and coastal habitats ■ Water pollution from agricultural and industrial sources ■ Fragmentation or destruction of natural tidal barriers and reefs ■ Invasion of nonnative species ■ Potential sea level rise	■ Population growth ■ Increasing demand for food and coastal tourism ■ Urbanization and recreational development, which is highest in coastal areas ■ Government fishing subsidies ■ Inadequate information about ecosystem conditions, especially for fisheries ■ Poverty and insecure tenure ■ Uncoordinated coastal land-use policies ■ Climate change
Forest Ecosystems	■ Conversion or fragmentation resulting from agricultural or urban uses ■ Deforestation resulting in loss of biodiversity, release of stored carbon, air and water pollution ■ Acid rain from industrial pollution ■ Invasion of nonnative species ■ Overextraction of water for agricultural, urban, and industrial uses	■ Population growth ■ Increasing demand for timber, pulp, and other fiber ■ Government subsidies for timber extraction and logging roads ■ Inadequate valuation of costs of industrial air pollution ■ Poverty and insecure tenure
Freshwater Systems	■ Overextraction of water for agricultural, urban, and industrial uses ■ Overexploitation of inland fisheries ■ Building dams for irrigation, hydropower, and flood control ■ Water pollution from agricultural, urban, and industrial uses ■ Invasion of nonnative species	■ Population growth ■ Widespread water scarcity and naturally uneven distribution of water resources ■ Government subsidies of water use ■ Inadequate valuation of costs of water pollution ■ Poverty and insecure tenure ■ Growing demand for hydropower
Grassland Ecosystems	■ Conversion or fragmentation owing to agricultural or urban uses ■ Induced grassland fires resulting in loss of biodiversity, release of stored carbon, and air pollution ■ Soil degradation and water pollution from livestock herds ■ Overexploitation of game animals	■ Population growth ■ Increasing demand for agricultural products, especially meat ■ Inadequate information about ecosystem conditions ■ Poverty and insecure tenure ■ Accessibility and ease of conversion of grasslands

Box 1.8 Invasive Species

No ecosystem is immune to the threat of invasive species. They crowd out native plants and animals, degrade habitats, and contaminate the gene pools of indigenous species. Island ecosystems are particularly vulnerable because of their high levels of endemism and isolation; many island species evolved without strong defenses against invaders. On Guam, for example, the brown tree snake from Papua New Guinea has eaten twelve of the island's fourteen flightless bird species, causing them to become extinct in the wild. In New Zealand, roughly two-thirds of the land surface is covered by exotic plants (Bright 1998:115). Half of Hawaii's wild species are nonnative (OTA 1993:234).

Invasive species are a costly problem:

- Leidy's comb jellyfish, native to the Atlantic coast of the Americas, was pumped out of a ship's ballast tank into the Black Sea in the early 1980s. Its subsequent invasion has nearly wiped out Black Sea fisheries, with direct costs totaling $250 million by 1993 (Travis 1993:1366). Meanwhile, the zebra mussel, native to the Caspian Sea, was similarly dumped into the United States' Great Lakes in the late 1980s. Controlling this invader, which colonizes and clogs water supply pipes, costs area industries millions of dollars per year—perhaps $3–$5 billion total to date (Bright 1998:182).
- The Asian tiger mosquito, now spreading throughout the world, is a potential transmitter of 18 viral pathogens (Bright 1998:169). One of those pathogens is the West Nile virus. In 1999, a director with the U.S. Geological Survey noted that recent crow die-offs in Wisconsin suggest that the West Nile virus could be more deadly to North American bird species than to species in Africa, the Middle East, and Europe, where the virus is normally found (USGS 1999:1).
- In South Africa's Western Cape, invasive trees threaten to cut Cape Town's water supply by about a third in the next century. (See Chapter 3, "Working for Water.")

Regulation and control are complicated by the many modes of invasion. Some species find their way to new habitats by accident: they hitchhike in ships or planes, on traded goods or travelers. Other species are intentionally introduced for hunting, fishing, or pest control. Still other invasives "escape" their intended confines, like the seaweed *Caulerpa taxifolia*, which was originally intended for aquariums in Europe but now also carpets thousands of acres of French and Italian coastlines (MCBI 1998).

Cumulative Number of Nonnative Species in U.S. Regions by Decade of Introduction

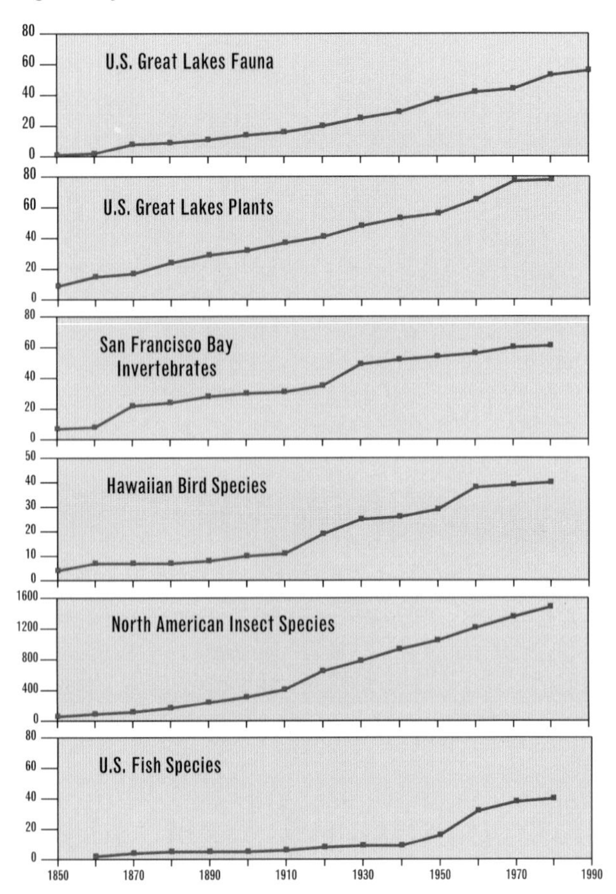

Source: Ruesink et al. 1995:466.

Native vs. Nonnative Plant Species in Selected Regions

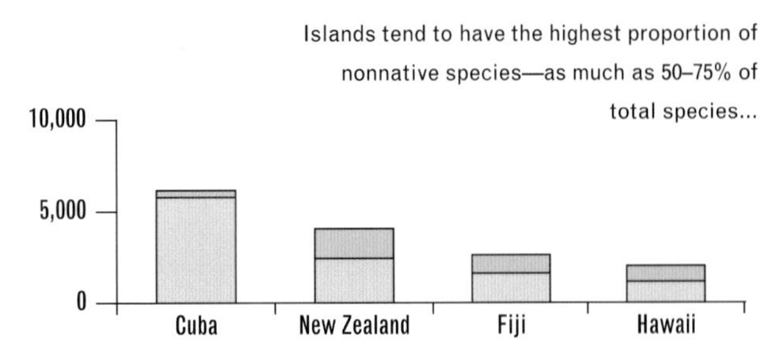

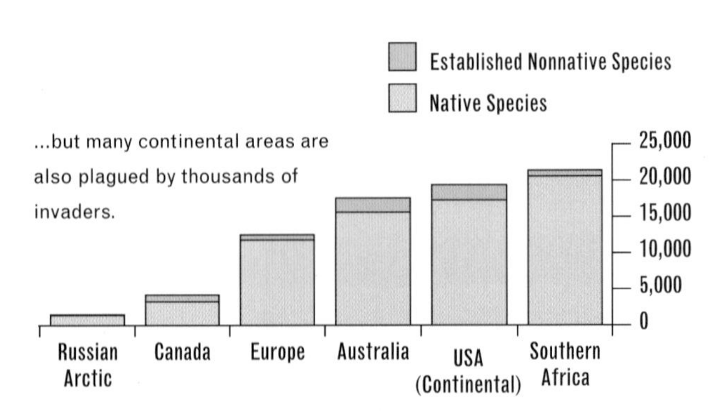

Sources: Vitousek et al. 1997; Vitousek et al. 1996.

Box 1.9 Trade-Offs: Lake Victoria's Ecosystem Balance Sheet

Trade-offs among various ecosystem goods and services are common in the management of ecosystems, although rarely factored into decision making. For example, farmers can increase food production by applying fertilizer or expanding the land they have under cultivation, but these strategies harm other goods and services from the land they farm, like water quality and biodiversity.

In very few cases do resource managers or policy makers fully weigh the various trade-offs among ecosystem goods and services. Why? In some cases, lack of information is the obstacle. Typically, not much is known about the likely impact of a particular decision on nonmarketed ecosystem services such as water purification or storm protection. Or, if such information does exist, it may not include estimates of the economic costs and benefits of the trade-offs. In other cases the obstacle is institutional. A government's Ministry of Agriculture naturally focuses primarily on its mission of food production and lacks the expertise or mandate to consider impacts of its actions on water quality, carbon sequestration, or coastal fisheries, for instance.

The example of Africa's Lake Victoria illustrates how profound and unpredictable trade-offs can be when management decisions are made without regard to how the ecosystem will react. Lake Victoria, bounded by Uganda, Tanzania, and Kenya, is the world's largest tropical lake and its fish are an important source of food and employment for the region's 30 million people. Before the 1970s, Lake Victoria contained more than 350 species of fish from the cichlid family, of which 90 percent were endemic, giving it one of the most diverse and unique assemblages of fish in the world (Kaufman 1992:846–847, 851). Today, more than half of these species are either extinct or found only in very small populations (Witte et al. 1992:1, 17).

The collapse in the lake's biodiversity was caused primarily by the introduction of two exotic fish species, the Nile perch and Nile tilapia, which fed on and outcompeted the cichlids for food. But other pressures factored in the collapse as well. Overfishing depleted native fish stocks and provided the original impulse for introducing the Nile perch and tilapia in the early 1950s. Land-use changes in the watershed dumped pollution and silt into the lake, increasing its nutrient load and causing algal blooms and low oxygen levels in deeper waters—a process called eutrophication. The result of all these pressures was a major reorganization of the lake's fishlife. Cichlids once accounted for more than 80 percent of Lake Victoria's biomass and provided much of the fish catch (Kaufman 1992:849). By 1983, Nile perch made up almost 70 percent of the catch, with Nile tilapia and a native species of sardine making up most of the balance (Achieng 1990:20).

Although the introduced fishes devastated the lake's biodiversity, they did not not destroy the commercial fishery. In fact, total fish production and its economic value rose considerably.

Trading Biodiversity for Export Earnings

Percentage Contribution to Lake Victoria Fish Catch (Kenya Only), 1968–1988

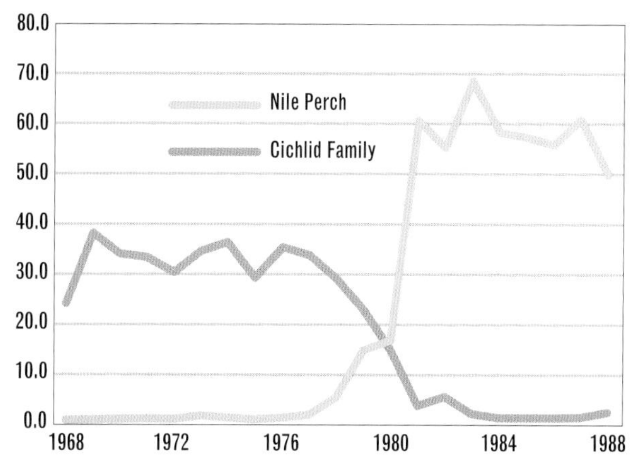

Source: Achieng 1990:20, citing Fisheries Department of Kenya, *Statistical Bulletin.*

Today, the Nile perch fishery produces some 300,000 metric tons of fish (FAO 1999), earning $280–$400 million in the export market—a market that did not exist before the perch was introduced (Kaufman 2000). Unfortunately, local communities that had depended on the native fish for decades did not benefit from the success of the Nile perch fishery, primarily because Nile perch and tilapia are caught with gear that local fishermen could not afford. And, because most of the Nile perch and tilapia are shipped out of the region, the local availability of fish for consumption has declined. In fact, while tons of perch find their way to diners as far away as Israel and Europe, there is evidence of protein malnutrition among the people of the lake basin (Kaufman 2000).

The sustainability of the Nile perch fishery is also a concern. Overfishing and eutrophication are major threats to the fishery, and the stability of the entire aquatic ecosystem—so radically altered over a 20-year span—is in doubt. The ramifications of the species introductions can even be seen in the watershed surrounding Lake Victoria. Drying the perch's oily flesh to preserve it requires firewood, unlike the cichlids, which could be air-dried. This has increased pressure on the area's limited forests, increasing siltation and eutrophication, which, in turn, has further unbalanced the precarious lake ecosystem (Kaufman 1992:849–851; Kaufman 2000).

In sum, introducing Nile perch and tilapia to Lake Victoria traded the lake's biodiversity and an important local food source for a significant—although perhaps unsustainable—source of export earnings. When fisheries managers introduced these species, they unknowingly altered the balance of goods and services the lake produced and redistributed the economic benefits flowing from them. Knowing the full dimensions of these trade-offs, would they make the same decision today?

Not all agricultural conversions are equal. Some may retain or carefully harbor aspects, and services, of the original ecosystem. In Sumatra, some traditional agroforestry systems (where trees and crops are mixed) contain as much as half the species diversity found in the neighboring forest. Traditional Central American coffee plantations raise their coffee plants in the shade of native trees that provide essential bird habitat and a range of secondary products. Even many modern agricultural systems include careful tillage practices aimed at preventing erosion and preserving the soil's water-holding properties and beneficial soil organisms.

URBAN AND INDUSTRIAL CONVERSION

Unfortunately, conversion to urban or industrial uses is usually not so benign. Radical changes in ecosystem benefits occur as structures and paved surfaces replace native plant and animal communities. As city dwellers cover permeable soil surfaces with concrete and asphalt, watershed functions decline. With few places to sink in, rainfall runs off quickly and local flooding can ensue. Still, the more simplified ecosystems in parks, backyards, and vacant lots do provide important services–shade, areas for relaxation, removal of air pollutants, and even some wildlife habitat–that city dwellers enjoy.

POLLUTION AND CLIMATE CHANGE

The effects of pollution put indirect pressures on ecosystems. Acid rain, smog, wastewater releases, pesticide and fertilizer residues, and urban runoff all have toxic effects on ecosystems–sometimes at great distances from the activities that gave rise to the pollution. For example, nitrogen releases from industry, transportation, and agriculture have seriously altered the global nitrogen cycle, affecting the function of both terrestrial and aquatic ecosystems.

Biologically active, or "fixed," nitrogen is an essential nutrient for all plants and animals. But nitrogen releases from human sources like fertilizers and fossil fuels now exceed those from natural sources, leaving ecosystems awash in fixed nitrogen. The impacts include an overgrowth of algae in waterways, caused by the fertilizing effect of excess nutrients; acidification of soils and loss of some soil nutrients; loss of plants adapted to natural low-nitrogen conditions; and more smog and greenhouse warming from higher levels of nitrogen oxides in the atmosphere (ESA 1997b:1–14).

Climate change from the buildup of greenhouse gases provides an even more profound example of the potential for pollution to inadvertently disrupt ecosystems on a global scale. Scientists warn that global ecosystems could undergo a major reorganization as Earth's vegetation redistributes itself to accommodate rising temperatures, changes in rainfall patterns, and the potential fertilizing effects of more carbon dioxide (CO_2) in the atmosphere. Computer models estimate that doubling atmospheric CO_2 levels from preindustrial levels, which will likely happen within the next century, could trigger broad changes in the distribution, species composition, or leaf density of roughly one-third of global forests. Tundra areas could also shrink substantially and coastal wetlands shift markedly, among many other effects. It is not at all clear how present ecosystems would weather such significant changes or how these changes might affect their productivity (Houghton et al. 1997:30).

What Drives Degradation?

Behind all the pressures impinging on ecosystems are two basic drivers: human population growth and increasing consumption. Closely related are a suite of economic and political factors–market forces, government subsidies, globalization of production and trade, and government corruption–that influence what and how much we consume, and where it comes from. Issues of poverty, land tenure, and armed conflict are also significant factors in how people treat the ecosystems they live in and extract goods and services from.

DEMOGRAPHICS AND CONSUMPTION

Population growth is in many ways the most basic of environmental pressures because everyone requires at least some minimum of water, food, clothing, shelter, and energy–all ultimately harvested directly from ecosystems or obtained in a way that affects ecosystems. Over the next 50 years, demographers expect the world's population to grow from the current 6 billion to 9 billion or so, with most of this growth taking place in developing nations (UN Population Division 1998:xv). Simple arithmetic dictates this will increase the demand for ecosystem products and increase the pressure on global food and water supplies.

Increasing pressure on ecosystems is not simply a matter of population growth, however. In fact, it is more a matter of how much and what we consume. Global increases in consumption have greatly outpaced growth in population for decades. From 1980 to 1997, the global economy nearly tripled to some US$29 trillion, yet the world population increased only 35 percent (World Bank 1999b:194; UN Population Division 1998:xv). Per capita consumption levels are rising quickly in many nations as their economies develop; and consumption levels in most industrialized nations are already remarkably high. This higher consumption of everything from paper to refrigerators to computers to oil is the result of greater wealth. Personal-income levels have climbed steadily in developed nations and a number of rapidly developing countries such as China, India, and Thailand; and consumption has increased accordingly.

At the same time, the world's economy has become more integrated. Trade has made consumer markets more global.

Industries have become more international and less tied to a single place or production facility. This "globalization" means that consumers derive goods and services from ecosystems around the world, with the costs of use largely separated from the benefits. This tends to hide the environmental costs of increased consumption from those doing the consuming.

For example, a housing contractor in Los Angeles installs copper plumbing but has no way of knowing whether the copper has come from the infamous Ok Tedi mine in Papua New Guinea. The giant mine, which is owned by an international consortium of companies, dumps 80,000 tons per day of untreated tailings into the Ok Tedi River, destroying much of the river's aquatic life and disrupting the subsistence lifestyle of the local Wopkaimin people. Globalization means the eventual homeowners who benefit from the copper have no knowledge of their link with the damaged Ok Tedi watershed and don't suffer the environmental costs (Da Rosa and Lyon 1997:223-226).

It's not surprising that those doing the most consuming live in developed countries, but the unevenness of consumption of ecosystem goods and services worldwide is striking. It takes roughly 5 ha of productive ecosystem to support the average U.S. citizen's consumption of goods and services versus less than 0.5 ha to support consumption levels of the average citizen in the developing world (GEF 1998:84). Annual per capita CO_2 emissions are more than 11,000 kg in industrial countries, where there are far more cars, industries, and energy-consuming appliances. This compares with less than 3,000 kg in Asia (UNDP 1998:57). On average, someone living in the developed world spends nearly $16,000 (1995 international dollars) on private consumption each year, compared with less than $350 spent by someone in South Asia and sub-Saharan Africa (UNDP 1998:50).

Of course, greater consumption of nutritious food, safe housing, clean water, and adequate clothing is absolutely necessary to relieve poverty in many nations, particularly in the developing world. In the words of the UN's 1948 Universal Declaration of Human Rights, "Everyone has the right to a standard of living adequate for the health and well-being of himself and of his family" (Article 25). Accommodating such basic human development, however, is far from the predominant pressure on ecosystems today. Even considering that almost four times as many people live in developing countries as in developed ones, the greatest burden on ecosystems currently originates with affluent consumers in developed countries, as well as wealthy elites in developing countries. It is the pattern of excessive consumption that often accompanies wealth that brings a disproportionate impact on ecosystems.

(continues on p. 30)

Increasing pressure on ecosystems is not simply a matter of population growth, however. In fact, it is more a matter of how much and what we consume.

Box 1.10 Domesticating the World: Conversion of Natural Ecosystems

Since the dawn of settled agriculture, humans have been altering the landscape to secure food, create settlements, and pursue commerce and industry. Croplands, pastures, urban and suburban areas, industrial zones, and the area taken up by roads, reservoirs, and other major infrastructure all represent conversion of natural ecosystems.

These transformations of the landscape are the defining mark of humans on Earth's ecosystems, yielding most of the food, energy, water, and wealth we enjoy, but they also represent a major source of ecosystem pressure.

Conversion alters the structure of natural ecosystems, and how they function, by modifying their basic physical properties—their hydrology, soil structure, and topography—and their predominant vegetation. This basic restructuring changes the complement of species that inhabits the ecosystem and disrupts the complex interactions that typified the original ecosystem. In many cases, the converted ecosystem is simpler in structure and less biologically diverse. In fact, habitat loss from conversion of natural ecosystems represents the primary driving force in the loss of biological diversity worldwide (Vitousek et al. 1997:495).

Historically, expansion of agriculture into forests, grasslands, and wetlands has been the greatest source of ecosystem conversion. Within the last century, however, expansion of urban areas with their associated roads, power grids, and other infrastructure, has also become a potent source of land transformation.

- Worldwide, humans have converted approximately 29 percent of the land area—almost 3.8 billion ha—to agriculture and urban or built-up areas (WRR calculations).

- Agricultural conversion to croplands and managed pastures has affected some 3.3 billion ha—roughly 26 percent of the land area. All totaled, agriculture has displaced one-third of temperate and tropical forests and one-quarter of natural grasslands. Agricultural conversion is still an important pressure on natural ecosystems in many developing nations; however, in some developed nations agricultural lands themselves are being converted to urban and industrial uses (WRR calculations).

- Urban and built-up areas now occupy more than 471 million ha—about 4 percent of land area. Almost half the world's population—some 3 billion people—live in cities. Urban populations increase by another 160,000 people daily, adding pressure to expand urban boundaries (UNEP 1999:47). Suburban sprawl magnifies the effect of urban population growth, particularly in North America and Europe. In the United States, the percentage of people living in urban areas increased from 65 percent of the nation's population in 1950 to 75 percent in 1990, but the area covered by cities roughly doubled in size during the same period (PRB 1998).

- Future trends in land conversion are difficult to predict, but projections based on the United Nations' intermediate-range population growth model suggest that an additional one-third of the existing global land cover could be converted over the next 100 years (Walker et al. 1999:369).

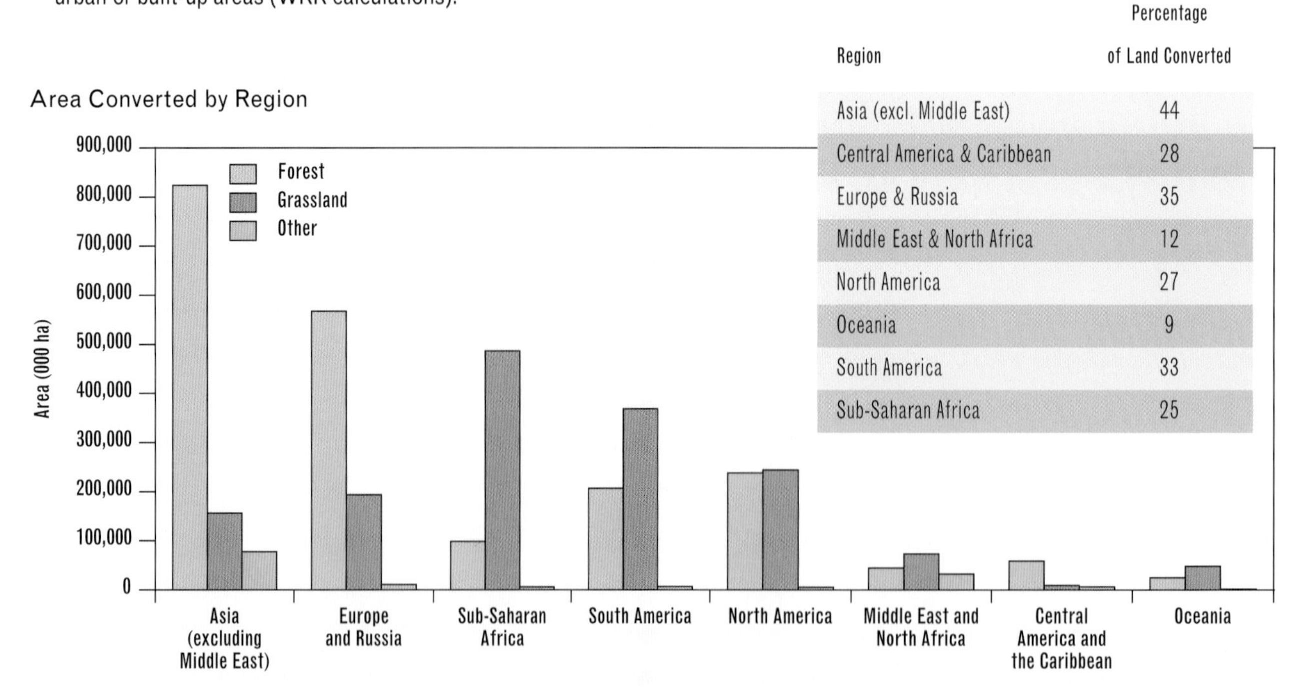

Region	Percentage of Land Converted
Asia (excl. Middle East)	44
Central America & Caribbean	28
Europe & Russia	35
Middle East & North Africa	12
North America	27
Oceania	9
South America	33
Sub-Saharan Africa	25

Source: WRR calculations.

Global Map of Converted Areas

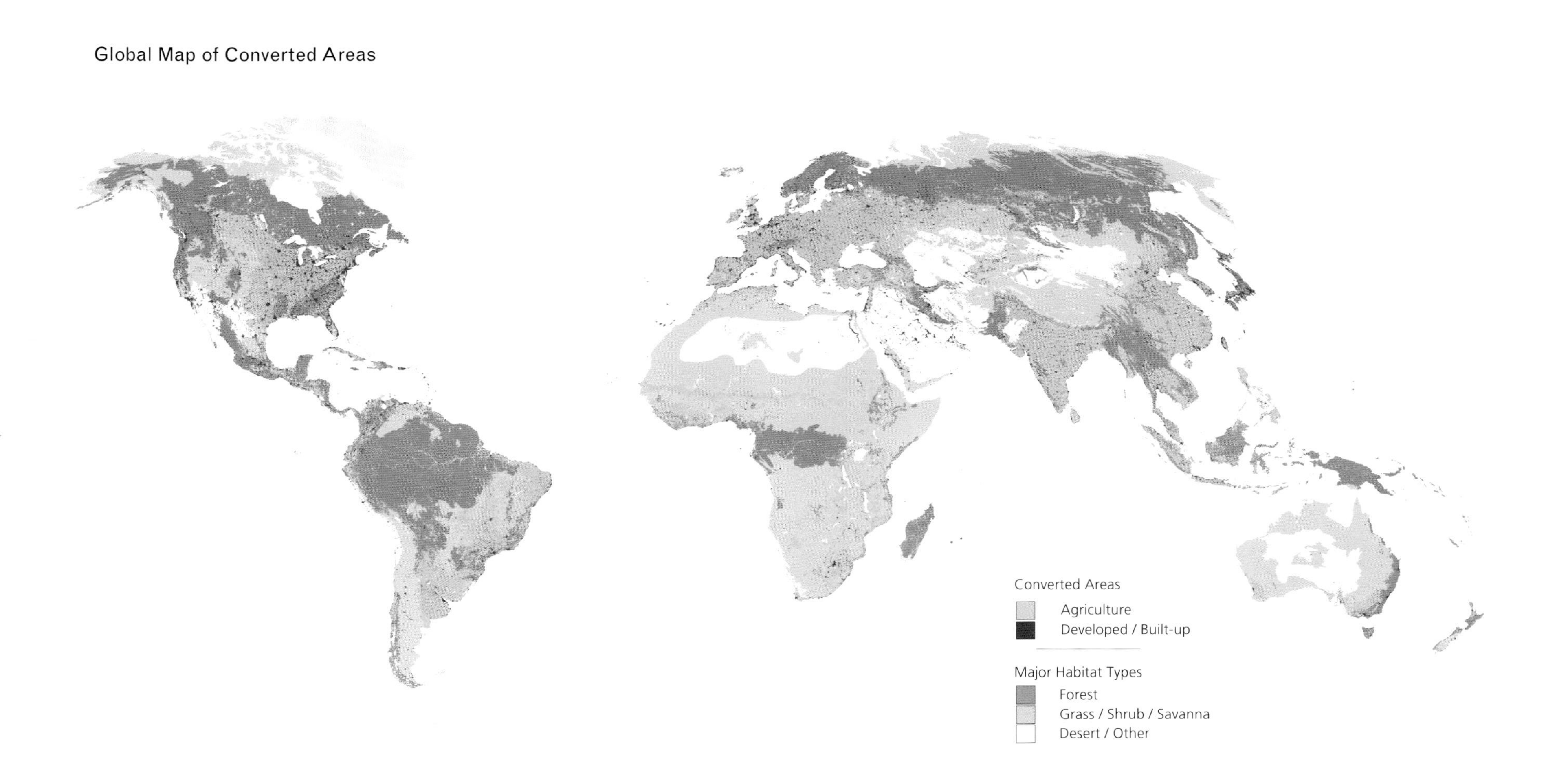

Source: Created for this publication by S. Murray [PAGE] based on data from Global Land Cover Characteristics Database Version 1.2 (Loveland et al. 2000); NOAA-NGDC (1998); WWF (1999).

Box 1.11 How Much Do We Consume?

Humans consume goods and services for many reasons: to nourish, clothe, and house ourselves, certainly. But we also consume as part of a social compact, since each community or social group has standards of dress, food, shelter, education, and entertainment that influence its patterns of consumption beyond physical survival (UNDP 1998:38–45).

Consumption is a tool for human development—one that opens opportunities for a healthy and satisfying life, with adequate nutrition, employment, mobility, and education. Poverty is marked by a lack of consumption, and thus a lack of these opportunities. At the other extreme, wealth can—and often does—lead to excessive levels of material and nonmaterial consumption.

In spite of its human benefits, consumption can lead to serious pressure on ecosystems. Consumption harms ecosystems directly through overharvesting of animals or plants, mining of soil nutrients, or other forms of biological depletion. Ecosystems suffer indirectly through pollution and wastes from agriculture, industry, and energy use, and also through fragmentation by roads and other infrastructure that are part of the production and transportation networks that feed consumers.

Consumption of the major commodities ecosystems produce directly—grains, meat, fish, and wood—increased substantially in the last 4 decades and will continue to do so as the global economy expands and world population grows. Plausible projections of consumer demand in the next few decades suggest a marked escalation of impacts on ecosystems (Matthews and Hammond 1999:5).

- Global wood consumption has increased 64 percent since 1961. More than half of the 3.4 billion m^3 of wood consumed annually is burned for fuel; the rest is used in construction and for paper and a variety of other wood products. Demand for lumber and pulp is expected to rise between 20 and 40 percent by 2010. Forest plantations produce 22 percent of all lumber, pulp, and other industrial wood; old-growth and secondary-growth forests provide the rest (Matthews and Hammond 1999:8, 31; Brown 1999:41).

- World cereal consumption has more than doubled in the last 30 years, and meat consumption has tripled since 1961 (Matthews and Hammond 1999:7). Some 34 percent of the world's grain crop is used to feed livestock raised for meat (USDA 2000). A crucial factor in the rise in grain production has been the more than fourfold increase in fertilizer use since 1961 (Matthews and Hammond 1999:14). By 2020, demand for cereals is expected to increase nearly 40 percent, and meat demand will surge nearly 60 percent (Pinstrup-Andersen et al. 1999:11).

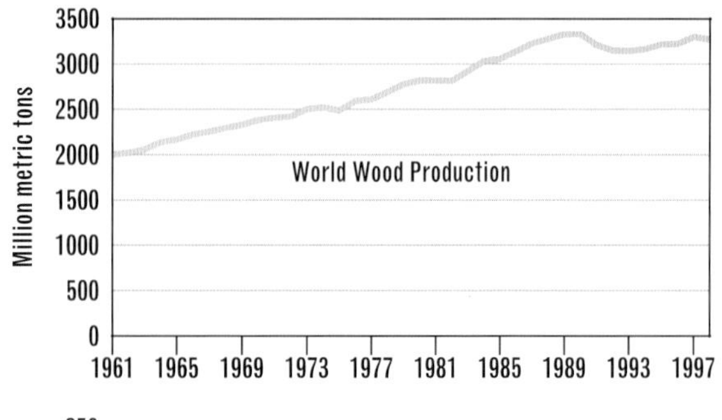

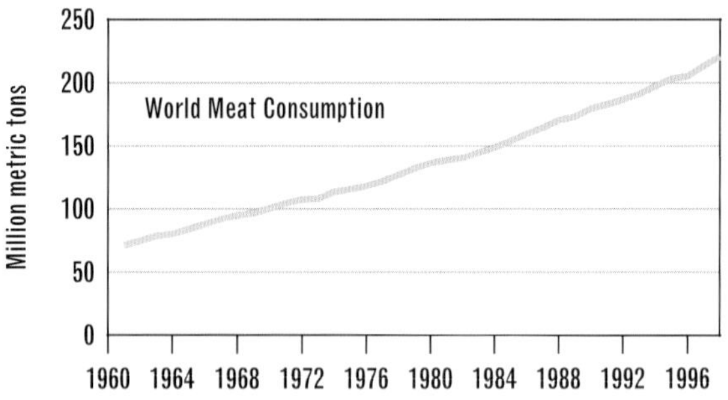

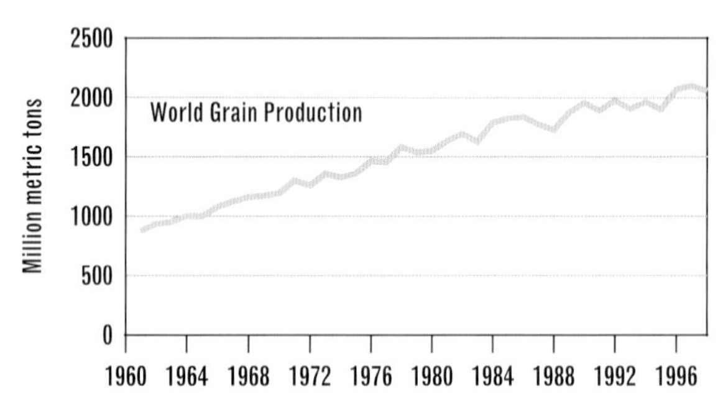

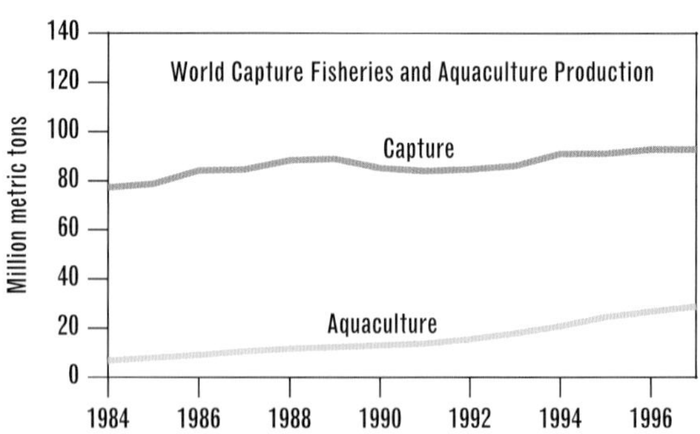

Sources: FAO 1999; FAO 2000.

- The global fish catch has grown more than sixfold since 1950 to 122 million metric tons in 1997. Three-fourths of the global catch is consumed directly by humans as fresh, frozen, dried, or canned fish and shellfish. The remaining 25 percent is reduced to fish meal and oil, which is used for both livestock feed and fish feed in aquaculture. Demand for fish for direct consumption is expected to grow some 20 percent by 2010 (FAO 1999:7, 82; Matthews and Hammond 1999:61).

The Unequal Geography of Consumption

While consumption has risen steadily worldwide, there remains a profound disparity between consumption levels in wealthy nations and those in middle- and low-income nations.

- On average, someone living in a developed nation consumes twice as much grain, twice as much fish, three times as much meat, nine times as much paper, and eleven times as much gasoline as someone living in a developing nation (Data Table ERC.3; Laureti 1999:50, 55).

- Consumers in high-income countries—about 16 percent of the world's population—accounted for 80 percent of the money spent on private consumption in 1997—$14.5 trillion of the $18 trillion total. By contrast, purchases by consumers in low-income nations—the poorest 35 percent of the world's population—represented less than 2 percent of all private consumption. The money spent on private consumption worldwide (all goods and services consumed by individuals except real estate) nearly tripled between 1980 and 1997 (World Bank 1999:44, 226).

Global Share of Private Consumption, 1997 (in billlions)

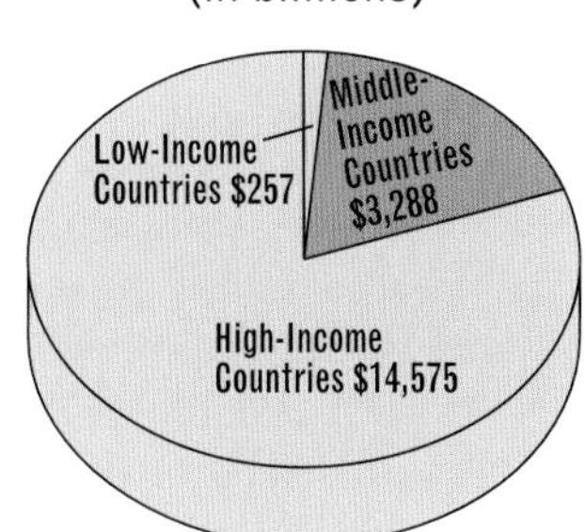

Disparities in Consumption: Annual per Capita Consumption in Selected High-, Medium-, and Low-Income Nations

Country	Total Value of Private Consumption* (1997)	Fish (kg) (1997)	Meat (kg) (1998)	Cereals (kg) (1997)	Paper (kg) (1998)	Fossil Fuels (kg of oil equivalent) (1997)	Passenger Cars (per 1,000 people) (1996)
United States	$21,680	21.0	122.0	975.0	293.0	6,902	489.0
Singapore	$16,340	34.0	77.0	159.0	168.0	7,825	120.0
Japan	$15,554	66.0	42.0	334.0	239.0	3,277	373.0
Germany	$15,229	13.0	87.0	496.0	205.0	3,625	500.0
Poland	$5,087	12.0	73.0	696.0	54.0	2,585	209.0
Trinidad/Tobago	$4,864	12.0	28.0	237.0	41.0	6,394	94.0
Turkey	$4,377	7.2	19.0	502.0	32.0	952	55.0
Indonesia	$1,808	18.0	9.0	311.0	17.0	450	12.2
China	$1,410	26.0	47.0	360.0	30.0	700	3.2
India	$1,166	4.7	4.3	234.0	3.7	268	4.4
Bangladesh	$780	11.0	3.4	250.0	1.3	67	0.5
Nigeria	$692	5.8	12.0	228.0	1.9	186	6.7
Zambia	$625	8.2	12.0	144.0	1.6	77	17.0

*Adjusted to reflect actual purchasing power, accounting for currency and cost of living differences (the "purchasing power parity" approach).

Sources: Total Private Consumption (except China and India): World Bank 1999: Table 4.11; (fish) Laureti 1999: 48–55; (meat) WRI et al. 2000a: Agriculture and Food Electronic Database; (paper) WRI et al. 2000b: Data Table ERC.5; (fossil fuels) WRI et al. 2000b: Data Table ERC.2; (passenger cars) WRI et al. 2000b: Data Table ERC.5.

Box 1.12 The Human Population

Population growth stresses ecosystems because it contributes to increases in both consumption and conversion. Each year, the human population grows by approximately 80 million. Although global fertility rates decreased since the 1950s from 5.0 to 2.7 births per woman (UN Population Division 1998b:514–515), the population will continue to grow. Past high fertility rates created today's pool of more than 1.5 billion people at the prime reproductive age—between 15 and 29 years old; another 1.9 billion are younger than 15 (UN Population Division 1998a). An adjunct to population growth is the significant decrease in mortality. Since the 1950s the global mortality rate has dropped from about 20 to fewer than 10 deaths per year per 1,000 people (UNFPA 1999). In contrast, the seven African countries hardest hit by the AIDS epidemic have actually experienced a decrease in life expectancy because of the high number of deaths caused by the disease (UN Population Division 1998a).

- Growth is fastest in less developed nations, among populations most dependent on ecosystems for a subsistence living. Demographers expect 97 percent of all population growth in the next 5 decades to occur in developing countries.

World Population Growth

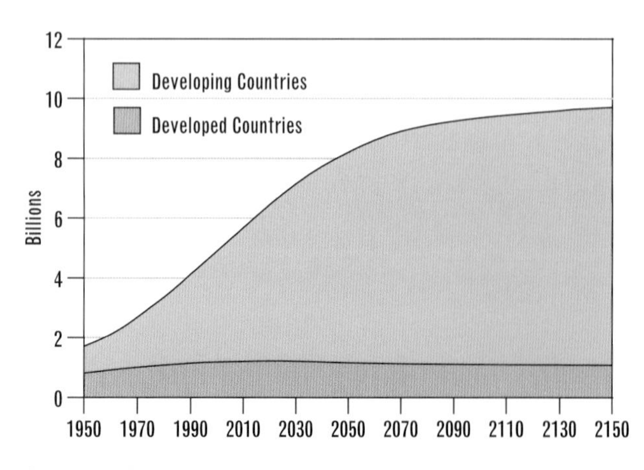

Source: UN Population Division (1998a).

- In both more and less developed nations, cities are drawing people into ever greater concentrations. Urban regions tend to offer more opportunities for economic development as well as better education and health resources. Although urban areas occupy only about 4 percent of the Earth's land area, they are home to nearly half the world's population (UNEP 1999:47; Wood et al. [PAGE] 2000). Currently cities are expansive consumers of ecosystem goods and services and prolific generators of ecosystem-damaging wastes—essentially concentrated centers of ecosystem pressures. By 2030, more than 60 percent of all people are likely to be living in urban areas. In industrial countries and Latin America, the share is expected to exceed 80 percent (UN Population Division 1998a).

Trends in Urbanization

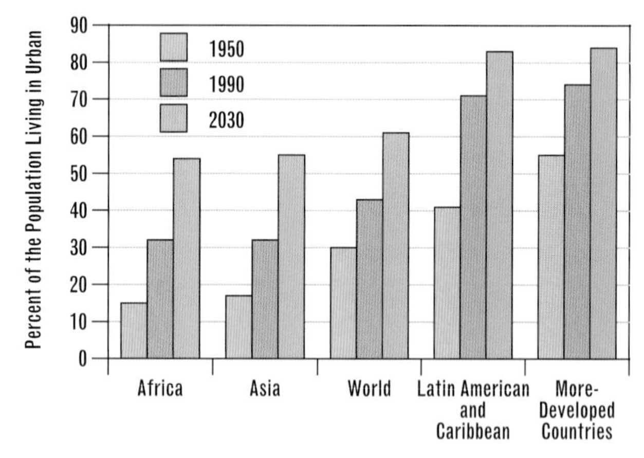

Source: UN Population Division 1998a.

- As the population grows in the next quarter century, pressures will increase, especially in countries where arable land is in short supply. In 14 countries, arable land per capita is expected to be less than 0.07 ha—equivalent to an area about 0.25 km^2—to sustain each human life (WHO 1997:59). Richer countries may supplement their food resources with imports, but poorer countries will have a more difficult time following such a strategy to feed their hungry populations.

Available Arable Land per Capita in 2025 for Selected Countries

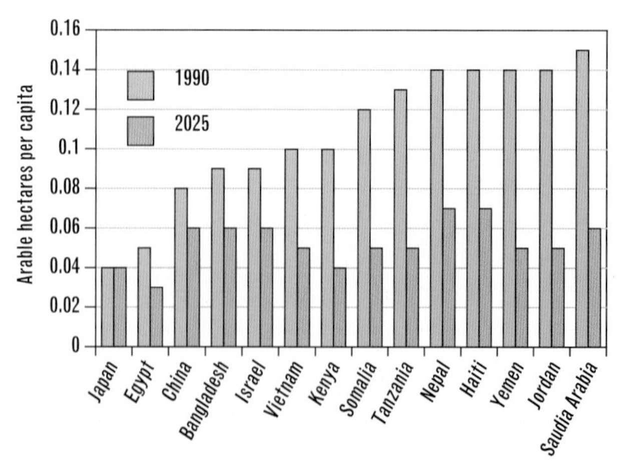

Source: WHO 1997:59.

Box 1.13 Pollution and Ecosystems

In the last century, a growing and rapidly industrializing world has produced greater quantities of common pollutants like household garbage and sewage, and more toxic and persistent contaminants like pesticides, polychlorinated-biphenyls (PCBs), dioxins, heavy metals, and radioactive wastes. The environmental costs of contemporary society's pollutant load are difficult to quantify, both because there is little comprehensive data on pollution emissions on a global scale and because the effects of pollutants on ecosystems are often hard to measure. But the problem is surely growing.

Pollutants affect ecosystems in a variety of ways. Pesticides and heavy metals may harm exposed orgnisms by being acutely toxic or by accumulating in plant and animal tissue through repeated exposures. Pollutants like acid rain can act at a system-wide level, disrupting soil acidity and water chemistry—both critical environmental factors that affect the nutrition and physical development of plants and aquatic life. Multiple pollutants can create a toxic synergy that weakens organisms and gradually reduces an ecosystem's productivity and resilience. All of these effects on ecosystems are much in evidence.

- Although there is greater awareness today of the dangers associated with toxic materials, toxic emissions continue to be significant. For example, the US$37 billion global pesticide market dispenses 2.6 billion kg of active ingredients (pesticides excluding solvents and dilutants) on the world's farms, forests, and household gardens, with a variety of collateral effects on wildlife and human health (Aspelin and Grube 1999:10).

- Accidental releases of toxic substances like mining wastes, or of oil or industrial chemicals, occur routinely and with devastating effect. In January 2000, 99,000 m^3 of cyanide-laden wastes escaped a Romanian gold mine when an earthen tailings dam collapsed; the toxic plume wiped out virtually all aquatic life along a 400-km stretch of the Danube and its tributaries (D'Esposito and Feiler 2000:1,4). In 1997, more than 167,000 tons of oil spilled from pipelines, storage vessels, tankers, and other carriers and sources to contaminate the world's marine and inland environments (Etkin 1998:5)

- Air pollution from sulfur dioxide (SO_2), nitrogen oxides (NO_x) and ground-level ozone still exceeds the "critical load"—the amount an ecosystem can absorb without damage—over wide areas of Europe, North America, and Asia, with documented effects on crops, forests, and freshwater ecosystems from acid rain. For example, the fraction of healthy Norway spruce, one of the most common conifers in European forests, decreased from 47 percent in 1989 to 39 percent in 1995—an indicator of the continued stress air pollution imposes on Europe's forest ecosystems (EEA 1999:144–145).

- Fertilizer runoff, human and animal sewage, and inadequately treated industrial wastes can add nutrients to freshwater and coastal ecosystems, stimulating algal blooms and depleting the water of oxygen—a process called eutrophication. Oxygen-depleted waters can't support aquatic life. Eutrophication is a growing problem worldwide. A roughly 18,000 km^2 "dead zone" of oxygen-depleted waters in the northern Gulf of Mexico stems from a tripling of the nutrient pollution carried to the coast by the Mississippi River over the last 40 years (Rabalais and Scavia 1999; NOAA 2000).

Total Waste Volumes Generated by Low-, Middle-, and High-Income Countries (per day)

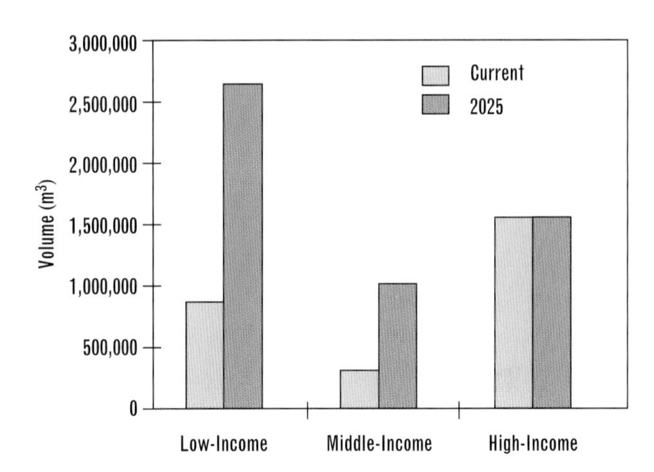

Source: Hoornweg and Thomas 1999:11.

Excess Nutrients Translate to Water Pollution

Country	Total Nitrogen Supply from Fertilizer and Manure (1,000 tons)	Nitrogen Uptake by Crops	Residual Nitrogen	Residual Equivalence per Hectare (kg)
Belgium and Luxembourg	580	211	369	240
Denmark	816	287	529	187
Netherlands	1255	285	970	480

Note: Because some nitrogen is lost to the atmosphere, only a part of the residual nitrogen stays in the soil for possible nitrate leaching.
Source: Matthews and Hammond 1999.

DISTORTED PRICES, UNDERVALUED SERVICES

People don't generally consciously decide to damage ecosystems, but many of the things we do have that effect. Given that ecosystems provide so many benefits, why do people do things that jeopardize these benefits?

Economic signals–reflected in prices and government policies–are one of the prime factors determining how we treat ecosystems. They are behind our choices of what to consume and how to manage our lands and our businesses. A farmer deciding what crops to plant and what farm chemicals to use, or whether to increase the cultivated area by clearing adjacent forests, is guided by calculating commodity and pesticide prices as well as many other farm costs. Similarly, a developer's choice of where to locate a tract of housing or a factory, or a fisher's decision on what type of fishing gear to use and how many days to spend at sea are driven largely by economic factors–the price of land or boat, of labor or fishing licenses, of the finished house or the harvested fish.

But prices all too frequently send us the wrong signals. In most cases, they don't reflect the real costs to the environment of harvesting ecosystem goods and services. The problem is, many of the less tangible aspects of ecosystems, particularly the services they provide, are not bought or sold in the marketplace and are therefore harder to assign a value. How much is carbon storage in a forest worth? What price tag can be put on flood protection provided by the wetlands along a river?

The connection between these services and the more tangible marketable goods–timber or fish or crops–is not always obvious to those exploiting these goods and services. The value of biodiversity to the future of food crops is, for example, of little immediate import to an individual farmer trying to maximize his or her profit. The result is that most ecosystem services have been undervalued in the past and neglected in decisions about whether to exploit or alter an ecosystem. The market has failed to register the real worth of these services in its price system–a "market failure."

Consider the case of deciding whether to clear native forests for a new agricultural settlement. The potential farmers will take into account the cost of the labor needed to clear land, the fertilizers used to increase yields, and the construction materials required to build houses or roads. They may even factor in some reductions in ecosystem services. For example, they may consider the cost of forgoing the benefits of using the forest as a source of fuelwood and the loss of wild animals and plants.

It is, nonetheless, very likely that they won't take into full account the many environmental costs of forest clearing. Cutting down forests might increase downstream flooding and sedimentation, for example, but since these costs are borne by people living far downstream, they will often be ignored by the upstream farmers. The result is that more forest is cleared than would make sense from an overall economic standpoint, and the forest ecosystem suffers needless damage, as may the downstream populations. Extending this argument to the global level, a better accounting of all the costs and benefits of forest conversion would not necessarily mean that all forest is preserved, but it would certainly result in a lower rate of deforestation than is occurring now.

SUBSIDIES AND OTHER POLICY FAILURES

Government policies often contribute to ecosystem decline through their effect on prices. Fiscal policies affect prices through taxes and subsidies. Tariffs increase the price of imported goods directly and import quotas increase them indirectly. Exchange-rate policies affect the value of all tradable commodities. Government agencies also actively buy and sell farm commodities, often at predetermined prices. All of these actions can influence the decisions of farmers, fishers, developers, timber and mining companies, and others who use the land and sea, harvest from it, or impact it through pollution.

Subsidies. Government subsidies contribute importantly to current pressures on ecosystems, often encouraging damaging activities–such as overfishing or the liberal use of coal or other fossil fuels–that would not otherwise be economically viable. Generous loans to build fishing boats, agricultural price supports, depletion allowances for timber and oil producers, and outright grants for road construction are just a few of the ways that governments subsidize activities that can damage ecosystems. One recent analysis reported that government expenditures on environmentally damaging subsidies in just four sectors–water, agriculture, energy, and road transportation–totaled some $700 billion per year worldwide (de Moor and Calamai 1997:1).

Subsidies often promote laudable social goals–employment, higher productivity, economic development–when first instituted, but these goals are often subverted over time through unintended consequences such as environmental impacts. For example, governments have subsidized the use of various farm inputs, such as pesticides and fertilizers, partly to boost agricultural production and partly to support the industries producing these chemicals. Pesticide subsidies, in particular, have been common in developing countries. In the mid-1980s, Indonesia was spending about $150 million annually on pesticide subsidies, mostly to protect the rice crop. This led to considerable overuse. Rather than reducing crop-damaging insects, however, this liberal pesticide use actually triggered periodic outbreaks by reducing natural predators and prompting pesticide resistance among target insects. It also caused substantial downstream pollution and adversely affected the health of farmers. When the government ended its subsidies, pesticide use dropped, the government saved money, and rice production continued to increase (World Bank 1997:26).

Subsidizing irrigation projects is another common practice that has seriously harmed aquatic ecosystems. Throughout the world, government support has typically allowed

Economic signals—reflected in prices and government policies—are one of the prime factors determining how we treat ecosystems. Subsidies often encourage damaging activities that would not otherwise be economically viable.

water utilities to sell irrigation water for far less than the cost of supplying it, which has inevitably led to overuse. In arid Tunisia, for example, farmers pay no more than one-seventh the cost of water they use to irrigate their fields. Similar practices of underpricing irrigation water in the western United States cost U.S. taxpayers an estimated US$2–$2.5 billion per year (de Moor and Calamai 1997:14–15). With water costs low, farmers have little incentive to use water efficiently or to restrict its use to high-value crops. Direct water diversions and overpumping from irrigation wells often rob streams of much of their normal flow. Too often pesticide and fertilizer runoff pollutes what flow remains.

Regulations. Beyond their effect on prices, government policies can also impact ecosystems more directly, through such mechanisms as zoning ordinances, pollution standards, or other regulations that affect land use and business practices. Programs to promote economic development may foster "grow now, clean up later" policies that encourage industrialization no matter what the environmental costs. China's dramatic industrialization after economic reforms in 1978 followed this pattern, and by the early 1990s, the nation was estimating that economic costs associated with ecological destruction and pollution had reached as high as 14 percent of its gross national product (WRI et al. 1998:115-116). Hoping to reverse its environmental losses and reduce the health impacts of polluted air and water, China has recently begun a costly effort to tighten and enforce its environmental regulations.

(continues on p. 33)

Box 1.14 Valuing the Invaluable

The economic values we assign to our work and the fruits of our labor are important factors in our behavior and the decisions we make about our assets. Similarly, the values we assign to *ecosystem* assets—goods and services like pollination, water purification, nitrogen fixation, and carbon storage—are an important factor in how we treat ecosystems. Yet because these services are not routinely bought and sold in markets, there's no easy way to calculate their worth. Too often, decision makers and traditional economists simply ignore their value, essentially treating ecosystem goods and services as though they will always be in profuse supply. A result is that loggers may harvest a patch of forest for the value of its timber alone, ignoring the value the forest provides in terms of flood control, water purification, or habitat for migratory songbirds.

How does one assign a monetary value to all the ecological amenities of an ecosystem? As the state of the art of economic analysis has improved, economists have identified a variety of tools to quantify direct—and even some indirect and intangible—ecosystem services.

Where possible, actual market values are used. For example, the price of fish and shellfish harvested in an estuary provides one value for direct goods provided by that ecosystem. Another way to estimate value is to calculate the cost of replacing an ecosystem service. For New York City, natural habitats in its upstate watershed were shown to provide the same water purification services as a new water filtration plant. The $3–$8 billion price tag (Ryan 1998) for the proposed filtration plant is a good base estimate of the value of the water purification service that the intact ecosystem provides—although it does not capture the value of the other watershed services including carbon sequestration, recreational opportunities, and support for biodiversity.

Similarly, the price difference between two comparable houses, one near a shoreline and one inland, is thought to capture the aesthetic value of the shore. Still another market-based method of calculating a lake's or a park's or a wilderness area's value, both as a scenic and a recreational site, is to calculate how much money and time visitors spend to travel there.

When market data are not available, or to supplement them, researchers resort to other means. They ask people what they'd pay, for example, to keep a wetland from being filled and developed or to prevent a wilderness area from being mined. Properly done, such "contingent valuation" surveys can go beyond measuring the practical benefits humans extract from nature to encompass the ethical and spiritual values they attach. But surveys can be unreliable and subject to bias, especially when people are queried about paying to minimize the effects of something as complex as climate change.

Valuation exercises can be a useful policy tool in educating audiences about the many ways we depend on and profit from ecosystem services. Ultimately, however, creating financial incentives for ecosystem conservation is more important than finding an accurate market value for any or all ecosystem services. Incentives for conservation may come from creating markets for ecosystem services where none exist, or finding other ways for landowners to gain financially from the services their land provides. Auctioning permits to emit carbon or compensating countries or companies that reforest land to sequester carbon are examples of ways to create such markets.

Ecotourism, where the beauty and unspoiled quality of an ecosystem is marketed directly, may be another incentive to conserve. In South Africa, a private enterprise called Conservation Corporation negotiated with farmers to return 168 km^2 of their land to its original habitat and stock it with big-game animals. Open for business as a safari destination, the land is now yielding $200–$300 per hectare annually from visitor fees instead of $21–$68 from ranching or farming, and providing a biologically diverse resource base to support the large game (Anderson 1996:207; Honey 1999:374). In the Maldives, a government study determined that a single live shark yields approximately US$33,500 annually in tourist revenue, compared to US$32 when caught and sold by a fisherman. This and other studies supplied the incentive for the Maldives to make sharks, turtles, and dolphins protected species (Sweeting et al. 1999:66, citing WTO 1997).

In some ways, "priceless" may be the most accurate value that we can ever place on intangible ecosystem goods and services such as a coastal area's beauty or a mountain range's spiritual importance. But used as one of many measures of an ecosystem's worth, and with recognition of its limitations, environmental economics offers a powerful ecosystem management tool in a political world. Until we fully understand ecosystem values, we are handicapped in deciding what to use and what to save.

Sectoral Divisions. Other government-related factors also affect the use of ecosystems. Government institutions, for example, are routinely divided along sectoral lines–the Ministry of Agriculture, the Forest Department, the Environment Agency, and so on. This works against adopting any integrated view of ecosystems or their management. The Ministry of Agriculture's prime concern, for instance, will be farm production. Like an individual farmer, the Ministry will likely see preserving biodiversity or minimizing forest conversion as peripheral to its mission. It may even see the Forest or Environment Departments as competitors for budget and administrative control, reducing the chances of cooperation between agencies that manage ecosystems. This limited focus makes it unlikely that agencies as now configured will recognize or account for the environmental trade-offs that their policies promote.

Corruption. Government corruption is another common institutional failure that allows unchecked exploitation of ecosystems–often by a small elite. Even when laws and management policies are sound, they may be undermined by government officials who turn a blind eye to illegal harvesting or themselves take part in the plunder through sweetheart deals or insider investments. The scale of corruption in the forest sector, for example, is staggering. In Indonesia, illegal logging accounts for more than half of the nation's timber production, with timber smuggling taking place in some national parks in full view of park authorities (EIA and Telepak 1999:4). As a result, the government loses an estimated US$1-$3 billion per year in timber royalties, and the forests suffer from haphazard cutting (WCFSD 1999:36). Similarly, the Russian government collected only a fraction–estimated at 3-20 percent–of the timber revenues it was due in 1994 (WCFSD 1999:36). The rest was lost to theft and fraud.

Who Owns Ecosystems?

Ownership is a crucial factor in how we manage ecosystems. The question of who owns the land or has the right to use its resources is key in determining what services or products are reaped from an ecosystem, how they are harvested, and who gains the benefits. Some patterns of ownership can work against good management of ecosystems, as when property rights are concentrated in the hands of those whose economic interests may favor unsustainable harvest levels or extensive development.

PROPERTY RIGHTS

In 1985, Maxxam Corporation acquired the locally based Pacific Lumber Company in Northern California, owner of the state's largest remaining tract of mature redwood forest. For years, Pacific Lumber had managed its forests to maintain their long-term productivity, emphasizing moderate harvest levels that could continue to feed its lumber mills indefinitely. Maxxam quickly abandoned Pacific Lumber's modest but sustainable harvest practices, more than doubling the harvest rate to help pay off its large corporate debt. Maxxam stockholders reaped the benefits of this short-term approach, with little regard for its long-term effects on the local economy or the health and productivity of the forest (Harris 1996:130-135, 170-171; LOE 1996:12-18).

Lack of ownership can also be a problem. Many of the world's poor lack legal property rights–tenure–over the lands they live on. A poor farmer without secure land tenure may not feel much incentive to consider long-term productivity because he or she has no assurance of being able to stay and capitalize on any investments in good soil or water management. In fact, lack of legal title tends to discourage some land uses, like agroforestry, that are relatively benign to ecosystems but require long periods to reach peak productivity (Scherr 1999). In addition, landless immigrants, often fleeing unemployment and poverty or civil strife in more populated regions, have been important contributors to deforestation in frontier areas as they clear forest plots for subsistence farming. In some instances, clearing forest areas is actually a means to gain land title, since it converts the land to agriculture–a legally recognized land use.

Sometimes, modern systems of private or state ownership can conflict with more traditional forms of group or community ownership, with the environment suffering as a consequence. Cultures around the world have developed systems of communal management of shared resources to control overharvesting. Forests in Indonesia, rangelands in Mongolia, and coastal fishing areas in the Philippines are all current examples. An extensive literature documents that these traditional systems of property rights and communal management can be very effective at preserving ecosystems over the long term even as they are routinely harvested. Nonetheless, governments often ignore these traditional forms of ownership, denying them legal recognition.

POVERTY

The question of who owns ecosystems and their benefits ultimately becomes a question of equity. Those with property rights or with the money to buy consumer items are most likely to control the goods and services that ecosystems produce and to influence how ecosystems are managed. Yet it is the poor who are most directly dependent on ecosystems for their immediate survival and therefore most vulnerable when ecosystems decline. Subsistence farmers and others who cannot afford fertilizers depend on natural soil fertility; and subsistence fishers depend on the continued productivity of lakes, rivers, estuaries, and coastal wetlands. When these systems are depleted, impoverished people can't insulate themselves from the effects as the wealthy can. They must bear the costs of lost ecosystem services directly.

(continues on p. 40)

Box 1.15 Ecotourism and Conservation: Are They Compatible?

From African wildlife safaris, to diving tours in the Caribbean's emerald waters and coral reefs, to guided treks in Brazil's rainforests, nature-based tourism is booming. The value of international tourism exceeds US$444 billion (World Bank 1999:368); nature-based tourism may comprise 40–60 percent of these expenditures and is increasing at 10–30 percent annually (Ecotourism Society 1998).

This burgeoning interest in traveling to wild or untrammeled places may be good news, especially for developing countries. It offers a way to finance preservation of unique ecosystems with tourist and private-sector dollars and to provide economic opportunities for communities living near parks and protected areas. For Costa Rica, tourism generated $654 million in 1996, and for Kenya $502 million in 1997, much of it from nature and wildlife tourism (Honey 1999:133, 296). Tourism has been influential in helping to protect Rwanda's mountain gorillas and their habitat in Volcanoes National Park. Prior to the outbreak of civil war, tourist visits provided $1.02 million in direct annual revenues, enabling the government to create antipoaching patrols and employ local residents (Gossling 1999:310).

But the reality of nature-based travel is that it can both sustain ecosystems and degrade them. Much nature-based tourism falls short of the social responsibility ideals of "ecotourism," defined by the Ecotourism Society as "travel to natural areas that conserves the environment and sustains the well-being of local people" (Ecotourism Society 1998). Destinations and trips marketed as ecotourism opportunities may focus more on environmentally friendly lodge design than local community development, conservation, or tourist education. Even some ecosystems that are managed carefully with ecotourism principles are showing signs of degradation.

Ecotourism's Costs and Benefits

At first glance, Ecuador's Galápagos Islands epitomize the promise of ecotourism. Each year the archipelago draws more than 62,000 people who pay to dive, tour, and cruise amidst the 120 volcanic islands and the ecosystem's rare tropical birds, iguanas, penguins, and tortoises. Tourism raises as much as $60 million annually, and provides income for an estimated 80 percent of the islands' residents. The tenfold increase in visitors since 1970 has expanded the resources for Ecuador's park service. Tour operators, naturalist guides, park officials, and scientists have worked together to create a model for low-impact, high-quality ecotourism (Honey 1999:101, 104, 107).

But closer examination reveals trade-offs: a flood of migrants seeking jobs in the islands' new tourist economy nearly tripled the area's permanent population over a 15-year period, turned the towns into sources of pollution, and added pressure to fishery resources (Honey 1999:115, 117). Only 15 percent of tourist income directly enters the Galápagos economy; most of the profits go to foreign-owned airlines and luxury tour boats or floating hotels—accommodations that may lessen tourists' environmental impacts, but provide little benefit to local residents (Honey 1999:108, citing Epler 1997). The hordes of tourists and immigrants have brought new animals and insect species that threaten the island's biodiversity (Honey 1999:54).

The Galápagos Islands well illustrate the complexities of ecotourism, including the potential to realize financial benefits nationally, even as problems become evident at the local or park scale. For example, to a government that is promoting ecotourism, more visitors means more income. But more visitors can translate into damage to fragile areas. Park officials often complain of habitat fragmentation, air pollution from vehicle traffic, stressed water supplies, litter, and other problems. In Kenya's Maasai Mara National reserve, illegal but virtually unregulated off-road driving by tour operators has scarred the landscape (Wells 1997:40).

These impacts can be minimized with investments in park management, protection, and planning. However, developing countries often lack the resources to monitor, evaluate, and prevent visitor impacts, and infrastructure and facilities may be rudimentary or nonexistent.

Low entrance fees are part of the problem; they often amount to just 0.01-1 percent of the total costs of a visitor's trip (Gossling 1999:309). Setting an appropriate park entry fee—one that covers the park's capital costs and operating costs, and ideally even the indirect costs of ecological damage—is one way that management agencies can capture a larger share of the economic value of tourism in parks and protected areas. Most parks have found that visitors are willing to pay more if they know their money will be used to enhance their experience or conserve the special area. To ensure broad affordable access to parks, Peru, Ecuador, Kenya, Jordan, Costa Rica, and several other countries have raised fees for foreigners while maintaining lower fees for residents.

Unfortunately, tourism revenues are not always reinvested in conservation. Of the US$3 million that Galápagos National Park generates each year, for example, only about 20 percent goes to the national park system. The rest goes to general government revenues (Sweeting et al. 1999:65). This is typical treatment of park income in many countries, but it undermines visitors' support for the fees and destroys the incentive for managers to develop parks as viable ecotourism destinations. Fortunately, some countries are using special fees and tourism-based trust funds to explicitly channel tourist dollars to conservation. Belize, for example, raises funds for conservation through a US$3.75 tourist tax levied on every foreign visitor as they depart the country, generating about US$750,000 per year (Sweeting et al. 1999:69).

Well-planned and -managed ecotourism offers greater potential to bolster local and rural economic development than traditional tourism, in which most of the economic benefits linked to tourist expenditures "leak" back to commercial tour operators in the richer countries (where most tourists originate) or are captured by large cities of the host countries (Wells 1997:iv). But increasing prices for land, food, and other products can coincide with the growing popularity of a tourist or ecotourist haven, to the detriment of local residents. In Zanzibar, villagers and townspeople have been enticed into selling their property to tourism investors who do not guarantee any profit sharing, joint ownership, or other form of sustained benefit (Honey 1999:287). In Tonga, tourism-driven inflation has caused shortages of arable land (Sweeting et al. 1999:29).

Some countries have introduced policies that help reimburse local residents for the direct and indirect costs of establishing a protected area. Kenya, for example, aims to share 25 percent of revenue from entrance fees with communities bordering protected areas (Lindberg and Huber 1993:106). Ecotourism planners also advocate sales of local handicrafts in gift stores, patronage of local lodges, use of locally grown food in restaurants and lodges, and training programs to enable residents to fill positions as tour guides, hotel managers, and park rangers. Both tour operators and visitors have a role to play by screening trips carefully and committing to ecotourist principles. Developers can choose sites based on environmental conditions and local support, and use sustainable design principles in building and resort construction.

Poorly planned, unregulated ecotourism can bring marginal financial benefits and major social and environmental costs. But with well-established guidelines, involvement of local communities, and a long-term vision for ecosystem protection rather than short-term profit by developers, ecotourism may yet live up to its promise.

Box 1.16 Uprooting Communal Tenure in Indonesian Forests

Many communities on the outer islands of Indonesia, and elsewhere in the developing world, use traditional systems of community-based, group tenure rights to manage forest resources. Many of these management systems are generations old and meet local economic needs while maintaining vital ecosystem functions, including protection of biodiversity (Lynch and Alcorn 1994:374, 381). Unfortunately, most of these systems are threatened by legal and development pressures.

In Indonesia, traditional community-based property rights are called *adat* rights. Across the Indonesian archipelago, communities adapt adat rights to their specific economic and environmental needs. Agroforests in Sumatra and Kalimantan, for example, are managed for rubber, durian fruits, illepe nuts, resins, and rattan.

Between 12 and 60 million people depend on Indonesia's forests, with a substantial proportion practicing traditional agroforestry (Poffenberger et al. 1997:22). Detailed information is lacking, but research suggests much of this land is managed under adat rights.

Threats to Group Tenure

Adat rights in Indonesia face four significant threats:

- Adat rights are not meaningfully recognized by the state, despite their widespread importance. The Indonesian Ministry of Forestry manages and claims exclusive ownership of 131 Mha of forest land—68 percent of Indonesia's land area, including 90 percent of the Outer Islands. Even though government planners admit knowledge of adat tracts is important in formulating sustainable resource management plans, the government does not know how much of this land is also claimed under traditional group tenure regimes (Fox and Atok 1997:32; Peluso 1995:390–391).

- State-sponsored development activities constantly override adat rights. Where 20-year timber concessions have been granted, forest-based communities find their traditional rights of use and access usurped (Lynch and Talbott 1995:52–54). Government-directed development plans—including mining, transmigration settlements, and conver-

sion of forests to timber or oil palm plantations—degrade or destroy these ecosystems (Michon and de Foresta 1995:103–104). In East Kalimantan province, 30 percent of Long Uli village land was lost to a government nature reserve, and 20 percent (including half of the village's cultivated land) was included in a timber concession, all without the consent of or consultation with the villagers (Sirait et al. 1994:416). Over the protests of villagers in eastern Maluku province, local government officials signed agreements with timber companies granting them access to the village's resin-producing agroforests, which were then destroyed without adequate compensation, thus undermining environmental sustainability and local economic stability (Zerner 1992:31–33).

- The imminent nature of state-sponsored development projects provokes communities to overexploit their resource base. Faced with irretrievably losing control of their lands and resources, some forest-dependent communities will incautiously reap maximum harvests and, in the process, destroy the resource base (Lynch and Talbott 1995:98; Sirait et al. 1994:416).

- Government policies that disproportionately reward agricultural production can also promote forest degradation. More favorable prices for agricultural commodities, relative to nontimber forest products, encourage farmers to pursue less sustainable forms of agriculture than those used by traditional agroforestry systems (Padoch and Pinedo-Vasquez 1996:113).

New Approaches

Many conflicts would be mitigated if adat rights were legally recognized and granted political legitimacy. In 1998, before the fall of the Suharto government, the Indonesian Ministry of Forestry issued a decree that created a new land-use category, the *kawasan dengan tujuan istemewa*, or "area of special/extraordinary objective," for 60 resin-producing agroforest villages in the vicinity of Krui, Sumatra. The decree established a process for granting official use and management rights to local villages covering 29,000 ha of forest. The regulation was the first ever to grant legally recognized management rights to community agroforesters.

Other important political and legal changes include President Habibie's emphasizing the importance of civil society and governmental accountability. The Basic Forest Law of 1999 acknowledges that local people have a key role in sustainable forest management; however, it fails to recognize adat rights. Within the Forestry Ministry, a new regulation currently being considered would authorize the demarcation of indigenous territories within areas designated as state forestland. The Ministry of Agrarian Affairs, in a related vein, has issued a decree providing for delineation and registration of community-based adat rights in some forested areas (Lynch 2000).

Wider legal recognition of traditional community rights of access to and management of forests in Indonesia could follow these important developments (Campbell 1998). Still needed, however, are clearer policies on adat rights that also define local and state rights and responsibilities (Bromley and Cernea 1989:52; Lynch and Alcorn 1994:376–377).

Current progress toward wider legal recognition of local tenure by the Indonesian government, however, is fragile in light of the country's recent economic and political turmoil. Similar efforts to promote legal recognition of group tenure in Thailand and the Philippines are also at precarious stages.

At current population growth rates, tensions between development and sustainability are sure to continue. An additional 15–33 Mha of forest in Indonesia is expected to suffer deforestation by 2020 (Lynch 2000). Plans are already under way to create more pulp, paper, and oil palm plantations, all of which replace natural forests (Barber 1997:74).

Logged-over areas of natural forest currently provide forest-dependent communities space for agriculture, grazing, and collection of forest products such as timber, rattan, and rubber. Converting these areas to intensively managed pulp and oil palm plantations will permanently exclude local populations; their claims to resources, which had tenuous legitimacy before, will be made irrelevant (Barber 1997:75). Securing the community-based property rights of Indonesia's forest-dependent communities would help to both protect the interests of Indonesia's rural inhabitants and promote environmental sustainability.

Box 1.17 Rural Poverty and Adaptation

Near a rural Bengali village, peasant families searching for firewood pick a local forest patch clean. A refugee from war-torn Rwanda flees to Tanzania where he poaches game in a national park to feed his family. A poor Kenyan family continues to cultivate their small farm plot in spite of severe erosion and exhausted soil. These are the typical images of the rural poor—people hugely dependent on ecosystems, unable to afford sound management practices, and caught in a vicious cycle of overusing already fragile and degraded resources.

A more nuanced view has emerged, however, that recognizes that the poor may have limited resources and great dependence on the environment, but they also have considerable ability to protect their ecosystems, when given the opportunity. Research is bringing to light abundant examples of *adaptation*—strategies that the poor use to lessen the impacts of environmental, economic, or social change on their resources. Adaptive measures include innovative land-use practices, the adoption of new technologies, economic diversification, and changes in social organization (Batterbury and Forsyth 1999:8).

Who Are the Poor?

Approximately 1.3 billion people, one-quarter of the world's population, live on about $1 a day (World Bank 1999:117). In addition to encompassing insufficient financial assets, poverty often means a lack of education, mobility, employment opportunities, or access to basic services such as safe water, and physical isolation in remote villages. Limited access to land is another key aspect of poverty; 52 percent of the rural poor have landholdings too small to provide an adequate income, and 24 percent are landless (UNCHS 1996:109).

The vulnerability of the poor is often exacerbated by a lack of political power to defend their rights to environmental resources or defend themselves against outright oppression. In South and Southeast Asian countries, for example, many governments consider forest-dependent people to be squatters who are illegally using state-owned resources. They can be arbitrarily displaced, often with state sanction, no matter how long they have occupied the forest (Lynch and Talbott 1995:21). War and civil conflict in Central and Eastern Europe, Somalia, the Congo, Lebanon, and other countries have torn people from their land and plunged them into poverty.

Urban poverty is a growing phenomenon, but the largest numbers of poor people in developing countries still live in rural areas—as much as 80 percent in 1988 (Jazairy et al. 1992:1). Many struggle to subsist on lands variously described as "poverty traps," "less favored," or "marginal." These tend to be areas of high ecological vulnerability (such as subtropical drylands or steep mountain slopes) or low levels of biological or resource productivity combined with high human demands. There may be almost twice as many poor living on marginal lands as on favored lands in developing countries—630 million compared to 325 million (CGIAR et al. 1997). If current trends in poverty and natural resource degradation persist, by 2020 more than 800 million people could be living on less favored lands, places like the upper watersheds of the Andes and the Himalayas, the East African highlands, and the Sahel (Hazell and Garrett 1996).

Protecting Their Ecosystems

It is increasingly evident that the poor can fight back against environmental degradation. In some places, they have been fighting back for centuries, using adaptive measures whenever ecosystem changes have demanded them.

One example of adaptation can be found in the highlands of Papua New Guinea, where the Wola people grow crops on slopes cleared of native forests by means of slash and burn techniques. Instead of accelerating soil exhaustion and furthering deforestation, as traditional models would predict, the Wola have maintained soil fertility by constructing mounds of soil using rotting vegetation as compost. They select strategically what crops to plant, using a variety of crops in the first years of cultivation when soils are rich. In later years when soil fertility declines, the Wola plant only sweet potatoes, a crop that can thrive without many nutrients (Batterbury and Forsyth 1999:8, citing Sillitoe 1998 and Sillitoe 1996).

The Mossi people in Burkina Faso offer other examples of successful adaptation. As rapid population growth and frequent droughts have degraded their soils, Mossi farmers have responded by creating compost pits and building *diguettes*—semipermeable lines of stone placed at right angles to the slope to prevent erosion (Batterbury and Forsyth 1999:9–10). The significant number of Mossi who have migrated to cities or the neighboring country of Cote d'Ivoire for wage employment during the dry season is also an adaptive response that reduces pressures on the land and food supply, provides remittances for families, and diversifies income sources. Like all adaptations, however, these local strategies have their limitations. Severe drought or a shortage of nonfarm job opportunities can undermine the Mossi's successes.

A third adaptation example comes from the forest-savanna zone of Guinea in West Africa. For 200 years, researchers erroneously blamed the Kissi and Kuranko people for the deforestation of a large forest in the Kissidougou province. Research into historical land-cover patterns eventually revealed that the Kissi and Kuranko had actually *created* patches of forest on relatively treeless savannas through targeted burning to reduce the risk of fire and to increase soil fertility, and by tethering animals and promoting fast-growing tree species (Batterbury and Forsyth 1999:10–11, citing Fairhead and Leach 1996).

Examples of Indigenous Soil and Water Conservation Techniques in Selected West African Countries

Country	Rainfall (mm)	Population Density (per km²)	Indigenous Soil and Water Conservation Techniques
Burkina Faso	1,000–1,100	35	Stone bunds in slopes network of earth bunds and drainage channels in lowlands
	1,000	35–80	Contour stone bunds on slopes, drainage channels
	400–700	29	Stone lines, stone terraces, planting pits
Cameroon	800–1,100	80–250	Bench terraces (0.5–3 m high), stone bunds
Cape Verde	400–1,200 (uplands)	>100	Dry stone terraces (walls 1–2 m high), rectangular basins (approx. 2 m x 4 m)
Chad	250–650	5–6	Water harvesting in drier regions: various earth bunding systems with upslope wingwalls and catchment area
Niger	300–500		Stone lines, planting pits
Nigeria	1,000–1,500	110–450	Stepped, level benched stone terraces, rectangular ridges, mound cultivation
Mali	400	20–30	Pitting systems
	500–650	13–85	Cone shaped mounds, planting holes, terraces square basins, stone lines, bunds or low walls
Sierra Leone	2,000–2,500	38	Sticks and stone bunding on fields and drainage techniques in gullies
Togo	1,400	80	Bench terraces and contour bunds, (rectangular) mound cultivation

Source: IFAD 2000.

Adaptation is not confined to rural areas. In cities the poor supplement their diets and income by transforming vacant lots, rooftops, and the lands along roadsides and other rights-of-way into highly productive plots of vegetables, fruits, and trees. As food and fuel are the largest household expenses for low-income urban populations, urban agriculture can be a first line of defense against hunger and malnutrition. Shantytown dwellers who mobilize to secure access to water and sanitation and improve their environments are engaging in another form of adaptation. But adaptation can be more difficult in cities, where a community's response may be more dependent on access to and support from local and state governments, corporations, or international agencies. In addition, many environmental risks are relatively new or beyond the experience of the urban poor, or difficult to detect, such as solvent or lead poisoning (Forsyth and Leach 1998:26).

How a community adapts to ecosystem decline depends on the knowledge that individuals have and the local biophysical environment, such as rainfall and soil conditions. Economic and political factors such as the availability of labor and access to markets also are crucial.

Governments, NGOs, and development agencies can help the poor respond positively to natural resource management challenges by working with local residents—supporting locally designed adaptations and community-based institutions, creating employment opportunities, and providing new knowledge, technical and marketing assistance, training, and credit. Those institutions also can hinder adaptations and progress against poverty. Limiting the voice of the poor in resource management decisions or denying local people security of tenure and rights of access to resources are among the most detrimental factors. Without recognition of traditional tenure rights and grants of control over resources, the poor have less incentive and capacity to adapt.

Experiences of the people of Sukhomajri, India, illustrate the difference that stable tenure systems can make in the health of an ecosystem. Twenty years ago, the forest department granted villagers the right to harvest the grass in the watershed for a nominal fee, rather than auctioning the grass to a contractor who, in turn, would charge the villagers high rates for the grass (Agarwal and Narain 1999:16). With the assurance that they would reap the benefits of increased biomass production, villagers identified ways to protect the watershed—regulating livestock grazing, investing in the construction of water tanks for increased crop production, and sustainably harvesting wood from the forest that lies within the catchment. By the mid-1980s, Sukhomajri was no longer importing food but exporting it. Between 1979 and 1984, household income increased from Rs 10,000 to Rs 15,000. The village also earns about Rs 350,000 annually from the sale of milk, and another Rs 100,000 from the sale of *bhabhar*—a fibrous grass that can be used as fodder and sold to paper mills (Agarwal and Narain 1999:16). The result—a once degraded watershed is today a wetter, greener, more productive and prosperous area.

The connection between poverty and the environment is complex. In many instances, poverty contributes to pressures on ecosystems. Roughly half of the world's poorest people live on marginal lands–arid areas, steep slopes, and the like–that are prone to degradation (UNDP 1998:66). Even when the slope erodes, or the fish harvest tapers off, the poor often have no choice but to keep depleting the resource or to convert other vulnerable areas for use.

But this isn't always the case. In fact, the poor can be a source of conservation and environmental protection as well (Scherr 1999). Many people around the world have learned to extract goods from marginal systems without further degradation. For instance, the Mien people of the northern highlands of Thailand center their cultivation on the least erosive slopes, allowing local forests to remain intact and even expand (Batterbury and Forsyth 1999:8). Similar successes, as a result of diversifying both crops and income-generating activities, are taking place in the Machakos region of Kenya (see Chapter 3, Regaining the High Ground: Reviving the Hillsides of Machakos), the drylands and forests of West Africa, and other areas.

Managing for Ecosystem Health

Well-managed ecosystems can provide a range of benefits over the long term. We can choose to emphasize one or a few benefits over others–timber production over scenery, more food over unbroken forests, hydropower over fish harvests–but each choice has a consequence. Poor management choices

The challenge for the 21st century is to understand the vulnerabilities and resilience of ecosystems, so that we can find ways to reconcile the demands of human development with the tolerances of nature.

in the past have often needlessly degraded ecosystems, yielding fewer goods and services today when demand is rising quickly. Retaining the productive capacity of ecosystems in the face of the trade-offs we make marks the difference between good and poor management.

But what does it take to manage ecosystems so that they remain resilient and productive, so that they retain–or recover–their health? We are still struggling to find out. There is no standard measure of ecosystem health or resilience. How much productivity should we expect from ecosystems, and how much degradation can we tolerate? How much can we repair what we have broken, and how much will it cost?

Certainly, answering these questions requires a fundamental knowledge of ecosystem processes and the relationship between various goods and services. Yet these are not scientific questions alone. They are also matters of societal judgment, of economics, and even of ethics. We may choose to forgo harvesting a tract of old-growth forest simply because it is a beautiful and rare habitat, or we may deem it more beneficial used as lumber for housing and left to regenerate as second growth. In either case, the forest may persist in a vital state, but deliver a very different complement of benefits.

Whatever we decide, our opportunities to improve our management of ecosystems are substantial. Our understanding of how ecosystems function, of the links between them and their biological limits, and of their total value has improved significantly in just a few short decades. Satellites and improved measurement techniques have heightened our ability to monitor ecosystems and measure the results of our management. Ecosystem restoration techniques have also advanced, giving the hope that some recovery of productivity is possible (Parrotta and Turnbull 1997). And, more and more, governments and communities have begun to understand the link between ecosystem health and their own economic prosperity and quality of life. Many have already started to define for themselves what sustainable ecosystem management might be–a regional approach to watershed management, perhaps, or land-use restrictions that seek to cluster suburban development rather than encourage sprawl.

The very process of global development, although it places greater pressures on ecosystems, can also be a positive force, changing the way we look at and manage ecosystems. As personal incomes rise and education and environmental awareness expand, the value we place on intact ecosystems will surely grow as well (Panayotou 1999). This is already in evidence in wealthier nations. The demand for nature-based tourism, for example, has started to increase sharply. Initiatives to preserve farmland and curb suburban sprawl have begun in many urban areas. Ambitious projects to restore threatened ecosystems such as the Rhine River or the Florida Everglades have garnered political and financial backing. These projects are evidence of a growing desire to experience and conserve ecosystems, and a willingness to pay for it.

Despite these positive signs, the challenge of defining equitable and sustainable ecosystem management at a global level should not be minimized. It includes asking ourselves such difficult questions as:

- How can we manage watersheds and water resources in the face of potential increases in demand of up to 50 percent for irrigation water and up to 100 percent for industrial water by 2025 (WMO 1997:19–20)?

- Even if irrigation water can be found, how can we intensify our agriculture enough to feed future populations without increasing the damage from nutrient and pesticide runoff or without continuing to convert forests and other ecosystems to croplands?

- How can we continue to supply the roughly 1 m^3 of wood products per year that the average person consumes without decimating existing forests? And what if wood demand doubles in the next 50 years, as some project (Watson et al. 1998:18)?

- How can we lessen the impact of climate change on ecosystems given that CO_2 emissions will likely increase as the global economy grows, at least in the short term?

- How can we reduce the impacts of urban areas–from sprawl to water use to air pollution and solid waste generation–on surrounding ecosystems as urban populations rise to an estimated 5 billion by 2025 (UNPD 1997)?

We have no option but to confront these and similar questions. Our dependence on ecosystems is growing, not diminishing. The productivity of ecosystems, once it is lost through poor management, is difficult and costly to replace.

Tackling these issues will require new strategies that reach across political boundaries without losing critical local support. These, in turn, will rely on an ever clearer understanding of the real state of global ecosystems–how much we have and how much we stand to lose without better management. As a first step, Chapter 2 presents the results of a comprehensive, albeit preliminary, assessment of the world's major ecosystems. The hope is that such background knowledge can help to reveal the trade-offs we have already made and crystallize the management choices that remain to us.

CHAPTER 2

TAKING STOCK OF ECOSYSTEMS

This chapter takes on the critical question: *What condition are the world's ecosystems in?* As Chapter 1 makes clear, the capacity of ecosystems to produce goods and services ranging from food to clean water is fundamentally important for meeting human needs and, ultimately, influences the development prospects of nations. Although policy makers have ready access to information about the condition of their nation's economy, educational programs, or health care system, comparable information about the condition of ecosystems is unavailable. In fact, no nation or global institution has ever undertaken a comprehensive assessment of how well ecosystems are meeting human needs.

We know a good deal about environmental conditions in many places, and we have a fair understanding of the pressures many ecosystems face. But this information lacks the coherence and global coverage needed to provide a clear picture of the state of major ecosystems worldwide.

To help fill this information gap, this chapter presents the results of a first-of-its-kind assessment: the Pilot Analysis of Global Ecosystems (PAGE). The PAGE study assessed five of the world's major ecosystem types.

- ***Agricultural ecosystems*** or "agroecosystems" cover 28 percent of the land surface (excluding Antarctica and Greenland) and account for $1.3 trillion in output of food, feed, and fiber and for 99 percent of the calories humans consume.

- ***Coastal ecosystems*** (including marine fisheries) cover approximately 22 percent of the total land area in a 100-km band along continental and island coastlines, as well as the ocean area above the continental shelf. The coastal zone is home to roughly 2.2 billion people or 39 percent of the world's population and yields as much as 95 percent of the marine fish catch.

- ***Forest ecosystems*** cover 22 percent of the land surface (excluding Antarctica and Greenland) and contribute more than 2 percent of global GDP through the production and manufacture of industrial wood products alone.

- ***Freshwater systems*** cover less than 1 percent of Earth's surface but they are the source of water for drinking, domestic use, agriculture, and industry; freshwater fish and mollusks are also a major source of protein for humans and animals.

- ***Grassland ecosystems*** (including shrublands) cover 41 percent of the land surface (excluding Antarctica and Greenland) and are critical producers of protein and fiber from livestock, particularly in developing countries.

Together these five ecosystem types, which overlap in some places, cover the bulk of Earth's land area and a significant portion of the ocean area. They are also home to much of the world's population. Other ecosystems, such as polar zones, high mountains, ocean areas beyond the continental shelves, and even urban ecosystems account for the remainder of the area and are important in their own right (see the Appendix to this Chapter). But the condition of the goods and services produced by these five major ecosystems will largely determine how well Earth's living systems meet human needs today and in the future.

A Unique Approach

The PAGE study is unique in that it evaluated the state of five ecosystems by examining the condition of a range of goods and services these ecosystems produce:

- food and fiber production,
- provision of pure and sufficient water,
- maintenance of biodiversity,
- storage of atmospheric carbon, and
- provision of recreation and tourism opportunities.

This "goods and services approach" makes explicit the link between the biological capacity of ecosystems and human well-being.

Notably, the PAGE analysis considered not just the current level of production of goods and services, but also the *capacity* of the ecosystem to continue to produce these goods and services in the future. For example, in evaluating food production in the coastal and marine assessment, PAGE researchers looked not only at the current marine fish catch, but also at trends in the condition of the fish stocks that contribute to this catch. In this way, the PAGE study–to the extent possible–addressed the question of the sustainability of current patterns of ecosystem use (Box 2.1 The Difficulty of Assessing Ecosystems).

A Global Synthesis of Current Information

The first objective of PAGE was to review existing environmental assessments and compile available data into a globally comprehensive package. PAGE researchers synthesized information from dozens of sources:

- national, regional, and global data sets on food and fiber production;
- sectoral assessments of agriculture, forestry, biodiversity, water, and fisheries;
- national state-of-the-environment reports;
- national and global assessments of ecosystem extent and change;
- biological assessments of particular species or environments.
- scientific research articles; and
- various national and international data sets.

For each of the five ecosystem types, PAGE researchers first assembled the best information available on the extent of the

Box 2.1 The Difficulty of Assessing Ecosystems

It is enormously challenging to measure the overall condition or health of an ecosystem. The ecosystem "indicators" most readily available, and that have shaped our current understanding of ecosystems, are far from complete. Each provides only a partial description of the bigger picture, like the parable of the five blind men giving different descriptions of the same elephant because each can feel only a small part of the whole animal. These indicators include:

- *pressures* on ecosystems, including such factors as population growth, increased resource consumption, pollution, and overharvesting;
- *extent* of ecosystems—their physical size, shape, location, and distribution; and
- *production* or output of various economically important goods by the system, such as crops, timber, or fish.

Each of these indicators is important, but collectively they provide only a narrow view of ecosystem condition and how well ecosystems are being managed. Indicators of pressure, for example, reveal little about the actual health of the system. With proper management, an ecosystem can withstand significant pressures without losing productivity. Indeed, some agroecosystems have withstood the pressure of intensive cultivation for generations, but have sustained productivity with the help of organic fertilizers and crop rotation. And although growing populations may increase pressures on forests or fisheries, examples abound of community-based management systems that maintained the productivity of ecosystems even in the face of significant population growth.

Similarly, changes in ecosystem extent—such as loss of forests and expansion of agriculture—may indicate that the form of land use and the predominant vegetation have changed, but don't reveal how well the remaining forest or agroecosystem is functioning. And information about the production or output of various ecosystem goods and services doesn't provide a complete picture because production information is rarely available for nonmarketed commodities such as water filtration or storm protection; and the nonmarketed commodities are sometimes the most valuable services ecosystems provide.

Most important, none of these traditional indicators provides information about the underlying capacity of ecosystems to continue to supply their life-sustaining goods and services. The history of the world's fisheries illustrates this problem well. Routinely in fisheries around the world, overfished stocks have collapsed after several years or decades of bountiful harvests. The high production in the good years thus revealed nothing about the health of the fishery; it merely foreshadowed the exhaustion of the resource. Similarly, food production statistics don't reveal evidence of the degradation of agroecosystems that might result from excessive soil erosion or nutrient depletion, since some degradation can be offset by increased fertilization and new crop varieties. With time, though, the diminished capacity of the agricultural lands will increase production costs and may ultimately take land out of production.

Indicators of ecosystem capacity are not easy to obtain. Such indicators must probe the underlying biological state of the ecosystem, including physical factors such as soil fertility or water's dissolved oxygen content that lie at the base of the ecosystem's ability to function. For example, data about the size and structure of some marine fish stocks are available. When these basic population data are combined with knowledge of breeding cycles, the availability of basic nutrients, and large-scale ocean trends like El Niño, the result can lead to an estimate of the maximum sustainable yield for the monitored fish stocks—in other words, the maximum amount of fish that can be harvested without risking depletion of the resource. If calculated carefully, this represents a true measure of the ecosystem's capacity to sustainably produce fish.

Unfortunately, the basic biological data needed to judge ecosystem capacity are often available only for limited areas or species. Even when these data are available, the complex interactions between the elements of the ecosystem and how they affect ecosystem capacity are often unclear. Capacity indicators thus represent the frontier of ecosystem assessment and one of its most problematic aspects.

ecosystem and any modifications to the ecosystem, such as conversion to agriculture or urban areas. PAGE researchers asked:

- Where is the ecosystem located?
- What are its dominant physical characteristics?
- How has it changed through time?
- What pressures and changes is it experiencing today?

They then concentrated on assembling the best indicators of production and condition of the various goods and services produced by each ecosystem:

- What is the quantity of the service being produced (and its value, where possible)?

An International Collaboration

Many organizations collaborated to produce the PAGE study:

- Centro Internacional de Agricultura Tropical (CIAT)
- Global Runoff Data Centre, Germany
- International Fertilizer Development Center (IFDC)
- International Food Policy Research Institute (IFPRI) (agroecosystem coordinator)
- International Institute for Applied Systems Analysis (IIASA)
- International Potato Center (CIP)
- International Soil Reference and Information Centre (ISRIC)
- Food and Agriculture Organization of the United Nations (FAO)
- MRJ Technologies, USA
- Ocean Voice International
- UN Environment Programme
- UN Development Programme
- US Geological Survey, EROS Data Center
- University of Maryland, USA
- University of New Hampshire, USA
- University of Umeå Sweden
- World Bank
- World Conservation Monitoring Centre (WCMC)
- World Resources Institute (PAGE coordinator)

- Is the capacity of the ecosystem to provide that service being enhanced or diminished through time?

Essentially, for each good and service, the PAGE study asked: *Why is it important?* and *What shape is it in?* To the extent possible, researchers also included information about the plausible future condition of the ecosystem.

The results of the PAGE study were subjected to a thorough peer review by more than 70 scientific experts around the world.

The "Big Picture," but with Limitations

The goal of PAGE was not only to provide "state of the art" information about the condition of global ecosystems, but also to help identify gaps in data and information. In addition, PAGE was designed to demonstrate, on a global level, the utility of an *integrated assessment approach*–one that simultaneously assesses the full range of both goods and services an ecosystem produces rather than focusing on just one or two, such as timber production or biodiversity.

The PAGE findings provide a "big picture" view of ecosystem condition and change at a global or continental scale and indicate how these ecosystem characteristics are linked to development prospects. PAGE did not attempt to produce the more detailed site-specific data and information needed at a national scale by resource managers. Nor did it examine specific trade-offs among various goods and services (except for a few illustrative cases), since that type of analysis is most meaningful at smaller scales, such as a nation or river basin, where these choices are actually made.

Although the PAGE study strove to be as integrated as possible in its approach, it is not, strictly speaking, an "integrated assessment." A truly *integrated* ecosystem assessment would focus not on categories such as "forests" and "grasslands," as PAGE has done, but instead on spatially contiguous regions, such as an entire nation, or even a river basin. The Amazon River Basin ecosystem, for example, includes agroecosystems, coastal areas, grasslands, forests, and freshwater habitats. An integrated assessment of the Amazon would examine the array of goods and services produced from this mosaic of land uses and land cover and the trade-offs among them, rather than examine each in isolation (see Box 4.3 The Need for Integrated Ecosystem Assessments).

Nonetheless, at a global scale, the broad ecosystem categories used by PAGE provide a useful way to present information. Moreover, these categories are useful to some of the environmental institutions charged with the conservation and sustainable use of ecosystems. For example, these are the categories used by the Convention on Biological Diversity, the treaty signed by the international community in 1992.

PAGE FINDINGS: The Ecosystem Scorecard

In spite of the narrowness of current ecosystem indicators, we must use them in judicious combination to assemble a picture of ecosystem status. Thus, the PAGE study has negotiated carefully through the various indicators available on ecosystem pressures, production, underlying biological condition, and physical extent to arrive at its findings.

For summary purposes, PAGE researchers chose to represent their findings as two separate "scores" for each of an ecosystem's primary goods or services (see the Ecosystem Scorecard). The *Condition* score (indicated by color) reflects how the ecosystem's ability to yield goods and services has changed over time by comparing the current output and quality of these goods and services with output and quality 20–30 years ago. It is drawn from indicators of production such as crop harvest data, wood production, water use, and tourism, as well as data on biological conditions, such as species declines, biological invasions, or the amount of carbon stored in the vegetation and soils of a given area.

The *Changing Capacity* score reflects the trend in an ecosystem's biological capacity—its ability to continue to provide a good or service in the future. It integrates information on ecosystem pressures with trends in underlying biological factors such as soil fertility, soil erosion and salinization, condition of fish stocks and breeding grounds, nutrient loading and eutrophication of water bodies, fragmentation of forests and grasslands, and disruption of local and regional water cycles.

In all cases, the ecosystem scores represent expert judgments that integrate a number of different variables, and accommodate gaps in the data sets. Although far from perfect, the Condition and Changing Capacity scores, when taken together, offer a reasonable picture of how ecosystems are serving us today, and their trend for the future, given current pressures.

Scorecard

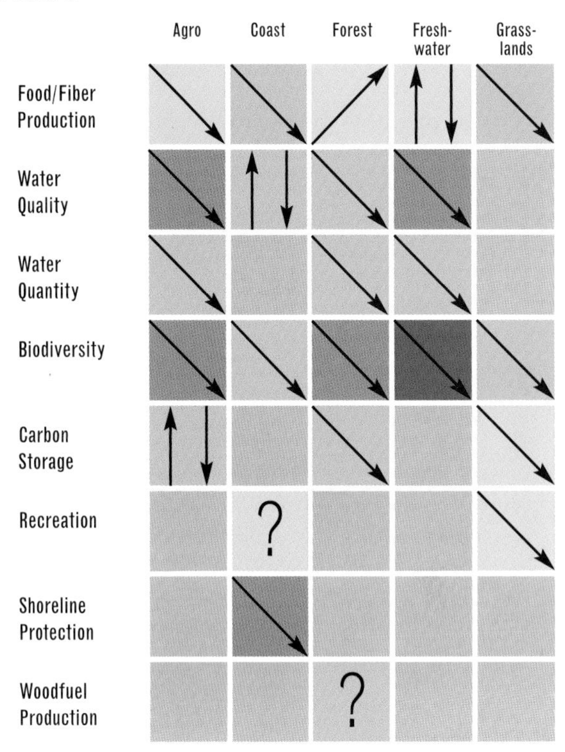

Key

Condition assesses the current output and quality of the ecosystem good or service compared with output and quality of 20–30 years ago.

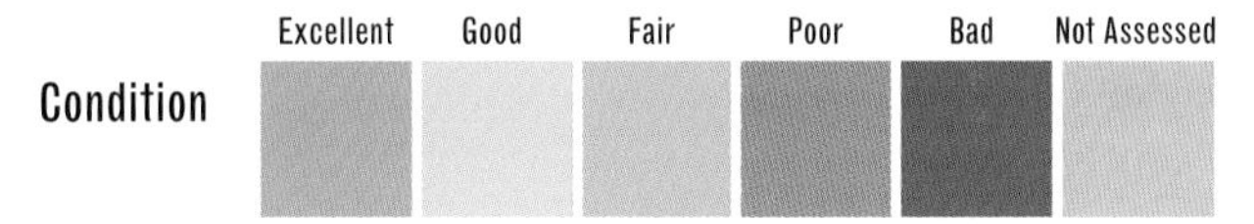

Changing Capacity assesses the underlying biological ability of the ecosystem to continue to provide the good or service.

Increasing Mixed Decreasing Unknown

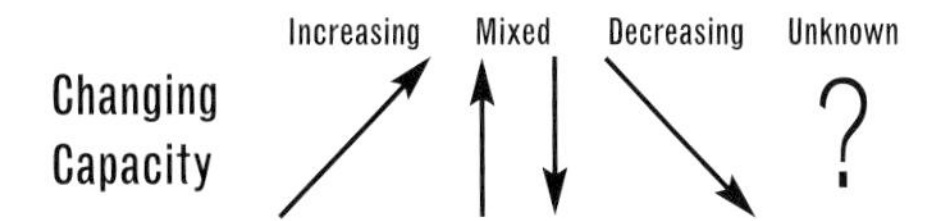

Scores are expert judgments about each ecosystem good or service over time, without regard to changes in other ecosystems. Scores estimate the predominant global condition or capacity by balancing the relative strength and reliability of the various indicators. When regional findings diverge, in the absence of global data, weight is given to better-quality data, larger geographic coverage, and longer time series. Pronounced differences in global trends are scored as "mixed" if a net value cannot be determined. Serious inadequacy of current data is scored as "unknown."

PAGE Findings: What Shape Are the World's Ecosystems In?

The results of the PAGE study confirm that humans have dramatically altered the capacity of ecosystems to deliver goods and services, with the most significant changes taking place over the past century. For some goods and services, such as food production, we have greatly increased the capacity of ecosystems to provide what we need, while for others, such as water purification and biodiversity conservation, we have greatly degraded their capacity. The balance sheet of the positive and negative impacts of our management of ecosystems is shown in the Ecosystem Scorecard and summarized below.

FOOD PRODUCTION

People have dramatically increased food production from the world's ecosystems, in part by converting large areas to highly managed agroecosystems–croplands, pastures, feedlots–that provide the bulk of the human food supply. The condition of agroecosystems from the standpoint of food production is mixed. Although crop yields are still rising, the underlying condition of agroecosystems is declining in much of the world. Soil degradation is a concern on as much as 65 percent of agricultural land. Historically, inputs of water, fertilizers, and technologies such as new seed varieties and pesticides have been able to more than offset declining ecosystem conditions worldwide (although with significant local and regional exceptions), and they may continue to do so for the foreseeable future. But how long can that kind of compensation continue? The diminishing capacities of agroecosystems will make that task ever more challenging.

The outlook for fish production–also a major source of food–is more problematic. The condition of coastal ecosystems from the standpoint of food production is only fair and becoming worse. Twenty-eight percent of the world's most important marine fish stocks are depleted, overharvested, or just beginning to recover from overharvesting. Another 47 percent are being fished at their biological limit and are, therefore, vulnerable to depletion. Freshwater fisheries present a mixed picture; we are currently overexploiting most native fish stocks, but introduced species have begun to enhance the harvest in some water bodies, and production from aquaculture ponds is growing steadily. Overall, the pattern of increasing dependence on aquaculture and the decline of natural fish stocks will have serious consequences for many of the world's poor who depend on subsistence fishing.

WATER QUANTITY

Dams, diversions, irrigation pumps, and other engineering works have profoundly altered the amount and location of water available for both human uses and for sustaining aquatic ecosystems. People now withdraw annually about half of the water readily available for use from rivers. Dams and engineering works have strongly or moderately fragmented 60 percent of the world's large river systems; they have so impeded flows, that the length of time it takes the average drop of river water to reach the sea has tripled. The changes we have made to forest cover and other ecosystems such as wetlands also have altered water availability and affected the timing and intensity of floods. For example, tropical montane forests, which play key roles in regulating water quantity in the tropics, are being lost more rapidly than any other tropical forest type. Freshwater wetlands, which store water and moderate flood flows, have been reduced by as much as 50 percent worldwide.

WATER QUALITY

Water quality is degraded directly through chemical and nutrient pollution and indirectly when the capacity of ecosystems to filter water is degraded and when land-use changes increase soil erosion. Nutrient pollution from fertilizer-laden runoff is a serious problem in agricultural regions around the world; it has resulted in eutrophication and human health hazards in coastal regions, particularly in the Mediterranean, Black Sea, and northwestern Gulf of Mexico. The frequency of harmful algal blooms, linked to nutrient pollution, has increased significantly in the past 2 decades. We have greatly exceeded the capacity of many freshwater and coastal ecosystems to maintain healthy water quality. And although developed countries have improved water quality to some extent in the past 20 years, water quality in developing countries–particularly near urban and industrial areas–has been degraded substantially. Decreasing water quality poses a particular threat to the poor who often lack ready access to potable water and are most subject to the diseases associated with polluted water.

CARBON STORAGE

The plants and soil organisms in ecosystems remove carbon dioxide (CO_2)–the most important greenhouse gas–from the atmosphere and store it in their tissues. This carbon storage process helps to slow the buildup of CO_2 in the atmosphere. Unfortunately, the steps we have taken to increase production of food and other commodities from ecosystems have had a net negative impact on their capacity to store carbon. This is principally the result of converting forests to agroecosystems; agroecosystems support less vegetation overall and therefore store less carbon. Such land-use changes are in fact an important source of carbon emissions, contributing approximately 20 percent of global annual carbon emissions.

Ecosystems nonetheless still store significant carbon (Box 2.2 Terrestrial Storage of Carbon). Of the carbon currently stored in terrestrial systems, 38–39 percent is stored in forests and 33 percent in grasslands. Agroecosystems, which overlap grasslands and forests somewhat, store 26–28 percent. How we manage these ecosystems–whether we promote afforestation and other carbon-storing strategies or increase the forest

Box 2.2 Terrestrial Storage of Carbon

Carbon stored in terrestrial ecosystems plays a large role in the global carbon cycle. To map the distribution of terrestrial carbon storage, PAGE researchers combined recent satellite maps of Earth's vegetation with estimates of how much carbon various types of vegetation and soil store. As the map shows, the highest quantities of stored terrestrial carbon are located in the tropics and in the boreal region. In the tropics, a larger portion of the carbon is found in the vegetation, while in boreal regions, especially peatlands, most carbon is stored in the soils. Boreal peatlands are especially important carbon storage areas. Unforested lands generally store less carbon than forested ecosystems.

Global Terrestrial Storage of Carbon

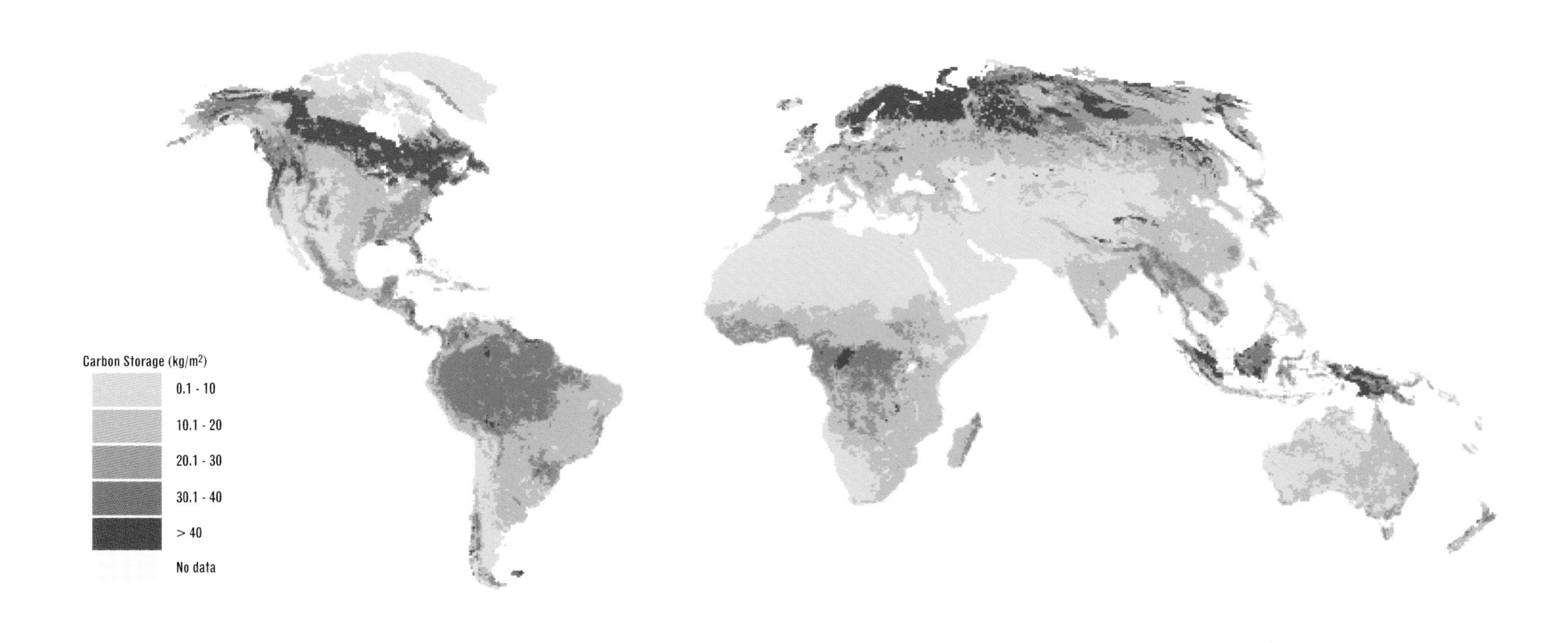

Sources: Matthews et al. [PAGE] 2000. The map is a combination of two maps: a map of carbon stored in above- and below-ground live vegetation based on USGS/EDC (1999b) and a map of carbon stored in soils based on Batjes (1996) and Batjes and Bridges (1994).

Box 2.3 Are We Altering Earth's Basic Chemical Cycles?

Tracking the changes in Earth's chemical cycles—carbon, nitrogen, and water cycles—is essential to understanding the condition of ecosystems. These cycles serve as the basic metabolism of the biosphere, affecting how every ecosystem functions and linking them all on a global level. Human-induced changes in these global processes can alter climate patterns and affect the availability of basic nutrients and water that sustain plant and animal life.

The Carbon Cycle

Carbon dioxide (CO_2) concentrations in the atmosphere rose 30 percent from 1850 to 1998, from 285 parts per million to 366 parts per million (IPCC 2000:4) (see Box 1.6 Carbon Storage, p. 15). This rise in atmospheric CO_2 levels is largely the result of increased CO_2 emissions from burning fossil fuels. However, changes in use and management of ecosystems have also played a major role by releasing carbon that had been stored in vegetation and soil. About 33 percent of the carbon that has accumulated in the atmosphere over the past 150 years has come from deforestation and changes in land use (IPCC 2000:4).

Climate models tell us that rising carbon concentrations in the atmosphere will alter Earth's climate, affecting precipitation, land and sea temperatures, sea level, and storm patterns. The extent and structure of ecosystems will change as they transform in response to these basic physical parameters. Changing climate will also affect the rate of greenhouse gas emissions from some ecosystems. For example, models suggest that a warmer climate in the Arctic will elevate the rate of decomposition of the vast peat reserves in tundra and taiga ecosystems, increasing the release of CO_2 into the atmosphere.

Elevated atmospheric CO_2 can, in turn, have more direct impacts on ecosystems. Because plants depend on CO_2 for growth, elevated CO_2 concentrations will have a "fertilizer effect," increasing the growth rate of some plants and changing some of the chemical and physical characteristics of their cells. Some species will benefit more than others, and this in turn will alter the composition of biological communities.

Climate change could also have a profound impact on growing patterns and yields in agriculture. PAGE researchers estimated that a warmer climate could raise cereal production by 5 percent in mid- to high-latitude regions (mostly developed countries) but might decrease cereal yields in low-latitude regions by 10 percent (particularly in African developing countries).

The Nitrogen Cycle

Although we are more familiar with the influence humans have had on the carbon cycle, human influence on the global nitrogen cycle is more profound and already more biologically significant. In most natural systems, lack of nitrogen is an important limiting factor for plant growth, which is what accounts for significant increases in crop yields in response to nitrogen fertilizers. However, as explained in Chapter 1, the production and use of fertilizers, burning of fossil fuels, and land clearing and deforestation also increase—far beyond natural levels—the amount of nitrogen available to biological systems (Vitousek et al. 1997:5). This added nitrogen has caused serious problems, particularly in freshwater and coastal ecosystems where excess nitrogen stimulates growth of algae, sometimes depleting available oxygen to the point where other aquatic organisms suffocate, a process known as eutrophication.

The Freshwater Cycle

The scale of human impact on freshwater cycles is also massive. Humans currently appropriate more than half of accessible freshwater runoff, and by 2025, demand is projected to increase to more than 70 percent of runoff (Postel et al. 1996:7, 787). A substantial amount—70 percent—of the water currently withdrawn from all freshwater sources is used for agriculture (WMO 1997:9). By shifting water from freshwater systems to agroecosystems, crop production increases, but at significant cost to downstream ecosystems and downstream users. Some of the water diverted from rivers or directly consumed does return to rivers but, typically, carrying with it pollution in the form of agricultural nutrients or chemicals, or human or industrial waste. But as much as 60 percent of water withdrawn from rivers is lost to downstream uses (Postel 1993:56; Seckler 1998:4).

Global Cycles, Global Impacts

The importance of these global cycles to the functioning of ecosystems cannot be overstated. There is no question that sound management of Earth's ecosystems will require changes in the use of resources at a local level; but it is not enough to only examine and assess the condition of ecosystems at the local level. Some of the most important features of Earth's ecosystems—with the most profound influence on the future role of ecosystems in meeting human needs—can only be fully understood on regional and even global levels. Thus, it is vital that we examine and assess the condition of ecosystems at those levels.

conversion rate–will have a significant impact on future increases or decreases in atmospheric carbon dioxide.

BIODIVERSITY

The erosion of global biodiversity over the past century is alarming. Major losses have occurred in virtually all types of ecosystems, much of it simply by loss of habitat area. Forest cover has been reduced by at least 20 percent and perhaps by as much as 50 percent worldwide; some forest ecosystems, such as the dry tropical forests of Central America, are virtually gone. More than 50 percent of the original mangrove area in many countries is gone; wetlands area has shrunk by about half; and grasslands have been reduced by more than 90 percent in some areas. Only tundra, arctic, and deep-sea ecosystems have emerged relatively unscathed.

Even if ecosystems had retained their original spatial extent, many species would still be threatened by pollution, overexploitation, competition from invasive species, and habitat degradation. In terms of the health of species diversity, freshwater ecosystems are far and away the most degraded, with 20 percent of freshwater fish species extinct, threatened, or endangered in recent decades. Forest, grassland, and coastal ecosystems all face major problems as well. The rapid rise in the incidence of diseases affecting marine organisms, the increased prevalence of algal blooms, and the significant decreases in amphibian populations all attest to the severity of the threat to global biodiversity.

Apart from the loss of medicines, useful genetic materials, and ecotourism revenues this erosion of biodiversity represents, it also threatens the basis of ecosystem productivity. The diversity of species undergirds the ability of an ecosystem to provide most of its other goods and services. Reducing the biological diversity of an ecosystem may well diminish its resilience to disturbance, increase its susceptibility to disease outbreaks, and thus threaten its stability and integrity.

RECREATION AND TOURISM

The capacity of ecosystems to provide recreational and tourism opportunities was assessed only for coastal and grassland ecosystems. It is likely that the demand for these services will grow significantly in coming years, but the condition of the service is declining in many areas because of the overall degradation of biodiversity as well as the direct impacts of urbanization, industrialization, and tourism itself on the ecosystems being visited.

The Bottom Line

Overall, there are numerous signs that the capacity of ecosystems to continue to produce many of the goods and services we depend on is decreasing. In all five ecosystem types PAGE analyzed, ecosystem capacity is decreasing over a range of goods and services, not just one or two. PAGE results confirm that major modifications of ecosystems–through deforestation, conversion, nutrient pollution, dams, biological invasions, and regional-scale air pollution–continue to grow in scale and pervasiveness. Furthermore, human activities are significantly altering the basic chemical cycles that all ecosystems depend on (Box 2.3 Are We Altering Earth's Basic Chemical Cycles?). This strikes at the foundation of ecosystem functioning and adds to the fundamental stresses that ecosystems face at a global scale.

This downward trend in global ecosystem capacity is not impeding high production levels of some goods and services today. Food and fiber production have never been higher, and dams have allowed unprecedented control of water supplies. But this wealth of production is, in many instances, the product of intensive management that threatens to reduce the productivity of ecosystems in the longer term. Our use of technology–whether it is artificial fertilizer, more efficient fishing gear, or water-saving drip-irrigation systems–has also helped mask some of the decrease in biological capacity and has kept production levels of food and fiber high. However, services like maintaining biodiversity and high water quality and carbon storage show reductions in output that technology cannot so easily mask. In sum, the PAGE findings starkly illustrate the trade-offs we have made between high commodity production and impaired ecosystem services, and indicate the dangers these trade-offs pose to the long-term productivity of ecosystems.

The remaining sections of this chapter present an ecosystem-by-ecosystem discussion of the conclusions of the PAGE study.

The Pilot Analysis of Global Ecosystems

Technical Reports Available in Print and On-Line at http://www.wri.org/wr2000

Agroecosystems
Stanley Wood, Kate Sebastian, and Sara Scherr, *Pilot Analysis of Global Ecosystems: Agroecosystems, A joint study by International Food Policy Research Institute and World Resources Institute*, International Food Policy Research Institute and World Resources Institute, Washington, D.C.
December 2000 / 100 pages / paperback / ISBN 1-56973-457-7 / US$20.00

Coastal Ecosystems
Lauretta Burke, Yumiko Kura, Ken Kassem, Mark Spalding, and Carmen Revenga, *Pilot Analysis of Global Ecosystems: Coastal Ecosystems,* World Resources Institute, Washington, D.C.
December 2000 / 100 pages / paperback / ISBN 1-56973-458-5 / US$20.00

Forest Ecosystems
Emily Matthews, Richard Payne, Mark Rohweder, and Siobhan Murray, *Pilot Analysis of Global Ecosystems: Forest Ecosystems*, World Resources Institute, Washington, D.C.
October 2000 / 100 pages / paperback / ISBN 1-56973-459-3 / US$20.00

Freshwater Systems
Carmen Revenga, Jake Brunner, Norbert Henninger, Richard Payne, and Ken Kassem, *Pilot Analysis of Global Ecosystems: Freshwater Systems*, World Resources Institute, Washington, D.C.
October 2000 / 100 pages / paperback / ISBN 1-56973-460-7 / US$20.00

Grassland Ecosystems
Robin White, Siobhan Murray, and Mark Rohweder, *Pilot Analysis of Global Ecosystems: Grassland Ecosystems*, World Resources Institute, Washington, D.C.
December 2000 / 100 pages / paperback / ISBN 1-56973-461-5 / US$20.00

The full text of each report will be available on-line at the time of publication. Paper copies may be ordered by mail from WRI Publications, P.O. Box 4852, Hampden Station, Baltimore, MD 21211 USA. Orders may be placed by phone by calling 1-800-822-0504 (within the United States) or 410-516-6963 or by faxing 410-516-6998. Orders may also be placed on-line at **http://www.wristore.com**.

The agroecosystem report is also available at http://www.ifpri.org. Paper copies may be ordered by mail from the International Food Policy Research Institute, Communications Service, 2033 K Street, N.W., Washington, D.C. 20006-5670 USA.

AGROECOSYSTEMS

Agroecosystems provide the overwhelming majority of crops, livestock feed, and livestock on which human nutrition depends. In 1997, global agriculture provided 95 percent of all animal and plant protein and 99 percent of the calories humans consumed (FAO 2000). Agroecosystems also contribute a large percentage of the fiber we use–cotton, flax, hemp, jute, and other fiber crops.

Globally, agroecosystems have been remarkably successful, when judged by their ability to keep pace with food, feed, and fiber demands (Box 2.4 Taking Stock of Agroecosystems). Per capita food production is higher today than 30 years ago, even though the global population doubled since then. However, agriculture faces an enormous challenge to meet the food needs of an additional 1.7 billion people–the projected population increase–over the next 20 years.

Historically, agricultural output has increased mainly by bringing more land into production. But the amount of land remaining that is both well suited for crop production (especially for annual grain crops) and not already being farmed is limited. A further limitation is the growing competition from other forms of land use such as industrial, commercial, or residential development. Indeed, in densely populated parts of India, China, Indonesia, Egypt, and Western Europe, limits to expansion were reached many years ago. Approximately 2.8 billion people live in or near agroecosystems (not including adjacent urban areas) (Wood et al. [PAGE] 2000).

Intensifying production–obtaining more output from a given area of agricultural land–has thus become essential. In some regions, particularly in Asia, farmers have successfully intensified production by raising multiple crops each year, irrigating fields, and using new crop varieties with shorter growth cycles. On high-quality, nonirrigated lands, farmers have intensified production mainly by abandoning or shortening fallow periods and moving to continuous cultivation, with the help of modern technologies. Agricultural intensification is widespread even on lower-quality lands, particularly in developing nations. Intensification has also been significant around major cities (and to an unexpected extent, within cities), principally to produce high-value perishables such as dairy products and vegetables for urban markets, but also to meet subsistence needs.

The unprecedented scale of agricultural expansion and intensification has raised concerns about the state of agroecosystems. First, there is growing concern about their productive capacity–can agroecosystems withstand the stresses imposed by intensification? These stresses include increased erosion, soil nutrient depletion, salinization and waterlogging of soils, and reduction of genetic diversity among major crops. There is also concern about the negative impacts of agriculture on other ecosystems–impacts that are often accentuated by intensification. Examples include the harmful effects of increased soil erosion on downstream fisheries

(continues on p. 56)

Box 2.4 Taking Stock of Agroecosystems

Highlights

- Food production has more than kept pace with global population growth. On average, food supplies are 24 percent higher per person than in 1961, and real prices are 40 percent lower.
- Agriculture faces an enormous challenge to meet the food needs of an additional 1.7 billion people over the next 20 years.
- Agroecosystems cover more than one-quarter of the global land area, but almost three-quarters of the land has poor soil fertility and about one-half has steep terrain, constraining production.
- While the global expansion of agricultural area has been modest in recent decades, intensification has been rapid, as irrigated area increased, fallow time has decreased, and the use of purchased inputs and new technologies has grown and is producing more output per hectare.
- About two-thirds of agricultural land has been degraded in the past 50 years by erosion, salinization, compaction, nutrient depletion, biological degradation, or pollution. About 40 percent of agricultural land has been strongly or very strongly degraded.

Key

Condition assesses the current output and quality of the ecosystem good or service compared with output and quality of 20–30 years ago.

	Excellent	Good	Fair	Poor	Bad	Not Assessed
Condition						

Changing Capacity assesses the underlying biological ability of the ecosystem to continue to provide the good or service.

	Increasing	Mixed	Decreasing	Unknown
Changing Capacity	↗	↑↓	↘	?

Scores are expert judgments about each ecosystem good or service over time, without regard to changes in other ecosystems. Scores estimate the predominant global condition or capacity by balancing the relative strength and reliability of the various indicators. When regional findings diverge, in the absence of global data weight is given to better-quality data, larger geographic coverage, and longer time series. Pronounced differences in global trends are scored as "mixed" if a net value cannot be determined. Serious inadequacy of current data is scored as "unknown."

Conditions and Changing Capacity

FOOD PRODUCTION

Since 1970, livestock products have tripled and crop outputs have doubled, a sign of rising incomes and living standards. Food production, which was worth US$1.3 trillion in 1997, is likely to continue to increase significantly as demand increases. Nonetheless, soil degradation is widespread and severe enough to reduce productivity on about 16 percent of agricultural land, especially cropland in Africa and Central America and pastures in Africa. Although global inputs and new technologies may offset this decline in the foreseeable future, regional differences are likely to increase.

WATER QUALITY

Production intensification has limited the capacity of agroecosystems to provide clean freshwater, often significantly. Both irrigated and rainfed agriculture can threaten downstream water quality by leaching fertilizers, pesticides, and manure into groundwater or surface water. Irrigated agriculture also risks both soil and water degradation through waterlogging and salinization, which decreases productivity. Salinization is estimated to reduce farm income worldwide by US$11 billion each year.

WATER QUANTITY

Irrigation accounts for fully 70 percent of the water withdrawn from freshwater systems for human use. Only 30–60 percent is returned for downstream use, making irrigation the largest net user of freshwater globally. Although only 17 percent of agroecosystems now depend on irrigation, that share has grown; irrigated area increased 72 percent from 1966 to 1996. Competition with other kinds of water use, especially for drinking water and industrial use, will be stiffest in developing countries, where populations and industries are growing fastest.

BIODIVERSITY

Agricultural land, which supports far less biodiversity than natural forests, has expanded primarily at the expense of forest areas. As much as 30 percent of the potential area of temperate, subtropical, and tropical forests has been lost to agriculture through conversion. Intensification also diminishes biodiversity in agricultural areas by reducing the space allotted to hedgerows, copses, or wildlife corridors and by displacing traditional varieties of seeds with modern high-yield but uniform crops. Nonetheless, certain practices, including fallow periods and shade cropping, can encourage diversity as well as productivity.

CARBON STORAGE

In agricultural areas the amount of carbon stored in soils is nearly double that stored in the crops and pastures that the soils support. Still, the share of carbon stored in agroecosystems (about 26–28 percent of all carbon stored in terrestrial systems) is about equal to the share of land devoted to agroecosystems (28 percent of all land). Agricultural emissions of both carbon dioxide and methane are increasing because of conversion to agricultural uses from forests or woody savannas, deliberate burning of crop stubble and pastures to control pests or promote fertility, and paddy rice cultivation.

Data Quality

FOOD PRODUCTION

Value, yield, input, and production data are from the Food and Agriculture Organization (FAO) national tables, 1965-97. Consistency and reliability vary across countries and years. Ecosystem analysis requires more spatially disaggregated information. Fertility constraints are spatially modeled from the soil mapping units of FAO's Soil Map of the World. Global and regional assessments of human-induced soil degradation are based primarily on expert opinion. Developing reliable, cost-effective methods for monitoring soil degradation would help to both mitigate further losses and target restoration efforts.

WATER QUALITY

There are no globally consistent indicators of water quality that relate specifically to agriculture. In agricultural watersheds, the quantity of pesticides and nutrients—nitrogen and phosphorus—are good indicators of pollution from leaching and surface runoff. In mixed-use catchments it is much more difficult to separate from other sources such as human effluents and pesticides applied in gardens and public recreation areas. Pesticide data are more expensive to monitor. Data on suspended solids from soil erosion are also scarce and difficult to interpret.

WATER QUANTITY

Irrigated area is assessed using the Kassel University global spatial data, which indicate the percentage and area of land equipped for irrigation but has some inconsistencies in scale, age, and reliability of source. Irrigation water use data are derived from country-specific tabular data sets on irrigated area, water availability and use, and water abstraction. Little crop-specific information is available on irrigated area and production. Global estimates of rainfall from the University of East Anglia are based on spatial extrapolations of monthly data from climate stations over a 30-year period. Even though the resolution of these data is coarse, it allows assessment of both spatial and temporal variability.

BIODIVERSITY

World Wildlife Fund for Nature (WWF) global spatial data describe potential natural habitats and ecoregions. These were developed from expert opinion and input maps of varying resolution and data, but the data do provide a general understanding of the spatial patterns of natural habitats. Genetic diversity data are compiled from major germplasm-holding institutions. Area adoption data for modern varieties of cereals are compiled from survey and agricultural census.

CARBON STORAGE

Storage capacity is modeled for vegetation and soils based on carbon storage capacity by land cover type at a resolution of half a degree for a single point in time. Data would be improved by better characterization of agricultural land-cover types and their vegetation content. Soil carbon data were derived for Latin America using FAO and the International Soil Reference and Information Centre's Soil and Terrain database.

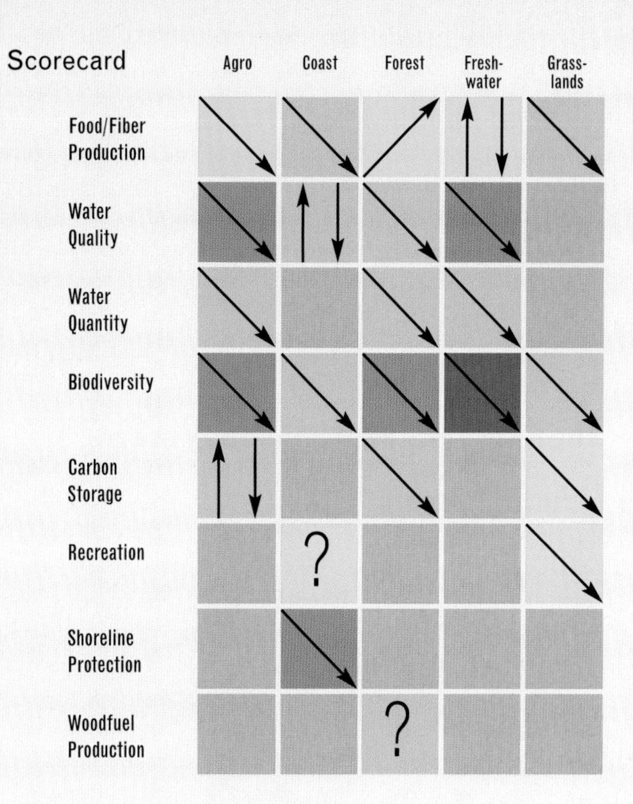

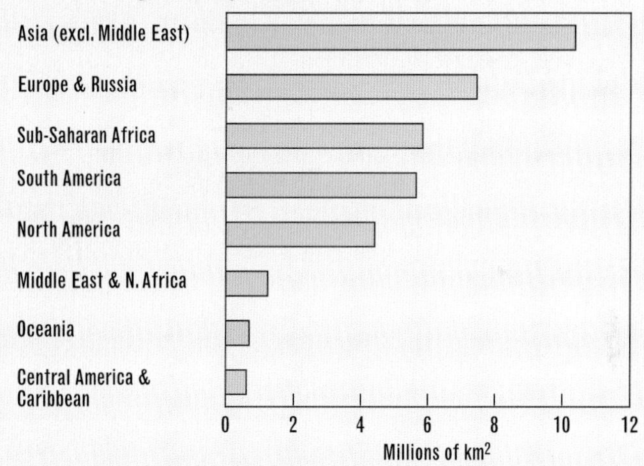

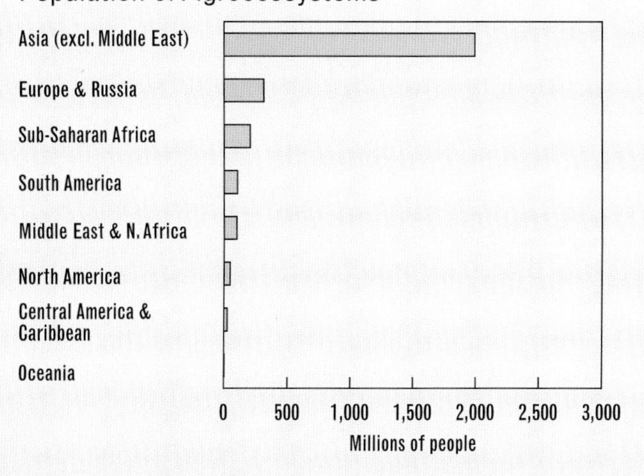

and reservoirs and the damage to both aquatic ecosystems and human health from fertilizer and pesticide residues in water sources, in the air, and on crops. Agricultural practices also have even broader consequences for biodiversity and for alteration of the global carbon, nitrogen, and hydrological cycles (Thrupp 1998; Conway 1997).

Characterizing Agroecosystems

EXTENT AND GROWTH

Agriculture is one of the most common land uses on the planet and agroecosystems are quite extensive. Determining their exact extent depends on how they are defined.[1] The PAGE study, making use of satellite imagery, defined agricultural areas as those where at least 30 percent of the land is used for cropland or highly managed pasture (Box 2.5 The Global Extent of Agriculture). Using this definition, agroecosystems cover approximately 28 percent of total land area (excluding Greenland and Antarctica). This includes some overlap with forest and grassland ecosystems because land-use is often quite fragmented spatially, with agricultural plots forming part of a mosaic of uses–agriculture alongside forest or grassland areas. The Food and Agriculture Organization of the United Nations (FAO) reports an even greater percentage of land in agriculture–37 percent (FAO 2000). FAO's figures are derived from national production statistics rather than from satellite data and include all permanent pasture.

The actual area of agroecosystems probably falls somewhere between these estimates. Since the satellite data are based on only 1 year of data, areas that were not cultivated that year but are still used for agricultural purposes (for example, an area under fallow or regions that alternate year to year between cropland and pasture) may be underestimated in the satellite images. It is also more difficult to detect extensive pastures and some perennial crops using satellite data because of their similarity to natural grasslands and forests.

According to FAO, 69 percent of agroecosystems consist of permanent pasture, with the remainder of the area under crops. However, this global average masks very large differences among regions in the balance between crops and pastureland. In some regions, pastureland predominates: pastures make up 89 percent of the agroecosystem area in Oceania, 83 percent in Sub-Saharan Africa, 82 percent in South America, and 80 percent in East Asia. In other regions, croplands occupy much larger areas: 92 percent of agroecosystem area in South Asia and 84 percent in Southeast Asia. In India, crops cover 94 percent of the agroecosystem area. On croplands, annual crops such as wheat, rice, maize, and soybeans occupy 91 percent of the area, with the remainder in permanent crops, such as tea, coffee, sugarcane, and most fruits (FAO 2000).

Most agricultural production, with the exception of dairy and perishable vegetable production, is derived from intensively managed croplands located away from major concentrations of population. However, since the 1980s, the growth of urban and periurban agriculture has accelerated, especially in developing countries. By the early 1990s, approximately 800 million people globally were actively engaged in urban agriculture, using a variety of urban spaces including homesites, parks, rights-of-way, rooftops, containers, and unbuilt land around factories, ports, airports, and hospitals (FAO 1999a). Urban residents, who would otherwise spend a high proportion of income on food, engage in agriculture to increase their own food security and nutrition or as an income source. An estimated 200 million urban dwellers produce food for sale (Cheema et al. 1996).

FAO statistics show that the total area in agriculture expanded slowly between 1966 and 1996, from 4.55 Bha to 4.92 Bha–about an 8 percent increase (FAO 2000). This low growth rate masks a more dynamic pattern of land-use changes, with land conversion to and from agriculture taking place at much higher rates. It is these aggregate changes, for which data are scarce, that are most relevant from an ecosystem perspective.

Despite global growth, agricultural area has actually decreased in many industrialized countries. Both the United States and Western Europe have progressively been taking land out of agriculture for the last 30 years, and Oceania for the last 20. During this period, these three regions have removed a total of 49 Mha from agricultural production. Agricultural land has also decreased significantly in Eastern Europe, largely because of liberalization of production and marketing and poor economic conditions. South Asia's total agricultural area has remained constant for more than 20 years at approximately 223 Mha. However, expansion of agricultural area is still significant in some regions. Agricultural land increased by almost 0.8 percent/year during 1986–96 in China and Brazil and by 1.38 percent/year in West Asia (FAO 2000).

INTENSIFICATION

Although the net global expansion of agricultural area has been modest in recent decades, intensification has been rapid. Irrigated area grew significantly over the past 3 decades, from 153 Mha in 1966 to 271 Mha in 1998. Globally, irrigated land accounts for only 5.5 percent of all agricultural land–17.5 percent of cropland–but in some regions irrigation is much more extensive. For example, China and India together contain 41 percent of the global irrigated area and Western Europe and the United States contain another 12.5 percent. In contrast, the arid and semiarid regions of Sub-Saharan Africa and Oceania (primarily Australia), contain only 3 percent of the world's irrigated land (FAO 2000) (Box 2.6 Intensification of Agriculture).

(continues on p. 60)

Box 2.5 The Global Extent of Agriculture

Agricultural lands cover about 36 Mha, 28 percent, of Earth's land area (excluding Greenland and Antarctica). Although agricultural area has increased worldwide in the past 30 years, it has decreased in many industrialized countries. Globally, about 31 percent of agroecosystems are croplands and 69 percent are pasture, but actual proportions of each vary widely among regions.

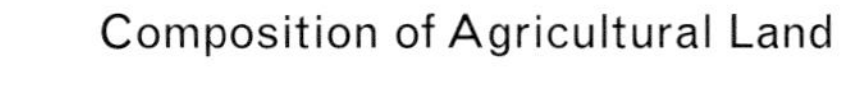

Composition of Agricultural Land

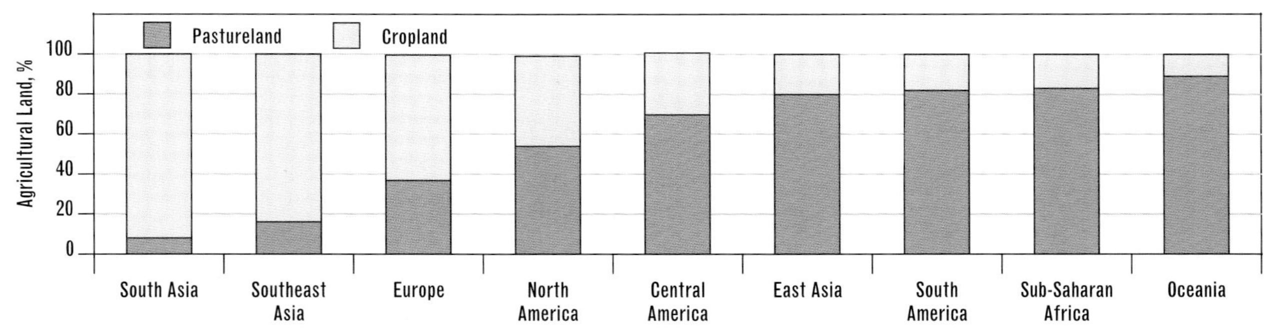

The Global Extent of Agriculture

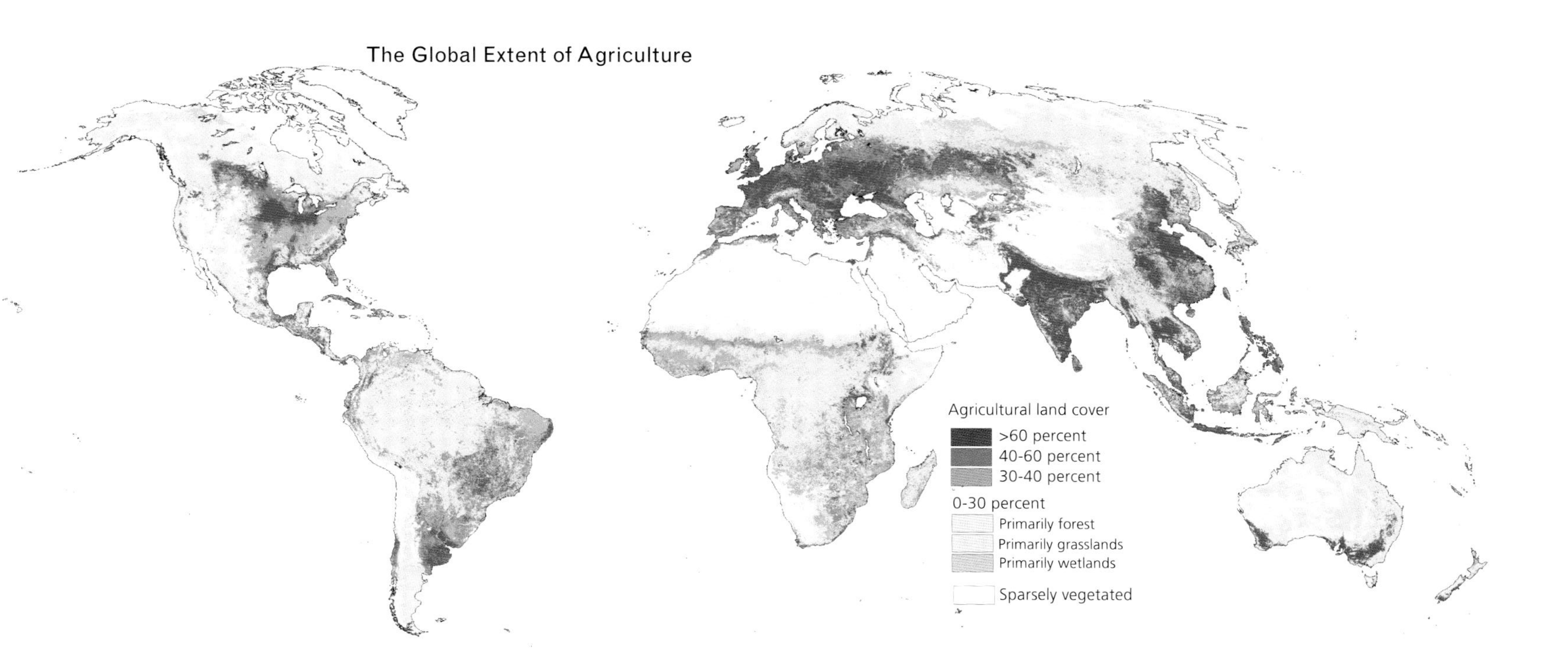

Sources: Wood et al. [PAGE] 2000. The map is based on Global Land Cover Characteristics Database Version 1.2 (Loveland et al. [2000]) and USGS/EDC (1999a). The figure is based on FAOSTAT (1999).

Box 2.6 The Intensification of Agriculture

As population has grown and good agricultural land has become scarcer, inputs such as water, fertilizer, pesticides, and labor have been applied more intensively to increase output. In Asia, where population pressures are greatest, virtually all of the cropland is harvested each year, sometimes two to three times a season, as the use of irrigation, new varieties of quick-growing seeds, and fertilizers has replaced traditional practices of leaving land fallow to restore

Wheat Yields, 1866–1997

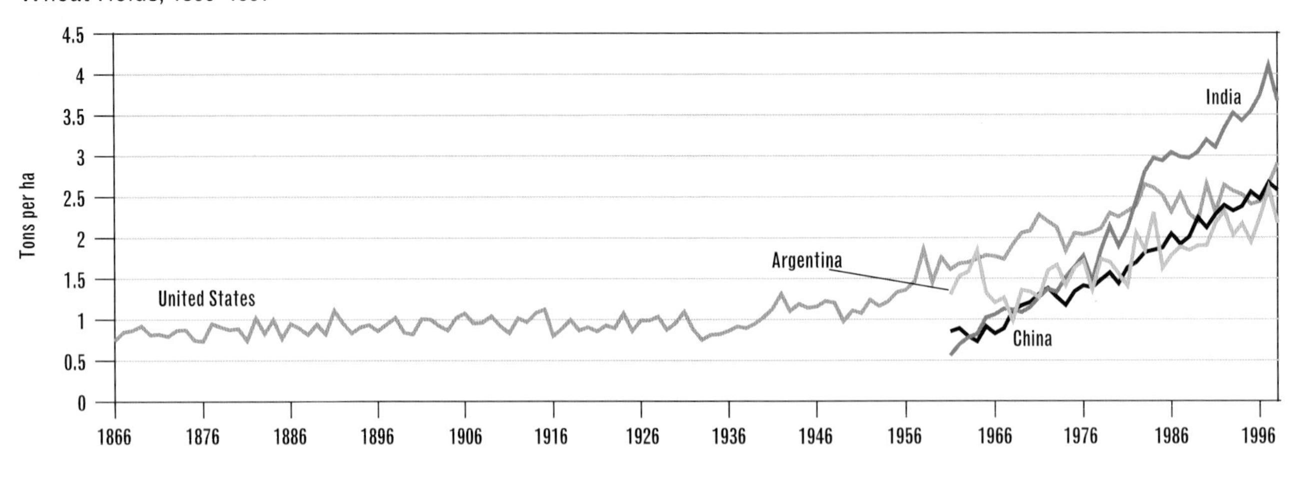

Intensification of Cropping, 1995–97

The cropping index is the harvested area of land planted in annual crops divided by the total area of such land. A value of more than 1 indicates more than one crop harvested per year per hectare.

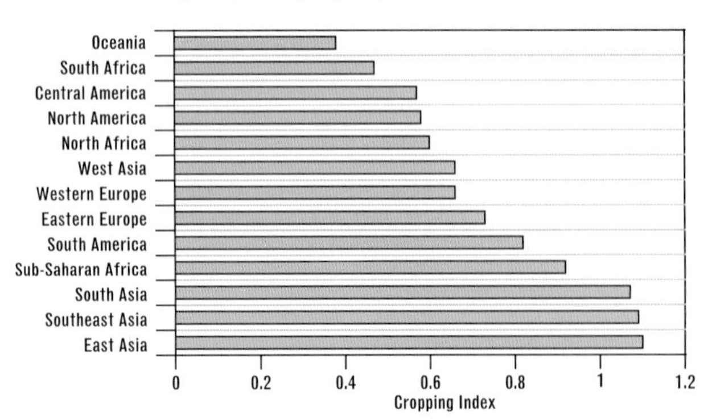

Use of Commercial Fertilizer, 1995–97

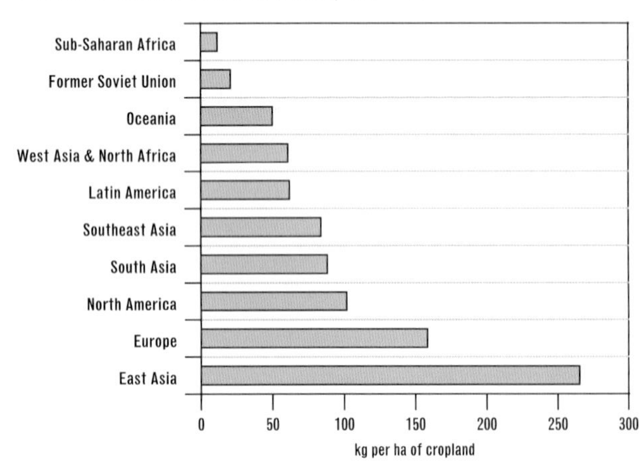

Intensification of Irrigation, 1995–97

The irrigation index is the irrigated area of cropland divided by the total area of cropland.

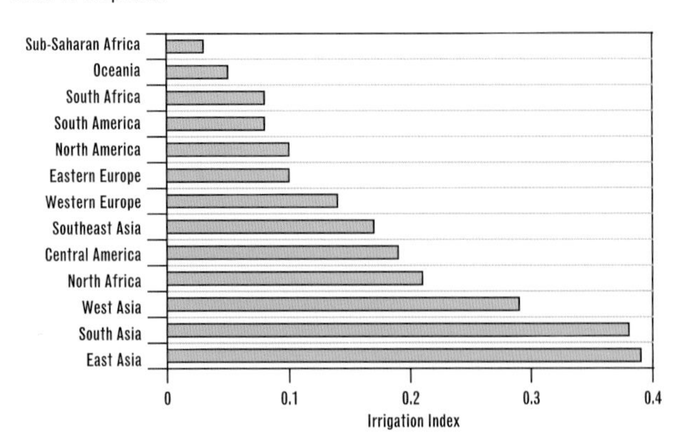

Irrigated Land Damaged by Salt, 1987

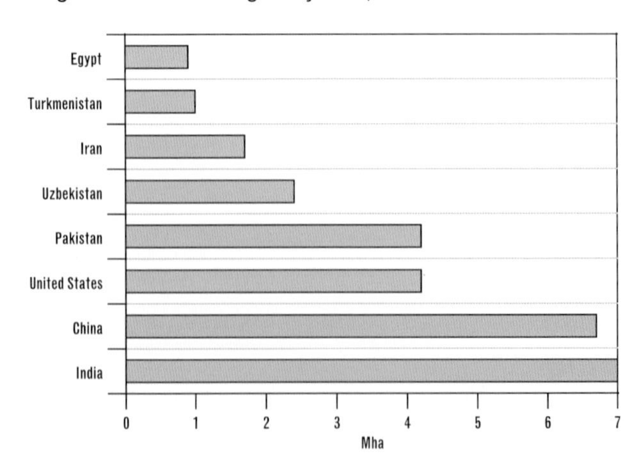

fertility. Even marginal lands in Africa are in continuous use to meet demands for food, although water and fertilizer inputs are much lower there.

Many agroecosystems are vulnerable to the stresses imposed on them by intensification. There is much local evidence of soil salinization caused by poorly managed irrigation systems, loss in soil fertility through overcultivation, compaction by tractors or livestock, and lowering of water tables through overpumping for irrigation.

Continued agricultural intensification need not lead inexorably to environmental degradation, however. Farming communities in all parts of the world have responded to degradation, particularly when it affects their livelihoods, with measures such as planting trees to control erosion, regulating cultivation around local water sources, restricting pesticides and other pollutants, rehabilitating degraded soils, and adopting new technologies. (See Chapter 3, Regaining the High Ground: Reviving the Hillsides of Machakos, Kenya.)

Commercial Application of Fertilizer to Cropland

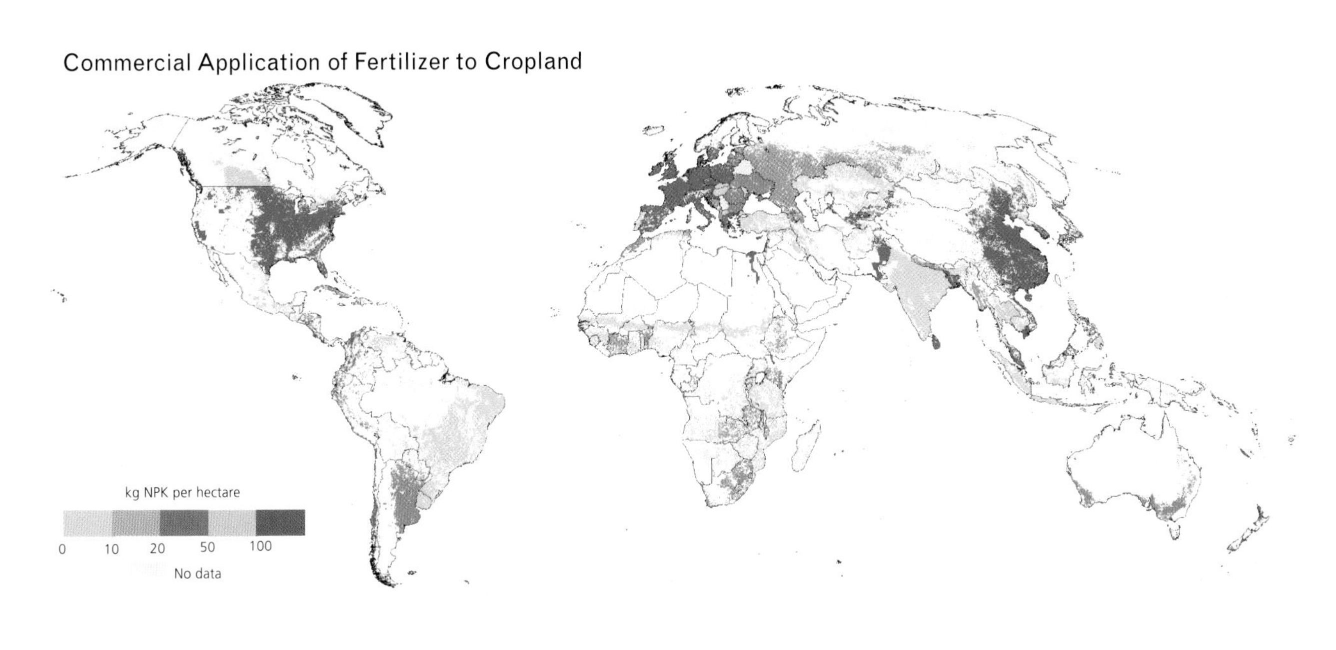

Distribution of Tractors on Cropland

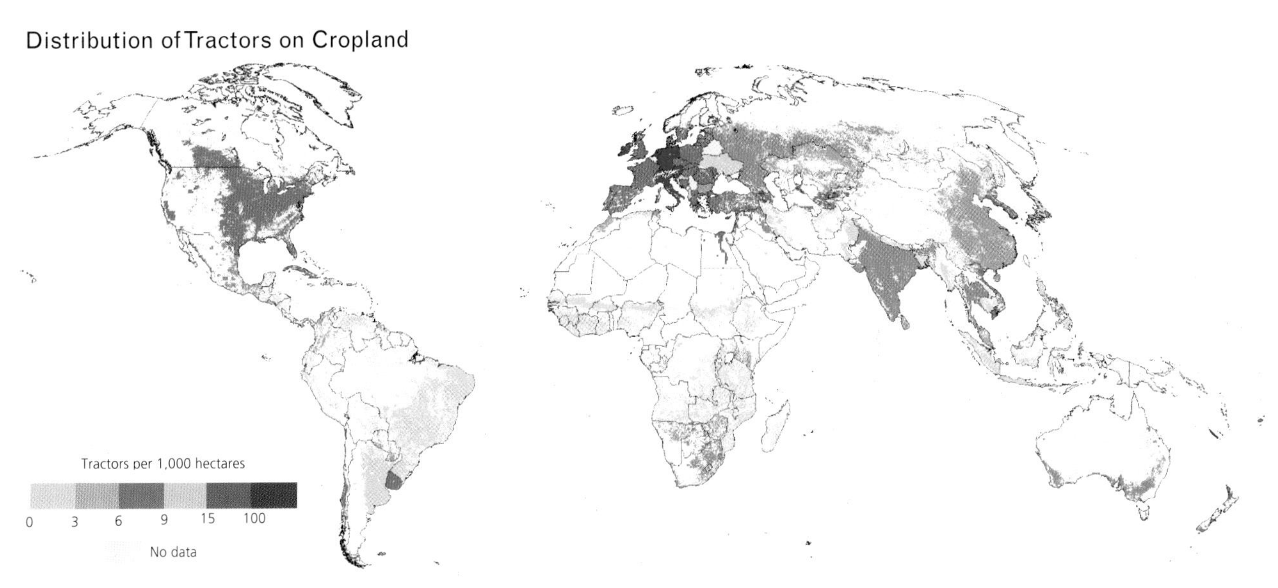

Sources: Wood et al. [PAGE] 2000. The maps are based on FAOSTAT 1999. They show national values within the global extent of agriculture, augmented by additional irrigated areas (Döll and Siebert 1999). Wheat yields are from USDA-NASS (1999). Irrigated land damaged by salt is based on Postel (1999:93). All other figures are based on FAOSTAT (1999).

Production intensity is also reflected in the use of inputs such as tractors and fertilizers. The current global consumption of fertilizer totals about 137 million tons/year (1997), representing a dramatic increase in consumption during the last 50 years (FAO 2000).

In recent years, irrigation growth rates have slowed considerably and growth in fertilizer consumption has moderated. Following a decline from the late 1980s to the mid-1990s, total fertilizer consumption is again increasing and is currently around 6 percent below its 1988 peak (FAO 2000).

SOIL AND SLOPE CONSTRAINTS

Despite the high productivity of global agriculture and the rapid intensification of production on some lands, many of the world's agricultural lands offer less than optimal conditions. Steep slopes (more than 8 percent incline) or poor soil conditions limit production on a significant portion of agricultural land. Soil fertility constraints include high acidity, low potassium reserves, high sodium concentrations, low moisture-holding capacity, or limited depth. If more than 70 percent of agricultural land in a particular region has one or more of these constraints, it is said to have "significant" soil constraints.

Using these definitions, 81 percent of agricultural land has significant soil constraints and around 45 percent of agricultural land is steep. Approximately 36 percent of agricultural land is characterized by both significant soil constraints and slopes of 8 percent or more. Areas with both steep slopes and significant soil constraints make up 30 percent of temperate, 45 percent of subtropical, and 39 percent of tropical agricultural land. Average agricultural yields are generally lower and degradation risks are generally higher in these areas than in more ecologically favored environments. Nonetheless, these marginal lands represent a significant share of global agriculture and support roughly one-third of the world's population (Wood et al. [PAGE] 2000).

Assessing Goods and Services

FOOD, FEED, AND FIBER

Economic Importance

The food, fiber, and animal feed that the world's agroecosystems produce is worth approximately $1.3 trillion per year[2] (Wood et al. [PAGE] 2000). Agriculture is most important to the economies of low-income countries, accounting for 31 percent of their GDP, and more than 50 percent of GDP in many parts of Sub-Saharan Africa. In middle-income countries, agriculture accounts for 12 percent of GDP. But in the high-income countries of Western Europe and North America, where other economic sectors dominate, the contribution of agriculture to GDP is just 1–3 percent, even though the value of the agricultural output in these countries represents 79 percent of the total market value of world agricultural products (Box 2.7 The Economic Value of Agricultural Production).

Conventional measures of agriculture's share of GDP actually understate agriculture's contribution to economies. For example, agricultural GDPs in the Philippines, Argentina, and the United States comprise 21 percent, 11 percent, and 1 percent of those countries' total GDPs, respectively; yet the total value of agriculture, including manufacturing and services further along the marketing chain, comprises 71 percent, 39 percent, and 14 percent of their respective total GDPs (Bathrick 1998:10).

Beyond the economic value of the food produced, agroecosystems also provide employment for millions. Agricultural labor represents the livelihood, employment, income, and cultural heritage of a significant part of the world's population. In 1996, of the 3.1 billion people living in rural areas, 2.5 billion–44 percent of the world population–were estimated to be living in households dependent on agriculture. The labor force directly engaged in agriculture is an estimated 1.3 billion people–about 46 percent of the total labor force. In North America, only 2.4 percent of the labor force is directly engaged in agriculture, while in East, South, and Southeast Asia as well as in Sub-Saharan Africa, agricultural labor accounts for 56–65 percent of the labor force (FAO 2000).

Human Nutrition

Agriculture was developed for a simple but fundamental purpose–to provide adequate human nutrition. Globally, agroecosystems produce enough food to provide every person on the planet with 2,757 kcal each day, which is sufficient to meet the minimum human requirement for nutrition (FAO 2000). However, many people do not have adequate access to that food, and an estimated 790 million people are chronically undernourished. In Sub-Saharan Africa, 33 percent of the population is undernourished; in the Caribbean 31 percent; and in South Asia 23 percent (FAO 1999b:29)

Global demand for food is still increasing significantly, driven by population growth, urbanization, and growth in per capita income. One of the most notable changes in demand is the dramatic increase in meat consumption, particularly in the developing world. This has been dubbed the "livestock revolution." Between 1982 and 1994, global meat consumption grew by 2.9 percent per year, but it grew five times faster in developing countries than in developed countries, where meat consumption is already high (Delgado et al. 1999:9–10).

Between 1995 and 2020, global population is expected to increase by one-third, totaling 7.5 billion people. Global demand for cereals is projected to increase by 40 percent, with 85 percent of the increase in demand coming from developing countries. Meat demand is projected to increase

Box 2.7 The Economic Value of Agricultural Production

The total value of output from agroecosystems is US$1.3 trillion per year. Worldwide, 46 percent of the total labor force works in agriculture, and almost half the total population lives in rural communities that depend on agriculture. Cropland generally has more valuable outputs per hectare than pasture, except in Europe, South Asia, and Southeast Asia, where pastures support intensive livestock production. Output per worker varies dramatically from region to region, reflecting difference in level of commercialization of agriculture and opportunities for off-farm employment.

Value of Production per Hectare, 1995–97

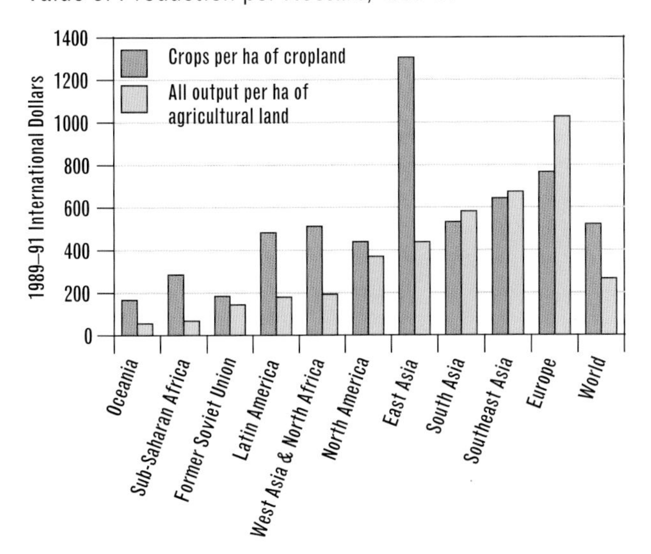

Value of Agricultural GDP per Agricultural Worker, 1995–97

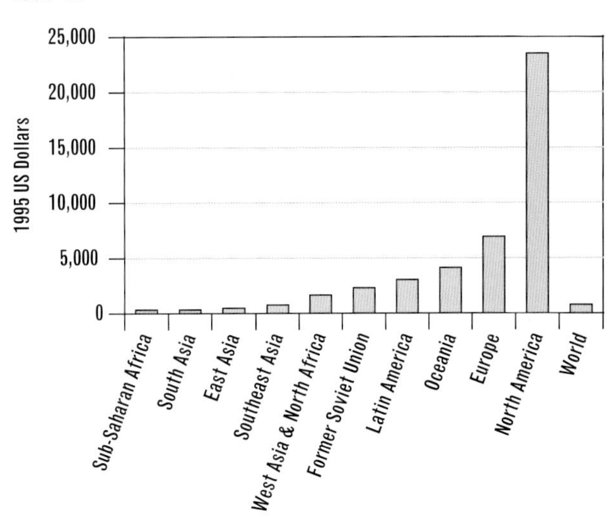

The Value of Crops per Hectare of Cropland, 1995–97

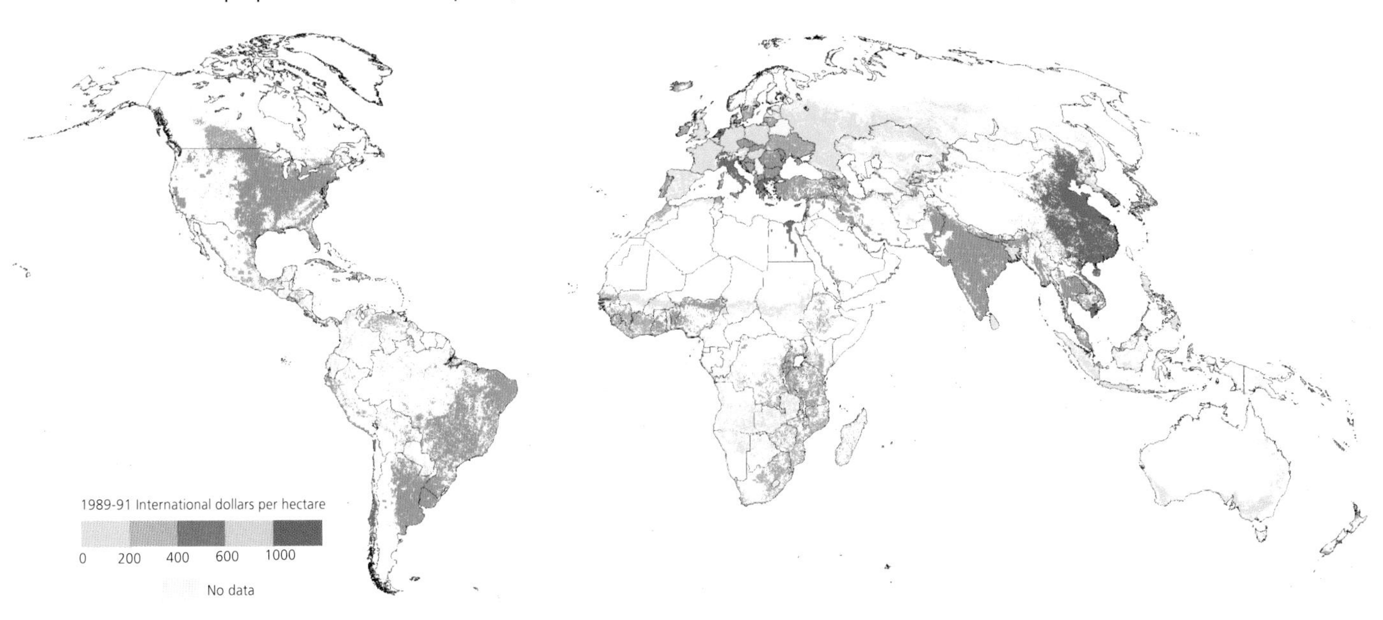

Sources: Wood et al. [PAGE] 2000. The map shows national values within the global extent of agriculture, augmented by additional irrigated areas (Döll and Siebert 1999). Value of production table and map are based on FAO (1997) and FAOSTAT (1999). Value of agricultural production weights the output of 134 primary crop and 23 primary livestock commodity quantities by their respective international agricultural prices for 1989–91. Value of crop production is based on the 134 primary crops only. Value of agricultural GDP per agricultural worker is based on World Bank (2000) and FAOSTAT (1999).

by 58 percent, with approximately 85 percent of the increase coming from developing countries. Demand for roots and tubers is expected to grow 37 percent, with 97 percent of this increase coming from the developing world (Pinstrup-Andersen et al. 1999:5–12). And, if significant progress is made in alleviating poverty during this period, there will be an additional increase in demand as the poor and malnourished use their increased income to buy food they previously could not afford.

Productive Capacity

Changes in Yield Growth. Rapid yield growth in most major crops has been instrumental in meeting the food needs of growing populations, particularly in the second half of this century. Recently, however, the growth of cereal crop yields has been slowing, raising concerns that future production may not be able to keep pace with demand. Moreover, there is evidence from some parts of the world that maintaining the growth in yields, or even holding yields at current levels, requires proportionately greater amounts of fertilizer input, implying that the quality of the underlying soil resource may be deteriorating.

These trends must be interpreted cautiously. Even if yields continue to grow rapidly, this does not necessarily indicate that agroecosystems are in good shape, since increased inputs like fertilizer and pesticides could mask underlying depletion of soil nutrients. Nor does a slowdown in the growth of crop yields prove agroecosystem conditions are worsening, since market factors such as falling commodity prices and high fertilizer prices may also account for slower production. Nonetheless, the declining rate of yield growth is worrisome in a world where the growth in food demand is not expected to slow.

Soil Degradation. One measure of the long-term productive capacity of an agroecosystem is the condition of its soil. Natural weathering processes and human management practices can both affect soil quality. Sustaining soil productivity requires that soil-degrading pressures be balanced with soil-conserving practices. The principal processes of soil degradation are erosion by water or wind, waterlogging and salinization (the buildup of salts in the soil), compaction and crusting, acidification, loss of soil organic matter and soil microorganisms, soil nutrient depletion, and accumulation of pollutants in the soil.

Different types of soil degradation are associated with different types of agricultural land use. For example, salinization is associated most often with intensification of irrigated land, and compaction with mechanized farming in high-quality rain-fed lands. Nutrient depletion is often associated with intensifying production on marginal lands but can occur on any soil if nutrients extracted by crops are not adequately replenished. Water erosion is also often associated with marginal lands that have been extensively cleared and tilled. Soil pollution is a particular problem in periurban agriculture (Scherr 1999).

The 1990 Global Assessment of Soil Degradation (GLASOD), based on a structured survey of regional experts, provides the only continental and global-scale estimates of soil degradation (Oldeman et al. 1991). The GLASOD study suggested that 1.97 Bha had been degraded between the mid-1940s and 1990 (Scherr 1999:17; Wood et al. [PAGE] 2000). This represents 15 percent of terrestrial area (excluding ice-covered Greenland and Antarctica).

To assess the extent and severity of soil degradation on agricultural lands in particular, PAGE researchers overlaid the GLASOD data on the map of agricultural land (land with more than 30 percent agricultural use). This revealed that 65 percent of agricultural lands have some amount of soil degradation. About 24 percent were classified as "moderately degraded" which, according to GLASOD, signifies that their agricultural productivity has been greatly reduced. A further 40 percent of agricultural land fell into the GLASOD categories of "strongly degraded" (lands that require major financial investments and engineering work to rehabilitate) or "very strongly degraded" (lands that cannot be rehabilitated at all) (Wood et al. [PAGE] 2000). Among the most severely affected areas are South and Southeast Asia,[3] where populations are among the densest and agriculture the most extensive (Box 2.8 Soil Degradation in South and Southeast Asia).

Soil Nutrient Balance. One indicator of soil condition–and productive capacity–is soil nutrient balance. One of the most common management techniques used to maintain the condition of agroecosystems, particularly intensively cultivated systems, is to replenish soil nutrients with organic manures or inorganic fertilizers containing nitrogen, phosphorus, and potassium. Too little replenishment can lead to soil nutrient mining–the progressive loss of nutrients as crops draw on them for growth. Too much replenishment (overfertilization) can lead to leaching of excess nutrients and the consequent soil and water pollution problems as these unused nutrients find their way into surrounding soils and freshwater systems.

An estimate of the nutrient balance of an agroecosystem can be obtained by measuring the nutrient inputs (inorganic and organic fertilizers, nutrients from crop residues, and nitrogen fixation by soybeans and other legumes) and outputs (nutrient uptake in the main crop products and the crop residue). PAGE researchers calculated these nutrient balances at the national level for individual crops in Latin America and the Caribbean (Henao 1999) and found that for most of the crops and cropping systems, the nutrient balance is significantly negative–in other words, soil fertility is declining (see Box 2.9 Hot Spots and Bright Spots).

The observed increases in production in recent decades must therefore be due to a combination of area expansion,

Box 2.8 Soil Degradation in South and Southeast Asia

South and Southeast Asia, where agricultural production systems are among the most intensive in the world, have soils that are among the most degraded. In these regions, soils are significantly steeper, more subject to erosion, and more likely to be salinized, acidic, depleted of potassium, and saturated with aluminum than the soils of most other regions.

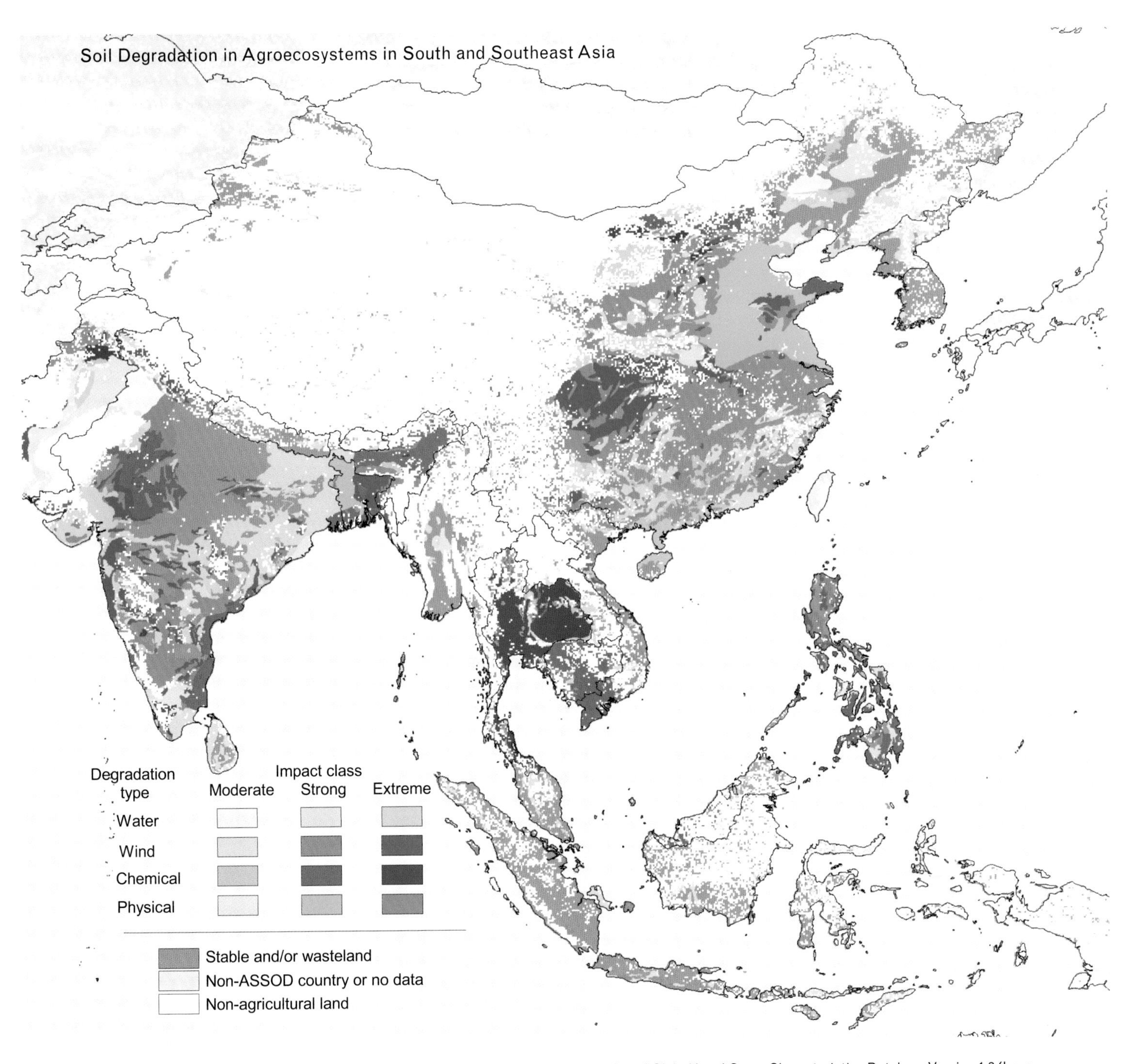

Sources: Wood et al. [PAGE] 2000. The map is based on Van Lynden and Oldeman (1997), and Global Land Cover Characteristics Database Version 1.2 (Loveland et al. [2000]). It shows soil degradation within the extent of agriculture.

improved varieties, and other factors that mask or offset the effects of soil degradation. By overlaying nutrient balance with trends in yields, it is possible to identify potential degradation "hot spots" where yield growth is slowing and soil fertility is declining. Areas where the capacity of agroecosystems to produce food appears most threatened include northeast Brazil and sections of Argentina, Bolivia, Colombia, and Paraguay.

Soil nutrient balances are also available for most of Sub-Saharan Africa at continental, national, and district levels (Smaling et al. 1997:47–62). Findings there also suggest widespread nutrient depletion.

Productivity Losses. The cumulative productivity loss from soil degradation over the past 50 years has been roughly estimated, using GLASOD figures, to be about 13 percent for cropland and 4 percent for pasture lands (Oldeman 1998:4). The economic and social impacts of this degradation have been far greater in developing countries than in industrialized countries. In industrialized countries, soil quality plays a relatively less important role in overall agricultural productivity because of the high level of fertilizer and other inputs used. Furthermore, the most important grain-producing areas in industrialized countries typically have deep, geologically "new" soils that can withstand considerable degradation without having yields affected.

Soil degradation has more immediate impacts on the food supply in developing countries. Agricultural productivity is estimated to have declined significantly on approximately 16 percent of agricultural land, especially on cropland in Africa and Central America, pastures in Africa, and forests in Central America. The GLASOD study estimates that almost 74 percent of Central America's agricultural land (defined by GLASOD as cropland and planted pastures) is degraded, as is 65 percent of Africa's and 38 percent of Asia's (Scherr 1999:18). Detailed studies based on predictive models for Argentina, Uruguay, and Kenya calculated yield reductions between 25 and 50 percent over the next 20 years (Mantel and van Engelen 1997:39–40).

Subregional studies have documented significant aggregate declines in crop yields due to degradation in many parts of Africa, China, South Asia, and Central America (Scherr 1999). Crop yield losses in Africa from 1970 to 1990 due to water erosion alone are estimated to be 8 percent (Lal 1995:666). Estimates of the economic losses associated with soil degradation in eight African countries range from 1 to 9 percent of agricultural GDP (Bøjö 1996:170). Total annual economic loss from degradation in South and Southeast Asia is estimated to be 7 percent of the region's agricultural GDP (Young 1994:75). Given that more than half of all land in this region is not affected by degradation, the economic effects in the degraded areas appear to be quite significant. Economic losses from erosion in different regions of Mexico vary from approximately 3 to 13 percent of agricultural GDP (McIntire 1994:124).

The Bottom Line for Food Production. At a global level there is little reason to believe that crop production cannot continue to grow significantly over the next several decades. That said, the underlying condition of many of the world's agroecosystems, particularly those in developing countries, is not good. Soil degradation data, while coarse, suggest that erosion and nutrient depletion are undermining the long-term capacity of agricultural systems on well over half of the world's agricultural land. And competition for water will further magnify the issue of resource constraints to food production. Although nutrient inputs, new crop varieties, and new technologies may well offset these declining conditions for the foreseeable future, the challenge of meeting human needs seems destined to grow ever more difficult.

WATER QUANTITY AND QUALITY

Agriculture is perhaps the most significant human influence on the world's water cycle, affecting quantity, timing, and quality of water available to freshwater systems. At a global scale, agriculture accounts for the greatest proportion of total freshwater withdrawals of any sector of human activity. Agriculture also has the highest consumptive use of water (use that results in returning water to the atmosphere, rather than back to streams or groundwater). Approximately 70 percent–2,800 km^3–of the 4,000 km^3 of water humans withdraw from freshwater systems each year (Shiklomanov 1997:69) is used for irrigation (WMO 1997:9).

This volume of water irrigates 271 Mha of croplands (FAO 2000). Although this number represents only 17 percent of total cropland, it produces 40 percent the world's crops (WMO 1997:9). Of the water used for irrigation, 50–80 percent is returned to the atmosphere or otherwise lost to downstream users (Shiklomanov 1993:19). As a consequence, irrigation can significantly decrease river flows and aquifer levels and can shrink lakes and inland seas.

The Aral Sea represents an extreme case of the ecological damage agricultural water diversions can inflict. Withdrawals to irrigate cotton and other crops shrank the sea to one-third of its original volume by the early 1990s, thus increasing its salinity. Fish species and fishing livelihoods were lost before steps were taken to restore some of the flows (WRI 1990:171; Gleick 1998:189).

For 82 percent of the world's agroecosystems, rainfall is the sole source of water for agricultural production. Although rain-fed agriculture has less sweeping impacts on freshwater flows than irrigated agriculture, it can still affect the quantity and timing of downstream flows. These impacts are highly site-specific, depending on the type of agriculture, the soil's slope and condition, and the patterns and intensity of rainfall.

Both irrigated and rain-fed agriculture can pose threats to water quality from the leaching of fertilizer, pesticides, and ani-

Box 2.9 Hot Spots and Bright Spots in Latin American Agroecosystems

Cereal yields have generally been increasing in Latin America over the past 20 years (left map), but at the expense of stocks of nutrients in the soils in which cereals and other crops are grown. In fact, most Latin American agricultural soils show a negative "nutrient balance," meaning that more nutrients are lost through plant growth and harvest than are replaced through additions of fertilizer, manure, or legume cover crops (center map). Combining these maps yields a picture of agricultural "hot spots"—areas where yield growth is slowing and soil fertility is declining (right map). Hot spots where agricultural capacity appears to be most threatened are in northeast Brazil and parts of Argentina, Bolivia, Columbia, and Paraguay. Some "bright spots"—where yields are stable or increasing and nutrient balances are positive—also appear, but cover a much smaller area.

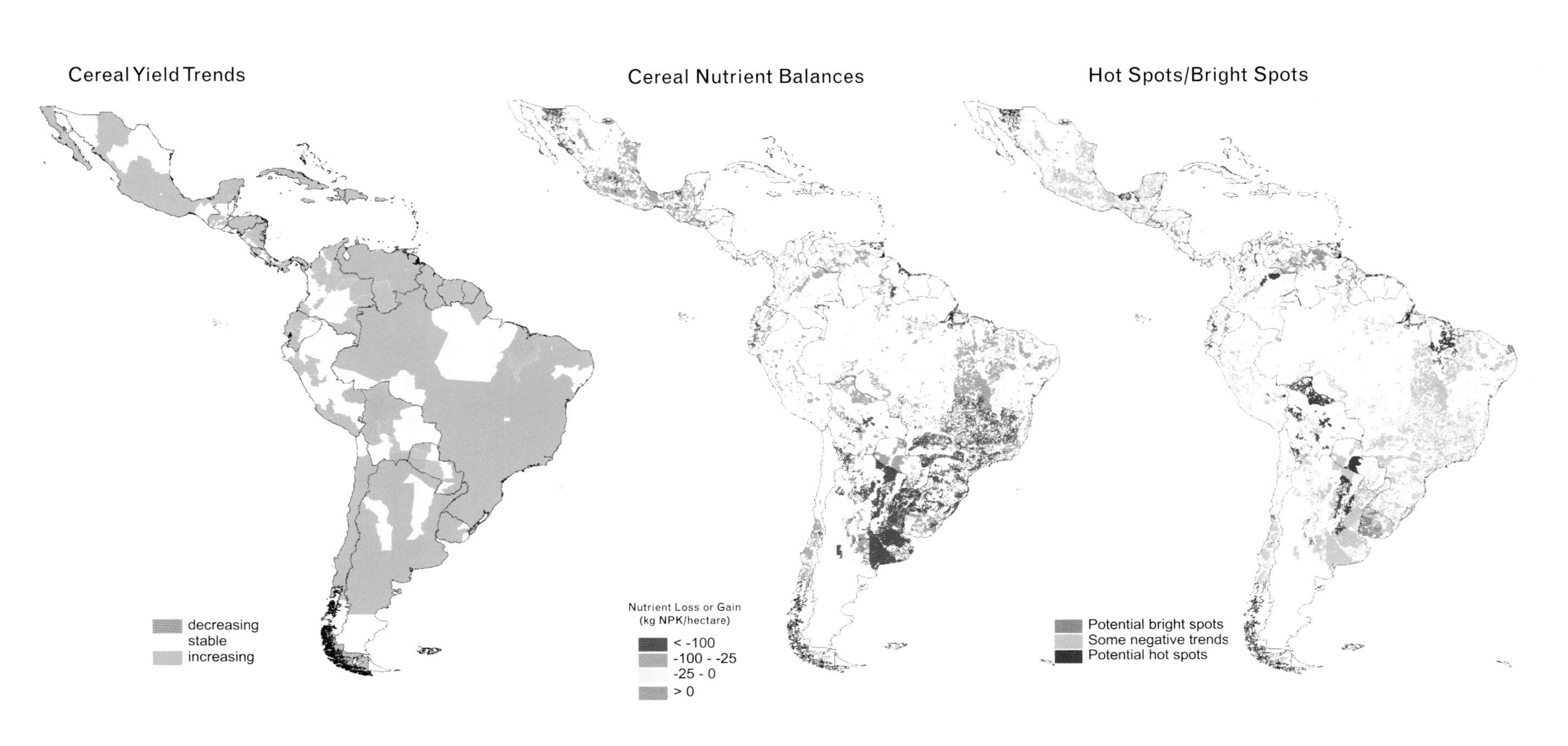

Sources: Wood et al. [PAGE] 2000. Cereal yield trends are based on subnational 1975-98 data for rice, wheat, maize, and sorghum. Nutrient balances are based on national balances of applied nutrients less extraction by cereal crops. They were allocated to specific geographic areas using subnational production statistics and information on climate, soil, and elevation. Map of hots pots and bright spots combines the map of cereal yield trends and the map of cereal nutrient balances.

mal manure into groundwater or surface water. Sediment from erosion can also greatly degrade surface water quality. Irrigated agriculture also creates problems associated with excess water in the soil profile: waterlogging and salinization. Both problems can decrease productivity and lead to abandonment of the affected land. In India, China, and the United States–countries that rely heavily on irrigation–an average of 20 percent of irrigated land suffers from salinization. According to one estimate, salinization costs the world's farmers $11 billion/year in reduced income–almost 1 percent of the total value of agricultural production (Postel 1999:92; Wood et al. [PAGE] 2000).

One measure of the relative impact of various agroecosystems on freshwater systems is their efficiency of water use. Seckler et al. (1998) calculated average irrigation efficiency–the proportion of irrigation water that is actually consumed by crops for growth, compared with the proportion that evaporates or is otherwise wasted. More efficient irrigation systems require less water to meet crop needs, often by delivering water more directly to plant roots, and they are better timed to meet plant growth requirements.

Globally, irrigation efficiency averaged 43 percent in 1990 (Seckler et al. 1998:25). In general, agroecosystems in arid regions have more efficient irrigation systems. Irrigation efficiency in the driest regions runs as high as 58 percent, whereas regions with abundant water supplies have efficiency as low as 31 percent. Thirty-one percent efficiency means more than two-thirds of irrigation water in these areas is wasted, although some water lost to underground leakage may become available for downstream use (Seckler et al. 1998:25). Irrigation efficiencies in China and India are intermediate–39 percent and 40 percent, respectively.

The increasing competition for water from other sectors poses a challenge for agriculture, especially in developing countries where urban populations and the industrial sector are growing quickly. Both industrial and domestic water demands generally take precedence over agriculture. Indeed, irrigated agriculture may increasingly have to rely on recycled water from industrial facilities and wastewater treatment plants to meet its needs. Many believe that water scarcity and its impact on water services such as irrigation is one of the most immediate natural resource concerns from the perspective of human welfare (Rosegrant and Ringler 1999). Certainly, current trends emphasize the critical importance of developing agroecosystems that use water more efficiently, and that minimize the salinization and waterlogging of soils and the leaching of pesticides, fertilizer, and silt into surface and groundwater.

The Bottom Line for Water Services. Overall, the capacity of agroecosystems to maintain the quantity and quality of incoming water resources, and deliver those to downstream users, is declining. Although the consumptive use of water to produce more food represents an important and legitimate water service within agroecosystems, the deterioration in water quality that accompanies this is an often significant penalty for other ecosystems. Irrigation inefficiency increases water withdrawals and contributes to unsustainable rates of groundwater extraction, reduced river flows, and damage to aquatic ecosystems. Downstream water quality is particularly at risk in areas where farmers apply agrochemicals and animal manure abundantly. Poorly managed irrigation can also directly reduce the productivity of agroecosystems through waterlogging and salinization. Improvements in the efficiency of agricultural water use are increasingly important as both food needs and competing water demands continue to grow.

BIODIVERSITY

Agricultural lands support far less biodiversity than the natural forests, grasslands, and wetlands that they replaced. Even so, the biodiversity harbored in agricultural regions is important in its own right. From a purely agricultural perspective, the diversity of naturally occurring predators, bacteria, fungi, and plants in a region can contribute to agricultural production by helping to control pest and disease outbreaks, improving soil fertility and soil physical properties, and improving the resilience of agroecosystems to natural disasters such as floods and droughts. Moreover, the genetic diversity found in traditional crop varieties and in wild species provides a reservoir of genetic material that breeders can use to develop improved crop and animal varieties.

The expansion of agricultural land has, nonetheless, had major impacts on biodiversity. Using maps of the potential habitat that would naturally occur in a region, based on climate and soil characteristics, PAGE researchers estimated the percentage of different habitat types that had been converted to agriculture. Among the most heavily affected natural habitats, 46 percent of the potential area of temperate broadleaf and mixed forests is now agricultural land, accounting for 24 percent of total agricultural land. Close behind, 43 percent of the potential area of tropical deciduous forest (similar to rainforest, but with distinct dry seasons and more open canopy) has been converted to agriculture, accounting for 10 percent of total agricultural land. These types of forest are far more biodiverse than agroecosystems.

Within agroecosystems, different management practices can further alter biodiversity. Intensification tends to greatly diminish the capacity of agroecosystems to support biodiversity by fragmenting and reducing the area of hedgerows, copses, wildlife corridors, and other refuges and natural habitats within the agricultural landscape. Pesticides and other agrochemicals can also be toxic to wildlife and soil microorganisms, including many beneficial birds, pollinators, and carnivorous insects. On the positive side, the increasing use of trees on agricultural lands can increase their biodiversity potential. In Latin America, Sub-Saharan Africa, and South and Southeast Asia, trees are a significant and often a growing part of the agricultural landscape (Wood et al. [PAGE] 2000).

In addition to on-farm tree planting, positive trends include the increasing adoption of "no tillage cultivation," where disturbance of the soil is greatly minimized, helping to preserve soil integrity and minimize erosion. The use of integrated pest management, where pesticides are used more sparingly and in combination with nonchemical pest controls to protect crops, is also expanding. Further, the growth of high-yielding, intensive production systems has a positive side, too, in that it has forestalled the conversion of at least 170 Mha of natural habitat in the tropics (Nelson and Maredia 1999) and perhaps as much as 970 Mha worldwide (Golkany 1999).

In terms of genetic diversity, global agriculture focuses on relatively few species and thus begins from a somewhat narrow base. More than 90 percent of the world's caloric intake comes from just 30 crops, and only 120 crops are economically important at a national scale (FAO 1998:14). Nonetheless, there has traditionally been immense genetic diversity within these crop species, and this diversity has historically helped to maintain the productivity of agroecosystems and is a source of genetic material for modern plant breeding.

Today, however, crop genetic diversity is tending toward decline. Modern crop varieties are taking on more uniform characteristics, and these varieties are planted over large areas in monocultures. This tendency is not limited to high-income countries where the commercialization of agriculture is most prevalent. Modern crop varieties are displacing traditional varieties throughout the world, threatening the loss of an enormous genetic resource and increasing the vulnerability of large areas of homogeneous crops to pest and disease attack. Across all developing countries, modern rice varieties were being grown on 74 percent of the planted area in 1991, modern wheat on 74 percent in 1994, and modern maize on 60 percent in 1992 (Morris and Heisey 1998:220).

The Bottom Line for Biodiversity. Through habitat conversion, landscape fragmentation, the specialization of crop species, and intensification, agriculture plays an important role in shaping global patterns of biodiversity. Currently, the capacity of agroecosystems to support biodiversity is highly degraded, particularly in areas of intensive agriculture. Approaches to enhance biodiversity in agricultural regions while still maintaining or increasing production are only now beginning to develop. Better agricultural practices will almost certainly constitute central elements in any strategy to preserve global biodiversity in the 21st century.

CARBON STORAGE

Carbon is of fundamental importance to the fertility of agroecosystems. The organic matter content of soil, and its stability over time, are key indicators of long-term soil quality and fertility. The level of soil organic matter affects the water retention and tilth of soils, as well as the richness of the soil biota.

Typically, when natural ecosystems such as forest or savanna are converted to agriculture, their soils quickly lose a significant percentage of their soil organic matter. Successful agriculture can arrest this decline and rebuild soil organic matter to its original levels through appropriate crop rotations and the application of nutrients (particularly from organic sources), or through such practices as zero or minimum tillage. On the other hand, excessive tilling, removing crop residues from fields, and practices that promote soil erosion will accelerate loss of organic matter.

Carbon in agroecosystems–in both soils and vegetation–also plays an important role in the global carbon cycle. Except for some production systems in the tropics, agricultural soils generally store more carbon than do the crops or pastures they support. Agricultural vegetation stores an average of 5-6 kg of carbon per square meter (kgC/m^2), while agricultural soils store an average of 7-11 kgC/m^2 (Wood et al. [PAGE] 2000). Together, the vegetation and soils in agroecosystems contain approximately 26-28 percent of all the carbon stored in terrestrial ecosystems.

Land-use change and land management practices, of which agricultural activities are an important part, emit an estimated 1.6 GtC to the atmosphere annually, about 20 percent of human-related greenhouse gas emissions (IPCC 2000:5). There are many distinct agricultural sources of carbon emissions. Prime sources of carbon dioxide include conversion of forests and woody savannas to agricultural land, and deliberate burning of crop stubble and pastures to control pests and diseases and promote soil fertility. Other activities produce methane–another carbon-based molecule that is a more powerful greenhouse gas than CO_2. Livestock rearing and paddy rice cultivation are both major methane sources.

Some researchers believe that the net release of carbon dioxide from agriculture could decrease between 1990 and 2020 (Sombroek and Gommes 1996), while emissions of methane will continue to climb, pushed by the continuing growth in the number of livestock. Emissions of nitrous oxide (N_2O), an even more potent greenhouse gas derived from nitrogen fertilizers, is also rising rapidly.

There is a growing belief that agriculture can play a much greater role in reducing global carbon emissions and in increasing carbon storage. For example, control of agricultural burning, improved diets for cattle and other livestock, and soil conservation can reduce emissions. Meanwhile, better cultivation practices, mixing trees into agricultural systems, and planting improved pasture grasses can help store more carbon. Recent studies show that conservation programs and the adoption of no tillage cultivation in the United States increased carbon storage in U.S. croplands by around 138 MtC/year during the 1980s (Houghton et al. 1999:577).

The Bottom Line for Carbon Storage. Agroecosystems store about 26–28 percent of total terrestrial carbon—mostly in the soil. Improved nutrient management, reduced soil erosion, and the widely adopted use of minimum tillage cultivation tend to increase soil organic matter and, hence, can play some role in increasing carbon storage capacity in agricultural soils. On the other hand, livestock rearing and rice cultivation are significant and growing sources of carbon emissions tied to agriculture, and agricultural burning and land conversion remain prime sources as well.

COASTAL ECOSYSTEMS

The continental margins, where coastal ecosystems reside, are regions of remarkable biological productivity and high accessibility. This has made them centers of human activity for millennia. Coastal ecosystems provide a wide array of goods and services: they host the world's primary ports of commerce; they are the primary producers of fish, shellfish, and seaweed for both human and animal consumption; and they are also a considerable source of fertilizer, pharmaceuticals, cosmetics, household products, and construction materials.

Encompassing a broad range of habitat types and harboring a wealth of species and genetic diversity, coastal ecosystems store and cycle nutrients, filter pollutants from inland freshwater systems, and help to protect shorelines from erosion and storms. On the other side of shorelines, oceans play a vital role in regulating global hydrology and climate and they are a major carbon sink and oxygen source because of the high productivity of phytoplankton. The beauty of coastal ecosystems makes them a magnet for the world's population. People gravitate to coastal regions to live as well as for leisure, recreational activities, and tourism.

Extent and Modification

Many different definitions of *coastal zone* are in use. For the purpose of the ecosystem analysis, PAGE researchers define coastal regions as "the intertidal and subtidal areas above the continental shelf (to a depth of 200 m) and adjacent land area up to 100 km inland from the coast." The PAGE analysis of coastal ecosystems also includes marine fisheries because the bulk of the world's marine fish harvest–as much as 95 percent, by some estimates–is caught or reared in coastal waters (Sherman 1993:3). Only a small percentage comes from the open ocean (Box 2.10 Taking Stock of Coastal Ecosystems).

EXTENT

Because the world's coastal ecosystems are defined by their physical characteristics (their proximity to the coast) rather than a distinct set of biological features, they encompass a much more diverse array of habitats than do the other ecosystems in the PAGE study. Coral reefs, mangroves, tidal wetlands, seagrass beds, barrier islands, estuaries, peat swamps, and a variety of other habitats each provides its own distinct bundle of goods and services and faces somewhat different pressures.

(continues on p. 72)

Box 2.10 Taking Stock of Coastal Ecosystems

Highlights

- Almost 40 percent of the world's population lives within 100 km of a coastline, an area that accounts for only 22 percent of the land mass.
- Population increase and conversion for development, agriculture, and aquaculture are reducing mangroves, coastal wetlands, seagrass areas, and coral reefs at an alarming rate.
- Fish and shellfish provide about one-sixth of the animal protein consumed by people worldwide. A billion people, mostly in developing countries, depend on fish for their primary source of protein.
- Coastal ecosystems have already lost much of their capacity to produce fish because of overfishing, destructive trawling techniques, and destruction of nursery habitats.
- Rising pollution levels are associated with increasing use of synthetic chemicals and fertilizers.
- Global data on extent and change of key coastal habitats are inadequate. Coastal habitats are difficult to assess from satellite data because areas are small and often submerged.

Key

Condition assesses the current output and quality of the ecosystem good or service compared with output and quality of 20–30 years ago.

	Excellent	Good	Fair	Poor	Bad	Not Assessed
Condition						

Changing Capacity assesses the underlying biological ability of the ecosystem to continue to provide the good or service.

Increasing Mixed Decreasing Unknown

Changing Capacity

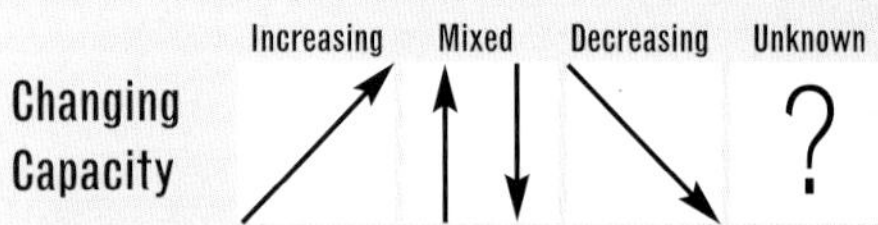

Scores are expert judgments about each ecosystem good or service over time, without regard to changes in other ecosystems. Scores estimate the predominant global condition or capacity by balancing the relative strength and reliability of the various indicators. When regional findings diverge, in the absence of global data weight is given to better-quality data, larger geographic coverage, and longer time series. Pronounced differences in global trends are scored as "mixed" if a net value cannot be determined. Serious inadequacy of current data is scored as "unknown."

Conditions and Changing Capacity

FOOD PRODUCTION

Global marine fish production has increased sixfold since 1950, but the rate of increase annually for fish caught in the wild has slowed from 6 percent in the 1950s and 1960s to 0.6 percent in 1995–96. The catch of low-value species has risen as the harvest from higher-value species has plateaued or declined, masking some effects of overfishing. Approximately 75 percent of the major fisheries are fully fished or overfished, and fishing fleets have the capacity to catch many more fish than the maximum sustainable yield. Some of the recent increase in the marine fish harvest comes from aquaculture, which has more than doubled in production since 1990.

WATER QUALITY

As the extent of mangroves, coastal wetlands, and seagrasses declines, coastal habitats are losing their pollutant-filtering capacity. Increased frequency of harmful algal blooms and hypoxia indicates that some coastal ecosystems have exceeded their ability to absorb nutrient pollutants. Although some industrial countries have improved water quality by reducing input of certain persistent organic pollutants, chemical pollutant discharges are increasing overall as agriculture intensifies and industries use new synthetic compounds. Furthermore, while large-scale marine oil spills are declining, oil discharges from land-based sources and regular shipping operations are increasing.

BIODIVERSITY

Indicators of habitat loss, disease, invasive species, and coral bleaching all show declines in biodiversity. Sedimentation and pollution from land are smothering some coastal ecosystems, and trawling is reducing diversity in some areas. Commercial species such as Atlantic cod, five species of tuna, and haddock are threatened globally, along with several species of whales, seals, and sea turtles. Invasive species are frequently reported in ports and enclosed seas, such as the Black Sea, where the introduction of Atlantic comb jellyfish caused the collapse of fisheries.

RECREATION

?

Tourism is the fastest-growing sector of the global economy, accounting for $3.5 trillion in 1999. Some areas have been degraded by the tourist trade, particularly coral reefs, but the effects of tourist traffic on coastal ecosystems at a global scale are unknown.

SHORELINE PROTECTION

Human modification of shorelines has altered currents and sediment delivery to the benefit of some beaches and detriment of others. Coastal habitats with natural buffering and adaptation capacities are being modified by development and replaced by artificial structures. Thus, the impact from storm surges has increased. Furthermore, rising sea levels, projected as a result of global warming, may threaten some coastal settlements and entire small island states.

Data Quality

FOOD PRODUCTION

Global data on fish landings are underreported in many cases or are not reported by species, which makes assessing particular stocks difficult. Data are fragmentary on how many fish are unintentionally caught and discarded, how many boats are deployed, and how much time is spent fishing, which obscures the full impact of fishing on ecosystems. Many countries fail to report data on smaller vessels and their fish landings.

WATER QUALITY

Global data on extent and change of wetlands and seagrasses are lacking, as are standardized and regularly collected data on coastal or marine pollution. Monitoring of nutrient pollution by national programs is uneven and often lacking. Current information relies heavily on anecdotal observation. Effective national programs are in place in some countries to monitor pathogens, persistent organic pollutants, and heavy metals, but data are inconsistent. No data are available on oil pollution from nonpoint sources.

BIODIVERSITY

Detailed habitat maps are available for only some areas. Loss of mangrove, coastal wetlands, and seagrasses are reported in many parts of the world, but little is documented quantitatively. Species diversity is not well inventoried, and population assessments are available only for some key species, such as whales and sea turtles. Data on invasive species are limited by difficulty in identifying them and assessing their impact. Few coral reefs have been monitored over time. Information on the ecological effects of trawling is poorly documented.

RECREATION

Typically, only national data on tourism are available, rather than data specific to coastal zones. Not all coastal countries report tourism statistics, and information on the impacts of tourism and the capacity of coastal areas to support tourism is very limited.

SHORELINE PROTECTION

Information on conversion of coastal habitat and shoreline erosion is inadequate. Information is lacking on long-term effects of some coastal modifications on shorelines. Predictions of sea level rise and storm effects as a result of climate change are speculative.

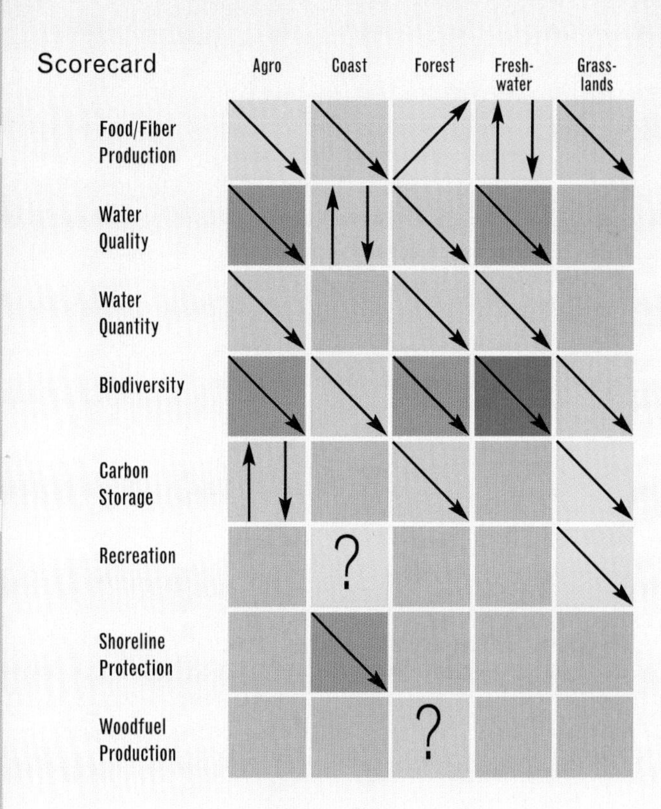

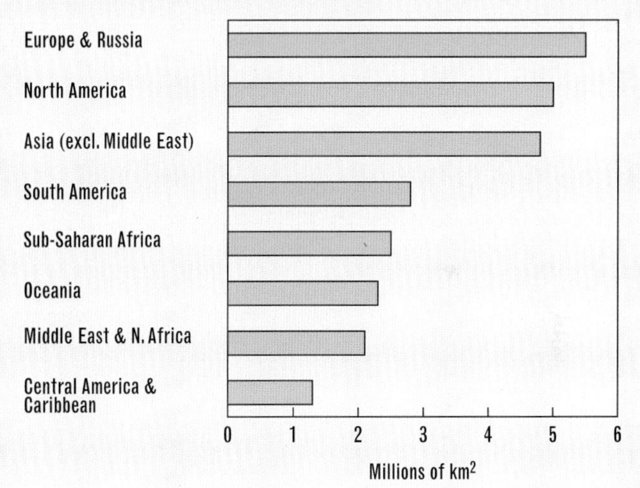

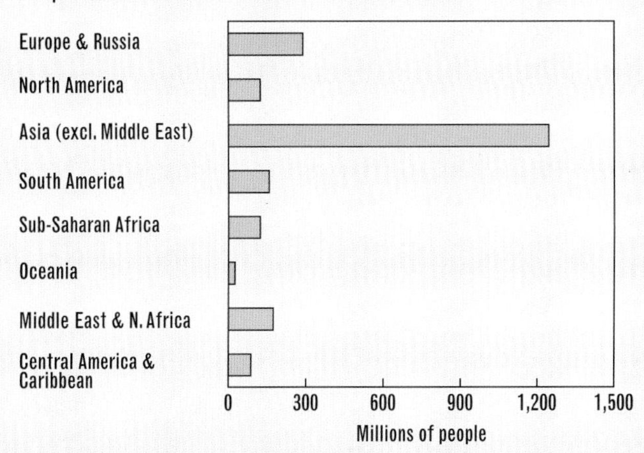

The extent of coastal ecosystems and how they have been modified over time is less well known than are the extents of the other ecosystems examined in the PAGE study. Individual coastal habitats such as wetlands or coral reefs tend to cover relatively small areas, and detailed mapping is necessary to accurately measure extent or change in these areas. Until the advent of satellite imagery, such mapping was beyond the capacity of most nations. Even today, high-resolution mapping of these systems is imperfect and expensive and has not been attempted at a global scale for the entire 1.6 million km of coastlines (Burke et al. [PAGE] 2000).

MODIFICATIONS

In the absence of such maps, PAGE researchers used satellite imagery to estimate how much coastal area remains in natural vegetation (dunes, wetlands, wooded areas, etc.) versus how much is now urban and agricultural land. Overall, 19 percent of all lands within 100 km of the coast is classified as highly altered, meaning they have been converted to agricultural or urban uses, 10 percent semialtered (involving a mosaic of natural and altered vegetation), and 71 percent are unaltered (Burke et al. [PAGE] 2000) (Box 2.11 Coastal Population and Altered Land Cover).

Mangroves and Coral Reefs

More detailed information is available about the extent and modification of a few coastal habitats, such as mangroves and coral reefs, than is known about the extent of coastal ecosystems. Mangroves line approximately 8 percent of the world's coastline (Burke et al. [PAGE] 2000) and about one-quarter of tropical coastlines, covering a surface area of approximately 181,000 km^2 (Spalding et al. 1997:23). Some 112 countries and territories have mangroves within their borders (Spalding et al. 1997:20). Although scientists cannot determine exactly how extensive mangroves were before people began to alter coastlines, based on historical records, anywhere from 5 to nearly 85 percent of original mangrove area in various countries is believed to have been lost. Extensive losses have occurred in the last 50 years. For example, much of the estimated 84 percent of original mangroves lost to Thailand were lost since 1975 (MacKinnon 1997:167; Spalding et al. 1997:66); Panama lost 67 percent of its mangroves just during the 1980s (Davidson and Gauthier 1993) (Box 2.12 Mangroves). Overall, it is estimated that half of the world's mangrove forests have been destroyed (Kelleher et al. 1995:30). Although the net trend is clearly downward, in some regions mangrove area is actually increasing as a result of plantation forestry and small amounts of natural regeneration (Spalding et al. 1997:24).

Knowledge of the extent and distribution of coral reefs is probably greater than for any other marine habitat. Rough global maps of coral reefs have existed since the mid-1800s because of the hazard they posed to ships. WCMC has compiled a coarse-scale (1:1,000,000) map of the world's shallow coral reefs; more detailed maps exist for many countries. Worldwide, an estimated 255,000 km^2 of shallow coral reefs exist, with more than 90 percent in the Indo-Pacific region (Spalding and Grenfell 1997:225, 227) (Box 2.13 Coral Reefs). Adding deep water reefs would make the total reef area much higher–perhaps more than double the area–but these deeper reefs are poorly mapped.

Both reef-building corals and coral reef fish show broadly similar patterns in the distribution of species richness, with highest species diversity in the Indo-Pacific region and lower diversity in the Atlantic. Currently, on a global basis, coral reef degradation is a more serious problem than outright loss of coral through, for example, land reclamation and coral mining. Nonetheless, coral area has been significantly reduced in some parts of the world.

Other Coastal Ecosystems

No comprehensive global information, and only limited reliable national information, is available to document change in seagrass habitats, peat swamps, or other types of coastal wetlands besides mangroves. Where data do exist, however, the habitat loss is often dramatic. For example, 46 percent of Indonesia's and as much as 98 percent of Vietnam's peat swamps are believed to have been lost (MacKinnon 1997:104, 175). Similarly, the extent of change in seagrass habitats is thought to be high. In the United States, more than 50 percent of the historical seagrass cover has been lost from Tampa Bay, 76 percent from the Mississippi Sound, and 90 percent from Galveston Bay because of population growth and changes in water quality (NOAA 1999:19).

PRESSURES ON COASTAL ECOSYSTEMS

Along with direct loss of area, a variety of other factors are significantly altering coastal ecosystems. Chief among these are population growth, pollution, overharvesting, and the looming threat of climate change.

Population

Globally, the number of people living within 100 km of the coast increased from roughly 2 billion in 1990 to 2.2 billion in 1995–39 percent of the world's population (Burke et al. [PAGE] 2000). However, the number of people whose activities affect coastal ecosystems is much larger than the actual coastal population because rivers deliver pollutants from inland watersheds and populations to estuaries and surrounding coastal waters. As coastal and inland populations continue to grow, their impacts–in terms of pollutant loads and the development and conversion of coastal habitats–can be expected to grow as well.

Pollution

A vast range of pollutants affects the world's coasts and oceans. These can be broadly classified into toxic chemicals (including organic chemicals, heavy metals, and radioactive

(continues on p. 76)

Box 2.11 Coastal Population and Altered Landcover

In 1990, 2 billion people lived within 100 km of the sea. By 1995, coastal areas were home to 200 million more, or 39 percent of the population.

Concentrated coastal populations are having a profound impact on marine coastal ecosystems. Much of the shoreline has been developed to meet needs for shelter, subsistence, commerce, and recreation. Even inland populations have an impact on coastal ecosystems. Coastal problems such as algal blooms and eutrophication can be attributed to added pollutants and nutrients from inland freshwater systems.

Overall, 29 percent of all lands within 100 km of a coastline is classified as altered—19 percent is highly altered, converted to agricultural and urban uses; and 10 percent is semialtered, with natural vegetation and cropland interspersed. Some 71 percent remains unaltered.

Population Living Near a Coastline, 1995

Proximity to Coastline	Population (cumulative totals in thousands)	Percentage of Total Population
Within 25 km	1,143,828	20
Within 50 km	1,645,634	29
Within 100 km	2,212,670	39

Natural and Altered Landcover within 100 km of a Coast

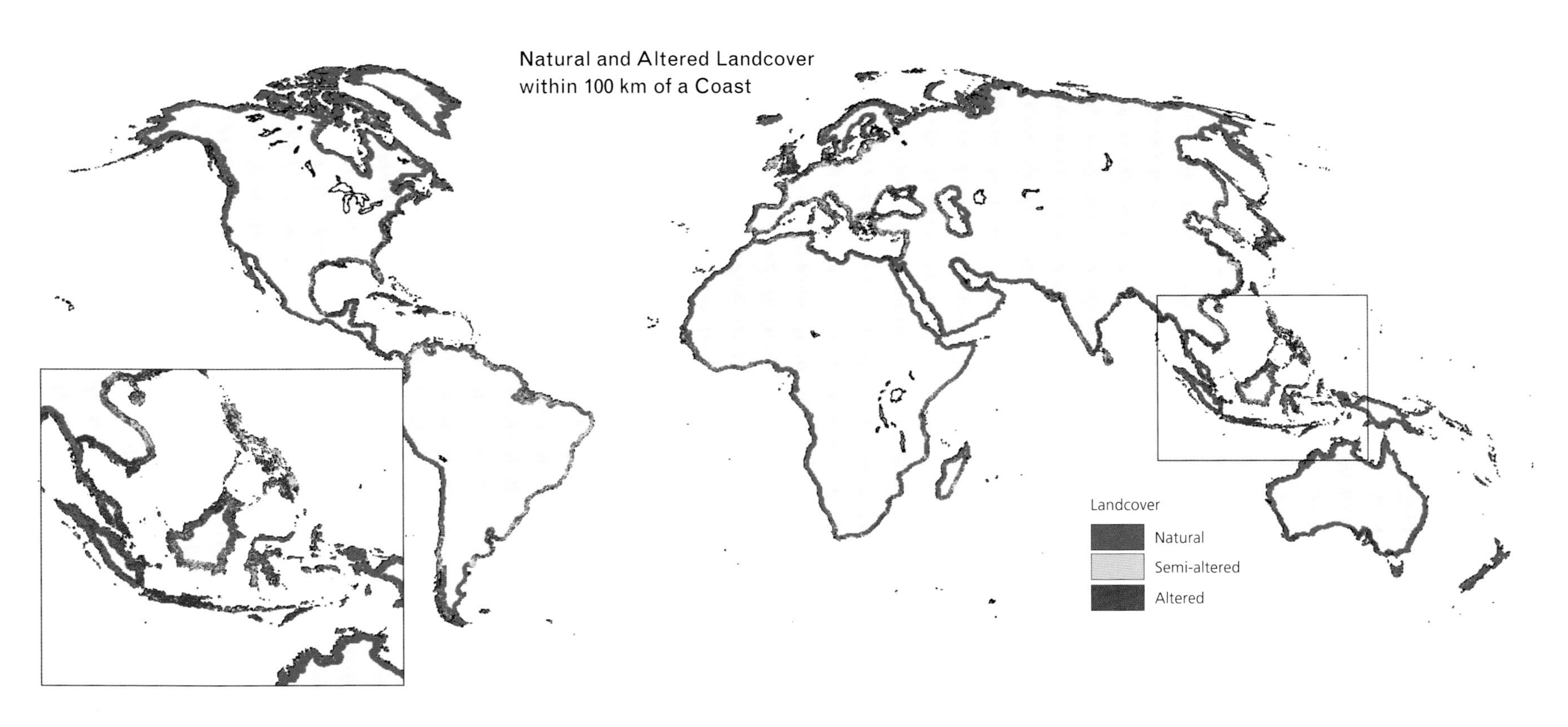

Sources: Burke et al. [PAGE] 2000. The map is based on Global Land Cover Characteristics Database Version 1.2 (Loveland et al. [2000]). The table is based on CIESIN (2000).

Box 2.12 Mangroves

Mangroves line 8 percent of the world's coasts and about one-quarter of the world's tropical coastlines, covering a surface area of approximately 181,000 km^2 (Spalding et al. 1997:23). Adapted to conditions of varying salinity and water level, they flourish in sheltered coastal areas, such as river estuaries.

Mangroves are crucial to the productivity of tropical fisheries because they act as spawning grounds for a wide range of fish species. They also provide local communities with timber and fuelwood and help stabilize coastlines.

Historical records indicate that the original extent of mangrove forests has declined considerably under pressure from human activity. National proportions of original mangrove cover lost vary from 4 to 84 percent, with the most rapid losses occurring in recent decades. Overall, as much as half of the world's mangrove forests may have been lost (Kelleher et al. 1997:30)

Excessive cutting for fuel and timber as well as clearance for agriculture and shrimp farming and for coastal development have all contributed to these high loss rates. In a few regions, however, mangrove area is actually increasing as a result of plantation forestry and natural regeneration.

Mangrove Area in Selected Countries

Region and Country	Current Extent (km^2)	Approximate Loss (%)	Period
Africa			
Angola	*1,100*	50	Original extent to 1980s
Cote d'Ivoire	640	60	Original extent to 1980s
Gabon	*1,150*	50	Original extent to 1980s
Guinea-Bissau	*3,150*	70	Original extent to 1980s
Kenya	*610*	4	1971–88
Tanzania	*2,120*	60	Original extent to 1980s
Latin America and the Caribbean			
Costa Rica	*413*	–6	1983–90
El Salvador	*415*	8	1983–90
Guatemala	161	31	1960s–90s
Jamaica	106	30	Original extent to 1990s
Mexico	5,315	65	1970s–90s
Panama	*1,581*	67	1983–90
Peru	51	25	1982–92
Asia			
Brunei	200	20	Original extent to 1986
Indonesia	*24,237*	55	Original extent to 1980s
Malaysia	*2,327*	74	Original extent to 1992–93
Myanmar	*4,219*	75	Original extent to 1992–93
Pakistan	*1,540*	78	Original extent to 1980s
Philippines	*1,490*	67	1918–80s
Thailand	1,946	84	Original extent to 1993
Vietnam	2,525	37	Original extent to 1993
Oceania			
Papua New Guinea	*4,627*	8	Original extent to 1992–93

Source: Burke et al. [PAGE] 2000. The table is based on *World Resources 1990–91*; UNEP Kenya Coastal Zone Database (1997), Spalding et al. (1997), Davidson and Gauthier (1993), MacKinnon (1997), World Bank (1989), and BAP Planning (1993). Current extent estimates in italics are not in agreement with recent estimates in the Data Tables in this volume, because of differences in year assessed and methodology.

Box 2.13 Coral Reefs

Coral reefs exist mostly in shallow tropical waters with minimal silt content. Shallow coral reefs occupy only 255,000 km² of the world's surface. Nonetheless, they support nearly 1 million species of plants and animals (Reaka-Kudla 1997; Spalding and Grenfell 1997:225). Besides harboring rich biodiversity, coral reefs provide an accessible area for small-scale fishing and help to protect coastlines from storm damage.

Coral reefs are most extensive around the islands and coasts of the Western Pacific and Southeast Asia, which together encompass two-thirds of the world's coral ecosystems. These areas are also the richest in species diversity.

Coral ecosystems are extremely vulnerable to the direct and indirect effects of human activity. In many parts of the world, reef area has been reduced by land reclamation, coastal development, and coral mining. Such direct threats can be combated by extending protected-area status, but the indirect effects of human activity such as increased siltation, pollution, and increases in sea level and temperature are broader in impact and harder to counter.

The mass bleaching of reefs that occurred during the 1997–98 El Niño was the most extensive such event yet recorded. If, as is generally thought, coral bleaching is caused by elevated sea temperatures, global warming is likely to make these events more severe and more threatening to the long-term survival of reefs.

Global and Regional Reef Areas, 1997

Region	Area (thousands of km²)	Percentage of Total Area
WORLD	255	100.0
Indo-Pacific	233	91.4
Western Pacific (including Hawaii)	105	41.2
Eastern Pacific	3	1.2
Red Sea	17	6.7
Arabian Gulf	3	1.2
Indian Ocean	36	14.1
Southeast Asia	68	26.7
Atlantic	22	8.6
Wider Caribbean	21	8.2
West Africa	1	0.4

Coral Bleaching Events in 1997–98

Sources: Burke et al. [PAGE] 2000. The map shows observations of coral bleaching from NOAA/NESDIS (2000) and WCMC (1999) and sea surface temperature data from NOAA/NESDIS (2000).

waste), nutrients (including agricultural fertilizers and sewage), sediments, and solid waste. The occurrence of bacterial contamination is a special case, often associated with nutrient pollution. Oil pollution (from spills and seepage) includes toxic, nutrient, and sediment-based pollutants.

Most pollution of coastal waters comes from the land, but atmospheric sources and marine-based sources such as oil leaks and spills from vessels also play a role. Approximately 40 percent of toxic pollution in Europe's coastal waters is thought to stem from atmospheric deposition; the percentage could be even greater in the open ocean (Thorne-Miller and Catena 1991:18; EEA 1998:213).

In some regions, such as North America and Europe, heavy metal and toxic chemical pollution has decreased in recent decades as the use of these compounds has decreased, but toxic chemicals continue to be a major problem worldwide (NOAA 1999:14; EEA 1998:216). Some progress has also been achieved in reducing the volume of oil spilled into the oceans. Both the number of oil spills and total amounts of oil spilled have decreased considerably since the 1970s (ITOPF 1999; Etkin 1998:10). Indeed, spills from vessels, although they can be catastrophic, are not the major source of oil pollution; runoff and routine maintenance of oil infrastructure are estimated to account for more than 70 percent of the total annual oil discharged into the ocean (National Research Council 1985:82).

Nutrient pollution, especially nitrates and phosphates, has increased dramatically this century. Greater use of fertilizers, growth in quantities of domestic and industrial sewage, and increased aquaculture, which releases considerable amounts of waste directly into the water, are all contributing factors (GESAMP 1990:96). Some local improvements in nutrient pollution have been achieved through sewage treatment and bans on phosphate detergents (NOAA 1999:iv; EEA 1999:155). However, the Joint Group of Experts on the Scientific Aspects of Marine Pollution (GESAMP) identified marine eutrophication, caused by these nutrients, as one of the most immediate causes of concern in the marine environment (GESAMP 1990:3) (Box 2.14 Pollution in Coastal Areas).

Overharvesting

Forty-five years of increasing fishing pressure have left many major fish stocks depleted or in decline. Yet overfishing is not a new phenomenon; it was recognized as an international problem as long ago as the early 1900s (FAO 1997:13). Prior to the 1950s, however, the problem was much more confined, since only a few regions such as the North Atlantic, the North Pacific, and the Mediterranean Sea were heavily fished and most world fish stocks were not extensively exploited. Since then, the scale of the global fishing enterprise has grown rapidly and the exploitation of fish stocks has followed a predictable pattern, progressing from region to region across the world's oceans. As each area in turn reaches its maximum productivity, it then begins to decline (Grainger and Garcia 1996:8, 42–44) (Box 2.15 Overfishing).

Overexploitation of fish, shellfish, seaweeds, and other marine organisms not only diminishes production of the harvested species but can profoundly alter species composition and the biological structure of coastal ecosystems. Overharvest stems in part from overcapacity in the world fishing fleet. Worldwide, 30–40 percent more harvest capacity exists than the resource can withstand (Garcia and Grainger 1996:5). A recent review of Europe's fisheries by the European Union indicates that the fishing fleet plying European waters would need to be reduced by 40 percent to bring it into balance with the remaining fish supply (FAO 1997:65).

Trawling. Not only is harvesting excessive, but many modern harvesting methods are destructive as well. Modern trawling equipment that is dragged along the sea bottom to catch shrimp and bottom-dwelling fish such as cod and flounder can devastate the seafloor community of worms, sponges, urchins, and other nontarget species as it scoops through the sediment and scrapes over rocks. Extent of damage to sea-bottom habitats that have been swept by trawling equipment may be light, with effects lasting only a few weeks, or intensive, with some impacts on corals, sponges, and other long-lived species lasting decades or even centuries (Watling and Norse 1998:1185–1190).

One global estimate puts the area swept by trawlers at 14.8 million km^2 of the seafloor (Watling and Norse 1998:1190). To better estimate the percentage of the continental shelf areas affected by trawling, PAGE researchers mapped the total area of trawling grounds for 24 countries for which sufficient data were available. These countries include about 41 percent of the world's continental shelves. The PAGE analysis shows that trawling grounds covered 57 percent of the total continental shelf area of these countries (Burke et al. [PAGE] 2000) (Box 2.16 Trawling).

Bycatch. Another destructive practice associated with commercial fishing comes from the "bycatch" or unintended catch of nontarget species as well as juvenile or undersized fish of the target species. Some of these fish are kept for sale, but many are discarded and eventually thrown back to the sea, where most die of injuries and exposure. Fisheries experts estimate that bycatch accounts for roughly 25 percent of the global marine fish catch–some 20 million metric tons per year (FAO 1999a:51). In certain fisheries, bycatch can outweigh the catch of target species. For example, in the shrimp capture fishery, discards may outweigh shrimp by a ratio of 5 to 1 (Alverson et al. 1994:24).

Climate Change

Global climate change may compound other pressures on coastal ecosystems through the additional effects of warmer ocean temperatures, altered ocean circulation patterns,

Box 2.14 Pollution in Coastal Areas

Marine nutrient pollution, especially from nitrates and phosphates, has increased dramatically this century largely because of increased use of agricultural fertilizers and growing discharges of domestic and industrial sewage (GESAMP 1990:96). Excessive nutrient concentrations in water can stimulate excessive plant growth—eutrophication. As the plant matter becomes more abundant, its decomposition can reduce oxygen concentrations in the water to less than the 2 parts per million needed to support most aquatic animal life. This not only jeopardizes native species, it also jeopardizes human health, livelihoods, and recreation.

Harmful algal blooms, which consist of algae that produce harmful biotoxins, can also be fueled by excessive nutrient runoff. More than 60 kinds of algal toxins are known today (McGinn 1999), and the number of incidents annually affecting public health, fish, shellfish, and birds has increased from around 200 in the 1970s to more than 700 in the 1990s (HEED 1998).

Hypoxia, the depletion of dissolved oxygen, is also related to nutrient pollution of coastal waters. Fish leave or avoid hypoxic areas and bottom-dwellers such as shrimp, crabs, snails, clams, starfish, and worms eventually suffocate. Current data suggest that hypoxic zones occur most frequently in enclosed waters adjacent to intensively farmed watersheds and major industrial centers off the coasts of Europe, the United States, and Japan.

Global Distribution of Hypoxic Zones

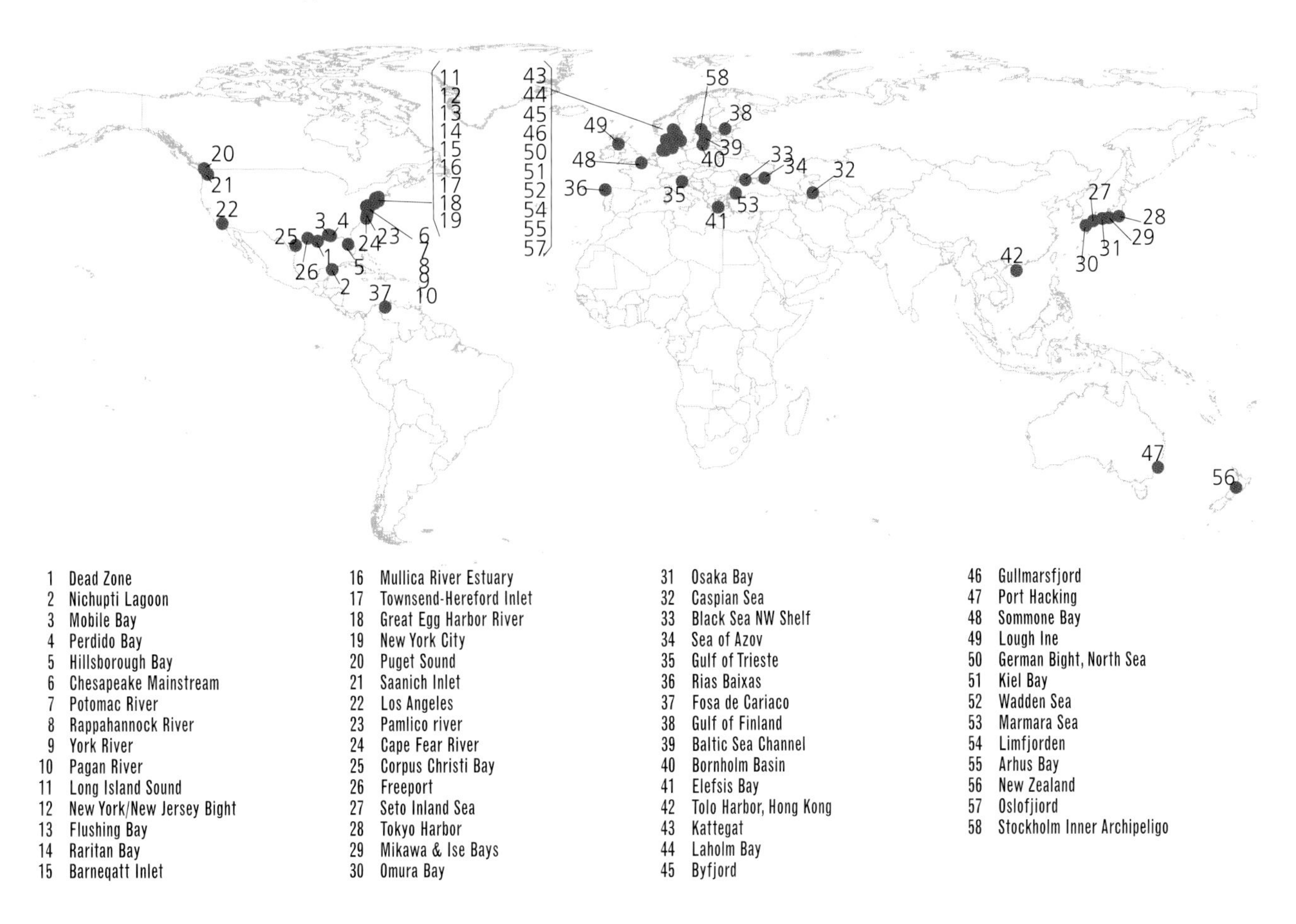

Source: Burke et al. [PAGE] 2000. The map is based on R.J. Diaz, Virginia Institute of Marine Science, personal communication (1999), updating Diaz and Rosenberg (1995).

Box 2.15 Overfishing

Prior to the 1950s, overfishing was confined to heavily fished regions in the North Atlantic, North Pacific, and Mediterranean Sea. Today overfishing is global, and current harvest trends put fishing, as both a source of food and a source of employment, at risk.

Fish account for one-sixth of all animal protein in the human diet, and around 1 billion people rely on fish as their primary protein source. As demand for fish has increased, many major stocks have declined or have been depleted. FAO reports that as of 1999, more than a quarter of all fish stocks are already depleted as a result of past overfishing or are in imminent danger of depletion from current overharvesting. Almost half of all fish stocks are being fished at their biological limit and are therefore vulnerable to depletion if fishing intensity increased.

Employment within fisheries is likely to change profoundly, especially for small-scale fishers who fish for the local market or for subsistence. Over the past 2 decades, these fishers, who number some 10 million worldwide, have been losing ground as competition from commercial vessels has grown. However, commercial fleets don't face bright prospects, either. Worldwide the fishing industry has 30–40 percent more harvest capacity than fish stocks can support, and the European Union recently estimated that the fleet working in Europe would need to be reduced 40 percent to bring it into balance with the remaining supply of fish.

A History of Decline: Peak Fish Catch vs. 1997 Fish Catch, by Ocean

Fishing Area	1997 Catch (thousand tons)	Maximum Catch (thousand tons)	Year of Maximum Catch
Atlantic			
Northeast	11,663	13,234	1976
Northwest	2,048	4,566	1968
Eastern Central	3,553	4,127	1990
Western Central	1,825	2,497	1984
Southeast	1,080	3,271	1978
Southwest	2,651	2,651	1997
Pacific			
Northeast	2,790	3,407	1987
Northwest	24,565	24,565	1997
Eastern Central	1,668	1,925	1981
Western Central	8,943	9,025	1995
Southeast	14,414	20,160	1994
Southwest	828	907	1992
Indian			
Eastern	3,875	3,875	1997
Western	4,091	4,091	1997
Mediterranean	1,493	1,990	1988
Antarctic	28	189	1971

Fishing Grounds Overfished or Fully Fished, 1994

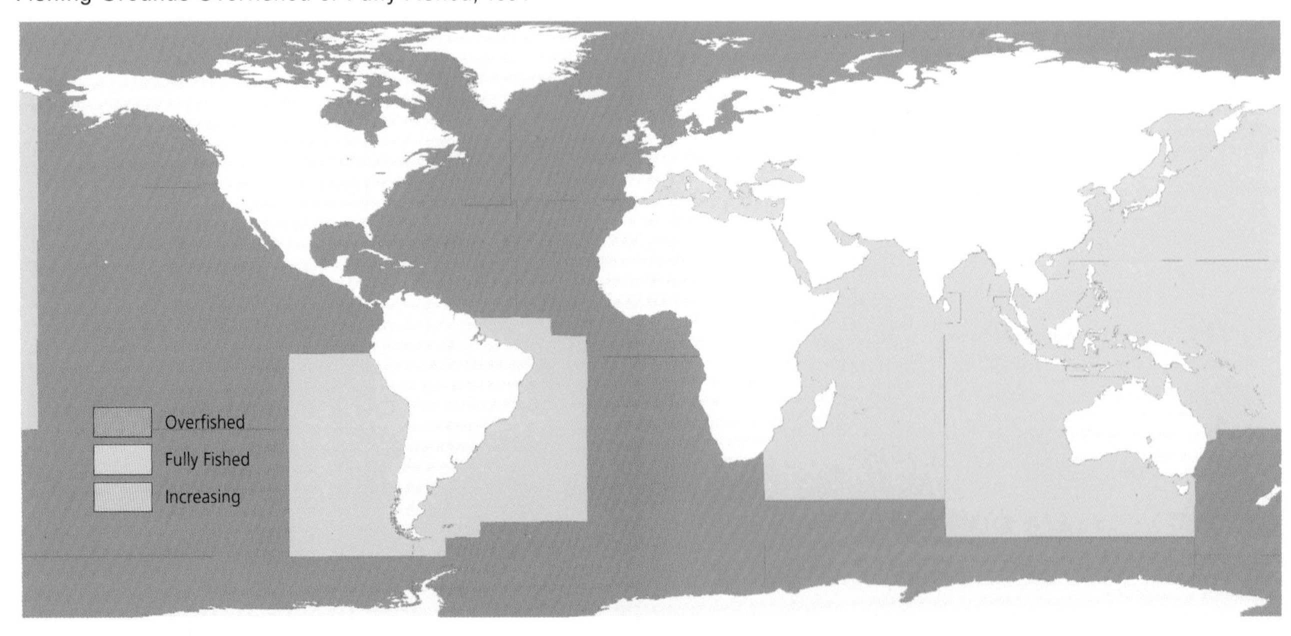

Source: Burke et al. [PAGE] 2000. The map is based on Grainger and Garcia (1996); analysis is based on landings data collected between 1950 and 1994 for the top 200 species-/fishing-area combinations, which represent 77 percent of the world's marine production, as explained in the technical notes for Data Table 4 in Coastal, Marine, and Inland Waters. Table is based on FAO (1999c, 1999d).

changing storm frequency, and rising sea levels. Changing concentrations of CO_2 in ocean waters may also affect marine productivity or even change the rate of coral calcification (Kleypas et al. 1999). The widespread coral bleaching observed during the 1997-98 El Niño is a dramatic example of the effect of elevated temperatures at the sea surface. Similarly, changes in ocean currents and circulation patterns could dramatically affect the biological composition of coastal ecosystems by changing both the physical characteristics of the habitat–the water temperature and salinity–and the pattern of migration of larvae and adults of different species.

Rising sea level, associated with climate change, is likely to affect virtually all of the world's coasts. During the past century, sea level has risen at a rate of 1.0-2.5 mm per year (IPCC 1996:296). The Intergovernmental Panel on Climate Change (IPCC) has projected that global sea level will rise 15-95 cm by the year 2100, due principally to thermal expansion of the ocean and melting of small mountain glaciers (IPCC 1996:22).

Some of the areas most vulnerable to rising seas are coastal lands whose highest points are within 2 m of sea level, in particular the so-called "lands of no retreat"–islands with more than half of their area less than 2 m above sea level. Rising sea levels will also increase the impact of storm surges. This, in turn, could accelerate erosion and associated habitat loss, increase salinity in estuaries and freshwater aquifers, alter tidal ranges, change sediment and nutrient transport, and increase coastal flooding. River deltas are at risk from flooding as a result of sea-level rise as are saltwater marshes and coastal wetlands if they are blocked from migrating inland by shoreline development (NOAA 1999:20).

Assessing Goods and Services

FOOD FROM MARINE FISHERIES

The forecast for world fisheries is grim despite the fact that fish provided 16.5 percent of the total animal protein consumed by humans in 1997 (Laureti 1999:63). On average this accounts for 6 percent of all protein–plant and animal–that humans eat annually. Approximately 1 billion people rely on fish as their primary source of animal protein (Williams 1996:3). Dependence on fish is highest in developing nations: of the 30 countries most dependent on fish as a protein source, all but four are in the developing world (Laureti 1999:v). In developing countries, production of fish products is almost equal to the production of all major meats–poultry, beef, sheep, and pork (Williams 1996:3).

Global marine fish and shellfish production has increased sixfold from 17 million tons in 1950 to 105 million metric tons in 1997 (FAO 1999c). This rapid growth–particularly in the last 20 years–has come partly from growth in aquaculture, which now accounts for more than one-fifth of the total harvest (marine and inland) (FAO 1999a:10). From 1984 to 1997, aquaculture production in marine and brackish environments tripled and continues to expand rapidly (FAO 1999c). Another 30 percent of the marine harvest consists of small, low-valued fish like anchovies, pilchard, or sardines, many of which are reduced to fish meal and used as a protein supplement in feeds for livestock and aquaculture. Over time, the percentage of the global catch made up by these low-value species has risen as the harvest of high-value species like cod or hake has declined, partially masking the effects of overfishing (FAO 1997:5).

Fish and shellfish production is of global economic importance and is particularly significant for developing countries, where more than half of the export trade in fish products originates (FAO 1999a:21). The value of fishery exports in 1996 amounted to US$52.5 billion, 11 percent of the value of agricultural exports that year (FAO 1999a:20).

Employment

Fishing and aquaculture are major sources of employment as well, providing jobs for almost 29 million people worldwide in 1990 (FAO 1999a:64). Some 95 percent of these fish-related jobs were in developing countries (FAO 1999b). The pattern of employment within the fisheries sector is likely to shift dramatically in coming years, especially for small-scale fishers harvesting fish for local markets and subsistence. Small-scale fishers have been losing ground over the last 2 decades as competition from commercial vessels has grown. Surveys off the west coast of Africa show that fish stocks in the shallow inshore waters where artisanal fishers ply their trade dropped by more than half from 1985 to 1990 because of increased fishing by commercial trawlers (FAO 1995:22). This trend is likely to intensify as fish stocks near shore continue to decrease under heavy fishing pressure.

Ecosystem Condition

The condition of coastal ecosystems, from the standpoint of fisheries production, is poor. Yields of 35 percent of the most important commercial fish stocks declined between 1950 and 1994 (Grainger and Garcia 1996:31). As of 1999, FAO reported that 75 percent of all fish stocks for which information is available are in urgent need of better management–28 percent are either already depleted from past overfishing or in imminent danger of depletion due to current overharvesting, and 47 percent are being fished at their biological limit and therefore vulnerable to depletion if fishing intensity increased (Garcia and DeLeiva 2000).

Another indicator of the condition of coastal fisheries is the relative abundance of fish stocks at different levels of the food web. In many fisheries, the most prized fish are the large predatory species high on the food web, such as tuna, cod, hake, or salmon. When these "top predators" are depleted through heavy fishing pressure, other species lower on the

Box 2.16 Trawling

Increasingly, trawling—dragging weighted nets across the sea floor to catch shrimp and bottom-dwelling fish—is taking place beyond the continental shelf. Harvesters are trawling at depths up to 400 m and, in some places, more than 1,500 m. An estimated 14.8 million km^2 of the sea floor is swept by trawlers (Watling and Norse 1998:1190). PAGE researchers mapped the total area of trawling grounds for 24 countries for which sufficient data were available. Trawling grounds in these countries encompass 8.8 million km^2. Extrapolating from these figures suggests that the world's trawling grounds total approximately 20 million km^2, nearly two and one-half times the size of Brazil.

Trawling sea floors is a major source of pressure on the biodiversity of coastal ecosystems. Modern trawling techniques are capable not only of rapidly depleting targeted fish stocks, but also of damaging or destroying nontarget species including corals and sponges. Because deep-living species tend to grow more slowly than shallow-water species, the long-term impact of trawling is magnified as trawl depths increase.

The thick natural carpet of bottom-dwelling plants and animals is important for the survival of the fry of groundfish such as cod, which find protection there (Watling and Norse 1998:1184). Thus, destruction of sea-floor habitats is one of the principal factors in the decline of fishing stocks in heavily trawled areas.

Trawled Areas of the World

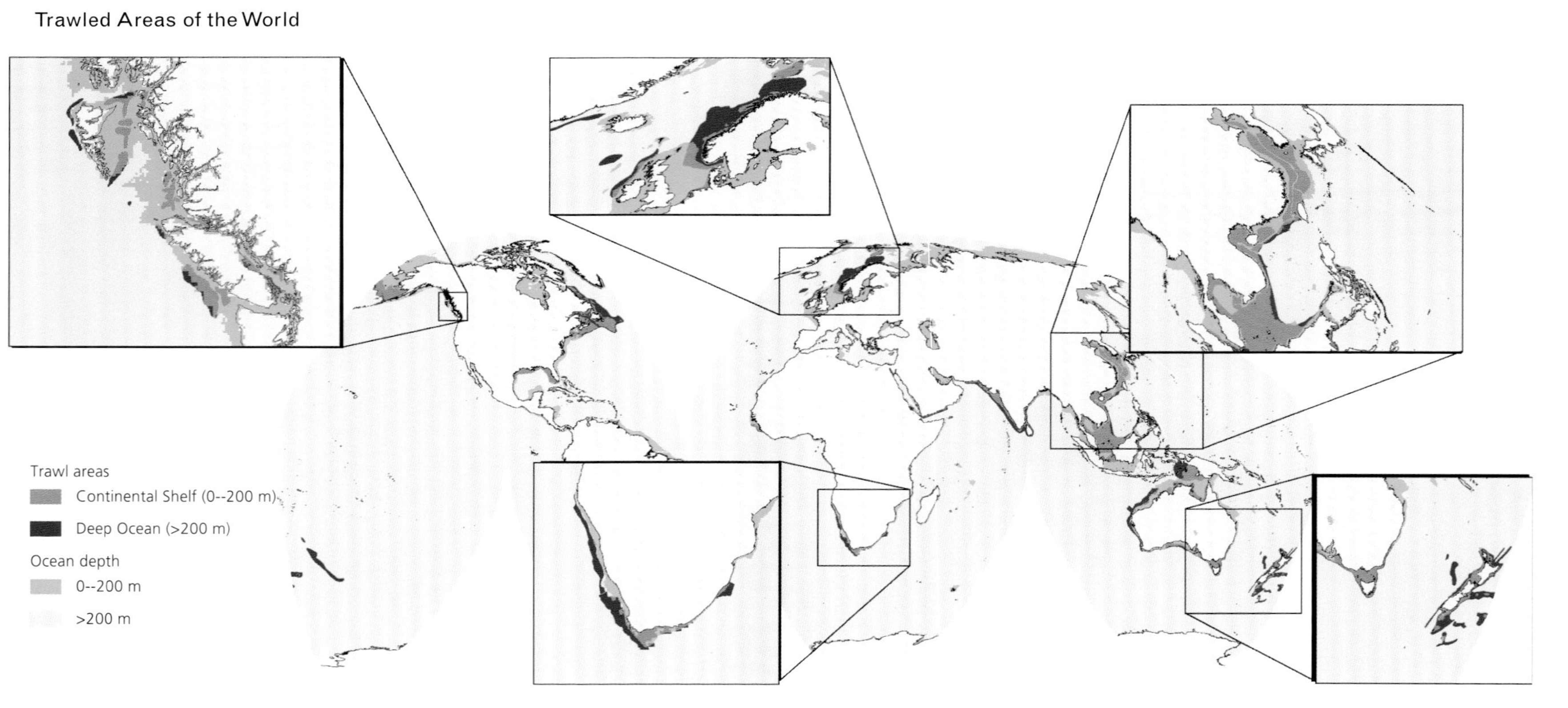

Source: Burke et al. [PAGE] 2000. The map is based on McAllister, D., et al. (1999). Data reflect preliminary results of a partial global trawling survey.

food web–plankton eaters–may begin to dominate the fish catch. This pattern of exploitation was described by Pauly et al. (1998) as "fishing down the food web," and it may signal a deterioration in the species structure of the ecosystem.

On behalf of the PAGE study, FAO analyzed global catch statistics for signs of ecosystem change, particularly for signs of "fishing down the food web." The results of the analysis show relatively strong evidence of this exploitation pattern only in the Northern Atlantic. Other regions show shifts in the relative abundance of species; but only in the North Atlantic did fishing practices seem to be the major influence causing this broad-scale ecosystem shift (Burke et al. [PAGE] 2000). In other areas such as the Mediterranean and Baltic Seas, an increase in plankton-eating fish low on the food web may indicate the presence of excess nutrients, which stimulates plankton growth and thus provides a larger food supply for plankton eaters (Caddy et al. 1998).

Continued deterioration of coastal ecosystems and the fish stocks they support could have serious implications for future fish consumption. FAO expects demand for fish and shellfish as a human food source to continue to increase well beyond today's consumption of 93 million tons per year. FAO warns that only under the most optimistic scenario–where aquaculture continues to expand rapidly and overfishing is brought under control so that fish stocks can recover–will there be enough fish to meet global demand (FAO 1999d). If the present deterioration continues, however, a substantial gap between supply and demand will likely develop, raising the price of fish and threatening food security in some regions (Williams 1996:14–15, 25–26).

The Bottom Line for Food Production. Global marine fish stocks still yield significant supplies of fish and shellfish, and marine aquaculture production is growing rapidly. However, current fishing practices show a global pattern of stock depletions and destructive fishing techniques that harm coastal ecosystems. Currently, nearly 75 percent of assessed fish stocks are either overfished or fished at their biological limit and susceptible to overfishing. Other factors, such as water pollution and loss of spawning habitat compound the harm. As a result, the capacity of the world's coastal and marine ecosystems to produce fish for human harvest is highly degraded and is continuing to decline. This could have a significant impact on nutrition and local and national economies in many countries.

WATER QUALITY

Coastal ecosystems provide the important service of maintaining water quality by filtering or degrading toxic pollutants, absorbing nutrient inputs, and helping to control pathogen populations. But the capacity of estuaries and coasts to provide these services can easily be exceeded in at least three ways. First, toxic pollutants can build to levels in fish and shellfish that are harmful to human health. Second, polluted coastal waters can harbor pathogens such as cholera and hepatitis A, which are also significant health hazards. Third, excessive nutrient inputs from agricultural and urban runoff, and sewage effluent, can cause eutrophication, whereby the additional nutrients stimulate rapid growth of algae. This in turn depletes the dissolved oxygen level in the water as it decomposes, which then harms or drives away all but the hardiest species.

Coastal pollution is most commonly measured by how much pollution is being discharged into the sea, such as the number of oil spills or the amount of sewage. However, this does not indicate what effect the pollution is having on coastal ecosystems. Consequently, the PAGE researchers examined several other indicators that better reflect biological changes in coastal ecosystems, although global data are available for relatively few of these indicators.

Oxygen Depletion

One such indicator is oxygen depletion in the water–a condition known as hypoxia. Hypoxia, which is often associated with more severe forms of eutrophication, can be quite harmful to marine organisms, especially sedentary organisms that live on the sea floor. Although historical information on hypoxia is limited, experts believe that the prevalence and extent of hypoxic zones have increased in recent decades (Diaz 1999; Diaz and Rosenberg 1995). One of the most well-known examples of hypoxic conditions is the so-called "Dead Zone" at the mouth of the Mississippi River in the northern Gulf of Mexico. Over the last 4 decades, the amount of nitrogen delivered to the coast by the Mississippi River–which drains the entire midsection of North America–has tripled, helping to create a hypoxic zone that covers 7,800–10,400 km^2 at mid-summer, when the zone is at its worst (Rabalais and Scavia 1999).

Somewhat better historical information exists for algal blooms, which also may be exacerbated by nutrient pollution.

Harmful Algal Blooms

Scientists have assembled information on harmful algal blooms (HABs)–rapid increases in the populations of algae species that produce toxic compounds. More than 60 harmful algal toxins are known today. They are responsible for at least six types of food poisoning, including several that can be lethal (McGinn 1999:21; NRC 1999:52). In the United States, HABs have caused nearly $300 million in economic losses since 1991 from fish kills, public health problems, and lost revenue from tourism and the seafood industry (McGinn 1999:25). From the 1970s to the 1990s, the frequency of recorded HABs has increased from 200 to 700 incidents per year (NRC 1999:52; HEED 1998). Some of this increase may be due to better reporting, since awareness of HABs has been heightened; but much of the increase is real, confirmed in areas with long-term monitoring programs.

Pathogens and Toxic Chemicals

Information about the ecosystem effects of pathogens, toxic chemicals, and persistent organic pollutants is less available than information about nutrient pollution. Limited data are available from some regions of the world–mostly industrialized countries–where programs have been established to monitor shellfish beds to guard against consumption of shellfish contaminated with pathogens. Data from the United States' shellfish monitoring program show gradually improving conditions; 69 percent of U.S. shellfish-growing waters were approved for harvest in 1995, up from 58 percent in 1985 (Alexander 1998:6).

Persistent Organic Pollutants

Persistent organic pollutants (POPs) include a number of chemicals that do not exist naturally in the environment, including polychlorinated biphenyls (PCBs), dioxins and furans, and pesticides such as DDT, chlordane, and heptachlor. POPs persist in the environment and can accumulate through the marine food web or in coastal sediments to a level that is toxic to aquatic organisms and humans.

"Mussel Watch" programs in North America, Latin America and the Caribbean, and France have provided a tool for monitoring changes in POPs (as well as other toxic compounds) in coastal ecosystems. These monitoring programs measure accumulations of toxic compounds in the tissues of mussels, which feed by filtering large quantities of sea water, and thus are prone to accumulate any available toxins. Mussel Watch data indicate that chlorinated hydrocarbons, though still high in coastal sediments near industrial areas and in the fat tissue of top predators such as seals, are now decreasing in some northern temperate areas where restrictions on their use have been enforced for some years (O'Conner 1998; GESAMP 1990:52). However, contamination appears to be rising in tropical and subtropical areas because of the continued use of chlorinated pesticides (GESAMP 1990:37).

The Bottom Line for Water Quality. Although there is relatively little monitoring of the actual condition of coastal waters (as opposed to the pollutants discharged into them), evidence indicates decreasing capacity of coastal ecosystems to maintain clean water in many regions of the world. In particular, the increased frequency of harmful algal blooms and hypoxia suggests that the capacity of ecosystems in these regions to absorb and degrade pollutants has been exceeded. Only within some of the OECD countries is there evidence of water quality improvements, which appear to be the result of reduced input of certain pollutants such as POPs.

BIODIVERSITY

Only 250,000 of the 1.75 million species cataloged to date in all ecosystems are found in marine environments, but experts believe that the majority of marine species have yet to be discovered and classified (Heywood 1995:116; WCMC In preparation). Life first evolved in the sea, and marine ecosystems still harbor an impressive variety of life forms. Of the world's 33 phyla (groups of related organisms), 32 are found in the marine environment, and 15 of these are found only there (Norse 1993:14–15). Coral reefs are one coastal marine ecosystem often singled out for their high biodiversity. Although coral reefs inhabit less than a quarter of 1 percent of the global sea bottom, they are the most diverse marine environment, with 93,000 species identified so far, and many more yet to be found (Reaka-Kudla 1997:88–91).

Evidence abounds of the significant pressures on coastal biodiversity. The loss of coastal habitats such as mangroves, seagrasses, and wetlands is one direct measure of declining condition of biodiversity in coastal habitats. Coral reefs face degradation at a global scale, with loss of area, overfishing of reef fish, and degradation of near-coastal water quality having inevitable consequences for reef biodiversity. A 1998 study that mapped pressures on coral reef ecosystems concluded that 58 percent of the world's reefs are at risk from human activities, with 27 percent at high risk (Bryant et al. 1998:20).

Invasive Species

One of the most significant changes in the condition of coastal biodiversity has been growth in the number and abundance of invasive species. For example, the marine ecosystems in the Mediterranean now contain 480 invasive species, the Baltic 89, and Australian waters contain 124 species (Burke et al. [PAGE] 2000). A principal source of biological invasion is from the ballast water of ships. On any one day, 3,000 different species are thought to be carried alive in the ballast water of the world's ocean fleets (Bright 1999:156).

The introduction of the Leidy's comb jellyfish from the western Atlantic into the waters of the Black Sea in 1982 provides one of the most dramatic examples of how a nonnative species can impact marine ecosystems. Unchallenged by natural predators in the Black Sea, the Leidy's comb jellyfish proliferated to a peak in 1988 of 0.9–1 billion tons wet weight (about 95 percent of the entire wet weight biomass in the Black Sea). These animals devastated the natural zooplankton stocks, which allowed the unleashing of massive algal blooms. Natural food webs were disrupted, ultimately contributing to the collapse of the Black Sea fish harvest (Bright 1999:157; Travis 1993:1366).

Other causes of biological invasion include intentional introduction of nonnative species for fisheries stocking or even for ornamental purposes, accidental introduction from aquaculture, and species migration through artificial canals, most notably through the Suez Canal from the Red Sea into the Mediterranean and vice versa.

Depletion

Another measure of direct change in the condition of coastal ecosystem biodiversity is the reduced abundance of various commercially important fish species. Excessive harvests of fish reduce their populations, sometimes to the point they become threatened with extinction, at least in substantial portions of their original range. The IUCN Red List of threatened species includes species such as the Atlantic cod, Atlantic halibut, five species of tuna, and yellowtail flounder–all species heavily exploited for food (IUCN 1996:70–88).

Disease

Additional evidence of declining condition of coastal biodiversity is found in the incidence of new diseases in coastal organisms (Harvell et al. 1999:1505). These diseases cause mass mortalities among plants, invertebrates, and vertebrates, including kelp, seagrasses, shellfish, corals, and marine mammals such as seals and dolphins. Better detection of new diseases may be a factor in the increase in reported incidents, but a careful review of the evidence shows that the number of new diseases is indeed rising (Harvell et al. 1999:1505).

Corals provide one of the best examples of the increase in disease incidence in marine ecosystems. A recent worldwide survey has documented more than 2,000 individual coral disease incidents from more than 50 countries. The earliest records date back to 1902, but the vast majority have occurred since the 1970s (Green and Bruckner In press). In Florida, for example, more than a fourfold increase in coral disease has been observed at 160 monitoring sites since 1996 (Harvell et al. 1999:1507). Although the exact causes of these diseases remain unclear, researchers have linked them to the increasing vulnerability of corals caused by stresses such as pollution and siltation.

Coral Bleaching

Coral bleaching provides a direct indicator of the condition of coral reefs. Reef-building corals contain microscopic algae (zooxanthellae) living within their tissues in a mutually dependent partnership. This partnership breaks down when corals are stressed, and one of the most common causes of such stress is exposure to higher-than-normal temperatures. When this happens, corals lose the algae from their tissues and become a vivid white color, as if they had been bleached. Although corals may recover from such an event, they may die if the cause of bleaching reaches particularly high levels or persists for a long period. Temperatures just 1–2 °C higher than average in the warm season are sufficient to cause bleaching.

Before 1979, there were no records of mass-bleaching of entire reef systems, but that changed in the last 2 decades. In 1987, 1991, and 1996, mass-bleaching was observed in 6 of the 10 major coral reef provinces of the world. The most recent and widespread bleaching event occurred from late 1997 until mid-1998, during one of the largest El Niño events of this century. Bleaching was recorded in all 10 provinces (Hoegh-Guldberg 1999:8). Coral death reached more than 90 percent in some locations; fortunately, many reefs have since recovered (Salm and Clark 2000:8). Experts believe high water temperatures caused the coral bleaching. There is no way of knowing whether human-induced climate change had any bearing, but researchers believe that the elevated sea temperatures associated with climate change could have this same detrimental effect.

Management Efforts

Evidence of the declining condition of coastal biodiversity has stimulated a number of actions by local communities, NGOs, and national governments to slow the rate of loss of particular habitats and to protect the species that remain. Although PAGE researchers did not attempt to survey the entire array of response measures, one important response has been the rapid growth in the number of marine protected areas. To date, more than 3,600 marine protected areas have been designated throughout the world (WCMC 2000). Even so, the total area under protection still falls well short of the minimum area that many marine scientists believe is necessary for the conservation of marine biodiversity.

The Bottom Line for Biodiversity. The variety of coastal habitats—from coral reefs to kelp beds—gives coastal ecosystems a wide array of species and complex communities. However, many indicators show a significant decline in this biodiversity. Degradation and area loss affect all major habitat types such as mangroves, seagrasses, coral reefs, and coastal wetlands. Invasive species have made significant inroads in many marine environments, especially near ports and other highly trafficked areas. Heavily exploited fish species such as cod and haddock have recently been listed as threatened species. Disease incidence among marine mammals and coral reefs has risen dramatically, as have coral bleaching events. Overall, the capacity of marine ecosystems to support their normal biodiversity has been greatly diminished.

SHORELINE PROTECTION

The economic and human costs of coastal storm damage are growing as more people expand into coastal settlements and put lives and property at risk. Economic losses in Europe from floods and landslides between 1990 and 1996 were four times greater than the losses suffered in the 1980s and more than twelve times those of the 1960s (EEA 1998:274). From 1988 to 1999, the United States sustained 38 weather-related disasters causing damage that reached or exceeded $1 billion each, for a total cost in excess of $170 billion (NCDC 2000). In both Europe and the United States, many of these weather-related natural disasters involved flooding in coastal areas or, in the case of the United States, hurricane impacts in coastal regions. Worldwide, more than 40 million people per year are currently at risk of flooding due to storm surges (IPCC 1996:292).

Healthy coastal ecosystems cannot completely protect communities from the impacts of storms and floods, but they do play an important role in stabilizing shorelines and buffering coastal development from the impact of storms, wind, and waves. For example, Sri Lanka spent US$30 million on revetments, groins, and breakwaters in response to severe coastal erosion that occurred in areas where coral reefs were heavily mined (Berg et al. 1998:630). Japan spent roughly 4.5 trillion yen (US$41 billion) on shoreline protection projects from 1970 to 1998 (Japanese Ministry of Commerce 1998).

For many countries, protection of coastal ecosystems is likely to be one of the most cost-effective means of protecting coastal development from the impact of storms and floods. Clearly, with the substantial loss in extent of various coastal ecosystems, the ability to provide this service of shoreline protection has significantly diminished in most nations.

The Bottom Line for Shoreline Protection. There is no doubt that the dramatic loss of coastal habitats around the world has diminished the capacity of coastal ecosystems to protect human settlements from storms. There are few estimates of how great the economic cost of the loss of this service might be, but losses from storm damage already cost billions of dollars annually. With intensive development of the world's coasts proceeding rapidly, the value of the coastal protection service will undoubtedly rise quickly, too.

COASTAL TOURISM AND RECREATION

Travel and tourism, encompassing transport, accommodation, catering, recreation, and services for travelers, is the world's largest industry and the fastest growing sector of the global economy. The World Travel and Tourism Council projected travel and tourism would generate US$3.5 trillion and account for more than 200 million jobs in 1999–about 8 percent of all jobs worldwide (WTTC 1999). In most countries, coastal tourism is the largest sector of this industry and in a number of countries, particularly small island developing states tourism contributes a significant and growing portion to GDP and foreign exchange. Travel and tourism in coastal zones can promote both conservation and economic development, if properly managed.

Most statistics related to tourism are aggregated by country, and agencies and organizations compiling statistics typically do not distinguish inland from coastal tourism. With this in mind, PAGE researchers chose the Caribbean–where the vast majority of tourism is coastal or marine in nature–to assess the condition of coastal ecosystems with regard to their potential to support the recreation and tourism industry.

In 1998, travel and tourism in the Caribbean accounted for more than US$28 billion or about 25 percent of the region's total GDP. The industry provided more than 2.9 million jobs in 1998 (more than 25 percent of all employment), with projections in excess of 3.3 million jobs by 2005 (WTTC/WETA 1998). The number of tourists arriving in the Caribbean is growing rapidly. Over the next decade, tourist arrivals are expected to increase by 36 percent (Caribbean Tourism Organization 1997).

Ecotourism

Different types of tourism differ in their benefits to local economies as well as in their environmental impacts. In the Caribbean, for example, most of the prosperous hotels are large resorts; nature-based tourism (ecotourism) is a small niche market. Worldwide, relatively few local communities have realized significant benefits yet from nature-based tourism on their own lands or in nearby protected areas. The participation of local communities in nature tourism has been constrained by a lack of relevant knowledge and experience, lack of access to capital for investment, inability to compete with well-established commercial operations, and simple lack of ownership rights over the tourism destinations (Wells 1997:iv).

Protected areas often supply the most valuable part of the nature tourism experience, but capture little of the economic value of tourism in return (Wells 1997:iv). Although many governments have successfully increased tourist numbers by marketing their country's nature tourism destinations, most have not invested sufficiently in managing those natural assets or in building the infrastructure needed to support nature tourism. Thus sensitive sites of ecological or cultural value have been exposed to risk of degradation by unregulated tourism development, too many visitors, and the impact of rapid immigration linked to new jobs and business opportunities (Wells 1997:iv–v) (see Box 1.15 Ecotourism, pp. 34–35).

Tourism Related Pressures

Tourism has a tremendous potential to bring economic prosperity and development, including environmental improvements, to the destinations in which it operates. However, poorly planned and managed tourism can harm the very resources on which it is based. Adverse impacts of tourism in the Caribbean include scarring mountain faces with condominium and road construction; filling wetlands and removing mangrove forests for resort construction; losing beach area and lagoons to pollution and to sand mining, dredging, and sewage dumping; and damaging coral reefs with anchoring, sedimentation, and marina development (UNEP/CEP 1994). A 1996 Island Resources Foundation study found that tourism was a major contributor to sewage and solid waste pollution in virtually every country in the Caribbean, as well as the prime contributor to coastal erosion and sedimentation (IRF 1996). Since the success of tourism in the Caribbean has been built on the appeal of excellent beaches and a high-class marine environment suitable for a range of outdoor activities, this inattention to the harmful impacts of tourism itself directly threatens the industry's growth in the region.

? **The Bottom Line for Tourism and Recreation.** Information is not available to accurately judge whether the capacity of coastal ecosystems to support tourism is being diminished at a global scale. However, in some areas, such as parts of the Caribbean region, there is clear evidence of degradation. Nonetheless, this industry has the potential—and indeed incentive—to bring long-term sustainable benefits to coastal communities without degrading the resource on which it depends.

FOREST ECOSYSTEMS

Forests, woodlands, and scattered trees have provided humans with shelter, food, fuel, medicines, building materials, and clean water throughout recorded history. In recent decades they have become a source of new goods and services including pharmaceuticals, industrial raw materials, personal care products, recreation, and tourism. Forests regulate freshwater quality by slowing soil erosion and filtering pollutants, and they help to regulate the timing and quantity of water discharge. In addition, forests harbor much of the world's biological diversity. Although scientists know that most of the world's species have not yet been identified, they think that at least half and possibly well over two-thirds of these species are found in forest ecosystems–in particular, in tropical and subtropical forests (Reid and Miller 1989:15).

Forests provided an important springboard for industrial and socioeconomic development for northern hemisphere countries. They were often recklessly used, but former forested lands usually became productive in new ways. For example, wide tracts of forest were converted permanently to agriculture. In some areas, such as parts of the eastern United States, forests that had been clear-cut have regrown. For now, the northern hemisphere and temperature zone industrialized countries–with the exception of Japan–are broadly self-sufficient in wood, though tropical woods must still be imported.

Forests are now playing a similar socioeconomic development role in many developing countries. That role is more critical in these nations because forests supply industrial wood both for domestic consumption and for export to obtain foreign currency. At the same time, traditional goods and services–woodfuels, food, and medicines–continue to support the livelihoods of many rural populations. Millions of people in tropical and subtropical countries still depend entirely on forest ecosystems to meet their every need.

From the range of goods and services provided by forest ecosystems, PAGE focused on five of the most important for human development and well-being: timber production and consumption, woodfuel production and consumption, biodiversity, watershed protection, and carbon storage.

(continues on p. 90)

Box 2.17 Taking Stock of Forest Ecosystems

Highlights

- Forests cover about 25 percent of the world's land surface, excluding Greenland and Antarctica. Global forest cover has been reduced by at least 20 percent since preagricultural times, and possibly by as much as 50 percent.
- Forest area has increased slightly since 1980 in industrial countries, but has declined by almost 10 percent in developing countries. Tropical deforestation probably exceeds 130,000 km^2 per year.
- Less than 40 percent of forests globally are relatively undisturbed by human action. The great majority of forests in the industrial countries, except Canada and Russia, are reported to be in "semi-natural" condition or converted to plantations.
- Many developing countries today rely on timber for export earnings. At the same time, millions of people in tropical countries still depend on forests to meet their every need.
- The greatest threats to forest extent and condition today are conversion to other forms of land use and fragmentation by agriculture, logging, and road construction. Logging and mining roads open up intact forest to pioneer settlement and to increases in hunting, poaching, fires, and exposure of flora and fauna to pest outbreaks and invasive species.

Key

Condition assesses the current output and quality of the ecosystem good or service compared with output and quality of 20–30 years ago.

	Excellent	Good	Fair	Poor	Bad	Not Assessed
Condition						

Changing Capacity assesses the underlying biological ability of the ecosystem to continue to provide the good or service.

Changing Capacity

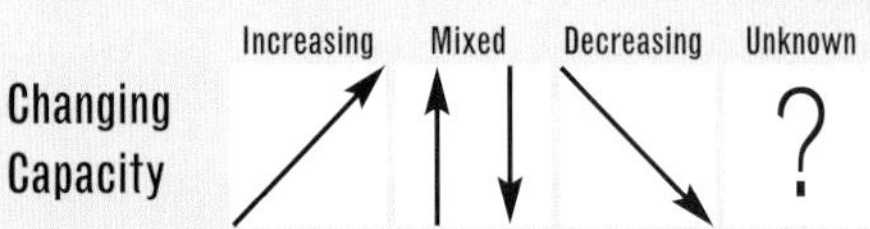

Scores are expert judgments about each ecosystem good or service over time, without regard to changes in other ecosystems. Scores estimate the predominant global condition or capacity by balancing the relative strength and reliability of the various indicators. When regional findings diverge, in the absence of global data, weight is given to better-quality data, larger geographic coverage, and longer time series. Pronounced differences in global trends are scored as "mixed" if a net value cannot be determined. Serious inadequacy of current data is scored as "unknown."

Conditions and Changing Capacity

FIBER PRODUCTION

Fiber production has risen nearly 50 percent since 1960 to 1.5 billion cubic meters annually. In most industrial countries, net annual tree growth exceeds harvest rates; in many other regions, however, more trees are removed from production forests than are replaced by natural growth. Fiber scarcities are not expected in the foreseeable future. Plantations currently supply more than 20 percent of industrial wood fiber, and this contribution is expected to increase. Harvesting from natural forests will also continue, leading to younger and more uniform forests.

WATER QUALITY AND QUANTITY

Forest cover helps to maintain clean water supplies by filtering freshwater and reducing soil erosion and sedimentation. Deforestation undermines these processes. Nearly 30 percent of the world's major watersheds have lost more than three-quarters of their original forest cover. Tropical montane forests, which are important to watershed protection, are being lost faster than any other major forest type. Forests are especially vulnerable to air pollution, which acidifies vegetation, soils, and water runoff. Some countries are protecting or replanting trees on degraded hillslopes to safeguard their water supplies.

BIODIVERSITY

Forests, which harbor about two-thirds of the known terrestrial species, have the highest species diversity and endemism of any ecosystem, as well as the highest number of threatened species. Many forest-dwelling large mammals, half the large primates, and nearly 9 percent of all known tree species are at some risk of extinction. Significant pressures on forest species include conversion of forest habitat to other land uses, habitat fragmentation, logging, and competition from invasive species. If current rates of tropical deforestation continue, the number of all forest species could be reduced by 4-8 percent.

CARBON STORAGE

Forest vegetation and soils hold almost 40 percent of all carbon stored in terrestrial ecosystems. Forest regrowth in the northern hemisphere absorbs carbon dioxide from the atmosphere, currently creating a "net sink" whereby absorption rates exceed respiration rates. In the tropics, however, forest clearance and degradation are together a net source of carbon emissions. Expected growth in plantation area will absorb more carbon, but likely continuation of current deforestation rates will mean that the world's forests remain a net source of carbon dioxide emissions and a contributor to global climate change.

WOODFUEL PRODUCTION

?

Woodfuels account for about 15 percent of the primary energy supply in developing countries and provide up to 80 percent of total energy in some countries. Use is concentrated among the poor. Woodfuel collection is responsible for much local deforestation in parts of Asia, Africa, and Latin America, although two-thirds of all woodfuel may come from roadsides, community woodlots, and wood industry residues, rather than forest sources. Woodfuel consumption is not expected to decline in coming decades, despite economic growth, but poor data make it difficult to determine the global supply and demand.

Data Quality

FIBER PRODUCTION

Generally good global data on industrial roundwood production by country are published annually by the Food and Agriculture Organization (FAO) and the International Tropical Timber Organization (ITTO). Production is recorded by value and by volume in cubic meters per year. Various studies forecast future production and consumption rates. Forest inventory data, recording annual rates of tree growth, tree mortality, size and age of stands, and harvest rates, are generally available for industrial countries but are incomplete and must be estimated for many developing countries. Information on plantation extent and productivity varies widely among countries.

WATER QUALITY AND QUANTITY

Global data on current forest cover and historic loss in major watersheds have been compiled by World Resources Institute (WRI). Data on water runoff, soil erosion, and sedimentation in deforested watersheds are available mostly at regional or local levels. Evidence of the importance of forest cover in regulating water quality and quantity is based on experience in forests managed primarily for soil and water protection in the industrial countries and on studies that value forests according to the avoided costs of constructing water filtration plants. Forest degradation by air pollution in Europe is surveyed by the UN Economic Commission for Europe (UN-ECE).

BIODIVERSITY

Global data sets are few, and evidence is often anecdotal. Forests with high conservation value are identified by field observation and expert opinion. More quantitative information on threatened species is available globally for forest trees and regionally for some birds, butterflies, moths, and larger mammals. Good-quality data on restricted-range birds are available, as are data on threatened birds in the neotropics. Identification of global centers of plant diversity is based on field observation and expert opinion.

CARBON STORAGE

Methodologies for estimating the size of carbon stores in biomass and soils are developing rapidly. This study relied on the estimates of carbon stored in above- and below-ground live vegetation developed by Olson. This data set was modified by updating carbon storage estimates to accord with the land-cover map from the International Geosphere-Biosphere Programme (IGBP), delineated by global ecosystems. Estimates of soil carbon stores were based on the International Soil Reference and Information Centre—World Inventory of Soil Emission Potentials (ISRIC-WISE) Global Data Set of Derived Soil Properties.

WOODFUEL PRODUCTION

The International Energy Agency (IEA) holds good recent data on wood energy production and consumption in industrial countries, where most wood energy is derived from industrial wood processing residues. Global time series data on woodfuel and charcoal production, available from FAO, are modeled or estimated from household surveys. Data on woodfuel plantations and nonforest sources of production (such as public lands) are patchy. Human dependence on woodfuel in developing countries is largely inferred from information on availability and price of other energy sources.

Scorecard

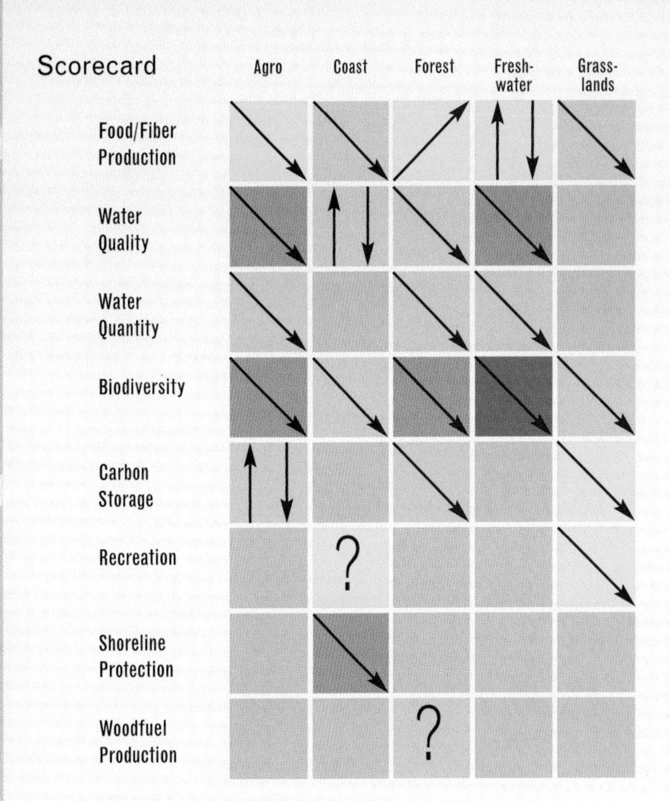

Area of Forest Ecosystems

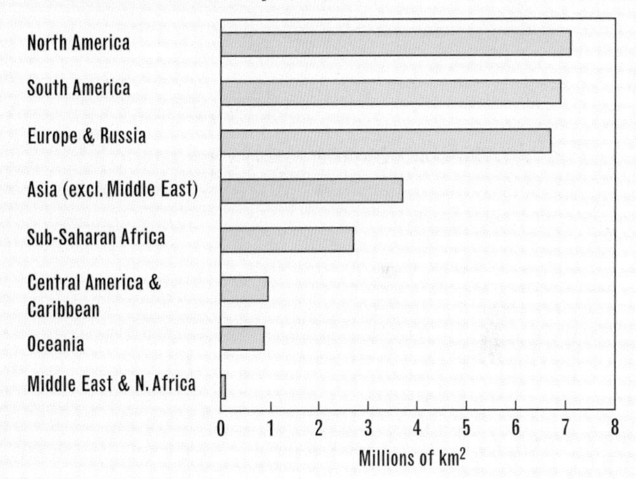

Population of Forest Ecosystems

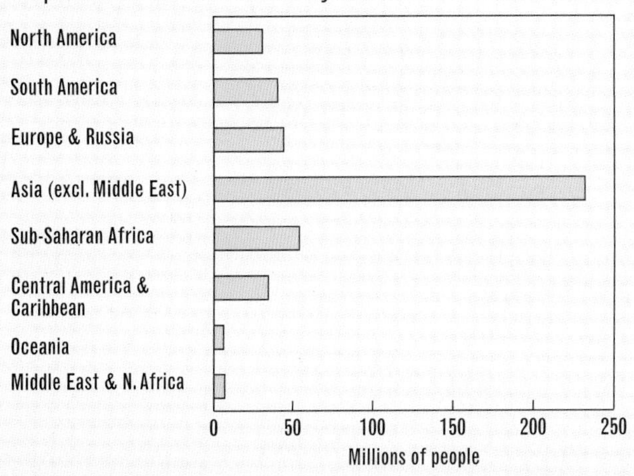

Forest Extent and Modification

More than 90 different definitions of "forest" are in use throughout the world, complicating the effort to measure and evaluate global forest ecosystems. PAGE researchers adopted the definition used by IGBP, which defines *forest ecosystems* as "the area dominated by trees forming a closed or partially closed canopy" (Box 2.18 The Changing Extent of Forests). Forest ecosystems include tropical, subtropical, temperate, and boreal forests as well as woodlands.

Using the IGBP definition, and using data from satellite imagery, the PAGE study calculated the total forest area in 1993 as 29 million km^2, approximately 22 percent of the world's land area (excluding Antarctica and Greenland). This estimate differs somewhat from that calculated by FAO, which is compiled from national forest inventories rather than satellite data and reflects a somewhat different definition. (FAO defines forests to be all areas having a minimum crown cover of 10 percent and minimum tree height of 5 m.) The FAO estimate puts global forest area in 1995 at 34.5 million km^2 (FAO 1997a:185), or 27 percent of the world's land area.

The area of transition between forest and other land cover is one of the most dynamic portions of forest ecosystems and makes up a significant percentage of forest ecosystems in many parts of the world. Nearly 4 million km^2 in Africa now qualifies as forest/cropland mosaics; cropland accounts for between 30 percent and 40 percent of the vegetation cover and forests account for some part of the remainder. Because these forest transition zones typically have at least 10 percent crown cover and still contain more than 30 percent agricultural land, PAGE researchers–as well as FAO and other researchers–included them in the analyses of both forest and agricultural ecosystems.

The change from closed forest to a forest-agriculture mosaic inevitably changes the goods and services that the "forest" provides. The transition zone could, in principle, be managed sustainably to provide timber, tree and fodder crops, and shelter for field crops, fuelwood, and habitat for wildlife. But without effective management, land-use change and ecosystem degradation in transition zones can proceed rapidly. Currently, neither national nor global forest inventories offer insight into how fast forest transition zones are expanding or how well they are functioning as ecosystems.

DEFORESTATION AND FOREST LOSS

Human actions have caused the world's forest cover to shrink significantly over the last several millennia, but it is difficult to specify exactly how much. Scientists can't precisely determine what the original extent of forest was prior to human impact. Forests are not static; their size and composition have evolved with changing climate. However, scientists can determine–by using knowledge of the soil, elevation, and climatic conditions required by forests–where forest could potentially exist if it were not for human actions. Comparing this "potential" forest area to today's actual forest cover gives a plausible estimate of historical forest loss.

Using this approach, Matthews (1983:474–487) estimated that as of the early 1980s, humans had reduced global forest cover about 16 percent. Updating this study with more recent deforestation data available from FAO brings the total loss of original forest cover to roughly 20 percent. Historical forest loss could be much higher, however. A 1997 study by WRI, which used a higher resolution map of potential forest than the Matthews study, estimates that original forest cover has been reduced by nearly 50 percent (Bryant et al. 1997:1).

Calculating current deforestation rates is every bit as challenging as estimating past forest loss. FAO estimates that forested area increased by 0.2 million km^2 (2.7 percent) in industrialized countries between 1980 and 1995 (Matthews et al. [PAGE] 2000; FAO 1997a:17), while it decreased by 2 million km^2 (10 percent) in developing countries (FAO 1997a:16–17). FAO also estimates that the rate of forest loss in developing countries decreased by 11 percent between 1980–90 and 1990–95, from 154,600 km^2 to 130,000 km^2 annually (FAO 1997a:18). However, the uncertainty in these estimates is high. Measuring deforestation on a global level is complicated by a scarcity of reliable direct measurements and the expense and difficulty of satellite measurements. As a result, estimates of the current deforestation rate vary widely, from about 50,000 km^2 to 170,000 km^2/year (Tucker and Townshend 2000:1461). Although the FAO estimate of 130,000 km^2/year is widely quoted, more recent studies–notably of Indonesia and Brazil–suggest that it underestimates actual forest loss.[4]

The underlying causes of forest loss have been the focus of many studies and reports over the past several decades. In its 1997 forest assessment, FAO attributes forest loss in Africa principally to the expansion of subsistence agriculture, under pressure from rural population growth (FAO 1997a:20). Forest loss in Latin America was due more to large-scale cattle ranching, clearance for government-planned settlement schemes, and hydroelectric reservoirs. FAO found forests in Asia to be subject about equally to pressure from subsistence agriculture and economic development schemes (FAO 1997a:20).

Historically, woodfuel collection was considered a leading factor in deforestation in some regions of the world; however, better information is undermining that conclusion. FAO does not consider woodfuel collecting to be an important cause of deforestation, although it can add to pressures that degrade forest quality and health. As much as two-thirds of woodfuel is obtained from nonforest sources such as woodlands, roadside verges, and wood industries (FAO 1997c:21).

FOREST FRAGMENTATION

Although change in actual extent clearly has an impact on the various goods and services that forests provide, fragmentation of forests can have just as great an impact. As part of the

Box 2.18 The Changing Extent of Forests

Since the 16th century, the forests of the northern temperate zone have suffered the most extensive losses as a result of human activity. In recent years, they have begun to recover. However, these gains have been more than offset by rapid decreases in the more extensive and species-rich forests of the developing world.

Many of the world's trees grow within areas that are only partially forested. These lands provide many of the goods associated with forests, especially woodfuel, species habitat, and soil protection. Such areas are particularly vulnerable to clearance, however, since they are often more accessible and less likely to be legally protected than forest areas with higher tree cover.

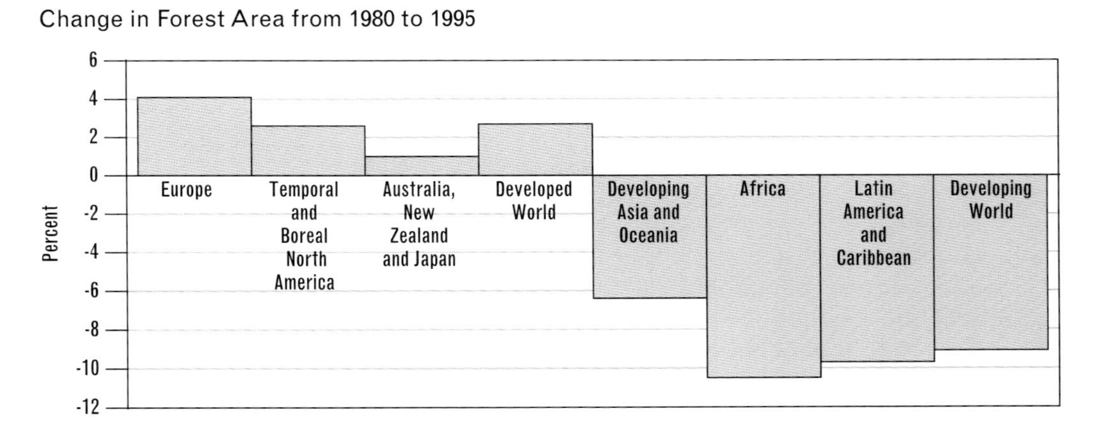

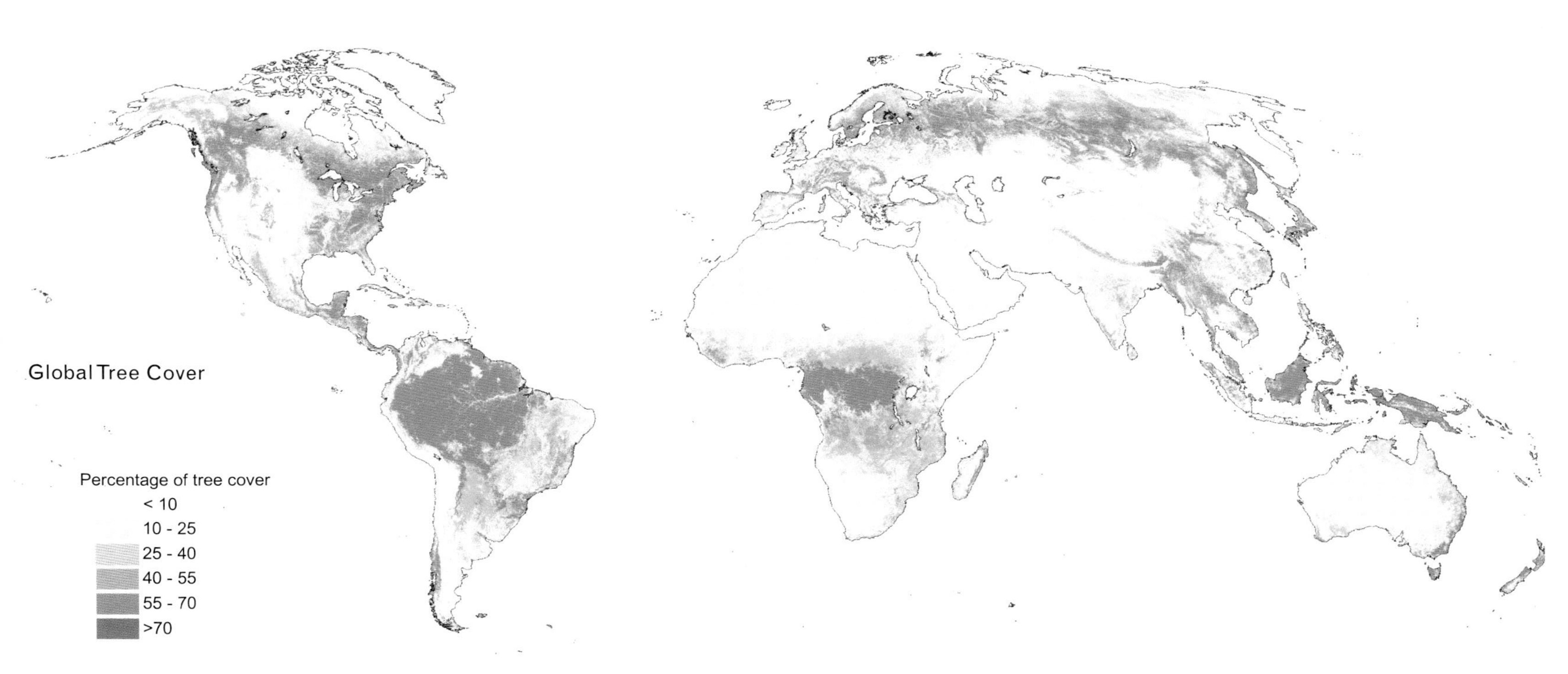

Sources: Matthews et al. [PAGE] 2000. Map is based on Defries et al. (2000). Figure is based on FAO (1997a).

characterization of the extent and change of forests, PAGE researchers developed an indicator of forest fragmentation based on the world's growing road network. Roads provide development benefits, but they also fragment otherwise continuous stretches of forest.

The impact of fragmentation is twofold. First, fragmentation directly affects species biodiversity by diminishing the amount of natural habitat available, blocking migration routes, providing avenues for invasion by nonnative species, and changing the microclimate along the remaining habitat edge. Second, roads provide access for hunting, timber harvest, land clearing, and other human disturbances that further change the characteristics of the local ecosystem.

Forests are naturally fragmented to some extent by such features as rivers, mountain ranges, natural fires, and storm damage. Road networks, however, provide a relatively unambiguous and globally applicable indicator of human-caused fragmentation, albeit a conservative indicator since human actions fragment forests in other ways as well. To demonstrate the potential use of such a fragmentation indicator, the PAGE study included a pilot analysis of forest fragmentation in Central Africa in which researchers documented the effect of road building in breaking up large forest blocks (Box 2.19 Fragmentation of Forests in Africa). In the absence of roads, large continuous blocks of habitat–more than 10,000 km^2–would naturally make up 83 percent of the forest area in Central Africa. However, in the presence of the existing road network, large forest blocks account for just 49 percent of the forest area (Matthews et al. [PAGE] 2000).

FOREST FIRES

In addition to outright conversion and fragmentation of forests, a third human-caused pressure is the frequency and intensity of fires. Wildfires are a natural and necessary phenomenon in many forest ecosystems, helping to shape landscape structure, improve the availability of soil nutrients, and initiate natural cycles of plant succession. In fact, some plant species can't reproduce without periodic fire.

The number of human-caused fires, however, greatly exceeds naturally occurring fires. Fires are set intentionally for timber harvesting, land conversion, or shifting agriculture, and also in the course of disputes over property and land rights. Tropical forest fires were unusually severe in 1997-98, following less-than-average rainfalls due to El Niño. The number of fires in Brazil increased dramatically between 1995 and 1998, spreading from agricultural areas into moist forest that traditionally had not burned (Elvidge et al. 1999). Brazilian fires increased 50 percent between 1996 and 1997, and another 86 percent between 1997 and 1998 (FAO 1999:3) (Box 2.20 Forest Fires).

Globally, humans initiate as much as 90 percent of total biomass burning (including savannas) (Levine et al. 1999:iv). Human-caused fires are thus already reshaping forest ecosystems and their impact could grow substantially. Recent studies indicate that fires in tropical moist forests create feedback loops that increase the forest's susceptibility to subsequent fires. The first fire serves to open up the canopy, allowing sun and air movement to increase drying of the forest. Previously fire-killed trees increase fuel availability, and invading grasses and weeds add combustible live fuels. Second and third fires are faster-moving, more intense, and of longer duration. Initial fires have been demonstrated to kill no more than 45 percent of trees more than 20 cm in diameter, whereas in recurrent fires, up to 98 percent of trees are liable to be killed (Cochrane et al. 1999:1832-1835). This enhanced fire cycle raises the risk that large areas of tropical forest could be transformed into savanna or scrub.

The social and economic costs of forest fires are also significant. An estimated 20 million people were at risk of respiratory problems from the recent fires in Southeast Asia (Levine et al. 1999:12), with economic damages (excluding health impacts) conservatively estimated at $4.4 billion (Economy and Environment Programme for Southeast Asia 1999, cited in Levine et al. 1999:14).

Despite the advent of satellite imagery and the growing significance of fires to the condition of global forests, no reliable global statistics are available for the total forest area burned annually. Within boreal forests, detailed records for the United States and Canada reveal that the annual area burned has more than doubled in the past 30 years (Kasischke et al. 1999:141, 147). Information about tropical forests is more uncertain. For example, estimates of the total area burned in Indonesia during 1997-98 range from 6,000 km^2 (official Indonesian estimates) to more than 45,000 km^2 (unofficial estimate based on analysis of satellite images) (Levine et al. 1999:8-10).

Assessing Goods and Services

FIBER

Commercial timber production is a major global industry. In 1998, global production of industrial roundwood–which includes all wood not used as fuel–was 1.5 billion m^3 (FAO 2000). In the early 1990s, production and manufacture of industrial wood products contributed about US$400 billion to the global economy, or about 2 percent of global GDP (Solberg et al. 1996:48). North America and Europe dominate production, but the timber industry is of greater economic importance to developing countries such as Cambodia, Solomon Islands, and Myanmar, where wood exports can account for more than 30 percent of international trade (FAO 1997a:36).

The three main sources of industrial roundwood are primary forests, secondary-growth forests, and plantations. Secondary-growth forests have replaced virtually all of the primary or original forests of eastern North America, Europe, and large parts of South America and Asia. Estimates of plantation area vary, partly because of differences in how plantations are defined. Plantations are generally defined as forests that have considerable human intervention in their establishment and management, but no clear line divides a "plantation" from an intensively managed "secondary forest."

FAO estimates that industrial roundwood plantations account for approximately 3 percent of total forest area, or about 1 million km^2. However, they provide about 22 percent of the world's industrial roundwood supply (Brown 1999:7, 41). Plantation forest area is highly concentrated. Five countries–China, Russia, United States, India, and Japan–account for 65 percent of global plantation forests (Brown 1999:15).

Assessing a forest's capacity to produce timber is difficult in part because the cycle of harvest and regrowth stretches over many decades. One clear indicator that a forest's capacity to produce timber is being degraded would be evidence of harvest rates greater than the rate of tree growth. According to preliminary data (FAO 1998), it appears that many countries are cutting more timber than grows each year.

In most European countries and the United States, the volume of wood felled is less than the volume of yearly growth (FAO 1998:Technical Annex 1). However, in some countries, like the United States, even though net removal is less than net growth, the rate of growth has diminished in recent years (Haynes et al. 1995:43). This imbalance suggests that current timber production may not be sustainable in the long term (Johnson and Ditz 1997:226). Moreover, information about the diameter of trees in the United States indicates a long-term trend toward smaller, younger trees, and a simplified forest structure, with less diversity of sizes and ages of trees. This could, in turn, reduce the diversity of plant and animal species the forest supports.

For most developing nations, there is a lack of reliable data on net annual forest growth and removal rates and the age of trees–information that is needed to accurately assess the long-term condition of forests. Even so, there is considerable evidence that in some regions, harvest rates greatly exceed regrowth. Typically, in such regions, once forest is cleared, the land is eventually converted to other uses. In other regions, overall harvest may be less than annual growth, but not for certain highly valued species such as mahogany, which are harvested at rates far in excess of their growth rate, which will lead to eventual depletion.

The Bottom Line for Fiber Production. Increasing demand for wood fiber has increased production and, in particular, increased the extent of plantations, which now provide 22 percent of the world's industrial wood. This has not reduced pressure on natural forests. Although forests that have been in timber production for decades show no distinct signs that their capacity to maintain that production is in doubt, some indicators give cause for concern. In developing countries, evidence exists of degradation of timber production capacity, and in these regions, after forests are harvested, the land is often converted to other uses.

WOODFUELS

Fuelwood, charcoal, and other wood-derived fuels (collectively known as woodfuels) are the most important form of nonfossil energy. Biomass energy, which includes woodfuels, agricultural residues, and animal wastes, provides nearly 30 percent of the total primary energy supply in developing countries. Rough estimates indicate that more than 2 billion people depend directly on biomass fuels as their primary or sole source of energy. Woodfuels are the dominant form of biomass energy for many countries, although the data are too sparse to know whether this is true for all countries (IEA 1996:II.289–308, III.31–187).

Available data show woodfuels account for more than half of biomass energy consumed in developing countries and, if China is excluded (where agricultural residues are a particularly important fuel), they account for about two-thirds (IEA 1996:II.289–308, III.31–187)(Box 2.21 Global Use of Woodfuels). Woodfuels are also significant sources of energy in some developed countries. Wood energy supplies nearly 17 percent of total energy consumption in Sweden and 3 percent in the United States (FAO 1997b:7, 11). Economic growth in developing countries has reduced the proportion of energy provided by woodfuel, but overall biomass energy consumption has continued to rise.

Will there be enough woodfuel in the future? Already, in some regions, particularly near urban centers, woodfuel availability has decreased significantly in recent decades. In some cases, production has been maintained even in the face of growing demand by tree planting programs and community woodlots. By 2010, an estimated 2.3–2.4 billion m^3 of fuelwood and charcoal will be available (Nilsson 1996), approximately 30 percent more than in 2000. However, woodfuel demand by 2010 is forecast to be 2.4–4.3 billion m^3 (Matthews et al. [PAGE] 2000). Whether a regional or even global woodfuel crisis will develop depends on a variety of factors such as the affordability of alternative fuels. Nevertheless, there is little doubt that growing woodfuel scarcity will increase the economic burden on the poor in some regions.

Perhaps the most striking feature of this information about woodfuels is how limited and imprecise the information actually is. Woodfuel is a critical energy source for a large percentage of the world's population but, despite the efforts of international institutions such as FAO and the International Energy Agency, the information needed to determine whether ecosystems will be able to meet the growing demand is largely unavailable.

(continues on p. 99)

Box 2.19 Fragmentation of Forests in Africa

Fragmentation can affect forest ecosystems as profoundly as changes in the total tree cover. In Africa and many other parts of the world, the effect of human encroachment on closed canopy forests has been to create forest "transition zones," in which forested land is interspersed with cropland to form an intricate mosaic, as shown on the facing page.

Road networks provide an unambiguous and easily measured, if conservative, indicator of the extent of human-induced fragmentation. When a road is built through a forest, it breaks up species habitats, sometimes into parcels too small to support viable breeding populations. It also provides avenues for invasion by nonnative species and alters the microclimate along the remaining habitat edge. Roads open up previously inaccessible areas of forest to hunting, timber cutting, and clearing for cultivation.

The maps below show the distribution of various sized blocks of forest in central Africa with and without roads. Without roads, continuous blocks of habitat of more than 10,000 km^2 make up 83 percent of the forested area's total extent. When roads are taken into account, this proportion drops to only 49 percent of the total.

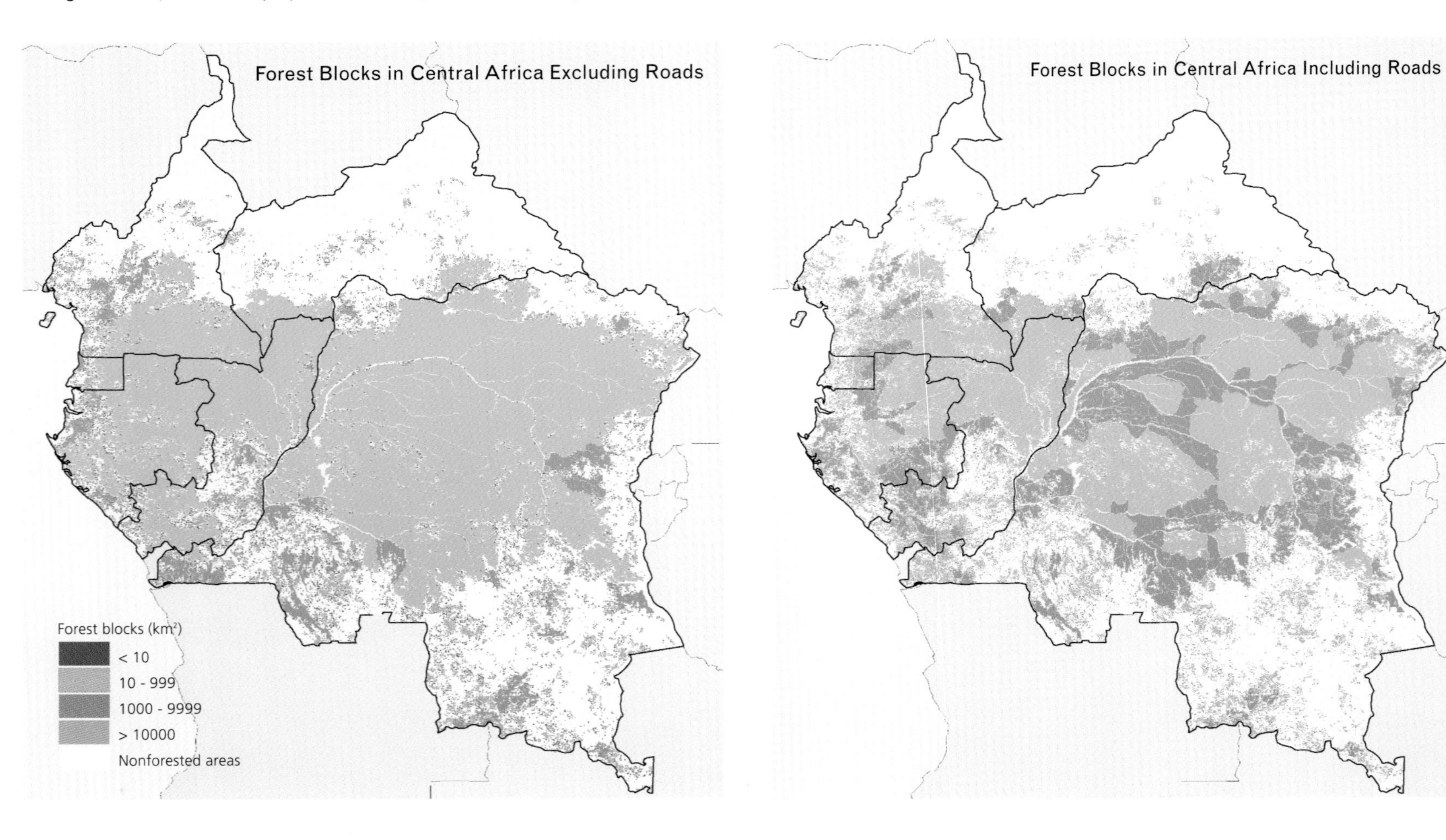

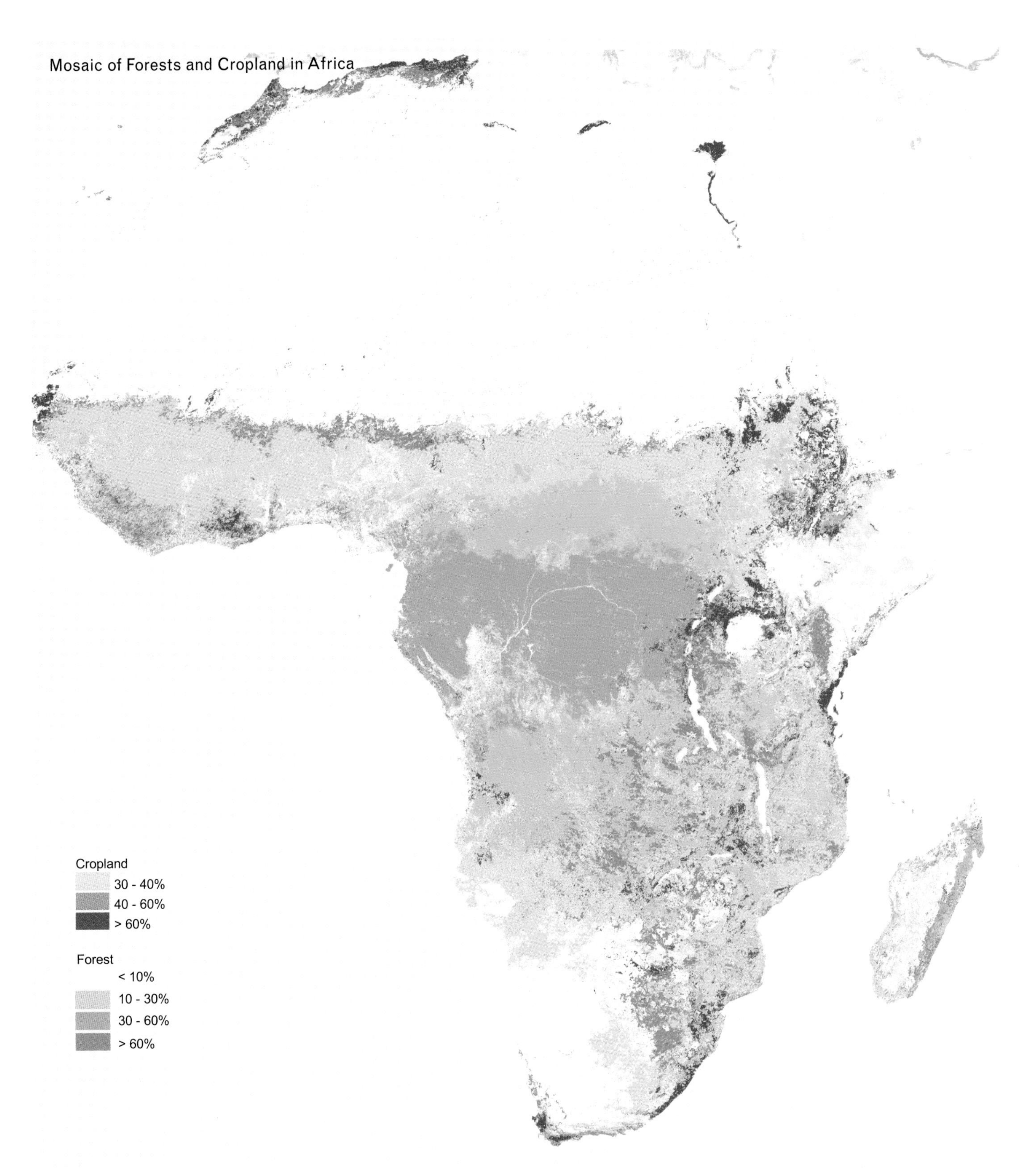

Sources: Matthews et al. [PAGE] 2000. The road fragmentation maps on the previous page are based on CARPE (1998) and Global Land Cover Characteristics Database Version 1.2 (Loveland et al. [2000]). The map above is based on Defries et al. (2000) and Global Land Cover Characteristics Database Version 1.2 (Loveland et al. [2000]).

Box 2.20 Forest Fires

Wildfires are a natural phenomenon in many forest ecosystems. They structure the landscape, improve the availability of soil nutrients, and initiate natural cycles of plant succession. Human-induced fires can have pervasive impact on the condition of forests and their capacity to produce goods and services.

Worldwide, forest fires were especially severe in 1997–98, when millions of hectares of tropical forest in Indonesia, Central America, and the Amazon went up in smoke. Tropical forests, which are normally too wet to sustain extensive fires, were especially susceptible then because of the dry conditions created by El Niño. Evidence suggests, however, that people opportunistically used the dry conditions to set fires to clear land for further development. The burn areas shown for the Amazon in 1998 are adjacent to areas burned to clear land in 1995. This suggests that routine burning of unusually dry fields or pastures may have gotten out of hand. Similar patterns were found in Indonesian forests (Barber 2000).

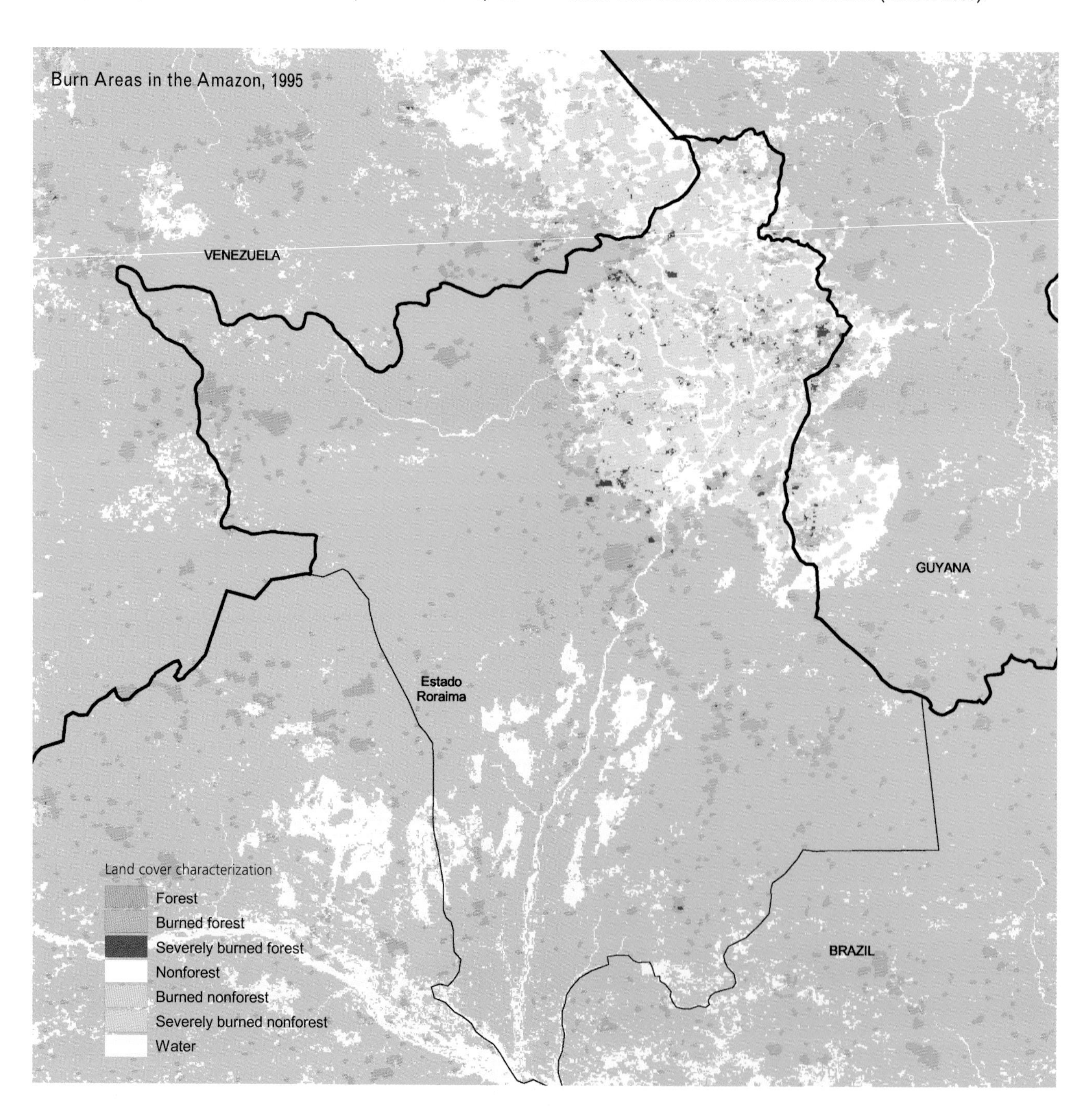

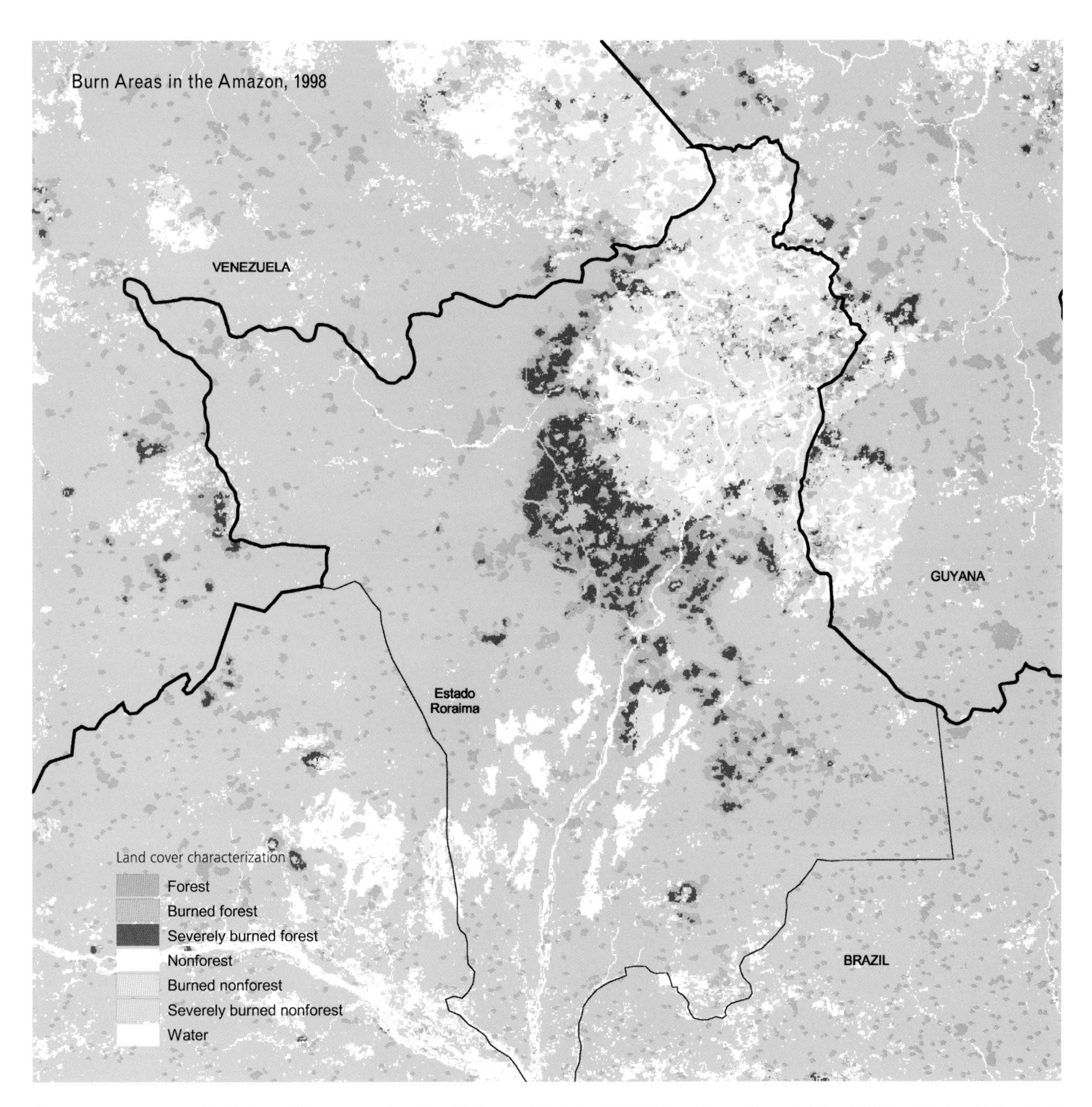

Sources: Matthews et al. [PAGE] 2000. The maps are based on Elvidge et al. (1999) and Global Land Cover Characteristics Database Version 1.2 (Loveland et al. [2000]). Fire data were collected between January and March 1995 and between the same months in 1998. Land-cover data were collected in 1992–93. Nonforested areas include grasslands, croplands, and some seasonal wetlands.

Box 2.21 Global Use of Woodfuels

Woodfuels are the most important source of nonfossil energy. Wood-derived fuels, including fuelwood and charcoal, account for approximately half the biomass energy used in developing countries (IEA 1996), while in some African countries, such as Tanzania, Uganda and Rwanda, woodfuel is the source of 80 percent of total energy consumed.

Although woodfuel collection was assumed to be a major cause of deforestation, recent studies show that up to two-thirds of all woodfuel is collected from nonforest sources such as dispersed woodland and roadside verges (FAO 1997c).

At present, data are insufficient to assess the global sustainability of woodfuel use. It is clear, however, that much of the world's population will continue to rely on wood energy for the foreseeable future and that total demand will increase significantly in coming decades. There is also evidence that in many densely populated areas of the developing world, such as the cities of Côte d'Ivoire, acquiring sufficient wood to meet energy needs is becoming increasingly arduous and costly as populations increase (Garnier 1997).

Share of Woodfuels in National Energy Consumption

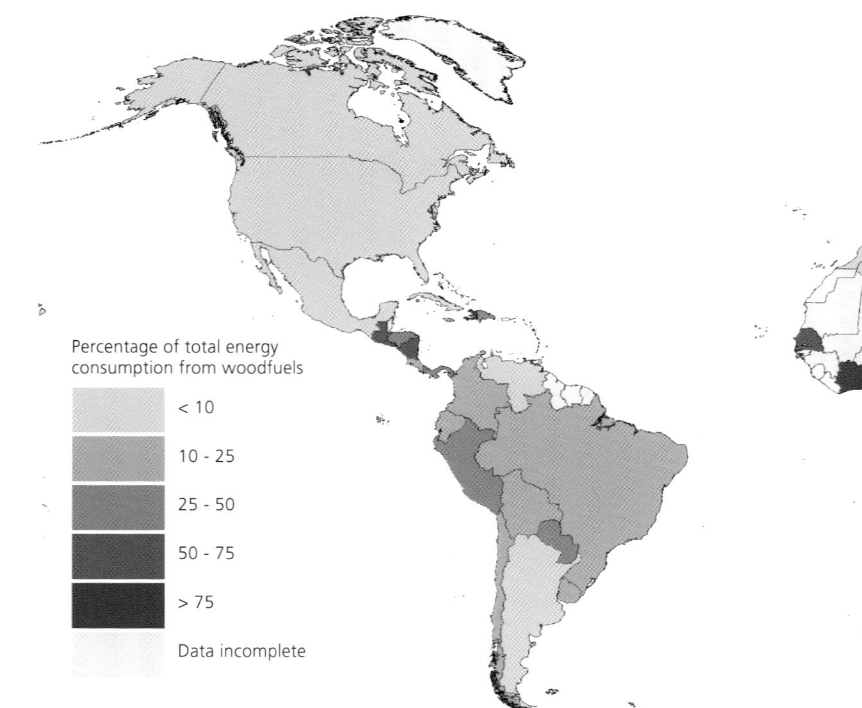

Sources: Matthews et al. [PAGE] 2000. The map is based on IEA (1996).

The Bottom Line for Woodfuel. Woodfuels are the primary source of energy for approximately 2 billion people and by far the most important of the biomass fuels. But we have inadequate information about actual consumption at the household level or the capacity of ecosystems to continue to provide this good. Woodfuels will remain of prime importance in the developing world for the foreseeable future. It is essential to put wood energy data collection and planning on an equal footing with commercial energy sources like oil, coal, natural gas, and hydroelectricity.

BIODIVERSITY

Forest biodiversity is a good in its own right. Diverse species found only in forest habitats are sources of new pharmaceuticals, genetic resources, and nontimber forest products such as resins, fruits, vines, mushrooms, and livestock fodder. Even more important, all other forest goods and services depend to some extent on the diversity of forest species. The condition of biodiversity is thus a useful indicator of the aggregate condition of the forest ecosystem.

Forests are particularly important ecosystems for biodiversity conservation. Two-thirds of 136 ecologically distinct terrestrial regions identified as outstanding examples of biodiversity are located in forested regions, according to WWF (Olson and Dinerstein 1998:509). Similarly, BirdLife International identified 218 areas containing two or more species of birds with restricted ranges. BLI reasoned that these "narrowly endemic" species were likely to be most susceptible to extinction. Eighty-three percent of these 218 areas occur in forests, mostly tropical lowland forests (32 percent) and montane moist forest (24 percent) (Stattersfield et al. 1998:31). Finally, of 234 centers of plant diversity worldwide identified by IUCN and WWF, more than 70 percent are found in forests (Davis et al. 1994, 1995:12-36).

The condition of forest biodiversity can be most directly measured by changes in the number of species found in the forest, including loss or extinction of native species or introductions of nonnative species. Any change in the number or relative abundance of different species represents ecosystem degradation from the standpoint of biodiversity. Because most species have not yet even been identified, it is possible to monitor threats to only the best-known species groups: in practice, this means birds and trees. Of an estimated 100,000 species of trees, WCMC reports that more than 8,700 (Oldfield et al. 1998) are now threatened globally (Box 2.22 Endangered Trees).

Similar global data for forest-dwelling birds have not been compiled, but BLI has mapped the locations of 290 threatened birds in the Neotropics (excluding the Caribbean), allowing comparison among different ecosystems to determine where threats are greatest. Of 596 key areas harboring threatened species, more than 70 percent were in forests (Wege and Long 1995:15-16).

Another direct measure of biodiversity condition is the extent to which invasive species have colonized an ecosystem. Invasions by nonnative species are now ranked by many ecologists as second only to habitat conversion as a threat to global biodiversity. Comprehensive global data on invasives is not yet available, but information compiled by WWF shows how invasive plants have changed the condition of biodiversity in North American forests. In northeastern coastal forests of the United States, up to 32 percent of total vascular plant species are nonnative, although it is not known how many of these species are harmful (Ricketts et al. 1997:82).

Although these direct measures of change in the number of species in forests are the best way to assess the condition of forest biodiversity, data are unavailable for much of the world. Consequently, most of what is known about the condition of forest species is only inferred from various measures of the pressures on forest biodiversity. Three such pressures–habitat fragmentation, logging, and loss of habitat area–are known to change the numbers and types of species found in forest regions. Areas with high levels of fragmentation or logging, or regions that have experienced significant loss of forest habitat, will not contain as many of the native species previously found in the region.

The relationship between habitat area and species diversity is well enough established that it is possible to estimate how many native species might ultimately be lost from a particular habitat as its area is reduced. The Global Biodiversity Assessment conducted in 1995 under the auspices of UNEP found that if recent rates of tropical forest loss continue for the next 25 years, the number of species in forests would be reduced by approximately 4-8 percent (Heywood 1995:235).

The Bottom Line for Biodiversity. Forests have the highest species diversity and endemism of any ecosystem. Pressure on this diversity is immense, as judged from forest loss and fragmentation, but direct information about condition is more limited. What evidence exists suggests that the number of threatened forest species is significant and growing, and species introductions are very high in certain regions. Not only is forest area shrinking, but the capacity of remaining forests to maintain biodiversity appears to be significantly diminished.

CARBON STORAGE

Forests play a central role in the global carbon cycle. Trees capture carbon from the atmosphere as they grow and store it in their tissues. Because of their great biomass, global forests comprise one of the largest terrestrial reservoirs or "sinks" of carbon. Forests store 39 percent (471-929 GtC) of the 1,213-2,433 GtC that PAGE researchers calculated are stored in all terrestrial ecosystems. By way of comparison, grass-

Box 2.22 Endangered Trees

Survival of the world's estimated 100,000 tree species is threatened by conversion of forest land to other uses, timber harvesting, fire, pest attack, and ecosystem simplification resulting from forest management. WCMC has compiled a list of threatened species, assessed according to the 1994 IUCN categories of threat. Altogether, more than 8,700 tree species, almost 9 percent of the world total, are at risk.

A major threat is posed by the deliberate or accidental introduction by humans of nonnative plants and animals to forest habitats. These can threaten the survival of native species by attacking them, competing with them for food and space, or altering local ecosystems to the point that they can no longer support indigenous tree populations. The number of nonnative species are, thus, an indicator of the degree of potential "assault" on native flora.

In North America, the highest concentrations of nonnative species are found around ports, along major transportation routes, and in fertile agricultural regions that have proved favorable to both introduced crops and their pests. Densely forested taiga regions away from major human settlements appear to be little affected, and the conifer forests of the Southeast have proved relatively resistant to invasive species.

Risk Categories for the World's Trees

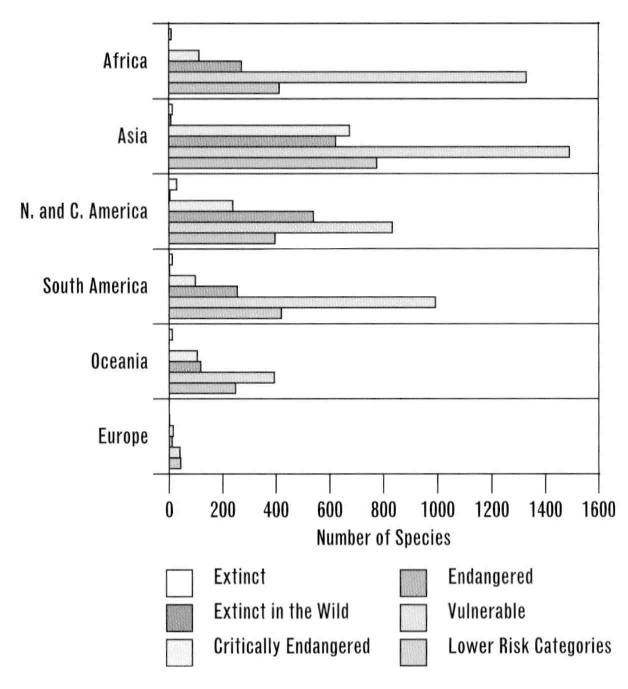

Nonnative Plants in Forest Ecoregions of North America

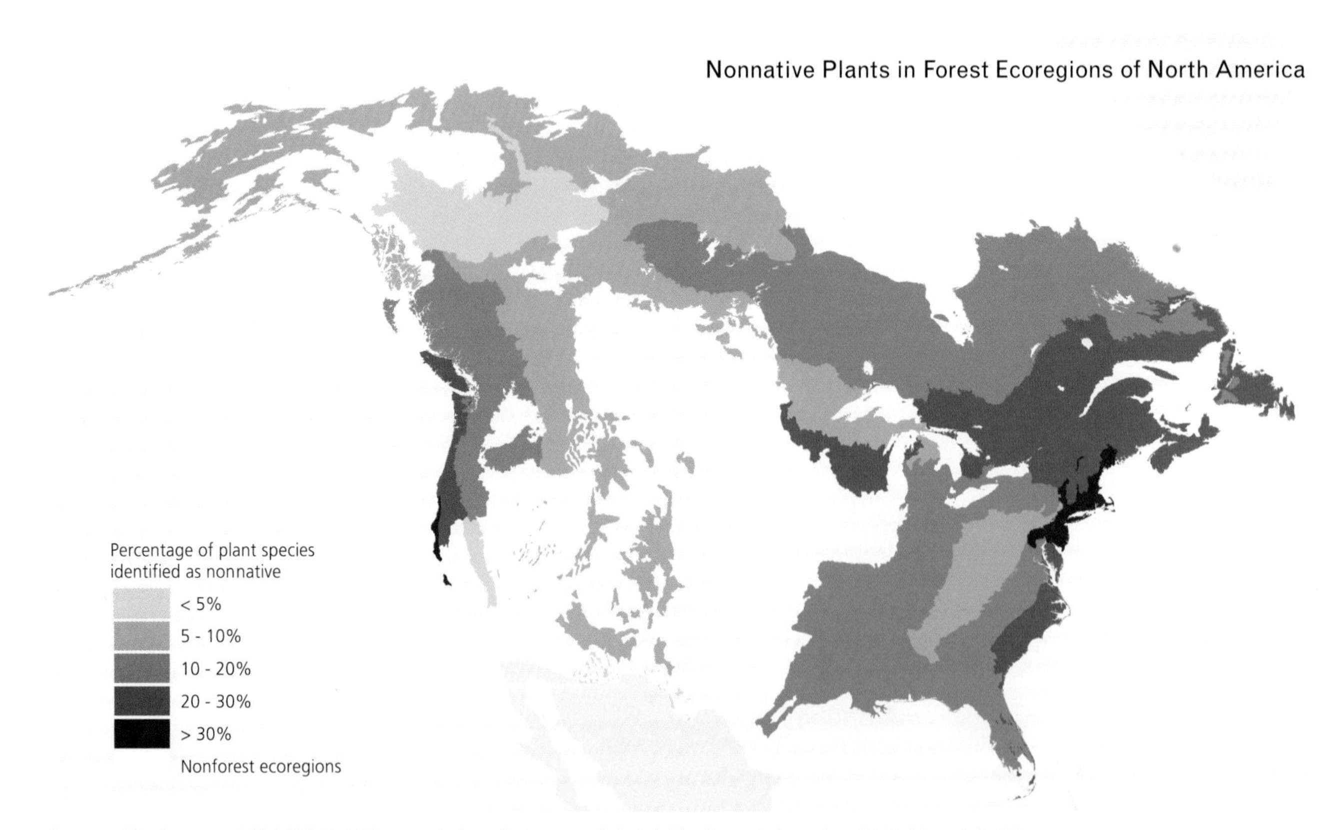

Sources: Matthews et al. [PAGE] 2000. The map is from Ricketts et al. (1997). The figure is based on Oldfield et al. (1998).

lands store about 33 percent of terrestrial carbon, yet cover nearly twice as much area as forested regions.

Land-use change is thought to release an average of 1.6 GtC to the atmosphere each year, or roughly 20 percent of all carbon emissions caused by human action (IPCC 2000:5). By far the most significant component of global land-use change is deforestation in the tropics (Houghton 1999:305, 310). Clearing forests and burning the debris releases large amounts of carbon stored in the vegetation back into the atmosphere. On the other hand, restoring degraded forests or changing their management can increase their carbon storing ability and thus increase the total carbon stored in world forests.

Loss of carbon storage in forests does not always take the form of large-scale clearance or outright deforestation. Logging and clearing small areas for agriculture can also degrade forests and significantly reduce their carbon-storing capacity. One recent study in tropical Asia reported that deforestation accounted for two-thirds of carbon loss in Asian forests, whereas one-third was due to degradation from logging and shifting cultivation (Houghton and Hackler 1999:486). Another study, in Africa, found that outright loss of forest accounted for 43 percent of carbon loss, while degradation of the forest was responsible for 57 percent (Gaston et al. 1998:110).

The Bottom Line for Carbon Storage. Forests store more carbon than any other terrestrial ecosystem—nearly 40 percent of total carbon stored. Deforestation and forest degradation are responsible for approximately 20 percent of annual carbon emissions. The condition of forest ecosystems from the standpoint of carbon storage is clearly declining, but with appropriate economic incentives, this trend could potentially be reversed. However, there are trade-offs to be borne in mind: more carbon is sequestered by young, fast-growing trees than by mature trees. Simply managing forests to store maximum carbon might encourage replacement of many existing old-growth forests with plantations, which would clearly jeopardize biodiversity, tourism, and other services that natural forests provide.

WATER QUALITY AND QUANTITY

Forests provide several valuable services in relation to watershed protection. They physically stabilize the upper reaches of watersheds. Tree roots "pump" water out of the soil to be used by the plant, thereby reducing soil moisture and the likelihood of mud slides; root structures increase the shear strength of soil and help prevent landslides. Forests also tend to moderate the rate of runoff from precipitation, reducing flows during flooding and increasing flows during drier times.

Forest cover also helps to maintain drinking water supplies. Within the United States, more than 60 million people in 3,400 communities rely on National Forest lands for their drinking water, a service estimated to be worth $3.7 billion per year (Dombeck 1999). Finally, forest cover affects the total amount of water available in a watershed. In many regions, forest loss will increase net water discharge because less water is transpired to the atmosphere. In other regions, however, forest loss can decrease net discharge. In cloud forests, for example, forests play a role in directly condensing or "stripping" water from moisture-laden air and making it available for discharge. In other regions, precipitation is dependent in part on the transpiration of water-laden air from the local forest. For example, climate researchers have estimated that temperatures are about 1°C higher and precipitation is 30 percent lower in large deforested patches in the Amazon (Couzin 1999:317).

Overall, forest loss has certainly impaired the world's watersheds to a significant degree. A 1998 analysis by WRI found that nearly 30 percent of the world's major watersheds have lost more than three-fourths of their original forest cover, and 10 percent have lost more than 95 percent of their original forest cover (Revenga et al. 1998:I-13) (Box 2.23 The Deforestation of Watersheds).

Perhaps a more revealing measure of the condition of forests for watershed protection today is the status of montane forests. These forests play an especially important role in the hydrological processes of watersheds by controlling soil erosion in steeply sloping mountains and sometimes "capturing" water in cloud forests.

In temperate regions, the extent of montane forest has increased in recent years, except in the mature old-growth coniferous forests of the Pacific Northwest of North America, Chile, Tasmania, and southern New Zealand. Highly prized for producing lumber, these forests may have been reduced to less than half their original extent by logging (Denniston 1995:32). In the tropics, montane forests are under even greater pressure. According to FAO, tropical montane forests were disappearing at a rate of 1.1 percent/year in the 1980s, which exceeded the rate of loss for all other tropical forest types (FAO 1993:28).

The Bottom Line for Water Quality and Quantity. Forests retain water in soil, regulate flow, influence precipitation, and filter drinking water. The water purification service alone has high economic value in certain regions. Forest loss in general has eroded the capacity of the world's forests to protect watersheds and provide water-related services, and this decline will likely continue as pressures on forests mount. Nearly 30 percent of the world's major watersheds have lost more than three-quarters of their original forest. Montane forests, which are particularly important in protecting watersheds, have suffered extensively. In spite of the importance of forests for vital water services, these services are rarely factored into land-management decisions.

Box 2.23 The Deforestation of Watersheds

Deforestation is a useful indicator of watershed degradation, because forests are often crucial for maintaining water quality and moderating water flow. The loss of original forest cover is estimated from the extent of forests that are believed to have existed 8,000 years ago assuming current climate conditions. Almost a third of all watersheds have lost more than 75 percent of their original forest cover, and seventeen have lost more than 90 percent. Most of these basins are relatively small. Large basins, such as the Congo and the Amazon, still have extensive original forest cover and have lost a relatively small percentage of their original forest. Nonetheless, the total area of original forest lost is large: nine large basins have lost more than 500,000 km^2 (Revenga et al. 1998:I-13).

Watersheds Losing the Greatest Share of Original Forest Cover

Region and Watershed	Percentage of Original Forest Lost
Africa	
Lake Chad	100
Limpopo	99
Mangoky	97
Mania	98
Niger	96
Nile	91
Orange	100
Senegal	100
Volta	97
Asia and Oceania	
Amu Darya	99
Indus	90
Europe	
Guadalquivir	96
Seine	93
Tigris & Euphrates	100
South America	
Rio Colorado	100
Lake Titicaca	100
Uruguay	92

Watersheds Losing the Greatest Area of Original Forest Cover

Region and Watershed	Area of Original Forest Lost (km^2)
Africa	
Congo	>1,000,000
Asia and Oceania	
Ganges	500,000–1,000,000
Mekong	500,000–1,000,000
Ob	500,000–1,000,000
Yangtze	>1,000,000
Europe	
Volga	500,000–1,000,000
North America	
Mississippi	500,000–1,000,000
South America	
Amazon	500,000–1,000,000
Paraná	500,000–1,000,000

Source: Revenga (personal communication, 2000) updating Revenga et al. (1998).

FRESHWATER SYSTEMS

Freshwater ecosystems in rivers, lakes, and wetlands contain just a fraction–one one-hundredth of one percent–of Earth's water and occupy less than 1 percent of Earth's surface (Watson et al. 1996:329; McAllister et al. 1997:18). Yet these vital systems render services of enormous global value–on the order of several trillion U.S. dollars, according to some estimates (Postel and Carpenter 1997:210).

The most important services revolve around water supply: providing a sufficient quantity of water for domestic consumption and agriculture, maintaining high water quality, and recharging aquifers that feed groundwater supplies. But freshwater ecosystems provide many other crucial goods and services as well: habitats for fish (for food and sport), mitigation of floods, maintenance of biodiversity, assimilation and dilution of wastes, recreational opportunities, and a transportation route for goods. Harnessed by dams, these systems also produce hydropower, one of the world's most important renewable energy sources.

Prior to the 20th century, global demand for these goods and services was small compared to what freshwater systems could provide. But with population growth, industrialization, and the expansion of irrigated agriculture, demand for all water-related goods and services increased dramatically, straining the capacity of freshwater ecosystems. Many policy makers are aware of the growing problems of water scarcity, but scarcity is only one of many ways in which these ecosystems are stressed today.

Extent and Modification

Freshwater systems have been altered since historical times; however, the pace of change accelerated markedly in the early 20th century. Rivers and lakes have been modified by altering waterways, draining wetlands, constructing dams and irrigation channels, and establishing connections between water basins, such as canals and pipelines, to transfer water. Although these changes have brought increased farm output, flood control, and hydropower, they have also radically changed the natural hydrological cycle in most of the world's water basins (Box 2.24 Taking Stock of Freshwater Systems).

RIVERS

Modification of rivers has greatly altered the way rivers flow, flood, and act on the landscape. In many instances, rivers have become disconnected from their floodplains and wet-

(continues on p. 106)

Box 2.24 Taking Stock of Freshwater Systems

Highlights

- Although rivers, lakes, and wetlands contain only 0.01 percent of the world's freshwater and occupy only 1 percent of the Earth's surface, the global value of freshwater services is estimated in the trillions of U.S. dollars.
- Dams have had the greatest impact on freshwater ecosystems. Large dams have increased sevenfold since the 1950s and now impound 14 percent of the world's runoff.
- Almost 60 percent of the world's largest 227 rivers are strongly or moderately fragmented by dams, diversions, or canals.
- In 1997, 7.7 million metric tons of fish were caught from lakes, rivers, and wetlands, a production level estimated to be at or above maximum sustainable yield for these systems.
- Freshwater aquaculture contributed 17 million metric tons of fish in 1997. Since 1990, freshwater aquaculture has more than doubled its yield and now accounts for 60 percent of global aquaculture production.
- Half the world's wetlands are estimated to have been lost in the 20th century, as land was converted to agriculture and urban areas, or filled to combat diseases such as malaria.
- At least 1.5 billion people depend on groundwater as their sole source of drinking water. Overexploitation and pollution in many regions of the world are threatening groundwater supplies, but comprehensive data on the quality and quantity of this resource are not available at the global level.

Key

Condition assesses the current output and quality of the ecosystem good or service compared with output and quality of 20–30 years ago.

	Excellent	Good	Fair	Poor	Bad	Not Assessed
Condition						

Changing Capacity assesses the underlying biological ability of the ecosystem to continue to provide the good or service.

	Increasing	Mixed	Decreasing	Unknown
Changing Capacity	↗	↑↓	↘	?

Scores are expert judgments about each ecosystem good or service over time, without regard to changes in other ecosystems. Scores estimate the predominant global condition or capacity by balancing the relative strength and reliability of the various indicators. When regional findings diverge, in the absence of global data weight is given to better-quality data, larger geographic coverage, and longer time series. Pronounced differences in global trends are scored as "mixed" if a net value cannot be determined. Serious inadequacy of current data is scored as "unknown."

Conditions and Changing Capacity

FOOD PRODUCTION

At the global level, inland fisheries landings have been increasing since 1984. Most of this increase has occurred in Asia, Africa, and Latin America. In North America, Europe, and the former Soviet Union, landings have declined, while in Australia and Oceania they have remained stable. The increase in landings has been maintained in many regions by stocking and by introducing nonnative fish. The greatest threat for the long-term sustainability of inland fisheries is the loss of fish habitat and the degradation of the aquatic environment.

WATER QUALITY

Even though surface water quality has improved in the United States and Western Europe in the past 20 years (at least with respect to phosphorus concentrations), worldwide conditions appear to have degraded in almost all regions with intensive agriculture and large urban or industrial areas. Algal blooms and eutrophication are being documented more frequently in most inland water systems, and water-borne diseases from fecal contamination of surface waters continue to be a major cause of mortality and morbidity in the developing world.

WATER QUANTITY

The construction of dams has helped provide drinking water for much of the world's population, increased agricultural output through irrigation, eased transport, and provided flood control and hydropower. People now withdraw about half of the readily available water in rivers. Between 1900 and 1995, withdrawals increased sixfold, more than twice the rate of population growth. Many regions of the world have ample water supplies, but currently almost 40 percent of the world's population experience serious water shortages. With growing populations, water scarcity is projected to grow dramatically in the next decades. On almost every continent, river modification has affected the natural flow of rivers to a point where many no longer reach the ocean during the dry season. This is the case for the Colorado, Huang-He (Yellow), Ganges, Nile, Syr Darya, and Amu Darya rivers.

BIODIVERSITY

The biodiversity of freshwater ecosystems is much more threatened than that of terrestrial ecosystems. About 20 percent of the world's freshwater fish species have become extinct, threatened, or endangered in recent decades. Physical alteration, habitat loss and degradation, water withdrawal, overexploitation, pollution, and the introduction of nonnative species all contribute to declines in freshwater species. Amphibians, fish, and wetland-dependent birds are at high risk in many regions of the world.

Data Quality

FOOD PRODUCTION

Data on inland fisheries landings are poor, especially in developing countries. Much of the catch is not reported at the species level, and much of the fish consumed locally is never reported. No data are systematically collected on the contribution to inland fisheries of fish stocking, fish introduction programs, and other enhancement programs. Historical trends in fisheries statistics are only available for a few well-studied rivers.

WATER QUALITY

Data on water quality at a global level are scarce; there are few sustained programs to monitor water quality worldwide. Information is usually limited to industrial countries or small, localized areas. Water monitoring is almost exclusively limited to chemical pollution, rather than biological monitoring, which would provide a better understanding of the systems' condition and capacity. For regions such as Europe, where some monitoring is taking place, differences in measures and approaches make the data hard to compare.

WATER QUANTITY

Statistics are poor on water use, water availability, and irrigated area on a global scale. Estimates are frequently based on a combination of modeled and observed data. National figures, which are usually reported, vary from estimates used in this study, which are done at the watershed or river catchment level.

BIODIVERSITY

Direct measurements of the condition of biodiversity in freshwater systems are sparse worldwide. Basic information is lacking on freshwater species for many developing countries, as well as threat analyses for most freshwater species worldwide. This makes analyzing population trends impossible or limited to a few well-known species. Information on nonnative species is frequently anecdotal and often limited to records of the existence of a particular species, without documentation of the effects on the native flora and fauna. Spatial data on invasive species are available for a few species, mostly in North America.

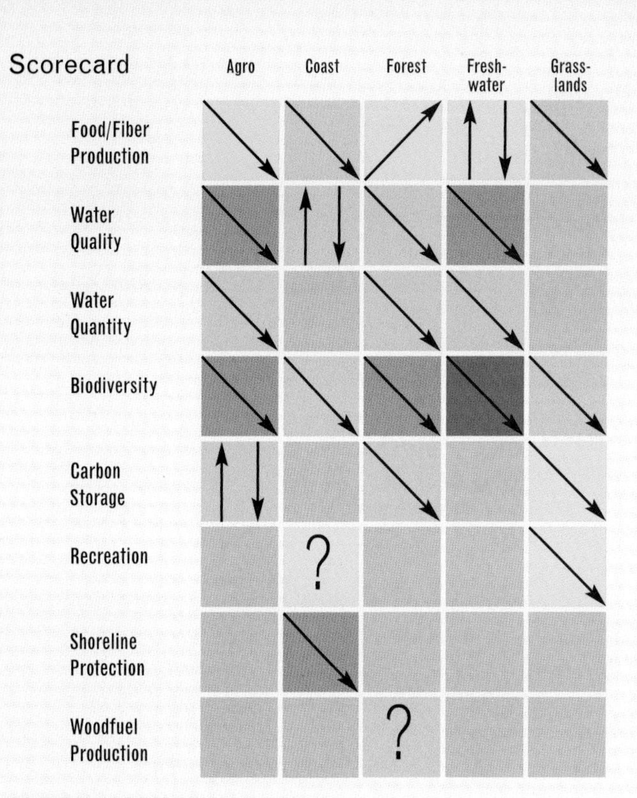

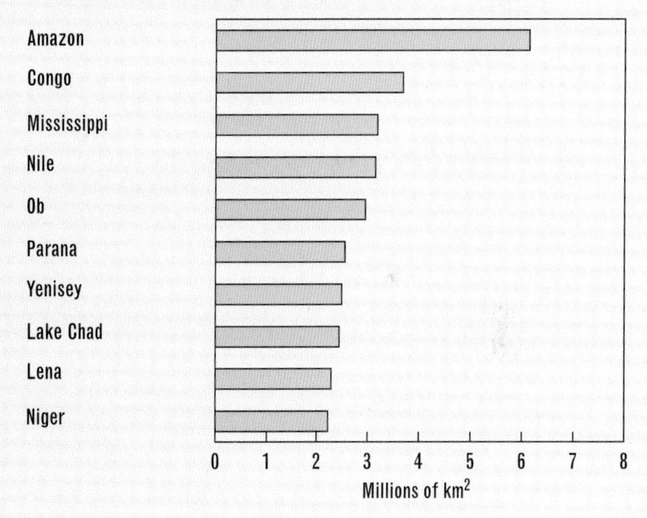

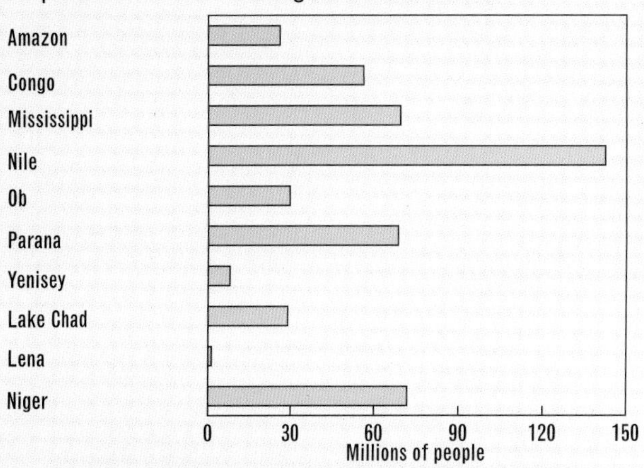

lands. Dams, the most significant physical impact on freshwater systems, have slowed water velocity in river systems, converting many of them to chains of connected reservoirs. This fragmentation of freshwater ecosystems has changed patterns of sediment and nutrient transport, affected migratory patterns of fish species, altered the composition of riparian habitat, created migratory paths for exotic species, and contributed to changes in coastal ecosystems.

Damming the World's Rivers

The number of large dams (more than 15 m high) has increased nearly sevenfold since 1950, from about 5,750 to more than 41,000 (ICOLD 1998:7, 13), impounding 14 percent of the world's annual runoff (L'vovich and White 1990:239). Even though dam construction has greatly slowed in most developed countries, demand and untapped potential for dams is still high in the developing world, particularly in Asia. As of 1998, there were 349 dams more than 60 m high under construction around the world (IJHD 1998:12–14). The regions with the greatest number of dams under construction are Turkey, China, Japan, Iraq, Iran, Greece, Romania, Spain, and the Paraná basin in South America. The river basins with the most large dams under construction are the Yangtze basin in China, with 38 dams under construction; the Tigris and Euphrates basin with 19; and the Danube with 11.

PAGE researchers assessed most of the world's large rivers (average annual discharge of at least 350 m^3/second) to quantify the extent to which dams and canals have fragmented river basins and to determine how water withdrawals have altered river flows. The PAGE analysis shows that, of the 227 major river basins assessed, 37 percent are strongly affected by fragmentation and altered flows, 23 percent are moderately affected, and 40 percent are unaffected (Dynesius and Nilsson 1994:753–762; Revenga et al. [PAGE] 2000) (Box 2.25 Fragmentation and Flow). "Strongly affected" systems include those with less than one-quarter of their main channel left without dams, as well as rivers whose annual discharge has decreased substantially. "Unaffected rivers" are those without dams in the main channel of the river and, if tributaries have been dammed, river discharge has declined no more than 2 percent.

In all, strongly or moderately fragmented systems account for nearly 90 percent of the total water volume flowing through the rivers in the analysis. The only remaining large free-flowing rivers in the world are found in the tundra regions of North America and Russia, and in smaller basins in Africa and Latin America.

Slowing the Flow

Clearly, water diversions and extractions have profoundly affected river flow on a global basis. On almost every continent, the natural flow of one or more major rivers has decreased so much that it no longer reaches the sea during the dry season; the Colorado, Huang He (Yellow), Ganges, Nile, Syr Darya, and Amu Darya, all run dry at the river mouth during the dry season (Postel 1995:10). The Amu Darya and Syr Darya used to contribute 55 billion m^3 of water annually to the Aral Sea prior to 1960, but diversions for irrigation reduced this volume to an annual average of 7 billion m^3–6 percent of the previous annual flow–during 1981-90 (Postel 1995:14–15).

By slowing the movement of water, dams also prevent large amounts of sediment from being carried downstream–as they normally would be–to deltas, estuaries, flooded forests, wetlands, and inland seas. This retention can rob these areas of the sediments and nutrients they depend on, affecting their species composition and productivity. Sediment retention also interferes with dam operations and shortens their useful life. In the United States, about 2 km^3 of reservoir storage capacity is lost to sediment retention each year, at a cost of $819 million annually (Vörösmarty et al. 1997:217). And retention eliminates or reduces spring runoff or flood pulses that often play a critical role in maintaining downstream riparian and wetland communities (Abramovitz 1996:11).

Water and sediment retention also affect water quality and the waste processing capacity of rivers–their ability to break down organic pollutants. The slower moving water in reservoirs is not well-mixed, but rather is stratified into layers, with the bottom layers often depleted of oxygen. These oxygen-starved waters can produce a toxic hydrogen sulfide gas that degrades water quality. In addition, oxygen-depleted waters released from dams have a reduced capacity to process waste for as far as 100 km downstream, because the waste-processing ability of river water depends directly on its level of dissolved oxygen.

An indicator of the extent to which dams have affected water storage and sediment retention at the global level is the change in "residence time" of otherwise free-flowing water–in other words, the increase in time that it takes an average drop of water entering a river to reach the sea. Vörösmarty et al. (1997:210–219) calculated the changes in this residence time, or "aging" of river water, at the mouth of each of 236 drainage basins (see also Revenga et al. [PAGE] 2000). Worldwide, the average age of river water has tripled to well over 1 month. Among the basins most affected are the Colorado River and Rio Grande in North America, the Nile and the Volta Rivers in Africa, and the Rio Negro in Argentina.

WETLANDS

Wetlands include a variety of highly productive habitat types from flooded forests and floodplains to shallow lakes and marshes. They are a key component of freshwater ecosystems, providing flood control, carbon storage, water purification, and goods such as fish, shellfish, timber, and fiber. Although wetlands are a significant feature of many regions, a recent review by the Ramsar Convention on Wetlands concluded that available data are too incomplete to yield a reliable estimate of the global extent of wetlands (Finlayson and Davidson 1999:3).

Because wetlands are valued as potential agricultural land or feared for harboring disease, they have undergone massive conversion around the world, sometimes at considerable ecological and socioeconomic costs. Without accurate global information on the original extent of wetlands, scientists can't say precisely how much wetland area has been lost; but based on a variety of historical records and sources, Myers (1997:129) estimated that half of the wetlands of the world have been lost this century. More detailed studies have tracked freshwater wetland loss in specific regions and countries. For example, experts estimate 53 percent of all wetlands in the lower 48 states of the United States was lost from the 1780s to the 1980s (Dahl 1990:5). In Europe, wetland loss is even more severe; draining and conversion to agriculture alone has reduced wetlands area by some 60 percent (EEA 1999:291).

Assessing Goods and Services

WATER QUANTITY

Water, for domestic use as well as use in agriculture and industry, is clearly the most important good provided by freshwater systems. Humans withdraw about 4,000 km^3 of water a year–about 20 percent of the normal flow of the world's rivers (their nonflood or "base flow") (Shiklomanov 1997:14, 69). Between 1900 and 1995, withdrawals increased more than sixfold, which is more than twice the rate of population growth (WMO 1997:9).

Scientists estimate the average amount of runoff worldwide to be between 39,500 km^3 and 42,700 km^3 per year (Fekete et al. 1999:31; Shiklomanov 1997:13). However, most of this occurs in flood events or is otherwise not accessible for human use. In fact, only about 9,000 km^3 is readily accessible to humans, and an additional 3,500 km^3 is stored by reservoirs (WMO 1997:7).

Given a limited supply of freshwater and a growing population, the amount of water available per person has been decreasing. Between 1950 and 2000, annual water availability per person decreased from 16,800 m^3 to 6,800 m^3 per year, calculated on a global basis (Shiklomanov 1997:73). However, such global averages don't portray the world water situation well. Water supplies are distributed unevenly around the world, with some areas containing abundant water and others a much more limited supply. For example, the arid and semiarid zones of the world receive only 2 percent of the world's runoff, even though they occupy roughly 40 percent of the terrestrial area (WMO 1997:7).

High Demand, Low Runoff

In river basins with high water demand relative to the available runoff, water scarcity is a growing problem. In fact, water experts frequently warn that water availability will be one of the major challenges facing human society in the 21st century and the lack of water will be one of the key factors limiting development (WMO 1997:1, 19). A 1997 analysis estimated that roughly one-third of the world's people live in countries experiencing moderate to high water stress–a number that will undoubtedly rise as population and per capita water demand grow (WMO 1997:1).

To get a better understanding of the balance of water demand and supply, and to better estimate the dimensions of the global water problem, PAGE researchers undertook a new analysis of water scarcity using a somewhat different method than the 1997 study. PAGE researchers calculated water availability and population for individual river basins, rather than on a national or state level,[5] with the object of identifying those areas where annual water availability per person was less than 1,700 m^3. Water experts define areas where per capita water availability drops below 1,700 m^3/year as experiencing "water stress"–a situation where disruptive water shortages can frequently occur. In areas where annual water supplies drop below 1,000 m^3 per person, the consequences are usually more severe: problems with food production, sanitation, health, economic development, and loss of ecosystems occur, except where the region is wealthy enough to use new technologies for water conservation or reuse (Hinrichsen et al. 1998:4).

According to the PAGE analysis, 41 percent of the world's population, or 2.3 billion people, live in river basins under water stress, where per capita water availability is less than 1,700 m^3/year (Revenga et al. [PAGE] 2000) (Box 2.26 The Quantity and Quality of Freshwater). Of these, 1.7 billion people reside in highly stressed river basins where annual water availability is less than 1,000 m^3/person. Assuming current consumption patterns continue, by 2025, PAGE researchers project that at least 3.5 billion people–or 48 percent of the world's population–will live in water-stressed river basins. Of these, 2.4 billion will live under high water stress conditions.

Even some regions that normally have water availability above scarcity levels may in fact face significant water shortages during dry seasons. The PAGE study identified a number of such river basins, particularly in northeast Brazil, southern Africa, central India, eastern Turkey, northwest Iran, and mainland Southeast Asia.

Groundwater Sources

Global concerns about water scarcity include not only surface water sources but groundwater sources as well. Some 1.5 billion people rely on groundwater sources, withdrawing approximately 600–700 km^3/year–about 20 percent of global water withdrawals (Shiklomanov 1997:53–54). Some of this water–fossil water–comes from deep sources isolated from the normal runoff cycle, but much groundwater comes from shallower aquifers that draw from the same global runoff that feeds freshwater systems. Indeed, overdrafting of ground-

(continues on p. 112)

Box 2.25 Fragmentation and Flow

For centuries, in all parts of the world, rivers and lakes have been modified to improve navigation, wetlands drained to make way for settlement, and dams and channels built to control the flow of water for human purposes. These changes have raised agricultural output by making more land and irrigation water available, easing transport, and providing flood control and hydropower.

But human modifications have also had far-reaching effects on hydrological cycles and the species that depend on those cycles. Rivers have been disconnected from their floodplains and wetlands, and water velocity has been reduced as river systems are converted into chains of connected reservoirs. These changes have altered fish migrations, created access routes for nonnative species, and narrowed or transformed riparian habitats. The result has been species loss and an overall reduction in the level of ecosystem services freshwater environments are able to provide.

The construction of dams has had an impact on most of the world's major river systems. There are more than 41,000 large dams in the world—a sevenfold increase in storage capacity since 1950 (ICOLD 1998, Vörösmarty et al 1997). The map at the top of the facing page shows the extent of fragmentation, or interruption of natural flow, caused by human intervention in 227 large river systems (Dynesius and Nilsson 1994; Nilson et al. 1999; Revenga et al. [PAGE] 2000). Almost all large river systems in temperate and arid regions are classified as highly or moderately affected, while all but a handful of the unaffected systems in which water still flows freely are located in Arctic or boreal regions. This trend will continue as new large dams are built throughout Asia, the Middle East, and Eastern Europe.

Dams slow the rate of natural flow, thereby increasing sedimentation and lowering levels of dissolved oxygen. The most affected river systems, in which length of water retention has risen by more than a year, include the Colorado River and Rio Grande in North America, the Nile and Volta Rivers in Africa, and the Rio Negro in Argentina.

Aging of Continental Runoff in Major Reservoir Systems

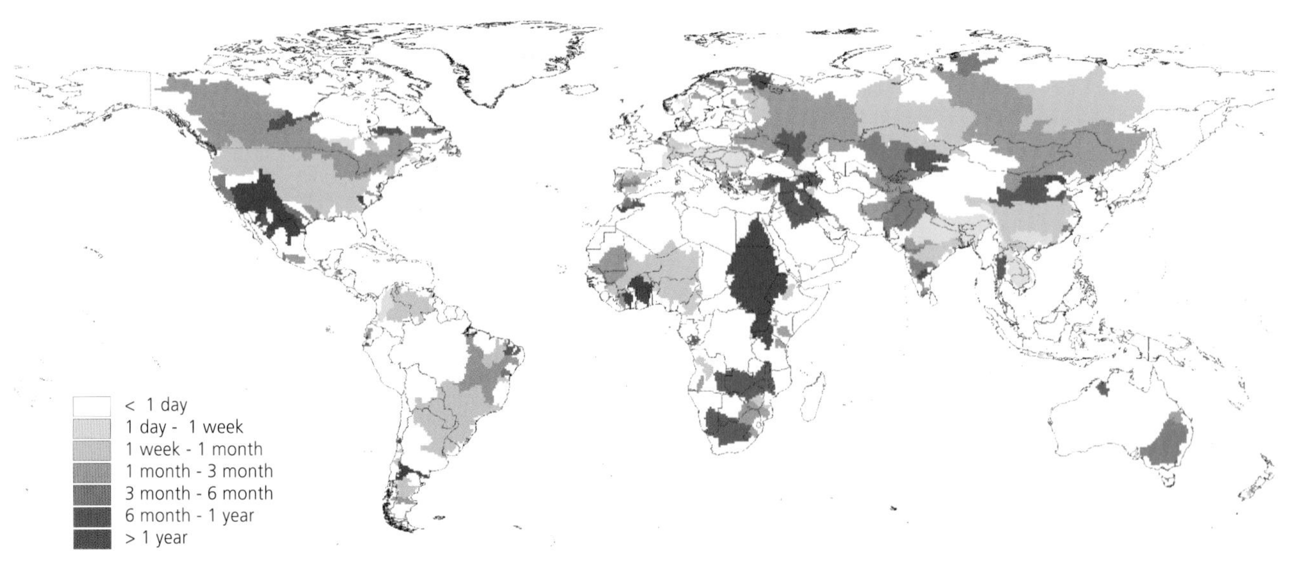

River Channel Fragmentation and Flow Regulation

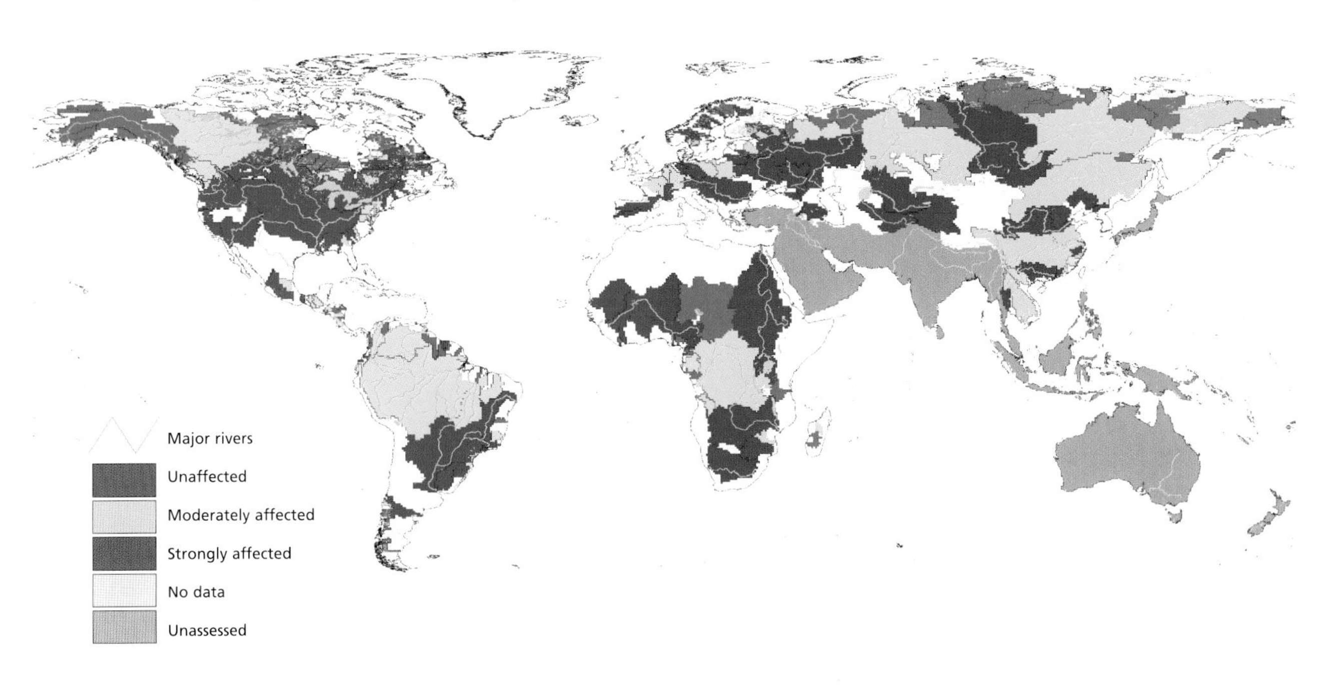

New Dams under Construction by Basin, 1998

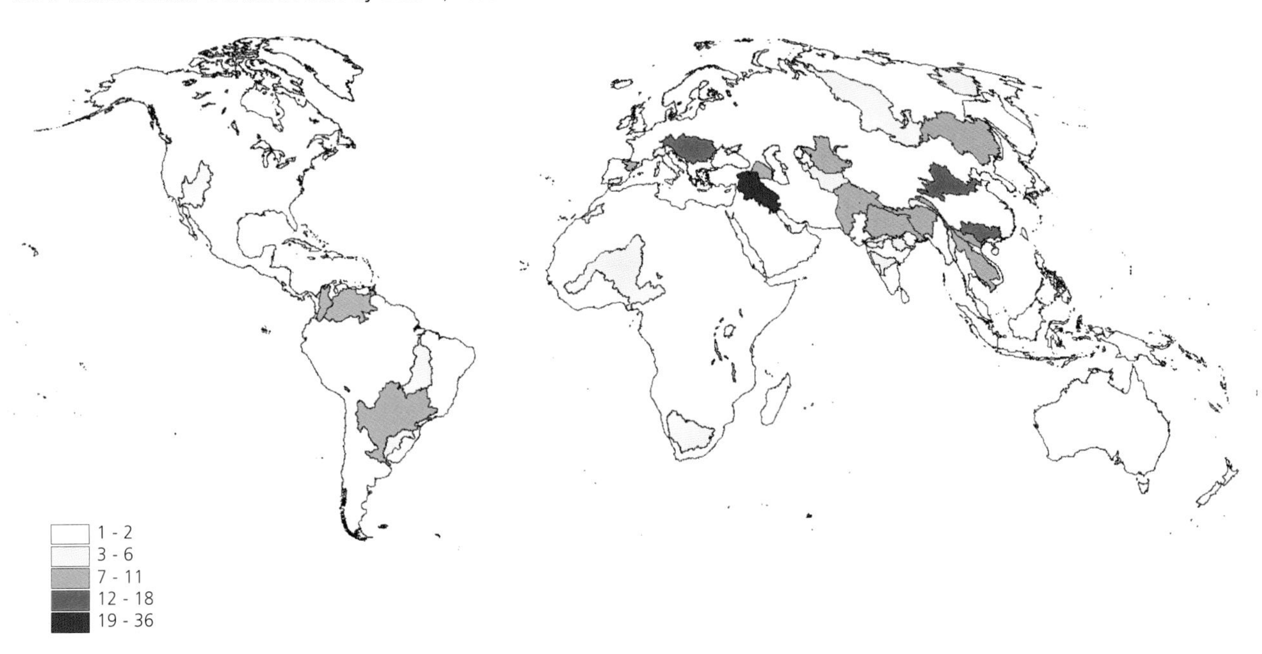

Sources: Revenga et al. [PAGE] 2000. The continental runoff map on the preceding page is from Vörösmarty et al. (1997.) The fragmentation map above is based on Revenga et al. (1998), Dynesius and Nilsson (1994), and Nilsson et al. (1999). The map showing dams under construction are based on data from IJHD(1998)

Box 2.26 The Quantity and Quality of Freshwater

Freshwater systems provide the single most essential good: water—for drinking, cooking, washing, rinsing, mixing, growing, processing, and countless other human uses. Increases in population, industrial production, and agricultural demand have caused the global rate of water consumption to grow twice as fast as the population rate (WMO 1997:9).

The quantity and quality of water available from freshwater systems is greatly influenced by land use within the watershed from which the water is drawn. The mix of cities, roads, agroecosystems, and natural areas affects transpiration, drainage, and runoff and often dictates the amount of pollution carried in the water. Natural waters have low concentrations of nitrates and phosphorous, but these levels increase in rivers fed by runoff from agroecosystems (especially in Europe and North America, where synthetic fertilizers are widely used) and urban areas. The excess nutrients stimulate plant growth, which can choke out local freshwater species, clog distribution systems, and endanger human health.

Just as clean water is often a victim of development, development, too, can be a victim of the lack of clean water. Many experts predict that the lack of clean water is likely to be one of the key factors limiting economic growth in the 21st century. As of 1995, more than 40 percent of the world's population lived in conditions of water stress (less than 1,700 m^3 of water available/person/year) or water scarcity (less than 1,000 m^3 of water available/person/year). This percentage will increase to almost half the world's population by 2025. River basins with more than 10 million people by 2025 that will move into situations of water stress are the Volta, Farah, Nile, Tigris and Euphrates, Narmada, and Colorado (Brunner et al. 2000).

Nutrient Pollution in Selected Rivers, 1994

Region	River	Area (millions of km^2)	Concentration (mg/l) Nitrates	Phosphates
Africa	Zaire	3.69	n.a.	n.a.
	Nile	2.96	0.80	0.03
Asia	Huang He	0.77	0.17	0.02
	Brahmaputra	0.58	0.82	0.06
Europe	Volga	1.35	0.62	0.02
	Seine	0.06	4.30	0.40
N. America	Mississippi	3.27	1.06	0.20
	St. Lawrence	1.02	0.22	0.02
Oceania	Murray Darling	1.14	0.03	0.10
	Waikato	0.01	0.30	0.10
S. America	Amazon	6.11	0.17	0.02
	Orinoco	1.10	0.08	0.01

Global Water Availability, 1995 and 2025

		1995		2025	
Status	Water supply (m^3/person)	Population (millions)	Percentage of Total	Population (millions)	Percentage of Total
Scarcity	<500	1,077	19	1,783	25
	500–1,000	587	10	624	9
Stress	1,000–1,700	669	12	1,077	15
Adequacy	>1,700	3,091	55	3,494	48
Unallocated		241	4	296	4
Total		5,665	100	7,274	100

Annual Water Availability per Person by River Basin 1995

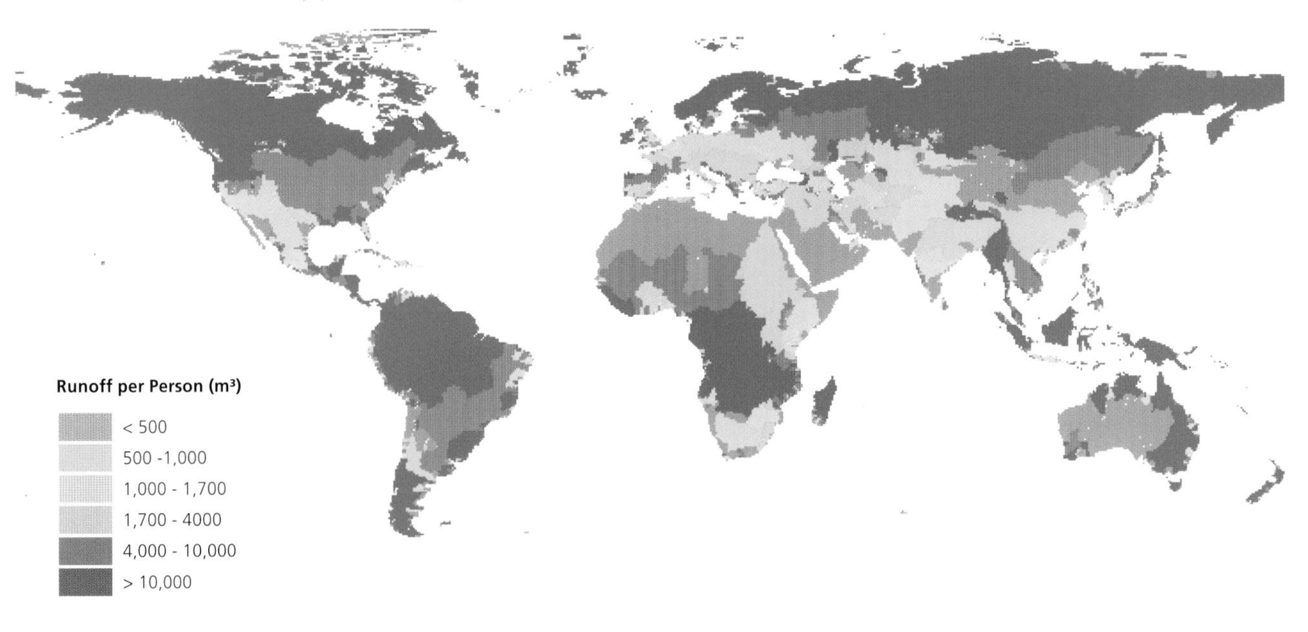

Annual Water Availability per Person by River Basin 2025

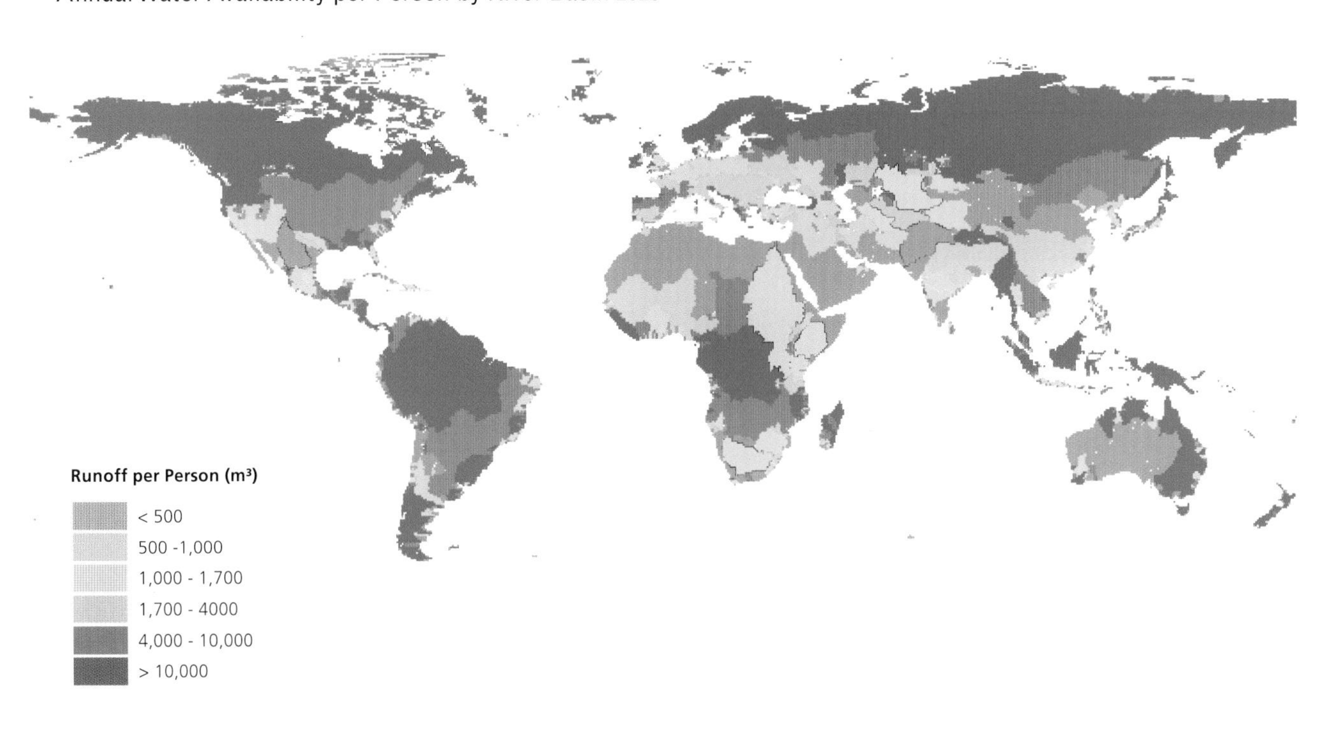

Sources: Nutrient pollution table is based on UNEP-GEMS (1995). The water availability table and maps are from Revenga et al. [PAGE] 2000, based on Brunner et al. (2000), Fekete et al. (1999), and CIESIN (2000). Water scarcity projections are based on the UN's low-growth projection of population growth or decline; they do not take into account effects of pollution and climate change.

water sources can rob streams and rivers of a significant percentage of their flow. In the same way, polluting aquifers with nitrates, pesticides, and industrial chemicals often affects water quality in adjacent freshwater ecosystems. Although overdrafting from and polluting groundwater aquifers are known to be widespread and growing problems (UNEP 1996:4-5), comprehensive data on groundwater resources and pollution trends are not available on a global level.

The Bottom Line for Water Quantity. Humans now withdraw annually about one-fifth of the normal (nonflood) flow of the world's rivers, but in river basins in arid or populous regions the proportion can be much higher. This has implications for all species living in or dependent on these systems, as well as for future human water supplies. Currently, more than 40 percent of the world's population lives in water-scarce river basins. With growing populations, water scarcity is projected to increase significantly in the next decades, affecting half of the world's people by 2025. Widespread depletion and pollution of groundwater sources, which account for about 20 percent of global water withdrawals, is also a growing problem for freshwater ecosystems, since groundwater aquifers are often linked to surface water sources.

WATER QUALITY

Freshwater systems, particularly wetlands, play an essential role in maintaining water quality by removing contaminants and helping to break down and disperse organic wastes. But the filtering capacity of wetlands and other habitats is limited and can be overwhelmed by an excess of human waste, agricultural runoff, or industrial contaminants. Indeed, water quality is routinely degraded by a vast array of pollutants including sewage, food processing and papermaking wastes, fertilizers, heavy metals, microbial agents, industrial solvents, toxic compounds such as oil and pesticides, salts from irrigation, acid precipitation, and silt.

Information about water quality on a global level is poor and difficult to obtain for a number of reasons. Water-quality problems are often local and can be highly variable depending on the location, season, or even time of day. In addition, monitoring for water quality is by no means universal, and water-quality standards often vary significantly from country to country.

Nonetheless, existing information makes it clear that there are many consistent trends in the contamination of water supplies worldwide. One hundred years ago, the main contamination problems were fecal and organic pollution from untreated human waste and the by-products of early industries. These pollution sources have been greatly reduced in most industrialized countries, with consequent improvements in water quality. However, a new suite of contaminants from intensive agriculture and development activities in watersheds has kept the clean-up from being complete. Meanwhile, in most developing countries, the problems of traditional pollution sources and new pollutants like pesticides have combined to heavily degrade water quality, particularly near urban industrial centers and intensive agriculture areas (Shiklomanov 1997:28; UNEP/GEMS 1995:6).

Increased use of manure and manufactured fertilizers–a major source of nutrients such as nitrates and phosphorous–has been a significant cause of pollution in freshwater systems. Nitrate and phosphorus concentrations are low in natural systems but increase with runoff from agroecosystems and urban and industrial wastewater. As a consequence, algal blooms and eutrophication are being documented more frequently in most inland water systems. The highest nitrate concentrations occur in Europe, but high levels are also found in watersheds that have been intensively used and modified by human activity in China, South Africa, and the Nile and Mississippi basins (UNEP/GEMS 1995:33-36). These high nitrate levels, in turn, are associated with extreme eutrophication caused by agricultural runoff in at least two areas: the Mediterranean Sea and the northern Gulf of Mexico at the mouth of the Mississippi River. Water pollution caused by agricultural runoff remains an intractable problem because of its extremely diffuse nature, which makes it hard to control even in industrialized countries.

Although water quality measurements that focus on levels of contaminants are useful, they do not directly tell us how water pollution affects freshwater ecosystems. To determine this, the aquatic community itself must be monitored. The Index of Biotic Integrity (IBI), which includes information about fish or insect species richness, composition, and condition, is one of the most widely used approaches for assessing the health of the aquatic community in a given water body or stretch of river (Karr and Chu 1999). A number of states in the United States now use various IBI approaches and it has been applied in France and Mexico; as yet its use is too limited to give an idea of global aquatic conditions (Oberdorff and Hughes 1992; Lyons et al. 1995).

The Bottom Line for Water Quality. Surface water quality has improved in the United States and Western Europe during the past 20 years, but nitrate and pesticide contamination remain persistent problems. Data on water quality in other regions of the world are sparse, but water quality appears to be degraded in almost all regions with intensive agriculture and rapid urbanization. Unfortunately, little information is available to evaluate the extent to which chemical contamination has impaired freshwater biological functions. However, incidents of algal blooms and eutrophication are widespread in freshwater ecosystems the world over—an indicator that these systems are profoundly affected by water pollution. In addition, the massive loss of wetlands on a global level has left the capacity of freshwater ecosystems to filter and purify water much impaired.

FOOD: INLAND FISHERIES

Fish are a major source of protein and micronutrients for a large percentage of the world's population, particularly the poor (Bräutigam 1999:5). Inland fisheries–stocks of fish and shellfish from rivers, lakes, and wetlands–are an important component of this protein source. The population of Cambodia, for example, gets roughly 60 percent of its total animal protein from the fishery resources of Tonle Sap, a large freshwater lake (MRC 1997:19). In Malawi, the freshwater catch provides about 70–75 percent of the animal protein for both urban and rural low-income families (FAO 1996).

Inland Fish Catch. Worldwide, the inland fisheries harvest totaled 7.7 million metric tons in 1997. Not counting the fish raised in aquaculture, this represents nearly 12 percent of all fish–freshwater and ocean-caught–that humans directly consume (FAO 1999a:7–10). The inland fisheries catch consists largely of freshwater fish, although mollusks, crustaceans, and some aquatic reptiles are also caught and are of regional and local importance (FAO 1999a:9) (Box 2.27 Changes in Inland Fisheries).

The inland fisheries harvest is believed to be greatly underreported–by a factor of two or three (FAO 1999b:4). Asia and Africa lead the world's regions in inland fish production. According to FAO, most inland capture fisheries (all fish except those raised in aquaculture) are exploited at or above their maximum sustainable yields. Globally, inland fisheries production (including aquaculture) increased at 2 percent per year from 1984 to 1997, although in Asia the rate has been much higher–7 percent per year since 1992. This growth in part results from deliberate fisheries enhancements such as artificial stocking or introduction of new species. Such enhancements are particularly important in Asia, which produces 64 percent of the world's inland fish catch (FAO 1999b:6). Another factor in increased production may, ironically, be the eutrophication of inland waters, which, in mild forms, can raise the production of some fish species by providing more food at the base of the food chain (FAO 1999b:7).

Aquaculture. As important as the inland fish catch is, production from freshwater aquaculture has now eclipsed it in size, value, and nutritional importance. Freshwater aquaculture production reached 17.7 million tons in 1997 (FAO 1999b:6). Marine and freshwater aquaculture together provided 30 percent of the fish consumed directly by humans in 1997, and more than 60 percent of this production is freshwater fish or fish that migrate between fresh and saltwater (FAO 1999a:7; FAO 1998). Asia, and China in particular, dominate aquaculture production (FAO 1999b:7).

Recreational Fishing. In Europe and North America, freshwater fish consumption has declined in recent decades and much of the fishing effort now is devoted to recreation. Recreational fishing contributes significantly to some economies. For instance, Canadian anglers spend $2.9 billion Canadian dollars per year on products and services directly related to fishing (McAllister et al. 1997:12). In the United States, anglers spent US$447 million on fishing licenses alone in 1996 (FAO 1999b:42). Recreational fisheries also contribute to the food supply since anglers usually consume what they catch, although recently there is a trend toward releasing fish after they are caught (Kapetsky 1999). The recreational catch is currently estimated to be around 2 million tons per year (FAO 1999b:42).

Condition of Inland Fisheries. The principal factor threatening inland capture fisheries is the loss of fish habitat and environmental degradation (FAO 1999b:19). In certain areas like the Mekong River basin in Asia, overfishing and destructive fishing practices also contribute to the threat (FAO 1999b:19). In addition, nonnative species introduced into lakes, rivers, and reservoirs–either accidentally or for food or recreational fishing–affect the composition of the native aquatic communities, sometimes increasing levels of production and sometimes decreasing them. Introduced species can be predators or competitors or can introduce new diseases to the native fauna, sometimes with severe consequences. (See Box 1.9 Trade-Offs: Lake Victoria's Ecosystem Balance Sheet, p. 21).

Assessing the actual condition of inland fisheries is complicated by the difficulty of collecting reliable and comprehensive data on fish landings. Much of the catch comes from subsistence and recreational fisheries and these are particularly hard to monitor, since these harvests are not brought back to centralized markets or entered into commerce (FAO 1999b:4).

Nevertheless, harvest and trend information exist for certain well-studied fisheries. Harvest information includes changes in landings of important commercial species and in the species composition of well-studied rivers. Without exception, each of the major fisheries examined has experienced dramatic declines during this century.

A somewhat different picture of the condition of inland fisheries is provided by data from FAO. By analyzing catch statistics over 1984–97, FAO found positive trends in inland capture fish harvests in South and Southeast Asia, Central America, and parts of Africa and South America. Harvest trends were negative in the United States, Canada, parts of Africa, Eastern Europe, Spain, Australia, and the former Soviet Union (FAO 1999b:9–18, 51–53).

(continues on p. 116)

Box 2.27 Changes in Inland Fisheries

Catches from inland fisheries account for nearly 12 percent of the total fish consumed by humans (FAO 1999a). In many landlocked countries, such as Malawi, freshwater fish make up a high proportion of total protein intake, particularly among the poor (FAO 1999b).

Globally, landings from inland capture fisheries (wildfish caught by line, net, or trap) have increased by an average of 2 percent per year from 1984 to 1996. Regional trends, however have diverged widely, with declines in Australia, North America, and the former Soviet Union and increases in much of Africa and Asia. Since 1987, aquaculture has outstripped capture fisheries as the major source of freshwater fish, with production dominated by Asian countries (FAO 1999a).

According to FAO, most inland capture fisheries are being exploited at above-sustainable levels. The effects of overharvesting are exacerbated by the loss or degradation of freshwater habitat caused by factors like dam building and pollution. The growth in total catch has been achieved only through reliance on restocking and the introduction of more productive species in major producing countries such as China.

Inland Capture Fisheries Landings, 1997

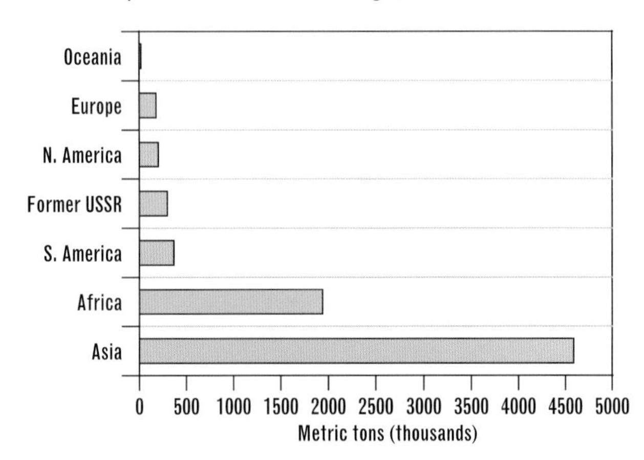

Inland Capture Fisheries Trends, 1984–97

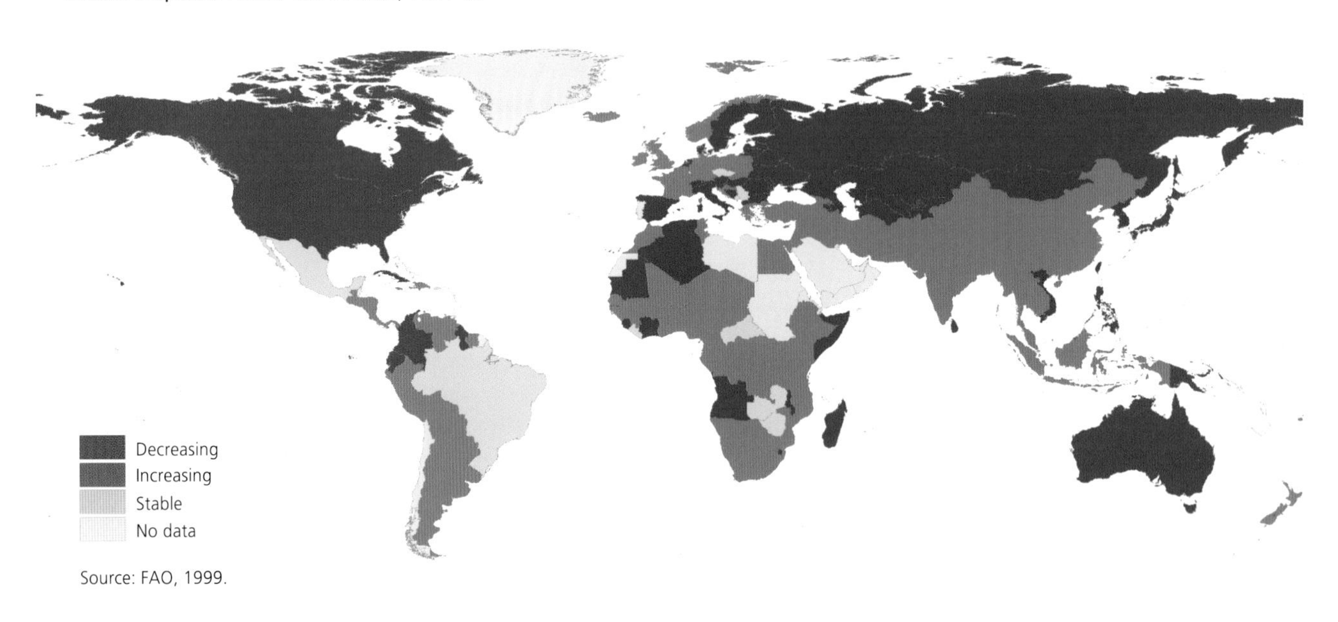

Source: FAO, 1999.

Sources: Revenga et al. [PAGE] 2000. The map is based on (FAO 1999b). The figure is based on FAO (1998). Table is derived from Carlson and Muth (1989), Bacalbasa-Dobrovici (1989), Postel (1995), Abramovitz (1996, citing Missouri River Coalition 1995), Hughes and Noss (1992), Sparks (1992), Kauffman (1992), and Liao et al. (1989).

Changes in Fish Species Composition and Fisheries for Selected Rivers

River	Change in Fish Species and Fishery	Major Causes of Decline	Main Goods and Services Lost
Colorado River, USA	Historically native fish included 36 species, 20 genera, and 9 families; 64 percent of these were endemic. Current status of species under the Endangered Species Act: 2 extinct, 15 threatened or endangered, 18 proposed for listing or under review	Dams, river diversions, canals, and loss of riparian habitat.	Loss of fisheries and biodiversity.
Danube River	Since the early 1900s, Danube sturgeon fishery has almost disappeared. Current fisheries are maintained through aquaculture and introduction of nonnative species.	Dams, creation of channels, pollution, loss of floodplain areas, water pumping, sand and gravel extraction, and nonnative species introductions.	Loss of fisheries, loss of biodiversity, and change in species composition.
Aral Sea	Of 24 fish species, 20 have disappeared. The commercial fishery that used to have a catch of 40,000 tons and support 60,000 jobs is now gone.	Water diversion for irrigation, pollution from fertilizers and pesticides.	Loss of important fishery and biodiversity. Associated health effects caused by toxic salts from the exposed lakebed.
Rhine River	Forty-four species became rare or disappeared between 1890 and 1975. Salmon and sturgeon fisheries are gone, and yields from eel fisheries have declined even though it is maintained by stocking.	Dams, creation of channels, heavy pollution, and nonnative species introductions.	Loss of important fishery, loss of biodiversity.
Missouri River	Commercial fisheries declined by 83 percent since 1947.	Dams, creation of channels and pollution from agriculture runoff.	Loss of fishery and biodiversity.
Great Lakes	Change in species composition, loss of native salmonid fishery. Four of the native fish have become extinct and seven others are threatened.	Pollution from agriculture and industry, non-native species introductions.	Loss of fishery, biodiversity, and recreation.
Illinois River	Commercial fisheries decreased by 98 percent in the 1950s.	Siltation from soil erosion, pollution, and eutriphication.	Loss of fishery and biodiversity.
Lake Victoria	Mass extinction of native cichlid fishes. Changes in species composition and disappearance of the small-scale subsistence fishery that many local communities depended on.	Eutrophication, siltation from deforestation, overfishing, and introduction of nonnative species.	Loss of biodiversity and local artisanal fishery.
Pearl River (Xi Jiang)	In the 1980s, yield levels in commercial fisheries dropped to 37 percent of 1950s levels.	Overfishing, destructive fishing practices, pollution, and dams.	Loss of fishery.

Depending on the region, the growth in harvests that FAO documented could stem from a variety of reasons: the exploitation of a formerly underfished resource, overexploitation of a fishery that will soon collapse, or enhancement of fisheries by stocking or introducing more productive species. FAO found that in every region, the major threat to fisheries was environmental degradation of freshwater habitat (FAO 1999b:19).

The Bottom Line for Food Production. Freshwater fish play an extremely important role in human nutrition as well as in local economies. Harvests have increased significantly in recent decades, reaching their current 7.7-million ton level for captured fish and 17.7 million tons for aquaculture-raised fish. Data are inadequate to determine sustainable yields for most wild populations, but where data exist, they show that the capacity of freshwater ecosystems to support wild fish stocks has declined significantly because of habitat degradation and overharvest. Production of freshwater aquaculture, however, has been increasing rapidly and is expected to continue to do so. The yield of some inland capture fisheries focused on introduced species has also increased, but sometimes to the detriment of native fish species.

BIODIVERSITY

Freshwater systems, like other major ecosystems, harbor a diverse and impressive array of species. Twelve percent of all animal species live in freshwater ecosystems (Abramovitz 1996:7) and many more species are closely associated with these ecosystems. In Europe, for example, 25 percent of birds and 11 percent of mammals use freshwater wetlands as their main breeding and feeding areas (EEA 1994:90).

Although freshwater ecosystems have fewer species than marine and terrestrial habitats, species richness is high, given the limited extent of aquatic and riparian areas. According to estimates from Reaka-Kudla (1997:90), there are 44,000 described aquatic species, representing 2.4 percent of all known species; yet freshwater systems occupy only 0.8 percent of Earth's surface (McAllister et al. 1997:5).

Some regions are particularly important because they contain large numbers of species or many endemic species (those that are found nowhere else) (Box 2.28 Biodiversity in Freshwater Systems). Many of the most diverse fish faunas are found in the tropics, particularly Central Africa, mainland Southeast Asia, and South America, but high diversity is also found in central North America and in several basins in China and India.

Physical alteration, habitat loss and degradation, water withdrawal, overexploitation, pollution, and the introduction of nonnative species all contribute directly or indirectly to declines in freshwater species. These varied stresses affecting aquatic systems occur all over the world, although their particular effects differ from watershed to watershed.

Threats and Extinctions

Perhaps the best measure of the actual condition of freshwater biodiversity is the extent to which species are threatened with extinction. Globally, scientists estimate that more than 20 percent of the world's freshwater fish species–of which some 10,000 have been described–have become extinct, are threatened, or endangered in recent decades (Moyle and Leidy 1992:127, cited in McAllister et al. 1997:38; Bräutigam 1999:5). According to the 1996 IUCN Red List of Threatened Animals, 734 species of fish are classified as threatened; of those, 84 percent are freshwater species (IUCN 1996:37 Introduction; McAllister et al. 1997:38). In Australia, 33 percent of freshwater fish are threatened, and in Europe, the number rises to 42 percent (Bräutigam 1999:4).

In the United States, one of the countries for which good data on freshwater species exist, 37 percent of freshwater fish species, 67 percent of mussels, 51 percent of crayfish, and 40 percent of amphibians are threatened or have become extinct (Master et al. 1998:6). In western North America, data from 1997 show that more than 10 percent of fish species are imperiled in most ecoregions (distinct ecological regions), with more than 25 percent imperiled in eleven ecoregions (Abell et al. 2000:75). Similar patterns are found for endangered frogs and salamanders. Based on recent extinction rates, an estimated 4 percent of freshwater species will be lost in North America each decade, a rate nearly five times that of terrestrial species (Ricciardi and Rasmussen 1999:1220).

It is not surprising that wetland species are often most threatened in arid areas, where there isn't enough water to meet the competing needs of humans and the environment. For example, of 391 "important bird areas" in the Middle East identified by BirdLife International, half are wetlands (Evans 1994:31). Moreover, these wetland sites were also judged to be the most threatened (Evans 1994:35).

Amphibian Declines

Population trends are one of the best ways to measure the condition of individual species and groups of species. Continental- or global-level data on population trends for extended time periods are not readily available for many freshwater-dependent species. But the availability of global population data for one taxonomic group–amphibians–has grown dramatically over the past 15 years as scientists have sought to ascertain the causes of an apparent world-wide decline of frogs and other amphibians (Pelley 1998). These data show significant declines in all world regions over several decades. For example, of nearly 600 amphibian populations studied in Western Europe, 53 percent declined beginning in the 1950s (Houlahan et al. 2000:754). In North America, 54 percent of the populations studied declined, while in South America, 60

Box 2.28 Biodiversity in Freshwater Systems

Despite their small area, compared with other ecosystems, freshwater systems are relatively rich in the number of species they support. Although 12 percent of all animal species live in freshwater systems (Abramovitz 1996:7), many more depend on them for survival. Physical alterations, habitat loss and degradation, water withdrawal, overexploitation, and introduction of nonnative species all contribute to declines in freshwater species. Globally, more than 20 percent of the world's freshwater fish species have become extinct, threatened, or endangered in recent decades (Moyle and Leidy 1992:127).

Freshwater biodiversity is not uniformly distributed around the world; some regions are particularly important because they contain large numbers of species or many endemic species (species occurring only in a restricted area). Endemism tends to correlate with overall species richness. Most of the highest concentrations of both endemism and species diversity are found in the tropics, particularly the Amazon, Congo, and Mekong watersheds.

Fish Species Richness and Endemism, by Watershed

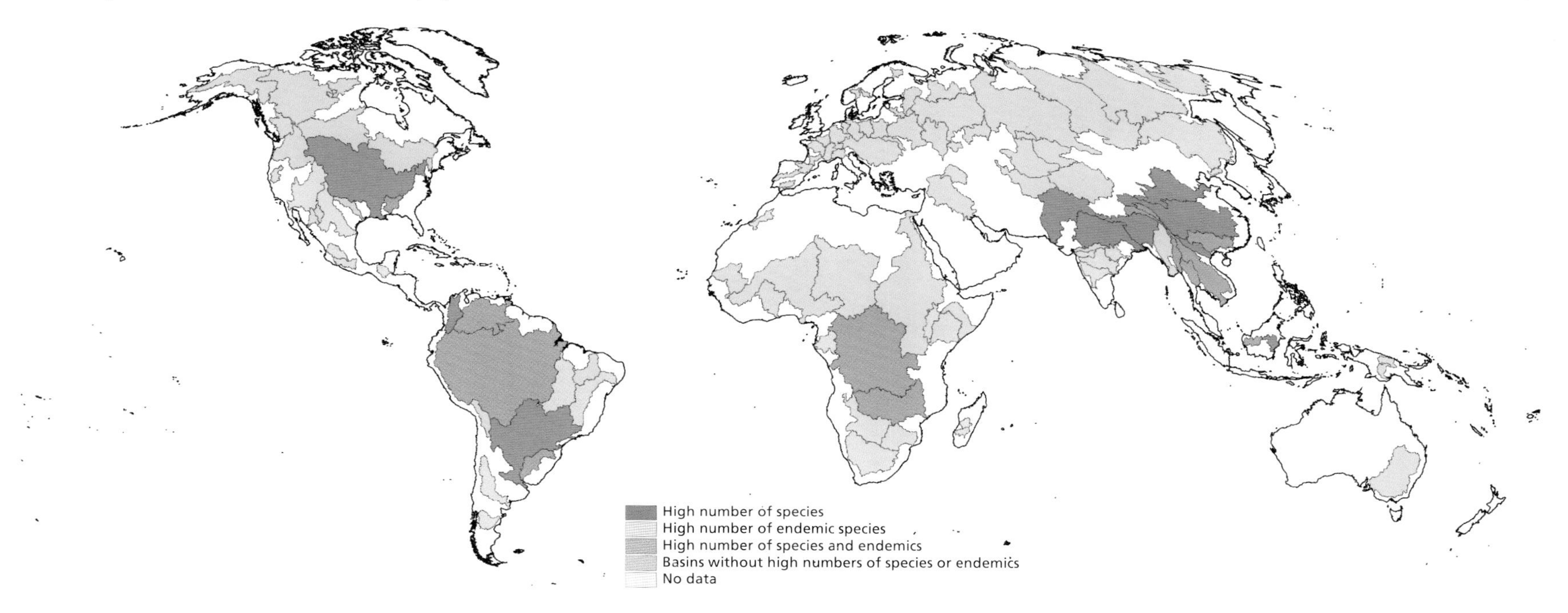

Sources: Revenga et al. [PAGE] 2000. The map is based on Revenga et al. (1998). Because there is a correlation between number of species and total area sampled, large watersheds tend to have more fish species than smaller ones (Oberdorff 1995). To reduce bias in size differences, basins were categorized as large (more than 1.5 million km^2), medium (400,000 to 1.5 million km^2), and small (less than 400,000 km^2). The map shows large basins with more than 230 fish species, medium basins with more than 143 species, and small basins with more than 112 species. For endemics, the map shows large basins with more than 166 species, medium basins with more than 29 species, and small basins with more than 15 species. Cut-off points for each category were determined by selecting the upper two-thirds within each range.

percent declined. In Australia and New Zealand, as much as 70 percent of studied populations declined, although far fewer populations were monitored. The mechanisms thought to be responsible for declines include increased exposure to ultraviolet-B rays, resulting from the thinning of the stratospheric ozone layer; chemical pollution from pesticides, fertilizers, and herbicides; acid rain; pathogens; introduction of predators; and global climate change (Lips 1998; Pelley 1998; DAPTF 1999).

Invasive Species

The number and abundance of nonnative species is another important indicator of the condition of freshwater biodiversity. Introduced species are a major cause of extinction in freshwater systems, affecting native fauna through predation, competition, disruption of food webs, and the introduction of diseases. Species introductions have been particularly successful in freshwater ecosystems. For example, two-thirds of the freshwater species introduced into the tropics have subsequently become established (Beveridge et al. 1994:500).

Nonnative fish introductions are common and increasing in most parts of the world. Fish are often deliberately introduced to increase food production or to establish or expand recreational fisheries or aquaculture. For example, introduced fish account for 97 percent of fish production in South America and 85 percent in Oceania (Garibaldi and Bartley 1998). However, nonnative fish introductions often have significant ecological costs. A 1991 survey of fish introductions in Europe, North America, Australia, and New Zealand found that 77 percent of the time, native fish populations decreased or were eliminated following the introduction of nonnative fish (Ross 1991:359). In North America, introduced species have played a large role in the extinction of 68 percent of the fish that have become extinct in the past 100 years (Miller et al. 1989:22).

The economic costs of accidental introductions can also be high. For example, the introduction of the sea lamprey (*Petromyzon marinus*) in the Great Lakes of North America was a factor in the crash of the lake trout fishery in the 1940s and 1950s. In 1991, efforts to control sea lampreys through chemical and mechanical means cost Canada and the United States $8 million, with an additional $12 million spent on lake trout restoration (Fuller et al. 1999:21). Similarly, between 1989 and 1995, the costs of zebra mussel (*Dreissena polymorpha*) eradication in the United States and Canada totaled well over $69 million, with some estimates as high as $300–$400 million (O'Neill 1996:2; O'Neill 1999). On the ecological front, zebra mussel infestation has dramatically reduced populations of native clams at 17 different sampling stations, leading to the near-extinction of many species.

Some of the most dramatic trade-offs between economic benefits and ecological costs involve introductions of species of tilapia (*Oreochromis niloticus* and *O. mossambicus*) and the common carp (*Cyprinus carpio*). These important aquaculture species have now been introduced around the world. In 1996, 1.99 million tons of common carp and 600,000 tons of Nile tilapia were produced through aquaculture (FAO 1999a:14). But in lakes and rivers where these species have been introduced, native species have suffered. By feeding at the bottom of lakes and rivers, carp increase siltation and turbidity, decreasing water clarity and harming native species (Fuller et al. 1999:69). They have been associated with the disappearance of native fishes in Argentina, Venezuela, Mexico, Kenya, India, and elsewhere (Welcomme 1988:101–109).

Water hyacinth (*Eichhornia crassipes*) is another example of a widespread invasive species that is causing considerable economic and ecological damage in many parts of the world. This plant, thought to be indigenous to the upper reaches of the Amazon basin, was spread widely across the planet for use as an ornamental plant beginning in the mid-19th century and is now distributed throughout the tropics (Gopal 1987:1). Water hyacinth poses practical problems for fishing and navigation, and is a threat to biological diversity, affecting fish, plants and other freshwater life. The plant spreads quickly to new rivers and lakes in the tropics, clogging waterways and causing serious disruption to the livelihood of local communities that depend on goods and services derived from these freshwater ecosystems (Hill et al. 1997). In addition, hyacinth and other aquatic plants act as vectors in the life cycles of insects that transmit diseases such as malaria, schistosomiasis, and lymphatic filariasis (Bos 1997).

The Bottom Line for Biodiversity. Physical alteration, water withdrawals, overharvesting, and the introduction of nonnative species have all taken a heavy toll on freshwater biodiversity. Indeed, of all the ecosystems examined in this report, freshwater systems by far are in the worst condition from the standpoint of their ability to support biological diversity—on a global level. More than 20 percent of the world's 10,000 freshwater fish species have become extinct, threatened, or endangered in recent decades. In the United States, where data are more complete, 37 percent of freshwater fish species, 67 percent of mussels, 51 percent of crayfish, and 40 percent of amphibians are known to be threatened or extinct. Increased global demands for food and water will increase the already considerable pressures on freshwater systems.

GRASSLAND ECOSYSTEMS

The goods and services provided by the world's grasslands have received far less attention than those supplied by, for example, tropical forests and coral reefs, although grasslands are arguably more important to a larger percentage of people. Grasslands are home to 938 million people–about 17 percent of the world's population (White et al. [PAGE] 2000). They are found throughout the world, in humid as well as arid zones, but grasslands are particularly important features of the world's drylands. Approximately half of the people living in grassland regions live in the world's arid, semiarid, and dry subhumid zones (White et al. [PAGE] 2000). Scant rains make these drylands particularly susceptible to damage from human management and slower to recover from degradation such as overgrazing or improper cultivation practices.

Grassland ecosystems have historically been crucial to the human food supply. The ancestors of nearly all the major cereal crops originally developed in grasslands, including wheat, rice, rye, barley, sorghum, and millet. Agroecosystems have replaced many grasslands, but grasslands still provide genetic resources for improving food crops and are a potential source of pharmaceuticals and industrial products.

Grasslands are important habitats for many species, including breeding, migratory, and wintering birds, and support many wild and domestic grazing animals. Grassland vegetation and soils also store a considerable quantity of carbon. Other grassland ecosystem goods and services include meat and milk; wool and leather products; energy from fuelwood and wind generated from windfarms; cultural and recreational services such as tourism, hunting, and aesthetic and spiritual gratification; and water regulation and purification. PAGE researchers examined four of these goods and services: food production, biodiversity maintenance, carbon storage, and tourism (Box 2.29 Taking Stock of Grassland Ecosystems).

Extent and Modification

PAGE researchers defined *grassland ecosystems* as "areas dominated by grassy vegetation and maintained by fire, grazing, and drought or freezing temperatures." Using this broad definition, grasslands encompass nonwoody grasslands, savannas, woodlands, shrublands, and tundra. Grassland ecosystems are found on every continent. Among the most extensive are the savannas

(continues on p. 122)

Box 2.29 Taking Stock of Grassland Ecosystems

Highlights

- Grasslands, which cover 40 percent of the Earth's surface, are home to almost a billion people, half of them living on susceptible drylands.
- Agriculture and urbanization are transforming grasslands. For some North American prairies, conversion is already nearly 100 percent. Road-building and human-induced fires also are changing the extent, composition, and structure of grasslands.
- All of the major foodgrains—corn, wheat, oats, rice, barley, millet, rye, and sorghum—originate in grasslands. Wild strains of grasses can provide genetic material to improve food crops and to help keep cultivated varieties resistant to disease.
- Grasslands attract tourists willing to travel long distances and pay safari fees to hunt and view grassland fauna. Grasslands boast some of the world's greatest natural phenomena: major migratory treks of large herds of wildebeest in Africa, caribou in North America, and Tibetan antelope in Asia.
- As habitat for biologically important flora and fauna, grasslands make up 19 percent of the Centers of Plant Diversity, 11 percent of Endemic Bird Areas, and 29 percent of ecoregions considered outstanding for biological distinctiveness.

Key

Condition assesses the current output and quality of the ecosystem good or service compared with output and quality of 20–30 years ago.

	Excellent	Good	Fair	Poor	Bad	Not Assessed
Condition						

Changing Capacity assesses the underlying biological ability of the ecosystem to continue to provide the good or service.

Changing Capacity: Increasing, Mixed, Decreasing, Unknown

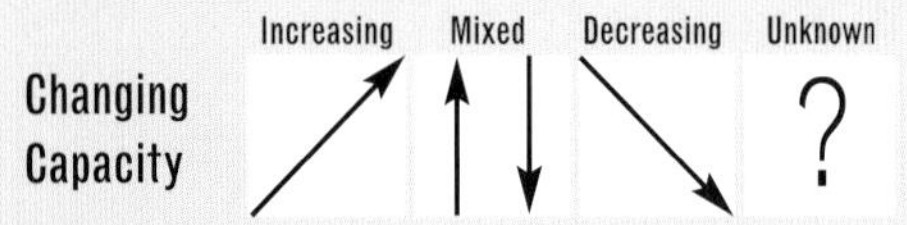

Scores are expert judgments about each ecosystem good or service over time, without regard to changes in other ecosystems. Scores estimate the predominant global condition or capacity by balancing the relative strength and reliability of the various indicators. When regional findings diverge, in the absence of global data weight is given to better-quality data, larger geographic coverage, and longer time series. Pronounced differences in global trends are scored as "mixed" if a net value cannot be determined. Serious inadequacy of current data is scored as "unknown."

Conditions and Changing Capacity

FOOD PRODUCTION

Many grasslands today support high livestock densities and substantial meat production, but soil degradation is a mounting problem. Soil data show that 20 percent of the world's susceptible drylands, where many grasslands are located, are degraded. Overall, the ability of grasslands to support livestock production over the long term appears to be declining. Areas of greatest concern are in Africa, where livestock densities are high, and some countries already show decreases in meat production.

BIODIVERSITY

Regional data for North America document marked declines in grassland bird species and classify 10–20 percent of grassland plant species in some areas as nonnative. In other areas, such as the Serengeti in Africa, In other areas, such as the Serengeti in Africa, population levels of large grassland herbivores have not changed significantly n the past 2 decades.

CARBON STORAGE

Grasslands store about one-third of the global stock of carbon in terrestrial ecosystems. That amount is less than the carbon stored in forests, even though grasslands occupy twice as much area. Unlike forests, where vegetation is the primary source of carbon storage, most of the grassland carbon stocks are in the soil. Thus, the future capacity of grasslands to store carbon may decline if soils are degraded by erosion, pollution, overgrazing, or static rather than mobile grazing.

RECREATION

People worldwide rely on grasslands for hiking, hunting, fishing, and religious or cultural activities. The economic value of recreation and tourism can be high in some grasslands, especially from safari tours and hunting. Some 667 protected areas worldwide include at least 50 percent grasslands. Nonetheless, as they are modified by agriculture, unbanization, and human-induced fires, grasslands are likely to lose some capacity to sustain recreation services.

Data Quality

FOOD PRODUCTION

Soil degradation can be determined globally, but assessment often relies on expert opinion, and the scale of the data is too coarse to apply to national policies. Data on livestock density in grasslands include global and some regional coverage, but only for domestic animals. We still lack corresponding studies of vegetation, soil condition, management practices, and long-term resilience. Data on meat production are available globally, but meat produced from livestock raised in feedlots cannot be separated from meat produced from range-fed livestock.

BIODIVERSITY

Long-term trends in grassland bird populations can be assessed from comprehensive regional data for the United States and Canada. Some long-term regional data within Africa show steady levels of major herbivore populations, but geographic coverage is limited. Other regional, national, and local data for grassland species lack long-term trends. Regional and local coverage of invasive species are more descriptive than quantitative.

CARBON STORAGE

Methods for estimating the size of carbon stores in biomass and soils continue to evolve. This study relied on previous global estimates for above- and below-ground live vegetation, updated to fit the current land cover map by the International Geosphere-Biosphere Programme, with the addition of soil carbon storage estimates. Models are needed to incorporate carbon storage modifications based on different management practices.

RECREATION

Regional information evaluates the exploitation of grassland wildlife but summaries are based primarily on expert opinion. Global country-level expenditures on international tourism provide estimates for all types of tourism but cannot be related specifically to grasslands. Regional data for tourism and safari hunting are good for some areas but rarely report long-term trends.

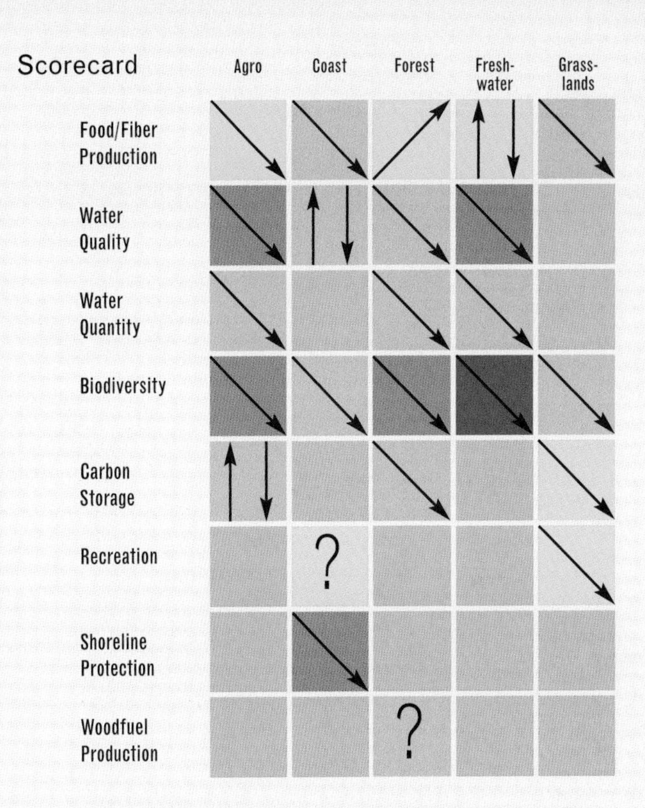

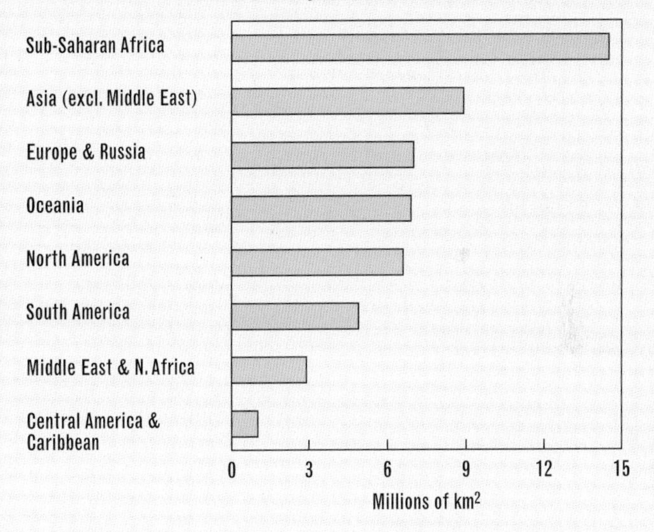

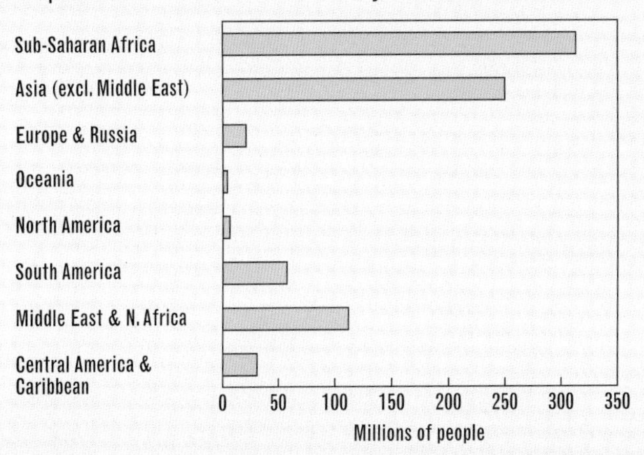

of Africa, the steppes of Central Asia, the cerrado and campo of South America, the prairies of North America, and the grasslands of Australia.

Extent

Estimates of the extent of the world's grassland ecosystems range from approximately 41 million km^2 to 56 million km^2, covering 31–43 percent of Earth's surface (Whittaker and Likens 1975:306, Table 15-1; Atjay et al. 1979:132–133; Olson et al. 1983:20–21). The differences among estimates are due, in part, to different definitions of grasslands; for instance, different researchers include more (or less) tundra or shrubland.

Using land-cover maps generated from recent satellite data, PAGE researchers produced a new map of the extent of the world's grasslands (Box 2.30 Global Extent of Grasslands). Some of the grasslands in this map are actually mosaics of grasslands and other land uses such as agriculture but are considered to be grasslands when those "other" land uses cover 40 percent or less of the area. Mapped this way, grassland ecosystems cover 52.5 million km^2–about 41 percent of the world's land area (excluding Antarctica and Greenland)–much more than forests or agroecosystems. Indeed, on a national basis, grasslands are one of the most common and extensive types of land cover. In 40 countries, grasslands cover more than 50 percent of the land area, and in 20 of these countries–most of them in Africa–grasslands make up more than 70 percent of the land area.

Grasslands are a significant ecosystem in many of the world's important watersheds as well. For example, grasslands comprise more than 50 percent of the land area in these watersheds: the Yellow River in China; the Nile, Zambezi, Orange, and Niger Rivers in Africa; the Rio Colorado in South America; and the Colorado and Rio Grande in North America (White et al. [PAGE] 2000). The extent of grasslands in these watersheds underscores the importance of managing grasslands so that they retain their watershed functions of absorbing rainfall to recharge aquifers, stabilizing soils, and moderating runoff. These essential watershed services are an often underappreciated aspect of grasslands.

Modifications

Like forests, the world's grasslands have lost much of their original extent through human actions–mostly conversion to agriculture. Scientists have no easy way to determine the extent of global grasslands prior to human disturbance, and thus no easy way to determine the exact amount of grasslands lost over time. However, PAGE researchers obtained a good rough estimate of historical loss by comparing current grasslands extent to "potential" grassland areas–those areas where grasslands would be expected to exist today (based on soil, elevation, and climate conditions) if humans had not intervened.

Using this approach, PAGE researchers examined in depth five regions for which the potential vegetation would likely be 100 percent grassland in the absence of humans disturbance. Among these regions, the Tallgrass Prairie in North America shows the greatest change. Croplands cover 71 percent of this region and urban areas cover 19 percent. In contrast, the grassland regions in Asia, Africa, and Australia each retain at least 60 percent of their area in grasslands with less than 20 percent in cropland and less than 2 percent in urban or built-up areas.

FIRE

Fire is a natural occurrence in most grassland ecosystems and has been one of the primary tools humans have used to manage grasslands. Fire prevents bushes from encroaching, removes dry vegetation, and recycles nutrients. Without fire the tree density in many of the world's grasslands would increase, eventually converting them to forests. In addition, fire helps hunters stalk grassland species and helps farmers control pests (Menaut et al. 1991:134).

Natural fires–typically caused by lightening–are thought to occur about every 1–3 years in humid areas (Frost 1985:232) and every 1–20 years in dry areas (Walker 1985:85). But today, the number of natural fires is insignificant compared to the number of fires started by humans (Levine et al. 1999:1). Humans have set fires in the savannas for at least 1.5–2 million years and continue to use fire as a low-cost and effective means to manage grasslands (Andreae 1991:4). Today, for example, in many African countries people use burning to maintain good forage conditions for grazing herds of livestock and to clear away dead debris (Box 2.31 Grassland Fires). Some 500 Mha of tropical and subtropical savannas, woodlands, and open forests now burn each year (Goldammer 1995, cited in Levine et al. 1999:4).

Although fire can benefit grasslands, it can be harmful too–particularly when fires become much more frequent than is natural. If too frequent, fire can remove plant cover and increase soil erosion (Ehrlich et al. 1997:201). Fires also release atmospheric pollutants. Because much of the biomass that is burned each year is from savannas, and because two-thirds of Earth's savannas are in Africa, UNEP reports that Africa is now recognized as the "burn center" of the planet (Levine et al. 1999:2). Burning of savannas is responsible for more than 40 percent of the carbon emissions from global biomass burning each year (Andreae et al. 1991:5).

FRAGMENTATION

Globally, grasslands have been heavily modified by human activities. Few large unaltered expanses remain (Box 2.32 Fragmentation of American Grasslands). Even many smaller grassland areas are extensively fragmented (Risser 1996:265). Fragmentation can affect the condition of grasslands in many ways, increasing fire frequency, degrading habitat, and damaging the capacity of the grassland to maintain biological diversity. Agriculture, urbanization, and road building are the biggest sources of grassland fragmentation, but livestock

Box 2.30 Global Extent of Grasslands

Grasslands are found on every continent and cover approximately 41 percent of Earth's land area (excluding Greenland and Antarctica). To gauge the impact of human activity on the extent of grasslands, PAGE researchers looked at five regions that could be expected to be entirely grasslands, based on current climate and geographic conditions. Of these the Tallgrass Prairie in North America shows the greatest change, with grasslands now accounting for only 9.4 percent of the total area. Only 21 percent of grasslands remains in South America. By contrast, more than 50 percent of the regions selected in Asia, Africa, and Australia remain as grasslands.

Estimated Grassland, Remaining and Converted (percent)

Continent and Region	Remaining in Grasslands	Converted to Croplands	Converted to Urban Areas	Total Converted
N. America Tallgrass Prairie in the United States	9.4	71.2	18.7	89.9
S. America Cerrado Woodland and Savanna in Brazil, Paraguay, and Bolivia	21.0	71.0	5.0	76.0
Asia Daurian Stepppe in Mongolia, Russia, and China	71.7	19.9	1.5	21.4
Africa Central and eastern Mopane and Miombo Woodlands in Tanzania, Rwanda, Burundi, Dem. Rep. Congo, Zambia, Botswana, Zimbabwe, and Mozambique	73.3	19.1	0.4	19.5
Oceania Southwest Australian shrublands and woodlands	56.7	37.2	1.8	39.0

The Global Extent of Grasslands

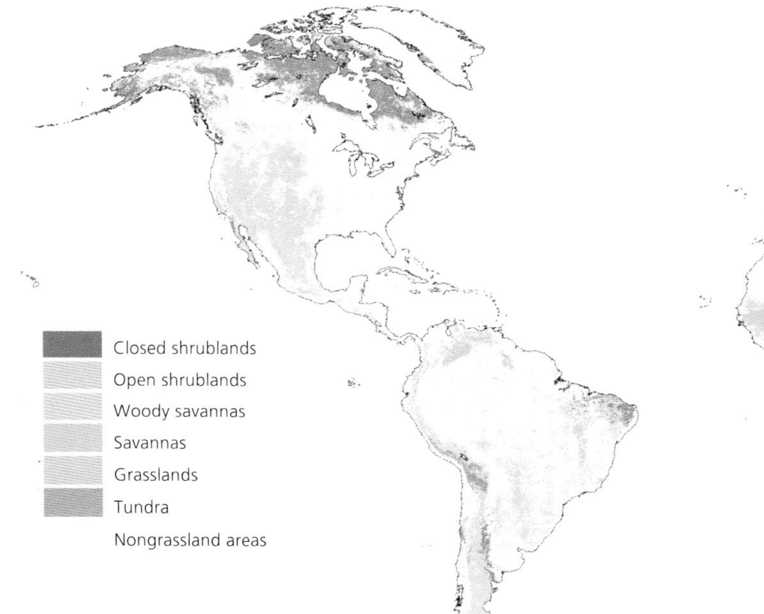

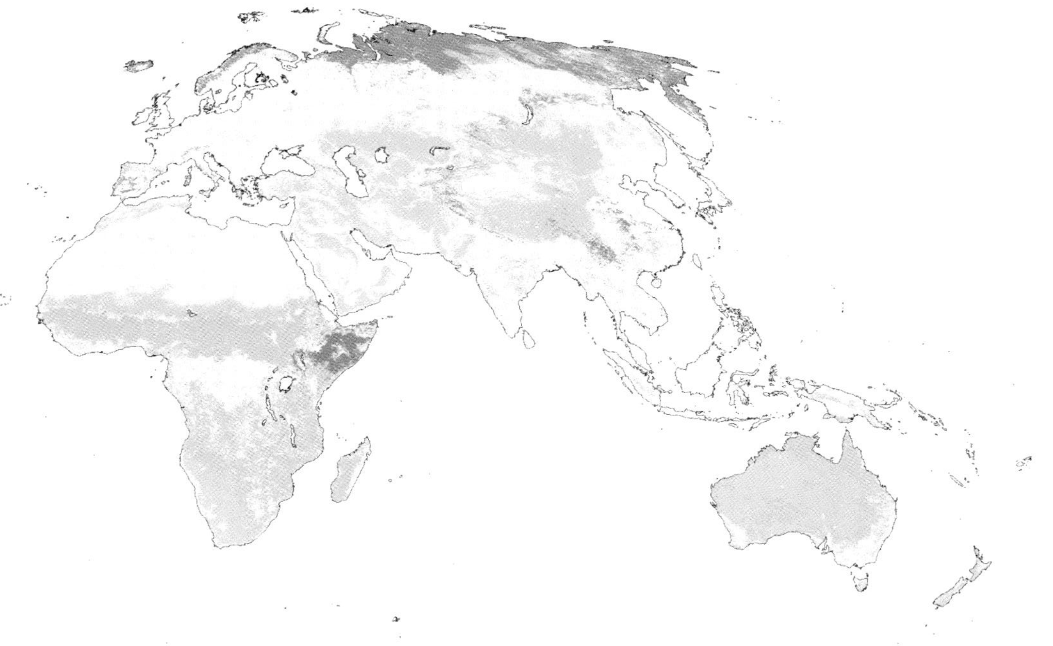

Sources: White et al. [PAGE] 2000. Map is based on the Global Land Cover Characteristics Database Version 1.2 (Loveland et al. 2000). The map shows all lands where grassland made up at least 60 percent of each 1 km^2 satellite mapping unit. Tundra areas are estimated using the Olson Global Ecosystem classification; all other areas are estimtaed from the International Geosphere-Biosphere Programme classification. Table is based on data from WWF and this map.

Box 2.31 Grassland Fires

Fire plays a vital role in determining the character and extent of the world's grasslands. Fires clear dry vegetation, prevent bush encroachment, and recycle nutrients. Without them, much of the world's grasslands would eventually become forested.

Today, the number of natural fires, typically caused by lightning, is insignificant compared with the number set by humans, who have used fire for millennia to hunt, clear land for cultivation and grazing, remove dead debris, and kill pests. Deliberate burning of grasslands is widely practiced in many African countries, with 25–50 percent of total land surface in the arid Sudan Zone and 60–80 percent in the humid Guinea Zone burned annually (Menaut et al. 1991:137).

Fires can be beneficial for grassland ecosystems, but if they become too frequent, they can remove vegetation cover and increase soil erosion (Ehrlich et al. 1997:201). In addition, fires are a significant source of atmospheric pollutants and carbon emissions, with savanna fires, mostly in Africa, accounting for a large proportion of the carbon released into the atmosphere as a result of biomass burning.

Fires Detected by Remote Sensing in Africa, South America, and Oceania, 1993

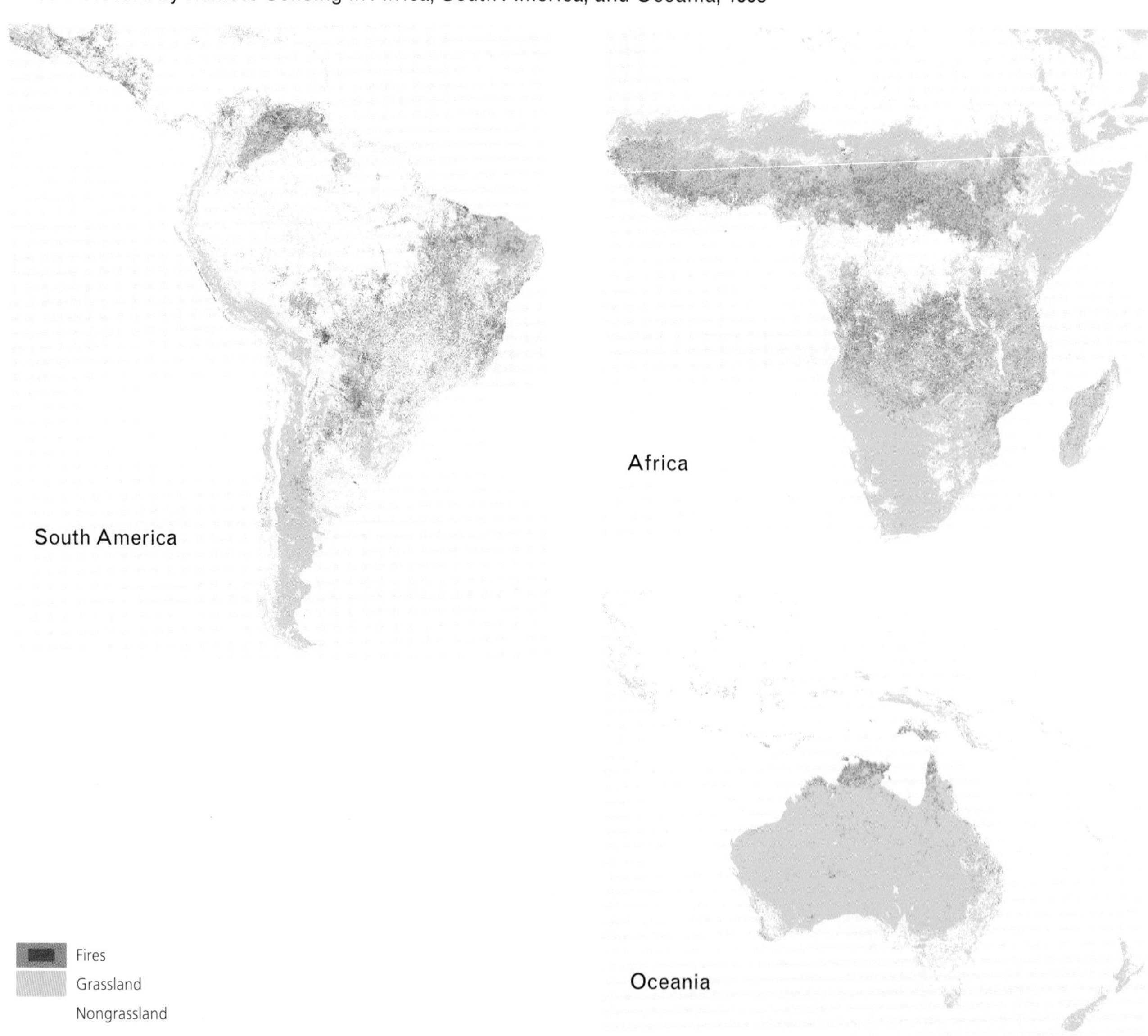

Source: White et al. [PAGE] 2000. Map is based on Arino and Melinotte (1998) and Global Land Cover Characteristics Database Version 1.2 (Loveland et al. 2000).

fencing and the spread of woody vegetation into grasslands also cause significant fragmentation and harm to native species.

One way to evaluate fragmentation is visually–using habitat maps and expert opinion to gauge the size of habitat blocks and the degree of fragmentation in an area. Using this approach, an analysis of 90 grassland regions in North and Latin America showed that the most heavily fragmented grasslands were in temperate and subtropical zones of North America, where there has been extensive agricultural development (Dinerstein et al. 1995:78–83; Ricketts et al. 1997:33, 147–150).

Another way to assess the pressure of fragmentation is to measure the extent to which road networks have contributed to the breakup of larger blocks of grasslands. PAGE researchers used this approach to measure fragmentation in two pilot regions: Botswana and the Great Plains in the United States. In Botswana, if the impact of roads is not considered, 98 percent of the grassland area is found in patches of at least 10,000 km^2. What little fragmentation researchers did observe is caused mainly by agricultural development or natural factors like rivers. When fragmentation by the road network is included, fragmentation increases somewhat, but 58 percent of the area still remains in 10,000 km^2 patches. In contrast, in the Great Plains of the United States, road fragmentation is pervasive. If the effect of roads is ignored, 90 percent of the grassland area is in patches of 10,000 km^2 or greater. But when roads are factored in, 70 percent of the area is in patches less than 1,000 km^2 and none larger than 10,000 km^2.

LIVESTOCK GRAZING

Grasslands and grazing animals have coexisted for millions of years. Large migratory herbivores–like the bison of North America, the wildebeest and zebra of Africa, and the Tibetan antelope of Asia–are integral to the functioning of grassland ecosystems. Through grazing, these animals stimulate regrowth of grasses and remove older, less productive plant tissue. Thinning of older plant tissues allows increased light to reach younger tissues, which promotes growth, increased soil moisture, and improved water-use efficiency of grass plants (Frank et al. 1998:518).

Grazing by domestic livestock can replicate many of these beneficial effects, but the herding and grazing regimes used to manage livestock can also harm grasslands by concentrating their impacts. Given the advantages of veterinary care, predator control, and water and feed supplements, livestock are often present in greater numbers than wild herbivores and can put higher demands on the ecosystem. In addition, herds of domestic cattle, sheep, and goats do not replicate the grazing patterns of herds of wild grazers. Use of water pumps and barbed wire fences has lead to more sedentary and often more intense use of grasslands by domestic animals (Frank et al. 1998:519, citing McNaughten 1993). Grazing animals in high densities can destroy vegetation, change the balance of plant species, reduce biodiversity, compact soil and accelerate soil erosion, and impede water retention, depending on the number and breed of livestock and their grazing pattern (Evans 1998:263).

Assessing Goods and Services

FOOD PRODUCTION

Grasslands are central to world food production. Historically, grasslands have been the ecosystem most extensively transformed to agriculture; they are the original source of many food crops and a continuing source of genetic material to improve modern crops. But grasslands are also major suppliers of food and income in the form of meat production from livestock. This is particularly important for rural populations. For example, in Africa, where rural populations are substantial, grasslands often support high livestock densities (the number of livestock raised per hectare) and are responsible for most of the continent's beef production (Box 2.33 Rangelands in Africa).

How much meat do grasslands currently produce? Global data on livestock production show more than 5 percent growth in world beef output in the last decade, to 54 million tons in 1998. Mutton and goat output increased even more–up 26 percent over the last decade to nearly 11 million tons. But such data do not provide a direct indicator of rangeland condition or its ability to support livestock. Meat production depends not only on grassland condition, but also on a range of other factors such as the availability of watering holes, dietary supplements, veterinary care, and the economic resources to acquire these things. In addition, some of the growth in meat production has come from the rapid rise in the use of feedlots (confined systems where animals cannot graze and are fattened on grain-based feeds to maximize weight gain). The popularity of intensive feedlot production is growing not only in developed countries where it is already common, but also in developing counties (Sere and Steinfeld 1996:40–41). It is not clear what implications the growing use of intensive livestock systems will have on grassland conditions, worldwide. Feedlots accounted for 12 percent of world beef and mutton production in 1996 (De Haan et al. 1997:53).

Information about livestock density is available for much of the world's grasslands and can provide a window on the grazing pressure grasslands face. However, like meat production, livestock density alone does not provide an accurate measure of the condition of the grassland system. Again, it is important to know how the livestock are managed–in particular, whether they are maintained in stable grazing systems, where livestock continuously graze a given parcel, or mobile grazing systems, where livestock are rotated over many different grazing lands. High livestock densities may indicate a highly productive system–one that effectively rotates cattle among grazing lands

(continues on p. 129)

Box 2.32 Fragmentation of American Grasslands

Fragmentation of grassland ecosystems can compromise their ability to provide goods and services and jeopardize their biodiversity. Agriculture, urbanization, and road building are the primary human-caused sources of grassland fragmentation, but fencing and encroachment by woody vegetation can also have significant impacts.

In the Western Hemisphere, the most fragmented grassland ecoregions are the intensively farmed areas of temperate and subtropical North America. The degree of fragmentation of the grasslands of the Great Plains region in the United States has been exacerbated by extensive road construction. If the road network is not taken into account, 90 percent of grassland area is composed of blocks 10,000 km^2 or more in extent. With roads factored in, however, no continuous blocks of this size remain, and 70 percent of the total area is made up of patches less than 1,000 km^2.

Grassland Blocks in the Great Plains, Excluding Roads

Grassland Blocks in the Great Plains, Including Roads

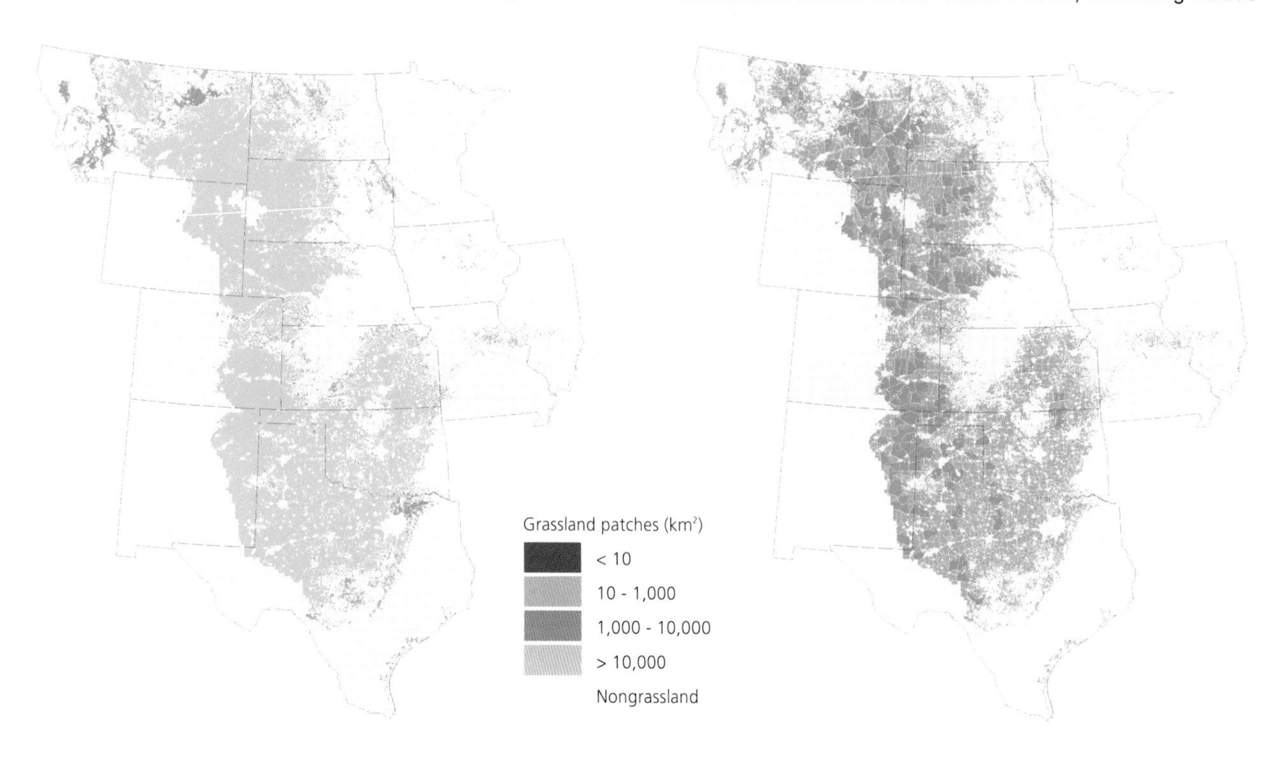

Fragmented Grassland Ecoregions of the Americas

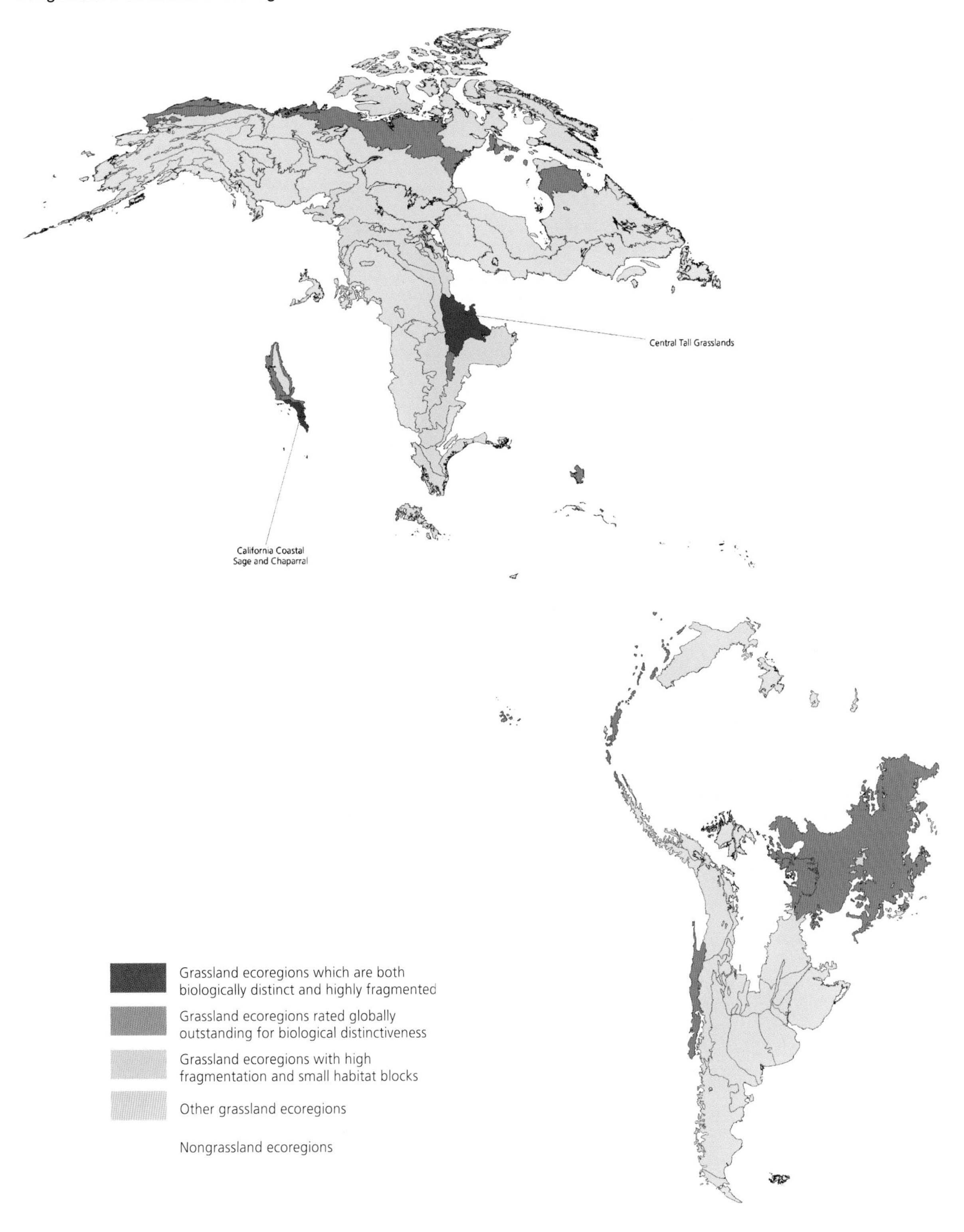

Sources: White et al. [PAGE] 2000. Maps of the Great Plains are based on Global Land Cover Characteristics Database Version 1.2 (Loveland et al. 2000). Map of the Americas is based on the WWF Conservation Assessment for North America, Latin America, and the Caribbean.

Box 2.33 Rangelands in Africa

Grasslands support some of the highest concentrations of cattle in Africa, where many rural populations depend on livestock for sustenance. High densities of livestock may indicate productive, well-managed systems or overstocked, poorly managed ones. Evidence of soil degradation often signals poor management because overstocking of herds diminishes vegetative cover and contributes to erosion. In Africa, a quarter of the susceptible drylands are now degraded, and much of that 320-Mha area is considered to be strongly or extremely degraded. The capacity of African grasslands to continue to support livestock production appears to be poor.

Beef and Veal Production in Sub-Saharan Africa, 1998

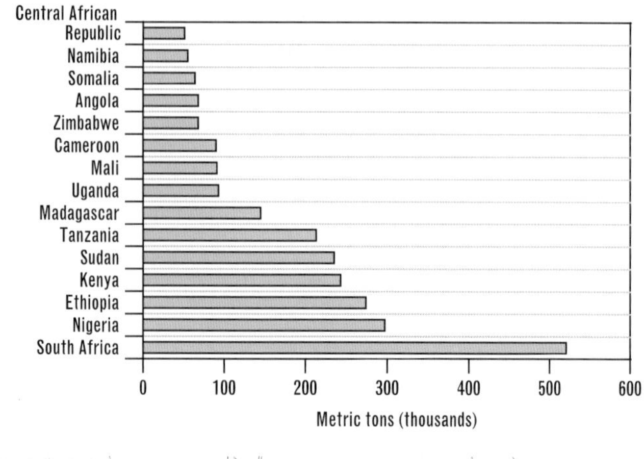

Cattle Density in Sub-Saharan Africa

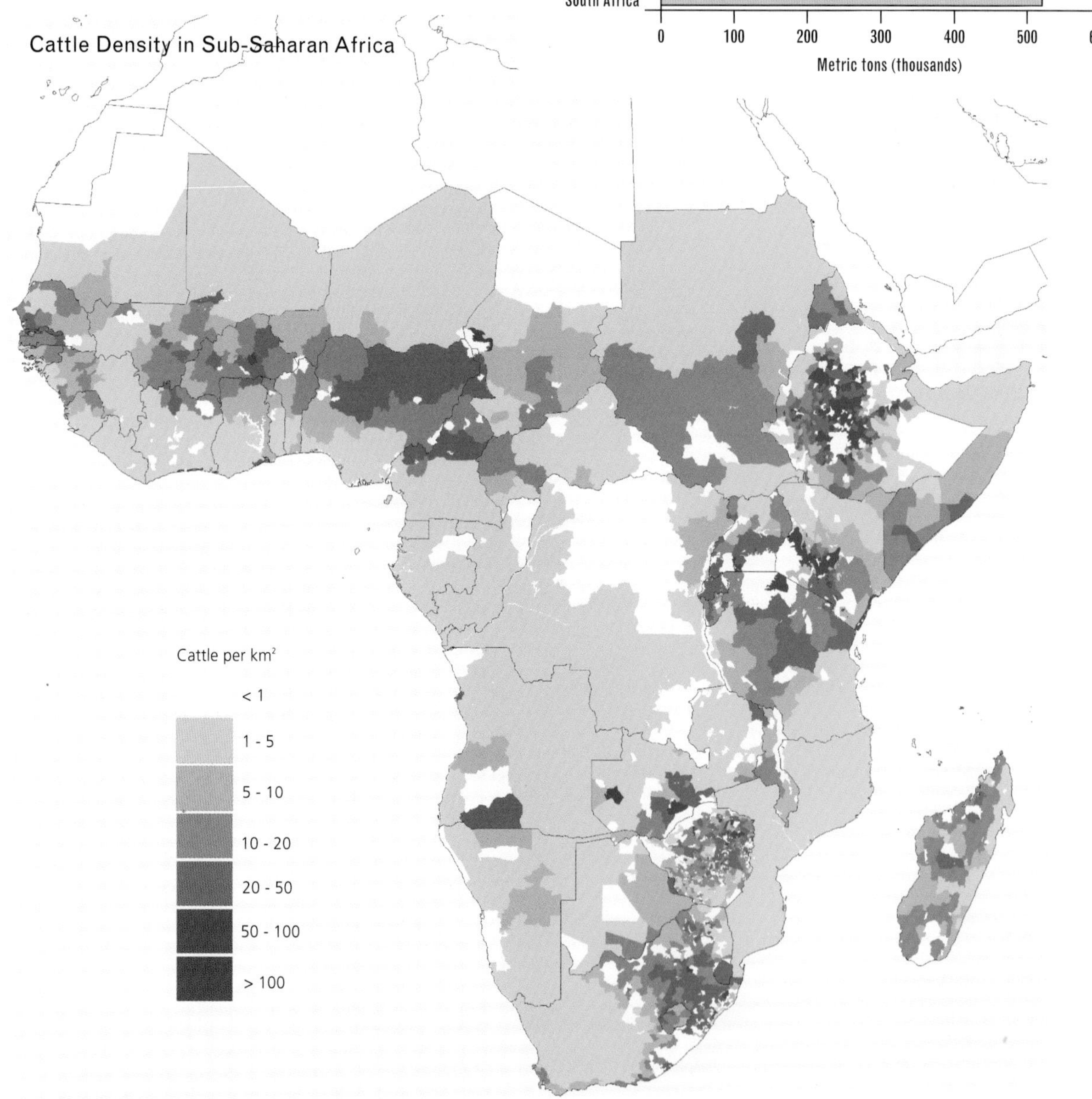

Source: White et al. [PAGE] 2000. Map is based on International Livestock Research Institute (1998). Table is based on FAOSTAT (1999).

and spreads the grazing pressure so that overgrazing does not occur. But high livestock densities could just as easily indicate an overstocked grassland, prone to overgrazing, and with production likely to decrease in subsequent years.

The importance of the livestock management system–mobile or static–is clear from a study of six grassland-rich regions of Mongolia, Russia, and China. In many parts of the study area, more recent sedentary methods of raising livestock using enclosed pastures have replaced older grazing systems more characterized by mobility, rotating livestock over multiple, sometimes widely separated, grazing sites. Comparisons among the regions indicate that the highest levels of grassland degradation are found where livestock mobility is lowest and static production systems have become the norm (Sneath 1998:1148) (see also Chapter 3 Sustaining the Steppe: The Future of Mongolia's Grasslands).

One of the most visible and useful indicators of degradation of grazing lands is soil erosion. High densities of livestock or poor management of herds diminish vegetative cover and contribute to erosion. This eventually will reduce the productivity of the grassland, although some areas with deep soils can withstand high rates of erosion for considerable time. Accordingly, information about soil condition provides a good indicator of the capacity of grassland ecosystems to sustain food production over the long term.

GLASOD provides the only source of comprehensive global information about soil loss for regions with extensive grasslands (Oldeman et al. 1991). The GLASOD study did not explicitly report on grassland areas as defined in the PAGE study; however, it did report data on the world's drylands, where grasslands are a major presence. Drylands in the arid, semiarid, and dry subhumid zones are considered particularly susceptible to soil degradation, and these susceptible drylands constitute 55 percent of grasslands as defined in PAGE. GLASOD found that slightly more than 1 Bha, or 20 percent, of all susceptible drylands globally have been degraded by human activity (Middleton and Thomas 1997:19). Water erosion is responsible for 45 percent of this damage and wind erosion 42 percent (White et al. [PAGE] 2000; Middleton and Thomas 1997:24).

Regionally, Asia has the largest area of degraded drylands: 370 Mha, or 22 percent of susceptible drylands. However, a larger fraction of Africa's susceptible drylands are degraded (25 percent, or 320 Mha) and–perhaps more critical–a higher proportion of these degraded areas are classified as "strongly degraded" and "extremely degraded"–GLASOD's severest degradation categories (Middleton and Thomas 1997:19). Elsewhere in the world, although the absolute area of degraded drylands is small, the proportionate area is sometimes large. In Europe, 99.4 Mha, or 32 percent, of the dryland area is degraded to some extent. North America, Australia, and South America have 11, 15, and 13 percent of susceptible dryland soils degraded, respectively (Middleton and Thomas 1997:19).

The Bottom Line for Food Production. Worldwide production of beef, mutton, and goat meat has never been higher. However, this reflects more the intensification of meat production into feedlots than an increase in grasslands' ability to support livestock. In fact, data on soil degradation in the world's susceptible drylands suggest that the capacity of grasslands to continue to support livestock production over the long term appears to be declining in many areas, with 20 percent of the world's susceptible drylands being degraded.

BIODIVERSITY

As in other ecosystems, grassland biodiversity supplies direct goods–game species, medicinal plants, tourism, and genetic material for breeding purposes, to name a few–and is also a critical factor underlying the capacity of grasslands to provide other goods and services. Many grasslands contain a rich assemblage of species–often species found in no other ecosystems. For example, PAGE researchers found that 19 percent of the world's recognized Centers of Plant Diversity (regions that contain large numbers of species, especially species found in only limited areas) are located in grasslands (White et al. [PAGE] 2000). Similarly, grassland areas contain 11 percent of the world's endemic bird areas (areas encompassing the ranges of two or more species that have relatively small breeding ranges).

The importance of grasslands for biological diversity is also evident from the biological distinctiveness index developed by WWF. This index considers species richness, species endemism, rarity of habitat type, and ecological phenomena, among other criteria. For North America, 10 of 32 regions rated as "globally outstanding" for biological distinctiveness are in grassland ecosystems. In Latin America, 9 of 34 of these regions are in grasslands (Dinerstein et al. 1995:21; Ricketts et al. 1997:33).

Information about the actual condition of grassland biodiversity is far less common than information about pressures threatening biodiversity, such as habitat loss and fragmentation. For this reason, the PAGE study does not include globally comprehensive measures of grassland biodiversity condition. However, PAGE researchers did draw on more restricted regional studies that can provide insight into grassland biodiversity trends.

For grasslands in North America, the North American Breeding Bird Survey provides 30-year population trends for a wide range of bird species. Survey data from 1966 to 1995 for bird species that breed in grasslands show declines throughout most of the United States and Canada. In contrast, a recent study of the Serengeti region of East Africa concluded that significant changes have not occurred in resident herbivore densities in the last 20 years. In areas close to protected area boundaries but less accessible to vehicle patrols, wildlife populations that were already low experienced declines (Campbell and Borner 1995:141).

The number and abundance of introduced species is also an indicator of biodiversity condition. Information about introduced species has never been assembled globally, but studies in North America are illustrative of nonnative species invasions in the grasslands there. The United States Congressional Office of Technology Assessment estimated that at least 4,500 nonnative species have been introduced into the United States, with approximately 15 percent causing severe harm (USCOTA 1993:3–5). A WWF study of the distribution of nonnative plant species in North America shows that at least 10 percent of the species in all ecoregions (ecologically distinct regions) within the Great Plains are nonnative, and more than 20 percent are nonnative in the California Central Valley Grasslands (Ricketts et al. 1997:83).

In the face of significant pressures on biodiversity and declining condition at a regional level, protected areas can play a pivotal role in maintaining at least samples of the natural diversity of species and habitats in grasslands. However, PAGE researchers determined that less than 15 percent of the world's protected areas consist of at least 50 percent grassland. Protected grasslands total 2.1 million km^2–about 4 percent of global grassland area (White et al. [PAGE] 2000).

The Bottom Line for Biodiversity. Direct measurements of biodiversity condition in grasslands are sparse. However, where information is available it shows that serious problems of species introductions are common and that populations of many native species are dropping. This suggests that, at least regionally, the capacity of grasslands to support biodiversity is decreasing. Indeed, the extensive conversion of grasslands to agriculture and urban areas and the growing degree of fragmentation suggest that many grassland ecosystems may already be unable to provide goods and services related to biodiversity. And, of the many areas that have been identified as still containing outstanding grassland biodiversity, few are monitored or protected by legislation or maintenance programs.

CARBON STORAGE

How the world's grasslands are managed will have a significant influence on atmospheric carbon concentrations. PAGE researchers calculated that the soil and vegetation in grasslands worldwide currently store 405–806 GtC–about 33 percent of the total carbon stored in terrestrial ecosystems. The amount of carbon stored in grasslands is about half the amount stored in forest ecosystems, even though the total area of grasslands is nearly twice as large.

Unlike tropical forests, where carbon is stored primarily in above-ground vegetation, soils store most of the carbon in grasslands (Middleton and Thomas 1997:141). In grasslands large amounts of carbon are deposited into the soil as organic litter and secretions from roots, and as nutrients for microbial organisms and insects. For example, in one savanna in South Africa, soil organic matter accounts for approximately two-thirds of the total carbon pool of about 9 kg C/m^2 (Scholes and Walker 1993:84).

A variety of human activities can disturb the carbon storage capacity of grasslands. When grasslands are converted to croplands, the removal of vegetation and subsequent cultivation reduces surface cover and destabilizes soil, leading to the release of organic carbon. Degradation of grass cover in drylands can also be a significant source of carbon loss in grasslands, as can the widespread practice of burning grasslands to improve their pasture value (Andreae 1991:5; Sala and Paruelo 1997:238). Even the growing threat of invasive species in grasslands may bode ill for carbon storage. For example, recent experiments suggest that crested wheatgrass–a shallow-rooted grass introduced to North American prairies from North Asia to improve cattle forage–stores less carbon than native perennial prairie grasses with their extensive root systems (Christian and Wilson 1999:2397).

On the other hand, programs aimed at curbing land degradation and rehabilitating grassland cover could increase carbon storage in the world's grasslands. Projections for carbon storage in the world's drylands from 1990 to 2040 show a difference of 37 gigatons in carbon emissions between a "business as usual" scenario where current degradation patterns continue, and a sustainable management scenario if programs for land rehabilitation are implemented (Ojima et al. 1993:108).

The Bottom Line for Carbon Storage. Although they store less carbon than world forests, grasslands do store approximately 33 percent of all carbon stored in terrestrial ecosystems, mostly in the soil. Thus the potential for soil degradation to decrease carbon storage in grasslands is significant. Current practices of grassland conversion and degradation of dry grassland areas are reducing the carbon storage potential in many regions of the world, especially the arid zones.

TOURISM

Grasslands provide important cultural, aesthetic, and recreational services. Many grasslands serve as choice hiking, hunting, and fishing areas, while other grasslands are sites of historical importance and religious and ceremonial activities. For example, Native American religious, ceremonial, and historical sites have been preserved in many places throughout the prairies of the United States (Williams and Diebel 1996:27).

The economic contribution of the recreational services provided by grasslands can be significant. For example, in Tanzania, gross earnings from tourism related to game hunting were $13.9 million in 1992–93, a threefold increase over 1988 (Planning and Assessment for Wildlife Management 1996:78). Similarly, total annual earnings in Zimbabwe's hunting industry grew from approximately $3 million in 1984 to close to $9 million in 1990 (Price Waterhouse 1996:85).

Other developing countries with extensive grasslands have also shown tremendous growth in international tourist receipts (income from visitors coming from out of the country) over the 10-year interval between 1985–87 and 1995–97. In Tanzania, for example, international tourist receipts rose 1441 percent, while in Ghana and Madagascar, receipts increased more than 800 percent (Honey 1999:368–369). Of course, not all this tourist growth necessarily corresponds to grassland tourism, but in some countries, such as Kenya, grasslands and their wildlife are clearly the most popular tourist destination (Honey 1999:329).

Given the growing importance of tourism as an income source, it is important to recognize that tourism also can become a pressure on ecosystems. Wildlife-seeking hunters and camera-wielding tourists can disturb wildlife, degrade grasslands with off-road excursions, pollute grasslands with a variety of pollutants including trash, and increase consumption of water and other resources in fragile areas. All these can impair the long-term ability of grassland ecosystems to provide the beauty and biodiversity that draws tourists in the first place. Analyses of tourist impacts in Kenya, Tanzania, and South Africa show mixed impacts in parks and other grassland areas, with damage mostly confined to heavily visited areas so far (Honey 1999:256).

Poaching is another modifying and degrading influence on grasslands that continues to be a problem in several African countries. In Kenya, elephant populations dropped 85 percent between 1975 and 1990 to approximately 20,000, and the rhinoceros population declined by 97 percent to less than 500 animals (Honey 1999:298).

The Bottom Line for Tourism. Growth in tourist numbers and tourism receipts in grassland-rich countries speaks to the significant economic contribution of grasslands tourism. But it is difficult to evaluate the present quality and long-term prognosis for grasslands tourism because of the lack of consistent, comprehensive data on wildlife exploitation, tourist impacts, and the size and quality of trophy animals, among other indicators. Nonetheless, the continued conversion of grasslands to agriculture and urban areas, increased fire frequency, the spread of invasive species, and the impacts of tourism itself suggest a potential decline in the capacity of grasslands to maintain tourism and recreational services over the long term.

APPENDIX: Although mountain, polar, and urban ecosystems were not included in the PAGE study, they are fundamentally important to human health and well-being. Mountain areas are the source of water for more than half of the world's population. Polar regions play a critical role in controlling global climate and sea level. Urban areas are home to half of all people, and urban populations are rising, especially in the developing world. This appendix gives brief profiles of each of these ecosystems.

MOUNTAIN ECOSYSTEMS

The grandeur of mountain ecosystems belies their delicacy. Weathering processes and gravity constantly pull rocks, soil, snow, and water downhill, inhibiting the development of soils. Thin soils and slope instability, in turn, limit plant growth, raise the vulnerability of mountains to human disturbance, and require lengthy recovery time once damaged. Mountain regions also have a long history of political neglect and economic exploitation.

Nevertheless, millions of people who live far beyond mountains' boundaries benefit from the water, timber, rich biodiversity, and awe-inspiring scenery that mountain ecosystems supply. Yet, it is the people who live in mountain and upland regions, about a tenth of the world's population, who depend most immediately on mountain ecosystems for subsistence (Grötzbach and Stadel 1997:17). Within mountainous regions of developing countries, transport links may be scarce, access to supplies and markets poor, population growth rates high, and employment opportunities limited. Mountain populations in Nepal, Ethiopia, and Peru, for example, rank among the world's poorest (FAO 1995).

Extent of Mountain Ecosystems

The definition of a *mountain region* can be based on numerous criteria–including height, slope, climate, and vegetation. A simple definition is "areas above 3,000 m"–a category that encompasses about 5 percent of the world's terrestrial surface and an estimated 120 million people. For simplicity, again, upland area is defined as the 27 percent of the world's surface above 1,000 m (Grötzbach and Stadel 1997:17; Ives et al. 1997:6–8). A total of about half a billion people live in uplands and mountains (Ives et al. 1997:8) Mountain ecosystems encompass a range of shapes, climates, and compositions of vegetation and animal species depending on elevation and latitude.

Goods and Services from Mountain Ecosystems

FOOD AND FIBER PRODUCTION

Mountains are not world centers of agriculture in terms of volume, but subsistence agriculture in mountains is the primary food source for most mountain inhabitants in developing countries–millions of people (Messerli and Ives 1997:10). Mountain agroecosystems also are valuable storehouses of food crop genes; many of the major food crops originated in uplands. Much of the world's remaining agricultural genetic diversity is believed to exist in the fields of subsistence mountain farmers or in still more remote areas.

Potatoes are a perfect example. Andean subsistence farmers have actively maintained the genetic diversity of potatoes.

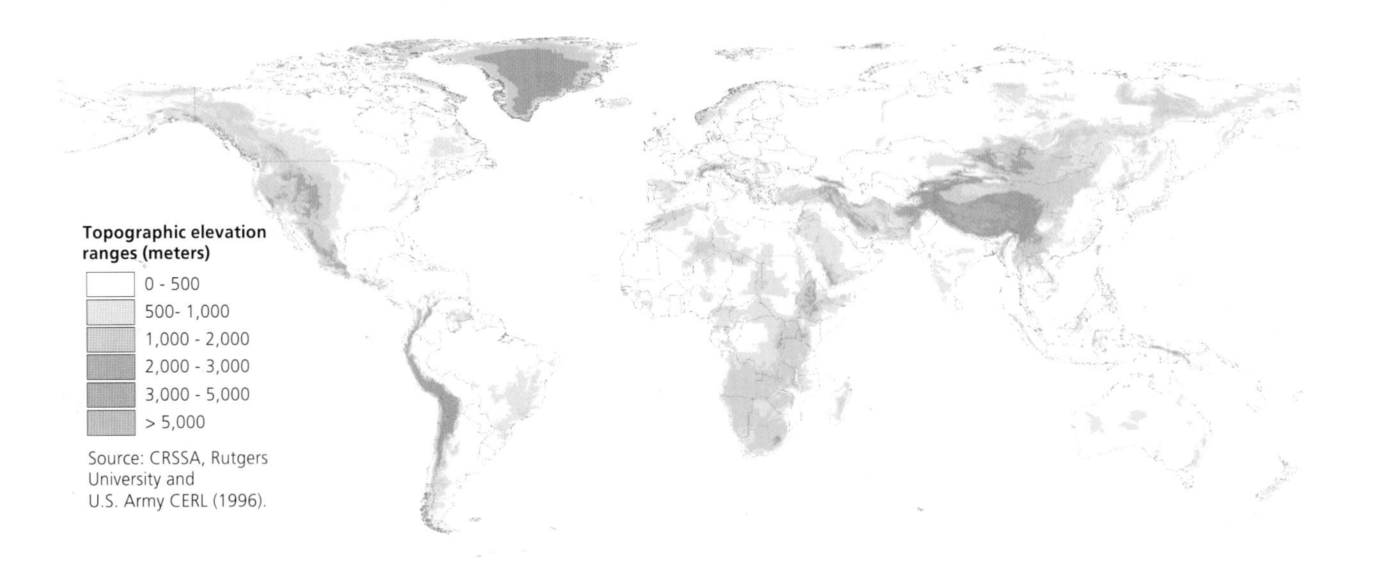

In Paucartambo, Peru, about 21 potato varieties are planted in each field, and the International Potato Center in Lima maintains the world's largest bank of potato germplasm, including some 5,000 distinct types of wild and cultivated potato and more than 160 noncultivated wild species (Tripp and van der Heide 1997; CIP 2000). By comparison, in most producer countries, a few commercial varieties dominate; and these monocultures are susceptible to epidemics of pests and diseases.

Mountains also have traditionally supplied timber resources to the world and fuel to local populations, but deforestation has reduced standing timber in many areas. In the tropics, mountain forests have had the fastest rates of loss over the last decade, compared with all types of lowland forests–about 1.1 percent a year (FAO 1993:ix).

WATER QUALITY AND QUANTITY

Half the world's population depends on mountain water. All the major rivers of the world originate in mountains, which receive high levels of precipitation as rain and snow that they store temporarily as ice, then release during spring and summer melt periods (Liniger et al 1998:5). Mountain forests help filter the water and protect its quality. On average, mountains in semiarid and arid environments provide 70–95 percent of downstream freshwater. In regions with higher rainfall, mountains provide 30–60 percent of the water supply (Liniger et al. 1998:18). High elevation water flows also power many of the world's hydroelectric plants.

Mountain watersheds will be expected to meet much of the projected increase in demand for freshwater by 2025. Will they be able to? Few assessments of the biological integrity of mountain rivers have been attempted, but trends in population growth, inadequate wastewater treatment, global warming, and increasingly extensive montane forest destruction and pollution all suggest that mountain ecosystems' ability to supply ample high-quality water is being degraded.

Mining is one of the greatest threats to the supply of clean water from mountains. Many countries have lax mining laws, regulatory controls, or enforcement, particularly in remote areas where citizens may be uninformed about mining impacts. Water drained or pumped directly from mines is often highly acidic and laden with cyanide and other heavy metals. Liquid wastes may be pumped directly into local waterways, or stored in ponds or behind earthen dams that are vulnerable to overflow or leaks. A partial survey of tailings dam failures by an NGO identified more than 70 spills and accidents in the last several decades, with considerable environmental damage (D'Esposito and Feiler 2000:5).

BIODIVERSITY

Mountains encompass numerous and varied habitats informed by altitude, soil and rock type, temperature, and sun exposure; their isolation has further enabled species diversity and endemism to flourish. The mountains of Central Asia, for example, are home to more than 5,500 species of flowering plants, with more than 4,200 species concentrated in Tajikistan alone (Jeník 1997:201). Mount Kinabalu in Sabah (Borneo) is estimated to harbor more than 4,000 plant species (Price et al. 1999:5).

Mountains also function as sanctuaries for plants and animals whose lowland habitats have been lost to conversion. Tropical montane forests, for example, are refuges for some of the world's rarest species including the mountain gorillas of Central Africa, the Quetzel of Central America, the red panda of the Eastern Himalaya, the Andean spectacled bear, and the European lynx found in isolated parts of Central

Europe. Ten percent of all bird species–already reduced to restricted ranges worldwide–are found solely or primarily in cloud forests, where the atmospheric environment is characterized by persistent, frequent, or seasonal cloud cover, usually on tropical or subtropical mountains exposed to oceanic climates.

Some protection of mountain biodiversity and other services is afforded by the designation of 141 biosphere reserves, 150 parks and reserves (above 1,500 m), and 39 World Heritage Sites in mountain and upland areas–more than in any other major landscape category. Still, numerous pressures–air and water pollutants, people–cross the boundaries of protected areas (Messerli and Ives 1997:20; Schaaf 1999).

Conversion

One sign of the potential decline in the capacity of some mountains to provide biodiversity is the reduction of unique mountain habitats, like tropical montane cloud forests, to just fragments of their original extent. Perhaps 90 percent of mountain forests have disappeared from the northern Andes (WCMC 1997, citing Weutrich 1993). Although half of the world's remaining montane cloud forests have some degree of protection, WCMC reports that many continue to be fragmented or cleared at a rapid rate for agriculture, fuel wood, grazing areas, mining, and road building, and as a result of fires that spread from adjacent cultivated areas (WCMC 1997:4).

Pollution

Air pollution is another pressure with documented impacts on mountain biodiversity. As high land masses, mountains intercept more air currents, and generally receive more precipitation, than other land forms. Most researchers believe that elevated ambient levels of sulfur and nitrogen oxides and ozone are responsible for the death or decline of extensive areas of montane forest in the northeastern United States and Canada. Long-range air pollutants also have damaged the mountain ranges along the border of the Czech Republic, Southeast Germany, and Southwest Poland (FRCFFP 1998:9).

RECREATION

Mountain tourism generates about US$70–$90 billion annually worldwide, about 15–20 percent of the global tourism industry. That total only begins to capture the value of mountains as sites of sacred rituals, sacrifice, and pilgrimage for all the major world religions, many minor ones, and as places for reverence of nature and wilderness (Price et al. 1999:4).

But mountains may have a difficult time sustainably accommodating further growth in tourist numbers. Tourism can significantly increase the employment and income levels of mountain communities, and sometimes provides funds for ecosystem protection. At the same time, tourism can be a primary degradation force. For example, mountains are heavily used by the 65–70 million downhill skiers worldwide (Price et al. 1999:36). They consume local supplies of food and water, generate solid waste and sewage, and require access to once pristine locales via roads, rail lines, airports, and hotels. Skiing also involves forest clearance and consumption of large volumes of water for snowmaking or watering.

High in the San Juan Mountains of Colorado, near the Continental Divide, the Summitville gold mine leaked contaminants into the Alamosa River in 1992, killing all aquatic life along a 27-km stretch. Clean-up is slated to cost $170 million (Carlson 2000:10).

The Bottom Line for Mountain Ecosystems. The demand for mountain areas' mineral resources, timber, scenic beauty, and water is growing. Yet there is a chronic lack of data regarding the state of mountain ecosystems and the extent and growth rates of activities damaging to mountain ecosystems. Agenda 21—the environmental blueprint crafted at the Rio Earth Summit in 1992—argued that mountains, as fragile areas, require integrated ecosystem treatment, like islands, polar regions, or tropical rainforests. Although acceptance of this viewpoint is growing, mountains are still low on the priority list of most national and international agendas. They remain vulnerable to exploitation by lowland populations through damaging extraction of natural resources and tourism development, for example, and by poorly designed government policies that contribute to the demise of traditional mountain farming systems and indigenous knowledge.

POLAR ECOSYSTEMS

The polar regions are the most remote places on Earth, yet their extreme conditions—cold, high, dry, windy, and largely removed from the public eye and political priority list—heighten their vulnerability. How the Arctic and Antarctica will respond to global environmental changes is a growing concern because these regions strongly influence the global climate system, hold a wealth of mineral and biological resources, and contain most of the world's freshwater as ice and permafrost. The fate of polar resources may signal dangers that will later become apparent in the rest of the world.

Managing the polar ecosystems requires cooperation. Eight countries share jurisdiction over the Arctic: Canada, Denmark/Greenland, Finland, Iceland, Norway, Russia, Sweden, and the United States. Antarctica is managed by interested countries on the basis of international agreements, although various countries have claims of sovereignty—some contested—over the continent, some sub-Antarctic islands, and adjacent territorial seas (UNEP 1999:327,329).

Extent of Polar Ecosystems

The areas surrounding the two poles have some things in common—cold climate, snow, and ice. Otherwise, their land and marine ecosystems are significantly different. A thick ice sheet covers the Antarctic continent; even during the summer season, only a few mountain and coastal areas are snow-free. The size of the ice sheet ranges from 4 to 19 million km^2, depending on the season; it is, on average, 2.3 km thick; and it represents 91 percent of the world's ice and the majority of the world's freshwater (GLACIER 1998; UNEP 1998:178). Surrounding Antarctica are open seas that have a productive shelf and upwelling areas where the shelf meets warmer waters. Other than about 4,000 researchers, Antarctica is uninhabited (Watson et al. 1998:89).

The Arctic, in contrast, consists of a large, deep ocean covered by drifting ice sheets a few meters thick. The land areas, which surround the ocean and are usually considered part of the Arctic region, are dominated by polar desert and tundra vegetation, although they include some prominent ice caps such as Greenland's inland ice. The Arctic's marine waters include the shallow and deep waters south and west of Alaska, the Barents Sea, and the northern Atlantic. The Arctic tundra is home to about 3.5 million people, many of whom make a living from marine and freshwater fishing, hunting, and reindeer husbandry (UNEP 1999:179).

Goods and Services from Polar Ecosystems

Although polar regions include some of the last large areas where human activity has not overtly altered the landscape, scientists have found solid evidence that human activities—often occurring in other parts of the world—are modifying polar environments and the goods and services they provide.

REGULATION OF GLOBAL CLIMATE, OCEAN CURRENTS, AND SEA LEVEL

Earth's vast polar ice sheets serve as a mirror, reflecting a large percentage of the sun's heat back into space, thus keeping the planet cool. Without the ice sheets, more heat from the sun would be retained in the ocean and more would be released into the atmosphere, feeding the warming process.

A warmer climate would also promote the release of more CO_2. For the past 10,000 years, tundra ecosystems in the Arctic have sequestered atmospheric carbon and stored it in the soil;

the tundra and boreal region store about 14 percent of the world's carbon (AMAP 1997:161). Some parts of the Arctic may now be sources of CO_2 emissions, however, because of the faster decomposition of dead plant matter in a warmer climate. If the permafrost under the tundra thaws, methane releases could also accelerate global warming (AMAP 1997:161).

The planet's weather patterns are driven largely by water circulation in the world's oceans, which is, in turn, driven by Arctic marine ecosystems. Warmer surface waters, including those from the nine major freshwater systems that drain into the Arctic Ocean, cool when they enter the North Atlantic (AMAP 1997:11). They become denser and sink to the bottom of the ocean—several million km^3 of water each winter–and slowly push water south along the bottom of the Atlantic. These water currents affect rainfall and climate worldwide (AMAP 1997:12).

The vast ice sheets in Antarctica and Greenland also control the world's sea level. If they shrink, sea level could rise, ocean currents could shift, and weather patterns could change and bring drought, severe storms, and the spread of tropical diseases.

Gradual disintegration and ice melt in polar regions are part of natural processes, but scientists are exploring the possibility that climate change may be altering those processes. Measures of ice thickness taken by U.S. submarines between the 1950s and 1970s compared with recent measurements indicate that the ice covering the Arctic Ocean may have thinned dramatically during the last few decades. The older submarine data showed an average thickness of 3.1 m, whereas data at the same sites in the 1990s show an average thickness of 1.8 m (Rothrock et al. 1999:3469). Satellite observations since the 1970s show the Arctic Sea cover to be shrinking at about 3 percent per decade (USGCRP 1999).

BIODIVERSITY

Hundreds of species are endemic to the Arctic, a place where organisms have adapted to the extremes of temperature, daylight, snow and ice found in polar regions. The Arctic also serves as habitat for several migratory bird species. Similarly, some islands of Antarctica have high levels of endemic species—some of New Zealand's southern islands are home to about 250 species, including 35 endemics. Still, much remains to be learned about the terrestrial fauna of the Antarctic, just as little is known about the fauna of the area's deep sea (UNEP 1999:183, 191, 192).

Pollution

Pollution may be the most immediate and evident threat to polar biodiversity. Airborne pollutants have turned the Arctic into a "sink" for contaminants from all over the world. Persistent organic pollutants (POPs) and other toxic chemicals travel on air, water, and wind currents until they settle in the Arctic, where they bioaccumulate in the food chain (AMAP 1997:viii). Radioactive materials have also accumulated in the Arctic; sources are fall-out from nuclear bomb tests, the accident at Chernobyl, and releases from European nuclear fuel reprocessing plants. For the general population in the Arctic and sub-Arctic, exposure to radioactive contamination is about five times higher than expected levels in a temperate area. Indigenous populations, who rely mainly on terrestrial food products, such as reindeer meat, have about 50 times higher exposure than other Arctic citizens (AMAP 1997:122–126).

The effects of POPs on wildlife are not fully understood, but it is clear that the biomagnification effects on certain species—birds, seals, polar bears, and others at the top of the food chain—are grave and will continue to worsen (UNEP 1999:184, 185). Polychlorinated biphenyls (PCBs), for exam-

Polar Pollution: Source regions for contaminated air

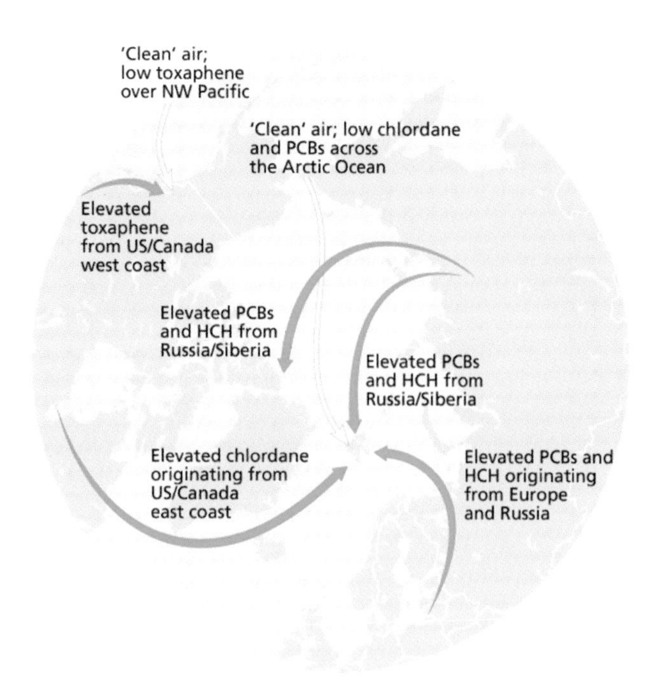

Source: AMAP 1997:79.

ple, are already found in polar bears in concentrations likely to affect their reproductive ability (AMAP 1997:89). People living in the polar regions exhibit similar high exposure to toxins with contaminant levels that can be 10–20 times higher than in most temperate regions (AMAP 1997:172). Numerous studies have linked even low-level or short-term exposure to dysfunction of the immune system, neurological deficits, endocrine disruption, and cancer.

Resource Extraction

Natural resource extraction is a growing threat to the biodiversity of polar ecosystems. Oil exploration is increasing, for example, and already its track record for pollution control includes 103 major pipeline failures in the Russian Federation between 1991 and 1993 (AMAP 1997:150). Natural resource extraction also causes damage to tundra, which is vulnerable to vehicular traffic. During the summer season, only the top few feet of soil melt, creating a layer of very wet soil between the permafrost and the thin vegetative cover. Erosion of the top vegetation easily leads to large-scale soil erosion that, because of Arctic ecological and climatic conditions, will take centuries to repair, while inducing further melting of the permafrost.

Ozone Depletion

It is not clear how ozone depletion in polar regions will affect biodiversity. Ozone depletion is more pronounced near the poles than elsewhere in the world. In 1985, a massive ozone hole was discovered over Antarctica in the spring. In recent years, ozone depletion over the Arctic has also been evident in smaller, less frequent holes (generally a few hundred kilometers in diameter, lasting a few days each), but the trend was clearly one of decreasing ozone levels through the 1990s in all seasons (Ferguson and Wardle 1998:8, 19; UNEP 1999:177). Ultraviolet (UV) radiation levels estimated in the spring, compared to the 1970s, are now about 130 percent higher in Antarctica and 22 percent higher in the Arctic (UNEP 1998:1). Polar ecosystems' heightened exposure to the sun's harmful UV-B rays could increase the incidence of cataracts and eye and skin cancer for humans, adversely affect plants and plankton accustomed to low-UV radiation, and perhaps harm algae at the base of the marine food web (UNEP 1998:xi–xiii).

Climate Change

The effect of climate change on polar biodiversity is another unknown. Warmer temperatures could convert tundra to boreal forests, change migration patterns of polar bears and caribou, alter the distribution of some small mammals whose food sources may be disrupted, and change fish species composition, among other effects (Watson et al. 1998:95–99).

FOOD PRODUCTION

The Arctic marine waters are among the richest fishing regions in the world and a major contributor to the world's fish catches. In much of Newfoundland, Greenland, Iceland, the Faroe Islands, and northern Norway, fishing is the primary livelihood (Hamilton et al. 1998:28). Local populations, particularly rural indigenous communities, are particularly reliant on hunting and fishing. Indigenous groups comprise about 50 percent of the population of Arctic Canada; and in some regions of the Yukon as much as one-third of the population lives off the land and another 30 percent support their families with activities that are not part of the cash economy (AMAP 1997:57). In much of Arctic Russia, reindeer meat is the primary food source and herding the main occupation. Secondary food sources may include moose, brown bear, bighorn sheep, alpine hare, ducks, geese, and other birds and fish.

Several polar fish stocks have been adversely affected in recent years, including salmon, cod, northern char, herring, and capelin. In the Faroe Islands, for example, cod landings decreased from about 200,000 tons to less than 70,000 tons between 1987 and 1993 after Faroese investments in catching and processing led to overfishing (Hamilton et al. 1998:30). Sometimes poaching is the biggest problem; Patagonian toothfish harvests have been driven to the brink of collapse in the Antarctic in the last 6–7 years because of illegal fishing

Polar Bears at Risk: Persistent organic pollutant (POPs) levels in polar bear tissues at several arctic locations

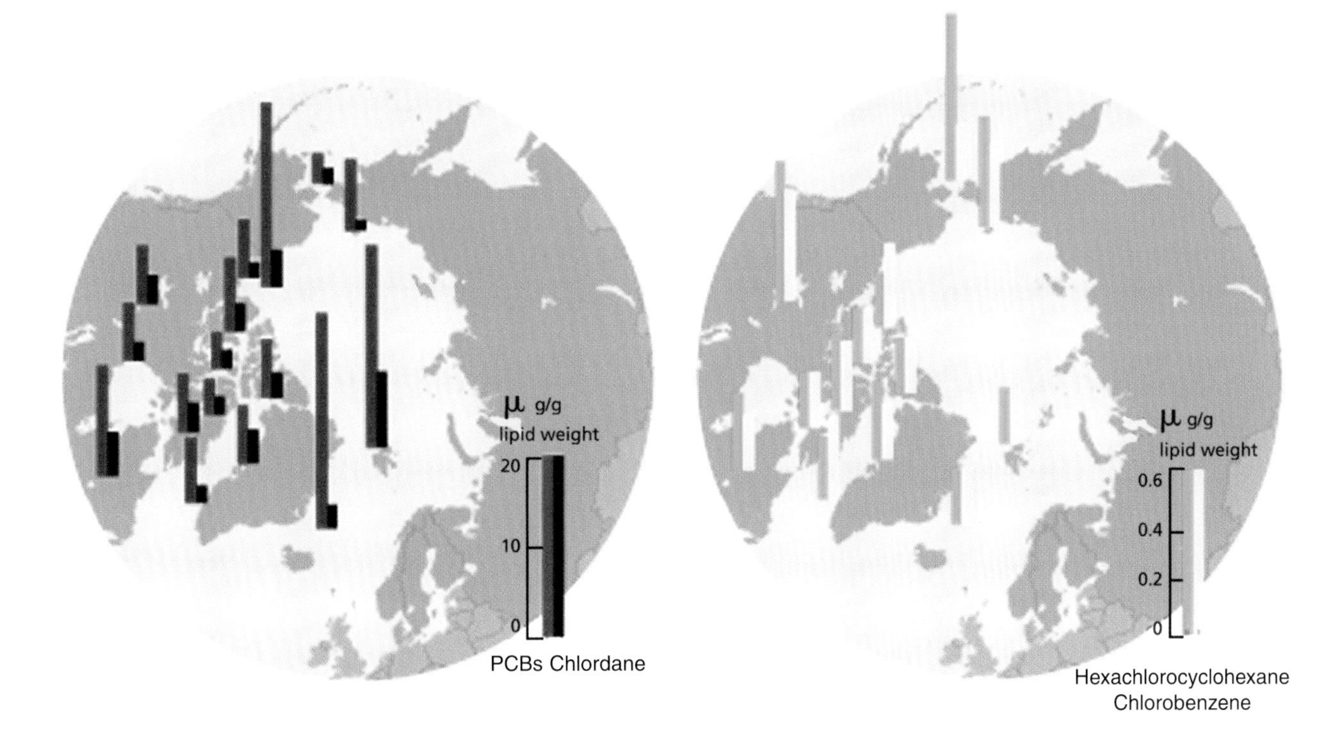

Source: AMAP 1997:89

and lax catch-limit enforcement. In 1997 the reported legal catch of Patagonian toothfish was 10,245 tons; the illegal catch was estimated at more than 100,000 tons in the Indian Ocean sector of the Southern Ocean alone (UNEP 1999:176).

RECREATION

There is a growing desire to explore polar areas. In the early 1990s, more than a million tourists were drawn to the Arctic (UNEP 1999:182). About 10,000 visited Antarctica in 1998–99, and a more than 50 percent increase to almost 16,000 was projected for 1999–2000 (IAATO 2000). Those may seem small numbers relative to the vast areas, but they have the potential for detrimental effects. Tourists are thought to frighten wildlife like breeding penguins in Antarctica, leave behind garbage, and create noise and pollution.

FEEDBACK

The poles are important to the world as early indicators of the pressures we are placing on global resources. For example, we can use analyses of the condition of the Arctic to better understand stratospheric ozone production, atmospheric cleansing, and pollution transport in northern latitudes. The massive ice sheets also serve as a kind of "time capsule" of information about volcanic activity, storminess, solar activity, and atmospheric composition (Stauffer 1999:412). Ice cores recently excavated from Vostok station in East Antarctica show that atmospheric concentrations of carbon dioxide and methane, two important greenhouse gases, are higher now than they have been in the past 420,000 years (Petit et al. 1999:429).

The Bottom Line for Polar Ecosystems. The polar ecosystems are still relatively unmodified when compared to other ecosystems, but their once-pristine condition already shows signs of climate change and other pressures. The effects of climate change are greater in polar regions than anywhere else on Earth. It is still unclear whether the ice thinning that has been observed in select areas is part of a natural climate variation or the result of human activities; nor is it clear whether the overall mass of the world's polar ice sheets is growing, shrinking, or fluctuating within normal parameters. But polar regions provide ample evidence of warming via ice cores and glacier retreat (Watson et al. 1998:90–91). Meanwhile, the immediate disruption caused by pollution and unsustainable levels of commercial fishing of some stocks is significant and growing.

URBAN ECOSYSTEMS

Urban areas are some of the most significant sectors on the planet in terms of human well-being, productivity, and ecological impact. Cities are centers of commerce, industrial output, education, culture, and technological innovation. As nexuses of the world's market economies and home to more than 2.7 billion people (World Bank 2000:152), cities are also centers of natural resource consumption and generators of enormous amounts of wastes, with environmental ramifications both locally and in distant ecosystems.

Urbanization's tremendous influence on humans and the environment will surely grow, as it is projected that global urban populations will nearly double by 2030 to 5.1 billion (UN Population Division 1996). But do urban areas–or portions of them–function as ecosystems? What defines an urban ecosystem?

Urban Ecosystems: Extent and Modifications

The concept of urban areas as ecosystems is new and controversial. There is no agreed-upon definition of an urban ecosystem, but the simplest and most useful one may be "a biological community where humans represent the dominant or keystone species and the built environment is the dominant element controlling the physical structure of the ecosystem." The physical extent of urban ecosystems is determined by the densities of both population and infrastructure. Administrative boundaries of cities generally are not reliable indicators of urban ecosystem boundaries for a number of reasons. For example, the U.S. Census Bureau defines urban areas as "areas where population density is at least 1,000 people/mi^2 (621 people/km^2)" (US Census Bureau 1995) but doesn't define a minimum infrastructure density. Another complicating factor is that urban areas are not sharply delineated but blend into suburbs and then rural areas. The PAGE estimate, however, is that urban ecosystems cover about 4 percent of the world's surface (see Box 1.10 Domesticating the World: Conversion of Natural Ecosystems, pp. 24-25).

Urban ecosystems, unlike natural ecosystems, are highly modified, with buildings, streets, roads, parking lots, and other artificial constructions forming a largely impenetrable

Built-Up Area of Selected European Cities

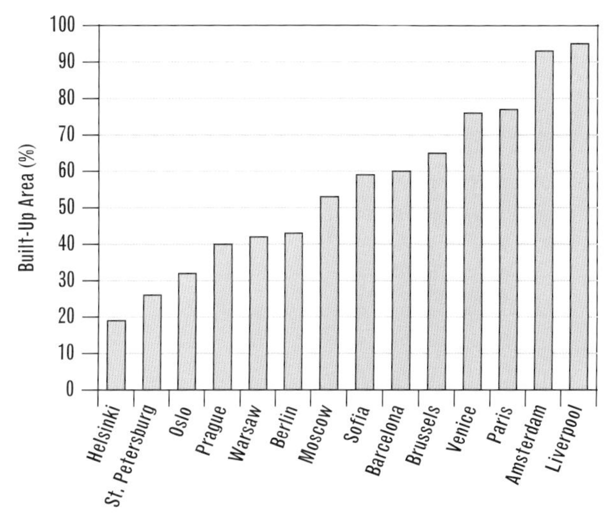

Source: Eurostat et al. 1995:202. 205.

Urban Tree Cover in Selected Cities

Tree cover in cities varies because of differences both in management and in the natural environment, particularly precipitation.

City	Tree Cover (%)
Baton Rouge, Louisiana (USA)	55
Waterbury, Connecticut (USA)	44
Portland, Oregon (USA)	42
Dallas, Texas (USA)	28
Denver, Colorado (USA)	26
Zurich, Switzerland	24
Windsor, Canada	20
Colima, Mexico	15–20
Hong Kong	16
Los Angeles, California (USA)	15
Chicago, Illinois (USA)	11
Ciudad Juarez, Mexico	4

Source: Nowak et al. 1996.

covering of the soil. Cities do contain natural and seminatural ecosystems–lawns and parks, forests, cultivated land, wetlands, lakes, streams–but the vegetation in those areas may be altered or highly managed, too.

Urbanization can change the structure and composition of vegetation of a region, whereby indigenous plants are replaced by nonnative species. For example, in the former West Berlin, approximately 40 percent of more than 1,400 plant species currently identified in the city are nonnative, and nearly 60 percent of native species are endangered (Kowarik 1990:47). In wooded areas, the ground leaf layer may be removed and replaced with shade-tolerant grass, disrupting the natural processes that create healthy soils and reducing an area's suitability as habitat for wildlife (Adams 1994:34).

Environmental stresses also modify the natural elements of urban ecosystems. Urban trees are subject to high levels of air pollutants, road salts and runoff, physical barriers to root growth, disease, poor soil quality, and reduced sunlight. Animal and bird populations are inhibited by the loss of habitat and food sources, toxic substances, and vehicles, among other intrusions.

Open space and tree cover vary widely in cities, depending on the natural environment and land use. In the United States, one analysis of more than 50 cities found that urban tree cover ranged from 0.4 percent in Lancaster, California, to 55 percent in Baton Rouge, Louisiana (Nowak et al. 1996:51).

Goods and Services Provided by Urban Ecosystems

The human elements of the city–its man-made infrastructure and economy–provide goods and services of enormous value, including human habitat, transportation networks, and a wide variety of income opportunities. But green spaces, which often form the vital heart of urban ecosystems, also contribute a wide range of goods and services. Just a few of them are focused on here.

AIR QUALITY ENHANCEMENT AND TEMPERATURE REGULATION

Temperatures in heavily urbanized areas may be 0.6–1.3 °C warmer than in rural areas (Goudie 2000:350). This "heat island" effect is the result of large areas of heat-absorbing surfaces, like asphalt, combined with a city's building density and high energy use. Higher temperatures, in turn, make cities incubators for smog. Air pollution levels in megacities like Beijing, Delhi, Jakarta, and Mexico City sometimes exceed WHO health standards by a factor of three or more (WRI et al. 1998:63).

Green space within cities significantly lowers overall temperatures and thus reduces energy consumption and air pollution (Lyle and Quinn 1991:106, citing Bryson and Ross 1972:106). A single large tree can transpire as much as 450 liters of water per day, consuming 1,000 megajoules (239,000 kcal) of heat energy to drive the evaporation process (Bolund and Hunhammer 1999:296). Urban lakes and streams also help moderate seasonal temperature variations. Urban trees and forests remove nitrogen dioxide, sulfur dioxide, carbon monoxide, ozone, and particulate matter. Trees in Chicago, for example, have been estimated to remove 5,575 tons of air pollutants per year, providing air cleansing worth more than US$9 million (Nowak 1994:71, 76). Urban forests in the Baltimore/Washington region remove 17,000 tons of pollutants per year, providing a service valued at $88 million (American Forests 1999:5). Even peripheral forests help urban air quality. Wind currents over the central city of Stuttgart, Germany draw cooler air from surrounding forest belts, cooling the downtown areas–one reason why Stuttgart has discouraged urban sprawl (Miller 1997:65, citing Miller 1983).

BIODIVERSITY AND WILDLIFE HABITAT

Cities support a relatively wide variety of plants and animals–both the native species that have specifically adapted to the urban landscape and its extreme ecological conditions and the numerous nonnative species humans have introduced.

Many of the animals, birds and fish that inhabit urban areas are valuable for the excitement and pleasure they bring to many urbanites, though some species are perceived as nuisances or dangerous. Almost a third of urban residents surveyed in the United States–more than 40 million people–report that they participate in wildlife watching activities

Changes in Tree Cover in the Baltimore-Washington Corridor, 1973–97

Overall tree cover has declined steadily in the rapidly growing Baltimore-Washington, D.C., urban corridor in the eastern United States. Urban and suburban expansion, as well as diminishing budgets for urban tree care, have shrunk tree cover from 51 percent of the land area in 1973 to 37 percent in 1997. Land with heavy tree cover (>50 percent wooded) declined by one-third, while land with little or no tree cover increased by nearly 60 percent.

Tree Cover Trends, 1973–97

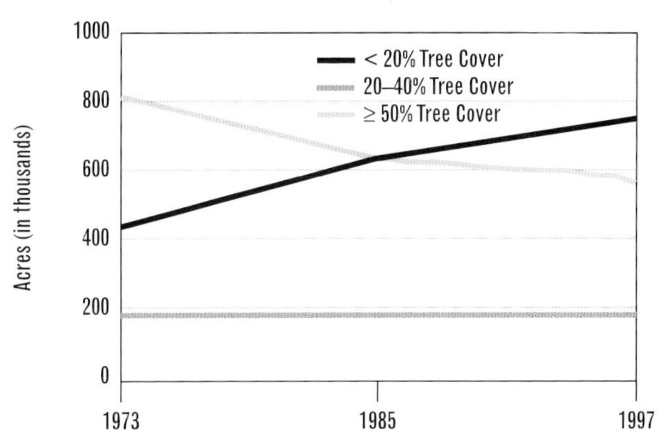

Source: American Forests 1999.

within 1 mile of their homes (U.S. Department of the Interior 1997:94).

Some urban wildlife also is valuable from the perspective of conservation and biodiversity. Urban parks and other green spaces are critical to migratory species and provide wildlife corridors, even though these corridors are often too fragmented to afford animals sufficient area to maintain diverse populations. Nevertheless, in many North American urban areas, deer and small herbivores such as squirrels are prevalent. Muskrats and beavers may be widespread in urban water areas, and some smaller predators like bats, opossum, raccoon, coyote, fox, mink, and weasels adapt well to the habitat changes wrought by development (Adams 1994:57-65). Rats, as scavengers, have adapted particularly well to crowded human living conditions.

Many urban streams are so polluted, littered, or channelized, or their riparian zone so substantially reduced and cleared of vegetation, that only the most pollution-tolerant species survive. Yet urban rivers also offer some of the greatest potential for restoration and the return of aquatic diversity. For example, in 1957 London's Thames was virtually devoid of fish in one stretch, but by 1975 efforts to improve the biological conditions were rewarded with the return of 86 different species of marine and freshwater fish (Douglas 1983:137).

Bird diversity in urban areas may provide a good indicator of urban environmental quality, since birds require differentiated habitat and are influenced by air and aquatic pollution through the food chain. For example, a 1993 survey of Washington, D.C., bird species richness identified 115 species–an

In Cuba in 1999, urban agriculture produced 800,000 tons of fresh organic produce and employed 165,000 people. Urban agriculture produced 65 percent of the nation's rice, 43 percent of the fruits and vegetables, and 12 percent of the roots and tubers.

estimate that agreed closely with totals from surveys decades earlier, and was almost as high as the number found in larger, surrounding counties. This suggests that Washington, D.C.–perhaps because parks and low to moderate density residential areas cover 70 percent of the metropolitan area–is providing diverse and good-quality habitat for birds. Unfortunately, such citywide studies are rare (U.S. National Biological Survey 2000).

STORM-WATER CONTROL

Urban forests, wetlands, and streamside vegetation buffer storm-water runoff, control pollution, help recharge natural groundwater reservoirs, and minimize flooding in urban areas. In contrast, buildings and roads cover much urban land with impervious surfaces and eliminate vegetation that provides natural water storage capacity.

Some studies have attempted to put a monetary value on the benefit of urban forests to storm-water control. Forests in the Baltimore/Washington area save the region more than $1 billion–money that would otherwise have to be spent on storm-water retention ponds and other systems to intercept runoff (American Forests 1999:2). Unfortunately, in most cities worldwide, urban trees are a resource at risk. Since the 1970s, three major U.S. metropolitan areas–Seattle, Baltimore/Washington, and Atlanta–have lost more than a third of their heavy tree cover (Smith 1999:35).

FOOD AND FIBER PRODUCTION

Many urban areas contribute substantially to their food supply. Urban agriculture includes aquaculture, orchards, and livestock and crops raised in backyards and vacant lots, on rooftops and roadsides, and on small suburban farms (UNCHS 1996:410). Urban and periurban agriculture is estimated to involve 800 million urban residents worldwide (FAO 1999). In Kenya and Tanzania, 2 of 3 urban families are engaged in farming; in Taiwan, more than half of all urban families are members of farming associations; in Bangkok, Madrid, and San Jose, California, up to 60 percent of the metropolitan area is cultivated (Smit and Nasr 1992:142; Chaplowe 1998:47). In Ghana's capital, Accra, urban agriculture provides the city with 90 percent of its fresh vegetables (The MegaCities Project 1994). Urban agriculture also provides subsistence opportunities and income enhancement for the poor and offers a way to recycle the high volumes of wastewater and organic solid wastes that cities produce.

RECREATIONAL OPPORTUNITIES AND AESTHETIC HAVENS

Trees provide visual relief, privacy, shade, and wind breaks. Trees and shrubs can also reduce cities' typically high noise levels; a 30-m belt of tall dense trees combined with soft ground surfaces can reduce noise by 50 percent (Nowak and Dwyer 1996:471). Parks provide urban dwellers with easy access to recreational opportunities and places to relax–an enormously valuable service where open space and escape from asphalt are often at a premium. Some urban parks, lakes, and rivers are also tourist attractions and enhance values of downtown areas. Furthermore, urban water bodies provide places for sportfishing, kayaking, sailing, and canoeing.

Managing Urban Areas as Ecosystems

One of the primary challenges to managing urban areas as ecosystems is the lack of information. Because the science of urban ecology is in its infancy, the knowledge base for urban areas as ecosystems is less comprehensive than for other ecosystems. In particular, there is a dearth of data concerning the "green" elements of cities. Air and water quality, sewerage connections, water withdrawals and solid waste per capita, and trends in the extent of urban forests and wildlife diversity are critical indicators of the condition and capacity of the more natural areas in urban spaces to provide environmental goods and services.

Another problem is lack of planning and budgeting for the care of green spaces; most budgets are geared toward removing dead trees. Many cities lack systematic tree-care programs, and little attention is paid to effects of soil conditions, restrictions to root growth, droughts caused of the channeling off of rain, the heat island effect, and the lack of undergrowth (Sampson 1994:165).

Managing urban consumption and its impact on neighboring ecosystems is perhaps the biggest challenge. Urban areas consume massive amounts of environmental goods and services–imported from ecosystems beyond their borders–and export wastes. It is estimated that a city with a population of 1 million in Europe requires, every day, an average of 11,500 tons of fossil fuels, 320,000 tons of water, and 2,000 tons of food, much of which is produced outside the city. The same city produces 300,000 tons of wastewater, 25,000 tons of CO_2, and 1,600 tons of solid waste (Stanners and Bordeau 1995:263). The total area required to sustain a city is called its "ecological footprint" (Rees 1992). In a study of the 29 largest cities in the Baltic Sea region, it was estimated that cities claim ecosystem support areas 500–1,000 times larger than the area of the cities (Folke et al. 1997:167). Any attempt to improve the sustainability of urban ecosystems must identify ways for cities to exist in greater equalibrium with surrounding ecosystems.

The good news is that urban areas present tremendous opportunities for greater efficiencies in energy and water use, housing, and waste management. Strategies that encourage better planning, mixed-use development, urban road pricing, and integrated public transportation, among other efforts, can dramatically lessen the environmental impacts of billions of people. The fact that land use changes rapidly in urban areas is a management and planning challenge, but also an opportunity as well. For example, the million or more brownfields (urban land parcels that once supported industry or commerce but lie abandoned or contaminated) that scar cities worldwide offer the chance to create new green spaces or lessen congestion and development pressure on remaining green areas (Mountford 1999). If well-managed, urban green spaces can add to the already proven health and education benefits of urban ecosystems.

The Bottom Line for Urban Ecosystems. Urban ecosystems are dominated by human activities and the built environment, but they contain vital green spaces that confer many important services. These range from removing air pollution and absorbing runoff to producing food through urban agriculture. Urban forests, parks, and yards also soften the urban experience and provide invaluable recreation and relaxation. The science of urban ecosystems is new and there is no comprehensive data showing urban ecosystem trends on a global basis. However, more localized data show that loss of urban tree cover, and the consequent decline of urban green spaces, is a widespread problem. The rapid growth in urban populations worldwide adds to the mounting stress on urban ecosystems. Continued decline in the green elements of urban ecosystems will erode the other values—economic, educational, and cultural. Urban population increases heighten the need to incorporate the care of city green spaces as a key element in urban planning.

CHAPTER 3

LIVING IN ECOSYSTEMS

This chapter traces the histories of several ecosystems and the people whose lives depend on them, whose actions have degraded them, and who hold the power to restore them. Included are the grasslands and traditions of pastoralism of Mongolia; a community-managed forest in India; mountain watersheds and downstream urban areas in South Africa; the agricultural plains of Machakos, Kenya; and the wetlands and croplands of southern Florida in the United States. These are places where the inhabitants are striving to safeguard their future, which depends so clearly on the health of their ecosystems.

Five brief stories from Cuba, the Caribbean, the Philippines, New York City, and the watershed of Asia's Mekong River complement the detailed case histories. Many of the cases and stories encompass multiple ecosystems, but for simplicity they are grouped in this chapter by the ecosystem most critical to the featured management challenge.

Together, the cases and stories capture diverse experiences from around the world–varying spatial scales, population sizes and densities, and ethnic groups. They illuminate the driving forces and impacts of degradation and the analyses of ecosystem condition presented in the earlier chapters. They also reflect the variety of trade-offs that we face as inhabitants and managers of ecosystems. For example, South Africans planted income-generating but invasive nonnative trees, then paid a high price in terms of diminished water supply to cities and towns. Drainage and conversion of parts of the Everglades to agriculture fueled the growth of the Florida sugar industry but reduced the ecosystem's water retention and filtration capacity and threatened biodiversity. The state government was able to intensify commercial cutting of timber in Dhani, India, from the 1950s through the 1970s but at the long-term expense of local livelihoods.

Individually, some of the cases and stories address many management issues, others just a few. None offers any ready-made "fixes" for ecosystems that have been degraded, but all can encourage an exploration of questions crucial to the future productivity of ecosystems:

- What causes an ecosystem to decline? Who gains the benefits of ecosystem use and who pays the costs of decline?
- What conditions increase recognition that ecosystem misuse or overuse must be supplanted by efforts to alleviate pressures and ensure long-term productivity? What circumstances move people to concern and action?
- How do we create the public and political will to take action to restore an ecosystem?
- What mechanisms and policies can help prevent ecosystem decline or ensure long-term sustainability?
- To what extent, and over what time frame, are an ecosystem and its services amenable to restoration?

The search for answers to these questions underscores the complexities of ecosystem change–the often-surprising natural dynamics of ecosystems as well as the human management challenges. Through case studies, we can examine ecosystems and the people who live in them as constituents in larger geographical regions and social contexts. No ecosystem, even an isolated Mongolian grassland or a forest in a small community like Dhani, is managed by a single person or institution that can act unilaterally. Ecosystem management is the sum of many individuals and institutions–public and private, formal and informal–and political and economic factors. A widening network of connections further complicates management. Many ecosystem problems have local roots and local or regional consequences. But the causes of problems such as acid rain, ozone depletion, invasive species, and global warming can originate in a neighboring country–or even half a world away–and affect us all.

AGROECOSYSTEMS

Regaining the High Ground: Reviving the Hillsides of Machakos

In Machakos, necessity is the mother of conservation. Because water is scarce and rainfall unpredictable in this mostly semiarid district southeast of Nairobi, farmers have learned to husband water. They collect water from their roofs, they channel road runoff onto their terraces, they scoop water out of seasonal streams or perennial rivers, and they dig ponds to collect rain. To minimize soil erosion, farmers have adopted a system of conduits, tree planting, and terraces found nowhere else in Kenya. "These [measures] are the lifeline of the people here in Machakos," said Paul Kimeu, soil and water conservation officer for the Machakos District.

Conservation efforts, plus persistence and hard work, have enabled the people of Machakos, the Akamba, to survive in the face of drought, poverty, and land degradation. In the 1930s, severe soil erosion plagued 75 percent of the inhabited area and the Akamba were described as "rapidly drifting to a state of hopeless and miserable poverty and their land to a parching desert of rocks, stones, and sand" (Tiffen et al. 1994:3, 101). Today, once-eroding hillsides are productive, intensively farmed terraces. The area cultivated increased from 15 percent of the district in the 1930s to between 50 and 80 percent in 1978, and the land supports a population that has grown almost fivefold, from about 240,000 in the 1930s to about 1.4 million in 1989 (Tiffen et al. 1994:5; Mortimore and Tiffen 1994:11). This environmental transformation has been called "the Machakos Miracle" (Mortimore and Tiffen 1994:14, citing Huxley 1960).

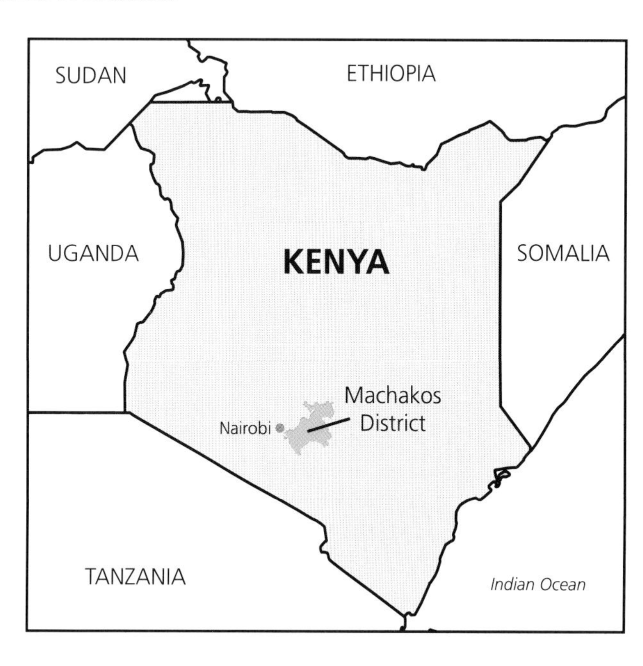

But the benefits of the "miracle" have not reached everyone. Those with the least fertile land often lack the financial

(continues on p. 152)

Box 3.1 Overview: Machakos

Through innovation, cultural tradition, access to new markets, and hard work, farmers in Kenya's Machakos District have turned once-eroding hillsides into productive, intensively farmed terraces. However, economic stagnation, population growth, increasing land scarcity, and a widening income gap raise the question: Is Machakos' agricultural transformation sustainable?

Ecosystem Issues

Agriculture

Since the 1930s, the Akamba people of Machakos have terraced perhaps 60–70 percent of arable fields to protect them from erosion. Land conditions and agricultural output have also benefited from penned livestock, tree planting, composting, and other measures. Yet with decreasing arable land per capita and sluggish economic development, poverty remains a problem for some, particularly during droughts. Poverty, in turn, decreases farmers' ability to invest in sustainable technologies and management.

Freshwater

Most streams in Machakos are seasonal, rainfall is variable, and groundwater limited. Water projects and conservation activities have expanded irrigation, reduced the risk of crop failure, cultivated higher-value crops, and freed labor from fetching water. But about half the population still lacks potable water and water availability constrains industrial and urban growth.

Forests

Contrary to expectations, aerial photos suggest that the District has become more, not less, wooded since the 1930s. Small-scale tree planting efforts have been beneficial; farmers plant trees to stabilize soils and supply fruits and timber. Akamba also minimize deforestation by using dead wood, farm trash, and hedge clippings for firewood.

Management Challenges

Equity and Tenurial Rights

Some of the most severe agroecosystem degradation in Machakos emerged in the decades when the colonial government divested the Akamba of their land rights and restricted market access. By contrast, greater Akamba control over farm techniques, lands, and livelihoods have coincided with self-led, often independently funded innovations in conservation.

Economics

Improved access to markets, the growth of urban areas like Nairobi and Mombasa, and the right to grow lucrative cash crops provided incentive for farmers to implement new technologies and maximize productivity. But market access remains difficult and economic growth sluggish; decreasing farm size and labor shortfalls are additional roadblocks to further agricultural intensification.

Stakeholders

For decades, government officials and farmers disagreed about farming objectives and methods. In an atmosphere of inequality and mistrust, officials promoted or regulated technologies that the Akamba did not accept or perceive as viable. Greater environmental progress has occurred since Akamba farmers have gained a more equal voice in the decisions about agricultural management and methods.

Information and Monitoring

NGOs, government extension workers, researchers, and self-help groups have vastly improved the information and resource base available to farmers, but improvements in the information base must be ongoing. For example, researchers have emphasized the weakness of data with which to analyze change in extent and condition of Machakos ecosystems, including data on soil health, changes in land use and vegetation, and production.

Timeline

1600s–1700s Akamba first occupy the Machakos uplands.

1889 Europeans arrive.

1895 British Protectorate of East Africa is established.

1897–99 Consecutive drought seasons result in devastating famine; 50–75 percent of Akamba die.

1906 British colonial government designates the most fertile Machakos lands as "White Highlands" for European settlers; Akamba are restricted to "Native Reserves." Only Europeans are allowed to grow high-value export crops like coffee and tea.

1928–29 Drought and famine strike.

1930s Growth of human and livestock populations without room for expansion cause farmlands on Native Reserves to deteriorate. Akamba migrate out of Reserve settlements in search of work or to occupy other lands illegally.

1933–36 Successive droughts occur. Officials acknowledge the "Machakos problem" when 75 percent of inhabited area is plagued by soil erosion.

1937–38 Colonial government creates the Soil Conservation Service and attempts to impose conservation measures on Akamba, including compulsory reductions of cattle. Akamba protest.

1940–45 Conservation funding and number of available male farm laborers are limited during WWII; famine relief is required.

1946 Government makes significant investments in land development and conservation in Africa—in Machakos in particular. Emphasis is on compulsory communal work, including government-selected systems of terracing.

1949–50 Consecutive drought seasons ensue.

1950s Growth of urban areas increases demand for agricultural products, making terracing and water conservation profitable and attractive.

1952 News spreads among Akamba that cultivators who use bench terraces, rather than government-mandated narrow terraces, are making big profits, sparking voluntary construction of bench terraces.

1954 Swynnerton Plan to revolutionize agriculture emphasizes production of crops for export. For the first time, Akamba are granted the right to grow coffee, another incentive to terrace land and a source of cash with which to purchase farm inputs.

1959–63 Akamba turn to political activity in build-up to Kenyan Independence (1963). Conservation efforts slow, as they are perceived as tainted by colonial authority.

1962 Akamba surge onto former Crown Lands. Population growth rates in some areas reach 10–30 percent per year, as people seek to escape land shortages in other areas.

c. 1965–70s Recognizing the potential for higher yields, farmers renew soil and water conservation efforts largely without government aid. New roads improve access to Nairobi, and growth of canning plants encourages fruit and vegetable production and, in turn, terracing.

1974–75 Drought returns.

1975–77 High prices for coffee inspire tripling of production and heavy investment in land conservation.

1978–80s Numerous church-led projects and national and international NGOs provide support for community development, including famine relief, food production, and water supply and irrigation.

1983–84 Drought strikes—called "dying with cash in hand" because of severe food shortages. After the drought, more terraces are rapidly constructed.

1996–98 Droughts followed by El Niño rains ruin subsistence crops and force farmers to sell livestock for food.

2000 Perhaps as much as 65 percent of farms are terraced, many farmers use additional conservation measures.

resources to tap the water below it. Higher living standards seem most achievable by those households with access to nonfarm income, but population growth and economic stagnation contribute to a shortage of jobs in towns and cities. For those farmers without access to nonfarm income, lack of capital or credit limits their ability to implement innovative agricultural practices.

On the one hand, then, Machakos offers a dramatic example of how knowledge, innovation, and respect for the vital services that soil and water provide have enabled people to restore and even increase the productivity of severely degraded lands. On the other hand, Machakos illustrates the continued vulnerability of both ecosystems and people in the face of cultural, economic, and environmental change.

A Land of Hills and Dry Plains

Machakos lies on a plateau that gradually slopes southeast from 1,700 to 700 m elevation, broken by groups of high hills. Rain has always been precious in Machakos; annual rainfall ranges from 1,200 mm in the highlands to less than 600 mm in the lowlands of the southeast and the dry plains of the extreme northwest (Mortimore and Tiffen 1994:12; Tiffen et al. 1994:18). Less than half the district has more than a 60 percent chance of getting enough rain to grow maize, the Akamba's preferred staple (Mortimore and Tiffen 1994:12, citing Jaetzold and Schmidt 1983). In most years the highlands are the only region that can support reliable agricultural harvests without irrigation.

The Akamba are believed to have settled the uplands of Machakos in the 17th and 18th centuries, when most of the area was an uninhabited thorny woodland. Evergreen forests crowned the wetter highlands and grasslands carpeted the drier plains. The Akamba raised cattle, goats, and sheep and cultivated grains, pulses, and sweet potatoes on wet hills. Close to water they irrigated small plots of vegetables, bananas, and sugarcane. They became skillful traders, providing ivory, honey, beer, ornaments, and weapons to the Kikuyu and Masai in exchange for food. Their lives changed dramatically in the late 1890s, however, after smallpox, cholera, and rinderpest decimated both human and animal populations and drought devastated the land. By 1900, 50–75 percent of the Akamba had perished in some areas; perhaps only 100,000 people were left in the district (Tiffen et al. 1994:44, citing Lindblom 1920; Tiffen 1995:4).

At about the same time, the new British colonial government gained sufficient power to impose boundaries on the Akamba and other native people in Kenya. They created several "Native Reserves" and claimed some of the best farmland for themselves in "Scheduled Areas" or "White Highlands." Though the Akamba retained most of their traditional lands, the government's policy blocked any expansion, with European ranches and farms on two sides and government-controlled "Crown Lands" on the other two.

Traditionally the Akamba had responded to drought, decreasing soil fertility, and population growth by moving to new fields or ranges. Without this mobility, shifting cultivation gave way to continuous cultivation. Although the population of both people and cattle in the Akamba reserve grew, the colonial government strictly enforced the reserve boundaries to maintain political control. By 1932, some 240,000 Akamba lived in Machakos, more than double the population at the turn of the century (Mortimore and Tiffen 1994:11). Within the reserves, soils became exhausted and crop yields fell.

For the already stressed ecosystem and its people, the return of severe drought in 1929 was catastrophic. The Akamba called the drought "*Yua ya nzalukangye*" or "looking everywhere to find food" (Tiffen et al. 1994:5). Then, from 1933 to 1936, droughts occurred during six of the eight semiannual growing seasons–the long rains from March to May, and the short rains from October to December. Locusts invaded the withering maize crops, and voracious quella birds ate the remains. Cattle denuded the parched brown hillsides, then began to starve, soon followed by the Akamba themselves. When the rains did come, the region's highly erodible red soil bled from the steep hillsides in torrents. Historical photographs reveal a landscape of treeless hillsides, deep gullies, denuded slopes, and fields stripped of topsoil.

Changing Attitudes: From Compulsory Conservation to Akamba Innovation

In reports written from 1929 to 1939, colonial agricultural officers argued that rapid population growth, surplus livestock, deforestation, and unscientific farming methods were leading to massive degradation of the region's natural resources. The Akamba recognized the worsening environmental crisis, too. "[T]his place was becoming a desert," reflected Joel Thiaka, a farmer from Muisuni, in 1938 (Tiffen et al. 1994:44).

Several factors prompted the colonial government to invest in land development: a global antierosion movement, catalyzed in part by the Dust Bowl in the United States; the increasing African populations; and the expense of providing emergency food aid to ward off massive starvation during times of drought (Tiffen et al. 1994:179). In 1937 the colonial government created a Soil Conservation Service led by Colin Maher. The Service's first efforts included the confiscation and slaughter of "excess" Akamba cattle. After Akamba protestors rallied in Nairobi, those initiatives were abandoned (Tiffen et al. 1994:181–182).

Maher next launched "compulsory conservation projects." These required Akamba to plant grass and build terraces–

structures used for centuries in Asia and Africa to cultivate steep hillsides. When these activities progressed too slowly, Maher mandated the building of conservation structures with government tractors and paid-labor gangs. The Akamba again protested, fearful of another government land grab; according to Akamba tradition, anyone clearing or cultivating land had permanent use-rights to the property. Some Akamba even threw themselves in front of the tractors. The Akamba finally agreed to send one family member two mornings a week to work on forced-labor gangs building terraces and water conservation projects and planting fodder crops.

The terraces that Maher required Africans to construct during this period were narrow-based terraces, also known as contour ditches. Building these small structures required workers to dig a shallow trench and throw the soil downhill to create a small berm to capture runoff. Though easy and relatively fast to construct, narrow terraces were also quick to wash away and required significant maintenance. They soon lost favor with Akamba farmers, but not with Maher.

Although soil conservation efforts languished during World War II (1940–45), they were renewed with vigor by an expanded Department of Agriculture after the war, as wide-scale erosion and famine returned to Machakos. There was much African opposition to many of these "betterment" projects. Yet several Akamba innovations emerged in the ensuing decades from these controversial programs, innovations which laid the foundation for the "Machakos miracle," though few recognized them at the time. One was workers' experimentation with the construction of a bench terrace called a *fanya juu.*

Fanya juu terraces are constructed by digging a trench along the contour of a slope and throwing the excavated soil uphill to form a gently sloping field with an earth embankment that collects rainfall and slows runoff. Though they require considerable labor to construct, such bench terraces soon become stable and require only periodic maintenance of the berm. Maher, however, thought they were too labor-intensive for the Akamba, and thus had mandated narrow terraces.

Maize, beans, and mango and banana trees are part of this well-designed hillside terrace.

Box 3.2 Machakos Agriculture

Results from a 1998–99 survey involving several hundred farmers and 484 plots of land suggest that the efforts put into conserving soil and water in Machakos have been well rewarded. The survey shows that terracing is by far the most popular conservation measure. Farmers who use terracing often use multiple conservation measures—adopting them as a package (Zaal 1999). Other research suggests that there was a substantial increase in productivity per hectare in the Machakos District between the 1930s and 1990s (Tiffen et al. 1994:95–96).

Land and Water Conservation Measures in Machakos

About half the terraced plots also incorporated another conservation measure.

Percent of Fields Given Over to...	
Terracing	65.7
Grass strips	14.0
Grass terrace border	10.7
Trash lines	8.5
Agroforestry	2.3
Cover crops	1.0
Open ridges	0.6
Stone terrace	0.4
Cut-off drain	0.2

Source: Zaal 1999.

Benefits of Terracing

The survey showed that farmers who use terraces reap numerous benefits.

Percent of Farmers Experiencing...	
Higher land value	97
Higher yield levels	94
Greater stability of yield	94
Less erosion	76
Less use of fertilizer	75
Less labor to plant	53
Less labor to weed	43

Source: Zaal 1999.

Crop and Livestock Production in Machakos

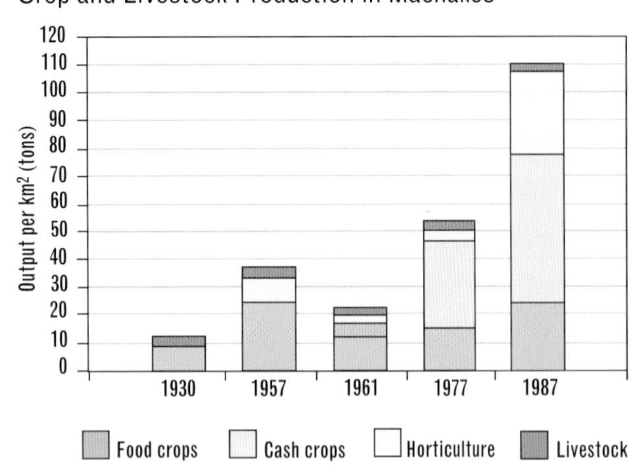

Source: Tiffen et al. 1994:95.

But the Akamba have a saying: "Use your eye, the ear is deceptive" (Tiffen et al. 1994:152). Many of the Akamba men fought as part of British forces overseas, where they saw other agricultural practices at work. In 1949, one veteran built a bench terrace patterned after one he had seen in India. He harvested a good crop of onions that he sold for a profit. Other farmers in the area soon followed his lead. After Maher's retirement in 1951, farmers were allowed to choose whether to have contour ditches or *fanya juu* in the compulsory betterment programs; more and more chose *fanya juu*.

During the 1950s, more than 40,000 ha was terraced in Machakos (Mortimore and Tiffen 1994:14, citing Peberdy 1958). One incentive for this large-scale shift to terraces was the government's decision in 1954 to allow Akamba farmers to grow coffee for the first time–a decision based on the Swynnerton Plan's emphasis on producing lucrative cash crops for export. The Akamba were eager to reap the economic benefits of growing coffee, but coffee can only be planted on steep slopes if they are terraced, to ensure that the nutrients and moisture essential to coffee's growth are retained. Other farmers used terraces to grow tomatoes and other vegetables for the expanding town of Nairobi.

Another breakthrough that would promote self-led Akamba innovation and conservation occurred in 1956. The new and mainly African-staffed community development service under a government-appointed chief replaced the hated compulsory work gang with the *mwethya*, or traditional work party, whose members chose each other and their own leaders. Normally Akamba families called a *mwethya* for a special project, such as building a hut; neighbors would help in exchange for food. With technical support from the government, *fanya juu mwethyas* were soon busy all over the district building terraces and undertaking other projects.

Since many Akamba men worked outside the district, most of the laborers who worked on the conservation projects and in the first *mwethya* were women. This was the first time in Akamba history that women were elected to leadership positions, providing them with increased status and political power and reinforcing the value of education for daughters. The traditional work group evolved, too, into self-help groups that today pool money as well as labor and are connected with organizations that provide community development, agricultural extension, and literacy services.

Kenya's independence from colonial rule in 1963 spurred a surge of Akamba families onto former Crown Lands. The new government ended all funding for soil conservation, and for a few years terracing fell out of favor with the Akamba, who saw conservation efforts as tainted by the colonial regime. But soon farmers who had seen the benefits of the *fanyu juu*–for yields of staple crops like grains and beans, cash-crop production, and survival during drought–began to build them again, on their own, either through *mwethyas* or hired labor. In fact, more terraces were built from 1961 to 1978 than were built during the 1950s, and without any government aid (Tiffen and Mortimore 1992:363). The period from 1960 to 1980 was also characterized by a phase of steep growth in land productivity in Machakos (Tiffen and Mortimore 1992:365). Another 8,500 km of terraces were built annually between 1981 and 1985, half of them by farmers with no outside assistance. By the mid-1980s, aerial surveys showed that 54 percent of Machakos' arable land was protected from erosion, with more than 80 percent protected in hilly areas (Tiffen et al. 1994:198). A 1998–99 survey of 484 fields in Machakos suggests that about 60 percent are terraced; many farmers also use additional conservation measures (Zaal 1999:5).

Overall, some 76 production technologies were introduced or expanded in the district between 1930 and 1990, including introduction of 35 crops varieties, 5 tillage practices, and 6 methods for managing soil fertility (Mortimore and Tiffen 1994:16). Many of these conservation and land development mechanisms were Akamba innovations.

The expansion of market opportunities clearly affected the popularity of conservation measures. The coffee boom in the 1970s, for example, increased demand for labor on the farms and in coffee processing factories and transport to markets. Coffee prices fell in the late 1980s, but large international horticultural firms in Nairobi began to encourage Machakos farmers to produce crops like French beans as export crops. Citrus, pawpaws, and mangoes have proved similarly successful with the rise of Kenya's canning industry and the growth of towns and tourist trade. According to a 1981–82 survey, 41 percent of rural income came from nonfarm businesses and wages (Mortimore and Tiffen 1994:16). For decades such income, usually earned by Akamba men with jobs outside the district, has been invested in farm improvements such as building terraces or water storage tanks and planting trees and hedges.

Farmers also began to invest in planting and protecting trees. Photographs comparing landscapes in 1937 and 1990 show a substantial increase in the density and average size of farm trees (Tiffen et al. 1994:218). Because farmers, particularly women, spent increasing time foraging for firewood after hillslopes were cleared, they developed the practice of planting woodlots to facilitate gathering. Often farmers planted trees at the bottom of their plot so as to minimize water uptake from their own crops and maximize that from their neighbors'; that location offered the added advantage of helping to keep hillside soil in place. Women farmers have favored fruit tree plantings because they offer household food supplies and an independent source of cash (Tiffen et al. 1994:221).

Adaptive changes in livestock management and the adoption of ox-drawn plows for weeding and cultivation have contributed to Akamba farmers' success. Since no communal grazing lands remain, animals are now fed on the farm. More than 60 percent of the district's livestock are stall-fed or tethered for all or part of the year, requiring fodder feeding, but also supplying manure for fields (Mortimore and Tiffen 1994:19, citing African Development and Economic Consultants 1986). Added advantages of "zero-grazing" systems are increased milk yield, reduced destruction of vegetation through overgrazing, decreased disease incidence, and labor savings. A transition to foddering cattle also brought the care of cattle into the female domain, further empowering women. Many women now derive useful income for themselves and the farm through milking, for example. Cutting of fodder by women, usually from napier grass on terrace edges, encourages their involvement in terracing.

Machakos' agricultural success didn't come without environmental costs. As the cultivated land in the district grew from 15 to nearly 80 percent, native plant and animal populations decreased dramatically, including some of Kenya's rarest species, such as the rhinoceros. Poaching and encroachment in Tsavo National Park and other protected areas remains a problem (Kenya Web 1999).

Small-scale, traditional irrigation in Machakos is based on seasonal streams.

Box 3.3 Ranking the Challenges in Machakos

At a 1999 conservation workshop sponsored by the World Resources Institute in Machakos, farmers unanimously agreed that lack of water was their most pressing concern, followed by farm size and land scarcity. As the population has increased, farms have been divided among heirs until the average farm size is little more than 1 ha. The high-potential lands have all been taken, so people are farming more marginal lands, either in the plains or on steep mountainsides where the government prohibits agricultural activities.

Lack of capital to invest in farm improvements and technologies and the lack of a ready labor pool were also at the top of this group's list of constraints to conservation. Because more children are in school and older children are migrating to cities to find work, women now provide most of the farm labor in Machakos—while still carrying out traditional responsibilities like raising children, keeping house, and fetching fuel and water.

Soil erosion didn't make their list of challenges. In fact, the largest contributor to soil erosion in the district today isn't farms but rather poorly constructed or unrepaired roads and sand mining from river beds by the concrete industry, which has flourished in conjunction with a building boom in Nairobi. Many roads are etched with deep gullies along steep roadsides, made worse by the El Niño rains, but repair requires public or community resources on a scale that the citizens of Machakos simply don't have. Poor roads also increase the cost of imported foods and the cost of transportation to get Machakos-produced goods to retail markets in places like Nairobi and Mombasa. Road conditions during the rainy season make it difficult for farmers to get their produce to markets before it spoils. Because the district is not completely supplied with electricity, food processing or refrigeration is not always feasible.

A road that connects Machakos Town to district hillsides. On the left is a roadside drain. Maize and bean crops and mango, banana, and eucalyptus trees are visible in the background.

An example of poorly maintained terraces near Machakos. Theses show only minimal management to reduce erosion of the unprotected terrace berms. Further up the slope this farmer has planted maize, beans, cassava, mango, and banana trees.

Machakos Today

"In Machakos today people are building soil conservation structures without being forced," says George Mbate, an economist with USAID (interview 19 February 1999). "They've come to relate production of crops with proper soil management."

The effect of drought is not as damaging today, thanks to investments in terraces; retention ditches, which encourage water seepage to the cropped area; and cut-off drains, which collect water and discharge it safely without causing erosion on the farm. The manure that farmers apply to fruit trees not only fertilizes the soil but improves water infiltration, lessening water runoff. Short-season maize varieties and early planting to allow enough time to prepare the land for the "long-rains" crops are also beneficial. These techniques, along with diversification of income from urban jobs, have made it possible to reduce food imports and famine relief, even during droughts (Tiffen and Mortimore 1992:373).

But even terraced crops are vulnerable, and the problems of Machakos are far from solved. Droughts in 1996 and 1997, followed by El Ninõ rains in 1998, ruined subsistence crops and forced some farmers to sell livestock to buy food. In the semiarid areas good harvests were achieved, but the heavy rains hit the hilly areas of Mwala division particularly hard, rotting crops, leeching nutrients from the soil, and destroying terraces, houses, and latrines.

"Most times, it's a food-deficient area," admits A.M. Ndambuki, agricultural officer for the district (interview 1 March 1999). "In a good year, there's enough food for that season. This year [1998] with the drought, we didn't harvest anything. Now almost all the food we're eating comes from outside the district." Importing food rather than producing it wouldn't be a problem if there were sufficient opportunity to earn money, but in Machakos, there is not. Many of the poorest farmers must search for alternative, often low-wage rural jobs in order to feed their families.

The farmers who fare the best are those like Samuel Milo, who grows tomatoes, maize, beans, and sugarcane on the sloping land of his 16-ha farm. He maximizes his terraces by planting napier grass for fodder on the terrace embankments, and a row of banana trees in the gullies to protect against erosion and to supply fruits. He plants trees as windbreaks between crops, too, and has a woodlot from which he sells timber and gets his firewood. His 4,200 coffee plants produce high-quality beans that he sorts, processes, and sells. By keeping his five cows penned and fed on napier grass harvested from the terrace, instead of allowing them to graze, he saves land space and has fertilizer for the soil.

But Mr. Milo is not just enterprising and conservation-minded, he is also fortunate. His farm is unusually large and a stream runs through his property. He has built an irrigation channel above the stream. Thanks to income-generating crops, he has been able to run a pipe from another stream into a large underground storage tank built on his property, ensuring a steady water supply.

Other farmers are not so lucky. For many, adaptations and conservation techniques like Mr. Milo's are too expensive or labor intensive. For the farmer with limited resources to hire help, for example, terracing can take years. In one Machakos village, researchers found that only 57 percent of farmers could afford the capital needed to produce cash crops for the market or to purchase farm inputs like fertilizer. Those were usually farmers with family members who earned money from off-farm jobs in urban areas (Murton 1999:40).

Another economic change that may undermine poor farmers' ability to apply best farm practices is a polarization of wealth and land. In 1965, the poorest 20 percent of the households in Mbooni owned 8 percent of the land; in 1996, they owned 3 percent. By contrast, the richest 20 percent owned 40 percent of the land in 1965 and 55 percent in 1996 (Murton 1999:41). This creates a group of viably large farms, but leaves very small farms struggling in poverty. Land concentration occurred as wealthier farmers, often those with a nonfarm income source, bought out farmers who sold their medium-sized or small farms. Some of the farmers who sold their farms migrated onto the former Crown Lands–the more fragile lands and drier frontier areas. There more acreage was available, but more inputs were needed to produce the same income.

Why do people bear the hardship of pioneering a new farm in difficult conditions or hang on to a tiny plot in the uplands? Because for the Akamba, owning land "is part of your identity, your value, your culture," according to Dr. Samuel Mutiso (interviewed 25 February 1999), a Kamba who heads the geography department at the University of Nairobi and is Kenya's representative to the UN Convention on Desertification. "We are torn between two worlds," he said.

Can the "Miracle" Continue?

"The changes in Machakos didn't come overnight," says Mutiso. Spurred by necessity and eventually freed from the constraints of dictatorial government land policies, the Akamba successfully intensified land use by selecting and adapting new technologies from a variety of places. They switched to more profitable crops, better staples, manure fertilizers, and systems of multiple cropping, reduced grazing, and tree cultivation. Community-level planning and leadership, such as the *mwethya* groups, and community preferences in technology and crops far more effectively increased fertility and decreased erosion than imposed conservation programs. When farmers have economic incentives to conserve soil–higher yields, the opportunity to grow more profitable crops, and access to markets–they are willing to invest more capital and labor in bench terraces. In a sample of five areas, the proportion of total area treated with soil conservation measures rose from about 52 percent in 1948 to 96 percent in the older settled areas in 1978. The areas also reflected substantial gains from soil erosion reduction and from rainfall infiltration and soil moisture retention (Tiffen and Mortimore 1992:368).

Migration to urban areas provided a flow of remittances that augmented capital for agricultural development. Income and experience from nonfarm jobs were combined with government extension efforts to dramatically facilitate the transfer of knowledge, technology, and capital to the farms.

Another important change was a shift from central government decision making about ecosystem issues to greater district-level participation, including direct engagement of local leaders in seminars. This approach afforded an opportunity to work with, rather than against, the Akamba's intimate knowledge of the land's problems and their culturally preferred agricultural methods. It also capitalized on their abiding attachment to the land. "It is not just economic," says Maria Mullei (interview 17 March 1999), an agricultural officer with USAID who also farms in Makueni, "you love the land so you protect it." In fact, much of the incentive and capital for the retreat from expected ecological disaster came from the people of Machakos themselves.

Decreasing farm size, growing land scarcity in the face of population growth, and loss of communal grazing lands also have pressured the Akamba to use their land and water as efficiently as possible. Yet no one has suggested that population growth might encourage further conservation, land intensification, and productivity. Today, population growth rates in Machakos are about 3 percent per year (Mortimore and Tiffen 1994:13). With increasing population density and high costs of raising children, however, birth rates are starting to fall.

Less encouraging are signs that without capital some erosion protection and water conservation technologies cannot be adopted even if they would improve the land. For example, more farmers would like to put in water storage tanks but face the problem of limited financial resources. On some upland farms there are too few bulls to haul plows, and terraces are too small to allow plows to turn easily.

Cyclical poverty may emerge, as Murton (1999) found in Mbooni, which was part of Machakos district prior to 1992. Those with an off-farm job, more fertile soils, or a water source fare better. Those that fare better and increase productivity are most able to switch to higher value crops, like citrus fruits and French beans, and tap commercial markets. But others abandon farming or migrate to marginal lands. Although all children complete primary school, the poorest families may not be able to afford to send their children to secondary school, which may deny them the opportunity to secure the off-farm jobs that lead to personal income.

The future of agricultural innovation and land productivity in Machakos also depends in no small part on the larger economy in which the district operates. The technologies to protect the land are in place, but the present greenness of the fields does not guarantee anyone a living. Economic and environmental sustainability also are determined by food prices, the availability of urban jobs, and external resources for improvement of roads or electrification to help farmers tap commercial markets.

Tempered by such challenges, Machakos remains an encouraging story, a place where the expected progression toward further environmental degradation has not occurred, a place where farms flourish in place of deserts. Whether such rewards and growth are sustainable will be determined in the decades to come.

Cuba's Agricultural Revolution: A Return to Oxen and Organics

The fall of the Berlin Wall in 1989 and the subsequent demise of communism in the Soviet Union occurred half a world away from Cuba. But the repercussions of that revolution directly affected Cuban soils: it transformed Cuba's agricultural lands by forcing a radical shift to organic inputs and farming methods on a scale unprecedented worldwide.

Cuban Agroecosystem Management from 1959 to 1989

From 1959 through the 1980s, being part of the socialist trade bloc significantly influenced Cuba's economic development and ecosystem management. Though a highly industrialized country that produced pharmaceuticals and computers as well as crops, sugar was the staple of the Cuban economy. By 1989 state-owned sugar plantations covered three times more farmland than did food crops (Rosset 1996:64). Sugar and its derivatives constituted 75 percent of the total value of Cuba's exports, purchased almost entirely by the Soviet Union, Central and Eastern Europe, and China (Rosset and Benjamin 1993:12). High crop yields were attained through agricultural methods that were more mechanized than in any other Latin American nation, in addition to extensive use of pesticides, fertilizers, and large-scale irrigation.

In return for its exports of sugar, tobacco, citrus, minerals, and other items, Cuba imported about 60 percent of its food as well as crude oil and other refined products, all from the socialist bloc at favorable terms of trade. Forty-eight percent of the fertilizer, 82 percent of the pesticides, and much of the fuel used to produce the sugar crops were imported as well, along with 36 percent of the animal feed for Cuban livestock (Rosset and Benjamin 1993:10, 15).

This trade regimen–though highly import-dependent–enabled Cuba's 11 million people to achieve economic equity, rapid industrialization, and advancements in quality of life. In the 1980s, Cuba exceeded most Latin American countries in nutrition, life expectancy, education, and GNP per capita. Sixty-nine percent of the population was urban, with virtually no unemployment (Rosset and Benjamin 1993:12). Ninety-five percent of Cubans had access to safe water and 96 percent of adults were literate (FAO 1999:20).

The Advent of Alternative Agriculture

The crumbling of the socialist trade bloc in 1989–91 brought upheaval to the Cuban economy and its conventional model of agricultural production. Cuba lost 85 percent of its trade (Murphy 1999). The United States tightened its already stringent economic blockade against Cuba, compounding the country's difficulties.

Cuba's access to basic food supplies was severely threatened. As food imports were halved, caloric intake dropped 22 percent, protein 36 percent, and dietary fats 65 percent (Bourque 1999). According to the FAO, Cuba endured the largest increase in undernourished people in Latin America in the 1990s–a jump from less than 5 percent to almost 20 percent (FAO 1999:8). Imports of pesticides, fertilizers, and feeds were reduced by 80 percent and petroleum supplies for agriculture were halved (Rosset 1996:64).

To avert widespread famine, Cuba had to find a way to produce twice the amount of food with just half of its previous agricultural inputs. The result is that Cuba is now in the midst of the largest conversion from conventional high-input chemical agriculture to organic or semiorganic farming in human history (Rosset 1996:64). Cuban farmers are attempting to produce most of their food supply without agrochemicals.

Cuba's prior investments in science, education, and agricultural research and development proved a great asset during these dire economic straits. In the 1980s, concerned by Cuba's vulnerability as the sugar plantation of the eastern bloc, government leaders had invested $12 billion in training scientists in biotechnology, health and computer sciences, and robotics

Cuba's Dependence on Imported Food, pre-1990

Imported foods accounted for 57 percent of Cubans' total caloric intake.

Food	Percentage of Food Imported
Beans	99
Oil and lard	94
Cereals	79
Rice	50
Milk and dairy	38
Animal feed	36
Meat	21
Fruit and vegetables	1–2
Roots and tubers	0
Sugar	0

Source: Rosset and Benjamin 1993:10.

(Rosset 1996:65). Although Cuba comprises only 2 percent of Latin America's population, it is home to 11 percent of the region's scientists (Rosset and Benjamin 1993:4).

Agricultural scientists influenced by the international environmental movement of the 1970s had begun to criticize Cuba's dependence on foreign inputs and the toll that conventional cultivation techniques were taking on the island's agroecosystems. As they noticed increasing pest resistance and soil erosion, many shifted their research in the 1980s to alternative methods of crop production, particularly the biological control of insect pests (Rosset and Benjamin 1993:21).

Most important, Fidel Castro gave his full support to the "alternative model" during this "Special Period." The government emphasized the importance of using Cuba's own scientific expertise instead of imported technology. "Cuban scientists will create resources that will one day be more valuable than sugarcane" Castro said in 1991. "Our problems must be resolved without feedstocks, fertilizers, or fuel" (Rosset and Benjamin 1993:24).

That was easier said than done. Cuban scientists had developed several alternative agricultural techniques during the 1980s but they were largely untried. Plus, the transition from chemical to organic agriculture takes time–roughly 3-5 years to regain soil fertility and re-establish natural controls of insect pests and diseases (Rosset and Benjamin 1993:25). Cuba did not have the luxury of 3-5 years.

Cuba's Access to Selected Imports in 1989 and 1992

Item	1989	1992	Percentage Decrease
Animal feeds	1,600,000 MT	475,000 MT	70
Fertilizer	1,300,000 MT	300,000 MT	77
Petroleum	13,000,000 MT	6,100,000 MT	53
Pesticides	US$80,000,000	>US$30,000,000	63

Source: Rosset and Benjamin 1993:17.

The first challenge was soil fertility. Fertilizer availability dropped 80 percent after 1989. To fill the void, Cuban farmers have employed a variety of "biofertilizers" and soil amendments, including composted animal wastes, cover crops, peat, quarried minerals, earthworm humus, and nitrogen-fixing bacteria. Though the *Rhizobium* bacterium has long been known to help legume crops obtain nitrogen from the atmosphere, Cuban scientists also have used *Azotobacter*, a free-living nitrogen-fixing bacterium, to supply nitrogen to many nonlegume crops. *Azotobacter* offers added advantages of shorter crop production cycles and reduces blossom drop, helping Cubans achieve a reported 30–40 percent increase in yields for maize, cassava, rice, and other vegetables (Rosset and Benjamin 1993:43). Similarly, the substitution of worm humus for chemical fertilizers increased yields of various crops by 12–46 percent (Monzote n.d.:9).

Intercropping, once rare in commercial scale farming, is being revived to diversify crop production and boost soil fertility. Another key component of Cuba's soil management efforts is reforestation; many forests were razed after the 1959 revolution to plant sugarcane and provide fuel for sugar manufacturing. In 1989-90, more than 200,000 ha were reforested (Rosset and Benjamin 1993:50).

The country is recycling its waste products on a massive scale, including household garbage and composted livestock and human waste. Wastewater is used to irrigate cane fields. Filter press cake, a by-product high in phosphorous, potassium, and calcium, serves as fertilizer. Bagasse, or dry pulp, is fed to livestock and burned to generate electricity for machinery in many sugar mills.

Cuba has a history of using biological controls for insect pests that dates back to 1928, when growers began releasing mass-reared parasitic flies (*Lixophaga diatraeae*) into sugarcane fields to control cane borers. Since the food crises, however, use of biological controls has intensified. Growers have been releasing predatory ants (*Pheidole megacephala*) to control the sweet potato weevil (*Cylas formicarious*), a method that has proven 99 percent effective (Rosset 1996:66).

Cuban researchers have focused also on the use of entomopathogens–bacteria, fungi, and viruses that infect insect pests but are nontoxic to humans. *Bacillus thuringiensis*, Cuba's first commercially produced biopesticide, is a soil bacterium widely used to control lepidopteran pests in pasture, cabbage, tobacco, corn, cassava, squash, and tomatoes, as well as mosquito larvae that transmit human diseases. The fungus *Beauveria bassiana* has also been used successfully against sweet potato and plantain weevils (Rosset 1996:67). In contrast, prior to 1989 the most common pesticide used in Cuba was methyl parathion, one of the most acutely toxic pesticides in the world (Gellerman 1996). By the end of 1991, an estimated 56 percent of Cuban cropland was treated with such biological controls, representing savings of US$15.6 million per year (Rosset and Benjamin 1993:27).

Overall, nonchemical weed control has been less successful than pest controls in Cuba, as elsewhere. Nevertheless, researchers continue to develop methods that hold promise–crop rotations based on mathematical modeling, methods involving weed densities, and traditional methods used by peasants before the advent of herbicides.

Perhaps the most striking change in the agricultural landscape was the return to the use of oxen in the fields while Russian tractors, lacking parts and fuel, were idle. Though more labor-intensive, ox traction actually provides advantages to Cuban farmers. Oxen are cheaper to operate, do not compact the soils, can be used in the wet season long before tractors, and their fodder provides much-needed organic fertilizer. New ox-powered plows, planters, and cultivators were developed, and the government encouraged oxen breeding programs to expand the herd.

Promotion of Small Farms and Urban Gardens

Alternative farming methods alone couldn't bring Cuba out of its agricultural slump. Huge Soviet-style state farms controlled 80 percent of the nation's agricultural land. The vast monocultures of sugarcane, pineapples, citrus and other crops they once produced with chemical fertilizers and pesticides were incapable of developing the natural pest controls or soil fertility produced by smaller, more dynamic organic systems. As a result, the state farms became extremely vulnerable to pests and disease (Rosset 1996:65, 69).

By contrast, *campesinos* were quick to adapt the new technologies, and their productivity soared. Many were descendents of generations of small farmers with long family and community traditions of low-input farming, and they remembered techniques that their parents and grandparents used

In the 1980s, Cuba used highly mechanized agricultural methods. After the economic crisis, oxen teams were substituted for tractors on both small and large farms. The number of oxen teams has tripled in the last decade. There is also a growing network of small workshops producing implements for farming with oxen teams.

such as intercropping and manuring. Even before the country-wide emphasis on organic agriculture in the 1990s, the small farmers had proven their efficiency: they worked only about 20 percent of the land but produced more than 40 percent of the domestic food supply (Rosset 1996:65, 68–69).

In 1993 the Cuban government broke up the unproductive state farms into Basic Units of Cooperative Production–worker-owned cooperatives that controlled about 80 ha each. Although the government still owns the land and sets production quotas for key crops, coop members own everything they produce above the quotas and can sell it in new farmer's markets. Sales at markets flourished and severe food shortages disappeared by mid-1995 (Rosset 1996:69–70).

Another factor that helped stave off hunger was the promotion of urban agriculture by the Cuban government on private and state land, which gardeners can use at no cost. Today, Havana alone has more than 26,000 self-provision gardens (Moskow 1999:127) that produced an estimated 541,000 tons of fresh organic fruits and vegetables for local consumption in 1998. Some neighborhoods were producing 30 percent of their food. Price deregulation provided another incentive, enabling urban farmers to earn two to three times as much as urban professionals (Murphy 1999).

Intensive, raised-bed agriculture is the model for urban agriculture in Cuba. These farms, called *organoponicos,* are approximately 1 ha and produce, on average, 20 kg of vegetables per square meter (Bourque 1999). Farmers rely on large applications of organic fertilizers from local sources and only use biologically based pest controls when absolutely necessary.

Will the Organic Revolution Be Overthrown?

In the 1996–97 growing season, Cuba recorded its highest-ever production levels for 10 of the 13 basic food items in the Cuban diet, largely because of small farms and backyard production (Rosset 1998). But FAO data suggest that total Cuban crop production in 1996–98 was still 40 percent lower than in 1989–91 (World Bank 2000:122), perhaps in part because sugar crop yields have not yet recovered. Furthermore, pest and disease outbreaks continue. Many of the biopesticides require critical timing of applications to work, and the quantity and quality of materials produced by the cooperatives vary widely. At one point a shortage of glass jars needed to grow fungal spores held up production (Rosset 1996:72).

Such stumbling blocks have led outside observers to speculate that the organic revolution in Cuba may dissolve after the economy improves and trade barriers come down. The topic is a subject of debate among Cuban agricultural scientists and farm managers, many of whom remain dedicated to high-input chemical agriculture common in the West (Mueller 1999).

Whatever the outcome, Cuba's ongoing experiment with alternative agriculture has left a powerful mark. Even though Havana now enjoys increased food availability, urban agriculture is stronger than ever (Murphy 1999). In a recent survey, 93 percent of gardeners interviewed affirmed their commitment to producing food in urban areas and once vacant lots even after the "Special Period" ends (Moskow 1999:133). Cuban scientists are already exporting their expertise, working with Mexico, Bolivia, Brazil, Laos, and other countries to develop and export biological controls for the coffee weevil and other pests (Bourque 1999). Moreover, Cuba has succeeded in feeding its people without the high inputs of conventional agriculture, providing a model that other countries can follow.

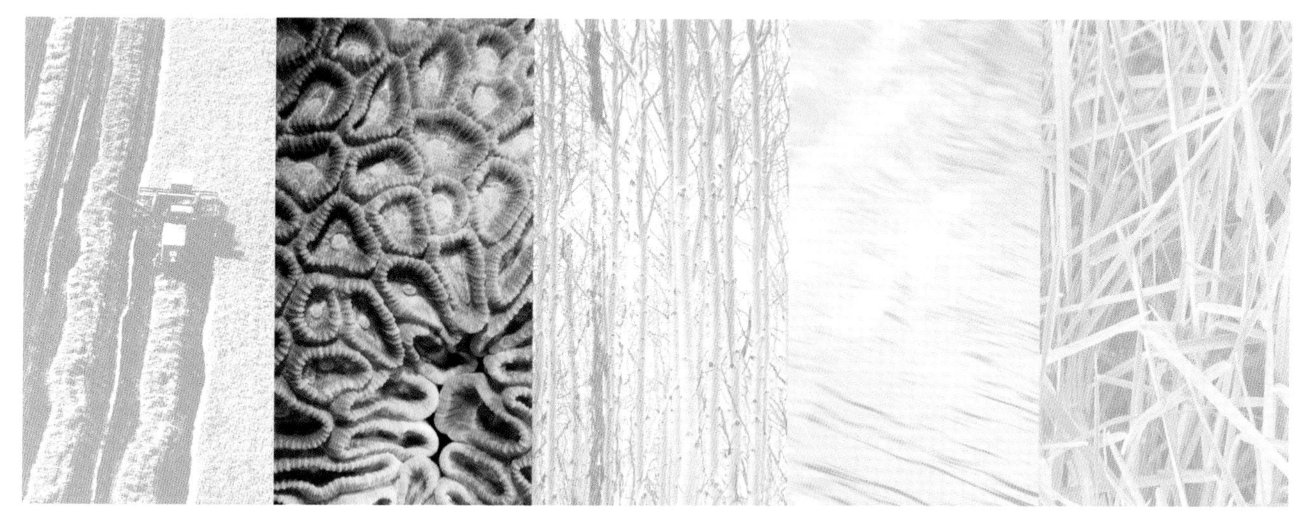

COASTAL ECOSYSTEMS

Replumbing the Everglades: Large-Scale Wetlands Restoration in South Florida

Look down on South Florida from a high enough altitude and the problem is obvious. Lake Okeechobee, the liquid heart of the giant watershed that covers the lower third of Florida, stands penned behind floodproof dikes. Massive changes in the landscape have clearly altered the flow of water through the area. Below Lake Okeechobee, the original shape of the Everglades is barely recognizable arcing south for 160 km from the Lake to the mangrove shallows of Florida Bay.

Water dominates the South Florida ecosystem like few other places in North America. This was once an unbroken marshland of sawgrass and small tree islands, fed by a shallow sheet of water flowing south from Lake Okeechobee. Now the marsh is a series of disconnected tracts separated by dikes, drained by a web of major and minor canals. Croplands–mostly sugarcane–have displaced the entire northern third of the Everglades; only the southern end remains in a relatively natural state as Everglades National Park and Big Cypress National Preserve.

The benefits of these changes–and the beneficiaries–are as clear as the changes themselves. To the east of the Everglades, safe behind a levee, lies the greater Miami area–a sea of tract houses and high-rise buildings, home to 6 million people and a burgeoning center of tourism, trade, international investment, and retirement living. The levees and canals protect the populated eastern corridor from floods and effectively turn most of the remaining tracts of Everglades into reservoirs for water supply. Agriculture, which represents the other major land use in the area, depends even more on the

(continues on p. 166)

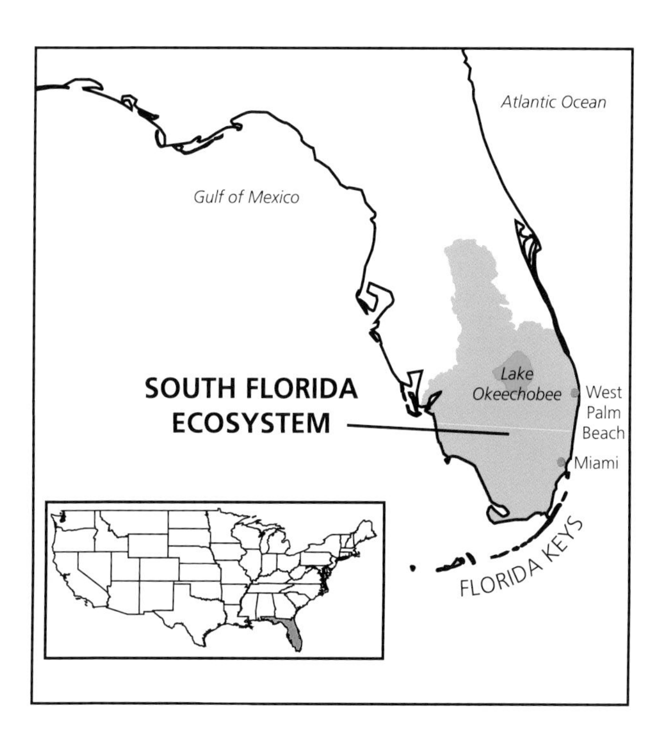

Box 3.4 Overview: Florida Everglades

In what may be the world's most ambitious effort to restore an ecosystem, U.S. government agencies, business interests, and environmentalists are combining forces—and US$7.8 billion—to reverse a century of draining and diking in the Florida Everglades. This vast inland marsh houses a rich assemblage of plants and wildlife and is the water source for the Miami area's 6 million residents and South Florida's lucrative farming sector.

Ecosystem Issues

Freshwater

The 23,000 km^2 Kissimmee-Okeechobee-Everglades watershed was once a single hydrologic system of rivers, lakes, and wetlands. Flood control and water supply structures have drastically reconfigured this once free-flowing water, reducing the water volume and disrupting the natural flooding and drying cycle. Nearly half of the wetlands have been lost; saltwater intrusion and pollution from intensive agriculture are additional problems.

Coastal

Changes in the natural water flow in the Everglades have greatly reduced the quantity of freshwater reaching the coast at Florida Bay, disrupting estuary salinity levels, and causing seagrass die-offs and turbidity in the bay. Traditional bird colonies have abandoned nearby mangrove forests and brackish marshes.

Agriculture

Croplands have displaced about one-third of the Everglades, but have made South Florida counties important producers of sugarcane, subtropical fruits, and winter vegetables. That output, however, now is threatened: agricultural acreage in Southern Florida is giving way to urban sprawl and soil subsidence.

Management Challenges

Economics

Although the restoration bill is daunting, the cost of allowing the Everglades' decline to continue could be far greater, particularly for local residents and businesses. For example, further declines in the health of Florida Bay could bring losses of more than $250 million/year in lost tourist dollars and reduced commercial fish catches. The area's $2 billion agriculture sector depends even more on the flood control and reliable water supply that the network of water control structures brings. No one has yet put an economic value on the many species whose lives hang in the balance of restoration.

Stakeholders

Sustaining the restoration effort will demand ongoing negotiations and commitment among an array of stakeholders—federal, state, and county governments; agribusinesses; environmental, sport, and recreation groups; and Native American tribes. Because restoration is intimately connected with regional patterns of land and resource use and economic expansion in Southern Florida, all of the area's 6 million residents are also ultimately affected.

Information and Monitoring

No restoration project of this magnitude has ever been undertaken; its effects on the social and biological aspects of the system are not entirely known. The many unknowns make ongoing monitoring of the ecosystem's health and productivity particularly essential: to ensure the maximum effectiveness of the $7.8 billion investment, to provide feedback to stakeholders, to guide changes in the restoration plan, and to inform similar efforts elsewhere.

Timeline

c. 0 AD Native Indian tribes—the Tequesta and the Calusa—migrate into South Florida.

1513 Spanish explorer Ponce de Leon claims Florida for Spain.

1820s Settlers from the United States begin to migrate south into Florida.

1821 U.S. purchases Florida territory from Spain.

1835–42 and 1855–58 The "Seminole Wars": Seminoles escape into the Everglades interior to avoid U.S. government troops.

1845 Florida territory is granted statehood in the United States of America.

1848 U.S. government first recommends draining Everglades for agriculture.

1855 Alligators begin to be hunted for their hides; at least 10 million killed from 1870 to 1965.

1881 Hamilton Disston finances first large-scale experiment in draining and farming in the Everglades.

1907 The Everglades Drainage District founded to fund major drainage canals.

1917 Four major canals completed from Lake Okeechobee to the Atlantic Ocean.

1926 and 1928 Hurricanes kill 2,500 people and cause more than $75 million in damages.

1928 Tamiami Trail (first road across the Everglades) is completed.

1947 Record rains flood 90 percent of southeastern Florida for 6 months. Everglades National Park is established.

1948 Central and South Florida (C&SF) Project is authorized.

1954–59 Everglades Agricultural Area created by diking and draining the northern Everglades.

1963–65 C&SF water managers stop water from flowing into Everglades National Park in order to fill new water conservation areas.

1970 Severe drought occurs.

1973 Construction complete on major elements of the C&SF Project.

1980–81 Severe drought occurs.

1983 Governor Robert Graham initiates Save Our Everglades program.

1986 Major algal bloom on Lake Okeechobee prompts state action to lower phosphorus pollution entering the lake.

1988 Seagrass die-offs and large algal blooms begin in Florida Bay. Federal government files suit against the South Florida Water Management District for releasing water polluted with agricultural runoff into the Everglades.

1991 Florida passes the Everglades Protection Act, mandating control of nutrient pollution of the Everglades.

1992 U.S. Army Corps of Engineers begins review of C&SF Project to determine how to reduce ecosystem damage.

1993 Federal government establishes the South Florida Ecosystem Restoration Task Force.

1994 Florida enacts the Everglades Forever Act to establish a comprehensive program to restore significant portions of the Everglades. Governor's Commission for a Sustainable South Florida is established.

1997 Restoration of the channelized Kissimmee River begins. Construction begins on the first of six filtering wetlands to remove phosphorus from agricultural runoff leaving the Everglades Agricultural Area.

1998 U.S. Army Corps of Engineers releases $7.8 billion plan to reconfigure the C&SF Project to restore a more natural water cycle.

flood control and reliable water supply that the network of water-control structures brings.

But the benefits that have come from bending the natural water cycle to human need have brought less welcome changes to the ecosystem. The Everglades and the whole of the South Florida ecosystem are uniquely dependent on the area's distinctive water flow pattern. When people began to disrupt this pattern, the health of the ecosystem began to deteriorate–at first slowly, but more rapidly in the last 2 decades. Wading bird populations have plunged, seagrass beds in Florida Bay have died back, sport and commercial fishing has suffered, and nonnative plants and fish have invaded, among other effects. Even the assurance of a plentiful water supply has evaporated as the urban population grew and the capacity of the Everglades to store water shrank.

Can the South Florida ecosystem be restored to health? Local powerbrokers and the public think so, and have already committed more than $2 billion to the effort over the last decade. Recently they have embraced a new $7.8 billion Everglades restoration plan proposed by the U.S. Army Corps of Engineers–the most ambitious and extensive ecosystem restoration effort in the world. With the goal of duplicating, as much as possible, the region's original water patterns, engineers are poised to rip out certain levees, refill some canals, and re-allocate water throughout the region. There are no guarantees of success, and even if some recovery occurs, scientists are not sure how much the total health of the ecosystem will improve over the long term, given that the Miami region is still developing rapidly. Yet the restoration effort has clearly generated local enthusiasm, as well as high-level support from the state and federal governments. How a contentious band of government agencies, business interests, and environmental and sporting groups came to agree on such an expensive and difficult program is a story of how convincing–and threatening–an ecosystem in distress can be.

Draining the Marsh, Stopping the Flood

Water had long been a barrier to human settlement of the Everglades region. Prior to the 19th century, a few Native Indian villages dotted the coast, but the marshy interior of the Florida Territory remained largely unpeopled until bands of Seminole and Miccosukee Indians fled to the Everglades to escape U.S. government troops in the 1830s.

Early white settlers regarded the Everglades and other seasonally flooded tracts as wasted land, inhospitable to commerce, food production, transportation, and personal safety, and fit only to be drained and "improved." At first, agriculture was the focus of these schemes. With a tiny population and no major cities or industrial base, Florida looked to its fertile muck soils for its future.

THE BEGINNING OF FLORIDA'S AGRICULTURE

In 1881, Philadelphia millionaire Hamilton Disston financed the first real attempts to drain and farm marshlands in South Florida on a 20,000 ha tract in the upper Kissimmee Basin. His success with rice and sugarcane crops on reclaimed land bore out the land's potential productivity. His canals–the area's first–opened a water route from Lake Okeechobee to the Gulf Coast. By the late 1920s, agriculture was well established around Lake Okeechobee and elsewhere in the basin and the rudiments of a drainage system–five major canals from Lake Okeechobee to the Atlantic–had been dug (Light and Dineen 1994:53–55; Light et al. 1995:120–122).

But these early canals and levees were not sufficient to protect the region from the disastrous floods that periodic hurricanes brought. Hurricanes in 1926 and 1928 claimed more than 2,500 lives and left an estimated $75 million in damages when flood waters breached the low levee protecting the farming areas south of Lake Okeechobee. These disasters intensified efforts to keep the lake safely within its bounds. The levee was raised and two flood bypass routes, to the east and the west, were created to help vent flood waters directly to the Gulf and Atlantic coasts rather than allowing the waters to flow south along their natural course (Light and Dineen 1994:55).

Unfortunately, when major hurricanes again hit the Everglades in 1947 and 1948, inundating 90 percent of southeastern Florida for 6 months, it was clear that flood protection was only partial at best. State and local representatives, backed by their powerful agricultural and urban constituents, pushed for the federal government to step in and fund a lasting solution to the area's flood problems (Light and Dineen 1994:58; USACE 1998:I-22).

THE CENTRAL AND SOUTH FLORIDA (C&SF) PROJECT

Federal officials responded with a major public works program–the Central and South Florida (C&SF) Project. It began in 1950 and took more than 20 years to complete. The C&SF Project is a large interlocked system of drainage canals, levees, pumps, water control gates, and water storage areas. The levees separated the Everglades from the urban eastern corridor to provide flood protection from Lake Okeechobee waters. As a by-product, the drainage canals and pumps allowed water tables in the area east of the levee to fall as much as 1.5 m, permitting suburban development to flourish (Light and Dineen 1994:58–76).

The intent of the C&SF Project was not just to tackle the flood threat, but also to secure an adequate water supply for both agricultural and urban users. Indeed, too little water was frequently as great a problem as too much. Drought years were not uncommon, bringing saltwater intrusion into local well fields and wildfires to the dry peat soils (USACE 1998:I-7).

To assure an ample water supply, C&SF Project engineers divided the central Everglades into three enormous tracts confined within perimeter dikes. These are the Water Conservation Areas. The Water Conservation Areas act as giant reservoirs to store water from the Kissimmee basin and Lake Okeechobee and serve as the principal recharge areas for the aquifer that supplies water to the urbanized eastern coastal strip.

A third major element of the C&SF Project was the creation of a special agricultural zone in the rich soils just south of Lake Okeechobee. The Everglades Agricultural Area, as it is called, converted about 20 percent of the original Everglades to intensive agriculture. Much of the 300,000 ha within the area is planted in sugarcane, making the sugar industry a significant economic force in the area (Light and Dineen 1994:60-66).

Providing Everglades National Park with sufficient water to keep it healthy was also on the list of project goals. In reality, this took a much lower priority than keeping human communities safe from floods and provided with water and became a sore point soon after the massive water project came on line. From the start, Everglades National Park supporters and conservationists were leery of the degree to which the C&SF plan would alter the natural water flow, but the fervor for flood control swept away their objections (Light et al. 1995:126-131).

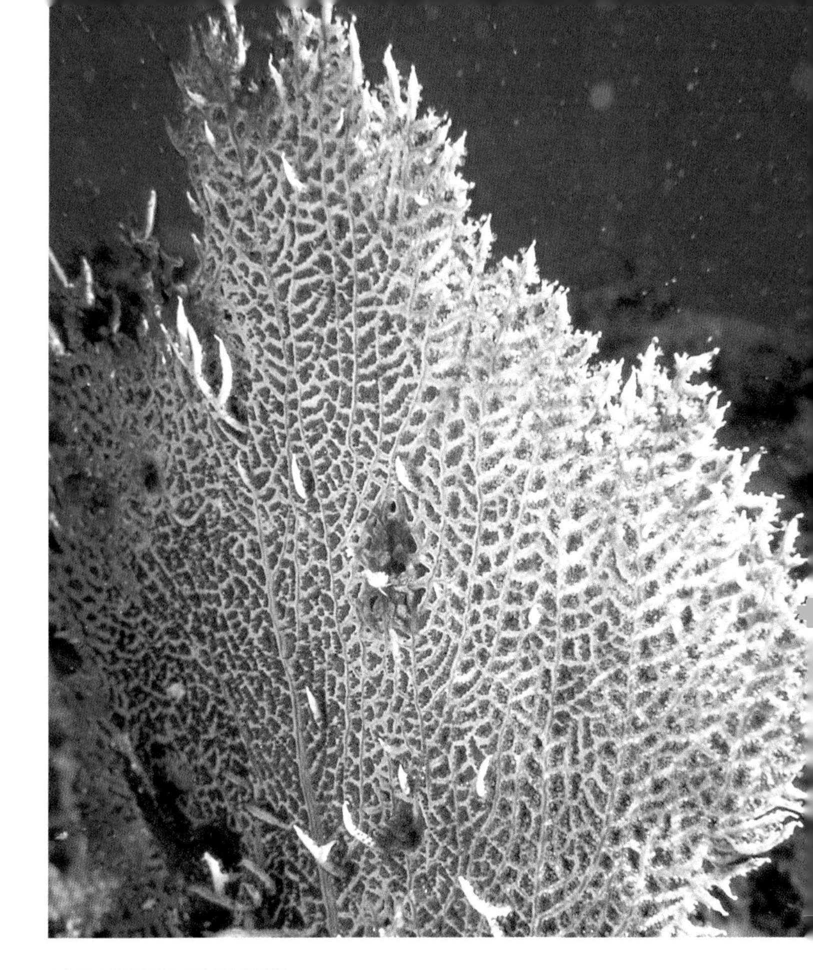

Trade-Offs: An Ecosystem in Transition

Overall, the C&SF Project has brought huge social and economic benefits to the region. Since the Project began in 1950, urban expansion in the Miami–Palm Beach corridor has brought new neighborhoods and livelihoods along with an additional 4.5 million people (USACE 1998:V-12). In the process, it has fueled the robust expansion of the service industries and international trade sector that currently account for more than half of the South Florida economy (GCSSF 1995: Regional Overview p.2).

Agriculture, which is largely the product of wetlands drainage and flood control works, contributes at least $2 billion annually to local coffers–a small but politically significant part of the local culture and economy (SFERTF 1998a:9). South Florida counties lead the nation in production of sugarcane, oranges, grapefruit, and snap beans and produce a variety of other important winter vegetables and tropical fruits that cannot be grown elsewhere in the United States. Even the lodging and resort industry, which is vital to the area's $14 billion tourist economy (1995), relies on the water supply that the C&SF Project assures (SFERTF 1998a:9-10).

But changes in the water cycle and land-use patterns in South Florida have impaired the natural functioning of the ecosystem in a number of important ways, degrading the services that it has traditionally supplied and threatening to undermine the region's economy.

LOST WATER CAPACITY

The most fundamental physical change in the ecosystem is that it no longer has the capacity to store and release enough water to meet all the demands of the region's wildlife and human communities, particularly in dry years. Conversion of large tracts of Everglades and other marshes to farmlands and suburbs has reduced the sponge-like capacity of the watershed to retain water in the wet season and release it during the dry season. By some estimates, nearly half of South Florida's original complement of wetlands has been lost, with a concomitant loss of storage capacity (SFERTF 1998a:3).

LOST SOIL CAPACITY

Draining and lowering the water tables over much of the watershed has caused widespread land subsidence and serious soil loss in many areas, threatening the future of the region's agriculture. In some parts of the Everglades Agricultural Area, topsoil loss from drying and oxidation of the peat soils exceeds 2 m–a loss of nearly half the original depth (Davis 1998). Topsoil loss has already brought a few fields perilously close to retirement and has convinced some observers that the area's future for agriculture is limited to only a few more decades (Snyder and Davidson 1994:107-108; Davis 1998).

LOST WATER QUALITY

Runoff from farm fields and urban areas has contaminated the water cycle with pollutants, lowering water quality throughout the region. Phosphorus contamination is the

(continues on p. 170)

Box 3.5 The South Florida Ecosystem

The South Florida ecosystem occupies a single large watershed—the Kissimmee-Okeechobee-Everglades watershed—that covers roughly the lower third of the state and its coastal waters, an area approximately 23,000 km^2 (McPherson and Halley 1996:16). Within this enormous region are several distinct environments, including freshwater marshes, wet prairies, cypress swamps, and pine forests in the interior; coastal prairies, beaches, and mangrove forests fringing the coasts; and coral reefs and seagrass beds in the warm waters of Florida and Biscayne Bays and the Straits of Florida.

Water flow across the region and into the coastal waters is the dynamic thread that weaves these communities into a single larger system—an interconnected tapestry of wetlands, uplands, and coastal and marine areas (USACE 1998:II–2).

At the center of the ecosystem are the Everglades, which originally stretched in a 11,650 km^2 swath from Lake Okeechobee to Florida Bay (McPherson and Halley 1996:16). Today, the Everglades have been nearly halved in extent, with Everglades National Park in the south preserving only a fifth of the native marshlands (USACE 1998:5–4).

The dynamics of the South Florida ecosystem were—and still are—driven by a seasonal cycle of flooding and drying. Most of the region's 100–165 cm of annual rainfall occurs from May through October and, under the natural regime, much of the land was flooded during this rainy season and gradually dried out during the late fall and winter (McPherson and Halley 1996:8). Natural water flow through the system is generally from north to south, but is very slow because of the flatness of the terrain. Water originating in the Kissimmee Basin in the north, where elevations are slightly higher, gradually flowed south through wetlands bordering the Kissimmee River and into Lake Okeechobee, which acted as a giant reservoir. Under high-water conditions during the wet season, the lake overflowed its southern banks, spilling water into the Everglades in a broad sheet just inches deep over much of the marsh. This sheet flow makes of the central Everglades a shallow, vegetation-covered river—a "river of grass," as the Everglades is frequently called. Because the slope is so gentle, with elevations falling just 6 m between Lake Okeechobee and Florida Bay, it takes the water flowing through the Everglades an average of 12 months to reach the coast (Jones 1999; USACE 1998:II–3).

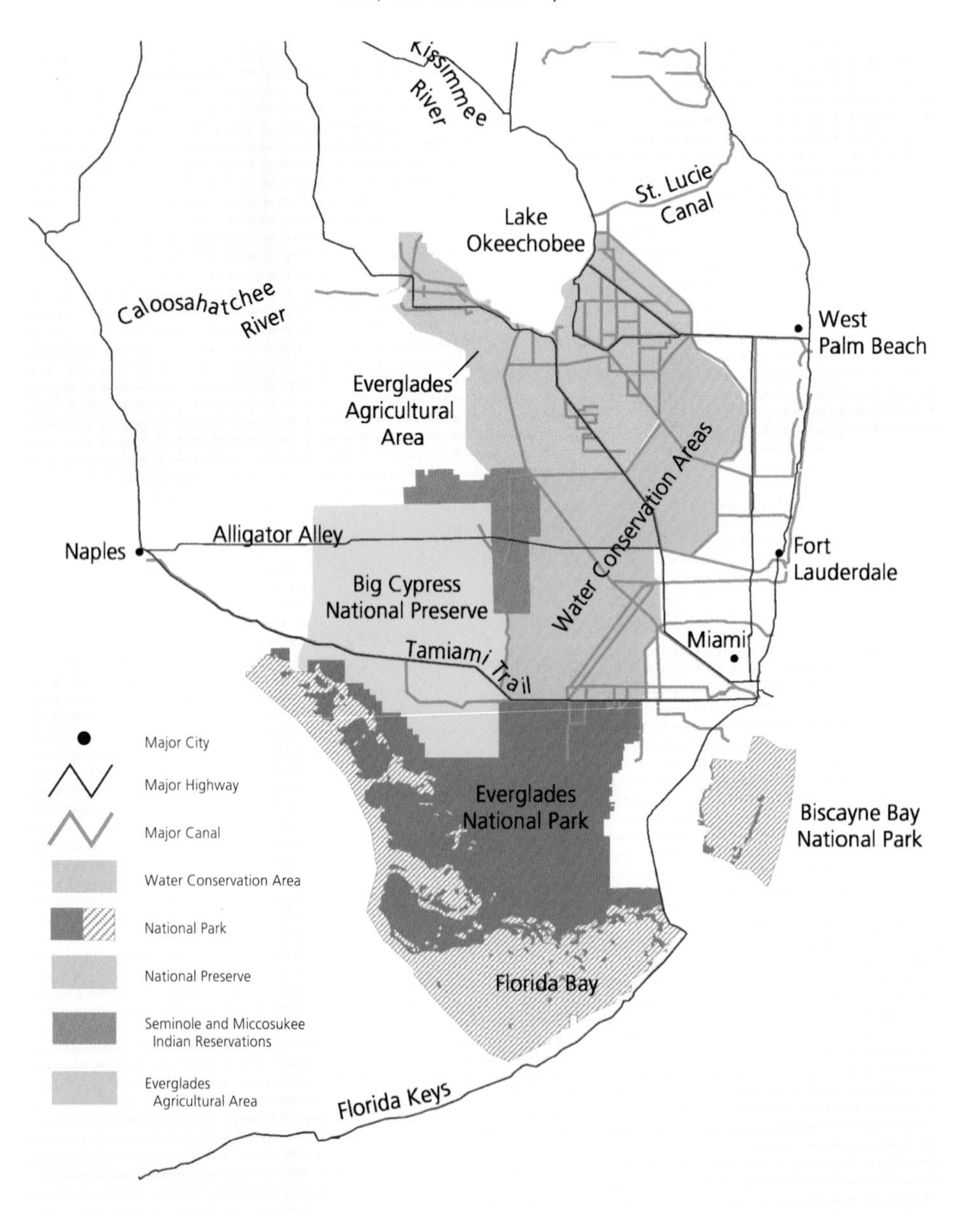

Sources: Birbeck 1990; Davis and Ogden 1994; ESRI 1993; Florida Department of Environmental Protection 1996a, 1996b.

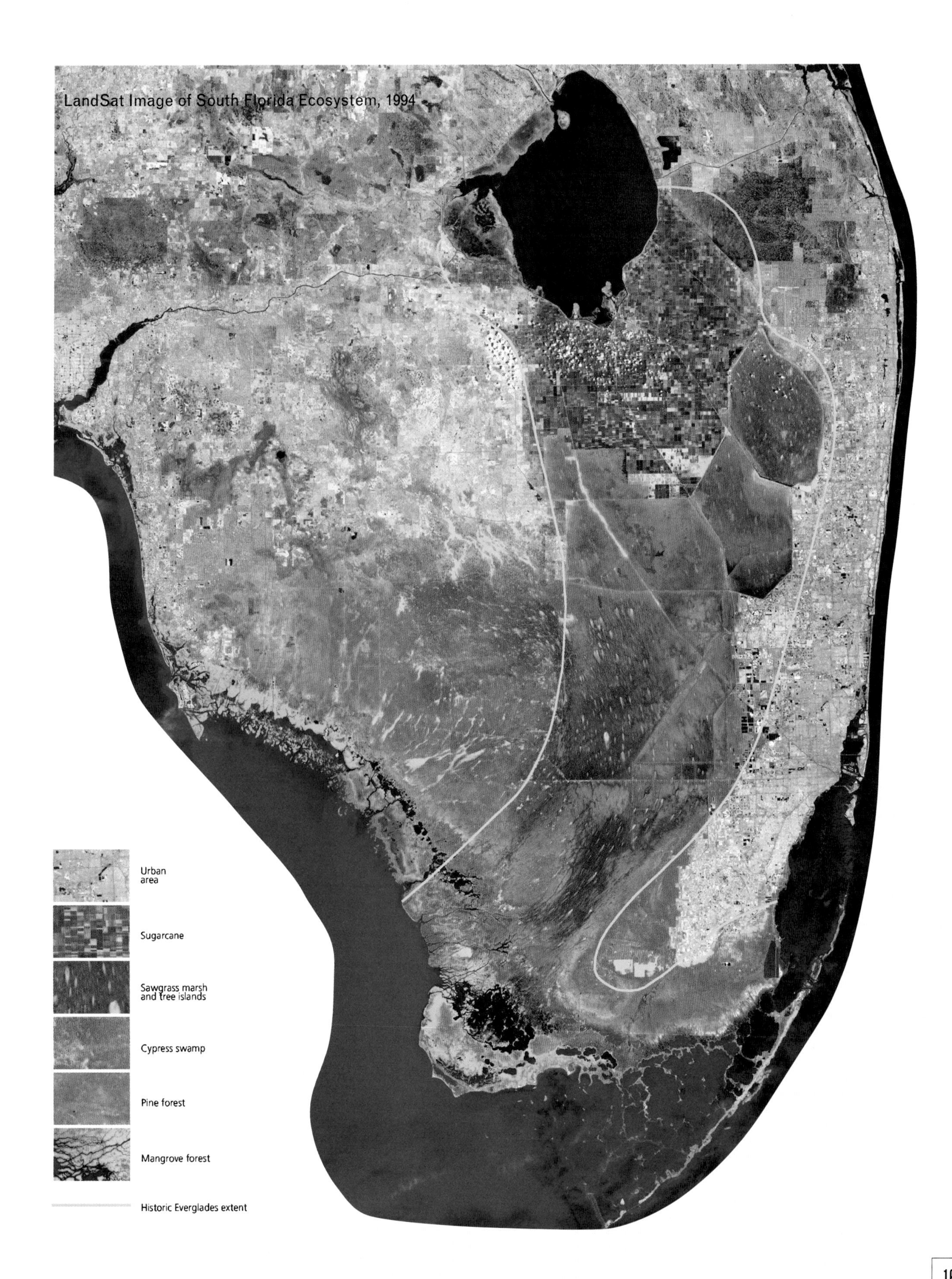
LandSat Image of South Florida Ecosystem, 1994
Urban area
Sugarcane
Sawgrass marsh and tree islands
Cypress swamp
Pine forest
Mangrove forest
Historic Everglades extent

most serious problem. The level of phosphorus in Lake Okeechobee and portions of the Everglades is now well above the natural tolerance of the ecosystem, throwing the biological community out of balance. For example, phosphorus levels have doubled in Lake Okeechobee in the last 20 years resulting from manure runoff from dairies and cattle ranches, causing repeated algal blooms and at least one significant fish kill in the 1980s (USACE 1998:III-21).

Phosphorus contamination of the Water Conservation Areas and Everglades National Park is just as worrisome as the situation in Lake Okeechobee, though the contamination comes from a slightly different source. Exposure of the peat soils in the Everglades Agricultural Area to air during cultivation naturally releases phosphorus as the soils oxidize. Phosphorus-enriched irrigation water pumped out of the Everglades Agricultural Area has already allowed cattails–which thrive under high-phosphorus conditions–to begin to displace the usually dominant sawgrass vegetation in some portions of the Water Conservation Areas. Scientists worry that too much phosphorus may next change the balance of plant and animal life in Everglades National Park (Armentano 1998; SFWMD 1998b:3-6).

LOST BIOLOGICAL DIVERSITY

Populations of many species of wildlife and fishes have dramatically declined as their food sources and nesting or spawning sites have degraded or disappeared. Disrupting the water cycle has also altered the seasonal pattern of flooding and drying on which the life cycles of many Everglades species depend. Sixty-eight species in the South Florida ecosystem are now listed by the U.S. Fish and Wildlife Service as endangered or threatened with extinction (SFERTF 1998a:3).

Populations of wading birds, including herons, egrets, storks, and spoonbills, have been particularly hard hit. Scientists estimate that in 1870, some 2 million wading birds crowded the marshes and estuaries of South Florida. By the 1970s that number had dropped to a few hundred thousand–about 10 percent of their historical level. The decline continues today (De Golia 1997:45).

The loss of biological diversity in the area is disturbing both from a conservation and an economic standpoint. Conservationists worldwide have recognized South Florida, and specifically Everglades National Park, for its biological richness. The Park is one of only three sites in the world to be designated a World Heritage Site, an International Biosphere Reserve, and a Ramsar Wetland of International Importance. The Park is also an important tourist destination, attracting 1 million visitors annually. If current patterns of damage continue in the Park, area officials have warned that the economic impact could be substantial. A government study calculated that if the recent declines in the health of Florida Bay at the southern end of the Park continue, economic losses could mount to more than $250 million/year in lost tourist dollars and reduced commercial catches of shrimp, lobster, snapper, and grouper (GCSSF 1995:Introduction p.2).

LOST NATIVE SPECIES

Exotic plant and animal species have invaded more than 3.7 Mha in South Florida and threaten to displace many of the native species, especially in Everglades National Park (SFERTF 1998a:3). Changes in the natural water cycle have fostered the spread of invasives such as *Melaleuca*, Brazilian pepper, and old world climbing fern, all of which thrive in dryer conditions (SFWMD 1998b:7). The system of canals,

Box 3.6 Indicators of Everglades Decline

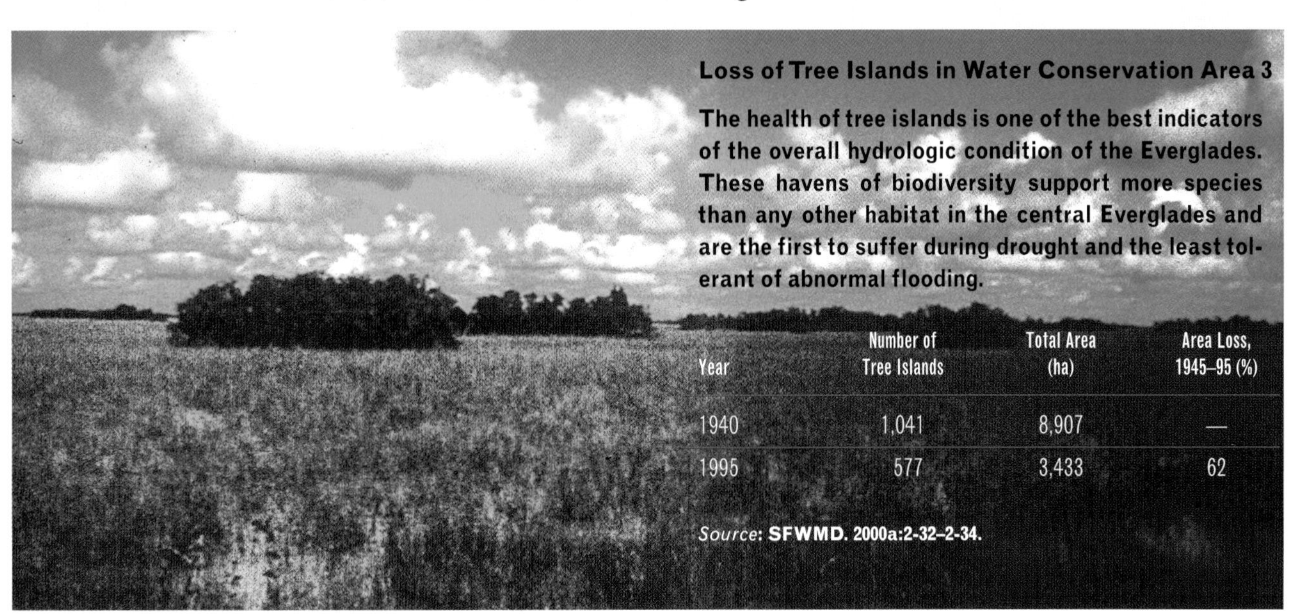

Loss of Tree Islands in Water Conservation Area 3

The health of tree islands is one of the best indicators of the overall hydrologic condition of the Everglades. These havens of biodiversity support more species than any other habitat in the central Everglades and are the first to suffer during drought and the least tolerant of abnormal flooding.

Year	Number of Tree Islands	Total Area (ha)	Area Loss, 1945–95 (%)
1940	1,041	8,907	—
1995	577	3,433	62

Source: **SFWMD. 2000a:2-32–2-34.**

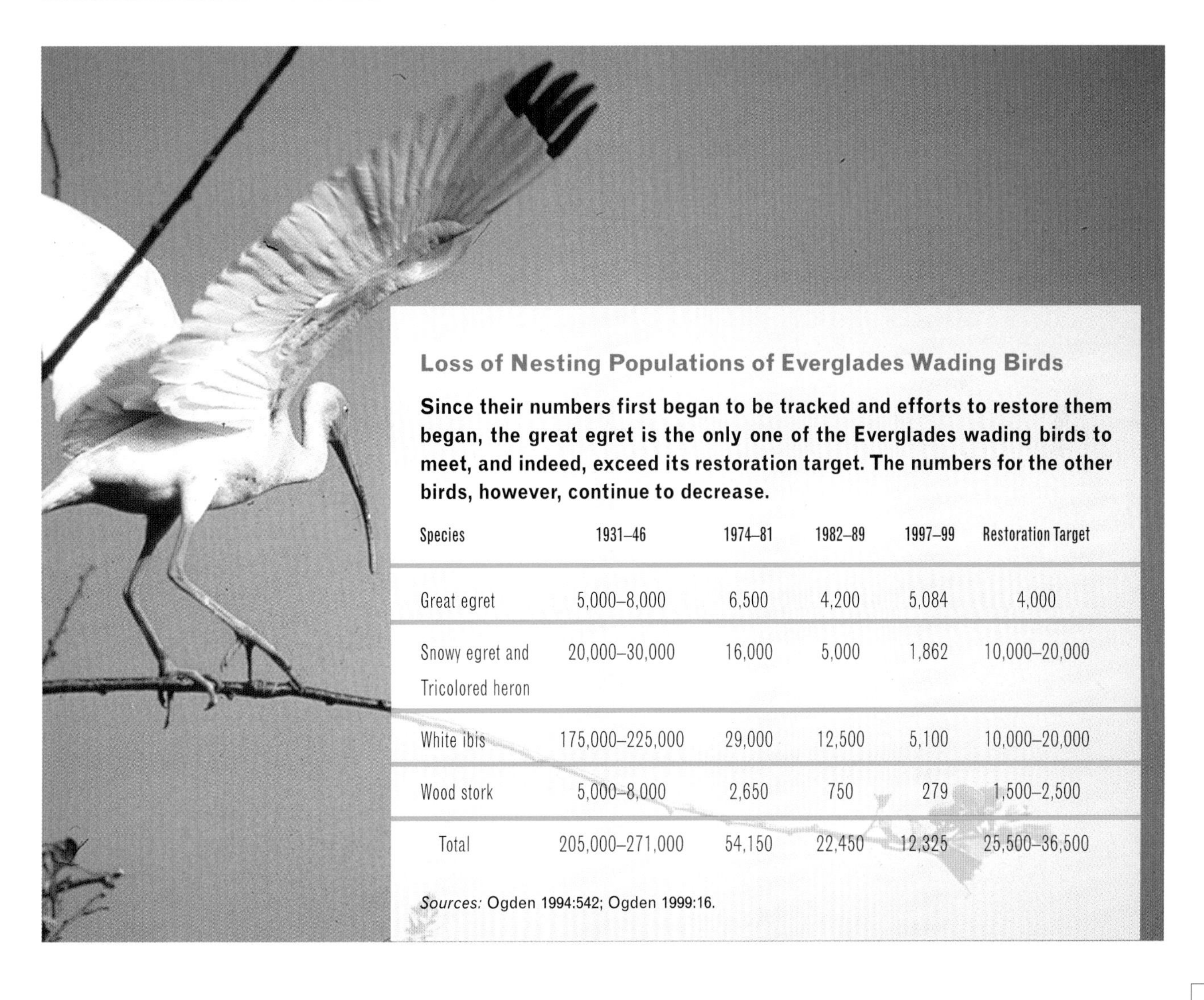

Loss of Nesting Populations of Everglades Wading Birds

Since their numbers first began to be tracked and efforts to restore them began, the great egret is the only one of the Everglades wading birds to meet, and indeed, exceed its restoration target. The numbers for the other birds, however, continue to decrease.

Species	1931–46	1974–81	1982–89	1997–99	Restoration Target
Great egret	5,000–8,000	6,500	4,200	5,084	4,000
Snowy egret and Tricolored heron	20,000–30,000	16,000	5,000	1,862	10,000–20,000
White ibis	175,000–225,000	29,000	12,500	5,100	10,000–20,000
Wood stork	5,000–8,000	2,650	750	279	1,500–2,500
Total	205,000–271,000	54,150	22,450	12,325	25,500–36,500

Sources: Ogden 1994:542; Ogden 1999:16.

Box 3.7 Restoration Means More Water and Clean Water

Currently, the C&SF project diverts much of the Everglades natural water flow for flood control. To prevent flooding, 3–4 times more water is released directly to the Atlantic Ocean than makes its way through the Everglades to Florida Bay. Water released to the Atlantic is lost for use by humans and wildlife. Restoration plans involve capturing some of this lost flow.

Restoration will also involve a major effort to remove phosphorus pollution from agricultural runoff by filtering it through 16,000 ha of artificial wetlands before releasing it into the Everglades. Filtering marshes reliably reduce phosphorus to 20 parts per billion (ppb) or less. Unfortunately, scientists believe that the ecosystem threshold where phosphorus begins to affect Everglades marshes is about 11 ppb, meaning an additional filtering step will be needed.

Discharge of Rainfall in the Everglades Region, 1980–89

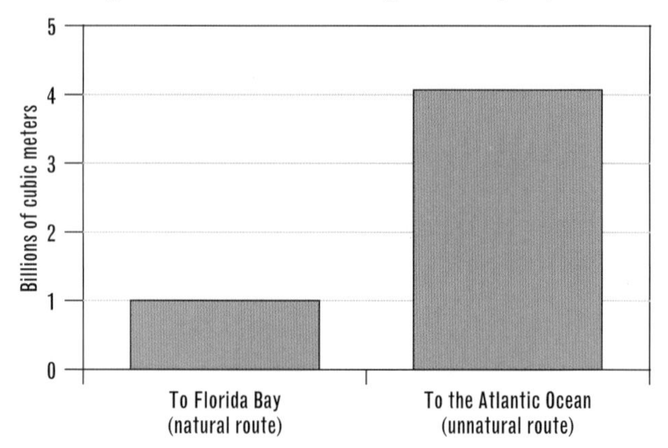

Source: Light and Dineen 1994:82.

Everglades Nutrient Removal Project

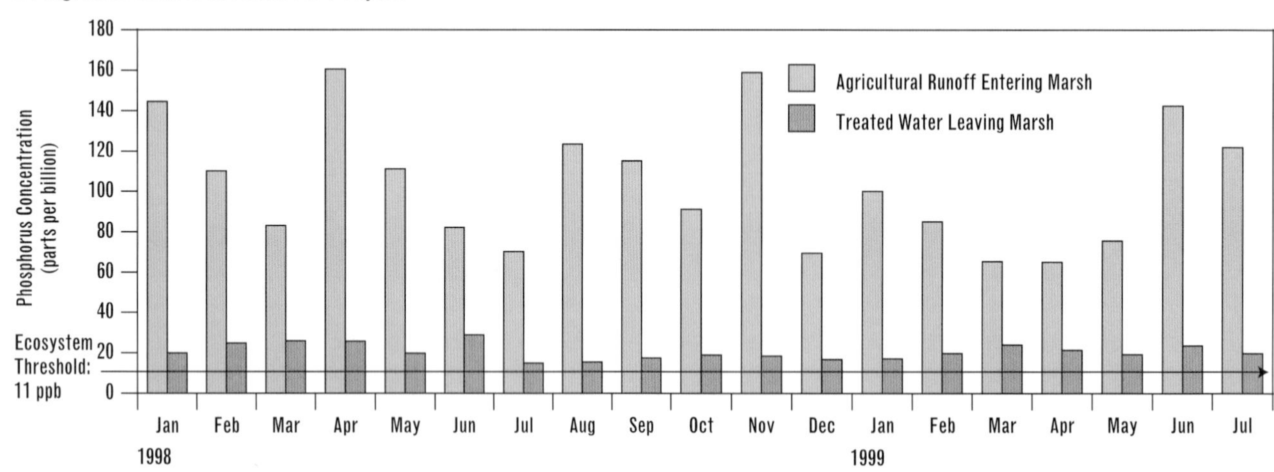

Source: SFWMD 2000b.

which provides unnatural routes into natural areas, has also been an important pathway for the spread of invasives such as the water hyacinth and the Asian swamp eel–a relatively new introduction whose voracious appetite may threaten native fishes (Armentano 1998; SFWMD 1998a:24).

A Change in Attitudes

The decline of key features of the ecosystem took time to be noticed, and even when environmental damage was obvious, a consensus on how to tackle the problem took years to evolve. But several key events and crises moved the process forward. As always, water–or lack of it–took center stage in convincing people that the alterations they had made in the natural system were anything but perfect.

From 1963 to 1965, C&SF water managers prevented water from flowing south into Everglades National Park in order to fill the newly constructed Water Conservation Areas. Drought conditions during those years meant the Park was water starved. Breeding colonies of ibis and egrets failed to form in their traditional spots for three consecutive years. Television cameras brought the Park's plight to a national audience and drove home the point that water conflicts were likely to become more common as water demand in the rapidly urbanizing area grew. The U.S. Congress subsequently ordered water managers to deliver adequate water to the Park, but the fight over how much was "adequate" would consume many more years and eventually direct the design of the restoration plan (Light et al. 1995:127, 129).

In 1970 drought struck again. The water shortages that plagued South Florida were so intense that state politicians took action, passing landmark legislation that mandated a regionwide approach to water management (Light et al. 1995:133). Governor Robert Graham launched the Save Our Everglades program in 1983–the first attempt to address the problems of the ecosystem at a regional scale, and the first public initiative to set out the goal of restoring the components of the ecosystem to something approaching their natural state (Light et al. 1995:142).

Rather than start to improve, conditions throughout the ecosystem continued to decline. In 1988, an ecological clarion call heralded the ecosystem's precarious health. Florida Bay is a shallow, tropical estuary at the southern tip of the Florida peninsula; a rapid die-off of seagrasses and a striking decline in water clarity occurred and continued for several years. Large, sustained algal blooms began to plague the waterway and commercial and sport fishing catches suffered (Armentano 1998; USACE 1998:III-23).

At about the same time, Dexter Lehtinen, a brash U.S. government attorney, filed suit against the regional water authority, the South Florida Water Management District, for releasing water polluted with agricultural runoff into the Everglades. The U.S. government suit–based on the water district's own studies–claimed that excess phosphorus from the Everglades Agricultural Area was threatening Everglades National Park and nearby Loxahatchie National Wildlife Refuge. The immediate intent of the suit was to force the District to require farmers to clean up their effluent before releasing it. But the larger effect of the suit was to highlight the inherent contradictions in the District's traditional service to the local agricultural community–to provide irrigation water and take away runoff–and its responsibility to provide clean water to Everglades National Park (Aumen 1998; Light et al. 1995:144-146).

At first the District fought the lawsuit; but in 1991, newly elected Governor Lawton Chiles directed the agency to admit that there was, indeed, a problem and begin to collaborate with federal authorities rather than continue to waste resources fighting the lawsuit. This began a process of redefining the Water District's mission to include stewardship of the South Florida ecosystem. The District eventually became a key promoter of the idea of ecosystem restoration (Aumen 1998).

In 1993, the federal government formed the South Florida Ecosystem Restoration Task Force, which has become a central player in developing a coherent restoration plan for the entire ecosystem. The Task Force has acted as the convening body to bring together all the parties with a legal interest in the restoration–a list that includes 10 federal and state agencies, several local county governments, the Miccosukee and Seminole Indian Tribes, and the South Florida Water Management District. Agribusiness interests, environmental groups, and sport and recreation groups also participate in the public hearings where decisions on restoration matters are made (SFERTF 1998a:7).

Just as significant, the state in 1994 created the Governor's Commission for a Sustainable South Florida, which has bluntly asserted that the problems with the South Florida ecosystem are intimately connected with the larger regional patterns of land and resource use and economic expansion. Without tackling these patterns, the Commission warns, restoration activities will not be effective in the long run (GCSSF 1995:Executive Summary p.1).

Restoring the Flow, Revitalizing the Ecosystem

What does restoring the South Florida ecosystem really mean? A decade of scientific study, debate, and negotiation has led to a broad consensus on what needs to be fixed and where to begin. Current plans already include 200 projects that restore habitat, manage urban growth, realign farming practices, and reconfigure the C&SF Project's water-control structures.

Three broad goals are behind these projects (SFERTF 1998a:1, 8–10):

- Restore the area's natural hydrological patterns as much as possible; the shorthand term for this is "getting the water right."

- Increase the health and extent of wildlife habitat so that depleted species can recover.

- Relieve pressure on the ecosystem by taming suburban growth and encouraging an economy that balances the needs of humans and the biological limits of the natural system.

GETTING THE WATER RIGHT

The first goal–restoring a more natural hydrological pattern–is the foundation on which all other aspects of ecosystem recovery are built. It forms the focus of the US$7.8 billion plan put forward in 1998 by the U.S. Army Corps of Engineers to revamp the C&SF Project. The basic strategy of this ambitious plan is to increase the capacity for storing water within the watershed. This will allow water managers to quit venting so much water directly to coastal estuaries from Lake Okeechobee during high water times and make it possible to direct water flows into the Everglades at the most appropriate times and in more sufficient quantities. It will also increase the water available for urban water supply and agriculture (SFERTF 1998a:8; USACE 1998:I-ix).

Computer models of the region's water flow predict that as population and industry continue to grow over the next 30 years, water shortages could occur, on average, every other year in most of the region's urban areas if the system is not reconfigured to store more water (USACE 1998:iv). This would strike hard at the area's economic stability and quality of life, and pit urban water users against farmers and both of these against the environment. Currently more than three times as much water is discharged directly to the coast than is allowed to continue its natural flow pattern through Everglades National Park and into Florida Bay (McPherson and Halley 1996:39). This water is essentially wasted for environmental and human purposes.

To create more storage in the system, the restoration plan calls for a combination of (a) new surface reservoirs, some created from existing rock quarries; (b) marshes; and (c) the use of an innovative technique of pumping water down wells into a shallow aquifer in the wet season for temporary storage and recovery during the dry season. These three elements will be combined into an interconnected system along the eastern flank of the Everglades that will also serve as a buffer against the encroachment of suburbs (USACE 1998:v–vi). In the Everglades Agricultural Area, converted cropland will also act as surface reservoirs. To implement this strategy, federal and state officials in 1999 bought a 259-km^2 tract of sugarcane fields that will be retired from production and eventually receive overflow flood waters (McClure 1999b). Elsewhere, advanced wastewater treatment plants will allow water managers to reuse wastewater to recharge coastal aquifers.

Restoration plans will also require that farmers discharge cleaner water into the Everglades. The legal settlement of the 1988 federal lawsuit against the water district directs farmers to use cultivation practices that reduce the phosphorus they release into their drainage water. At the same time, farmers in the Everglades Agricultural Area must pay one-third of the cost of constructing some 16,000 ha of special phosphorus-scrubbing marshes–the largest constructed wetlands in the world–through which they will send their drainage water before it goes into the Everglades. Ultimately, farmers will have to extract even more of the remaining phosphorus from their effluent in order to meet new water quality restrictions due to take effect in 2003. Researchers still haven't decided how this can be done at a reasonable cost (Aumen 1998).

Removing barriers to the sheetflow of water through the Water Conservation Areas and into Everglades National Park is also an essential part of restoring a more traditional hydrological pattern in the region. Current plans call for removing approximately 800 km of canals and levees within the Water Conservation Areas and revamping a portion of a major road that cuts through the Everglades; gates and culverts are to be installed along the road to restore the sheetflow interrupted by the road since its completion in 1928 (USACE 1998:vi).

RECOVERY OF WILDLIFE

Reconfiguring the C&SF Project to restore a more natural hydrological cycle should help with the second major restoration goal–improving habitat quality and recovering wildlife populations. The original system was huge and hydrologically interconnected. Animals could typically find appropriate food supply and breeding grounds somewhere within the system under a range of natural conditions. Draining and diking the watershed broke up the system's connectivity and disrupted the ability of many animals to find suitable habitats timed to their life cycle (USACE 1998:vii–viii).

By removing internal levees and allowing the delivery of more water, more appropriately timed and directed, water managers hope to recreate many conditions that favored wildlife. They anticipate that species at every level of the food chain–from small minnows and crayfish to alligators, herons, and otters–will start to recover their original population density and distribution. Water district biologists have particular hopes that wading bird populations will rebound; these birds are perhaps most sensitive to the habitat conditions over the entire watershed (USACE 1998:vi–ix).

But just how much and how fast the living elements of the ecosystem will recover is still very much in question. Scientists have drawn up biological criteria to judge whether the system is truly recovering; but there is still controversy and concern over what to expect, especially given its high price tag. Some critics feel that the recovery plan will not recreate the original hydrological pattern sufficiently to allow large-scale recovery and will yield far smaller benefits to wildlife than advertised (McClure 1999a; Santaniello 1998; Santaniello 1999; Stevens 1999). Even government biologists are cautious. They have labored hard to draw up an integrated strategy to ensure that the restoration plan benefits as many of the area's endangered species as possible, but do not expect all of the beleaguered species to survive.

CURBING DEVELOPMENT

Modifying development and economic activities in the Miami urban corridor so that they are less environmentally destructive is probably the most challenging of all restoration goals. Biologists and water planners know that without progress on this front, their efforts to restore the South Florida ecosystem will eventually be drowned in the flood of new development still surging into the Miami urban corridor. Each year, some 29,000 people relocate to the area to take advantage of the climate, natural beauty, and expanding economy (SFERTF 1998b:iii). By 2010, officials expect the region's population to expand to 8 million; by 2050, some forecasters think the population could nearly triple to more than 15 million (GCSSF 1995:Regional Overview p.1).

Plans to manage this expected influx include a number of steps to curb the proliferation of urban sprawl. A regional program called "Eastward Ho!" is encouraging local governments to establish urban development boundaries and to redirect new growth back into already developed areas by building on unused urban parcels, redeveloping run-down sites, and cleaning up brownfields. Modifying building regulations to require higher housing density in new suburban developments is a second essential step that restoration advocates are pressing on area governments. Upgrading the area's transportation system so that it encourages denser, less automobile-dependent development is also considered an important part of the overall effort to reduce the impact of future growth.

None of these steps will be easy; they involve land-use decisions by a large number of local governments whose land-use plans currently lack much regional coordination and are subject to intense local political pressure (GCSSF 1995:Executive Summary pp.1-7).

Beyond the Everglades

It is impossible to know yet whether the effort to rejuvenate the South Florida ecosystem will ultimately succeed. On one level, the Everglades restoration effort has made an impressive start and boasts a list of accomplishments and advantages that paint a hopeful picture: it enjoys widespread popular and political support that comes from a basic understanding of the current state of the system, its vulnerability to further decline, and an acceptance of the tenet that some minimum of ecosystem health is required to support the local economy and the quality of life that people enjoy. That alone is a tremendous step forward. But the difficulty of actually bringing back healthy populations of wading birds, returning full productivity to Florida Bay, or recovering even one of the 68 endangered species whose survival hangs in the balance cannot be underestimated.

Yet regardless of the outcome, the Everglades effort has already offered many lessons. First, it shows how vulnerable ecosystems are to single-purpose management, especially when managers are ignorant of the basic workings of the ecosystem. Without knowledge of how changes in area hydrology were likely to affect the South Florida ecosystem, it was impossible for the Army Corps of Engineers to foresee the trade-offs they were making when they built the C&SF Project. And even if they had had such knowledge, it was probably outside their mandate to act on it, given their primary goals of flood control and improved water supply.

The Everglades experience also provides a thoroughly convincing economic argument for taking care to not degrade a critical ecosystem in the first place. The $7.8 billion price tag for what is just the first stage of the overall restoration effort leaves no doubt that large-scale ecosystem restoration requires a huge investment–often many times the expense of altering the system in the first place. Still, this may be inexpensive compared to the benefits that will be lost if the ecosystem continues to degrade or fails completely. The tourist trade alone is worth $14 billion annually to the South Florida economy and the ecosystem's health is directly tied to that industry's overall success.

Perhaps the most important lesson is that the idea of ecosystem restoration is extremely compelling. The public's and politicians' acceptance of a restoration program of such magnitude and expense shows that a well-articulated vision of a restored ecosystem can be a potent force for consensus and change. At the same time, the Everglades experience leaves no doubt that following through on this vision requires patience and commitment. It takes time to learn how and why an ecosystem is failing and how to put it right again; time to negotiate the inevitable controversies about how best to spend the precious dollars available to attain maximum recovery. Efforts to restore the Everglades have taken nearly 3 decades to mature to their present state, and it will undoubtedly require much longer than 3 more decades for the Everglades to heal.

Ultimately, even attaining some level of ecosystem recovery will not be enough. Keeping the restored ecosystem from failing again will be the ultimate test and will require making good on the much more ambitious vision of a regional economy that does not, through its impacts, smother the renewed life so carefully nurtured.

Managing Mankòtè Mangrove

Some people call mangroves "the roots of the sea." Mangroves are gnarled, salt-tolerant trees that grow in intertidal zones and estuaries where the ocean, land, and freshwater meet; they cling to the loose soils, sands, and muds with a maze of roots that can withstand waves and erosion. These unique, adaptable plants, of which there are about 60 species, are found along the majority of the world's subtropical and tropical coastlines.

Some coastal residents might also call mangroves "the roots of their community." The forests, swamps, and wetlands where mangroves thrive are ecosystems of great biodiversity and productivity. Coastal residents use mangroves for fuel, construction materials, food, medicines, and tannins. For fishers the mangroves' networks of roots provide breeding grounds for many kinds of sea life. The leaves, small branches, propagules, and fruit that fall from the trees contribute to production of detritus that supply the fish and other wildlife with an abundant food supply. Mangroves are also prime nesting and migratory sites for hundreds of bird species. By serving as a buffer along the coastline, mangrove forests protect coastal areas, crops, and towns from flooding during storms, shelter fishers' boats, and protect coral reefs from suspended solids. Plus, mangroves control sedimentation and coastal erosion.

But a mangrove's natural resilience and value affords it little protection against a growing number of anthropogenic threats, as communities and institutions on St. Lucia's southeast coast came to understand in the 1980s. That realization inspired an innovative program to enable local residents to reap the benefits of Mankòtè, St. Lucia's largest mangrove forest, without degrading its ecosystem services and long-term viability.

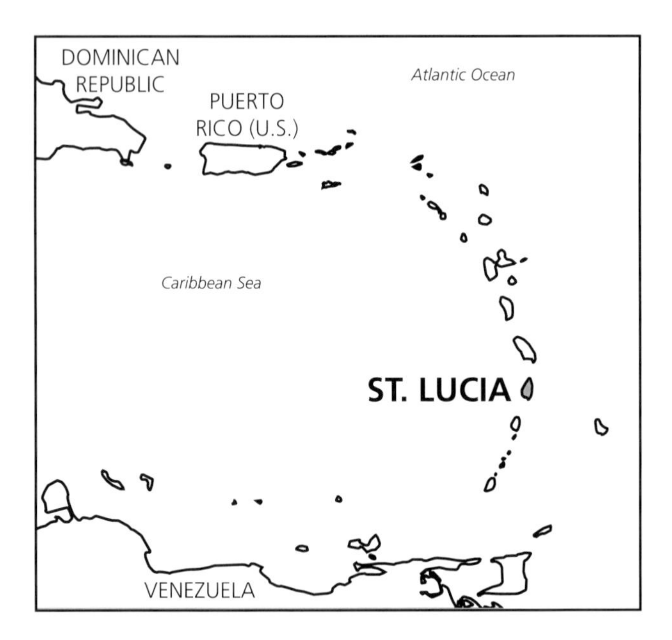

Changing Community Practices

Mankòtè was part of a U.S. military base during World War II. When the base closed and the area became public land in 1960, the 63-ha mangrove–20 percent of the total mangrove area of the country–was still covered with well-developed trees (Geoghegan and Smith 1998:1). As an open-access resource, it was soon subjected to varied and often destructive uses ranging from seasonal fishing, bird hunting, and crab catching to waste dumping and spraying of pesticides for mosquito eradication (Smith and Berkes 1993:123–124).

The greatest stress on the mangrove, however, was the extensive tree cutting by local citizens for commercial charcoal production. By the early 1980s, charcoal production had become a major source of subsistence income and an important cottage industry. The use of mangrove wood for charcoal is popular because it is cheap relative to petroleum-based fuels, can be easily transported, and is slow burning. Mankòtè became the main supply of charcoal for about 15,000 residents of Vieux Fort, a nearby community, and others in the southeast portion of the island. Although no data are available, older residents of the area observed that during those years, smaller trees in the mangrove were being harvested and large trees were becoming scarce (Smith 2000).

At about the same time, a regional NGO, the Caribbean Natural Resources Institute (CANARI), identified the Mankòtè mangrove as a priority area for conservation. CANARI soon realized that the charcoal producers themselves were key to Mankòtè's protection. Although charcoal producers' harvests were putting pressure on Mankòtè, they practiced a number of sound management measures. For example, they cut on a rotational basis, allowing time for the trees to regenerate before recutting, and left uncut the species of mangroves that make poor charcoal but provide cover to impede the evaporation of the swamp.

CANARI proposed a management strategy that was innovative and controversial for its time. They advocated that the mangrove be managed in collaboration with the harvesters–a landless, poor group with no legal right to the resource, but also the people most dependent on the mangrove and most damaging to it. With the government's tacit approval, CANARI launched what has become an ongoing effort to test ways to save the mangrove and maintain the charcoal producers' incomes (Geoghegan and Smith 1998:4, 7).

Among CANARI's key steps was to organize the harvesters into an informal cooperative of about 15 people; the cooperative is called the Aupicon Charcoal and Agricultural Produc-

The woodlot, as originally conceived, was to be managed by and benefit the group as a whole. Members would be organized for harvests and other activities. Similarly, pole production in the mangrove was meant to be a group activity. However, it has proven easier for people to continue using the mangrove and the woodlot without strict coordination of activities. Extractions are made by individuals or small teams and recorded each month.

ers Group (ACAPG). CANARI works with the group to monitor and track trends in charcoal production and the status of the mangroves. ACAPG committed to a set of sustainable harvesting practices, including a ban on cutting trees that line waterways, preservation of large trees, and cutting on a slant to preserve the tree's stump.

To reduce pressures on the mangrove, government agencies, local NGOs, and the harvesters sought to create a new wood supply for charcoal production. Between 1983 and 1985, the Department of Forest and Lands planted a 62-ha woodlot close to Mankòtè with fast-growing hardwoods, mainly *Leucaena*, and with a palm species that ACAPG members can harvest to make brooms. The government also loaned the producers a large plot of land and encouraged the producers to plant it with marketable products.

There have been significant communal harvests of plantation wood recently, although initial efforts in plantation and agricultural endeavors were plagued with problems, from fires to the charcoal producers' inexperience with agriculture, marketing, and working together. The woodlot is still far from a replacement for mangroves, but management strategies and income-diversifying opportunities continue to evolve. For example, in 1993 the harvesters began leading tourists and school groups on tours of the mangrove as an income-generating opportunity. Local NGOs have provided guide training; technical assistance grants to build interpretive signs, a boardwalk, and a viewing tower; and assisted with tour promotion and organization (Smith 2000; Brown 1996).

To limit outside threats to the mangrove, local institutions successfully protested the Department of Health's mosquito eradication program that was damaging the mangrove's fauna and hydrological functions, and secured the designation of Mankòtè as a marine reserve in 1986. That designation affords the mangrove complete protection from any extractive use without written permission of the Chief Fisheries Officer, ending years of illegal waste dumping. The charcoal producers have sole rights of use of timber resources (Smith 1999).

Like most participatory approaches to ecosystem management, the Mankòtè strategy has taken more than a decade to achieve many of its objectives. By the 1980s, the overall trend of degradation of the tree cover had been reversed. Monitoring four species of trees in each of four transects between 1986 and 1992 showed a significant increase in the number of mangrove stems larger than 25 mm/m^2–from 0.10 to almost 2 (Smith and Berkes 1993:126-127). The basal area, or total area of stems, increased fourfold. Because 1991 was a year of particularly high charcoal production, the increased regeneration of mangroves noted in the 1992 survey is especially noteworthy. Field observations and interviews indicate that preservation methods are still used rather than clear-cutting (Smith and Berkes 1993:126-127). Although the data are still limited, research in the last several years suggests that density and size of trees have continued to increase, while charcoal production has averaged 2 tons/month in early 2000, slightly less than the average in the past 15 years (Smith 2000).

Mankòtè's future is still uncertain. An economic downturn in St. Lucia could bring new pressures to the mangrove. The government continuously receives proposals for the development of the mangrove and surrounding land; fortunately, key agencies are concerned about identifying what kind of development would be possible without encroaching on the mangrove and its functions. Research is under way to ascertain other potentially significant pressures on the mangrove, including the impacts of crab hunting and fishing, and to test the effectiveness of some silviculture practices in the mangrove, with the hope of improving yields from regeneration. Nevertheless, there is agreement among all parties that the informal, collaborative arrangement at Mankòtè currently provides greater protection to the mangrove than any government agency or other institution can do on its own. The arrangement has also allowed rural families to continue to reap economic benefits.

Bolinao Rallies Around Its Reef

With its cascading waterfalls, rolling hills, white beaches, and spectacular sunsets, Bolinao has been called nature's masterpiece. But the most valuable asset in this northern Philippines municipality may be its 200 km^2 of coral reefs. About one-third of Bolinao's 30 villages and 50,000 people depend on fishing to make a living (McManus et al. 1992:43), and the Bolinao-Anda coral reef complex serves as the spawning ground for 90 percent of Bolinao's fish catch. More than 350 species of vertebrates, invertebrates, and plants are harvested from the reef and appear in Bolinao's markets each year (Maragos et al. 1996:89).

Imagine, then, the dismay among local residents, marine researchers, and NGOs who learned in 1993 that an international consortium intended to build what was claimed to be the world's largest cement factory right on Bolinao's coral reef-covered shoreline. The cement industry ranks among the three biggest polluters in the Philippines (Surbano 1998), and the plans for the Bolinao complex included a quarry, power plant, and wharf. It can take 3,500 pounds of raw materials to produce 1 ton of finished cement; pollutants commonly emitted from this energy-intensive industry include carbon dioxide, sulphur dioxide, nitrous oxide, and dust–about 360 pounds of particulates per ton of cement produced. Another by-product is highly alkaline water that is toxic to fish and other aquatic life (Environmental Building News 1993).

The ensuing debate over the plant's construction brought a new urgency and focus to local efforts to ensure the long-term viability of Bolinao's coastal resources. Pitted against a politically and economically powerful business consortium, residents successfully challenged the idea that a cement plant's short-term economic benefits would offset the risk of long-term ecosystem ruin. That outcome is an unusual and significant achievement, particularly in developing countries, where citizen advocacy and broad-based participation in natural resource management is likely to face daunting obstacles, including limited access to both environmental information and the political process.

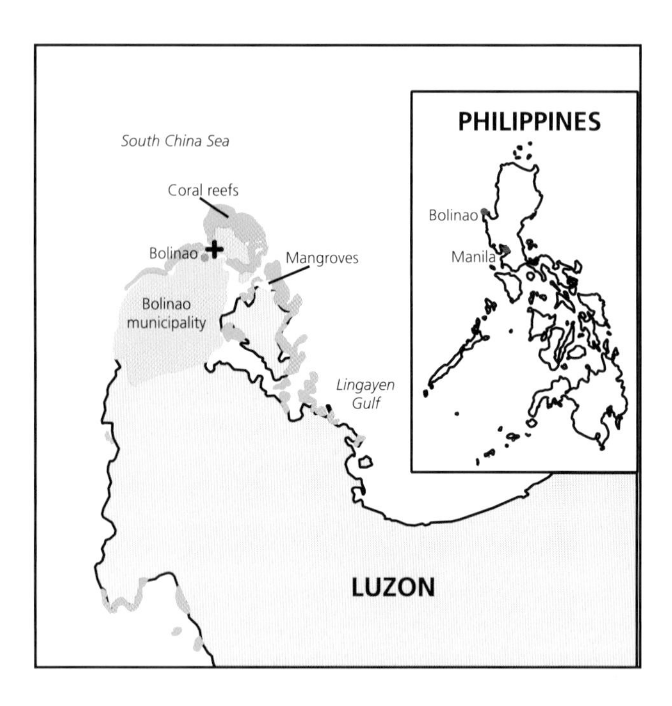

Bolinao's Threatened Marine Ecosystem

Bolinao's environmental fragility had been recognized, in some quarters, long before a Taiwanese business group called Tuntex announced its plans to build a mammoth cement complex. A 1986 study by the Marine Science Institute at the University of the Philippines, for example, documented significant damage to Bolinao's coral reef system. Researchers found that about 60 percent of the region's corals had been killed, mostly through destructive fishing practices that relied on dynamite and cyanide to enhance catches (McManus et al. 1992:44). In 1992, Bolinao's once-booming sea urchin industry was shut down indefinitely after the urchins had been exploited nearly to extinction to satisfy export demand for roe (Talaue-McManus and Kesner 1995:229). Fishers, fish vendors, and shell craftspeople had noted diminished catches, changes in dominant species, and decreases in the size of mature fish.

But it took the possibility that a cement factory would cause further deterioration of the area's marine resources to galvanize widespread action on behalf of the ecosystem. "We launched a vigorous education campaign focused on the cement plant's potential environmental impacts," explains Liana Talaue-McManus, a researcher from the Marine Science Institute (Talaue-McManus 1999). For many, this was the first time that they fully understood the extent and richness of their community's natural resources, as well as its vulnerability.

The plant complex would be located in the middle of the reef system, within 3 km of the municipal center. This was an ideal spot from investors' perspectives, given its abundance of limestone, the deep channel for marine transport, and Bolinao's proximity to Taiwan. Investors argued that the cement production complex would not cause any pollution, but local residents soon began to suspect otherwise.

With support from the University of Philippine's researchers, a local NGO–the Movement of Bolinao Concerned Citizens–challenged the Tuntex consortium. They played a critical role in the 2-year struggle against the cement

plant, rallying opposition and raising awareness of the complex's potential impacts. Those impacts, as their research revealed, could include air pollution, erosion from the quarrying of limestone, damage to the reefs from the widening of the shipping channel, oil pollution from shipping, and the threat to their limited freshwater supply.

Their efforts were rewarded. In August 1996, the Philippines Department of Environment and Natural Resources (DENR) denied "with finality" the application for an environmental permit, citing the unacceptable environmental risks the cement plant would pose to aquatic life and coral reefs, and the conflicts that would arise with existing land and marine uses (Ramos 1996).

Crafting a Long-Term Management Plan

The hard work of ecosystem protection didn't end with the cement plant fight. In fact, for Bolinao residents and NGOs, the toughest part of ecosystem management was ahead. Local NGOs are still working toward a larger goal: developing a coastal resource management plan that empowers fishers and other community members to participate in long-term decisions about the management and health of their resources.

Consensus on how to conserve and protect the marine areas has long been elusive. Since the early 1990s, a coastal planning team composed of representatives from the Haribon Foundation and from the Marine Science Institute and College of Social Work and Development (both at the University of the Philippines) sought to mobilize Bolinao's villages on behalf of marine protection. But many issues polarized the community:

- Most of Bolinao's fish harvesters are poor, with the reefs serving as their sole source of food and income. As farmlands deteriorated, many farmers migrated to reef areas, exacerbating competition for marine resources. Increased population in the coastal areas increased the amount of organic pollution; the pollution, in turn, reduced the resilience of Bolinao's coral reef ecosystems. Because of poverty, resource depletion, tradition, and lax enforcement of bans, fishing methods known to be destructive were sometimes still used.

- The town leadership lacked adequate information about the marine ecosystem and needed technical assistance to make sound resource decisions.

- Access to milkfish fry and siganid fishing in Bolinao was governed by an inequitable but ingrained system. Those who won concessions from the local government–through a sometimes corrupt bidding process–garnered exclusive privileges to fish in an area. Subsistence fishers were banned from the area or forced to sell their catch to the concession holders at below-market prices. The result was illegal fishing and minimal incentive to regulate the harvest, but significant income for the local government.

- One survey found that the number of aquaculture pens in the Caquiputan Channel between the Bolinao mainland and the islands of Santiago and Anda had increased from 330 in December 1996 to 3,100 in July 1997 (Talaue-McManus et al. 1999). Although they produced revenues for the town's political and economic elite, they reduced fishing grounds and navigation areas, causing water quality declines and fish kills.

- Resort owners wanted the shorefront left open and free of activity, while subsistence and deep sea fishers needed navigation and docking areas.

The challenge of finding a balance between these actors and between the different uses of the coastal resources made it all the more impressive when, in 1997, NGOs successfully crafted "a collective vision for the long-term viability of Bolinao's coastal living resources" (Talaue-McManus et al. 1999). This coastal development plan drew on more than 2 decades of scientific research by investigators from the Marine Science Institute and was drafted by 21 representatives of the municipal government, the religious sector, members of the fishing industry, ferryboat operators, and environmental advocates through community workshops and meetings.

The plan divides the municipal waters of Bolinao into four zones with different use designations–"reef fishing," "ecotourism," "multiple use" (which includes milkfish pens and fish cages), and "trade and navigation." One zone includes a marine protected area. The next steps were to determine exactly what activities were to be allowed or prohibited in each zone, to ensure that the marine protected area remains truly protected, and, of course, to implement the plan. Implementation is still under way.

Most of those involved agree that local input has been a hallmark of Bolinao's ecosystem management process. They credit the participatory process with winning much greater public acceptance for Bolinao's coastal development plan than a traditional plan could have secured; most often, plans are drawn up quickly by outside consultants with little or no local input. Plus, by including direct resource users–subsistence fishers and fish vendors as well as the local government–in the zoning process, there is a greater chance of achieving conservation goals. Local stakeholders are, after all, the people who will ultimately either respect the new rules and regulations or ignore and evade them. An ongoing research program, such as that conducted by the Marine Science Institute, is an important complement to the planning effort. It serves as a source of knowledge and data that public representatives can draw on to make informed decisions.

Perhaps the best news is that Bolinao is part of a growing number of communities, organizations, and sectors of government in the Philippines that are using a "bottom up" rather than "top down" approach to natural resource management, building on a long tradition of strong citizen advocacy. And although Bolinao's coastal development plan is still very much a work in progress, one thing appears certain: more and more people will get involved as the plan is implemented. As word has spread in the Philippines about the Bolinao experience, other municipalities have turned to the University of the Philippines-Haribon team. They seek help in formulating their own coastal development plans, offering the promise of more research and monitoring on the status of coral reef ecosystems, and generating new strategies and models for reef protection and new management abilities within local communities.

FOREST ECOSYSTEMS

Up from the Roots: Regenerating Dhani Forest through Community Action

Dhani Forest has reincarnated itself from the roots up. The stubbled, degraded slopes of a decade ago have regenerated more rapidly than many thought possible. Protected from uncontrolled grazing and harvest, root stumps have sprouted new branches, grasses have flourished, streams have recharged, and wildlife have returned. So, too, have the livelihoods of local villagers who traditionally made their living harvesting forest products, such as fuelwood and siali leaves used in making leaf plates. Under the supervision of a committee of local villagers, limited harvesting of forest products has resumed, steadily increasing the flow of benefits from Dhani to the five communities that flank the forest.

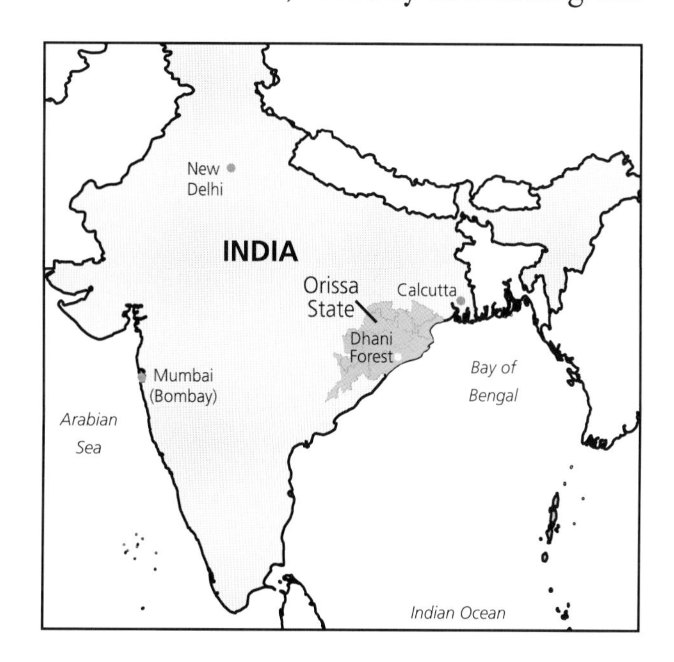

The rebirth of this mixed deciduous forest in the state of Orissa in India marks a new approach to managing the State's depleted forests–one that returns limited control to local communities. In fact, the State has had little to do directly with the forest's regeneration. The five villages surrounding the forest initiated the restoration effort. They crafted a detailed plan to regulate forest use, to carefully husband what remained of the forest and enhance it where they could, to distribute the forest benefits fairly, to educate their children in forest conservation, and to resolve disputes arising from their plan. They nursed the forest back to health because it had stopped giving them what they needed. In doing so, they became leaders in a trend toward community forest management that has spread across Orissa State and all of India.

(continues on p. 184)

Box 3.8 Overview: Dhani Forest

Twenty years ago, Dhani Forest in Orissa State was badly degraded. Commercial harvesters had removed much of the forest canopy; local residents had cleared slopes for crops, gathered fuelwood relentlessly, and allowed cattle to graze the forest floor heavily. Today, this mixed deciduous forest is reborn, thanks to a five-village effort to ensure its survival. These villages have become leaders in a trend toward community forest management that is spreading across India.

Ecosystem Issues

Forests

The 2,200 ha Dhani Forest is a primary source of food, fuel, building materials, fibers, and medicines for local people. Their dependence makes Dhani both extremely vulnerable to overuse and critical to protect.

Agriculture

At various times, villagers have cleared lower slopes of the forest to expand agricultural areas and feed their families. Clearing forest, however, decreased their supplies of leaves that serve as farm fertilizer and food and other resources that cushion the effects of drought and crop failure.

Freshwater

Local stream flows and water tables are vulnerable to changes in Dhani's forest cover and soils. Diminished water flows, in turn, affect the health of soils and crops in adjacent agroecosystems.

Management Challenges

Equity and Tenurial Rights

Today, villagers' rights to manage and use part of Dhani Forest's output is legally recognized—a far cry from the 1950s when the Orissa Forest Department ignored villagers' use rights and granted permits to contractors to harvest timber there. Yet some people argue that the State still does not treat the villages' forest protection committee as an equal, and some believe that the State should completely surrender title to Dhani Forest.

Economics

Dhani Forest's renewed health is essential to both local subsistence and local market economies. The State also reaps economic benefits; local management has lowered its forest protection expenses and is creating an asset from land that might otherwise be unproductive.

Stakeholders

Dhani's restoration and protection require collective decision making among the five villages who crafted the forest's protection plan, plus the cooperation of other neighboring villages who might infringe on this open-access forest. Restoration also depends on the State's willingness to respect community management and the value of nontimber ecosystem goods and services.

Information and Monitoring

Dhani Forest's successful restoration has largely depended on folk knowledge, wisdom, and commitment; the same is true of many similar projects in India. Orissa State has contributed some technical expertise, but more scientific analysis to complement local management is needed—guidance and research that are beyond the resources of the Dhani community.

Timeline

Pre-1799 Most forests in India are managed sustainably at the community level.

1799 British rule of India introduces commercial timber production and soon exhausts many forests.

1865 The British colonial government asserts state monopoly over forests with the Indian Forest Act.

1878 Purview of the Indian Forest Act is expanded and local control is further diminished. Dhani Forest remains under the control of Orissa's Raja until 1947 and is generally well managed.

1914–18 World War I massively increases demand for Indian timber.

1920s Railway lines reach Orissa, providing easier commercial access to Orissa's forests.

1940–45 India serves as the sole supplier of timber to Allied forces in the Middle East and Persian Gulf during World War II; forests are also under siege for fuelwood to offset the loss of coal to the war effort.

1947 Indian independence and state socialism put an emphasis on industrialization and use of forests for timber production and commerce rather than local use.

1940–50s Population in villages near Dhani begins to increase notably, intensifying pressure on the forest.

1950s Land Reform Bill declares forests on the boundary of a village to be village forests. Villages begin protecting and regenerating these tracts. National Forest Policy reinforces the state's exclusive control over forest protection, production, and management.

Late 1950s Tribal groups mount a sustained challenge to the continual denial of their rights to use forests.

1960 Orissa's Forest Department takes control of Dhani Forest and begins to permit commercial timber harvests; traditional conservation and community management systems decline.

1971 Beginnings of Joint Forest Management in Arabari in West Bengal and other districts.

1979 State permits a second major timber harvest in Dhani Forest.

1987 The villages closest to Dhani form a forest protection and management system to protect about one-third of the forest.

1988 Orissa becomes the first state to formally recognize local forest protection committees like Dhani's.

1991 Several other villages begin protecting another section of Dhani Forest.

1993 Orissa enters into a Joint Forest Management agreement with the villages surrounding Dhani Forest.

1997 Orissa awards the Dhani villages the *Prakriti Mitra* (Nature's Friend) award.

1998 Dhani Forest's canopy has filled out and the forest supplies increased goods and services.

1999 A cyclone severely damages Dhani Forest and the livelihoods of forest-dependent groups.

2000 A total of 400,000 ha is now under the protection and management of some 10,000 local villages throughout Orissa. The Dhani villages are active in the local federation of forest-protecting villages.

From Restricted Use to Overuse

Traditionally, local village folk did not own or manage the 2,200 ha of Dhani Forest. Nonetheless, they accrued many of the forest's benefits to augment their subsistence through a well-regulated system of forest harvesting.

Until Indian independence in 1947, the Dhani Forest lay within the domain of the Raja of Ranpur, one of 30 feudal states in Orissa that maintained a semi-independent status during the British colonial period. In Ranpur, as in other nearby feudal states, the Raja, or king, regulated access to forests and all forest products. During British rule, the Raja acted like a landlord, paying taxes on the forest estate to the colonial government. Some forests were essentially off-limits to local use. In others, villagers were permitted to meet their needs for timber and other forest products in exchange for modest royalty payments to the Raja or in exchange for free labor. Sometimes special considerations were given to the poor and to local tribal peoples with particularly high dependence on the forest.

After obtaining the required permit, villagers could gather a variety of products for personal use, from bamboo and wood for housing and agricultural tools, to fruits, fibers, leaves, and flowers. The forest rules banned cutting of selected "reserved" trees, and it was forbidden to sell or export trees without a permit from the ruler. The royal family also retained the privilege of hunting all wildlife within the forest.

The Raja maintained a separate administration of rangers, foresters, and guards to manage the "reserve forests," as forests like Dhani were known. The rangers strictly enforced the forest rules, both to prevent overuse by locals and to capture any commercial revenues from timber sales. Even without free access, villagers faced no shortage of forest products. During the Raja's tenure, the picture was one of a generally healthy forest with an abundance of resources.

In the early 1950s this picture began to change. Population was increasing rapidly, and agricultural land to meet local food requirements came into greater demand. Villagers cleared some of the forests on the lower slopes for planting using traditional swidden cultivation methods. More important, the era of the Raja's strict control had ended and the states of the newly independent India struggled to forge a "modern" forest policy–one that favored commercial uses of timber over meeting local needs. In 1960, the State Forest Department, which now controlled Dhani Forest, began permitting commercial contractors to harvest timber and remove much of the canopy in Dhani's low-lying areas. Villagers pressed some of the cut areas into crop production, and the State tried to establish teak plantations in other sections.

Over the next 2 decades, commercial cutting continued and local use intensified. Village cattle grazed the forest floor intensively and villagers gathered fuelwood relentlessly. Some came from more distant villages where forests were already exhausted. Sometimes even rootstocks were extracted for sale. Illegal timber cutters also took from the forest, smuggling out timber to meet growing urban lumber demands.

In 1979, the State allowed a second major timber harvest that left the forest devoid of large trees. Alarmed by the access given to outsiders, local villagers accelerated their own timber cutting in a rush to claim some of the forest goods and associated income for themselves. By the mid-1980s, the whole of Dhani Forest was degraded, much of it badly.

A Time for Action

The degradation of Dhani Forest had far-reaching impacts on the lives of local people. Materials from the forest on which they had always depended fell into short supply. People had to traverse long distances to collect fuelwood and to obtain small amounts of timber for house construction and farm tools. Firewood for traditional cremations dwindled. Fruits, tubers, herbs, and leafy vegetables that had long augmented food supplies during lean times gradually disappeared. The lack of forest productivity removed the cushion that the forest had always provided during dry periods and crop failures.

With the forest canopy removed, the forest soils dried out, reducing stream flows and decreasing local water tables. Because agriculture is the main occupation in the surrounding villages, soil moisture and water availability were prime concerns. Soil erosion also became a problem, affecting fertility in some neighboring fields. Loss of forest canopy also meant loss of the leaves and other sources of "green manure" that farmers had depended on for fertilizer.

Dhani Forest's worsening condition struck directly at the local economy, too. Without sales of products collected from the forest, many villagers had no source of cash. Selling fuelwood was the primary commercial activity, but the sale of leaves from kendu trees and siali vines was also important, particularly for women and poorer families. Approximately 50 Harijan families (the lowest castes and those with little land and high daily use of forest products) depend on the income from siali leaf collection in Dhani Forest. During peak season after the rains, one person working all day can collect as many as 3,000 leaves, which can then be stitched together into leaf plates or sold in bulk in Chandpur, the nearest town. Mats woven from date palm leaves were also sold locally; tubers like tunga, karba, and pichuli, as well as medicinal plants and vines, brought substantial local income. As these products dwindled, the pressure to migrate out of the nearby villages to urban areas for wage labor increased.

By the mid-1980s, villagers were convinced that Dhani Forest's poor condition was a serious community matter. They had begun to realize that it was they who were losing the most–not the private logging contractors or the State Forest

Department. It also disturbed them that future generations would inherit a depleted ecosystem. In early 1987, a respected village elder, Kanduri Pradhan, organized a meeting among the five villages that lay closest to Dhani Forest–Barapalli, Arjunpur, Panaspur, Balarampur, and Kiyapella. In ensuing meetings, a group of residents from all five villages discussed their options for collectively protecting Dhani Forest. A few villages in the Ranpur area had already begun to protect their forests, and this encouraged the group to commit to a joint program of action to guard and manage more than one-third, or 840 ha of Dhani Forest.

The decision to jointly manage Dhani Forest was a significant social and political event for the villages. Close cultural ties already linked the villages–they shared the observance of some local festivals, for instance, and a common school. Prior to their decision to protect the forest, they had formed an inter-village committee to coordinate collective activities. Yet they were also socially diverse, comprised of an assortment of tribal peoples and Hindu castes, including Brahmins (the most influential caste), Khandayats (farmers), and Harijans (the least powerful castes). Each of these groups lived in its own enclaves. Indigenous tribal people, the Saora and the Kandha tribes, populated Kiyapella and Panaspur villages. Balarampur village had a mixed tribal and Harijan community. In Barapalli and Arjunpur villages, Khandayats and Brahmins dominated. Dependence on the forest, however, linked them all, and village representatives realized that any hope of real forest protection lay in joint action.

A Plan for Life

By September 1987, the five villages had formalized their commitment to protect Dhani Forest. They formed a forest protection committee called the Dhani Panch Mauza Jungle Surakhya Committee. Out of lengthy discussions on the causes of the forest's poor condition and the possible ways to relieve pressures on the forest came a plan to restrict human uses of the forest.

From the beginning, the effort to protect and rejuvenate Dhani Forest was a true community affair. The elders of all households in each of the villages sat on the general body of the forest protection committee, which made all policy and budgetary decisions. A smaller executive committee included two members from each village to help implement the general committee's decisions. Community members were also required to take turns serving on the 25-person patrol squad that kept a daily vigil at the forest, restricting public access and preventing further degradation.

At first, the protection plan was simple: keep people and cattle out except for very restricted uses. Gradually, as the community's experience with protection evolved, so did the protection plan. The forest protection committee drew up an elaborate set of regulations and a schedule of fines. Cutting a valuable timber species like teak, for example, drew a fine of 1,001 rupees–a stiff penalty in the context of local incomes. In essence, the committee forbade any unsupervised cutting or collection of forest materials and set strict limits on those goods that could be harvested. The committee banned anyone entering the forest from carrying an ax or other sharp implement that could be used to cut woody material. It also banned grazing during the rainy season (July-September) to encourage regrowth of ground vegetation and restricted human access during the summer months to prevent fires. To help restore the lower slopes of the forest, the committee negotiated with local farmers to end the practice of periodically cultivating these areas.

It did not take long for Dhani Forest to rebound. Although they had lost much of their foliage, many of the trees and shrubs still had intact root systems and a number of these species were naturally fast growing; simple protection from defoliation allowed them to spring back. Still, Dhani is not the forest it once was. Some valuable species that were once abundant, like Sissoo, mango, Kendu, and Harida, are now scarce. The original forest species composition has been altered further with the planting of nonnative species like eucalyptus.

But even casual observers can see the improvements in the forest's condition. By mid-1999, the forest canopy had filled out and Dhani Forest boasted more than 250 plant species and 40 bird species. Other wildlife had begun to return as well. Soil erosion had diminished and stream volumes had increased, benefiting the agricultural fields that border the forest.

However, nature dealt the Dhani restoration a setback in October 1999 when a powerful cyclone battered Orissa, uprooting some 90 million trees in its path (Watts 1999). Although Dhani Forest is about 60 km inland, its forest canopy sustained considerable damage, losing many large teak, eucalyptus, and other valuable trees. Fierce winds uprooted bamboo bushes as well and destroyed many siali vines, ruining the siali leaf crop for the year (Singh 2000). In spite of the damage, Dhani Forest remains a functioning forest–testimony to the careful management that in just a little more than a decade transformed a degraded forest patch into a living community resource.

Sharing the Benefits

Conflicts with villagers who were harvesting against the rules were fairly frequent in the initial days of forest protection. But as the protection scheme gained acceptance within and beyond the local villages, cooperation increased. Soon the patrolling squad dropped to 10 people–two from each village–and in 1992 a professional watchman was appointed. At first the community paid the watchman with households' contributions of rice or cash donations. Gradually, revenues from sales of bamboo from the forest increased enough to fund the watchman's salary.

Locals' acceptance of the protection plan has been reinforced by a steady increase in the benefits they reap from the fast-regenerating forest. The forest protection committee has capitalized on the fact that short-term benefits demonstrate progress and breed long-term community support. As the forest has grown healthier, the committee has gradually raised the allowable harvest of different forest products, while taking care to make sure these uses are sustainable and do not impede long-term forest recovery.

Today, local villagers enjoy a much-increased supply of traditional forest products. Firewood from an annual cleaning and thinning operation is shared equally among the five villages, and locals can enter the forest any time to collect fallen branches, leaves, fruits, berries, and tubers at no cost. They also can collect green wood for cremations. With a permit, villagers can obtain poles and timber for a nominal fee, but they must appear before a committee and justify their need and the exact amount they require. Likewise, they can purchase up to 100 bamboo stalks for a fee. All materials are for personal use only and cannot be bartered or sold.

The forest protection committee has also taken care to extend the benefits of their management beyond the five villages. With permission and payment of a higher fee, neighboring villages can obtain many of the same forest goods as local villagers. Special concessions are made for community festivals if a village does not have access to any other forest. Victims of house fires can get timber for repairs at no cost.

Beyond Timber and Fuel: Pursuing Social Goals

The community effort to restore Dhani Forest has always been motivated as much by social as by biological goals. The community's forest management plan has grown to include much more than simple protective measures and rules for distributing benefits.

The Committee's local economic development efforts are perhaps its most ambitious work. The Committee has focused on improving the incomes of local people–mostly tribal peoples and Harijans–who are most dependent on the forest for a living and who effectively lost their livelihoods when the forest was closed to unrestricted use in the early days of Dhani's protection. At the Committee's urging, the State Forest Department has donated two leaf-plate stitching machines and trained local women's groups in siali leaf processing. The Committee was also instrumental in bringing a State-supported dairy program to the area; 40 forest-dependent families each have received one cow to provide a small income from milk.

The community also has decided to augment the natural growth in the forest by interplanting fruit trees, like cashews, that produce a crop that can be consumed locally or sold for cash. Other trees that produce collectible products are planted to help diversify the products that local people can harvest and to increase their production and dependability.

To fund the forest augmentation work and other community development activities, the forest protection committee aims to market any excess bamboo that remains after villagers' needs are met. A state survey of bamboo stocks (pre-cyclone) in the forest suggests that this can be a significant and sustainable source of revenue.

A related activity is the forest protection committee's efforts to pass on the traditional values of this forest-based community to the next generation of forest managers. Once every few months, the village children accompany the forest guard in his rounds. The guard familiarizes them with the plants, and teaches the children their common uses and local religious significance. The children also take part in raising seedlings and planting them to augment the forest stand. Children from Dhani visit various schools in the region to share their understanding of the forest and its importance with children whose villages are not yet involved in forest protection.

Equity and Other Challenges

Community forest management efforts like those in Dhani Forest have become quite common in Orissa and elsewhere throughout India. More than 6,000 rural communities in Orissa alone have made some attempt to protect local forest parcels for common use (Nayak and Singh 1999:8); 120 of these are in the Ranpur area (Panagrahi and Rao 1996:2). Like the Dhani villages, many of these communities have shown remarkable ingenuity, sophisticated planning, and success. But as with any group endeavor, forest protection by rural communities faces many obstacles. In some cases, the protection effort breaks down after a few years because of conflict within or between villages over how to manage the site. The problem becomes more acute once the forest regenerates and trees become larger and more valuable, increasing the temptation to harvest.

One source of internal conflict stems from the social structure of the community itself. Local forest protection programs have evolved in the same social context that has traditionally given rise to caste, class, and gender inequalities. An elite group often dominates the village decision-making process, which may marginalize women and lower-status sections of the community.

Also, the very act of protecting forests by limiting access to them tends to adversely affect the poorer and more forest-dependent members of the village, who have few other options for fuel and livelihood.

There are approximately 2,000,000 villagers living in some 10,000 villages across Orissa. More than 400,000 ha of forest is under JFM by village communities, but what they want is sole rights over the forests they protect and manage. They have formed a state-level forum to fight for ownership.

Dhani reflects both of these problems. The impetus for forest protection–and control of the forest protection process–has always been strongest in the villages populated with higher castes that owned land and had less absolute dependence on the forest. Conversely, the villages populated by tribal people and Harijans have shown greater reluctance to participate and have complained of less power over the forest's management. The forest protection committee's attempt to provide more income sources for the poorest members of the community has evolved as a response to this tension.

Likewise, the Dhani villages have wrestled with gender issues. Until 1995, the general committee (the main body of the forest protection committee) consisted of family elders, usually men. Since then it has included two members–one man and one woman–from each family in the five villages. The executive committee, a group of 21 villagers who implement the decisions of the larger general committee, has also included women since 1995, but only three and they are not routinely consulted when important decisions are made. Including women in the forest management makes sense because women are the predominant forest users, collecting most of the firewood, leaves, and other plants that enter local commerce.

Conflict with outside villages is another typical complication in forest protection efforts. Villages that have traditionally made use of a forest, yet have not been part of the effort to protect it, sometimes resist when a community group tries to limit free access to the forest. The conflict may remain latent as long as the forest is degraded, but once the forest regrows, neighboring villages may want a share. This was the case in Dhani. Kadamjhola, another village bordering Dhani Forest, declined to participate in the original forest protection plan but now wants to share in the project. The five original Dhani villages have agreed to involve Kadamjhola in the protection and management scheme.

Other neighboring villages have also sought a share of the replenished flow of forest products. In earlier years, these villages regularly infringed on the protected forest patch, causing many disputes. But in 1991, with the encouragement and advice of the forest protection committee, several of these villages joined together to protect their own piece of Dhani Forest–a section adjacent to the parcel that the five Dhani villages have under management. The efforts of the two groups will reinforce each other and reduce pressure on both parcels.

The Dhani Forest protection committee also has helped other community forest management groups resolve conflicts through their role in the recently formed regional federation of forest-protecting villages that has sprung up in the Ranpur area.

(continues on p. 190)

Box 3.9 History of Indian Forest Management

Although overuse of Dhani Forest did not begin until the 1950s, Indian forests have been systematically exploited for centuries. Many of the policies and inequities in wealth and political power that permitted historical forest destruction still influence the use and restoration of forests like Dhani.

British rule in India (1799–1947) left an indelible imprint on Indian forests, both through the outright destruction of forests for commercial timber and by dismantling centuries of local traditional forest governance systems. Certainly Indian forests had been altered prior to the arrival of the Europeans—for settled agriculture, for example—but in 1799 most were relatively unpressured. Pepper, cardamom and ivory were the only forest products for which there was significant commercial demand, and land for subsistence hunting and gathering was ample. Many forests in India were managed locally, with village systems and cultural traditions that carefully regulated members' harvesting practices.

But in the 19th century, the British turned to Indian timber for the royal navy's ships, for gun carriages, and to construct and fuel an expanding railway network. Large landowners, called *zamindars,* also promoted the conversion of forests to agriculture to make money and meet the tax demands of the colonial administrators.

By the mid-1800s, the British were concerned about rapidly dwindling supplies of teak, sal, and deodar—the best timbers for railway construction—and the government sought to expand its legal purview over Indian forests. They criticized villagers' customary use of forests as random and unscientific; colonials complained that rural Indians had become accustomed to grazing cattle and cutting wood wherever they wished. Although some colonials recognized that there were, in fact, complex systems of local forest governance that warranted praise and strengthening, their voices were overwhelmed by the assertion of the proprietary rights of the colonial government to India's forests.

The 1878 Forest Act dismantled the last vestiges of rural community control and instituted new classifications for forests: the compact and most valuable areas were labeled "reserved" or exclusively claimed for the state, others were classified as "protected"–places where the local people were given certain privileges but no formal rights. Eventually the colonial government converted many protected areas into reserve forests. Large areas of forest under the control of India's princes were also drawn into the colonial Act. Leases with local landlords and rajas divested surrounding populations of their forest rights. By World War II, the Forest Department's instructions were to produce the maximum output possible.

Traditional conservation and community management systems went into decline. In some areas, sale or bartering of forest produce was prohibited. New laws restricted small-scale hunting by tribes and British foresters. Indian princes sought to ban the traditional use of *jhum*—the shifting clearing and cultivation of forest in rotation—with the hope of enhancing the commercial value of their forests. Even in the few places, such as Madras, where the classification of *panchayat*, or village forests, lingered, bureaucratic government rules impaired their functioning. Loss of control induced a sense of helplessness among villagers, and protected areas became vulnerable to exploitation by both residents and outsiders.

With Indian independence in 1947, the domain of the Forest Department grew and the scope for local community management shrank still more. The Indian government took over extensive forests owned by landlords. But before surrendering their lands, many landlords cut as many trees as possible.

Clarifying the Forest Classifications

The terms used to describe forest-use rights and access privileges have specific connotations in the context of Indian forestry laws.

- **Reserve Forests** are those for which all rights are recorded and settled by the state. They represent the highest degree of state control—the state grants privileges but not rights to people.
- **Protected Forests** represent a lesser degree of state control, whereby rights are recorded but not yet settled.
- **Village Forests** constitute a fuzzier category. These are forests under management of representative village bodies, but the nature of these bodies and the kind of control they have varies. In the 1930s, for example, the state granted to village bodies some isolated, unprofitable (for the state) forest patches in western India in a bid to renew and bolster traditional management practices; these, for example, are referred to as village forests.
- **Common Property Lands** are lands with no individual ownership where resources are shared according to some established social norms. Grazing lands traditionally used by village communities are an example. Village forests can also be thought of as common property lands.

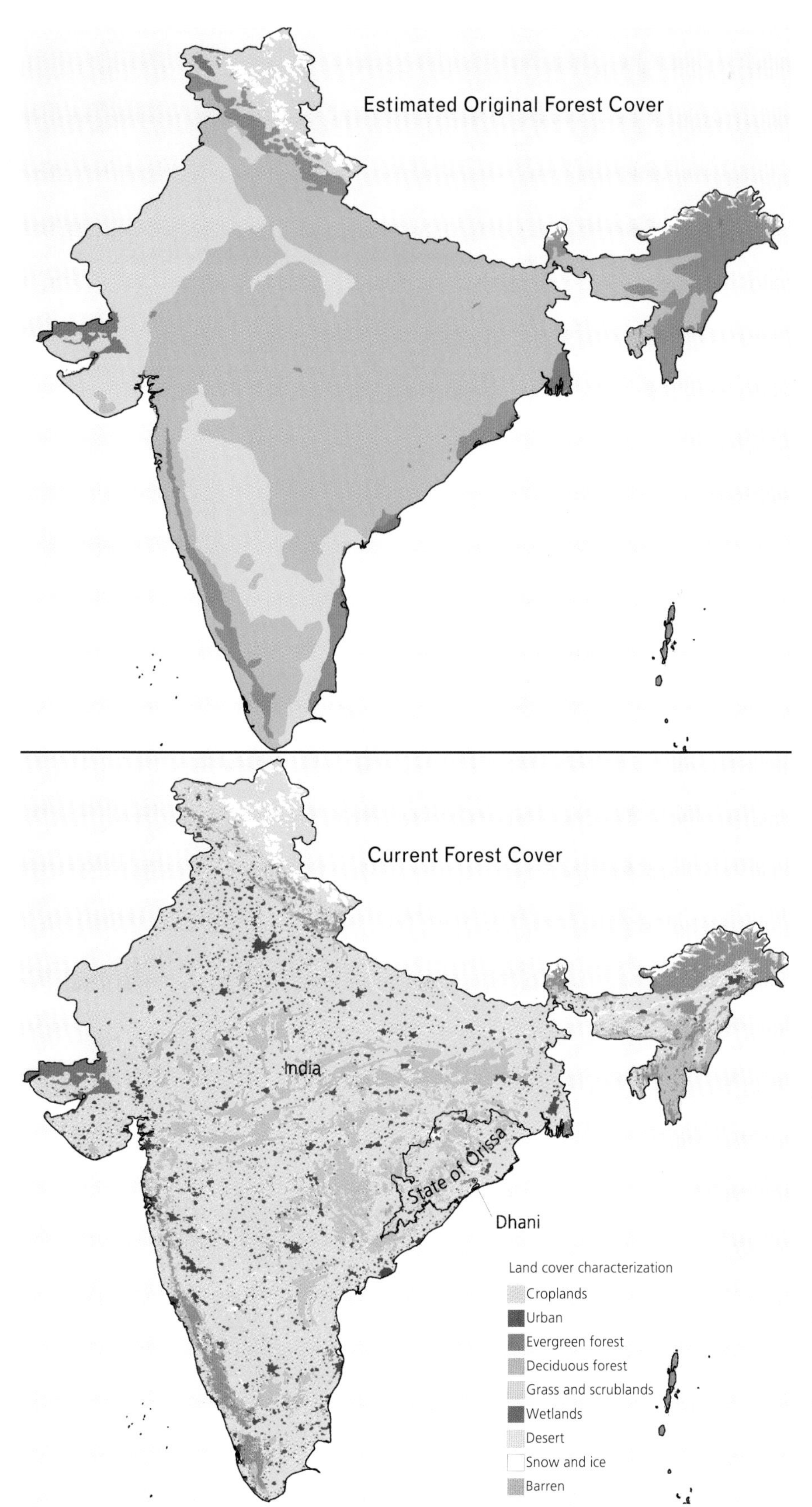

Sources: MacKinnon (1997) and Global Land Cover Characteristics Database Version 1.2 (Loveland et al. 2000).

Industrialization was an important objective of the newly independent Indian government and state timber plantations and production of paper and other wood-based industries were subsidized.

By the 1970s, when government forests were largely exhausted, some of the best tree stocks in India were what remained of locally managed village forests—like Dhani. The forest industry turned to some of these village forests and attempted to extract timber without the consent of local leaders.

At the same time, a growing population put those remaining forests under extreme pressure to be converted to other uses or to produce more wood, fuel, timber, and nonwood products. One survey found that between 1950 and 1980, the number of people supported by a single hectare of common property went from 4.9 to 13.7, with poor families deriving 77 percent of their fuel and fodder from such lands (Pachauri and Sridharan 1998:126, citing Jodha 1990).

In the early 1970s, however, experiments in Joint Forest Management were initiated, and would lead to a new era of forest co-management.

The pressures of population growth and forest conversion continue, yet Dhani and other forests are beginning to regenerate. Villagers are testing their rights to manage, reap, and perhaps even gain title to the lands they have restored. And governments at all levels are starting to realize the economic benefits of managing a forest for its nontimber goods and services—from leaves to healthy soils—as well as for its commercial timber potential.

State vs. Local Control: Who Should Reap the Benefits of Regeneration?

Title to Dhani Forest–both the land and the trees themselves–rests with the State of Orissa, yet it is only through the efforts of the Dhani villagers that a functional forest exists on the formerly degraded site. A similar situation exists on most of the forests in Orissa that have been regenerated through local community forest management–a total of approximately 400,000 ha, or about 7 percent of the State's forest lands (Mahapatra 1999:34). This tension between legal state control and de facto local control has been a source of local dissatisfaction and political friction for years.

In 1988, responding to pressure from a rapidly growing number of forest-protecting communities, Orissa became the first state to formally recognize the legitimacy of local forest protection committees. Soon after, it established a joint forest management (JFM) program through which it allows villages to co-manage local forests while sharing forest products with the state. Under the JFM formula, local communities are entitled to 100 percent of minor or intermediate harvests of commodities like fuelwood and nontimber products like leaves, grass, and fruits, and 50 percent of major harvests of timber.

Although the state maintains this is an equitable division, many local villagers throughout Orissa disagree. The State, they argue, has shown little interest in local forest management until now, when forests have begun to regrow and their value has risen. They complain that the State treats them like junior partners in the management effort, even though they have done the bulk of the restoration work. Many of these villages believe the State should surrender title to forests entirely to the local communities that protect them. Local activism over the subject of forest ownership has increased steadily in recent years, and the question of the State's role and right to harvest weighs heavily in the future of local forests like Dhani (Mahapatra 1999:32–42).

Dhani's own experience with the State has been more positive than most. Orissa State showed little interest, interference, or involvement in the beginning of the protection effort. In 1993, however, the State entered into a JFM agreement with the Dhani villages and has since been forthcoming with support. Lately, the State has cleared up one of the gray areas in the JFM rules: how to share the bamboo harvest. The state has also actively supported economic development initiatives of the Dhani community and offered technical help in improving the forest stand.

Even while it has maintained good relations with the State, the Dhani community has been active in the regional federation of forest-protecting villages. It has also taken a more visible role beyond the borders of Orissa, becoming a major learning center for those who want to study community forest management. In recognition of the Dhani villages' success in protecting and restoring the forest, Orissa State awarded them the *Prakriti Mitra* (Nature's Friend) award in 1997.

Forest Regrowth, Community Renewal

For the past 15 years, Dhani Forest has served as an 840-ha classroom. It has offered the community–and the world–some basic lessons in the value, degradation, and restoration of forest ecosystems.

The forest has always been a central feature–both spiritual and economic–in the lives of the communities around Dhani. It has been a source of livelihoods, a place for ritual, and the tangible abode of nature. As the forest condition degraded and these forest benefits dwindled, the fabric of the community began to fray. Both local subsistence and the cash economy suffered. Food supplies became less stable. Periodic migration out of the community for wage labor increased.

But the years of forest scarcity had a positive effect as well. Desperate to regain the benefits of the forest, the Dhani villagers came to a collective decision to act on their own–a grassroots campaign that provided a common rallying point among villagers and helped renew their traditional link to nature in the form of "Mother Forest."

Their efforts have brought tangible and significant financial reward to the communities. They have added money to the common village fund. They have also brought economic opportunities to the poorest and most forest-dependent villagers, the residents hardest hit by the original decision to limit access to the forest and an essential element in the long-term success of the restoration effort.

On another level, the Dhani experience emphasizes the importance of granting local residents a voice in how the ecosystems they live in are managed. Annexation of Orissa's forest lands by the State left locals with little control and stripped them of most of the forest's benefits. This set up the conditions for Dhani's demise. In contrast, when locals reasserted their control, they quickly established a workable management plan that garnered the community's and eventually the State's support. In this instance, and in many villages throughout India, community forest management has been far more effective than state management. Although Orissa State has acknowledged this truth in the form of its JFM program, there are indications that it still is unprepared to relinquish the level of control that local communities feel they deserve.

The Dhani example nonetheless demonstrates that the state can play a useful role in supporting community forest management. By lending financial and technical support to the community's forestry and community development goals, Orissa State improved the Dhani's prospects for success over the long term (Singh 2000). Experience here and in many other villages shows that community institutions such as the Dhani Forest protection committee tend to get stronger and more effective once they achieve financial and institutional independence. To the extent that the state has helped hasten that independence, it has nourished the roots of Dhani's restoration.

Box 3.10 The People of Dhani's Villages

Harijan women stitching siali leaf plates.

The five villages that manage Dhani Forest are home to 1,244 people in 212 households. Twenty-four percent of the households are families of the lower castes of Indian society, 29 percent are tribal, and 46 percent are upper caste families. Since 1935, the number of households has increased from 28 to 224—an increase of 700 percent. The economies of these villages are heavily forest dependent—75 percent of their income comes from a combination of forest resources and agriculture. Populations increased most in villages where families in the upper castes predominate, but lower caste and tribal families are the most dependent on forest products.

Caste Composition

Caste refers to the hereditary social classes of Hinduism; it governs the occupations members can aspire to and their associations with members of other castes. The division is based on wealth, inherited rank or privilege, or profession.

Villages	Number of Households: Upper Castes	Lower Castes	Tribals	Total
Arjunpur	52	21	—	73
Balarampur	4	11	18	33
Barapalli	43	19	—	62
Kiyapalla	—	—	30	30
Panaspur	—	—	14	14
Total	99	51	62	212

—, Data not available.

Source: Nayak and Singh 1999.

Population Trends in the Five Villages that Manage Dhani Forest, 1935–96

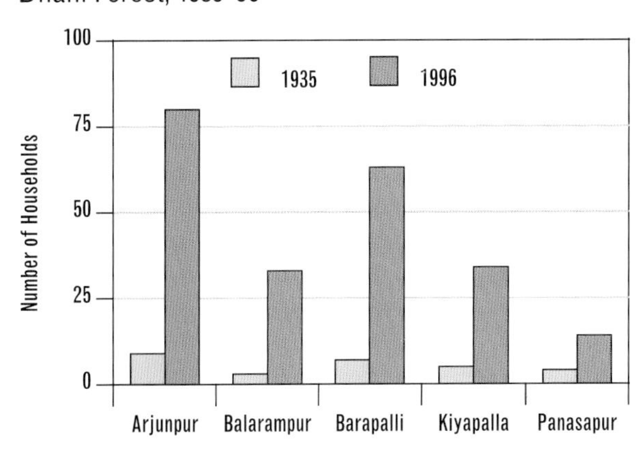

Sources of Primary Income in the Dhani Villages

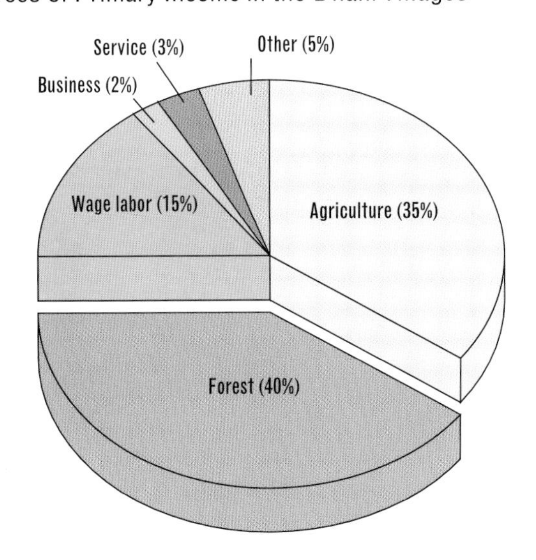

Box 3.11 Joint Forest Management in India

India's Joint Forest Management (JFM) initiatives are based on the concept of collaboration between local people and state authorities. Local people participate in forestry activities on land that remains, essentially, under state control; the Forest Department provides financial assistance and technical advice.

Joint Forest Management grew out of the tension in the 1970s and 1980s between Forest Department staff and local communities. This was an era of political upheaval in many states. Villages had increasing need for forest resources but decreasing access to them, as the government aggressively promoted state plantations in barren and degraded forest lands that had always been used by local people. In fact, by 1980 nearly 23 percent of India's land area had been placed under state management; the majority of the affected rural population were denied access to their traditional resource bases. Nonetheless, Indian forests were losing ground, converted to other uses. For example, during 1959–76, Indian forests lost 2.5 Mha to agriculture, mostly to encroachment by the people living on forest peripheries.

A woman carries a "head load" of wood from rejuvenated Dhani Forest.

During this period, Dr. Ajit Banerjee, a young Forest Service officer posted at a small research station in West Bengal, was exploring alternative methods of forest management. In 1971 Banerjee initiated an experiment in Arabari in which local villagers would work with Forest Department staff to jointly manage forest patches adjacent to their settlement. The idea was to provide residents with a supply of biomass and sources of income through the sale of nontimber forest products—fruit, leaves, mushrooms, twigs, and fodder grass—and in exchange the communities would help restore and protect the forests. Soon, 618 families from 11 villages were working with the West Bengal Forest Department to restore more than 1,200 ha of forest, salvaging sal trees where good rootstocks remained and planting barren patches with fast-growing species like cashews. Some of the deforested areas were cultivated with rice, jute, and maize. The produce was sold to member families at a nominal price. The members could get firewood and fodder free for their own use.

By the early 1980s, jointly managed forests in Arabari were flourishing. Today, West Bengal, Orissa, and other states have formally endorsed the "Arabari experiment" as a general model for jointly managing forests. Widespread replication of the JFM model—with corresponding regeneration of forests—offers strong evidence that the recognition of traditional rights of local people to use forest resources could be the most important condition for managing a forest sustainably.

There remain several challenges to the further success of JFM. Marketing of nontimber forest products is still under the control of an organized lobby of large merchants. The state-run corporation responsible for marketing timber remains vulnerable to a group of contractors who keep prices low at auctions. Moreover, the efficient functioning of forest protection committees still depends on, in many cases, the personal efficiency and willingness of concerned Forest Department officials.

Community Managed Forests in 15 of 30 Orissa Districts

District	Villages (no.)	Land Under Protection (ha)	District	Villages (no.)	Land Under Protection (ha)
Angul	630	6,000	Mayurbhanj	750	35,000
Balesore	450	7,000	Nabrangpur	150	1,000
Baudh	25	2,500	Nayagarh	650	110,000
Bolangir	600	24,000	Puri	250	6,000
Debgarh	110	4,500	Raigada	75	8,000
Dhenkanal	732	8,000	Sambalpur	650	80,000
Ganjam	80	2,500	Sundargarh	125	5,000
Koraput	125	12,250			

Source: Mahapatra 1999.

FRESHWATER SYSTEMS

Working for Water, Working for Human Welfare in South Africa

South Africa is waging a new sort of turf battle. Beginning at dawn each day, thousands of citizens wield scythes, axes, and pesticides against a rapidly advancing and thirsty enemy: the alien trees, shrubs, and aquatic plants that thrive in South Africa's mountain watersheds, drainage basins, and riparian zones. These invading nonnative plants are literally drinking the water that people desperately need in this semiarid country.

Imported for aesthetic and economic reasons and unchecked by natural enemies, alien plants have infested 10 Mha, or 8 percent of the country (Versveld et al. 1998:32). Their noxious spread creates a chain reaction of ecological and economic disasters. In addition to depriving South Africans of needed water, these plants obstruct rivers, exacerbate the risk and damage of wildfires and floods, and reduce biodiversity by crowding out native vegetation.

Destroying trees and aquatic plants may seem counterintuitive to basic concepts of watershed protection and ecosystem management. Watershed conservation is most often associated with the prevention of deforestation. But South Africa is a country naturally dominated by grasslands and fire-prone fynbos shrub vegetation that, because of its low biomass, requires little water–unlike an infestation of large alien trees and woody weed species.

Common invader species such as wattle (*Acacia*), silky hakea (*Hakea sericea*), and pine (*Pinus*) increase the aboveground biomass of fynbos ecosystems by 50–1,000 percent. The invaders dramatically decrease runoff from watersheds

(continues on p. 196)

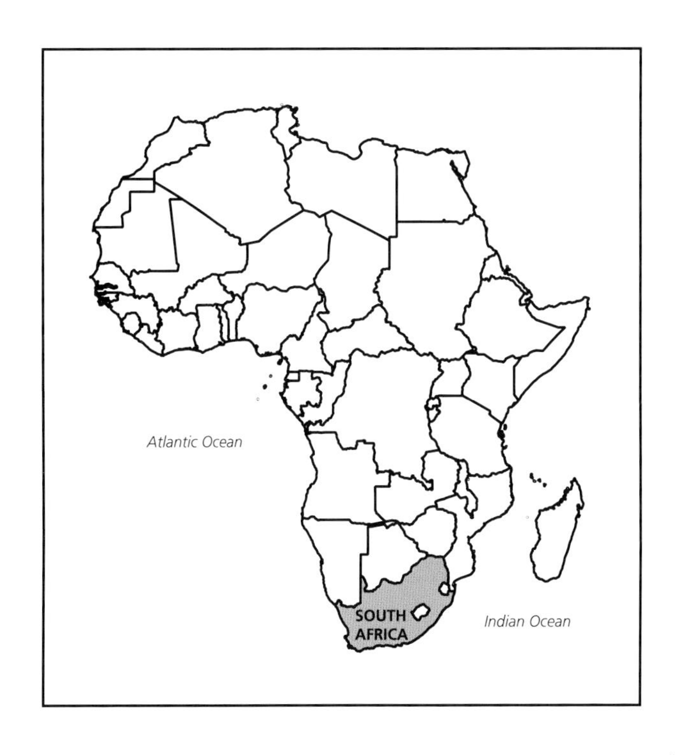

Box 3.12 Overview: South Africa's Invasives

Nonnative plants have invaded 10 Mha of South Africa. Though they provide valuable timber and other benefits, invasive plants deprive the country of precious water, reduce biodiversity, obstruct rivers, and increase risk and damage of wildfires and floods. South Africa's response, a multiagency effort called the Working for Water Programme, has hired thousands of poor, disadvantaged citizens to remove invasive species while acquiring a living wage and new skills.

Ecosystem Issues

Freshwater

Since the invasion of South Africa by nonnative plants, the water quantity provided by the country's freshwater ecosystems to downstream areas has dramatically decreased—by as much as 82 percent in some watersheds.

Forests

Converting grasslands and native forests to nonnative plantations made it possible for South Africa to increase fiber production. Today, timber contributes R1.8 billion to the national economy and forest-based industries another R10 billion. The trade-off: the nonnative trees drink almost 7 percent of water that would otherwise flow into rivers—far more than native species.

Grasslands

Already one-third of South Africa's Cape Floral Kingdom, a grassland and fynbos shrubland ecosystem, has been lost to urbanization, agriculture, and forestry. Invasives now threaten biodiversity in the remaining 90,000 km^2 of fynbos, home to 45 percent of the subcontinent's plant species. Invasives also increase soil erosion after wildfires and floods.

Agriculture

Conversion of lands to agriculture and habitat disturbance from road building and other developments promotes the spread of nonnative plants.

Management Challenges

Equity and Tenurial Rights

The end of apartheid began to return a voice to black citizens, whose control over land and water had previously been drastically limited. This era also brought a new commitment to supplying sufficient water to all. If that commitment is sustained, it provides impetus for the Working for Water Programme and other restoration efforts that promise to provide more water at minimal cost.

Economics

Once almost free, the government now charges citizens for water to discourage overuse and waste. Charges for other major water consumers like the forestry and agriculture sectors are critically needed, too, but hotly contested.

Stakeholders

The Working for Water Programme has found some common ground with stakeholders, but more difficult policy negotiations are ahead. For private landowners and commercial foresters, many invasives are valuable crops or decorative elements of yards; controlling them brings higher costs than benefits.

Information and Monitoring

Research on the impacts of invasives on water supply helped generate interest in today's integrated invasive plant control effort. More economic studies that illustrate the impacts of invaders and the financial benefits of control are essential to help justify the increasingly large-scale funding that the Working for Water Programme requires.

Timeline

c. 1000 Traders and nomads introduce plant and animal species to Southern Africa, but none significantly impact native vegetation.

1652 The Dutch colonize South Africa's Cape. They soon import more than 50 crop plants from Europe, Asia, and South America; some are present-day invaders.

1820–1870 A large influx of settlers from around the world introduces 11 of the 12 invasive species that now cause the greatest problems in fynbos.

1880s–1890s Botanists begin to note the spread of nonnative plants over mountain slopes and losses of endemic species in Cape fynbos vegetation. At the same time, foresters promote mountain plantations of nonnative trees.

1920s Controversy about effects of forest plantations on water supplies begins, even as demand for commercial timber and related products drives high rates of afforestation with nonnative hardwoods that continue for the next 60 years.

1930s Rapid spread of prickly pear (*Opuntia aurantiaca*) in the succulant Karoo sparks awareness of the threat invasives pose in arid areas as well as fynbos. Threats to biodiversity in grasslands and savanna are not fully understood for another 50 years.

1934 The South African parliament appoints an interdepartmental committee to assess water preservation options.

1937 The Weeds Act is passed, one of the first major legislative attempts to deal with invasives, but a lack of field staff and resources makes it difficult to enforce.

1940s–1970s Hydrological studies show that plantations have a negative effect on streamflow. Efforts to control invasives are launched, but they are uncoordinated, erratic, and hampered by limited follow-up after clearing.

1948 Apartheid designates 83 percent of South African land "whites only." Rural land and water laws in ensuing decades mainly serve white interests. Blacks are denied access to the political process.

1970 The Mountain Catchment Act gives the Department of Forestry management responsibility for high-lying areas; invasives there are tackled in earnest, with plants cleared from tens of thousands of hectares. The Plant Research Institute conducts vital research on biological controls for invasive plants.

1983 Conservation of Agricultural Resources Act grants government wider power to control invasive species and introduces the idea that landowners are obliged to manage their land sustainably.

1986 International program on biological invasions focuses attention and research on plant invasions in South Africa. A review of catchment experiments provides unequivocal evidence of the detrimental effect of nonnative plants on stream flow.

Late 1980s Responsibility for management of mountain catchments is passed from the Department of Forestry to the provinces; lack of funding ends momentum for integrated invasive plant control programs. Plants re-invade cleared areas.

1989 International SCOPE program on biological invasions focuses attention and research on South African plant invasions. A review of catchment experiments provides unequivocal evidence of the effect of nonnative plants on streamflow.

1993 Further government-sponsored research determines that clearing invasive vegetation can improve runoff from catchments.

1994 Apartheid ends. South Africa becomes a constitutional democracy.

1995 The Working for Water Programme is founded by South Africa's Minister of Water Affairs and Forestry, hires 7,000 people, and clears 33,000 ha in its first 8 months.

1998 The National Water Act recognizes water as a common resource; commits to protecting its quantity, quality, and reliability; and grants each South African a right of access to 25 l of water per day. Meeting that commitment to 14 million people without access to sufficient water is a daunting challenge.

2000 The Working for Water Programme employs tens of thousands of people and has successfully cleared more than 450,000 ha of land of invasive species, yet millions of hectares still require attention.

through greater water uptake from soil and subsequent transpiration (van Wilgen et al. 1996:186, citing Versfeld and van Wilgen 1986). Currently, invasive species in South Africa consume about 3.3 billion m^3 of water each year, almost 7 percent of the water that would otherwise flow into rivers (Versveld et al. 1998:iv). That's nearly as much water as is used by people and industries in South Africa's major urban and industrial centers (Basson 1997:10).

South Africa's response to the invasion may be the largest and most expensive program of alien-plant control ever undertaken. It is also an effort to address the impoverishment of black South Africans–poverty being one of the legacies of apartheid, the system of white rule that ended in 1994. Through a multiagency effort called the Working for Water Programme, the government has hired thousands of citizens to hack away the thirsty invasive plants and to turn the byproducts of their labors into saleable goods such as fuelwood, furniture, and toys. Since its inception in 1995, the Programme has offered men and women opportunities to acquire a living wage and new skills. In some project areas, the Programme provides childcare, community centers, and health and national water conservation education.

By uniting social goals with ecosystem restoration, and by capitalizing on public pressure to provide more water to millions of people, Working for Water has mustered political will, public support, and funding at a time of fierce competition among the many social welfare projects visualized by South Africa's new democratic government. Still, success is far from assured and the stakes are high. If the Programme fails, many pervasive invaders could double in extent over the next 10–20 years (Versveld et al. 1998:vi), jeopardizing the water supply to cities, industries, and agriculture. The Programme's high cost, conflicts of interest with landowners, and management and safety problems cannot be ignored. But the multiple dividends that Working for Water pays are substantial: a healthier ecosystem, more water at less cost, and employment for thousands in a country where opportunities to escape poverty are rare.

The Plant Invaders

Today, invasive plants and animals are considered one of the gravest threats to the biodiversity of natural ecosystems worldwide. That awareness, however, has come relatively recently. For centuries alien plants were seen as desirable; their cultivation offered immediate economic returns and social benefits, although their costs were usually slower to manifest. Alien plants can spend decades living innocuously in nonnative settings before some subtle adaptation or shift in ecological dynamics triggers an invasion. Even after years of study, it is not always clear which organisms will aggressively invade new ranges, where invasions will occur, when, or why.

A ribbon of invasive alien pines *(Pinus pinaster)* on the horizon; these pines spread from a plantation just over the mountain. They radically alter the structure of the fynbos and reduce streamflow from rivers.

IMPORTING THE INVADERS

Nonnative plants certainly seemed harmless to the Dutch, who introduced more than 50 plants within the first few years of their settlement at South Africa's Cape in 1652 (Wells et al. 1986:29). For the next 150 years, colonists from all over the world continued to import species that would provide firewood, timber, food, and shade, and would stabilize sand drifts, enhance gardens, and remind them of home.

In total, about 8,750 plant species have been introduced into South Africa. Fortunately, only 2 percent have become seriously invasive, mainly trees and shrubs that mature quickly, multiply prolifically, spread easily, and fare well in disturbed conditions (van Wilgen and van Wyk 1999:566). Species imported from southern continents and other fire-prone ecosystems, like Australia, took hold particularly readily in the fynbos, where fires trigger seed release and create conditions conducive to germination.

Some of the most problematic species took root in the late 19th century when forest authorities began to promote afforestation of the mountains around Cape Town. Imported pines, eucalyptus, and wattles were promoted to supply tannin and timber, since the extent of South Africa's natural forests is limited by climate and the fire regime. Officials believed also that alien plants would increase the water supply and provide aesthetic relief; they called the naturally bare and stony slopes of the Cape's mountains "a reproach and an eyesore." Government foresters provided private growers with free seeds and transplants of the alien species and awarded prizes for the best plantations (Shaughnessy 1986:41).

The nonnative trees proved fast growing and able to take root on all kinds of marginal lands. South Africa soon trans-

Box 3.13 Most Widespread Plant Invaders in South Africa

Species	Origin	Reason for Introduction	Approx. Area and System Invaded	Water Use (millions of cubic meters)
Syringa (*Melia azedarach*)	Asia	Ornamental, shade	3 Mha; savanna, along riverbanks, disturbed areas, roadsides, urban open spaces	165
Pines (*Pinus species*)	North America and Europe	Timber, poles, firewood, shade, ornamental	3 Mha; widespread in mountain catchments, forest fringes, grasslands, fynbos	232
Black wattle (*Acacia mearnsii*)	Australia	Shelter, tanbark, shade, firewood	2.5 Mha; widespread, except in arid areas	577
Lantana (*Lantana camara*)	Central and South America	Ornamental, hedging	2.2 Mha; forest and plantation margins, water courses, savanna	97

Sources: Versveld et al. 1998:75; Working for Water Programme n.d.:4.

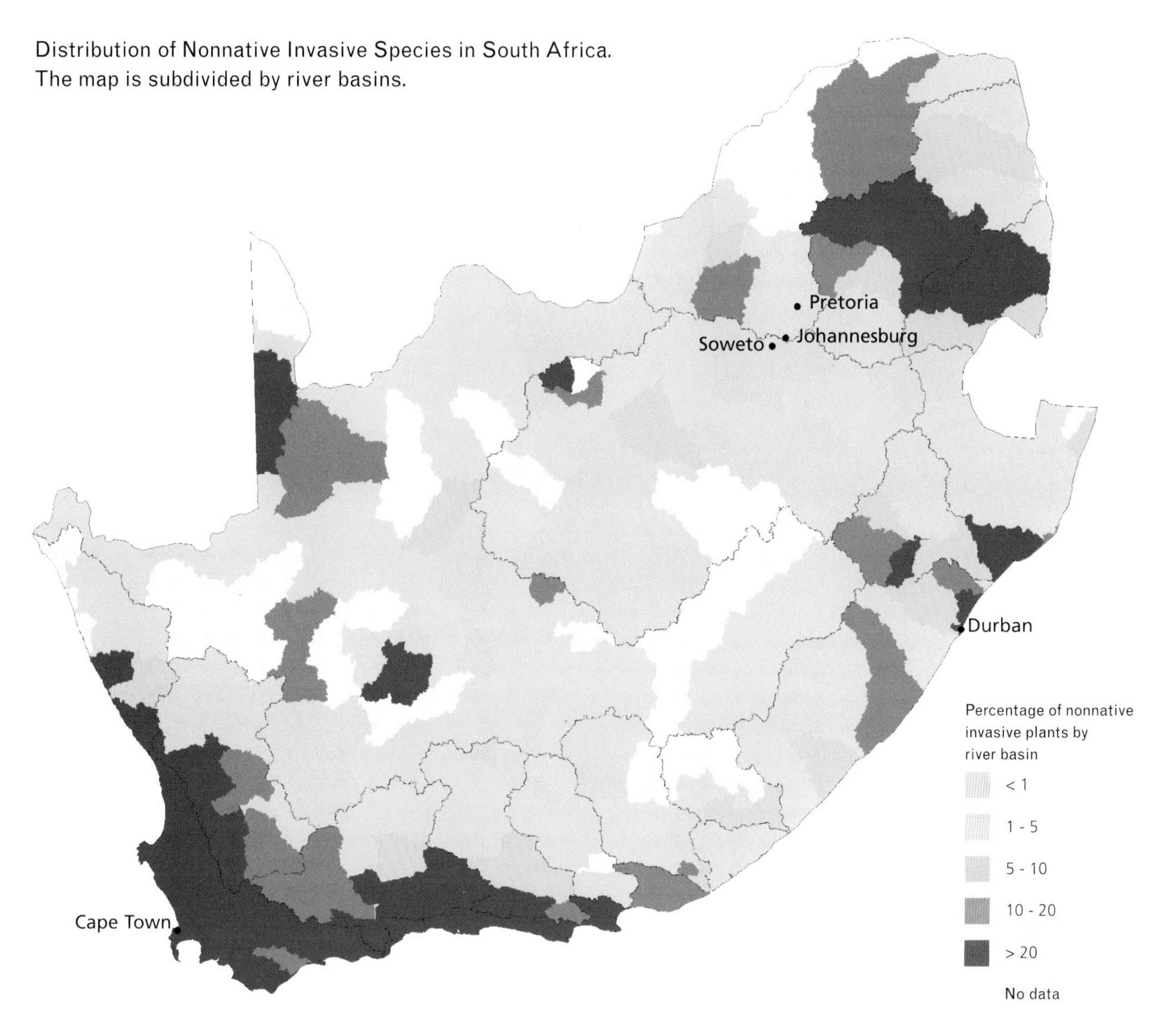

Distribution of Nonnative Invasive Species in South Africa. The map is subdivided by river basins.

Sources: Versfeld et al. 1998; USGS 1997.

formed grasslands and scrub-brushland habitats–largely unsuitable for agriculture and grazing though very rich in native biodiversity–into state-owned and private plantations to feed the burgeoning timber industry and pulp and paper mills. Today, plantations of alien trees cover 1.52 Mha. Natural forests cover less than 7,177 km^2–about 0.25 percent of South Africa (Le Maitre et al. Forthcoming).

Unfortunately, in riparian zones fast-growing aliens drink almost twice the amount of water that the same trees consume in areas away from rivers (van Wilgen and van Wyk 1999:567). And, plantations can only grow in the higher rainfall areas, like South Africa's mountain catchments. There they garner "first take" on some of the key water supplies for South Africa's lowlands. Although mountain catchments encompass just 8 percent of the land surface, they provide 49 percent of the total annual freshwater runoff for the country (van der Zel 1981:76).

LOSING WATER, GAINING AWARENESS

As early as the 1800s, South African botanists expressed concern that introduced plants might suppress and replace natural vegetation, eventually turning the species-rich fynbos into a biological desert. But among land managers and policy makers, there was little interest in alien plant control for almost another 100 years.

The threat of water shortages–more than the potential loss of biodiversity–is what eventually motivated a reevaluation of South Africa's land management practices. Suspicions that the proliferation of alien plants might be linked to water supply problems arose in the 1920s when farmers' associations petitioned the government to investigate why South Africa's rivers were drying up. The government initiated a series of experiments to assess the impact of commercial forestry on water resources in mountain areas. In study catchments, fynbos shrublands and grasslands were heavily planted with alien pines and eucalyptus, and the impact on stream flow was monitored and compared to untreated control catchments. In the following decades, researchers found stream flow sensitive even to small changes in catchment vegetation cover. In KwaZulu-Natal Drakensberg, for example, there was an 82 percent reduction in stream flow in grassland catchments 20 years after planting with pines, a 55 percent reduction in fynbos catchments in the Western Cape 23 years after planting with pines, and a total drying up of streams in Mpumalanga Province 6–12 years after completely replacing grassland catchments with pines and eucalyptus (van Wilgen and van Wyk 1999:x). Despite these findings, until the 1990s, efforts to protect watersheds and combat the spread of invasive plants were small and sporadic, petering out when funding waned.

Finally ecologists were able to galvanize support for change with a critical body of evidence that water losses to unchecked invasives could be economically disastrous. Advances in technology enabled the development of computer models that simulated the growth, spread, and water use of alien plants in a fire-prone landscape. The results were eye-opening. Even sparsely infested areas are likely to become dense with invasives over the next half century, resulting in reductions in streamflow of 30–60 percent (van Wilgen et al. 1997:406). During the dry months when water needs are greatest, runoff in some invaded catchments could be reduced to zero, converting perennial streams to seasonal ones.

Unchecked alien plants would have dire implications for the Cape region's native wildflower, foliage, and dried flower harvests and for the 1.3 Mha of irrigated croplands that produce 25 percent of the country's agricultural output (IWMI 1999:4). The Western Cape's harvests of apples, peaches, and pears, for example, depend entirely on water derived from adjoining mountain catchments; and the deciduous fruit industry generated gross export earnings of more than US$560 million and employment for 250,000 people in 1993 (van Wilgen et al. 1996:185).

The impetus for invasives control gained further momentum from a political transformation–the end of apartheid in 1994. A democratically elected government brought a new national focus to equitable water access, a radical departure from a history in which water was seen as the property of the person whose land it ran through, usually white farmers. Now, under South Africa's 1998 Water Law, all water is a common resource. Each South African has a right of access to sufficient water for basic needs, an amount provisionally set at 25 l/person/day.

Since 14 million South Africans have inadequate or no water supplies (Koch 1996:12), translating this new "right" into practice will make prior water shortages seem trivial. South Africa is already water stressed, and rapid population expansion in metropolitan areas like Cape Town threaten to create regional water crises. Studies have predicted that in parts of the Cape, water demand in the year 2010 could be 70–106 percent higher than in 1990 (Marais 1998:2, citing Spies and Barriage 1991).

A New Kind of Turf Battle

Watershed protection and poverty alleviation are dual goals paired effectively in South Africa's Working for Water Programme. In 1995 Kader Asmal, Minister of Water Affairs and Forestry, was convinced by the arguments of scientists and conservationists that clearing invading plants could supply water and other ecological benefits. He proposed that the government use Poverty Relief funds to hire disadvantaged citizens to remove invasive trees, shrubs, and aquatic plants.

The first year of the plant-clearing effort had a budget of R25 million and employed more than 6,100 people (van Wilgen 1999). Now in its fifth year, Working for Water's 1999–2000 budget is eight times larger–R202 million (van Wilgen 1999)

A Working for Water team clears a dense stand of *Pinus pinaster* in the mountains above the coastal town of Kleinmond, about 120 km east of Cape Town.

and funding 240 projects in eight heavily infested provinces. At times, employment has risen to 42,000 people, many of whom have never been employed before or only labored as migrant workers (Working for Water 1998, 1999). Priority is given to clearing invasives from riparian zones and areas with the greatest number of disadvantaged citizens.

PROTECTING THE WATERSHEDS

The Programme has cleared in excess of 450,000 ha of infested land. In some places streams have flowed again for the first time in decades (van Wilgen 1999). The clearing of a dense stand of pines and wattles from 500 m of river bank in Mpumalanga Province, for example, soon resulted in a 120-percent increase in stream flow. Removing pines for 30 m on either side of a stream (just 10 percent of the catchment) in the Western Cape resulted in a 44-percent increase in stream flow a year later–more than 11,000 m^3 of water gained per cleared hectare (Scott 1999:1151-1155; Dye and Poulter 1995:27-30).

Twelve to 18 months after clearing an area, workers must eliminate alien seedlings with herbicide treatments or burning and replant the land with indigenous species. Follow-up also may require the use of biological controls such as species-specific insects and diseases from the alien plant's home country. Examples include the tiny gall wasp that prevents the long-leafed wattle from flowering and producing seeds, or leaf-feeding insect species that damage the leaves and stems of lantana, another aggressive invader. In most cases, biological methods cannot control alien plant species on their own–they cannot remove existing established stands of trees, for example–but they can provide a cost-effective means of minimizing the invaders' future spread and an alternative to herbicide applications near water.

ALLEVIATING POVERTY

Working for Water's momentum comes as much from the jobs it creates as the water that flows anew from project areas. Employment is a powerful lever for change in a country with 37 percent unemployment (in 1997) (UNEP 1999, citing South African Institute for Race Relations 1998); 50 percent of all households are classified as "poor," earning less than R353/ adult/month (May 1998). In many project areas, citizens lack reliable sources of clean water, electricity, and permanent homes. Few have the education or skills to take on available jobs, especially those in an increasingly technological labor market.

Programme workers are paid a daily wage of R22-R55–on par with local wages for similar jobs (Marais 1999). Most workers spend the day removing invasives with scythes and chain saws. Some employees trained in mountaineering start the week with a helicopter flight to parts of Mpumalanga and Western Cape provinces that are inaccessible by foot. There they clear alien vegetation from peaks and gorges, camping until a return flight home on Friday.

The Programme's social welfare benefits are expanding along with the water supply. By supporting child daycare centers, Working for Water has built a workforce that is more than 50 percent female, including many single mothers. The Programme also strives to create jobs for youths, rural residents, and the disabled. Worker training and education, provided in collaboration with government agencies, schools, and nonprofit organizations, complements hiring programs. Topics include environmental awareness and health education–from first aid, to family planning, to HIV/AIDS prevention.

TEMPERING THE TAP

While striving to restore the mountain watersheds to a state of uninvaded abundance, the Working for Water Programme serves to awaken citizens to a new appreciation of the limits of South Africa's precious water resources. A combination of incentives is spurring the adoption of conservation measures and providing Programme income.

A major impetus comes from South Africa's new Water Law, which explicitly recognizes the need to protect "the quantity, quality, and reliability of water required to maintain the ecological functions on which humans depend" (see next page). Some municipalities where Working for Water operates

(continues on p. 202)

Box 3.14 South Africa's New Water Law: Managing Water for Equity, Economic Growth, and Ecosystem Resilience

Reforming the way water is managed is central to South Africa's economic and political reconstruction. Since the democratic elections of 1994, the nation has crafted a suite of water policies, including the Water Services Act of 1997 and the National Water Act of 1998 (NWA), to redress past inefficiencies, inequities, and environmental degradation. These new policies are considered among the most progressive in the world.

Like other countries, South Africa's has crafted water-sector reforms that emphasize a decentralized approach to water management, encourage local participation in decision making, and use innovative water pricing practices (Saleth and Dinar 1999:iii). What sets South Africa's approach apart are its far-sighted and ecologically grounded commitments to manage water efficiently, while ensuring equity of access and the sustainability of the resource. These goals have required radical departures from the nation's old practices.

Protecting Ecosystem Integrity

South Africa's new water policy is based on the principle that the nation must maintain the natural ecosystems that underpin its water resources if it expects to meet its ambitious water provision goals. To this end, the NWA requires that the country maintain an environmental "reserve"—the amount of water that its freshwater ecosystems require to remain robust (NWA No. 36, Chap. 3, Parts 2 and 3). The law also encourages an integrated, watershed-based approach to water management; actions that could fall under the law's purview include modifications of land-use practices along stream corridors, the clearing of nonnative vegetation, and measures to reduce the production of pollutants.

Water Allocations to Satisfy Basic Needs

The NWA establishes a "basic needs reserve" for humans, too—an allocation of water for drinking, food preparation, and personal hygiene. This reserve, provisionally targeted at 25 l/person/day, is guaranteed as each citizen's right (DWAF 1994:15; Water Services Act No. 108). To ensure that everyone has access to the reserve, the law directs the Department of Water Affairs and Forestry (DWAF) to oversee the provision of water and sanitation across the provinces.

After a supply of water to meet basic human needs and the environmental reserve is assured, South African law requires that remaining water be allocated so that: (a) all people have equitable access to the resource for productive purposes, especially within the agricultural sector; and (b) all people have equitable access to the benefits that flow from water use, such as jobs. For example, under law, the country would seek to remedy such inequities as the distribution of irrigation water; currently, irrigation accounts for more than half the water used in South Africa, but black farmers have access to less than 10 percent. The NWA also specifies that the government can implement water charges (described below) for certain regions or groups to further the goal of equitable access.

Water as Public Property

The 1998 law makes all water public property, repealing the previous statute that assigned water rights based on property ownership (NWA No. 36, Ch.4). For example, a landowner now needs permission to make large-scale water withdrawals from water that crosses his or her property. Other regulated water uses include storing water, impeding or diverting the flow of water in a watercourse, engaging in activities that can reduce stream flows such as plantation forestry, irrigating land with waste water, or altering the banks of a watercourse.

Individuals who want to use water beyond reasonable amounts for domestic use, livestock, emergencies, and recreation must apply for temporary licenses (NWA No. 36, Chap. 4, Part 1 and Schedule 1). Water authorities grant licenses for specific uses, like irrigation, and for specific periods of time. The maximum grant of water rights is 40 years, but all licenses of any length are subject to review at least every 5 years to ensure equitable distribution in a watershed. Reviews are conducted to maintain water quality, to redress situations where water has been over-allocated, or to address situations in which socioeconomic demands have changed. Licenses can be traded or auctioned.

New Governance Structures

The scope for local participation in water management in South Africa has been vastly broadened while the capacity to coherently plan and integrate water management at national and watershed levels has been retained.

At the national level, DWAF is charged with establishing the details of the national water strategy, making decisions about water transfers among watersheds, meeting the terms of international agreements in shared river basins, and determining water quality standards. But the responsibility for actually allocating water to users within an individual watershed rests with local "Catchment Management Agencies" (CMAs) (NWA No. 36, Chap. 7, Part 1). The CMAs and other institutions are expected to operate with broad participation from all interested parties—for example, they must make all applications for water licenses public and judge all water users' responses.

It is also worth noting that South Africa's water laws are among the first in the world to grant water rights to a person who farms a given piece of land, whether the person is the formal owner or merely the user of the plot. This arrangement is substantial help to holders of communal land (International Water Management Institute 1999:8).

Water Fees for Equity and Efficiency

The NWA relies on water fees as the main tool for financing the provision of water and encouraging efficient use (NWA No. 98, Chap. 5, Part 1). The law requires the DWAF to develop water pricing strategies and gives the agency considerable discretion in varying water prices by location, depending on circumstances. For example, the agency can apply a given water charge on a national or regional basis, or simply within a specific water management area. The DWAF can use three types of water fees:

- A charge to cover the full financial costs of providing access to water, including the costs of developing, operating, and maintaining the water infrastructure.

- A watershed management charge, which can apply to the use of rivers and other water bodies for waste disposal as well as to water consumption. Funds generated can be used to support water management, conservation, and research.

- A resource conservation charge that can be applied where a particular water use significantly affects others in the watershed. These charges are intended to reflect the scarcity value of water in a water-stressed area.

Implementation Challenges

South Africa's water reforms are lauded internationally, and people across South Africa recognize the merits of the changes outlined in the new water policies. Nevertheless, implementing the new policies is challenging. Weak management and inadequate training have plagued many water delivery projects in the past 5 years, and some communities have resisted paying the new water charges. These early experiences demonstrate that, no matter how lofty the goals, instituting profound changes in the management of a resource as basic as water takes time, both to build support among the wide array of water users and to build the capacity and professionalism of local water institutions.

An equally great challenge posed by the new water policies is the need for the South African government to take a multidisciplinary approach to water management issues. Hydrological and engineering considerations—for decades, the water department's focus—now are merely pieces of a larger management framework that gives equal consideration to economic, social, and ecosystem issues.

use water conservation campaigns to help implement that law. Prepaid meters encourage citizens to pace their water use and "save" water. Citizens use "grey water" (wastewater) in the garden, water-efficient toilets, and low-flow showerheads. They refrain from irrigation between 11 a.m. and 2 p.m., when 60 percent of the water applied evaporates.

Another conservation incentive is an increase in what had been some of the cheapest water prices in the world. Sliding scales for household water use make the first 5 m^3 of water just R0.007 each, but each additional cubic meter has a higher price–as much as R0.14/kl for use of more than 60 kl/household/month (van Wilgen 2000).

The results are striking. In Hermanus, for example, water use decreased by 25 percent, while revenue from the sale of water increased by 20 percent, helping to fund a local Working for Water project. Conservation measures have allowed Hermanus can delay building expensive additional water supply capacity–like a new dam (Working for Water 1998:17).

CALCULATING THE BOTTOM LINE

Currently, Working for Water is spending R200–R250 million/year, mainly on worker wages. Financial support comes principally from the government's Reconstruction and Development Programme and Poverty Relief funds, and about 40 percent from water tariffs (van Wilgen 1999). Substantial training, materials, and staff for the social welfare programs are provided by many partner agencies. In Walker Bay near Hermanus, landowners are paying half the clearing costs and the full maintenance costs. In Cwili-Kei Mouth/Komga on the Eastern Cape, farmers are paying 60 percent of the cost to clear their land (Marais 2000; Working for Water 1998:17). Programme leaders hope to replicate these models.

Yet at current rates of work and efficiency, the plants are still spreading faster than the Programme is removing them. Assuming an alien expansion rate of 5 percent/year, watershed restoration and plant control will require about 20 years of work–an annual investment of about R600 million. That's a total cost of about R5.4 billion, plus long-term maintenance of about R30 million/year (Versveld et al. 1998:iv–vi).

Still, put in the context of other water supply options, plant-clearing programs and watershed protection may be the best buy. One study suggests that the additional water generated by clearing aliens from catchments in the Western Cape would cost just over R0.06/m^3. By comparison, it would cost, per cubic meter, R5.70 to secure water from the best dam option in the Western Cape, R1.50 for treating sewage water, and R4.80 cents for desalination (van Wilgen et al. 1997:409; van Wilgen 2000). The studies also showed that early investment in clearing is financially prudent. The spatial cover of invasives in fynbos regions appears to spread and intensify from light to dense within four to six fire cycles (50–80 years). To clear lightly infested areas costs about R825/ha compared to R5,875/ha to clear a densely invaded area (Versveld et al. 1998:vi).

WINNERS AND LOSERS

Not only does the government face steep plant-clearing and weed-control costs, so do private companies and landowners. Many of the species targeted as "pests" sustain one of the country's fastest growing economic sectors: plantation forestry contributes 2 percent to South Africa's GDP, about R1.8 billion/year; and products from pines, eucalyptus, and wattles contribute another R10 billion/year. Yet forestry is a major source of invaders. Thirty-eight percent of South Africa's invaded areas are occupied by nonnative species used in commercial forestry, and nearly 80 percent of invasive pines occur within 30 km of plantation forestry (Nel et al. 1999:i,1,19). Many rural landowners are reluctant to finance the restoration of invaded areas for which they are responsible–areas where species like wattle and eucalyptus have escaped from intended use on farms as windbreaks, shade trees, and wood lots. Plant nurseries, too, have been targeted for tighter regulations on sales of invasive plants.

Private landowners and Working for Water have found some common ground. Working for Water proponents do not propose banning the use of invasives on plantations, and many landowners are eager to control weeds like lantana, bugweed, and chromolaena, which obstruct plantation operations and increase the fire hazard. The forest industry has committed to a code of conduct that requires riparian zones and nonafforested areas in their estates to be kept clear of alien plants. Some forestry companies have helped plant-control efforts by clearing weeds and commercial species from riverine areas or assisting with planning, mapping, vehicle donations, and worker training.

But broader consensus on the financial responsibility of the forest companies and the thousands of small independent farmers for clearing and controlling invasives is elusive. Not all agree

The Expansion of Forestry over the Past Century

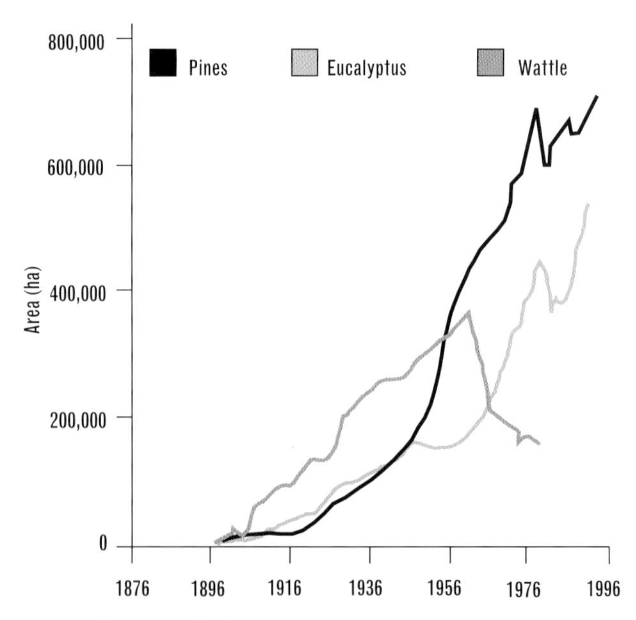

Source: Nel et al. 1999: 20.

Box 3.15 Valuing a Fynbos Ecosystem

The ability to estimate the value of South Africa's ecosystems with and without invasives has proved key to securing support for clearing programs. For example, a 1997 analysis valued a hypothetical 4-km^2 fynbos mountain ecosystem at R19 million with no management of alien plants and at R300 million with effective management of alien plants. The analysis was based on the value of just six major goods and services provided by the ecosystem: water production, wildflower harvest, hiker and ecotourist visitation, endemic species, and genetic storage (Higgins et al. 1997:165). The authors also determined that the cost of clearing alien plants was just 0.6–5 percent of the value of mountain fynbos ecosystems. That may be a very conservative estimate, given the extraordinary species richness and endemism in South Africa's eight biomes and the fact that invading plants threaten to eliminate about 1,900 species (van Wilgen and van Wyk 1999, citing Hilton-Taylor 1996).

In fact, South Africa's biodiversity is perhaps the strongest long-term justification for limiting the extent of invasives, but the most difficult ecosystem service to value. It is possible, for example, to estimate a "market worth" for fynbos plants when developed as food and medicines or horticultural crops. However, it is more difficult to put a value on a species like the Cape Sugarbird, whose habitat is endangered by invasions in the Western Cape, or the oribi antelope, threatened by invaders that disrupt grasslands habitats.

Benefits and Costs Associated with the Black Wattle *(Acacia mearnsii)* in South Africa

The black wattle, an aggressive invader, provides significant commercial benefits and is an important resource for rural communities. But one recent analysis suggests that its costs may be more than twice as high as its benefits.

Wattle Benefits	Net 1998 Value (R6 = US$1)	Wattle Costs and Negative Impacts	Net 1998 Value (R6 = US$1)
Timber and other commercial wood by-products, including tannins, pulp, woodchips	$363 million	Reduction of surface streamflow estimated at 577 million cm^3 of water annually	$1,425 million
Firewood	$143 million	Loss of biodiversity	Unknown, but believed to be significant
Building materials	$22 million	Increases in the fire hazard	$1 million
Carbon sequestration	$24 million	Increase in erosion	Unknown
Nitrogen fixation	Unknown	Destabilization of river banks	Unknown
Medicinal products	Unknown	Loss of recreation opportunities and aesthetic costs	Unknown
Combating erosion	Unknown		
Total	>$552 million		>$1,426 million

Source: de Wit et al. Forthcoming.

with proponents of Working for Water who advocate more clearing near and downstream from plantations and fines for illegal plantings within 20–30 m of riparian zones. Plus, the Programme advocates a polluter-pays approach to seed pollution, which would hold those who use invasives responsible for the costs if the plants spread. Private landowners question the practicality of trying to measure seed pollution. They fear being blamed for impacts caused by others, including the backlog of removal to be done in riverine areas–at least some of which were likely infested by the government before plantation forestry was privatized. Unless these disputes are overcome and the stakeholders work cooperatively, Working for Water's efforts will be crippled.

Foresters also oppose Working for Water's advocacy of water tariffs on "stream flow reduction activities"–effectively, a tax on the water consumed by their trees to help fund the clearing of alien-infested catchments. These tariffs will force the forest industry to come to grips with a system in which water is no longer a free service; the industry fears that such water controls will inhibit its global competitiveness. Singling out the forest industry for user fees complicates the dispute. Sectors like agriculture and mining pump more water from rivers than forestry but are not likely to be charged for several years. Detailed knowledge of their impact on water use lags far behind that of forestry, making it difficult to issue permits and bills.

Working for Water also poses problems for the many rural communities that depend on invasive plants for firewood, shelter, and food such as honey, prickly pears, and guava. So far, the Programme has avoided clearing where invasive plants are a major fuel source for impoverished communities, or has sold or donated felled species as firewood, charcoal, or barbecue wood. Eventually, though, it may be necessary to develop locally managed woodlots of species with minimum invasive potential or of fast-growing indigenous species.

The Programme's Future

Securing the buy-in and support of landowners is only one of a gamut of daunting obstacles faced by Working for Water. Living up to its promise of creating empowerment and alleviating poverty for local communities may prove harder than plant removal. The scope for employment in catchment clearing is massive if Programme funding is sustained, but it is less clear whether the Programme can provide meaningful and sustainable livelihoods for a significant number of people.

Success may depend on the Programme's ambitious aim of shifting many of the 92 percent of its participants who currently remove plants into higher-paying, permanent jobs in fire management, ecotourism, and "secondary" industries (Fynbos Working for Water Allied Industries 1998:4). Secondary industries are businesses that turn cleared invasives into profitable products like firewood, treated processed timbers, and crafts. Through a partnership between the Green Charcoal Company and Working for Water, for example, a factory is manufacturing charcoal processed from harvests of invasive alien trees. This partnership lowers the Programme's clearing costs and simplifies follow-up treatment of the cleared areas by removing the felled wood. In Mpumalanga Province, the Programme is producing wood chips that can be mixed with cement to create panels for inexpensive, insulated home construction. A possible partner is the Homeless People's Federation, a network of savings and credit collectives that help disadvantaged citizens secure loans to build homes or start businesses. Perhaps the most poignant example of the secondary industry concept is the mills that Working for Water is building to produce, from invasive biomass, low-cost coffins. There is no shortage of buyers. The devastating spread of HIV/AIDS in South Africa has forced thousands of impoverished families to spend precious funds to bury relatives in expensive coffins.

But running a successful secondary industry requires management and business acumen and a labor force with solid technical skills. That is one reason why Working for Water seeks to sign contracts with established businesses–to gain managerial, marketing, and product development experience for workers and establish outlets for the felled wood or finished products. Programme workers also gain critically needed training. An assessment of Working for Water found that about 70 percent of laborers lack the skills for furniture building, saw-milling, industrial woodworking, or ecotourism (Fynbos Working for Water Allied Industries 1998:8). That relegates the bulk of untrained laborers to lower-paying firewood, bark, and chip industries.

The management deficit identified in the secondary industries also hinders Working for Water as a whole. The idea and vision for the Programme were implemented quickly by Programme founders eager to begin "doing" rather than "planning." The rapid Programme expansion appears to have short-changed worker training. Thirty-six percent of the Western Cape projects reported problems, such as removal of the wrong species, use of the wrong extraction methods, or failure to carry out the required follow-up prescriptions (Raddock 1999). Some projects are led by managers who lack experience, training, mentoring, and supervisory skills. Worker productivity flags under the daily-pay system, and poor management exacerbates the problem.

To improve quality control and productivity, Working for Water is shifting from the daily wage to a contract system. The best workers are promoted to "contractors" who identify people with initiative and form a labor team. After training, the contractors can bid on plant removal and restoration jobs that fall under the auspices of the Programme and can contract with private industries to clear invasives from railway and utility easements or other large land holdings. In test contract system areas, productivity is up 30–50 percent, and in some places more than 65 percent of the clearing is achieved by self-employed teams (Marais 1999; Botha 1999).

Fynbos vegetation is a shrubland characterized by a mixture of three main growth forms: proteoids, ericoids, and restioids.

The environmental goals of the Programme present challenges as well. Some allege that Working for Water is too politically driven, leading to an emphasis on labor initiatives rather than research, monitoring, and conservation practices such as careful rehabilitation of cleared areas. The return of a full complement of ecosystem services in cleared areas mandates that topsoil be replaced followed by mulching and plantings of indigenous vegetation to prevent soil erosion; that nutrient cycling be initiated; and that the provision of a clean water supply be promoted. If felled trees are not removed, wildfires can burn very hot (invaded grassland and shrubland sites have 10 times more fuel than noninvaded ecosystems), killing indigenous seed banks and causing soil to become water repellent. In subsequent rainfalls, sheet and gully erosion may result. Prevention of further invasions through careful management of primary infestation routes and sources–roads, railways, rivers, and actions of private landowners–requires more attention, too.

Programme success also depends on overcoming financial problems. Until the government's recent commitment to provide funding in 3-year cycles, varying levels of income meant labor contracts could be as short as 1 month. Also, the timing of cash flows does not always correspond with optimal seasonal work plans. For example, the ideal time to cut wattles is in the winter when cold temperatures would help kill trees, but funding has sometimes only been available in the summer when regrowth is strongest. Another problem is that sudden infusions of cash from the Poverty Relief Fund might necessitate surges in hiring and clearing efforts without adequate management.

A Complex Fabric of Solutions

Without its tangible social welfare benefits, few democratic governments would embrace an investment of public resources on the scale of the Working for Water Programme. In a country with poverty as widespread as in South Africa, it would be hard to convince public leaders that limiting the spread of alien plants–even with compelling evidence that biodiversity or water is at risk–outweighs the need to provide a living wage.

But Working for Water relates ecosystem protection to local residents' lives, viewing social context not as a static background but as a promising avenue for ecosystem restoration. Rather than cordoning off one problem from another, the Programme weaves a solution around all of them. A surplus of unemployed citizens is tailored into a resource, not a drawback. Felled wood is an input, an opportunity for entrepreneurs, and a source of Programme funding, not waste. Clearing trees in a community offers a chance to provide education programs.

Many hands weave Working for Water's complex fabric of solutions. The Programme benefits immeasurably from a savvy public relations campaign and the support of myriad government agencies. Programme promoters have garnered international recognition and R23 million in foreign aid (Gelderblom 2000). Programme managers capitalize on marketing opportunities, such as outfitting workers in bright-colored T-shirts printed with the Programme logo and the names of financial sponsors. Partnerships with government agencies, nonprofit organizations, and the private sector yield management advice, research, ideas, and staff and materials. Perhaps most important, the tacit buy-in of those many partners has transformed Working for Water from an idea to a multimillion-dollar project in just 5 years. The high levels of recognition that the Programme has gained among national and international publics and policy makers also offers insurance against cutbacks in tough budgetary times.

Whether Working for Water can grapple comprehensively and cogently with invasive plants, water conservation, poverty, and even worker health remains to be seen. There is the strong possibility that the Programme will fall short of its goals. Controlling invasives completely may not be possible, but partial success will still warrant acclaim. Even if invasives' spread continues to outpace Working for Water's efforts, the Programme's expenditures have already translated into more water. The Programme's social welfare strategies have brought about greater public understanding of the value of ecosystem services, better health education, and worker skills training. These investments cannot be lost.

Persistence is critical to what must be an ongoing process of watershed restoration and biodiversity protection in South Africa. Sustaining the necessary public and political interest, sufficient to ensure millions in annual funding, is no small task. But the need for water–mandated for all by law and essential for economic growth–plus the need for jobs may be the ultimate insurance that the Working for Water Programme will succeed.

Managing the Mekong River: Will a Regional Approach Work?

The Mekong River represents a last chance of sorts–the last chance to tap a large, relatively pristine river basin's potential to supply energy and water without destroying its environmental integrity. The Mekong is the world's 12th longest river, stretching 4,880 km from its source on the Tibetan plateau to its outlet on the coast of Vietnam. It is the 8th largest river in terms of annual runoff and perhaps the world's least exploited major waterway in terms of dams and water diversions. But the Mekong's 795,000 km^2 watershed includes six of Southeast Asia's richest and poorest nations–Cambodia, China, Lao PDR, Myanmar, Thailand, and Vietnam. All these governments are eager to promote economic development using the Mekong's water resources (MRC 1997:14-15).

The drive to dam and divert the Mekong threatens the traditional uses of the river–as a source of fish and a barrier to salt water penetration into the rich Mekong delta soils. Ideally, a new model of coordinated regional water management will preserve those benefits while sharing new ones. The Mekong River Commission (MRC), originally known as the Mekong Committee, was established among the basin countries in 1957 to address potential conflict over hydropower development. The MRC provides a vehicle for joint management of the river and for the coordination of development strategies for the lower Mekong basin. In 1995, after almost 4 decades of political turmoil had hampered the Commission's effectiveness, the basin countries reaffirmed their interest in working together. Cambodia, Lao PDR, Thailand, and Vietnam signed the Agreement on Cooperation for the Sustainable Development of the Mekong River basin, which acknowledges the need for regional action. China and Myanmar have observer status.

Yet the MRC lacks any real power to develop or enforce a unified vision of sustainable water use in the basin, and each of the riparian countries is pursuing its ambitious development plans largely independently at this time. Can a truly regional approach to Mekong management evolve in time to influence the basin's environmental future?

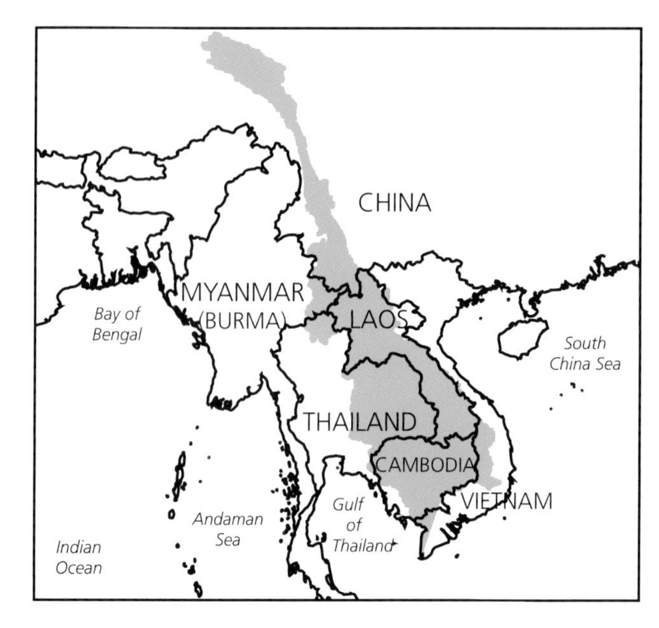

Damming the Mekong

The Mekong River and its tributaries have a potential hydroelectricity generating capacity of 30,000-58,000 MW (MRC 1997:5-19). Although plans to construct major hydroelectric dams have been afoot for years, as of 1997 less than 5 percent of this potential had been exploited.

Now, however, scores of large dams are under serious consideration in response to both the growing regional demand for electricity and the desire of the nations in the basin to earn foreign exchange from international sales of hydropower. The financial crisis that erupted in Asia in 1997 shook Thailand's economy particularly hard, slowing electricity consumption and delaying power purchase agreements and dam start-ups, but energy demand is expected to pick up again quickly as the recession recedes (EIA 1999). By 2020, electricity demand in the Mekong region could be six times greater than in 1993 (MRC 1997:5-9).

Hydropower potential varies greatly among the riparian nations. Highland countries like China and Lao PDR possess the greatest share, while countries like Vietnam and Cambodia–along the slower-moving, lower reaches of the Mekong–possess relatively little. Currently, major pressures on the Mekong include:

- China's Yunnan province at the top of the watershed is planning a cascade of up to 14 dams on the upper Mekong–known locally as the Lancang River. These dams would have a total installed capacity of 7,700 MW, equivalent to 20 percent of China's current energy consumption. Because of Yunnan's remoteness from China's more developed areas and the chance to earn export dollars, Yunnan authorities are likely to export electricity to Thailand. China has also proposed plans to divert water from the Mekong into the Yellow River to meet Northeast China's growing demand for water.

- Many of the tributaries feeding the Mekong in Thailand have already been dammed to provide power and irrigation water to its arid eastern provinces. However, Thailand has

Box 3.16 How the Mekong's Hydropower Resources Are Divided

The Mekong Basin at a Glance

Country	Average Flow from Catchment Area (m^3/sec)	Percentage of Total Flow	Population: National (millions)	Population: Basin (millions)	GDP ($ billions)	Consumption: Electricity (KWh/ person/yr)	Consumption: Fish (kg/ person/yr)
China	2,410[a]	16	1,278.0	5.9	902.0	260[a]	—
Cambodia	2,860	18	11.2	8.7	3.0	55	13
Lao PDR	5,270	35	5.4	4.6	1.8	55	7
Thailand	2,560	18	61.4	22.1	153.9	900	15–27[b]
Vietnam	1,660	11	79.8	14.0	24.8	140	21–30[c]
Myanmar	300	2	45.6	0.4	—	60	—

Note: —, data not available.

[a]Yunnan Province only. [b]Northeast Thailand only. [c]Mekong delta in Vietnam only.

Sources: UN 1998; CIESIN 1999; World Bank 1999; MRC 1997:5–11, 5–20.

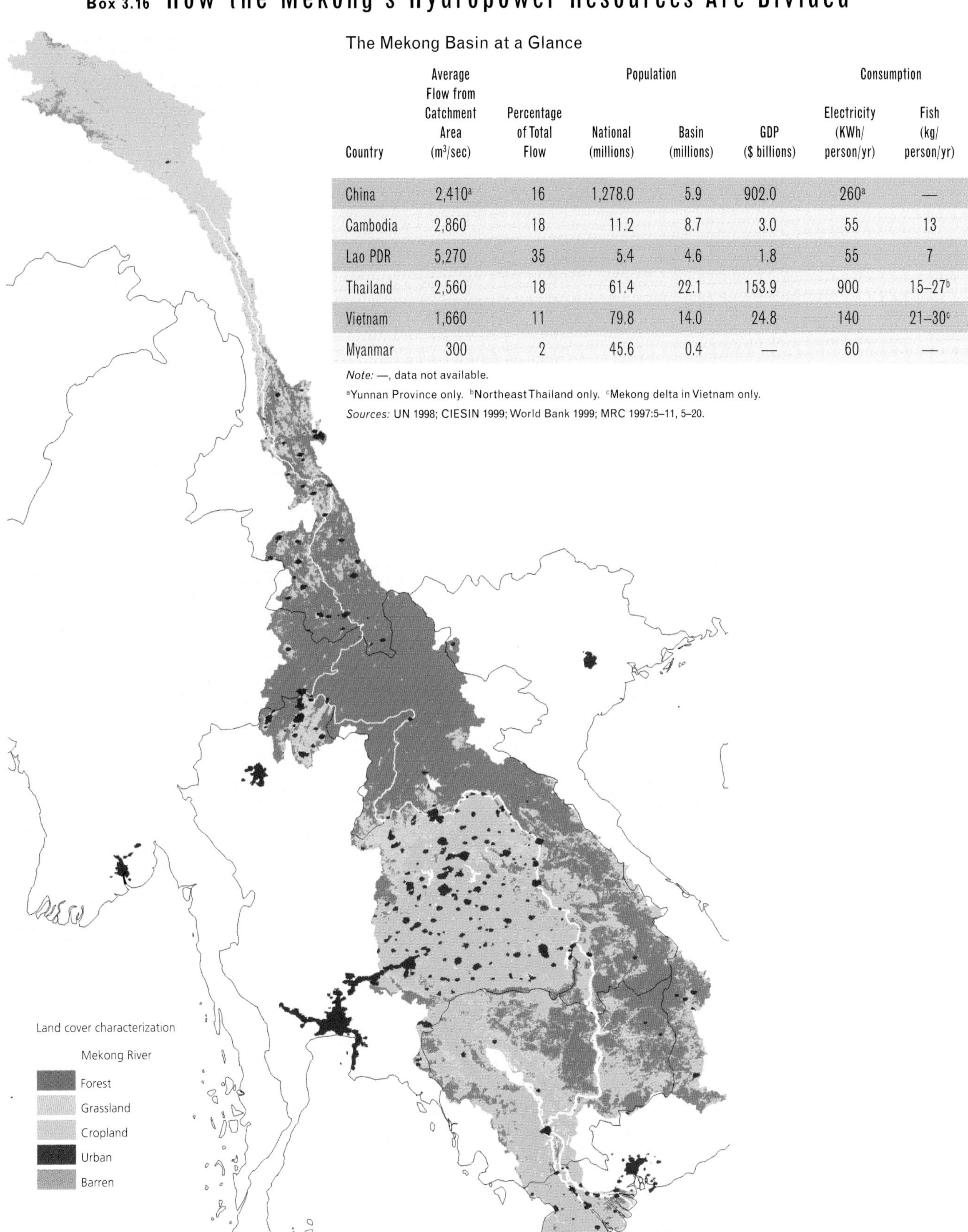

long-standing plans to divert water from the Mekong into the water-scarce Chao Phyra River, the main source of water for Thailand's economic heartland.

- One-third of the total flow of the Mekong originates in Lao PDR. Given its abundant rainfall and rugged topography, estimates of the country's hydropower potential reach 7,000 MW, of which only a fraction is currently exploited. Laos has prepared plans to construct as many as 17 new dams during the next decade to reduce the country's poverty. Most of the hydroelectricity will be sold to Thailand and Vietnam. Thailand already buys electricity from Lao PDR's Nam Ngum dam and is negotiating to buy power from the planned Nam Theun II dam.

Not all the proposed projects will be developed, however. Only a handful are both technically feasible and economically viable, and public and NGO outcry against some–like Nam Theun II–may stall construction. For those hydropower plans that do hold economic promise, the private sector stands ready to invest. Often the funding comes through "build-own-operate-transfer" (BOOT) projects, in which foreign investors finance, construct, and operate a dam, recouping their investment and sharing risk during a concession period, then transfer ownership of the project to the government.

Vulnerability Downstream

Although dams and diversion projects dominate the official development discourse, the Mekong has long provided many other environmental benefits to the basin's 55 million inhabitants. Approximately 30 percent of households in the Mekong delta are below the poverty line and most of the rural population depends on the river and its tributaries for their survival (MRC 1997:4-6).

For example, the fish caught in the Mekong are the source of 40-60 percent of the animal protein consumed by the population of the lower basin, and fish sustain an even higher percentage of people in much of Cambodia (Institute for Development Anthropology 1998:87-88). The 900,000 tons of fish harvested annually (Friederich 2000) and the Mekong's extraordinary fish species richness are threatened by dams, which interfere with spawning cycles by preventing fish migrations.

Dams also reduce the seasonal floods that sustain fish spawning and nursery grounds in the wetlands upstream and the delta region. The flood cycle, keyed to the monsoon rains, is a critical factor in the life cycle of many of the area's aquatic species. Even slight changes in peak flood flow could threaten the region's fish production and food security (MRC 1997:3-8). Impacts observed at dams already constructed on Mekong tributaries illustrate the area's vulnerability. At Nam Pong reservoir in Northeast Thailand, the number of fish species found in the river dropped from 75 to 55 after impoundment. Fishermen upstream of Thai dams at Tuk Thla and Kompol Tuol saw their catches decline from 5-10 kg/day to 1-2 kg/day after the dams were built (MRC 1997:5-14).

Altering the annual flood cycle, reducing the silt load of the water, or diverting the Mekong's flow could also have serious impacts on agriculture in the Mekong delta. Flood waters deposit 1-3 cm of fertile silt each year on the lowland floodplains in Vietnam and Cambodia, sustaining these intensively farmed areas (MRC 1997:2-17). In addition, river flows during the dry season are important for controlling salinity penetration into interior areas from the coast. According to the Vietnam Water Resources Sector Review, seawater penetrates up to 70 km inland during the dry season. If current trends in water abstraction in the delta continue, the area affected by salinity could increase from 1.7 to 2.2 Mha (Xie 1995:10). Increased salinity was cited as the primary cause of rice yield declines of 50-90 percent in Tra Vinh province over the last 30 years (Nguyen 1998:4).

The dangers that dams could pose to the biodiversity of the Mekong must also be considered in the context of the environmental degradation that the region has already suffered. A combination of deforestation, increasing conversion to intensive, chemical-dependent agriculture, continued population growth, and mangrove clearance for shrimp aquaculture in the delta region has compromised the basin's environmental health. Vietnam, for example, has already lost approximately 85-90 percent of its forest cover, largely because of decades of war and reconstruction. In Thailand, perhaps 55-65 percent of forests has been cleared for agriculture and tree plantations (WCMC 1994:106-107). Some of the highest rates of deforestation in the world continue to plague the riparian countries (FAO 1999:132). Many remaining forests are of poor quality, affecting water retention in the basin and promoting land degradation and soil loss in the uplands (MRC 1997:3-5). Disrupting flood cycles or decreasing base flows during dry times through water diversions could add significantly to these existing stresses.

Furthermore, where will countries resettle the thousands of people who will be displaced by dams? Just the nine proposed mainstream dam projects could displace 60,000 people (MRC 1997: 5-24).

Conflict Brewing?

With all its mighty waters, the Mekong ecosystem is finite and fragile. The array of current demands and future plans for the river has already led to increasing competition among the basin countries. The MRC was established to minimize the conflicts inherent in managing a river that crosses many international borders, but its efforts at regional coordination have been largely unsuccessful (China Environment Series 1998). Although it collects hydrological data from the basin, the MRC has done little to analyze the data, promote debate among the partners on the cumulative effects of their water developments, or craft a common vision of how water should be shared. As a result, the governments of Cambodia, Vietnam, Lao PDR, and Thailand are competing for international funding for their dam-building projects and have "... adopted a rhetoric of cooperation and sustainable development to mask underlying conflicts and competition" (China Environment Series 1998).

Complicating the equation is the fact that China is not a member of the MRC, although it controls the upper reaches of the river and has an ambitious dam-building program in place. China is reluctant to join the MRC until water-use rules are clarified and it is assured that restrictions on dam building and water diversions will not interfere with its upper Mekong development plans. The agreement specifies that the watershed nations have neither the right to veto the use nor the unilateral right to use the water of the Mekong. This implies that dam construction on the river's mainstream would only proceed by consensus, a system unacceptable to China.

In reality, compromise will be difficult for all the basin countries, whose negotiating powers vary greatly as a function of their location within the river basin and their wealth. Based on the size of its economy, China has by far the greatest capacity to mobilize funding and technology to exploit its "share" of the Mekong. Because its portion of the river runs through sparsely populated territory, China also has a relatively small population that depends on the river for irrigation and fish production. China, therefore, has much to gain and little to lose from dam construction. Cambodia and Vietnam, on the other hand, are extremely vulnerable because of their downstream location, relative poverty, and the large number of people that depend directly on the Mekong for their livelihoods. Lao PDR, one of the poorest nations in the world, is desperate to develop its hydropower resources to spur economic growth. Thailand is in an intermediate position. It has the largest within-basin population among the riparian countries, but has the economic and human resources to withstand potentially negative changes in the river upstream.

A Regional Vision

Despite the current imbalance of power among the riparian countries and the potential for conflict, the benefits of a regional approach are compelling. Development of a regional electricity transmission grid, for example, would benefit from a coordinated plan to develop the basin's hydropower potential. A regional grid would facilitate China's ability to market hydropower to other energy users in the region, offering advantages all around. In addition, a regional growth plan that helps expand the economies of the lower Mekong basin countries and promotes open markets in the region provides a longer-term inducement for Thailand and China to cooperate.

A basin-wide approach to water management would also offer clear environmental advantages. It would, by definition, force the riparian countries to examine how dams on the upper reaches of the river would affect flow conditions downstream. Currently, upstream countries can pursue water withdrawals and hydropower production while ignoring repercussons such as salt water intrusion, decreased catches for subsistence fishing, and soil depletion.

Since the governments in the region unanimously favor developing the region's hydropower potential, a regional approach to water management would not necessarily mean less power generation, but it would offer a chance to distinguish between environmentally "good" dams and "bad" dams. The challenge is to select dams that meet strict environmental and economic standards. Some have argued, for instance, that dams on the Lancang and in the uplands of Lao PDR are "good" because they generate a lot of power without displacing many people and flooding large areas. Thus, the social and environmental costs are relatively small. It is also possible that dams could actually benefit the local environment in some ways. Planners of Lao PDR's Nam Theun II dam have proposed earmarking a portion of the hydropower revenue for forest conservation in the surrounding watershed. Protecting forests around dams is desirable because it reduces sedimentation, lowers maintenance costs, and prolongs dam life.

But capitalizing on the benefits of a regional approach to water development and use in the Mekong region will take quick action, given the rapid changes under way. Water experts warn that now is the time to rethink basin-wide water management, not after the dams and diversion schemes have been built and the environmental and geopolitical repercussions are felt.

The MRC has a critical role to play in promoting regional cooperation. It has been criticized for failing to seriously address the potential negative environmental impacts of proposed dams and diversion schemes, and it has failed to build the predictive modeling capacity that is needed to assess the trade-offs between river basin development options. But the MRC reaffirmed its commitment to environmental analysis and assessment in 1995 and to serving as a regional information center on environment and development in the Mekong River basin. These developments could help basin nations to better visualize the benefits of a regional approach to managing the Mekong watershed and to quantify the damage–environmental and social–that may occur if they pursue an uncoordinated approach.

New York City's Watershed Protection Plan

To safeguard the city's drinking water, in 1997 New York City chose to launch an ambitious environmental protection plan, rather than build an expensive water filtration plant. By protecting its watershed the city would employ nature's ability to purify water while preserving open space and saving money. But as this widely heralded example of watershed protection is implemented, many question whether it will, in fact, deliver all that it promises.

For more than a century, New York City residents have enjoyed drinking water of such purity that it has been dubbed "the champagne of tap water." That water–about 1.3 billion gallons per day–flows from an upstate watershed that encompasses 1,970 mi^2 and three reservoir systems: the Croton, Catskill, and Delaware (NRC 1999:3, 17). Until relatively recently, undisturbed soil, trees, and wetlands provided natural filtration as the water traveled through the Catskill Mountains and the Hudson River Valley before reaching 9 million residents of the city and its suburbs. The only regular treatment needed was standard chlorination to control waterborne diseases such as cholera and typhoid.

But in the last several decades, development has brought increasing numbers of people and pollutants to the watershed, straining the land's buffering and filtering capacities. More than 30,000 on-site sewage treatment and disposal systems and 41 centralized wastewater treatment plants discharge wastewater into the upstate watersheds (NRC 1999:358). Runoff from roads, dairy farms, lawns, and golf courses contains fertilizers, herbicides, pesticides, motor oils, and road salts.

The need to attend to the development-pressured upstate watershed became clear in 1990. The U.S. Environmental Protection Agency (EPA) put New York City on notice: protect the source for the Catskill and Delaware reservoirs–the watershed, nature's own treatment plant–or construct and operate a water filtration system. Filtration would cost \$3–\$8 billion, according to various estimates, potentially doubling the average family residential water bill (Ryan 1998). By comparison, the City determined that the price tag for watershed protection would be just \$1.5 billion, increasing the average water bill of a New York City resident by about 1–2 percent, or \$7 per year (Revkin 1995, State of New York 1998).

The EPA's warning was compelled by the 1989 Surface Water Treatment Rule, which requires that surface water supplies for public water systems be filtered unless stringent public health criteria are met and extensive watershed protection strategies minimize risks to the water supply. The rising levels of bacteria and nutrients in the watershed, plus the risks posed by antiquated sewage treatment plants and failing septic systems, put New York City's Catskill and Delaware supplies in danger of violating the Rule. The Croton supplies

east of the Hudson River were in bigger trouble already: because of that area's greater pollution pressures, filtration was mandated. Even though the Croton system supplies just 10 percent of the City's water, compared to the 90 percent that flows from the Delaware and Catskill systems, the cost to build and maintain that plant is still expected to be at least \$700 million (Gratz 1999).

The cost savings from protecting the Delaware and Catskill supplies were clear, but crafting and implementing a major ecosystem protection plan is no small undertaking. Nationwide, less than 2 percent of municipalities whose drinking water systems are supplied by surface water have demonstrated to the EPA that they can avoid filtration by instituting aggressive watershed protection programs (Gratz 1999). The vast majority are far smaller than New York, less populated, and own substantially more of the critical watershed lands. When the protection agreement was crafted, New York City owned just 85,000 acres of the watershed, less than 7 percent of the total critical area, including the land beneath the reservoirs (Ryan 1998); another 20 percent was owned by the state (NRC 1999).

With so little watershed land under its direct control, but millions of water users dependent on it, New York City needed to obtain the support of upstate landowners for open-space conservation and stronger land-use protection. But from the perspective of upstate communities, watershed restrictions such as land acquisitions, limits on where roads and parking lots can be constructed, and strict standards for sewage treatment systems amounted to outsiders threatening local taxpayers' economic viability. Still, after years of contentious negotiations, city, state, and federal officials, some environmentalists, and a coalition of upstate towns, villages, and counties forged a 1997 watershed management agreement that convinced the EPA to extend its filtration waiver until 2002.

Perhaps the most crucial element of the program is the state's approval of New York City's plan to spend $250 million to acquire and preserve land in the watershed, with priority given to water-quality sensitive areas (NRC 1999:213). A local consultation process helps protect the interests of watershed communities. Other plan elements include new watershed regulations, direct city investments in upgrades to wastewater treatment plants to minimize contamination, city funding of voluntary farmer efforts to reduce runoff, and payments to upstate communities to subsidize sound environmental development (State of New York 1998).

In addition to economic savings, the ecosystem protection program offers some additional advantages that filtration cannot. It lowers health risks that are present even with filtration–for example, the risk that a sewage plant will malfunction or an incidence of the disinfectant-resistant pathogen *Cryptosporidium* will occur. Land acquisition and development controls also mean more land for parks, recreation, and wildlife habitat.

Ownership of Critical Watersheds

Only a handful of major U.S. cities have unfiltered water supply systems—mostly those that can ensure long-term water protection because significant portions of the critical watershed lands are owned by the water utility or are designated as protected open space under state or federal ownership and management. New York City is an exception—and accordingly, it must rely heavily on the cooperation of private upstate landowners to help protect its drinking water.

City	Ownership (percent) Public	Ownership (percent) Private	Watershed Area (acres)	Population Served (millions)
Seattle, WA	100	0	103,885	1.2
Portland, OR	100	0	65,280	0.8
New York, NY	26	74	1,279,995	9.0
Boston, MA	52	48	228,100	2.4
San Francisco, CA	100	0	475,000[a]	2.3

[a]Supplies 85 percent of the city's water; 15 percent is filtered and comes from other publicly owned watersheds.

Sources: NRC 1999; personal communications.

But whether this dramatic effort will prove to be a bargain remains to be seen. Among the unknowns are the effectiveness of voluntary pollution protection commitments by farmers, and still-evolving knowledge of best management practices to control roadway, lawn, farm, and other runoff. Environmental organizations are concerned that the negotiated settlement contains serious loopholes in the watershed rules and land-buying requirements. For example, the agreement provides no limits on the number of new sewage treatment plants that can be built in the City's cleanest reservoir basins.

Nor does the agreement specify an absolute acreage requirement that the city must purchase in the watershed, only that the city must *solicit* the purchase of 350,000 acres. The City projects that this approach could lead to its acquisition of about 120,000 acres, allowing it to increase its holdings to 17 percent of the critical land area in the next 10 years (Gratz 1999). However, the City's solicitation efforts might yield far less land, since the plan relies on the cooperation of upstate residents–and even 17 percent ownership gives the City limited watershed control. Another problem is that the plan sets criteria for types of land to be acquired but no assurance that the "best" lands from the perspective of water quality will be purchased, since land is obtained on a willing buyer/seller basis. From the perspective of the Natural Resources Defense Council, the plan may allow too much development to take place on sensitive watershed lands and the scientific aspects of water management were given insufficient attention by negotiators under pressure to craft a politically acceptable plan (Izeman 1999, Revkin 1997). Other concerns include inadequate requirements for buffers–zones of vegetation where discharge of pollutants, and development, cannot take place (NRC 1999:14)–and the agreement's failure to emphasize pollution prevention as much as pollution control.

Only years of extensive water quality monitoring will prove whether the watershed protection program is sufficient to protect public health. At the moment, the water is still deemed safe to drink, but some still think filtration ultimately will be required.

Shortcomings aside, the agreement is laudable. It formally acknowledges the interests of watershed residents and stresses the need to implement watershed protection plans fairly and equitably. Elements of the New York City watershed agreement may serve as a model for other communities. There is a growing recognition that filtration, by itself, is no panacea. It can reduce the threat of waterborne pathogens, but it cannot completely eliminate the threat, especially if the source water is poor. Watershed protection offers a cost-effective approach to clean drinking water, and benefits the environment as a whole. The challenge in the case of New York City is the need to compel many people and communities to work together, putting aside self-interest, toward the twin goals of saving the watershed and saving money.

GRASSLAND ECOSYSTEMS

Sustaining the Steppe: The Future of Mongolia's Grasslands

For thousands of years, most of central Asia's high steppe has been the realm of nomadic herders and their horses, camels, goats, sheep, and cattle. Today, this expanse of grasslands–the largest remaining natural grasslands in the world (WCMC 1992:287)–is divided, politically, between Russia, China, and the Republic of Mongolia. This entire region is sometimes called "Inner Asia."

For Mongolia, with a human population of just 2.4 million in a land area the size of Western Europe, there would seem to be an abundance of pasture for its 30 million head of livestock. But natural conditions make the grasslands of Inner Asia highly vulnerable to damage from human activities and slow to recover. The growing season is just 4 months long. Annual precipitation ranges from just 100 mm in the most arid regions to 500 mm in limited northern areas, and in much of the region is less than 350 mm. The steppe is subject to intense winds, snow can cover the ground 8 months of the year, and in the dry season grass and forest fires are common. These ecological and climatic factors inhibit the growth of vegetation and increase the severity of erosion in areas with unprotected soils (Palmer 1991:55).

In an environment of extremes, herders have recognized the merits of moving their herds seasonally or more frequently. Herd mobility seems to sustain the fertility of rangelands, and thus benefits livestock health and food security. In the feudal period, herders would rotate animals over pastures where they had access to abundant seasonal grasses or shelter from harsh weather–usually pastures to which use rights were coordinated by local authorities, such as lords or monasteries and their officials. Occasionally herders would use a technique called *otor*–movement of livestock to even more distant and lesser-used pastures. *Otor* helped to intensively feed the animals and prepare them for severe, grass-scarce winter and

spring seasons and could be used to relieve pastures when a shortage of forage or degradation became evident.

Important aspects of these coordinated, large-scale, highly mobile systems endured in Mongolia even through the socialist government campaigns that organized livestock herders into collectives in the 1950s. Since 1990, however, Mongolia has reoriented its economy from central planning toward privatized land and free markets. This has brought new opportunities to some, but it has also created social and economic conditions that are undermining the long-standing mobile herding culture and perhaps threatening its continued existence. Systems of wide pastoral movement, in many cases, broke down when the collectives ended and have been replaced with lower-mobility, small-scale pastoralism. This trend may pose a significant threat to the sustainability of Mongolia's grassland ecosystems.

A similar shift from mobile herding to more sedentary livestock rearing mixed with farming systems had already occurred in the Chinese and Russian regions of Inner Asia, and the environmental effects are discouraging. Like Mongolia, these countries experimented first with organizing herders into collectives–Russia in the 1930s and China in the late 1950s. Then, decades later, they privatized livestock operations in a bid to modernize and increase production. Meat and wool production increased but with costs to the ecosystems, including pasture degradation. Estimates vary widely, but local studies in Buryatia and Chita in Russia and in Inner Mongolia in China suggest that as much as 75 percent of grasslands has suffered some degree of degradation (Humphrey and Sneath 1999:52; Gomboev 1996:21). According to Chinese government figures, just 44 percent of Inner Mongolia's grasslands are considered usable and in good condition (Neupert 1999:426).

By comparison, Mongolia's grasslands are in relatively good condition. Officials have calculated that moderate or severe degradation affects 4–20 percent of pasture lands (Government of Mongolia 1995:28).

The ecosystem problems in parts of China and Russia underscore for Mongolia the merits of preserving elements of the mobile herding practices. Incorporating mobile herding into the modern Mongolian economy may be essential to local livelihoods and national prosperity. Grasslands cover about 80 percent of Mongolia's 1.567 million km^2 land area and agriculture–mainly livestock herding–supplied 33 percent of Mongolia's GDP in 1998. Approximately half the national workforce works in the agricultural sector, mostly as pastoralists (herders) (National Statistical Office of Mongolia 1999:45, 54, 95; Statistical Office of Mongolia 1993:6). Mongolian exports of livestock products have collapsed since the end of the socialist trade bloc in 1989–91, but in better economic times, pastoralism supplied substantial raw materials such as wool and hides for Mongolia's export trade and fledgling industrial sector. And Mongolia's future economic growth depends at least in part on livestock production. Economic growth is a priority for Mongolia, whose per capita GNP of US$380 (1998) makes it one of the poorest countries in Asia (World Bank 2000:11).

At individual and local levels, the meat, milk, and transport that livestock provide are vital to the many herders and their families living in remote, inaccessible places. Price inflation and fuel and commodity shortages during the current transition to a market economy make livestock even more essential to households' food security.

"Following the Water and Grass"

Large-scale, highly mobile herding operations have ancient roots. From the 17th until the 20th century, Mongolia was divided into administrative districts called *hoshuu* or "banners" ruled by a hereditary lord or a Buddhist monastery. The commoners were bound to particular geographic areas and required to work for local authorities. Buddhist monasteries, nobility, and the imperial administration owned millions of animals that were herded by subjects and servants who generally received a share of the animal produce in return.

The pastoral movement systems could be sophisticated. The herder groups were flexibly organized, consisting of one or more families. Herders and their families might move large groups of horses, sheep, goats, and other domesticated or semidomesticated animals to selected seasonal pastures in an annual cycle (Simukov 1936:49–55). Because different animals have different grazing habits, animals were segregated by species for efficient pasture use. Sheep, for example, crop so close that horses and cattle cannot get at what is left, forcing horses to dig up grass roots to eat. Some members of the herder group might specialize in working with a particular species. Others might cut wool, milk animals, make felt for tents, or help the group move to a new camp.

There was enormous variation in frequency and distance of moves. In better-watered northern regions, herders might move livestock twice a year. In other areas, herders might make three to four long-distance moves; in some places, more. The ancient Chinese description for these pastoral activities was "following the water and grass" (Hasbagan and Shan 1996:26).

With local lords and monasteries to coordinate general access to pastures and to support pastoral movement, herding families usually could share seasonal pastures efficiently and avoid pasture overuse. These flexible herding systems and collective-use arrangements also ensured that water sources or the best pastures were not controlled by a few herders to the detriment of the whole herding system (Mearns 1991:31).

Such herding principles and techniques have been passed down through the ages with remarkable continuity. Some pas-

(continues on p. 216)

Box 3.17 Overview: Mongolia's Grasslands

Nomadic herders have grazed livestock on Mongolia's vast but fragile grasslands for thousands of years. By rotating animals over shared pastures in collaborative seasonal and species-segregated patterns, herders have anchored their country's economy without degrading its ecosystems. Recent political and economic changes, however, may be eroding these sustainable practices. Analyses of neighboring grassland regions in China and Russia warn of the degradation possible when large-scale mobile herding practices decline and small-scale static systems expand.

Ecosystem Issues

Grasslands

Estimates of grassland degradation are much debated and range from 4 to 33 percent, but the clear potential for further degradation is cause for alarm. Grasslands are the basis of livestock production and approximately half of Mongolia's workforce depends on pastoralism or agriculture for their livelihoods and food security. Overgrazing, mining, vehicular traffic on the steppe, and other pressures threaten grassland biodiversity. Among the mammals at risk are Mongolia's gazelles, wild camels and horses, and the Asiatic wild ass.

Agriculture

Much of Inner Asia is not well suited for growing crops; half of all cultivated land in Mongolia is considered degraded. Sedentary livestock will require conversion of more land to agriculture to supply food and fodder for animals and people.

Freshwater

Mongolian herding practices are dictated in part by the uneven and irregular distribution of water in Mongolia. Growing concentrations of herders and settlements near water sources intensify pressure on natural resources in those areas. Those same water sources supply irrigation water for agriculture; agricultural water use in 2000 is projected to triple its 1970 amount.

Forests

Forests, found primarily in Mongolia's wetter, mountainous areas, are critical to the protection of soil, grasslands, water resources, and wildlife diversity. However, reduction of forests by logging, use for fuelwood, and forest fires is accelerating.

Management Challenges

Equity and Tenurial Rights

For centuries a variety of collective tenure arrangements have helped sustain grasslands and produce healthy livestock in Mongolia. The recent transition to private land and herd ownership, however, has decreased flexible systems such as rotational grazing and access to shared grazing lands. In some areas land tenure is ambiguous; in others wealthier pastoralists have fenced large areas of high-quality grasslands.

Economics

Reorientation from a centrally planned to a market economy may spark environmental problems and widen income inequality; poorer pastoralists may not be able to capitalize on economies of scale and access large areas of high-quality pastures. The government has cut supportive services to herders since the breakup of collectives, and few pastoralists can afford the fuel or other inputs necessary to sustain mobile herding operations.

Stakeholders

Privatization is bringing divisive elements to herding communities. The influx of new herders with limited experience in animal husbandry, the widening gap between rich and poor herders, and absentee herd ownership all weaken the system of shared beliefs and preferences for mobile herding that once helped protect grassland condition. Sustainable management suggests the need for government policies that facilitate and encourage mobility rather than sedentary production.

Information and Monitoring

Pastoralists' ecological knowledge, understanding of local geography, and animal husbandry skills need to be incorporated into management policies. There also is room for scientific analysis and research to help guide a transition to privatization without losing the best aspects of mobile herding. Assessments of pasture condition, arable land, and livestock use, and identification of pastures that are of strategic importance to mobile herders would greatly aid the transition.

Timeline

1691–1911 Mongolia becomes a frontier province of China. Herders move livestock for Buddhist monasteries, high lamas, and aristocratic lords in rotations over common lands; pasture rights are regulated by the local institutions and among clans and families according to customary law.

1911 Expulsion of Manchus in northern Mongolia brings a decade of Mongol autonomy.

1921 Bolshevik uprising in Russia inspires revolution in Mongolia.

1924 Mongolian People's Republic is founded in northern Mongolia, creating the world's second communist state after the Soviet Union (USSR). The southern part of Mongolia remains under Chinese control and becomes the Inner Mongolian Autonomous Region in 1947, though it lacks real political autonomy.

1929–32 The Mongolian government attempts to forcibly collectivize herding households. Thousands of Buddhist lamas are killed and private property is confiscated. Herders slaughter 6–7 million head of livestock in protest.

1932 The Mongolian government shifts to a more gradual organization of collectives; cooperation among herding households is encouraged. Russia has already collectivized most rural residents at this time.

1949 The communist People's Republic of China is founded. Rangelands in Xinjiang, Inner Mongolia, and other areas are nationalized, removing them from the control of landlords, Mongol princes, lamaseries, and clans.

1950s–60s Chinese and Russian governments emphasize agricultural expansion and highly mechanized farming methods.

1950s Socialist government campaigns in Mongolia increase momentum for the organization of pastoralists into collectives. Expansion of area under cereal and fodder crop production begins.

1950s Russia and China encourage use of foreign breeds of sheep and other livestock to increase productivity; these "improved" breeds eventually prove weaker and decrease herd mobility.

1955 A ceiling is placed on private livestock holdings in Mongolia to encourage the emergent collectives.

1957 China begins to establish large collectives (People's Communes) in rural districts and eradicates customary use-rights for pastures. Grasslands become pressured as livestock herds and cultivated area expand.

1960s Virtually all of Mongolia's herding households are members of collectives and all land is owned by the state. Households look after a share of the collectives' herd, although they are also permitted to own some private stock. Mongolia begins expanding its cultivated area.

1980s China begins shift from a centrally planned to free-market economy. Agricultural communes are dissolved and livestock distributed to pastoral households. Farmers and pastoralists have leases for lands, but uncertainty over pasture rights and location discourages mobility. Fenced areas emerge in the once-unbounded steppe. The communist era ends in Russia. Influenced by political change in the USSR and Eastern Europe, Mongolia begins a transition to a democratic government and market economy.

Early 1990s Farms in Russia retain communal structure despite the new central government policies; many farm leaders are reluctant to hand over land and livestock to individual private farmers.

1991 Prices are freed from state control. Constitution of Mongolia acknowledges the principle of private land ownership, but pastureland is specifically excluded from private ownership and lease systems are developed. Mongolia begins to dissolve collectives; herd numbers soon increase more than 20 percent.

1994 More than 90 percent of Mongolia's animals have been transferred to private ownership. Many are owned by "new" herders who were allocated animals in the dissolution of the collectives; some opt for more sedentary herd management. Land degradation is perceived around herders' settlements.

2000 Severe economic crisis that began with the breakup of the USSR continues to limit economic growth and reconstruction in Mongolia. Government resources to support mobile herding are scarce and the gap between wealthy and poor herders grows.

toralists still shift their herds 150-200 km between summer and winter pastures. Others shift their herds 25-50 km, and some less than 10 km depending on social and economic conditions (Humphrey and Sneath 1999:221-222). But many pastoral systems are, fundamentally, still mobile, and pastoralists continue to stress the benefits of mobility and cooperative grazing for pasture and livestock health.

Science tends to support what herders have observed for generations. Ecological studies show that continuous grazing of livestock in the same pastures can be much more damaging than systems of pasture rotation (Tserendash and Erdenebaatar 1993:9-15). Dense populations of sedentary livestock can impair grass regrowth. Some plant species may gradually disappear and be replaced by poorly palatable weeds or poisonous plants that can sicken or kill livestock. Once a pasture's soil is severely damaged, wind can cause desertification.

A New Era in Mongolia: 1921-90

The pastoral culture experienced major new influences in the 20th century. After only a decade of Mongol autonomy, following the collapse of the Chinese Qing Dynasty, struggles for power led to the 1921 Bolshevik-inspired revolution. Socialist central planning emerged under the leadership of the Mongolian People's Revolutionary Party in 1924. This era introduced technologies like irrigated agriculture and farm machinery. It also introduced state-controlled pastoralism and brought the beginnings of industrialization. Mobile herding techniques generally endured–even improved in some ways–during this period.

One of the first steps of the Soviet-style government was to organize herders into collectives. Early attempts at collectivism were so unpopular they had to be abandoned. However, in the 1950s, Mongolian pastoralists were organized as wage workers employed by about 250 *negdels* or collective farms and about 50 state farms, each managing pastoral or agricultural activity in a rural district or *sum*. A *sum* consisted of a central settlement of a few hundred households and a large area of grassland used as pasture by the herder households, most living in mobile felt *yurts* and herding the collective or state farm livestock and a few personal animals. Although the new *sum* districts were generally smaller than the earlier *banners*, most pastoralists continued to rotate pastures throughout the year and make use of *otor*. However, in some regions the distance of seasonal moves was reduced (Humphrey and Sneath 1999:233-264).

This "collective" system actually enhanced mobile pastoralism in some ways. The collectives maintained machinery

Box 3.18 Land Use in Inner Asia

The Asian steppe, including Mongolia and parts of China and Russia, support the most extensive natural grasslands in the world (WCMC 1992:280-292). The climate is harsh; on some regions of the steppe, snow can cover the ground for 5–8 months of the year. Extreme heat and drought are possible, too, particularly in the southern desert regions that cut off Mongolia from Tibet. In effect, much of Inner Asia is not readily adaptable to most economic activities; large areas of the Russian Federation, for example, consist mostly of high mountain ranges.

But livestock have thrived on the steppe for centuries. In fact, most of Inner Asia that is accessible is used for livestock grazing. Agriculture is also a significant land use, although less than 1 percent of Mongolia's land area is classified as arable (Mearns 1991:26). Thus, the way of life for many is rural, and the importance of herds as sources of food, wool, and transportation is paramount.

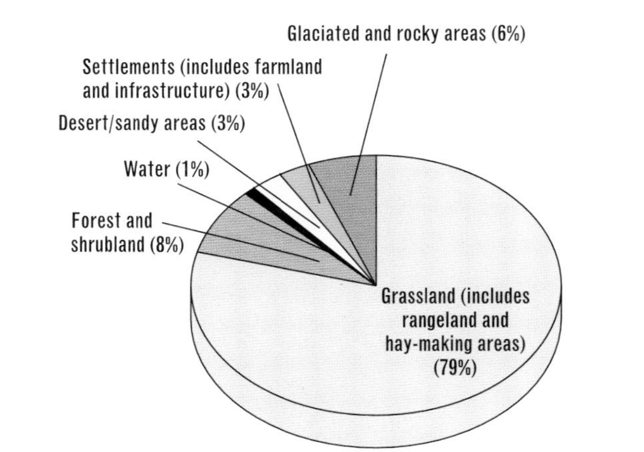

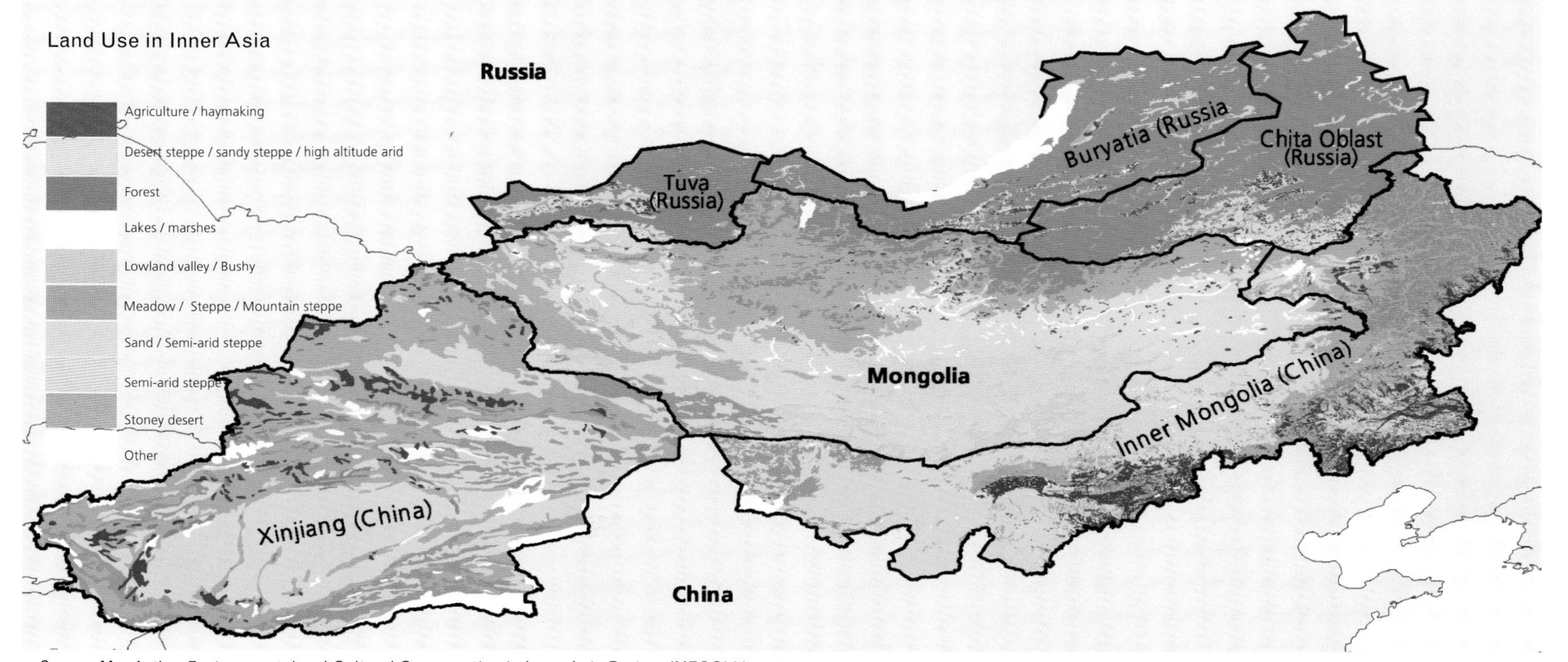

Source: MacArthur Environmental and Cultural Conservation in Inner Asia Project (MECCIA) 1995.

Box 3.19 Pastoral Movements

A variety of pastoral systems are practiced in herd movements in Inner Asia, depending on environmental, social, and economic conditions. In one area of Mongolia (Hovd *sum,* Uvs *aimag*), for example, most pastoralists use pastures that are high in the mountains in the summer—areas above 2,400 m. In autumn pastoral households move down near the lakes, at around 1,600 m. Winter is spent higher on the mountain slopes, at around 2,200 m, and the spring pastures are at a slightly lower elevation—2,000 m. In another, less mountainous area of Mongolia (Dashbalbar *sum*, Dornod *aimag*), the pastoral population generally spends the winter and spring in low areas in river or stream valleys and move to pastures in higher altitudes in the summer and autumn. The average movement in this area is about 25 km (Humphrey and Sneath 1999:236–247).

Cross-section Showing Pastoral Movement

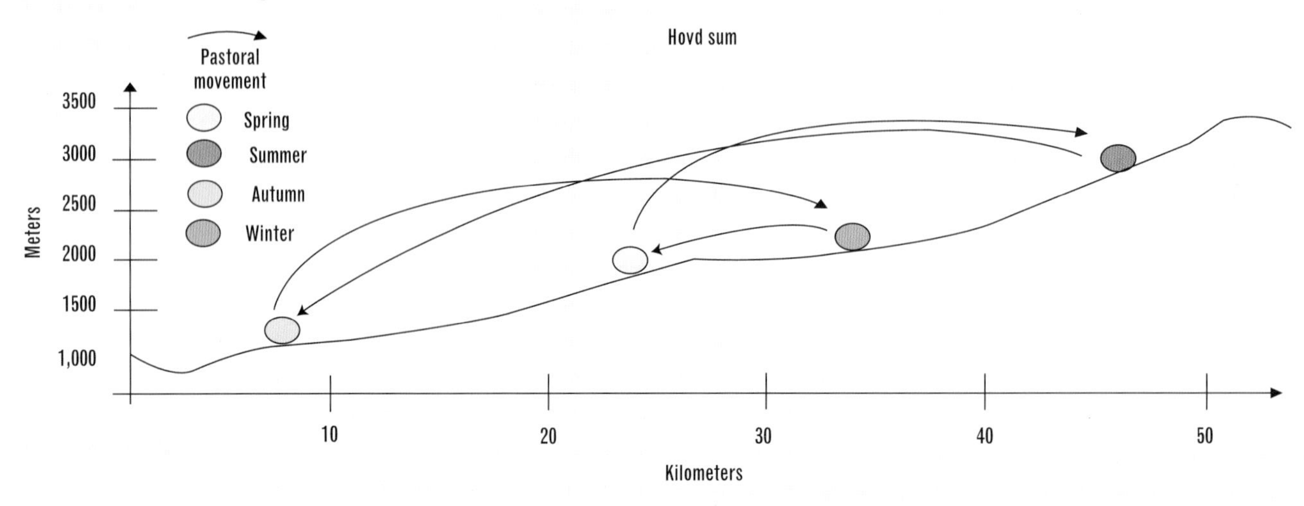

Source: Figure is adapted from Humphrey and Sneath 1999:237.

for transportation and hay-cutting services. Herding households were moved on long legs of the annual migration by collective trucks; and hay deliveries helped feed livestock during the winter and early spring. Recalled one herder, "In the collective period . . . *otor* was very good. The services provided to the herdsmen were excellent. Also, the making of hay [for fodder] and the repair of *hashaa* [enclosures and sheds] was done well" (Humphrey and Sneath 1999:39). Herding households were encouraged to work together. State loans were supplied for infrastructure improvements that would benefit pastoralists, such as boring wells, purchasing hay-making equipment, and constructing winter animal shelters.

But collectivism discouraged individual initiative. Noted the same herder, "Herdsmen had hay and so forth provided for them, and were instructed where and when to move, so they did not choose places to pasture the livestock themselves. They worked only at the command and direction of their leaders . . . cutting and making hay, shearing sheep . . . dipping the animals, all these things the brigade or groups did together. So [during collectivism] people . . . just followed instructions and waited to be told what to do" (Humphrey and Sneath 1999:39–40).

Still, Mongolia basically retained its mobile herding system and a relatively low livestock-to-pasture ratio. This pattern of land use does not appear to have caused much pasture degradation (Asian Development Bank/PALD 1993).

Chinese and Russian Experiences with Grassland Management

A comparison of Mongolia's grasslands to neighboring Chinese and Russian grasslands during roughly the same period (1920–90) underscores the pitfalls of abandoning large-scale, mobile herding techniques. Even in areas of Mongolia where livestock densities are comparable to neighboring regions of China and Russia, the Mongolian regions tend to be far less degraded, according to estimates and herders' perceptions. This may be because Chinese and Russian central governments placed more emphasis on settled pastoralism. Russia also relied heavily on highly mechanized farming methods.

In Russia, most herders were organized into collectives by the 1930s. Within a few decades, livestock in some parts of Russia were kept relatively immobile on fenced pastures. Heavy machinery and chemical fertilizers were used to cultivate fodder crops and grain.

In China's Inner Mongolia in the 1950s, families were similarly settled into "People's Communes." The communes centered on a village in a district with local government facilities, while herding families on the steppe were organized into production "brigades." The brigades retained some mobility and herded the commune livestock on seasonal pastures as directed by officials, along with the small number of personal livestock that households were allowed to own. The decrease in pasture rotation, however, required an increase in hay-making facilities and winter animal sheds.

China, like Russia, dictated a drastic expansion of agriculture in the 1950s and 1960s. Large-scale irrigation projects enabled fodder to be grown, so pastoralists no longer had to move livestock to different seasonal pastures.

Even the remnants of the former specialized herding systems in China's Inner Mongolia disappeared by the 1990s. The new post-Maoist government, as part of its economic reforms, dissolved the communes. Because the government's recent experience in allocating agricultural land to farming families in the rest of China had been relatively successful, the administration sought to apply a similar policy to pastoral regions. Livestock were distributed to pastoral households and quotas for animal production were phased out. Hay-making fields also were allocated to households. By the 1990s grazing land was divided and allocated to individuals and groups of households using long-term leases (Humphrey and Sneath 1999:165).

These 20th century political and economic changes brought benefits to Chinese and Russian pastoralists, but also introduced new inequalities and ecosystem problems. Growth in production was one benefit. In China's Inner Mongolia, the number of livestock rose from about 17 million head in 1957 to more than 32 million in 1980 (Inner Mongolian Territorial Resources Compilation Committee 1987:519–520). These increases were largely the result of a shift to fast breeding sheep and goats and away from larger livestock such as horses, cattle, and camels. Herders also gained rudimentary electrical service, roads, and wells provided by the central government. In Buryatia, Chita Oblast, and Tuva in Russia, farms provided members with guaranteed wages, living accommodations, pensions and insurance, medical facilities, kindergartens and schools, shops, central heating, fuel and firewood, clubs, libraries, and recreational facilities (Humphrey and Sneath 1999:79).

With economic reforms and the beginning of a market economy in the 1980s, living standards in China rose from the extremely low levels that had prevailed in the People's Communes. Some herders became wealthy; those who had better access to markets or who were able to buy machinery and vehicles usually were those who could obtain low-interest government loans through ties to the local administration. Those households could hire labor to look after large herds and could invest in hay-cutting machinery and other assets. Some could pay for special access to high-quality areas of pasture in addition to the minimal pasture allocated to each herding household. Those with the financial means fenced these formerly common lands, limiting the mobility of others to use or move across them.

Thus, benefits were brought at high cost to cultural traditions and ecosystems. Large-scale pastoral movements between seasonal pastures have been largely eliminated by the land allocations, and there has been a corresponding decline in the use of the pastoral technique of *otor*. The effect has been to increase the amount of hay cut to feed livestock, to increase the tendency for livestock to graze in one location all year, and to intensify the concentrations of animals in certain areas. Individual herders can no longer graze different species of livestock on a range of accessible, suitable territories. For example, riverside pastures that had been available to cattle from the whole district might today be divided among different households. Locals have identified deterioration of pasture in intensively grazed areas in Russia and China's Inner Mongolia, especially around water sources and households.

Where static herds do not have access to natural water sources year round, water must be trucked to those pastures; and vehicular traffic damages the fragile surface of those pastures. The need to increase production of hay and fodder to feed the settled livestock also damages the thin steppe soils. In the substantial areas of Inner Asia where soil cover is weak and the climate harsh, converted pastures supply low crop yields while exacerbating erosion and desertification (Humphrey and Sneath 1999:91); plowed grasslands rapidly lose topsoil to strong winds and soil moisture decreases.

Other problems include reduced production of grass in hay-making pastures each year, since people routinely cut in the same places. Herders in China's Inner Mongolia have been known to plow the spring pastures to plant hay and grain because they cannot afford the high price of grain sold in markets. Grassland specialists in Xinjiang estimate that it takes 15-20 years for plowed land to regain its previous productivity as pasture (Humphrey and Sneath 1999:106) because plowing destroys the extensive root system that supports perennial grasses.

Another issue is the introduction of foreign livestock breeds. Merino sheep, for example, were crossbred with Mongolian sheep starting in the 1950s to increase the productivity and quality of livestock products. Many of the "improved" breeds were weaker and slower moving than indigenous breeds, thus requiring heated sheds to survive the winter, further reducing herd mobility (Humphrey and Sneath 1999:239). In Buryatia in Russia, researchers noted that foreign breeds indirectly affected forest ecosystems. Building winter sheds and supplying fuel and housing for newly settled herders requires timber. As a result, forest areas along the Russian border have been heavily exploited. By comparison, most Mongolian herders still use *yurts* for shelter and burn dried dung for fuel; wooden houses are generally found only in central villages. Thus, forest pressures from Mongolia's pastoralists are lower (Humphrey and Sneath 1999:12).

A decline in nomadic practices brings cultural advantages and disadvantages. Interviews with herders from various parts of Inner Asia suggest that many still prefer a mobile life, particularly middle-aged and older herders. Others recognize that nomadism is essential for pasture health but can be a hard life. Time spent in *otor* is time cut off from other people and, often, from social services like formal education, health care, and postal services. Static farming and livestock rearing let families cultivate vegetables, drink water from wells, and access markets more readily (Yenhu 1996:21).

Mongolia after Socialism: Parallels to China and Russia

In 1990, Mongolia began a transition toward a free-market economy. In some ways, the lives of its herders and its economic climate show parallels to China and Russia. There are more sedentary living complexes, divided pastures, and pressures on grasslands and other ecosystems. As a consequence, overgrazing and soil degradation have increased. Records show that the number of dust storms in Ulaanbaatar, the Mongolian capitol, have increased from 16 per year on average during 1960-69 to 41 per year during 1980-89 (Whitten 1999:11). Mongolia's National Environmental Action Plan warns that desert in the country's southern region may be advancing northward by as much as 500 m per year (Government of Mongolia 1995:27–28).

INCREASE IN LIVESTOCK NUMBERS

Mongolia has dissolved its collectives, and most of the livestock and other agricultural resources have become the members' property. As in China's Inner Mongolia in the 1980s, this move toward privatization and markets has promoted rapid growth in Mongolian livestock numbers. That growth occurred as herders first sought prosperity through larger herds, then as they sought to at least earn subsistence income as the economy took a downturn. From 1990 to 1998, Mongolia's national herd increased by more than 20 percent, from 26 to 32 million head (Statistical Office of Mongolia 1993:28; Ministry of Agriculture and Industry of Mongolia 1998:2).

DECREASE IN COMMON PROPERTY GRASSLANDS

To date, the Chinese have progressed farthest in the transition from collective use of pastures to individual use, though Russian Buryatia and Chita are not far behind (Humphrey and Sneath 1999:97). Now Mongolia is following suit. All pastureland remains "common" land under the jurisdiction of provincial and district-level authorities, suggesting that Mongolia still has some of the largest areas of common grazing land in the world (Mearns 1996:308–309). In practice, however, access to and control of common grasslands is not clearly defined. Ownership and use of public land is a controversial topic in Mongolia, with active debate in the Mongolian parliament about the merits of private rights to land and how to ensure that the rich do not acquire all the best pastures. With ambiguous use rights and declining use of collective management, some herding families have begun to rotate their herds less, fearing that others may use the best pastures if they vacate them.

Furthermore, the dissolution of the motor pools of the old collectives and the increase in the cost of gasoline is making seasonal movement difficult for many pastoral families. Where they once used trucks, they now rely on animal transport. The organization of *otor* movement and the regulation of access to pasture, which had been overseen by collective and state farm officials, have declined.

INCREASING DEPENDENCE ON PASTORALISM

During the breakup of the state collectives, livestock were allocated to its former members–to herders and to those who performed other jobs, like veterinarians, drivers, and canteen workers. In some districts the majority of the population became directly dependent on their allocation of livestock for subsistence. The number of registered herders nationwide was 135,420 in 1989–less than 18 percent of the national workforce. Since the economic reforms of the 1990s, that total has more than tripled to 414,433 in 1998 (National Statistical Office of Mongolia 1999:95,45; Statistical Office of Mongolia 1993:6).

Many of these "new herders" maintain permanent dwellings in the district center and are less familiar with or guided by the traditional mobile grazing systems than the households who were part of the specialized herding brigades of the collectives. Some have part or all of their livestock herded by relatives or friends with access to more distant pastures. Others who have migrated from urban areas to take up herding are treated as outsiders and resented for what locals see as increased grazing pressures on local pastures. The presence of these migrants weakens the potential to successfully manage common grazing areas (Mearns 1996:328).

ECONOMIC CRISIS

In the collective era, Mongolia exported 25,000–40,000 tons of meat, 25,000–30,000 tons of livestock, and more than 60,000 horses each year. The vast majority of these products went to the Soviet Union and other members of the socialist trade bloc. With the collapse of the socialist trade bloc, those export markets almost disappeared. Mongolia's meat exports in 1998 amounted to just 7,500 tons, and livestock and horse exports were insignificant (National Statistical Office of Mongolia 1999:144). At the same time, Mongolia's access to affordable imports was undermined; pre-1990, Mongolia spent one-third of its GDP on imports from the Soviet Union, including all petroleum products, 90 percent of imported machinery and capital goods, and 70 percent of consumer goods (Mearns 1991:30).

Accordingly, there has been a collapse in living standards and a declining level of public services like veterinary services and provision of farm machinery. The economic crisis also has lowered agricultural output. The area under cultivation, yields per hectare, and overall production for staple crops like wheat and cereals all have decreased since the end of central planning. Many farmers cannot afford to buy machinery, seeds, and fertilizers (Economic and Social Commission for Asia and the Pacific 1999:336).

In retrospect, many herders stress the relative wealth, security and convenience that the collective period offered, in comparison with the shortages and uncertainty of the current transition to a market economy. Some pastoralists have tried to establish "cooperatives" by pooling their shares of the old collectives to take ownership of its assets, or to share transportation and other costs. However, most of these cooperatives have gone bankrupt as the economy has failed to improve.

INCOME INEQUALITY

Although economic liberalization has enabled some individuals to make money, those in the agricultural sector have struggled to realize any profit. Similar to China's Inner Mongolia, Mongolia is experiencing a growing difference between the living conditions of rich and poor herders. Today, about 37 percent of livestock-owning households struggle to subsist

Box 3.20 Livestock Density in Inner Asia

Densities of livestock in Inner Asia are significantly higher in parts of Inner Mongolia and Xinjiang compared to neighboring Mongolia. But it is not necessarily the case that high livestock densities mean reduced grassland productivity. In fact, researchers studying pastoralism in Inner Asia found that the mobility of the herd and the herd structure seem to be stronger determinants of degradation. For example, records from the 1930s suggest that Inner Mongolia supported about the same quantity of livestock (when calculated in terms of a standard unit of livestock) as it has in the 1990s—the equivalent of about 70 million sheep (Sneath 1998, citing Chang 1933). But in the 1930s, the herds contained a much smaller proportion of sheep and goats and the system of pastoralism was much more mobile. Environmental problems are perceived where herders have shown a tendency to graze their herds year round in specific areas. Pressure on grasslands is exacerbated when some of the best natural pastures are converted to hay making and agriculture.

Livestock Density in Inner Asia

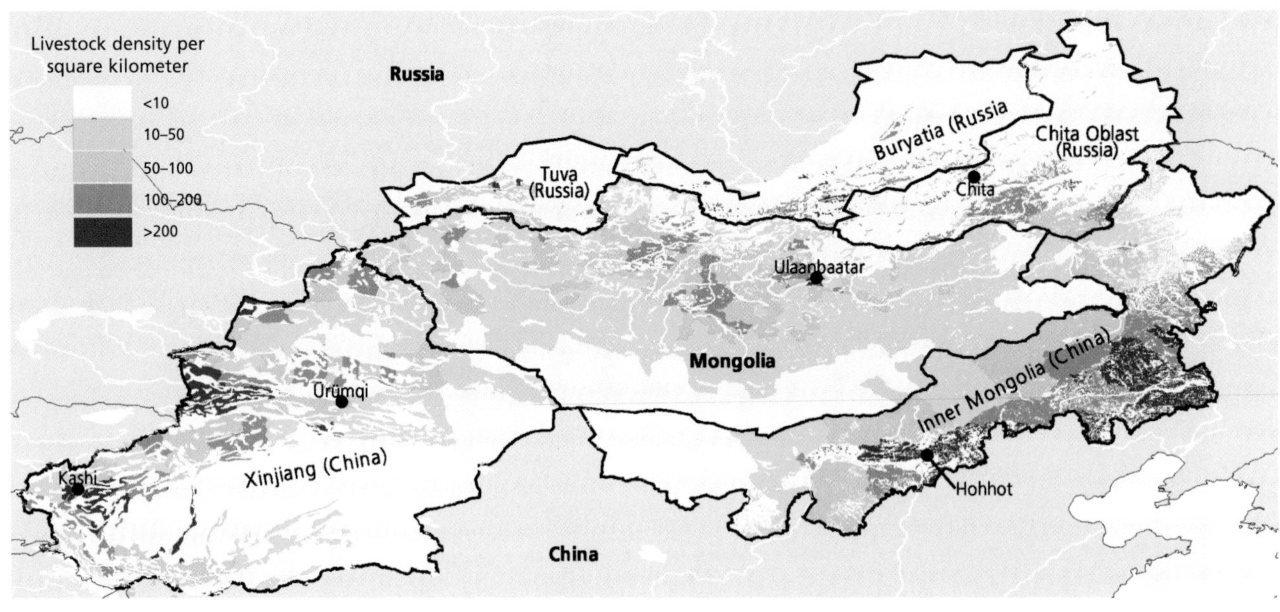

Source: MECCIA 1995.

on the income from less than 50 animals, and 11.5 percent had less than 10 animals in 1998 (National Statistical Office of Mongolia 1999:96). This situation is likely to have worsened during the harsh winter of 1999–2000 when more than 2.2 million livestock died of starvation (UNDP 2000).

One benefit of the emergence of a small stratum of wealthy livestock owners is the potential for them to reestablish some larger pastoral operations that can benefit from economies of scale and the old systems of extensive pastoral movement. The number of households in Mongolia that owned more than 1,000 animals rose from seven in 1992 to 955 in 1998; 33 of these owned more than 2,000 head of livestock (National Statistical Office of Mongolia 1998:96; *Zasagyn Gazar Medeel* 1992). The richest employ neighboring households to help herd livestock and can maintain trucks, jeeps, and wider systems of pastoral movement than most other households. Poor herders cannot afford such moves and, with smaller herds, have less incentive to do so. Their more meager flocks can survive on pastures around their fixed dwellings (Humphrey and Sneath 1999:254).

Poor herders also face more labor and education challenges now than they did under collective systems. For many it has become more economical to remove children from school to stay home and help with herding rather than employ laborers to look after herds (Ward 1996:33).

RELIANCE ON HAY AND FODDER CROPS

Unlike neighboring China and Russia, Mongolia has largely continued to use local breeds that can graze on natural pastures year round. But hay supplies are still critical in winter and early spring (Humphrey and Sneath 1999:236). In fact, the loss of the hay provision the government once supplied to Mongolian

Population and Livestock Density in Selected Districts

Country/Village	Population Density (person/km²)	Livestock Density (SSU[a]/km²)	Percentage of Useful Land[b] Cultivated	Percentage of Pasture[c] Considered Degraded
China				
Chinggel Bulag	0.70	54	0	54.4
Hosh Tolgoi	2.10	56	0.3	?
Handgat	3.25	54	0.44	12
Hargant	1.40-	36	0	22.9
Russia				
Argada	11.30	270	33	88.3
Gigant	4.00	125	18.8	76.9
Sholchur	1.80	65	0.9	1.5
Mongolia				
Hovd sum	0.96	48	0.008	0.07
Dashbalbar	0.40	22	0.17	0.03
Sumber[d]	1.56	36	1.2	2

[a]SSU, standard stocking unit: sheep = 1, goat = 0.9, cattle = 5, horse = 6, camel = 7.

[b]"Useful land" is all land not specifically unusable for farming economy as a whole. It includes arable and hay-making land.

[c]"Pasture" is land specifically designated for pasture.

[d]Data do not include the administratively separate town or Choir.

Source: Humphrey and Sneath 1999:77.

Growth in Mongolian Livestock Populations

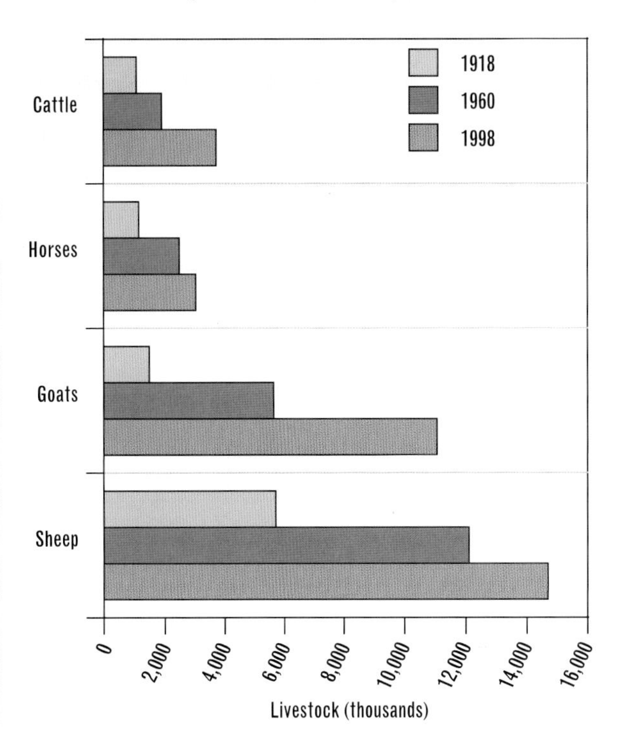

Source: Humphrey and Sneath 1999:44–45.

collectives seems to be harming livestock nutrition, especially as pastoralists make shorter and less frequent moves.

The lack of adequate hay production leaves flocks vulnerable to starvation, as evidenced during the winter of 1999-2000. Thousands of hectares of pasture were buried under heavy snow into the spring, yet the government was unable to provide supplementary feed because of limited funds, lack of hay stocks resulting from prior drought, and transportation problems (FAO 2000).

Another problem is that some of the pasture used for hay production is not ecologically suited for it. Perhaps 10 percent of the 1.34 million ha under cultivation in 1990 is now affected by erosion (Whitten 1999:14).

Mongolian herders have noted the negative impacts of recent trends. Remarked one man, "In the 1970s all the households used to go on *otor*, and the households were spread out at a distance from one other. But now most of the households do not move from their winter camps, so in the winter and autumn pastures the animals have eaten all the vegetation. So there has been significant pasture damage and reduction in vegetation" (Sneath 1993).

Modernization and Mongolia's Future

Looking at China's Inner Mongolia, some already foresee the passing of the era of mobile pastoralism. Economics could encourage production systems in which calves and lambs are shipped to farming areas for fattening, rather than raised on grass. For some herders, benefits of such a transition could include increased income, more leisure time, and greater economic security (Humphrey and Sneath 1999:93, citing Li et al. 1993).

It is too soon to tell if such a scenario is inevitable for Mongolia, or if the country can find a way to balance the old herding techniques of pastoral mobility with the new forces of urbanism and market economics. On one hand, old techniques of pastoral mobility still exist even in China's Inner Mongolia, with livestock raised to full weight on the steppe. On the other hand, the herding patterns that collectives used had retained some aspects of the older systems of land use, but the dissolution of these institutions brought a decline in large-scale pastoral operations and expanded the herds kept for use by individual families.

Currently, grazing land in Mongolia remains a public resource despite attempts to introduce legislation for its private ownership. However, without support, the poorer households with small numbers of livestock and limited domestic labor will have difficulty maintaining systems of wide pastoral movement, even where pasture land is not divided among individuals. A more sedentary life does not inevitably lead to pasture degradation, but the movement of the herds in relation to available pasture does appear to matter to herders. For example, in Dashalbar, Mongolians have a relatively settled way of life, with houses in the district center, but herders with a vast area of pasture at their disposal still make use of seasonal movement and occasional *otor* (Humphrey and Sneath 1999:212).

Other complicating influences include a tripling of the human population in Mongolia in the last 60 years and projected high growth rates for several more decades. This adds pressure to expand the pastoral economy and animal herds, although the number of livestock may be approaching the maximum level that Mongolia can support with the resources currently available to the pastoral sector. The desire to live near roads, markets, schools, and modern services also will draw people and their herds to populated areas where degradation is already a problem.

With current high inflation, debt, and depressed trade, it seems unlikely that local or central governments will be able to encourage large pastoral enterprises by renewing the government-supported motor pools and machinery for hay production. Yet such investments and government leadership may be essential if large-scale pastoral movement systems that include the majority of herders are to be retained. District governments might be able to coordinate labor for the maintenance of public resources such as wells and hay production, for example. Or, small farms and associations could be combined in scaled-down versions of collectives for more specialized and mobile livestock herding, even if households are more settled.

It is possible that wealthy Mongolian herd owners will accumulate sufficiently large livestock holdings to establish intermediate-scale pastoral operations, using labor from poorer households. However, decades may pass before such operations become large enough to encompass the majority of grazing land, and there would still be need for district authorities to coordinate herding and land use.

Significant investment in improved transportation services for herders could bolster environmentally sustainable systems of large-scale pasture rotation and might also benefit livestock processing industries by facilitating their purchase of livestock products at competitive prices. In China, at least, the close presence of markets and relatively high demand for pastoral products has enabled some herders to make a good living. But in Russia and Mongolia, the distance to markets, the high cost of production inputs like fuel, and low demand all depress the livestock economy. In Russia and Mongolia, the prices for livestock products like meat, cheese, and wool are very low; sugar, tea, flour, and other foods are expensive (Humphrey and Sneath 1999:75).

Market failures may cloud Mongolia's ability to see the short-term benefit of preserving large-scale herding patterns. This is especially true in the face of some farmers' increased wealth and the lack of policies that support and encourage mobile herding and collective action. But where herders' lives become highly settled, the grasslands appear to be overused. Pastoralists recognize the threat to the future productivity of their livestock operations. Herding populations from Tuva to western Mongolia and Mongol-inhabited parts of Xinjiang are deeply concerned about the environment. Whether that local awareness will translate into political change and sensitivity to ecological vulnerability, or what path "modernization" will take, is difficult to gauge.

CHAPTER 4

ADOPTING AN ECOSYSTEM APPROACH

Adopting an "ecosystem approach" means we evaluate our decisions on land and resource use in terms of how they affect the capacity of ecosystems to sustain life, not only human well-being but also the health and productive potential of plants, animals, and natural systems. Maintaining this capacity becomes our passkey to human and national development, our hope to end poverty, our safeguard for biodiversity, our passage to a sustainable future.

–from the Foreword to this volume

Just as ecosystems sustain us, we must sustain them. We exist with them in a worldwide web–a fraying web of life. The scientific evidence described in Chapter 2 and the practical experience recounted in Chapter 3 underscore the need to weave a different future.

The Pilot Analysis of Global Ecosystems (PAGE) shows that the overall capacity of ecosystems to deliver goods and services is decreasing. Yet human demand for ecosystem products–from water to food to timber–continues to increase. Globally, we have managed agriculture, forests, and freshwater systems to achieve remarkable growth in the output of food and fiber. But when PAGE researchers examined the full range of goods and services produced by five major ecosystems, they found that the increased output of some goods and services has resulted in steep declines

Box 4.1

What Is an Ecosystem Approach?

An ecosystem approach broadly evaluates how people's use of an ecosystem affects its functioning and productivity.

- *An ecosystem approach is an integrated approach.* Currently, we tend to manage ecosystems for one dominant good or service such as fish, timber, or hydropower without fully realizing the tradeoffs we are making. In doing so, we may be sacrificing goods or services more valuable than those we receive—often those goods and services that are not yet valued in the marketplace such as biodiversity and flood control. An ecosystem approach considers the entire range of possible goods and services and attempts to optimize the mix of benefits for a given ecosystem. Its purpose is to make tradeoffs efficient, transparent, and sustainable.

- *An ecosystem approach reorients the boundaries that traditionally have defined our management of ecosystems.* It emphasizes a systemic approach, recognizing that ecosystems function as whole entities and need to be managed as such, not in pieces. Thus it looks beyond traditional jurisdictional boundaries, since ecosystems often cross state and national lines.

- *An ecosystem approach takes the long view.* It respects ecosystem processes at the micro level, but sees them in the larger frame of landscapes and decades, working across a variety of scales and time dimensions.

- *An ecosystem approach includes people.* It integrates social and economic information with environmental information about the ecosystem. It thus explicitly links human needs to the biological capacity of ecosystems to fulfill those needs. Although it is attentive to ecosystem processes and biological thresholds, it acknowledges an appropriate place for human modification of ecosystems.

- *An ecosystem approach maintains the productive potential of ecosystems.* An ecosystem approach is not focused on production alone. It views production of goods and services as the natural product of a healthy ecosystem, not as an end in itself. Within this approach, management is not successful unless it preserves or increases the capacity of an ecosystem to produce the desired benefits in the future.

in virtually all others–from water quality and quantity to biodiversity and carbon storage. In many cases these trade-offs were unconcious. Nonetheless, even with a new awareness of the value of traditionally overlooked ecosystem services like biodiversity or carbon storage, we can't simply reverse the trade-offs we've made. We can't, for example, make do with less food in order to protect biodiversity or improve water quality. The poor and disadvantaged would pay the human consequences of such a strategy.

The case studies in Chapter 3 further underscore our dependence on ecosystems. The villagers who live near Dhani Forest in India have no ready replacement for the food and fiber that Dhani provides, any more than the residents of southern Florida–even with their greater financial means–can find an alternative supply for the plentiful water that the Everglades offers.

Fortunately, the case studies give reasons for optimism. The groundswell of political concern over the deterioration of the Everglades is one sign that awareness of the importance of ecosystems is growing. The community's response to Dhani Forest's degradation assures us that–at least in some places–we are changing our behavior for the better. With its Working for Water Programme, the South African government is simultaneously fighting invasive plants, rising water demand, and poverty. The Programme examines impacts and pressures across ecosystems, challenges political interest groups and perverse economic influences, and forges alliances with the private sector.

Nonetheless, most of the management approaches presented in Chapter 3, as innovative as they are and as difficult as they were to implement, still fall short of a true "ecosystem approach." Some focus only on facets of an ecosystem's health. They include reparative actions, but not always preventive ones. From Mongolia to Bolinao to New York City, none encompasses the broad-scale changes needed to cope with current environmental degradation and inevitable increases in consumption.

What Should We Do to Adopt an Ecosystem Approach?

The principles of the ecosystem approach, described in Box 4.1, are slowly gaining recognition among resource managers. For more than a decade, the concept of ecosystem management has been growing in theory and application. In 1992, the U.S. Forest Service officially adopted an ecosystem orientation to managing U.S. National Forests. Since then, it has struggled to articulate what this means for its timber harvest policies, grazing practices, recreation activities, and management of roadless and wilderness areas. Box 4.2 provides examples of the differences between a traditional approach and an ecosystem approach in forestry.

Box 4.2 Differences Between Traditional Forest Management and an Ecosystem Approach to Forest Management

	Traditional Forest Management	Forest Ecosystem Management
Objectives	■ Maximizes commodity production	■ Maintains the forest ecosystem as an interconnected whole, while allowing for sustainable commodity production
	■ Maximizes net present value	■ Maintains future options
	■ Aims to maintain harvest or use of forest products at levels less than or equal to their growth or renewal	■ Aims to sustain ecosystem productivity over time, with short-term consideration of factors such as forest aesthetics and the social acceptability of harvest practices
Scale	■ Works at the stand level within political or ownership boundaries	■ Works at the ecosystem and landscape level
Role of Science	■ Views forest management as an applied science	■ Views forest management as combining science and social factors
Role of Management	■ Focuses on outputs (goods and services demanded by people), such as timber, recreation, wildlife, and forage	■ Focuses on inputs and processes, such as soil, biological diversity, and ecological processes, since these give rise to goods and services
	■ Strives for management that fits industrial production	■ Strives for management that mimics natural processes and productivity
	■ Considers timber is the most important forest output (timber primacy)	■ Considers all species—plant and animal—important and considers services (protecting watersheds, recreation, etc.) are on an equal footing with goods (timber)
	■ Strives to avoid impending timber famine	■ Strives to avoid biodiversity loss and soil degradation
	■ Views forests as a crop production system	■ Views forests as a natural system, more than the sum of its parts
	■ Values economic efficiency	■ Values cost-effectiveness and social acceptability

Source: Adapted from Bengston 1994

Our dominance of Earth's productive systems gives us enormous responsibilities, but great opportunities as well.

The European Union likewise has begun to frame its environmental problems in terms of large-scale ecosystem effects such as forest loss, widespread nutrient pollution of rivers, and loss of biodiversity. Thus, in its periodic assessments of the environment, the European Environment Agency reports on such indicators as air pollution in excess of ecosystem "critical loads," trends in defoliation of European forest ecosystems, and the effects of fragmentation on Europe's ecosystems (EEA 1999).

At an international level, the ecosystem approach has also gained greater visibility and endorsement. At their biennial meeting in May 2000, the nations that signed the 1992 Convention on Biological Diversity formally spelled out 12 principles that define an ecosystem approach and called for governments to adopt these principles to manage their land, water, and living resources. In their declaration, the nations noted there is no single way to implement the ecosystem approach in all nations, but that the general framework for management must focus on ecosystem processes rather than political jurisdictions and sectoral divisions (COP-5 2000:103–109).

Although these steps toward incorporating an ecosystem approach into land-management decisions represent progress, the wide-scale reorientation of business practices, government policies, and personal consumption habits around an ecosystem approach is still far from reality. In most nations, and in most local practices, the idea of ecosystems as essential biological elements that touch daily life and business remains foreign. At an international level, there is little use of an ecosystem approach when shaping agreements on trade, agriculture, forests, or water use.

Lessons drawn from the PAGE findings and the case studies offer practical guidance for adopting an ecosystem approach. Our recommendations are grouped in four broad areas:

- Tackle the science and information gap.
- Recognize and measure the value of ecosystem services.
- Engage in a public dialogue on goals, policies, and trade-offs.
- Involve all stakeholders in ecosystem management.

These are not a series of sequential steps, but an on-going

dance in which we can progress in all areas simultaneously. By following the practical guidance from PAGE and the case studies, we will move more agilely in each area. We already have enough knowledge and experience to get the dance under way.

TACKLE THE SCIENCE AND INFORMATION GAP

Managing ecosystems holistically and sustainably requires a detailed understanding of their function and condition. Without a stronger base of scientific knowledge and indicators at local, national, and global levels, we are ill-prepared to judge ecosystems' productive capacity, to recognize the trade-offs we are making, or to assess the long-term consequences of these trade-offs.

Underlying all of our efforts to tackle the science and information gap is the need for more applicable scientific knowledge. For example, experimental evidence shows that the loss of biological diversity will reduce the resilience of an ecosystem to external perturbations such as storms, pest outbreaks, or climate change. But scientists are not yet able to quantify how much resilience is lost as a result of the loss of biodiversity in a particular site nor even how that loss of resilience might affect the long-term sustainability of the production of goods and services. Better scientific understanding of ecosystems' carrying capacity and thresholds for change would greatly benefit our management efforts.

In some cases, our scientific understanding of ecosystems is improving enough to allow us to build models that will help determine what resources are most at risk and forecast their future. In South Africa, for example, sophisticated computer modeling revealed that allowing invasive trees to spread would severely disrupt water supplies. In the Everglades, modeling of the entire watershed showed just how distorted the water cycle in the region had become. Fifty years earlier, when people were making decisions about altering waterflow in the Everglades, they didn't have such powerful scientific tools at hand.

But more than simply building a better scientific base and honing our understanding of ecology, we must develop and consistently measure indicators of ecosystem extent, condition, and performance. PAGE underscores how sorely our indicators of ecosystem condition are lacking. Often PAGE assessments had to be based on data measured in different periods, governed by inconsistent definitions, or riddled with blanks in coverage. Even for agroecosystems, for which studies of conditions and production abound, there are no globally consistent measurements of the impact of agriculture on water quality and little crop-specific information about the size and production of irrigated areas. In our era of supposed information overload, the PAGE results show that consistent, reliable measures of ecosystem conditions are difficult to ascertain both on a global scale and on a local or national scale where most land use decisions are made.

The case studies, too, clearly illustrate the need for improved indicators, consistent monitoring, and reporting on ecosystem condition. The longer cases chronicle the gradual transformation of ecosystems through physical alteration or overuse, a period when individuals and institutions sometimes failed to recognize early warnings of ecosystem decline or were unable to assess the long-term repercussions of their choices. Part of the challenge is that ecosystem decline may begin gradually, then manifest quickly as pressures increase. Florida Bay degraded slowly in the first two decades after the Central and South

(continues on p. 232)

Box 4.3 The Need for Integrated Ecosystem Assessments

How can we judge whether an ecosystem is in good condition? Scientists have taken several approaches:

- *Measuring against natural systems.* Some scientists have suggested that the condition of an ecosystem could be measured by comparing one or more of an ecosystem's properties (such as biomass, number of species, or the flow of nutrients through the ecosystem) to those of a "natural" or "undisturbed" ecosystem. This would effectively define the condition of an ecosystem to be its degree of "non-naturalness." But the shortcomings of this approach for policy and management decisions are clear. Judging condition with such an indicator of "naturalness" would mean, for example, that all agroecosystems or forest plantations would be defined as being in poor condition since they are quite different from the natural ecosystems that they replaced. Moreover, given the pervasive influence of human action on the global environment, it is increasingly difficult to define what a "natural" or "undisturbed" ecosystem would be like.

- *Measuring sectoral conditions.* Many reports have been written about the state of agriculture in various countries focusing only on food production, without considering the potential negative effects of that food production on biodiversity, water quality, or carbon sequestration. Or forest assessments have examined only timber production, without evaluating the potential impact of timber harvest on regional rainfall, energy production from downstream hydro-facilities, or biodiversity loss. This strictly sectoral approach made sense when trade-offs among goods and services were modest or unimportant. But it is insufficient today, when ecosystem management must meet conflicting goals and take into account the linkages among environmental problems. A nation can increase food supply by converting a forest to agriculture, but in so doing decreases the supply of goods that may be of equal or greater importance such as clean water, timber, biodiversity, or flood control. Both local resource managers and national policy makers need some means of weighing these trade-offs, which requires a more integrated view of just what those trade-offs might entail.

- *Measuring for optimization.* An integrated assessment determines the condition of an ecosystem by assessing separately the capacity of the system to provide each of the various goods and services and then evaluating the trade-offs among those goods and services. Even if the trade-offs are conscious choices, an integrated assessment will show whether the capacity of the system to provide a *combination* of the services is optimized. For example, in an acceptably productive agroecosystem that relies on chemical inputs, separate assessments could show whether the addition of a rotation of a green manure crop could greatly reduce nutrient inputs, dramatically increase water quality, or affect agricultural yield. Thus, it could be determined whether the ecosystem was being managed to optimize the provision of a combination of food and clean water or whether these goods might have been achieved through an alternative management approach.

This approach to ecosystem assessments is called an "integrated assessment" because it examines not just a single ecosystem product, such as crop production, but an entire array of products that the ecosystem might provide. The principal benefit of an integrated ecosystem assessment is that it provides a framework for examining the linkages and trade-offs among various goods and services. The opportunity to increase the aggregate benefits from the bundle of goods and services produced by an ecosystem would be hidden in an assessment of each sector in isolation. The goal of management of the ecosystem may well be to favor one service, say, food production, over the others, but by looking at the production and condition of the entire array of services, trade-offs among various services become apparent.

Box 4.4 Using Information to Support an Ecosystem Approach

In collaborating on this report and supporting a global assessment of ecosystems, the United Nations Development Programme, the United Nations Environment Programme, the World Bank, and the World Resources Institute confirm their commitment to use information to motivate actions that will maintain and restore ecosystems. Governments, businesses, organizations, and individuals everywhere have many opportunities to match that commitment:

- Governments can use their access to information to drive decisions on ecosystem use, protection, and restoration. Government agencies and officials now have more and better data than ever before, through advancements in science and technology, and they are in the best position to integrate satellite habitat imagery, air and water quality readings, biological data, demographic information, and transportation and land-use maps. For example, government regulators can incorporate scientific findings on ecosystem thresholds, such as the "critical load" of pollutants like SO_X and NO_X, in regulations that govern automobile and powerplant emissions or water quality standards.

- Businesses can improve their environmental performance in relation to ecosystems by collecting and disseminating information about the environmental aspects of their processes, products, and services. Although government regulations are powerful means of requiring businesses to manage and report on their performance, increasing numbers of businesses around the world are voluntarily adopting environmental management systems and publicly disclosing information on their performance. Many businesses do so to save money, to increase shareholder value, to benchmark their performance, and to monitor their compliance with external commitments.

- Industry associations can develop policies and codes that respect the need to keep ecosystems viable. One model for how such ecosystem-friendly business practices can be promulgated is the International Organization for Standardization's ISO 14000 standards, which provide guidance to companies that want to improve their environmental management in a number of areas, including environmental auditing, labeling, and product life-cycle assessment. As of July 2000, 14,106 companies in 84 countries have adopted the ISO 14000 standards. Another model is the Global Reporting Initiative (GRI), which was established in 1997 by the Coalition for Environmentally Responsible Economies and the UN Environment Programme, with the mission of designing globally applicable guidelines for preparing enterprise-level sustainability reports. The GRI guidelines are available online at http://www.globalreporting.org.

- Universities, environmental groups, and civic associations can help interpret the wealth of raw data that is already available—presenting data in user-friendly, indexed, nontechnical formats that allow anyone to navigate lots of information quickly. Such organizations can compile risk-ranked lists of facilities or production methods, integrate data sets, or create rankings of popular consumer products based on the presence of suspected toxins, for example. They can also "watchdog" ecosystem management, ensuring that we truly take an ecosystem approach by promoting open planning processes, organizing and informing constituents, and demanding accountability from governments, multilateral banks, and corporations.

- Consumers can seek product information and use purchasing power to drive businesses to better practices on behalf of ecosystems. Certification of sustainable management practices or "ecolabeling" already enables us to choose the timber, agricultural products, and fish products that are produced and harvested with the fewest ecological impacts. For example, the Forest Stewardship Council assesses forest management practices against a set of 10 environmental, social, and economic principles and has certified more than 15.8 Mha of productive forestland worldwide (Parker et al. 1999:12). Business leaders such as IKEA, the largest furniture manufacturer worldwide, are turning to those forest products both to gain a marketing advantage and to respond to consumer interest in more environmentally sensitive products. Similar certification processes, such as Energy Star ratings, are already in place to help consumers evaluate the energy consumption of appliances, and others could be developed for environmentally sensitive goods and services, such as community-based lodging and guides for ecotourism.

- Citizens everywhere can make a point of learning more about the environmental conditions and issues in their surroundings. Those with access to the Internet can readily get information to help them make decisions about voting, using local land and resources, recycling, and disposing of household wastes, for example. They also gain the means to share the information with friends and colleagues, or voice their opinions—sometimes just by sending a message with another click on the keyboard.

Florida Project altered the Everglades water flow, then rapidly in the last decade. In South Africa, the connection between imported plants and water supply took almost a century to identify with certainty. The years that it took to recognize the damage and change course amplified the repercussions of degradation–both on the ecosystem and on those dependent on the goods and services that had been compromised.

Not all information is equal, however, when it comes to supporting an ecosystem approach. Integrated assessments are the most effective means to encourage stakeholders to manage ecosystems for more than their immediate commercial value(Box 4.3 The Need for Integrated Assessments). Such assessments separately determine the capacity of an ecosystem to provide various goods and services and then evaluate the trade-offs among those goods and services. Narrower sectoral measures, which have been the principal sources for most decision making, focus on a single outcome, rather than consequences across the ecosystem. Thus, the government agencies that replumbed the Everglades judged their success on the basis of agricultural production and flood control. The agencies that forested South African mountains with pines had their sights set on maximum timber output, as did the government in Dhani, which permitted commercial contractors to harvest the forest canopy. Only at crisis points–when the supply of critical goods like food or water was interrupted–did serious interest develop in analyzing other indicators of the health of these ecosystems. Perhaps the crises would never have occurred if more integrated information had been available at the outset.

Of course, that's a wishful thought. No matter how sophisticated our scientific understanding, computer models, and original statistics, we are still likely to be surprised by ecosystem outcomes unless we monitor them continuously. Just as our knowledge of ecosystem dynamics is rapidly changing, so is the scale of pressures–demographic, economic, and biological–that will alter ecosystems. Periodically assessing ecosystems is key to avoiding unexpected outcomes. In Bolinao, only years of monitoring a variety of environmental indicators will show whether the new four-zone coastal management plan is helping fish stocks rejuvenate, or whether other factors outside the purview of the plan are more critical. The New Yorkers who drink unfiltered water must rely on extensive water quality monitoring to determine whether their ecosystem protection plan is adequate or whether an investment of billions of dollars in a filtration plant is necessary. A careful record of monitoring may verify suspicions that new ecosystem management is needed–and can help the largest and most expensive efforts, like the Everglades restoration plan, withstand inevitable public and legal challenges.

Sound scientific analysis, modeling, assessment, and monitoring can increase the wisdom of ecosystem management decisions. The scope of action for tackling the science and information gap is large indeed, and it spans governments, businesses, organizations, and individuals (Box 4.4. Using Information to Support an Ecosystem Approach). But it is not the only requirement for an ecosystem approach.

RECOGNIZE AND MEASURE THE VALUE OF ECOSYSTEM SERVICES

Undervaluing ecosystem services has contributed to many shortsighted management practices. The PAGE study of freshwater systems, for example, argues that heavily subsidized water prices, especially for agriculture, have promoted the inefficient use of water. The study documents the sixfold increase in water consumption since 1900 worldwide, more than twice the rate of population growth. The PAGE study of forest ecosystems shows that old-growth forests in Canada–where logging companies' operations are subsidized–are harvested far in excess of their rate of growth, despite the forests' value in terms of biodiversity, carbon storage, and watershed protection. Market mechanisms have generally failed to assign monetary values to such public goods, but market failure is not solely responsible for the exploitation of ecosystem services. Tax breaks, trade incentives, tariffs, public-investment strategies, and other economic policies have distorted the price of water, land, and other ecosystem inputs and outputs.

The case studies, too, provide a wealth of examples of economic policies that, despite good intentions, have aggravated declines in ecosystem condition and capacity by undervaluing essential ecosystem services. For example, government funds subsidized the drainage of nearly one-fourth of the Everglades south of Lake Okeechobee to create the Everglades Agricultural Area. In addition to the direct damage this drainage inflicted on wildlife habitats, it also set the stage for indirect injury to the Everglades through water withdrawals, polluted runoff, and soil subsidence from agricultural production.

An essential element of an ecosystem approach is recognizing and measuring the value of ecosystem services, so that governments, industries, and communities can factor these values into their production and consumption choices. A first step toward setting these values is calculating the cost of economic policies that subsidize the use of resources, either by comparing subsidized to market prices or by summing the cost of government subsidy programs. Worldwide, subsidies supporting environmentally unsound practices in the use of water, agriculture, energy, and road transport are estimated to total US$700 billion, with almost half that amount supporting farm production and farm income in OECD countries (UNEP 1999:207). Refining and disaggregating this amount into national, local, or sectoral components is feasible and, even if imprecise, would provide some empirical basis for adjusting distorted prices. Going further to remove subsidies and set explicit prices on ecosystem services may be politically difficult but would lead directly to more efficient resource use.

South Africa's water law is an example of explicit pricing to encourage efficiency(see Box 3.14, pp. 200–201). South Africa allows the Department of Water Affairs and Forestry to

We can do better at managing ecosystems than we have in the past, and we can do better *today*.

levy watershed management charges on those sectors that use rivers and other water bodies for waste disposal and water consumption. Those fees are expected to discourage waste, promote conservation, and provide funds to improve watershed health. Some sectors and communities have resisted new water charges, but others have instituted municipal conservation practices that reduced water use by 25 percent.

For ecosystem services that are not explicitly subsidized, other methods of valuation need to be developed or improved (see Box 1.14, p. 32). Environmental economists should continue to hone our abilities to guage the value of ecosystem goods and services, and such values should be transmitted to those making decisions on landuse and industrial production methods. An example of how such valuation can be brought into more common use is the Environmental Valuation Reference Inventory, compiled by Environment Canada. This database of valuation studies allows corporations and government agencies to quickly call upon accepted research on monetary values for a variety of environmental services. These values, in turn, can be used to estimate the effects of projects or developments that may degrade these services (EVRI 2000).

Ultimately, creating financial incentives for ecosystem conservation is more important than setting an accurate price on ecosystem services. The price of many ecosystem services may prove to be incalculable from any supply-and-demand equation. Nonetheless, we should not lose sight of the fact that subjective judgment is at work in every valuation. The aesthetic appreciation or spiritual significance of a given landscape depends on the values of the beholders, just as the price of a particular good depends on the buyers' willingness to pay. In a debate that has focused on scientific and economic measures of value, community and religious leaders have a unique opportunity to raise the ethical considerations that should guide our use of ecosystems. Thus the valuation of ecosystem services–like the ecosystem approach as a whole–is most effective when it engages a public dialogue on goals, policies, and trade-offs.

ENGAGE IN A PUBLIC DIALOGUE ON GOALS, POLICIES, AND TRADE-OFFS

With an ecosystem approach, knowledge of ecosystem processes and conditions serves as the foundation for public discourse on what we want and need from ecosystems, how the benefits should be distributed, what the ecosystems can tolerate in terms of degradation, and what we can tolerate in terms of costs. The discourse is itself a foundation for concensus about what actions need to be taken. Even a tenuous consensus among competing interests in the New York watershed or the Bolinao reefs or the Everglades wetlands is a powerful facilitator of change, often more powerful that any engineer's technology, government's mandate, or consultant's report.

The story of New York City's watershed management plan is an example of an effort to bring together all those who have a stake in the health of an ecosystem and identify a common theme around which they could unite–in this case, water. Although the negotiated outcome in cases like New York City may not be ideal from a scientific perspective (the protection plan has been criticized as inadequate), it represents progress over interminable wrangling or inaction. Plus, when all interest groups are part of the solution, the results are usually more sustainable than those achieved without broad stakeholder participation.

When governments fail to broaden the dialogue on ecosystem management to include all stakeholders, nongovernmental organizations with ties to the local community can be powerful agents of change. The value of NGOs stands out in stories like the restoration of the Mankòtè mangrove and coastal management in Bolinao. There, NGOs persisted with countless consultations to forge alliances among the stakeholders and to elicit wider participation in decision making.

Many public dialogues on resource use are not only about the present–the relocation of a levee in the Everglades or the area for work crews to fight invasives in South Africa–they are implicitly about the future. Discussions about the best course

(continues on p. 236)

Box 4.5 Filling the Information Gap

Ecosystem	Characteristic	Principal Information Needs
All ecosystems	Extent and land use	Satellite imagery has improved knowledge of the extent of various ecosystems, but available data are rarely precise enough to use at the national or subnational levels or to support all the needs of international environmental conventions. More frequent interpretations, improved data resolution, more systematic classification processes, and innovative approaches to ground-truthing are needed.
	Soil degradation	The only comprehensive global source of information on soil degradation (GLASOD) was undertaken in the late 1980s; a supplemental study, using more detailed information, only covered Asia (ASSOD). Needs include longer-term monitoring of soil organic matter, more detailed data on soil nutrient balances, and more work on indicators that show the link between soil quality and ecosystem goods and services.
	Biodiversity	Information on biodiversity is poor across ecosystems. Only an estimated 15-20 percent of species have been identified, although the Global Taxonomy Initiative is trying to address this issue. Even for known species, information on population trends and invasions is lacking. The Global Invasive Species Programme and the World Conservation Union are assembling databases on invasives, and considerable data exist among scientists, museums, or plant collections in all countries, but effort is needed to assemble them into a form that can inform national planning.
	Water quantity and quality	Better information on water resources can immediately benefit nations because of its direct link to human health and well-being. In most parts of the world (except OECD countries), water quality monitoring is rudimentary, and most efforts leave out important biological information. Groundwater data are not readily available at a global or continental scale.
Agroecosystems	Condition	Food production and yield statistics are copious, but less is recorded about the underlying condition of agricultural systems, much less about differences in farming systems and land management practices. Reasonably detailed land use data are needed to predict the impact of agriculture on soil fertility, water quality, and habitats. Current data on soil degradation, water quality, and biodiversity are qualitative and often controversial.

Ecosystem	Characteristic	Principal Information Needs
Coastal Ecosystems	Biodiversity	Availability of global biodiversity data for coasts and oceans remains limited; even data on the distribution of habitat types are lacking for most areas, except for coral reefs and mangroves. Because most coastal habitats are small and submerged, local surveys, such as the Global Coral Reef Monitoring Network, are still more reliable than remote sensing in determining extent and condition.
	Fisheries	Outside of North Atlantic fisheries, only 50-70 percent of landings are now reported by species, which precludes efforts to evaluate the impact of fishing on specific species. Population information on fish stocks, which is needed to assess whether harvests exceed sustainable levels, is still more fragmentary.
	Water quality	Remote sensing can help to fill information gaps about occurrence and duration of algal blooms, oil spills, turbidity, and sea surface temperature, but on-site monitoring is needed to evaluate many coastal water quality parameters, such as eutrophication, coliform bacteria, and persistent organic pollutants, as well as to monitor disease outbreaks among marine organisms. The Global Ocean Observing System established by the United Nations could compile such data.
Forest Ecosystems	Condition	Extraordinarily poor data on woodfuel production and consumption will be difficult to supplement, since monitoring will be costly in most developing countries. Key data needs related to timber production are the relative rates of growth and harvest in production forests. Improved deforestation estimates will require both better satellite coverage and corroboration on the ground.
Freshwater Ecosystems	Water quantity	Rain and stream gauges around the world are disappearing, victims of loss of funding for monitoring programs. Better basic hydrological information about river discharge, flood frequency, dry season flows, condition of wetlands, and location of dams would help planners meet the growing human demand for water.
	Fisheries	Improved data on inland fisheries, essential to ensure their sustainability, will require improved or new monitoring networks, since much of the catch is consumed locally and unrecorded.
Grassland Ecosystems	Condition	High resolution satellite data measuring the productivity of grasslands, combined with on-the-ground measures of rainfall, livestock densities, and management systems could greatly increase our understanding of desertification and help national governments better manage rangelands.

of growth in a crowded area or about the rationale for allocating scarce resources or even about the nature of sustainability itself can mold a common sense of value among diverse participants. Thus, public dialogue can help the community make judgments about the relative importance of different ecosystem services. The dialogue also promotes public awareness and education; it encourages participants to learn more about the social, economic, and physical trends that are likely to affect their best-laid plans in the future.

Thus, it is essential that the stakeholders now trying to ensure the viability of ecosystems–like the Mekong River Basin or Bolinao's coastal resources–strive to incorporate projected future social and ecological changes. In the Mekong, the extraordinary pace of economic and population growth will inexorably drive intertwined demands for irrigation, drinking water, hydropower, fish production, salinity control, and transport. Bolinao's new coastal management plan may suffice for the municipalities' current population of 50,000, but the area's long-term health will depend in part on the plan's ability to incorporate a potential doubling of the population in 30 years (McManus et al. 1995:195).

Governance systems that encourage community decision making create powerful incentives for local conservation. But local solutions may not always be sufficient to keep up with rapidly accelerating, rapidly changing stresses. In those circumstances, more enduring efforts have to involve the widest possible range of stakeholders not only in dialogue but in implementation.

INVOLVE ALL STAKEHOLDERS IN MANAGING ECOSYSTEMS

Local communities can be the most pernicious violators or the most prudent managers of ecosystems. Often motivated by poverty or short-term gains, they have the greatest opportunity to overuse ecosystem goods and services. At the same time, their knowledge of their ecosystem and their direct stake in its health are important assets that improve the chances for long-term stewardship.

Similarly, national agencies, multinational businesses, and international organizations have all demonstrated their powers of destruction, as well as capacities for broad vision and enlightened policies on the use of ecosystems. National or multinational goals may conflict with–and dominate–local ones, as they did in Dhani during the period of greatest local degradation. But the growing environmental sensitivity of internationally financed demonstration projects, such as some of the best ones undertaken by the World Bank and the United Nations, can encourage local and national interests to adopt an ecosystem approach.

Involving all essential local, national, and even international interests in ecosystem management thus produces better outcomes. Inclusion of all stakeholders brings more knowledge and experience to bear on problems. The process of inclusion can balance interests that may be legitimate but divergent and can yield a more equitable distribution of the benefits and costs of ecosystem use.

Local stakeholders, however, often have the most to gain or lose in managing ecosystems. Dhani provides the quintessential example of how community concern and action can revive a local ecosystem. Driven by their dependence on the forest and their understanding of how it had been degraded, the villagers of Dhani forest crafted an effective forest protection plan. When the state, which owns the forest land, later blessed the plan, it made the local community partners in the restoration rather than adversaries. Likewise, in Machakos, the demise of government-instigated compulsory work groups in the 1950s enabled the Akamba to return to the traditional clan-based *mwethya* and to undertake–on their own initiative–the conservation techniques and work styles that rejuventated their agroecosystems.

The case studies also underscore how local communities with secure rights of resource use tend to manage ecosystems more sustainably. By contrast, consider how Dhani residents abandoned carefully crafted rules of forest access and use in favor of hastened harvesting of fuelwood when state and commercial cutting in the 1960s–70s undermined their tenure. Similarly, pastoralists in Mongolia who are uncertain about their rights to common property grasslands are less likely to use the sustainable practice of pasture rotation, for fear of losing access to lands to another herder and his flocks.

Sadly, ecosystem mismanagement continues as a result of government policies that displace local people, exploit natural resources for quick capital, and fail to recognize the role that ecosystems play in the development of sustainable livelihoods, especially for the poor. Tenure remains in question for millions of people, even as experience has repeatedly shown that secure tenure and the authority to manage resources promote long-term investments in land improvements and careful stewardship.

What Does the Future Hold?

The case studies suggest that people do learn and adapt and that ecosystems do have some natural resilience. But they also warn that there are limits to how much an ecosystem can recover. It is possible for a forest that has lost biomass and habitat quality, like Dhani Forest, to rebound in just a few years once overuse is controlled. It is less likely that wetlands, as in Florida, can be restored to health in areas already converted to suburbs, roads, and malls. Meanwhile, restoration will demand expensive financial investments in places like South Africa and Florida, and significant human capital in places like Dhani, Machakos, and Cuba–outlays that depend on strong public and governmental will.

Box 4.6 The Call for a Millennium Ecosystem Assessment

It is impossible to devise effective environmental policy unless it is based on sound scientific information. While major advances in data collection have been made in many areas, large gaps in our knowledge remain. In particular, there has never been a comprehensive global assessment of the world's major ecosystems. The planned Millennium Ecosystem Assessment, a major international collaborative effort to map the health of our planet, is a response to this need. It is supported by many governments, as well as UNEP, UNDP, FAO and UNESCO. I call on Member States to help provide the necessary financial support for the Millennium Ecosystem Assessment and to become actively engaged in it.

— UN Secretary General Kofi A. Annan
From *We the Peoples: The Role of the United Nations in the 21st Century* (April 2000)

Also endorsing the Millennium Ecosystem Assessment as of September 2000:

- Conference of parties to the Convention to Combat Desertification
- Conference of parties to the Convention on Biological Diversity
- Conference of parties to the Ramsar Convention on Wetlands
- Consultative Group on International Agricultural Research and the International Agricultural Research Centers
- Millennium Assessment Steering Committee, representing 30 international agencies and research
- Ministers of the Environment meeting in Elmina, Ghana, September 1999, representing 20 countries
- Third World Academy of Sciences
- Third World Network of Scientific Organizations
- *World Resources* partners UNDP, UNEP, World Bank, and WRI

The case studies do not end here. Only time will reveal the level of health that any of these degraded ecosystems regain. We know the "restored" Everglades system will be different in species composition and functioning than the original system. South Africa will never entirely be rid of its invading plants, despite the best efforts of the Working for Water Programme.

Climate change, globalization, and urbanization are pressures that could undermine the long-term successes of even the most informed, carefully constructed management and restoration plans. Increasing global carbon emissions are already affecting ecosystems. Warmer temperatures and changes in rainfall patterns could encourage migrations and invasions of nonnative species, and rising sea levels could submerge many low-lying areas, from coral atolls to parts of the Everglades ecosystem. Globalization and industrialization are likely to destabilize many traditional economic patterns that focus on subsistence and local resource use. Suburban sprawl, habitat fragmentation, air pollution, and the sheer scale of resource demand and waste generation will take a toll before better urban planning begins to minimize these stresses.

Successful ecosystem management will increasingly require the cooperation of neighbors–sometimes people with widely divergent goals. Dhani residents had only to work with adjacent villages, but South Africa must work with Botswana and Zimbabwe to control dense infestations of nonnative plants like rose cactus, the distribution of which is accelerated by elephants and donkeys moving freely across borders. Even that is a relatively local problem compared with the transboundary issues raised by efforts to develop and manage the Mekong River sustainably. There, the wishes and needs of six nations all threaten the quantity and quality of the water in the Basin, and the livelihoods of the fishers and farmers in the Lower Mekong.

The international agreement to stem stratospheric ozone depletion (the Montreal Protocol) suggests that we can–aided by sound science–formulate a shared vision and commitment to manage a problem, once we understand its severity. But for some ecosystem services, like biodiversity and carbon storage, a shared understanding of their importance may not be enough to bring about cooperative global management. International markets do not value ecosystem services, such as biodiversity or carbon storage, as the public assets they are. Yet they are essential assets of global importance; thus, the global community may need to bear some of the costs of sustaining them. International efforts to supply public capital and leverage private-sector

investment will be a crucial factor in changing how countries value and conserve their ecosystems.

Perhaps the most important message in the case studies is that we can do better at managing ecosystems than we have in the past, and we can do better today. We often tout technology's promise of solving problems: making restoration cheaper or increasing the productivity of our ecosystems. These cases don't undermine technology's promise, but they remind us that we already have much of the knowledge and technology we need. Many of these "fixes" are simple and nontechnical. In South Africa, people are restoring the ecosystem by uprooting invasive trees by hand. In Dhani, a community employs watchmen and patrols, uses simple harvest plans and bans cattle grazing, and promotes alternative local employment. In Machakos, the Akamba collect rainwater and construct terraces–a practice dating back to ancient times in many parts of the world.

Put simply, we already know enough to begin to manage ecosystems more soundly and to restore some of the natural productivity we have lost. Mustering the local, national, and global commitment to use and expand that knowledge is the challenge.

A Millennium Ecosystem Assessment

Our failure to think in terms of ecosystems has been rooted in the profound lack of information about how ecosystems affect us and what condition they are in. The Pilot Analysis of Global Ecosystems begins to address this information issue. But one of the most important conclusions of the PAGE study is that we currently lack much of the baseline knowledge we need to assess ecosystem conditions adequately on a global, regional, or sometimes even a local scale. PAGE researchers noted the absence of dozens of critical data sets–from the level of fuelwood use to the impacts of livestock on grassland forage conditions (Box 4.5 Filling the Information Gap).

Considering our technological advances, it is surprising that the availability of information for assessing the condition of ecosystems has not improved in recent years and may actually be decreasing. On the one hand, remote sensing has made information available about certain features of ecosystems, such as their extent. On the other hand, on-the-ground information for such indicators as freshwater quality and river discharge is less available today than 20 years ago (Stokstad 1999:1199).

Gathering this kind of information and making it available in a form that governments, businesses, and local residents can easily understand and use will require a much larger, more comprehensive effort than PAGE. Such an effort, the Millennium Ecosystem Assessment (MEA), scheduled to begin in 2001, is organized and supported by an array of governments, UN agencies, and leading scientific organizations (Box 4.6 The Call for a Millennium Ecosystem Assessment). The PAGE study itself provided a demonstration of some of the methods and approaches the MEA will use, but the MEA will develop and expand these methods for global application by a diverse group of researchers acting at several scales, from local to global.

The MEA, like the PAGE study, will focus on the capacity of ecosystems to provide goods and services important to human development. Thus, it will consider the underlying ecosystem processes on which these goods and services depend. Furthermore, it will explicitly consider social and economic attributes such as employment and economic value. The MEA will consist of a global assessment more comprehensive than the PAGE study and approximately 10 assessments undertaken at regional, national, and local scales. It will also help nations develop more capacity to do their own assessments in the future:

- *The global component of the MEA* will establish a baseline for future assessments, help meet information needs of the international environmental treaties, like the Convention on Biological Diversity, establish methodologies for integrated ecosystem assessments, and raise public awareness about the importance of ecosystem goods and services. The global component will be uniquely suited to assessing change in global chemical cycles of carbon, nitrogen, and water.

- *The regional, national, and local components of the MEA* will cover only a small portion of the globe but will help to catalyze more widespread use of integrated assessments and help to develop the methodologies and modeling tools needed for those assessments. These components will also provide information that bears directly on management and policy decisions in the regions where they are conducted, and they will be uniquely suited to assessing tradeoffs and linkages among various goods and services. The development of scenarios describing plausible future conditions of ecosystem goods and services will also take place at a regional level and be synthesized at the global level.

- *Capacity building* will also be a central objective of the MEA process. The regional, national, and local components of the MEA will directly strengthen the institutions involved. The information, methodologies, and modeling tools developed through the MEA will be of use to national and subnational assessment processes around the world. Finally, the MEA will help to promote the data collection and monitoring efforts needed to meet information needs at all scales.

The MEA is just one of many steps necessary to reorient our view of ecosystems and how to manage them. Yet it is one of the first and most elemental. If the MEA is successful, it could provide a foundation of knowledge about ecosystems that would offer immediate utility and guidance for policy makers tackling such basic issues as water use, coastal development, agricultural policies, and biodiversity conservation. At a more fundamental level, it would mark an important step toward an ecosystem approach by beginning to frame the environmental information that decision makers use in terms of ecosystem goods and services. In time, this basic reorganization of how we measure and analyze environmental change will embed the concept of ecosystems into how we talk about and manage our impacts on the Earth.

What Better Time Than Now?

Our dominance of Earth's productive systems gives us enormous responsibilities, but great opportunities as well. Human demands on ecosystems have never been higher, and yet these demands are likely to increase dramatically, especially in developing countries, as rising populations mean more and more people are seeking better lives. Human understanding of ecosystems has never been greater, and yet even amid an abundance of data we are often confronted with our own ignorance about the world around us. Most important, human intervention in ecosystems is evident everywhere, yet so little has been done to protect them that we must not delay our actions.

The challenge for the 21st century, then, is to reconcile the demands of human development with the tolerances of nature. For this we have to understand the vulnerabilities and resilience of ecosystems. From the Foreword to this volume:

> *At the dawn of a new century, we have the ability to change the vital systems of this planet, for better or worse. To change them for the better, we must recognize that the well-being of people and ecosystems is interwoven and that the fabric is fraying. We need to repair it, and we have the tools at hand to do so. What better time than now?*

DATA TABLES

- Biodiversity and Protected Areas
- Forests and Grasslands
- Coastal, Marine, and Inland Waters
- Agriculture and Food
- Freshwater
- Atmosphere and Climate
- Energy and Resource Use
- Population and Human Development
- Basic Economic Indicators
- Small Nations and Islands

Information about the World Resources 2000–2001 Data Tables

Country groupings are based on lists developed by FAO (developed and developing countries), UNICEF (industrialized, developing, and least developed countries, and other regions), UNPD (populations in developed and developing countries and other regions), World Bank (high-, medium-, and low-income countries), and WRI (regions). (See pp. 309–313.)

Several general notes apply to all of the data tables in the report (except where noted otherwise):

- "X" in a data column signifies that data are not available or are not relevant (for example, country status has changed as with the former Czechoslovakia and the Czech Republic).
- Negative values are shown in parentheses.
- Zero is either 0 or less than one-half the unit of measure; (0) indicates a value less than 0 and greater than negative one-half.
- Except where identified with a footnote, regional totals are calculated using regions designated by the World Resources Institute (WRI). These regions are as listed at the back of this report in the "WRI Regions" table on page 309.
- Except where identified with a footnote, world totals are presented using totals as calculated by the original data source (which may include countries not listed in this database); for original source see Sources and Technical Notes.
- Except where identified with a footnote, developed and developing country totals are calculated using designations from the Food and Agriculture Organization of the United Nations (FAO). These designations are listed at the back of this report in the "FAO Regions" table.
- Additional technical notes to the Data Tables begin on page 315.
- Updates may be added to some tables such as the Freshwater Chapter. Please check our website for revisions.

Watch the website http://www.wri.org

World Resources data tables and analysis of global environmental trends are moving to a new searchable format on line. Beginning in January 2001, these critical components of *World Resources* will become part of a new portal for environmental information on the World Resources Institute's website. Data and trends will be continuously updated and expanded as new statistics become available.

As in previous editions, the information contained in the data tables in this volume is also downloadable and printable in .pdf format at the website and is available on CD-Rom by order from http://www.wristore.com.

Part II: Data Tables

TABLE OF CONTENTS

	National Protection Systems								Protected Areas that Are Part of Global Agreements					
	Protected Areas (IUCN Management Categories I-V)					Number of Marine Protected Areas {a} (IUCN Categories I-VI)			Biosphere Reserves {b}		World Heritage Sites {c}		Wetlands of International Importance {d}	
	Number	Area (000 ha)	Percent of Land Area	No. of Areas at least: 100,000 ha in Size	No. of Areas at least: 1 million ha in Size	Total	Littoral	Marine	Number	Area (000 ha)	Number	Area (000 ha)	Number	Area (000 ha)
WORLD {e}	**28,442**	**851,511**	**6.4**	**1,250**	**154**	**3,636**	**2,825**	**1,685**	**368**	**263,897**	**149**	**142,209**	**1,019**	**73,012**
ASIA (EXCL. MIDDLE EAST)	**2,421**	**148,692**	**6.0**	**210**	**20**	**831**	**545**	**419**	**45**	**13,350**	**25**	**5,154**	**59**	**3,121**
Armenia	5	213	7.6	1	0	X	X	X	0	0	0	0	2	492
Azerbaijan	34	478	5.5	0	0	3	3	0	0	0	0	0	N/A	N/A
Bangladesh	10	98	0.7	0	0	6	6	0	0	0	1	595	1	596
Bhutan	9	998	21.2	3	0	X	X	X	0	0	N/A	N/A	N/A	N/A
Cambodia	20	2,863	15.8	10	0	X	X	X	1	1,481	0	0	3	55
China	407	59,840	6.2	40	10	56	41	28	16	2,645	7	352	7	588 f
Georgia	18	195	2.8	0	0	1	1	0	0	0	0	0	2	34
India	493	14,312	4.4	22	0	115	113	12	0	0	5	292	6	193
Indonesia	331	19,253	10.1	35	5	102	93	40	6	1,329	3	2,845	2	243
Japan	96	2,561	6.8	8	0	187	52	140	4	117	2	28	11	84
Kazakhstan	73	7,337	2.7	18	1	1	1	0	0	0	0	0	N/A	N/A
Korea, Dem People's Rep	31	316	2.6	1	0	X	X	X	1	132	0	0	N/A	N/A
Korea, Rep	30	684	6.9	1	0	7	5	6	1	39	0	0	2	1
Kyrgyzstan	78	694	3.5	0	0	X	X	X	1	24	0	0	N/A	N/A
Lao People's Dem Rep	0	0	0.0	0	0	X	X	X	0	0	0	0	N/A	N/A
Malaysia	143	1,507	4.6	5	0	111	108	12	0	0	0	0	1	38
Mongolia	42	17,991	11.5	18	4	X	X	X	3	6,139	0	0	6	631
Myanmar	3	174	0.3	1	0	3	3	1	0	0	0	0	N/A	N/A
Nepal	12	1,112	7.6	4	0	X	X	X	0	0	2	208	1	18
Pakistan	81	3,727	4.7	11	0	3	3	0	1	37	0	0	8	62
Philippines	19	1,454	4.8	4	0	159	67	126	2	1,174	2	53	4	68
Singapore	5	3	4.7	0	0	3	3	2	0	0	N/A	N/A	N/A	N/A
Sri Lanka	110	869	13.3	0	0	16	12	6	2	9	1	9	1	6
Tajikistan	19	587	4.1	1	0	X	X	X	0	0	0	0	N/A	N/A
Thailand	158	7,077	13.8	19	0	18	18	17	4	56	1	622	1	<1
Turkmenistan	23	1,977	4.1	5	0	1	0	1	1	35	0	0	N/A	N/A
Uzbekistan	11	818	1.8	1	0	X	X	X	1	57	0	0	N/A	N/A
Viet Nam	54	995	3.0	1	0	2	2	2	1	76	1	150	1	12
EUROPE	**12,356**	**109,297**	**4.7**	**212**	**12**	**760**	**667**	**246**	**139**	**103,149**	**28**	**19,292**	**632**	**19,838**
Albania	48	84	2.9	0	0	7	7	0	0	0	0	0	1	20
Austria	695	2,451	29.2	3	0	X	X	X	4	28	0	0	10	116
Belarus	903	1,304	6.3	0	0	X	X	X	2	305	1 g	88	1	19
Belgium	72	86	2.8	0	0	2	2	0	0	0	0	0	6	8
Bosnia and Herzegovina	21	27	0.5	0	0	X	X	X	0	0	0	0	N/A	N/A
Bulgaria	127	500	4.5	2	0	2	2	0	17	25	2	41	5	3
Croatia	195	421	7.4	1	0	18	13	6	1	200	1	19	4	80
Czech Rep	1,789	1,247	15.8	2	0	X	X	X	6 g	435	0	0	10	42
Denmark {h}	220	1,380	32.0	2	0	76	58	41	1 i	70,000	0	0	38	2,283 j
Estonia	219	500	11.1	0	0	3	3	3	1	1,560	0	0	10	216
Finland	260	1,867	5.5	4	0	3	3	1	2	770	0	0	11	101
France {k}	1,341	7,437	13.5	29	0	126	102	42	8 g	832	2 g	23	15	579
Germany	1,398	9,620	26.9	26	0	49	48	15	14 g	1,559	1	<1	31	673
Greece	88	469	3.6	1	0	15	14	1	2	9	2	<1	10	164
Hungary	186	649	7.0	0	0	X	X	X	5	129	1 g	X	19	150
Iceland	79	981	9.5	2	0	8	8	0	0	0	0	0	3	59
Ireland	72	65	0.9	0	0	9	9	0	2	11	0	0	45	67
Italy	422	2,204	7.3	2	0	59	36	23	5	204	0	0	46	57
Latvia	158	807	12.5	1	0	2	2	1	1	398	0	0	3	43
Lithuania	79	645	9.9	0	0	3	2	3	0	0	0	0	5	50
Macedonia, FYR	26	181	7.1	0	0	X	X	X	0	0	1	X	1	19
Moldova, Rep	63	47	1.4	0	0	X	X	X	0	0	N/A	N/A	N/A	N/A
Netherlands {k}	82	232	5.7	0	0	19	18	5	1	260	0	0	18	325
Norway	180	2,093	6.5	7	0	11	8	7	0	0	0	0	23	70
Poland	522	2,929	9.1	4	0	6	5	1	8 g	320	1 g	5	8	90
Portugal	58	603	6.6	1	0	29	25	13	1	1	1	X	10	66
Romania	157	1,089	4.6	1	0	11	11	0	3 g	679	1	679	1	647
Russian Federation	219	52,907	3.1	92	12	16	12	10	20	23,386	5	17,343	35	10,324
Slovakia	1,039	1,085	22.1	1	0	X	X	X	4 g	241	1 g	X	11	37
Slovenia	32	120	5.9	0	0	1	1	0	0	0	1	<1	2	1
Spain	328	4,240	8.4	7	0	34	32	5	16	1,008	4 g	110	38	158
Sweden	348	3,645	8.1	8	0	51	51	4	1	97	1	940	30	383
Switzerland	257	1,063	25.7	0	0	X	X	X	1	169	0	0	8	7
Ukraine	28	944	1.6	1	0	5	3	2	5 g	294	0	0	22	716
United Kingdom {k}	515	5,000	20.4	15	0	192	189	62	13	47	4	11	140	664
Yugoslavia	104	339	3.3	0	0	2	2	1	1	183	1	32	4	40
MIDDLE EAST & N. AFRICA	**518**	**25,863**	**2.1**	**25**	**5**	**126**	**90**	**54**	**21**	**15,019**	**5**	**10,772**	**48**	**1,748**
Afghanistan	7	219	0.3	0	0	X	X	X	0	0	0	0	N/A	N/A
Algeria	18	5,891	2.5	2	2	8	6	3	3	7,294	1	8,000	3	5
Egypt	16	794	0.8	1	0	18	16	6	2	2,456	0	0	2	106
Iran, Islamic Rep	78	8,303	5.1	8	1	6	5	1	9	2,610	0	0	20	1,433
Iraq	8	1	<0.1	0	0	X	X	X	0	0	0	0	N/A	N/A
Israel	188	326	15.5	1	0	25	18	8	1	27	0	0	2	<1
Jordan	11	298	3.3	1	0	1	0	1	1	31	0	0	1	7
Kuwait	5	27	1.5	0	0	4	2	2	0	0	N/A	N/A	N/A	N/A
Lebanon	3	5	0.5	0	0	1	1	0	0	0	0	0	3	1
Libyan Arab Jamahiriya	8	173	0.1	1	0	5	2	3	0	0	0	0	N/A	N/A
Morocco	12	317	0.7	1	0	10	9	3	1	2,569	0	0	4	14
Oman	3	3,428	16.1	1	0	2	1	2	0	0	1	2,750	N/A	N/A
Saudi Arabia	71	4,973	2.3	8	2	4	3	4	0	0	0	0	N/A	N/A
Syrian Arab Rep	0	0	0.0	0	0	X	X	X	0	0	0	0	1	10
Tunisia	7	45	0.3	0	0	7	5	4	4	32	1	13	1	13
Turkey	64	985	1.3	1	0	21	15	8	0	0	2	10	9	159
United Arab Emirates	2	X	X	0	0	4	3	1	0	0	N/A	N/A	N/A	N/A
Yemen	0	0	0.0	0	0	X	X	X	0	0	0	0	N/A	N/A

	National Protection Systems								Protected Areas that Are Part of Global Agreements					
	Protected Areas (IUCN Management Categories I-V)					Number of Marine Protected Areas {a} (IUCN Categories I-VI)			Biosphere Reserves {b}		World Heritage Sites {c}		Wetlands of International Importance {d}	
	Number	Area (000 ha)	Percent of Land Area	No. of Areas at least: 100,000 ha in Size	1 million ha in Size	Total	Littoral	Marine	Number	Area (000 ha)	Number	Area (000 ha)	Number	Area (000 ha)
SUB-SAHARAN AFRICA	**1,005**	**146,904**	**6.0**	**202**	**33**	**150**	**128**	**93**	**40**	**39,757**	**32**	**29,373**	**70**	**14,348**
Angola	13	8,181	6.6	9	2	4	4	3	0	0	0	0	N/A	N/A
Benin	2	778	6.9	2	0	X	X	X	1	623	0	0	2	139
Botswana	12	10,499	18.0	7	3	X	X	X	0	0	0	0	1	6,864
Burkina Faso	12	2,855	10.4	6	1	X	X	X	1	186	0	0	3	299
Burundi	13	146	5.3	0	0	1	1	1	0	0	0	0	N/A	N/A
Cameroon	18	2,098	4.4	8	0	2	1	2	3	850	1	526	N/A	N/A
Central African Rep	13	5,110	8.2	11	2	X	X	X	2	1,640	1	1,740	N/A	N/A
Chad	9	11,494	9.0	9	2	X	X	X	0	0	0	0	1	195
Congo	9	1,545	4.5	5	0	1	1	1	2	246	0	0	1	439
Congo, Dem Rep	15	10,191	4.3	9	4	1	1	0	3	283	5	6,855	2	866
Côte d'Ivoire	11	1,986	6.2	4	1	3	3	1	2	1,480	3 g	1,504	1	19
Equatorial Guinea	0	0	0.0	0	0	4	4	0	0	0	N/A	N/A	N/A	N/A
Eritrea	3	501	4.3	2	0	X	X	X	0	0	N/A	N/A	N/A	N/A
Ethiopia	21	5,518	5.0	14	0	X	X	X	0	0	1	22	N/A	N/A
Gabon	5	723	2.7	1	0	4	4	1	1	15	0	0	3	1,080
Gambia	6	23	2.0	0	0	5	5	3	0	0	0	0	1	20
Ghana	10	1,104	4.6	3	0	X	X	X	1	8	0	0	6	178
Guinea	3	164	0.7	1	0	1	0	1	2	133	1 g	13	6	225
Guinea-Bissau	0	0	0.0	0	0	2	2	1	1	110	N/A	N/A	1	39
Kenya	50	3,507	6.0	7	1	14	13	10	5	891	2	300	2	49
Lesotho	1	7	0.2	0	0	X	X	X	0	0	N/A	N/A	N/A	N/A
Liberia	1	129	1.2	1	0	X	X	X	0	0	N/A	N/A	N/A	N/A
Madagascar	40	1,121	1.9	1	0	3	3	1	1	140	1	152	2	53
Malawi	9	1,059	8.9	3	0	X	X	X	0	0	1	9	1	225
Mali	13	4,532	3.7	8	2	X	X	X	1	2,349	1	400	3	162
Mauritania	9	1,746	1.7	3	1	5	5	2	0	0	1	1,200	2	1,216
Mozambique	11	4,779	6.0	7	1	7	7	6	0	0	0	0	N/A	N/A
Namibia	20	10,616	12.9	7	3	4	4	0	0	0	N/A	N/A	4	630
Niger	6	9,694	7.7	4	2	X	X	X	2	25,128	2	7,957	1	220
Nigeria	27	3,021	3.3	9	0	X	X	X	1	<1	0	0	N/A	N/A
Rwanda	6	362	13.8	1	0	X	X	X	1	15	N/A	N/A	N/A	N/A
Senegal	12	2,181	11.1	3	0	7	6	6	3	1,094	2	929	4	100
Sierra Leone	2	82	1.1	0	0	X	X	X	0	0	N/A	N/A	1	295
Somalia	2	180	0.3	1	0	2	2	2	0	0	N/A	N/A	N/A	N/A
South Africa	390	6,619	5.4	6	1	20	17	8	1	104	1	X	16	493
Sudan	11	8,642	3.4	6	2	2	0	2	2	1,901	0	0	N/A	N/A
Tanzania, United Rep	39	13,817	14.6	19	3	9	8	8	2	2,338	4	6,860	N/A	N/A
Togo	9	429	7.6	2	0	1	0	1	0	0	0	0	2	194
Uganda	37	1,913	7.9	6	0	X	X	X	1	220	2	132	1	15
Zambia	35	6,366	8.5	11	1	X	X	X	0	0	1 g	7	2	333
Zimbabwe	48	3,071	7.9	6	1	X	X	X	0	0	2 g	733	N/A	N/A
NORTH AMERICA	**6,146**	**213,822**	**11.1**	**255**	**46**	**525**	**357**	**263**	**52**	**22,350**	**18**	**20,405**	**53**	**14,229**
Canada	3,083	90,702	9.1	102	20	139	102	76	8	1,512	8 g	10,664	36	13,051
United States {k}	3,063	123,120	13.1	153	26	386	255	187	44	20,838	12 g	9,741	17	1,178
C. AMERICA & CARIBBEAN	**813**	**16,450**	**6.1**	**35**	**1**	**369**	**324**	**177**	**29**	**14,972**	**11**	**2,997**	**47**	**2,239**
Belize	32	479	20.9	2	0	19	18	8	0	0	1	96	2	X
Costa Rica	85	723	14.2	1	0	18	18	11	2	729	3 g	592	9	246
Cuba	81	1,909	17.2	4	0	51	44	25	6	1,384	1	42	N/A	N/A
Dominican Rep {l}	34	1,523	31.3	6	0	12	10	8	0	0	0	0	N/A	N/A
El Salvador	2	5	0.2	0	0	2	2	0	0	0	0	0	1	2
Guatemala	38	1,827	16.8	4	0	4	4	0	2	3,141	1	58	4	503
Haiti	8	10	0.3	0	0	X	X	X	0	0	0	0	N/A	N/A
Honduras	70	673	6.0	1	0	25	25	20	1	800	1	500	4	172
Jamaica	5	2	0.1	0	0	6	4	6	0	0	0	0	1	6
Mexico	166	6,637	3.4	12	1	56	48	24	11	5,393	2	899	6	1,095
Nicaragua	70	908	7.0	1	0	4	4	1	1	2,182	0	0	1	44
Panama	30	1,422	18.8	4	0	14	14	3	2	1,253	2 g	804	3	111
Trinidad and Tobago	25	31	6.0	0	0	14	14	2	0	0	N/A	N/A	1	6
SOUTH AMERICA	**1,116**	**131,663**	**7.4**	**202**	**26**	**189**	**172**	**57**	**29**	**50,205**	**13**	**9,125**	**45**	**11,551**
Argentina	147	4,909	1.8	11	0	32	30	4	7	2,235	3 g	861	7	1,000
Bolivia	32	15,601	14.2	19	6	X	X	X	3	735	0	0	2	805
Brazil	519	37,513	4.4	62	5	82	73	25	2	29,699	3 g	1,974	5	4,537
Chile	87	14,142	18.7	19	5	28	24	4	7	2,407	0	0	7	100
Colombia	94	9,363	8.2	18	2	11	10	9	3	2,514	1	72	1	400
Ecuador {l}	22	12,077	42.6	12	1	4	4	3	2	1,446	2	1,038	3	95
Guyana	1	59	0.3	0	0	X	X	X	0	0	0	0	N/A	N/A
Paraguay	20	1,401	3.4	3	0	X	X	X	0	0	0	0	4	775
Peru	21	3,463	2.7	6	1	4	4	1	3	2,701	4	2,180	7	2,932
Suriname	13	736	4.5	3	0	5	5	0	0	0	0	0	1	12
Uruguay	13	48	0.3	0	0	4	4	0	1	200	0	0	1	435
Venezuela	144	32,245	35.4	48	6	18	17	11	1	8,266	1	3,000	5	264
OCEANIA	**4,056**	**60,784**	**7.1**	**109**	**11**	**519**	**412**	**275**	**13**	**5,094**	**17**	**45,091**	**60**	**5,882**
Australia {l}	3,727	54,250	7.0	94	10	372	280	226	12	5,093	13	42,297	53	5,249
Fiji	15	20	1.1	0	0	6	6	1	0	0	0	0	N/A	N/A
New Zealand	233	6,334	23.4	15	1	88	81	13	0	0	3	2,756	5	39
Papua New Guinea	6	7	<0.1	0	0	12	10	9	0	0	0	0	2	595
Solomon Islands	0	0	0.0	0	0	6	6	0	0	0	1	37	N/A	N/A
DEVELOPED	**23,397**	**405,509**	**7.2**	**617**	**71**	**1,983**	**1,477**	**905**	**212**	**130,955**	**65**	**84,778**	**776**	**40,458**
DEVELOPING	**5,045**	**446,002**	**5.8**	**633**	**83**	**1,486**	**1,218**	**679**	**156**	**132,941**	**84**	**57,431**	**238**	**32,498**

Notes: a. Includes areas with substantial terrestrial components that reach the shore. An area can be both, marine and littoral (see technical notes). b. Biosphere reserves shared by several countries are counted only once in regional and world totals. c. Countries that have not signed the World Heritage Convention are designated by N/A. Sites shared by several countries are counted only once in regional and world totals. d. Countries that have not signed the Convention on Wetlands of International Importance are designated by N/A. e. World totals except for Marine Protected Areas and Wetlands of International Importance were calculated by WRI. World total for National Protection Systems excludes protected areas in Antarctica and Greenland. f. Includes one site in Hong Kong. g. Includes sites shared by two or more neighboring countries. h. Categories under National Protection Systems exclude protected areas in Greenland. i. Corresponds to a single site located in Greenland. j. Includes 11 sites in Greenland. k. Excludes protected areas and sites protected under global agreements located in overseas territories (i.e., Guadeloupe, Cayman Islands, etc.).
l. Extent of protected areas may include marine components that may lead to some unexpectedly high figures for percent of land area protected.

Sources: World Conservation Monitoring Centre; United Nations Educational, Scientific, and Cultural Organization; and Ramsar Convention Bureau

	Mammals				Birds				Reptiles			
	Total Number of Known Species			No. of Species per 10,000 km² {a}	Total Number of Known Species			Breeding Bird Species per 10,000 km² {a}	Total Number of Known Species			No. of Species per 10,000 km² {a}
	All Species	Endemic Species	Threatened Species		Breeding Species	Endemic Species	Threatened Species		All Species	Endemic Species	Threatened Species	
WORLD {b}	4,629 c	X	1,096	X	9,672 d	X	1,107	X	6,900 e	X	253	X
ASIA (EXCL. MIDDLE EAST)	X	X	X	X	X	X	X	X	X	X	X	X
Armenia	84	3	4	59	242	0	5	169	51	0	3	36
Azerbaijan	99	0	11	49	248	0	8	122	54	0	3	26
Bangladesh	109	0	18	45	295	0	30	122	119	2	13	49
Bhutan	99	0	20	59	448	0	14	269	19	2	1	11
Cambodia	123	0	23	47	307	0	18	118	82	1	9	32
China {f}	400	83	75	41	1,103	70	90	114	340	81	15	35
Georgia	107	2	10	56	X	0	5	X	52	0	7	27
India	316	44	75	47	926	58	73	137	390	188	16	58
Indonesia	457	222	128	81	1,530	408	104	271	514	305	19	91
Japan	188	42	29	57	250	21	33	75	87	33	8	26
Kazakhstan	178	4	15	28	396	0	15	62	49	0	1	8
Korea, Dem People's Rep	X	0	7	X	115	1	19	51	19	2	0	8
Korea, Rep	49	0	6	23	112	0	19	53	25	1	0	12
Kyrgyzstan	83	1	6	31	X	0	5	X	33	0	1	12
Lao People's Dem Rep	172	0	30	61	487	1	27	171	66	1	7	23
Malaysia	300	36	42	95	508	18	34	160	350	71	14	110
Mongolia	133	0	12	25	426	0	14	80	22	0	0	4
Myanmar	251	6	31	62	867	4	44	216	203	37	20	51
Nepal	181	2	28	75	611	2	27	252	100	1	5	41
Pakistan	151	4	13	36	375	0	25	88	172	23	6	41
Philippines	158	102	49	51	196	186	86	64	190	159	7	62
Singapore	85	1	6	213	118	0	9	295	140	0	1	350
Sri Lanka	88	15	14	47	250	24	11	134	144	77	8	77
Tajikistan	84	1	5	35	X	0	9	X	44	0	1	18
Thailand	265	7	34	72	616	2	45	168	298	37	16	81
Turkmenistan	103	0	11	29	X	0	12	X	82	0	2	23
Uzbekistan	97	0	7	28	X	0	11	X	64	0	0	18
Viet Nam	213	9	38	67	535	10	47	168	187	46	12	59
EUROPE	X	X	X	X	X	X	X	X	X	X	X	X
Albania	68	0	2	48	230	0	7	162	31	0	1	22
Austria	83	0	7	41	213	0	5	106	14	0	1	7
Belarus	74	0	4	27	221	0	4	81	7	0	0	3
Belgium	58	0	6	40	180	0	3	125	8	0	0	6
Bosnia and Herzegovina	72	0	10	42	218	0	2	127	27	0	0	16
Bulgaria	81	0	13	37	240	0	12	108	33	0	1	15
Croatia	76	0	10	43	224	0	4	126	29	0	0	16
Czech Rep	81	0	7	41	199	0	6	101	10	0	0	5
Denmark	43	0	3	27	196	0	2	121	5	0	0	3
Estonia	65	0	4	40	213	0	2	130	5	0	0	3
Finland	60	0	4	19	248	0	4	78	5	0	0	2
France	93	0	13	25	269	1	7	72	32	1	3	9
Germany	76	0	8	23	239	0	5	73	12	0	0	4
Greece	95	3	13	41	251	0	10	107	56	8	6	24
Hungary	83	0	8	40	205	0	10	98	15	0	1	7
Iceland	11	0	1	5	88	0	0	41	0	0	0	0
Ireland	25	0	2	13	142	0	1	75	1	0	0	1
Italy	90	3	10	29	234	0	7	76	40	1	4	13
Latvia	83	0	4	45	217	0	6	117	7	0	0	4
Lithuania	68	0	5	37	202	0	4	109	7	0	0	4
Macedonia, FYR	78	0	10	57	210	0	3	154	31	0	1	23
Moldova, Rep	68	0	2	46	177	0	7	119	9	0	1	6
Netherlands	55	0	6	35	191	0	3	120	7	0	0	4
Norway	54	0	4	17	243	0	3	77	5	0	0	2
Poland	84	0	10	27	227	0	6	72	9	0	0	3
Portugal	63	1	13	30	207	2	7	100	29	3	0	14
Romania	84	0	16	29	247	0	11	87	25	0	2	9
Russian Federation	269	22	31	23	628	13	38	54	58	0	5	5
Slovakia	85	0	8	50	209	0	4	124	20	0	0	12
Slovenia	75	0	10	59	207	0	3	164	25	0	0	20
Spain	82	4	19	22	278	5	10	76	53	11	6	15
Sweden	60	0	5	17	249	0	4	71	6	0	0	2
Switzerland	75	0	6	47	193	0	4	121	14	0	0	9
Ukraine	108	1	15	28	263	0	10	68	21	1	2	5
United Kingdom	50	0	4	17	230	1	2	80	8	0	0	3
Yugoslavia	96	0	12	45	224	0	8	104	70	0	1	33
MIDDLE EAST & N. AFRICA	X	X	X	X	X	X	X	X	X	X	X	X
Afghanistan	123	2	11	31	235	0	13	59	103	4	1	26
Algeria	92	2	15	15	192	1	8	32	81	4	1	13
Egypt	98	7	15	21	153	0	11	33	83	0	6	18
Iran, Islamic Rep	140	6	20	26	323	1	14	60	167	29	8	31
Iraq	81	2	7	23	172	1	12	49	81	1	2	23
Israel	116	4	13	91	180	0	8	141	97	1	5	76
Jordan	71	0	7	34	141	0	4	68	73	0	1	35
Kuwait	21	0	1	17	20	0	3	17	29	0	2	24
Lebanon	57	0	5	56	154	0	5	152	42	1	2	41
Libyan Arab Jamahiriya	76	5	11	14	91	0	2	17	56	1	3	10
Morocco	105	4	18	30	210	0	11	60	90	11	2	26
Oman	56	2	9	20	107	0	5	39	64	8	4	23
Saudi Arabia	77	0	9	13	155	0	11	26	84	4	2	14
Syrian Arab Rep	63	2	4	24	204	0	7	78	X	2	3	X
Tunisia	78	1	11	31	173	0	6	69	62	1	2	25
Turkey	116	2	15	28	302	0	14	72	105	7	12	25
United Arab Emirates	25	0	3	12	67	0	4	33	37	1	2	18
Yemen	66	1	5	18	143	8	13	39	77	30	2	21

	Mammals				Birds				Reptiles			
	Total Number of Known Species			No. of Species	Total Number of Known Species			Breeding Bird Species	Total Number of Known Species			No. of Species
	All Species	Endemic Species	Threatened Species	per 10,000 km² {a}	Breeding Species	Endemic Species	Threatened Species	per 10,000 km² {a}	All Species	Endemic Species	Threatened Species	per 10,000 km² {a}
SUB-SAHARAN AFRICA	X	X	X	X	X	X	X	X	X	X	X	X
Angola	276	7	17	56	765	12	13	156	X	19	5	X
Benin	188	0	9	85	307	0	1	138	X	1	2	X
Botswana	164	0	5	43	386	1	7	101	157	2	0	41
Burkina Faso	147	0	6	49	335	0	1	112	X	3	1	X
Burundi	107	0	5	76	451	0	6	322	X	0	0	X
Cameroon	409	14	32	114	690	8	14	193	183	21	3	51
Central African Rep	209	2	11	53	537	1	2	137	129	0	1	33
Chad	134	1	14	27	370	0	3	75	5	1	1	1
Congo	200	2	10	62	449	0	3	140	X	1	2	X
Congo, Dem Rep	450	28	38	74	929	24	26	153	377	35	3	62
Côte d'Ivoire	230	0	16	73	535	2	12	170	X	3	4	X
Equatorial Guinea	184	1	12	131	273	3	4	194	X	4	2	X
Eritrea	112	0	6	50	319	0	3	141	85	1	3	38
Ethiopia {g}	255	31	35	54	626	28	20	133	188	9	1	40
Gabon	190	3	12	64	466	1	4	157	X	3	3	X
Gambia	117	0	4	112	280	0	1	269	47	1	1	45
Ghana	222	1	13	78	529	0	10	186	X	1	4	X
Guinea	190	1	11	66	409	0	12	142	X	3	3	X
Guinea-Bissau	108	0	4	71	243	0	1	159	X	2	3	X
Kenya	359	23	43	94	847	9	24	222	190	18	5	50
Lesotho	33	0	2	23	58	0	5	40	X	2	0	X
Liberia	193	0	11	87	372	1	13	168	62	2	3	28
Madagascar	141	93	46	37	202	105	28	53	363	259	17	95
Malawi	195	0	7	86	521	0	9	230	124	7	0	55
Mali	137	0	13	28	397	0	6	81	16	4	1	3
Mauritania	61	1	14	13	273	0	3	59	X	1	3	X
Mozambique	179	2	13	42	498	0	14	117	167	5	5	39
Namibia	250	3	11	58	469	3	8	109	250	25	3	58
Niger	131	0	11	27	299	0	2	60	X	0	1	X
Nigeria	274	4	26	62	681	2	9	153	135	7	4	30
Rwanda	151	0	9	110	513	0	6	373	X	1	0	X
Senegal	192	0	13	72	384	0	6	144	100	1	7	37
Sierra Leone	147	0	9	77	466	1	12	243	X	1	3	X
Somalia	171	12	18	43	422	11	8	107	193	49	2	49
South Africa	255	35	33	52	596	8	16	122	315	97	19	65
Sudan	267	11	21	43	680	1	9	110	X	8	3	X
Tanzania, United Rep	316	15	33	70	827	24	30	184	289	61	4	64
Togo	196	0	8	110	391	0	1	220	X	1	3	X
Uganda	338	6	18	118	830	3	10	290	149	2	1	52
Zambia	233	3	11	56	605	2	10	145	145	3	0	35
Zimbabwe	270	0	9	81	532	0	9	159	153	2	0	46
NORTH AMERICA	X	X	X	X	X	X	X	X	X	X	X	X
Canada	193	7	7	20	426	5	5	44	41	0	3	4
United States	432	105	35	45	650	67	50	68	287	79	28	30
C. AMERICA & CARIBBEAN	X	X	X	X	X	X	X	X	X	X	X	X
Belize	125	0	5	95	356	0	1	271	107	2	5	81
Costa Rica	205	7	14	120	600	6	13	350	214	38	7	125
Cuba	31	12	9	14	137	21	13	62	105	83	7	47
Dominican Rep	20	0	4	12	136	0	11	81	117	34	10	69
El Salvador	135	0	2	106	251	0	0	196	73	4	6	57
Guatemala	250	3	8	114	458	1	4	208	235	24	9	107
Haiti	3	0	4	2	75	1	11	54	108	35	6	77
Honduras	173	2	7	78	422	1	4	190	162	22	7	73
Jamaica	24	2	4	23	113	26	7	110	36	27	8	35
Mexico	491	140	64	86	772	92	36	135	704	368	18	123
Nicaragua	200	2	4	86	482	0	3	207	161	6	7	69
Panama	218	16	17	112	732	9	10	376	226	25	7	116
Trinidad and Tobago	100	1	1	125	260	1	3	324	70	3	5	87
SOUTH AMERICA	X	X	X	X	X	X	X	X	X	X	X	X
Argentina	320	49	27	50	897	19	41	140	234	78	5	37
Bolivia	316	16	24	67	X	18	27	X	211	20	3	45
Brazil	417	119	71	45	1,500	185	103	162	491	201	15	53
Chile	91	16	16	22	296	16	18	71	82	43	1	20
Colombia	359	34	35	75	1,700	67	64	356	593	115	15	124
Ecuador	302	25	28	100	1,388	37	53	460	380	120	12	126
Guyana	193	1	10	70	678	0	3	246	X	2	8	X
Paraguay	305	2	10	90	556	0	26	164	120	3	3	35
Peru	460	49	46	93	1,541	112	64	310	360	96	9	73
Suriname	180	2	10	72	603	0	2	240	151	0	6	60
Uruguay	81	1	5	31	237	0	11	92	X	1	0	X
Venezuela	323	19	24	73	1,340	40	22	302	283	66	14	64
OCEANIA	X	X	X	X	X	X	X	X	X	X	X	X
Australia	260	206	58	29	649	350	45	72	748	641	37	83
Fiji	4	1	4	3	74	24	9	61	25	11	6	20
New Zealand	2	2	3	1	150	74	44	51	52	48	11	18
Papua New Guinea	222	65	57	63	653	94	31	184	280	80	10	79
Solomon Islands	53	21	20	37	163	43	18	115	61	11	4	43

Notes: a. Values are standardized using a species-area curve. b. World totals include countries not listed. c. Includes cetaceans. World total comes from: D.E. Wilson and D.M. Reeder (eds.), 1993. d. World total comes from C.G. Sibley and B.L. Monroe, 1993. e. World total was estimated by the World Conservation Monitoring Centre. f. Total number of species, number of endemics, and the species per 10,000 square kilometers calculations include Hong Kong, SAR. g. Total number of species and endemics for mammals and birds are for the Former People's Democratic Republic of Ethiopia, which included Eritrea.

Sources: World Conservation Monitoring Centre, IUCN-The World Conservation Union, Food and Agriculture Organization of the United Nations, and other multiple sources.

	Amphibians				Freshwater Fish		Plants					
	Total Number of Known Species			No. of Species per 10,000 km² {a}	Total Number of Known Species		Total Number of Known Species			Endemic Species of Higher Plants {c}	Threatened Species of Higher Plants {c}	No. of Plant Species per 10,000 km² {a}
	All Species	Endemic Species	Threatened Species		All Species	Threatened Species {b}	Higher Plants {c}	Flowering Plants	Ferns			
WORLD {d}	**4,522 e**	**X**	**124**	**X**	**25,000 f**	**734**	**270,000 g**	**X**	**X**	**X**	**25,971 h**	**X**
ASIA (EXCL. MIDDLE EAST)	**X**	**X**	**X**	**X**	**X**	**X**	**X**	**X**	**X**	**X**	**X**	**X**
Armenia	7	0	0	5	41	0	3,300	3,300	X	X	19	X
Azerbaijan	10	0	0	5	61	5	4,300	X	X	240	12	2,109
Bangladesh	19	0	0	8	260	0	5,000	5,000	X	X	18	2,074
Bhutan	24	0	0	14	44	0	5,468	5,446	X	75	5	3,281
Cambodia	28	0	0	11	215	5	X	X	X	X	0	X
China {i}	290	158	1	30	686	28	32,200	30,000	2,000	18,000	113	3,340
Georgia	13	0	0	7	84	3	4,350	X	X	380	12	2,292
India	209	122	3	31	748	4	16,000	15,000	1,000	5,000	785	2,363
Indonesia	285	115	0	50	1,400	60	29,375	27,500	1,875	17,500	184	5,196
Japan	61	45	10	18	186	7	5,565	4,700	630	2,000	537	1,679
Kazakhstan	12	0	1	2	104	5	X	X	X	X	36	X
Korea, Dem People's Rep	14	2	0	6	X	0	2,898	2,898	X	107	0	1,274
Korea, Rep	14	0	0	7	130	0	2,898	2,898	X	224	52	1,359
Kyrgyzstan	4	0	0	1	75	0	3,786	70	X	X	7	1,412
Lao People's Dem Rep	37	2	0	13	262	4	X	X	X	X	1	X
Malaysia	189	70	0	60	449	14	15,500	12,500	1,100	3,600	371	4,890
Mongolia	6	0	0	1	75	0	2,823	2,272	X	229	0	533
Myanmar	75	10	0	19	300	1	7,000	7,000	X	1,071	6	1,742
Nepal	43	11	0	18	185	0	6,973	5,160	380	315	7	2,871
Pakistan	17	4	0	4	156	1	4,950	4,929	X	372	5	1,168
Philippines	92	73	2	30	230	26	8,931	8,000	900	3,500	320	2,907
Singapore	24	0	0	60	41	1	2,282	2,000	170	2	2	5,713
Sri Lanka	39	20	0	21	65	8	3,314	3,000	314	890	431	1,781
Tajikistan	2	1	0	1	49	1	X	X	X	X	25	X
Thailand	112	21	0	31	600	14	11,625	11,000	600	X	355	3,170
Turkmenistan	5	0	0	1	107	5	X	X	X	X	13	X
Uzbekistan	2	0	0	1	83	3	4,800	X	X	400	11	1,369
Viet Nam	80	27	1	25	450	3	10,500	7,000	X	1,260	297	3,306
EUROPE	**X**	**X**	**X**	**X**	**X**	**X**	**X**	**X**	**X**	**X**	**X**	**X**
Albania	13	0	0	9	39	7	3,031	2,965	45	24	17	2,139
Austria	20	0	0	10	60	7	3,100	2,950	66	35	6	1,537
Belarus	12	0	0	4	40	0	2,100	1,590	28	X	0	772
Belgium	17	0	0	12	X	1	1,550	1,400	50	1	0	1,073
Bosnia and Herzegovina	16	0	1	9	56	6	X	X	X	X	1	X
Bulgaria	17	0	0	8	X	8	3,572	3,505	52	320	59	1,615
Croatia	20	0	1	11	87	20	X	X	X	X	0	X
Czech Rep	19	0	0	10	56	6	X	X	X	X	7	X
Denmark	14	0	0	9	41	0	1,450	1,200	50	1	1	895
Estonia	11	0	0	7	30	1	1,674	1,630	40	X	1	1,018
Finland	5	0	0	2	66	1	1,102	1,040	58	X	0	345
France	32	3	2	9	53	3	4,630	4,500	110	133	86	1,233
Germany	20	0	0	6	71	7	2,682	2,600	72	6	3	824
Greece	15	2	1	6	98	16	4,992	4,900	71	742	446	2,131
Hungary	16	0	0	8	81	11	2,411	2,343	60	38	8	1,155
Iceland	0	0	0	0	7	0	377	340	36	1	1	175
Ireland	3	0	0	2	25	1	950	892	56	X	0	499
Italy	41	12	4	13	45	9	5,599	5,463	106	712	202	1,820
Latvia	13	0	0	7	109	1	1,205	1,153	48	X	0	651
Lithuania	13	0	0	7	X	1	1,796	1,328	21	X	0	967
Macedonia, FYR	13	0	0	10	55	4	3,500	X	X	X	X	2,563
Moldova, Rep	13	0	0	9	82	9	1,752	X	X	X	0	1,173
Netherlands	16	0	0	10	X	1	1,221	1,170	48	X	0	767
Norway	5	0	0	2	31	1	1,715	1,650	61	1	4	544
Poland	18	0	0	6	104	2	2,450	2,300	62	3	1	778
Portugal	17	0	1	8	28	9	5,050	2,500	65	150	186	2,428
Romania	19	0	0	7	87	11	3,400	3,175	62	41	34	1,194
Russian Federation	41	0	0	4	290	13	X	X	X	X	129	X
Slovakia	20	0	0	12	78	7	3,124	X	X	92	11	1,849
Slovenia	20	0	1	16	95	5	3,200	X	X	22	3	2,535
Spain	28	4	3	8	50	10	5,050	4,916	114	941	822	1,383
Sweden	13	0	0	4	51	1	1,750	1,650	60	1	3	498
Switzerland	18	0	0	11	48	4	3,030	2,927	87	1	2	1,898
Ukraine	17	0	0	4	90	12	5,100	X	X	X	20	1,318
United Kingdom	7	0	0	2	36	1	1,623	1,550	70	16	13	565
Yugoslavia	21	0	0	10	110	13	4,082	3,905	68	X	50	1,896
MIDDLE EAST & N. AFRICA	**X**	**X**	**X**	**X**	**X**	**X**	**X**	**X**	**X**	**X**	**X**	**X**
Afghanistan	6	1	1	2	84	0	4,000	3,500	X	800	0	1,008
Algeria	10	0	0	2	X	1	3,164	3,100	46	250	125	520
Egypt	6	0	0	1	70	0	2,076	2,066	6	70	59	454
Iran, Islamic Rep	11	5	2	2	269	7	8,000	X	X	X	0	1,489
Iraq	6	0	0	2	X	2	X	X	X	X	X	X
Israel	7	0	0	5	36	0	2,805	2,780	25	X	19	2,194
Jordan	X	0	0	X	26	0	2,200	2,200	X	X	4	1,069
Kuwait	2	0	0	2	X	0	234	234	X	X	X	193
Lebanon	8	0	0	8	60	0	3,000	2,863	35	X	3	2,961
Libyan Arab Jamahiriya	3	0	0	1	X	0	1,825	1,800	15	134	41	331
Morocco	11	1	0	3	X	1	3,675	3,600	56	625	182	1,049
Oman	X	0	0	X	3	3	1,204	1,182	14	73	19	439
Saudi Arabia	X	0	0	X	8	0	2,028	1,729	X	X	1	345
Syrian Arab Rep	X	0	0	X	X	0	3,000	X	X	X	7	1,145
Tunisia	7	0	0	3	X	0	2,196	2,150	36	X	6	873
Turkey	18	3	2	4	152	18	8,650	8,472	85	2,675	1,610	2,059
United Arab Emirates	X	0	0	X	5	1	X	X	X	X	X	X
Yemen	X	1	0	X	5	0	1,650	X	X	135	124	446

	Amphibians				Freshwater Fish		Plants					
	Total Number of Known Species			No. of Species per 10,000 km² {a}	Total Number of Known Species		Total Number of Known Species			Endemic Species of Higher Plants {c}	Threatened Species of Higher Plants {c}	No. of Plant Species per 10,000 km² {a}
	All Species	Endemic Species	Threatened Species		All Species	Threatened Species {b}	Higher Plants {c}	Flowering Plants	Ferns			
SUB-SAHARAN AFRICA	**X**	**X**	**X**	**X**	**X**	**X**	**X**	**X**	**X**	**X**	**X**	**X**
Angola	X	22	0	X	X	0	5,185	5,000	185	1,260	20	1,055
Benin	X	0	0	X	150	0	2,201	2,000	200	X	2	990
Botswana	38	0	0	10	92	0	2,151	X	15	17	0	563
Burkina Faso	X	0	0	X	X	0	1,100	1,100	X	X	X	369
Burundi	X	2	0	X	X	0	2,500	2,500	X	X	0	1,783
Cameroon	190	66	1	53	354	26	8,260	8,000	257	156	67	2,310
Central African Rep	47	0	0	12	X	0	3,602	3,600	X	100	1	921
Chad	X	0	0	X	X	0	1,600	1,600	X	X	5	322
Congo	X	1	0	X	X	0	6,000	4,350	X	1,200	2	1,870
Congo, Dem Rep	80	53	0	13	X	1	11,007	2,867	381	1,100	69	1,818
Côte d'Ivoire	X	3	1	X	X	0	3,660	3,517	143	62	42	1,163
Equatorial Guinea	X	2	1	X	X	0	3,250	3,000	250	66	6	2,312
Eritrea	19	2	0	8	X	0	X	X	X	X	X	X
Ethiopia {j}	62	24	0	13	150	0	6,603	6,500	100	1,000	125	1,398
Gabon	X	4	0	X	X	0	6,651	6,500	150	X	78	2,248
Gambia	30	0	0	29	79	0	974	966	8	X	0	935
Ghana	X	4	0	X	X	0	3,725	3,600	124	43	22	1,308
Guinea	X	3	1	X	X	0	3,000	3,000	X	88	29	1,043
Guinea-Bissau	X	1	0	X	X	0	1,000	1,000	X	12	X	655
Kenya	88	10	0	23	X	20	6,506	6,000	500	265	130	1,703
Lesotho	X	0	0	X	8	1	1,591	1,576	15	2	0	1,103
Liberia	38	4	1	17	X	0	2,200	2,200	X	103	1	993
Madagascar	179	155	2	47	121	13	9,505	9,000	500	6,500	255	2,479
Malawi	69	3	0	31	X	0	3,765	3,600	161	49	46	1,665
Mali	X	1	0	X	X	0	1,741	1,741	X	11	5	355
Mauritania	X	0	0	X	X	0	1,100	1,100	X	X	2	239
Mozambique	62	1	0	15	X	2	5,692	5,500	183	219	57	1,340
Namibia	51	2	1	12	114	3	4,040	3,978	61	687	14	942
Niger	X	0	0	X	X	0	1,178	1,170	8	X	X	238
Nigeria	109	1	0	24	260	0	4,715	4,614	100	205	16	1,059
Rwanda	X	1	0	X	X	0	2,290	2,288	X	26	0	1,664
Senegal	2	1	0	1	79	0	2,086	2,062	24	26	15	780
Sierra Leone	X	2	0	X	X	0	2,090	2,090	X	74	8	1,091
Somalia	27	3	0	7	X	3	3,028	3,000	26	500	57	768
South Africa	108	49	9	22	94	27	23,420	23,000	380	X	1,875	4,797
Sudan	X	1	0	X	X	0	3,137	3,132	X	50	2	507
Tanzania, United Rep	133	49	0	30	X	19	10,008	10,000	X	1,122	326	2,231
Togo	X	3	0	X	X	0	3,085	2,484	99	X	0	1,739
Uganda	50	1	0	17	291	28	5,406	5,000	400	X	8	1,891
Zambia	65	1	0	16	106	0	4,747	4,600	146	211	5	1,141
Zimbabwe	120	3	0	36	112	0	4,440	4,200	234	95	73	1,325
NORTH AMERICA	**X**	**X**	**X**	**X**	**X**	**X**	**X**	**X**	**X**	**X**	**X**	**X**
Canada	41	0	1	4	177	13	3,270	2,920	65	147	40	335
United States	263	152	24	28	822	123	19,473	16,302	549	4,036	2,449	2,036
C. AMERICA & CARIBBEAN	**X**	**X**	**X**	**X**	**X**	**X**	**X**	**X**	**X**	**X**	**X**	**X**
Belize	32	1	0	24	63	4	2,894	2,750	134	150	10	2,200
Costa Rica	168	39	1	98	130	0	12,119	11,000	1,110	950	296	7,074
Cuba	55	50	0	25	28	4	6,522	6,004	495	3,229	834	2,949
Dominican Rep	35	15	1	21	16	0	5,657	5,000	650	1,800	62	3,354
El Salvador	23	0	0	18	16	0	2,911	2,500	400	17	8	2,277
Guatemala	107	34	0	49	220	0	8,681	8,000	652	1,171	170	3,948
Haiti	56	27	1	40	16	0	5,242	4,685	550	1,623	36	3,743
Honduras	81	34	0	36	46	0	5,680	5,000	650	148	34	2,559
Jamaica	24	21	4	23	6	0	3,308	2,746	558	923	681	3,207
Mexico	310	194	3	54	506	86	26,071	25,000	1,000	12,500	911	4,569
Nicaragua	59	2	0	25	50	0	7,590	7,000	576	40	29	3,256
Panama	164	22	0	84	101	1	9,915	9,000	900	1,222	1,018	5,088
Trinidad and Tobago	26	3	0	32	76	0	2,259	1,982	277	236	11	2,816
SOUTH AMERICA	**X**	**X**	**X**	**X**	**X**	**X**	**X**	**X**	**X**	**X**	**X**	**X**
Argentina	153	45	5	24	410	1	9,372	9,000	359	1,100	83	1,463
Bolivia	122	28	0	26	389	0	18,316	17,000	1,300	4,000	107	3,885
Brazil	581	375	5	63	3,000	12	56,215	55,000	1,200	X	751	6,058
Chile	49	34	3	12	44	4	5,292	5,125	150	2,698	268	1,269
Colombia	684	230	0	143	1,500	5	51,220	50,000	1,200	1,500	429	10,735
Ecuador	426	162	0	141	706	1	19,362	18,250	1,100	4,000	642	6,421
Guyana	X	14	0	X	X	0	6,409	6,000	407	X	54	2,329
Paraguay	85	3	0	25	X	0	7,851	7,500	350	X	38	2,311
Peru	376	152	1	76	855	0	18,245	17,121	1,100	5,356	653	3,674
Suriname	95	8	0	38	300	0	5,018	4,700	315	X	33	1,997
Uruguay	X	4	0	X	X	0	2,278	2,184	93	40	2	882
Venezuela	245	122	0	55	1,270	5	21,073	20,000	1,059	8,000	252	4,752
OCEANIA	**X**	**X**	**X**	**X**	**X**	**X**	**X**	**X**	**X**	**X**	**X**	**X**
Australia	205	183	25	23	216	37	15,638	15,000	400	14,074	1,871	1,741
Fiji	2	2	1	2	X	0	1,628	1,307	310	760	64	1,334
New Zealand	3	3	1	1	29	8	2,382	2,160	200	1,942	165	802
Papua New Guinea	225	128	0	63	282	13	11,544	10,000	X	X	66	3,257
Solomon Islands	17	9	0	12	X	0	3,172	2,780	370	30	21	2,235

Notes: a. Values are standardized using a species-area curve. b. Includes some marine fish species. c. Higher plants include flowering plants, conifers and cycads, and ferns and fern-allies. d. World totals include countries not listed. e. World total is from D.R. Frost (ed.), 1985. f. World total includes both marine and freshwater fish; freshwater fish make up around 40-45 percent of this estimate. World total comes from W.N. Eschmeyer et al., 1998. g. World total is from K.S. Walter and H.J. Gillett (eds.), 1998. h. This is a subset of the 31,195 species listed as threatened in the "1997 IUCN Red List of Threatened Plants." This subset of 25,971 records refer to full species (i.e., taxa listed as threatened at the intraspecific level are not included) that occur in a single country. "Country" is used in a general sense; for example, figures are provided for overseas dependencies. i. Total number of species, number of endemics, and the species per 10,000 square kilometers calculations include Hong Kong, SAR. j. Total number of fish, total number of higher plants, and endemic higher plants are for the Former People's Democratic Republic of Ethiopia, which included Eritrea.

Sources: World Conservation Monitoring Centre, IUCN-The World Conservation Union, Food and Agriculture Organization of the United Nations, and other multiple sources.

	CITES Entered into Force (year)	CITES Reporting Requirements Met as of 1997 {a} (percent)	International Legal Net Trade Reported by CITES, 1997 {b}									
			Number of:						Number of:			
			Live Primates	Live Parrots	Live Tortoises	Live Lizards	Live Snakes	Wild Orchids	Cat Skins	Crocodile Skins	Lizard Skins	Snake Skins
WORLD {c}	**X**	**X**	**25,733**	**235,336**	**76,079**	**948,497**	**258,715**	**343,801**	**21,864**	**850,198**	**1,637,973**	**1,457,767**
ASIA (EXCL. MIDDLE EAST)	**X**	**X**	**(9,787)**	**20,748**	**26,799**	**79,529**	**(19,677)**	**(52,312)**	**(8,013)**	**448,884**	**(397,934)**	**(794,741)**
Armenia	N/A	N/A	X	X	X	X	X	X	X	X	X	X
Azerbaijan	1999	X	4	X	X	X	2	X	X	X	X	X
Bangladesh	1982	75	X	39	X	X	X	X	X	X	X	(4,301)
Bhutan	N/A	N/A	X	X	X	X	X	X	X	X	X	X
Cambodia	1997	0	X	X	X	X	X	X	X	X	X	X
China	1981	100	(5,966)	4,731	(1,271)	103	46,222	(100,210)	(17,999)	105,946	13,236	867
Georgia	1996	0	X	X	X	X	X	X	X	X	X	X
India	1976	100	9	7	(1)	X	(1)	1	X	16	15	(510)
Indonesia	1979	95	(3,955)	(18,334)	(1,784)	(10,296)	(76,862)	(102)	X	(260)	(511,000)	(441,902)
Japan	1980	94	3,556	9,413	30,670	39,255	4,772	128,911	(354)	82,166	318,159	120,999
Kazakhstan	2000	X	1	4	X	X	X	X	X	X	X	X
Korea, Dem People's Rep	N/A	N/A	X	X	10	21	72	X	X	X	X	X
Korea, Rep	1993	100	62	661	31	45,448	279	100,300	9,550	70,332	20,140	36,416
Kyrgyzstan	N/A	N/A	2	X	X	X	X	X	X	X	X	X
Lao People's Dem Rep	N/A	N/A	X	(1)	X	X	X	X	X	X	X	(2,657)
Malaysia	1978	90	74	1,603	(877)	1,408	(8,708)	(2,837)	X	2,702	(237,993)	(432,512)
Mongolia	1996	100	X	X	X	X	X	X	(40)	X	X	X
Myanmar	1997	0	X	X	X	X	X	X	X	X	X	X
Nepal	1975	78	X	574	X	X	X	X	X	X	X	X
Pakistan	1976	82	14	(702)	X	X	1	X	X	1	X	X
Philippines	1981	82	(2,809)	772	10	475	15	197	X	X	X	0
Singapore	1987	100	17	9,277	60	99	2,376	5,043	X	5,327	(53,097)	10,082
Sri Lanka	1979	68	0	364	X	X	X	X	X	X	X	X
Tajikistan	N/A	N/A	X	X	X	X	X	X	X	X	X	X
Thailand	1983	73	(6)	3,449	1	1,809	X	(207,579)	X	146,613	(7,367)	(427,079)
Turkmenistan	N/A	N/A	X	1	X	X	X	X	X	X	X	X
Uzbekistan	1997	0	31	12	X	X	2	X	X	X	X	X
Viet Nam	1994	50	(819)	(246)	(394)	(81)	(7,416)	(1,951)	X	X	X	(55,718)
EUROPE	**X**	**X**	**10,607**	**140,449**	**(8,283)**	**182,643**	**30,422**	**16,880**	**9,210**	**270,814**	**353,031**	**322,871**
Albania	N/A	N/A	X	X	X	X	X	X	X	X	X	X
Austria	1982	100	16	342	X	175	104	2,778	1	(331)	6,219	2
Belarus	1995	67	X	9	3,430	X	(1)	X	X	X	X	X
Belgium	1984	100	670	6,094	1,147	24,399	2,676	270	1	23	72	X
Bosnia and Herzegovina	N/A	N/A	X	X	X	X	X	X	X	X	X	X
Bulgaria	1991	71	1	17	(1)	6	19	X	X	X	X	X
Croatia	2000	X	5	15	(1)	X	X	X	X	0	30	30
Czech Rep	1992	100	66	(3,927)	987	9,802	(161)	541	X	X	X	X
Denmark	1977	95	0	45	607	5,130	476	202	(246)	81	133	0
Estonia	1992	83	7	X	X	1	1	X	(12)	X	X	X
Finland	1976	82	X	6	295	129	44	142	X	X	X	X
France	1978	100	3,332	7,508	4,730	13,870	6,767	1,791	150	143,202	58,271	3,040
Germany	1976	100	812	8,182	788	23,809	4,135	3,157	112	(17,362)	35,051	(25,092)
Greece	1993	100	2	5,900	428	150	2	X	672	5	5	542
Hungary	1985	85	8	(580)	(140)	445	(26)	X	2	X	X	143
Iceland	2000	X	X	X	X	X	X	X	X	X	X	X
Ireland {d}	N/A	N/A	(1)	(133)	8	(10)	X	50	X	X	X	X
Italy	1979	100	384	6,528	606	9,892	2,357	280	3,198	108,544	75,014	293,169
Latvia	1997	100	0	(1)	X	4	0	X	10	X	X	X
Lithuania	N/A	N/A	10	5	X	X	3	X	X	X	X	X
Macedonia, FYR	N/A	N/A	X	X	X	X	X	X	X	X	X	X
Moldova, Rep	N/A	N/A	X	X	X	X	X	X	X	X	X	X
Netherlands	1984	100	412	(11,706)	2,081	20,164	3,087	6,100	X	0	(3)	(24)
Norway	1976	91	5	730	10	3	(4)	2	54	X	X	X
Poland	1990	88	48	254	64	1,068	62	105	4	1	X	50
Portugal	1981	71	16	37,429	54	2,127	2	X	X	4	23	4
Romania	1994	50	5	3	X	X	4	X	X	X	X	X
Russian Federation	1976	82	2,144	1,641	(28,547)	(1,098)	(312)	X	(1,627)	5	0	X
Slovakia	1992	100	(7)	(682)	556	689	308	104	X	X	X	X
Slovenia	2000	X	11	10	X	X	7	X	X	62	588	3
Spain	1986	100	88	78,753	2,720	50,654	5,543	(38)	4	13,422	167,021	36,810
Sweden	1975	100	111	(181)	48	1,805	28	303	1	4	3	X
Switzerland	1975	100	21	162	(455)	1,953	1,561	640	(126)	28,778	9,857	(19,872)
Ukraine	2000	X	14	3	3	1,123	54	X	X	X	X	X
United Kingdom	1976	100	2,424	3,297	1,684	16,326	3,652	453	7,000	(5,624)	747	34,066
Yugoslavia	N/A	N/A	3	79	X	X	X	X	X	X	X	X
MIDDLE EAST & N. AFRICA	**X**	**X**	**(31)**	**8,552**	**(945)**	**904**	**1,131**	**1**	**0**	**469**	**3,590**	**4,112**
Afghanistan	1986	0	X	X	X	X	X	X	X	X	X	X
Algeria	1984	64	X	5	(169)	(2)	10	X	X	X	X	X
Egypt	1978	15	0	147	(2)	(40)	46	1	X	X	X	X
Iran, Islamic Rep	1976	68	77	(3)	X	3	4	X	X	X	X	X
Iraq	N/A	N/A	X	X	X	X	X	X	X	X	X	X
Israel	1980	44	(121)	3,011	1	3,734	944	X	X	(17)	10	X
Jordan	1979	32	2	37	X	X	X	X	X	X	X	X
Kuwait {d}	N/A	N/A	3	1,046	X	X	2	X	X	X	X	X
Lebanon	N/A	N/A	X	194	(46)	(5)	2	X	X	483	3,540	4
Libyan Arab Jamahiriya	N/A	N/A	X	211	(669)	X	X	X	X	X	X	X
Morocco	1976	59	(30)	99	(160)	75	3	X	X	X	X	X
Oman	N/A	N/A	1	1	(5)	X	X	X	X	X	X	X
Saudi Arabia	1996	0	1	1,406	(9)	17	49	X	X	X	40	(2)
Syrian Arab Rep	N/A	N/A	12	(2)	(2)	X	X	X	X	X	X	X
Tunisia	1975	100	X	132	(7)	X	8	X	X	X	X	X
Turkey	1996	50	1	113	158	34	3	X	X	X	X	4,110
United Arab Emirates	1990	63	24	919	(33)	(1,136)	58	X	X	3	X	X
Yemen	1997	0	X	1	(2)	(1,776)	X	X	X	X	X	X

	CITES Entered into Force (year)	CITES Reporting Requirements Met as of 1997 {a} (percent)	International Legal Net Trade Reported by CITES, 1997 {b} Number of: Live Primates	Live Parrots	Live Tortoises	Live Lizards	Live Snakes	Wild Orchids	Number of: Cat Skins	Crocodile Skins	Lizard Skins	Snake Skins
SUB-SAHARAN AFRICA	**X**	**X**	**(9,902)**	**(102,051)**	**(38,501)**	**(201,723)**	**(123,359)**	**(7,026)**	**0**	**(60,482)**	**(253,211)**	**(25,700)**
Angola	N/A	N/A	(8)	5	X	X	X	X	X	X	X	(4)
Benin	1984	29	X	X	(5,177)	(15,488)	(56,591)	X	X	X	X	(2)
Botswana	1978	90	X	11	X	X	X	X	X	(337)	X	X
Burkina Faso	1990	63	X	(2)	X	X	X	X	X	X	X	X
Burundi	1988	30	X	(1)	X	(200)	X	X	X	X	X	X
Cameroon	1981	71	(85)	(1,083)	X	(5,781)	(39)	(50)	X	X	(40,410)	(9,900)
Central African Rep	1980	50	X	(18)	X	X	(1)	X	X	X	X	X
Chad	1989	56	X	(1)	X	X	X	X	X	X	(36,051)	X
Congo	1983	93	X	(498)	2	X	X	X	X	X	X	X
Congo, Dem Rep	1976	77	(8)	(13,431)	X	X	X	X	X	X	X	X
Côte d'Ivoire	1995	0	X	(28)	(1)	X	1	41	X	X	X	(2)
Equatorial Guinea	1992	17	X	(2)	X	X	X	X	X	X	X	X
Eritrea	1995	0	X	X	X	X	X	X	X	X	X	X
Ethiopia	1989	89	(144)	2	X	X	X	X	X	X	X	X
Gabon	1989	78	20	(6)	X	X	X	X	X	X	X	X
Gambia	1977	43	X	X	X	X	X	X	X	X	X	X
Ghana	1976	86	(70)	(6)	(5,176)	(17,809)	(25,481)	X	X	X	X	(10)
Guinea	1981	59	X	(14,279)	X	(46)	(576)	X	X	X	X	(1)
Guinea-Bissau	1990	0	(1)	(9)	X	X	X	X	X	X	X	X
Kenya	1979	58	(436)	(320)	3	X	X	35	X	(782)	X	X
Lesotho {d}	N/A	N/A	X	X	X	X	X	X	X	X	X	X
Liberia	1981	47	X	X	X	X	X	X	X	X	X	X
Madagascar	1975	91	0	(3,633)	(9)	(83,029)	X	(7,070)	X	(6,748)	12	73
Malawi	1982	81	X	1	X	X	X	X	X	(400)	X	X
Mali	1994	100	X	(4,326)	(720)	(14,480)	(2,101)	X	X	X	(132,820)	(8,754)
Mauritania	1998	X	X	X	X	X	X	X	X	X	X	X
Mozambique	1981	82	(82)	(102)	(17,534)	(8,333)	(309)	X	X	(1,430)	X	X
Namibia	1991	86	0	416	(26)	5	1	X	X	(120)	X	(1)
Niger	1975	52	X	(68)	(46)	188	(5)	X	X	X	X	X
Nigeria	1975	43	X	(157)	(20)	X	X	X	X	X	X	(4)
Rwanda	1981	18	X	X	X	310	25	X	X	X	X	X
Senegal	1977	86	(1)	(17,188)	129	X	X	X	X	20	X	(2)
Sierra Leone	1995	33	X	(98)	X	X	X	X	X	X	X	X
Somalia	1986	8	X	X	X	X	X	X	X	X	X	X
South Africa	1975	96	(148)	(46,078)	(84)	(58)	109	789	X	(13,351)	3,894	935
Sudan	1983	60	(17)	X	X	X	X	X	X	X	(62,010)	(8,019)
Tanzania, United Rep	1980	89	(2,413)	3	(1,383)	(30,834)	(36)	X	X	(650)	(1)	(4)
Togo	1979	79	(15)	(470)	(4,771)	(26,356)	(38,358)	X	X	12	8	(5)
Uganda	1991	57	(2)	4	X	X	X	X	X	X	(9)	X
Zambia	1981	65	2	5	(3,655)	X	X	X	X	(4,936)	X	X
Zimbabwe	1981	76	(1)	(1,038)	3	X	X	(901)	X	(35,230)	X	X
NORTH AMERICA	**X**	**X**	**10,877**	**(350)**	**23,633**	**651,939**	**152,004**	**61,923**	**(1,195)**	**(101,653)**	**401,997**	**108,818**
Canada	1975	100	1,016	997	(122)	11,146	1,772	2,805	261	280	42	573
United States	1975	91	9,861	(1,347)	23,755	640,793	150,232	59,118	(1,456)	(101,933)	401,955	108,245
C. AMERICA & CARIBBEAN	**X**	**X**	**(370)**	**(20,325)**	**(100)**	**(351,631)**	**(13,318)**	**(22,579)**	**(5)**	**27,640**	**438,368**	**402,754**
Belize	1981	71	X	1	X	X	X	(21,014)	X	X	X	X
Costa Rica	1975	83	3	773	4	(11,591)	14	(240)	X	(18)	X	X
Cuba	1990	88	1	(22,802)	X	(19)	X	(159)	X	X	X	X
Dominican Rep	1987	100	42	476	13	53	X	109	X	10	X	X
El Salvador	1987	36	X	36	X	(295,723)	X	X	X	2	(500)	X
Guatemala	1980	89	5	608	X	(38,154)	(545)	(172)	X	X	X	X
Haiti	N/A	N/A	X	X	X	X	X	21	X	X	X	X
Honduras	1985	23	X	37	X	(240)	(206)	X	X	11,682	X	X
Jamaica	1997	0	X	10	X	X	X	(1,054)	X	X	X	X
Mexico	1991	100	210	6,208	0	10,216	497	X	(5)	79,662	438,866	402,683
Nicaragua	1977	90	X	(8,022)	X	(16,255)	(13,074)	(338)	X	11,244	X	X
Panama	1978	85	4	33	X	53	1	(54)	X	(74,738)	X	X
Trinidad and Tobago	1984	71	X	88	(42)	9	(5)	X	X	X	X	X
SOUTH AMERICA	**X**	**X**	**(1,065)**	**(44,941)**	**(2,520)**	**(358,813)**	**(27,051)**	**1,068**	**0**	**(580,056)**	**(547,191)**	**(18,465)**
Argentina	1981	94	(22)	(18,834)	(5)	3,567	262	917	X	1	(377,055)	(21,579)
Bolivia	1979	63	X	8	X	X	X	X	X	(14,806)	X	X
Brazil	1975	57	(92)	2,861	(160)	12	(2)	134	X	(13,505)	1,942	3,119
Chile	1975	74	4	5,431	24	1,703	211	X	X	X	X	X
Colombia	1981	82	2	130	X	(340,840)	(23,227)	19	X	(517,712)	5,000	(1)
Ecuador	1975	74	X	X	0	X	X	28	X	16	X	(2)
Guyana	1977	76	(513)	(13,138)	(748)	(3,221)	(3,025)	X	X	(910)	X	X
Paraguay	1977	71	X	(10)	(2)	2,510	138	(32)	X	(514)	(180,924)	(1)
Peru	1975	74	(300)	(3,269)	X	(5,900)	(65)	X	X	X	X	(1)
Suriname	1981	100	(137)	(11,413)	(1,046)	(16,844)	(1,343)	2	X	X	X	X
Uruguay	1975	65	(2)	(7,600)	X	200	X	X	X	11	X	X
Venezuela	1978	80	(5)	893	(583)	X	X	X	X	(32,637)	3,846	X
OCEANIA	**X**	**X**	**26**	**(1,815)**	**(43)**	**(2,834)**	**(151)**	**1,520**	**2**	**(6,253)**	**(1)**	**154**
Australia	1976	95	26	(422)	X	2	83	1,498	2	(5,819)	(1)	144
Fiji	1997	0	X	(10)	(43)	X	X	X	X	X	(1)	X
New Zealand	1989	100	(22)	(770)	X	X	X	X	X	24	3	12
Papua New Guinea	1976	77	X	(28)	X	X	X	X	X	(458)	(2)	(1)
Solomon Islands	N/A	N/A	X	(593)	X	(2,836)	(234)	X	X	X	X	(1)

Notes: N/A indicates that a country has not signed CITES. a. Indicates the number of annual reports submitted by a country to the CITES Secretariat as a percentage of the number of reports expected since it became a party to the Convention. Values are for reports submitted for years up to 1997. "X" under reporting requirements met indicates that a country ratified CITIES after 1997. b. Balance of imports minus exports. Exports are shown as a negative balance (in parentheses). c. World totals reflect the total number of animals traded; in this case, the values represent total imports. d. These countries have signed, but have not ratified the Convention.

Source: World Conservation Monitoring Centre

	Forest Area									Forests Certified by Forest Stewardship Council-Accredited Certification Bodies		
	Total Forest			Natural Forest {a}			Plantations {a}					
	Extent 1990 (000 ha)	Extent 1995 (000 ha)	Average Annual % Change 1990-95	Extent 1990 (000 ha)	Extent 1995 (000 ha)	Average Annual % Change 1990-95	Extent 1990 (000 ha)	Extent 1995 (000 ha)	Average Annual % Change 1990-95	Natural Forests (hectares)	Plantations (hectares)	Mixed Forests {b} (hectares)
WORLD {c}	**3,510,728**	**3,454,382**	**(0.32)**	**X**	**X**	**X**	**X**	**X**	**X**	**5,621,629**	**1,002,230**	**10,907,593**
ASIA (EXCL. MIDDLE EAST)	**503,969**	**490,266**	**(0.55)**	**394,877**	**375,606**	**(1.00)**	**55,787**	**52,821**	**(1.09)**	**55,083**	**76,074**	**10,000**
Armenia	292	334	2.69	X	X	X	X	X	X	0	0	0
Azerbaijan	990	990	0.00	X	X	X	X	X	X	0	0	0
Bangladesh	1,054	1,010	(0.85)	819	700	(3.14)	235	443	12.68	0	0	0
Bhutan	2,803	2,756	(0.34)	2,799	2,748	(0.37)	4	12	21.47	0	0	0
Cambodia	10,649	9,830	(1.60)	10,642	9,823	(1.60)	7	X	X	0	0	0
China {d}	133,756	133,323	(0.06)	101,925	99,523	(0.48)	31,831	21,373	(7.97)	0	0	0
Georgia	2,988	2,988	0.00	X	X	X	X	X	X	0	0	0
India	64,969	65,005	0.01	51,739	50,385	(0.53)	13,230	20,252	8.52	0	0	0
Indonesia	115,213	109,791	(0.96)	109,088	103,666	(1.02)	6,125	4,956	(4.24)	0	62,278	10,000
Japan	25,212	25,146	(0.05)	X	X	X	X	X	X	0	1,070	0
Kazakhstan	9,540	10,504	1.93	X	X	X	X	X	X	0	0	0
Korea, Dem People's Rep	6,170	6,170	0.00	4,700	4,700	0.00	1,470	X	X	0	0	0
Korea, Rep	7,691	7,626	(0.17)	6,291	6,226	(0.21)	0	X	X	0	0	0
Kyrgyzstan	730	730	0.00	X	X	X	X	X	X	0	0	0
Lao People's Dem Rep	13,177	12,435	(1.16)	13,173	12,431	(1.16)	4	22	33.64	0	0	0
Malaysia	17,472	15,471	(2.43)	17,391	15,371	(2.47)	81	170	14.83	55,083	0	0
Mongolia	9,406	9,406	0.00	9,406	9,406	0.00	0	X	X	0	0	0
Myanmar	29,088	27,151	(1.38)	28,853	26,875	(1.42)	235	543	16.75	0	0	0
Nepal	5,096	4,822	(1.11)	5,040	4,766	(1.12)	56	140	18.33	0	0	0
Pakistan	2,023	1,748	(2.92)	1,855	1,580	(3.21)	168	840	32.19	0	0	0
Philippines	8,078	6,766	(3.54)	7,875	6,563	(3.65)	203	590	21.34	0	0	0
Singapore	4	4	0.00	4	4	(1.45)	0	X	X	0	0	0
Sri Lanka	1,897	1,796	(1.09)	1,758	1,657	(1.19)	139	138	(0.14)	0	12,726	0
Tajikistan	410	410	0.00	X	X	X	X	X	X	0	0	0
Thailand	13,277	11,630	(2.65)	12,748	11,101	(2.77)	529	868	9.90	0	0	0
Turkmenistan	3,754	3,754	0.00	X	X	X	X	X	X	0	0	0
Uzbekistan	7,989	9,119	2.65	X	X	X	X	X	X	0	0	0
Viet Nam	9,793	9,117	(1.43)	8,323	7,647	(1.69)	1,470	2,475	10.42	0	0	0
EUROPE	**930,726**	**933,320**	**0.06**	**X**	**X**	**X**	**X**	**X**	**X**	**2,187,542**	**1,050**	**10,676,900**
Albania	1,046	1,046	0.00	X	X	X	X	X	X	0	0	0
Austria	3,877	3,877	0.00	X	X	X	X	X	X	0	0	0
Belarus	7,028	7,372	0.96	X	X	X	X	X	X	0	0	0
Belgium {e}	709	709	0.00	X	X	X	X	X	X	0	0	1,890
Bosnia and Herzegovina	2,710	2,710	0.00	X	X	X	X	X	X	0	0	0
Bulgaria	3,237	3,240	0.02	X	X	X	X	X	X	0	0	0
Croatia	1,825	1,825	0.00	X	X	X	X	X	X	0	0	0
Czech Rep	2,629	2,630	0.01	X	X	X	X	X	X	0	0	10,441
Denmark	417	417	0.00	X	X	X	X	X	X	36	0	0
Estonia	1,913	2,011	1.00	X	X	X	X	X	X	0	0	0
Finland	20,112	20,029	(0.08)	X	X	X	X	X	X	0	0	0
France	14,230	15,034	1.10	X	X	X	X	X	X	0	1,050	0
Germany	10,740	10,740	0.00	X	X	X	X	X	X	2,209	0	82,971
Greece	5,809	6,513	2.29	X	X	X	X	X	X	0	0	0
Hungary	1,675	1,719	0.52	X	X	X	X	X	X	0	0	0
Iceland	11	11	0.00	X	X	X	X	X	X	0	0	0
Ireland	500	570	2.62	X	X	X	X	X	X	0	0	0
Italy	6,467	6,496	0.09	X	X	X	X	X	X	11,000	0	0
Latvia	2,757	2,882	0.89	X	X	X	X	X	X	0	0	0
Lithuania	1,920	1,976	0.57	X	X	X	X	X	X	0	0	0
Macedonia, FYR	989	988	(0.02)	X	X	X	X	X	X	0	0	0
Moldova, Rep	357	357	0.00	X	X	X	X	X	X	0	0	0
Netherlands	334	334	0.00	X	X	X	X	X	X	0	0	69,064
Norway	7,938	8,073	0.34	X	X	X	X	X	X	0	0	0
Poland	8,672	8,732	0.14	X	X	X	X	X	X	2,129,849	0	612,937
Portugal	2,755	2,875	0.85	X	X	X	X	X	X	0	0	0
Romania	6,252	6,246	(0.02)	X	X	X	X	X	X	0	0	0
Russian Federation	763,500	763,500	0.00	X	X	X	X	X	X	0	0	32,712
Slovakia	1,977	1,989	0.12	X	X	X	X	X	X	0	0	0
Slovenia	1,077	1,077	0.00	X	X	X	X	X	X	0	0	0
Spain	8,388	8,388	0.00	X	X	X	X	X	X	0	0	0
Sweden	24,437	24,425	(0.01)	X	X	X	X	X	X	18,012	0	9,026,683
Switzerland	1,130	1,130	0.00	X	X	X	X	X	X	4,252	0	0
Ukraine	9,213	9,240	0.06	X	X	X	X	X	X	0	0	0
United Kingdom	2,326	2,390	0.54	X	X	X	X	X	X	22,184	0	840,202
Yugoslavia	1,769	1,769	0.00	X	X	X	X	X	X	0	0	0
MIDDLE EAST & N. AFRICA	**20,412**	**19,420**	**(1.00)**	**9,732**	**8,741**	**(2.15)**	**1,581**	**1,857**	**3.22**	**0**	**0**	**0**
Afghanistan	1,990	1,398	(7.06)	1,982	1,390	(7.09)	8	X	X	0	0	0
Algeria	1,978	1,861	(1.22)	1,493	1,376	(1.63)	485	972	13.90	0	0	0
Egypt	34	34	0.00	0	0	0.00	34	X	X	0	0	0
Iran, Islamic Rep	1,686	1,544	(1.76)	1,607	1,465	(1.85)	79	X	X	0	0	0
Iraq	83	83	0.00	69	69	0.00	14	X	X	0	0	0
Israel	102	102	0.00	X	X	X	X	X	X	0	0	0
Jordan	51	45	(2.50)	28	22	(4.82)	23	X	X	0	0	0
Kuwait	5	5	0.00	0	0	0.00	5	X	X	0	0	0
Lebanon	78	52	(8.11)	65	39	(10.31)	13	X	X	0	0	0
Libyan Arab Jamahiriya	400	400	0.00	190	190	0.00	210	X	X	0	0	0
Morocco	3,894	3,835	(0.31)	3,573	3,514	(0.33)	321	565	11.31	0	0	0
Oman	0	0	0.00	0	0	0.00	0	X	X	0	0	0
Saudi Arabia	231	222	(0.79)	230	221	(0.79)	1	X	X	0	0	0
Syrian Arab Rep	245	219	(2.24)	118	92	(4.98)	127	X	X	0	0	0
Tunisia	570	555	(0.53)	369	354	(0.80)	201	320	9.30	0	0	0
Turkey	8,856	8,856	0.00	X	X	X	X	X	X	0	0	0
United Arab Emirates	60	60	0.00	0	0	0.00	60	X	X	0	0	0
Yemen	9	9	0.00	9	9	0.00	0	X	X	0	0	0

	Forest Area											
	Total Forest			Natural Forest {a}			Plantations {a}			Forests Certified by Forest Stewardship Council-Accredited Certification Bodies		
	Extent 1990 (000 ha)	Extent 1995 (000 ha)	Average Annual % Change 1990-95	Extent 1990 (000 ha)	Extent 1995 (000 ha)	Average Annual % Change 1990-95	Extent 1990 (000 ha)	Extent 1995 (000 ha)	Average Annual % Change 1990-95	Natural Forests (hectares)	Plantations (hectares)	Mixed Forests {b} (hectares)
SUB-SAHARAN AFRICA	**532,001**	**513,551**	**(0.71)**	**528,602**	**510,021**	**(0.72)**	**3,168**	**4,955**	**8.94**	**581,571**	**323,516**	**1,458**
Angola	23,285	22,200	(0.95)	23,265	22,080	(1.05)	120	160	5.75	0	0	0
Benin	4,923	4,625	(1.25)	4,909	4,611	(1.25)	14	30	15.24	0	0	0
Botswana	14,271	13,917	(0.50)	14,270	13,916	(0.50)	1	X	X	0	0	0
Burkina Faso	4,431	4,271	(0.74)	4,411	4,251	(0.74)	20	28	6.73	0	0	0
Burundi	324	317	(0.44)	232	225	(0.61)	92	120	5.31	0	0	0
Cameroon	20,244	19,598	(0.65)	20,228	19,582	(0.65)	16	23	7.26	0	0	0
Central African Rep	30,571	29,930	(0.42)	30,565	29,924	(0.42)	6	X	X	0	0	0
Chad	11,496	11,025	(0.84)	11,492	11,021	(0.84)	4	8	13.86	0	0	0
Congo	19,745	19,537	(0.21)	19,708	19,500	(0.21)	37	64	10.96	0	0	0
Congo, Dem Rep	112,946	109,245	(0.67)	112,904	109,203	(0.67)	42	60	7.13	0	0	0
Côte d'Ivoire	5,623	5,469	(0.56)	5,560	5,403	(0.57)	63	98	8.84	0	0	0
Equatorial Guinea	1,829	1,781	(0.53)	1,826	1,778	(0.53)	3	X	X	0	0	0
Eritrea	282	282	0.00	233	233	0.00	X	X	X	0	0	0
Ethiopia {f}	13,891	13,579	(0.45)	13,751	13,439	(0.46)	189	220	3.04	0	0	0
Gabon	18,314	17,859	(0.50)	18,293	17,838	(0.50)	21	30	7.13	0	0	0
Gambia	95	91	(0.86)	94	90	(0.91)	1	2	8.11	0	0	0
Ghana	9,608	9,022	(1.26)	9,555	8,969	(1.27)	53	19	(20.52)	0	0	0
Guinea	6,741	6,367	(1.14)	6,737	6,363	(1.14)	4	11	19.49	0	0	0
Guinea-Bissau	2,361	2,309	(0.45)	2,360	2,308	(0.44)	1	X	X	0	0	0
Kenya	1,309	1,292	(0.26)	1,191	1,174	(0.28)	118	240	14.20	0	0	0
Lesotho	6	6	0.00	0	0	0.00	7	6	0.00	0	0	0
Liberia	4,641	4,507	(0.59)	4,635	4,501	(0.58)	6	8	5.75	0	0	0
Madagascar	15,756	15,106	(0.84)	15,539	14,889	(0.85)	217	335	8.68	0	0	0
Malawi	3,612	3,339	(1.57)	3,486	3,213	(1.63)	126	122	(0.65)	0	0	0
Mali	12,154	11,585	(0.96)	12,140	11,571	(0.96)	14	40	21.00	0	0	0
Mauritania	556	556	0.00	554	554	0.00	2	8	27.73	0	0	0
Mozambique	17,443	16,862	(0.68)	17,415	16,834	(0.68)	28	45	9.49	0	0	0
Namibia	12,584	12,374	(0.34)	12,584	12,374	(0.34)	0	1	X	54,420	0	0
Niger	2,562	2,562	0.00	2,550	2,550	0.00	12	60	32.19	0	0	0
Nigeria	14,387	13,780	(0.86)	14,236	13,629	(0.87)	151	250	10.08	0	0	0
Rwanda	252	250	(0.16)	164	162	(0.25)	88	247	20.64	0	0	0
Senegal	7,629	7,381	(0.66)	7,517	7,269	(0.67)	112	216	13.14	0	0	0
Sierra Leone	1,522	1,309	(3.02)	1,516	1,303	(3.03)	6	8	5.75	0	0	0
Somalia	760	754	(0.16)	756	750	(0.15)	4	6	8.11	0	0	0
South Africa	8,574	8,499	(0.18)	7,279	7,204	(0.21)	965	1,429	7.85	502,301	275,862	1,458
Sudan	43,376	41,613	(0.83)	43,173	41,410	(0.83)	203	425	14.76	0	0	0
Tanzania, United Rep	34,123	32,510	(0.97)	33,969	32,356	(0.97)	154	200	5.23	0	0	0
Togo	1,338	1,245	(1.44)	1,317	1,224	(1.47)	17	34	14.04	0	0	0
Uganda	6,400	6,104	(0.95)	6,380	6,084	(0.95)	20	26	5.25	0	0	0
Zambia	32,720	31,398	(0.82)	32,677	31,355	(0.83)	48	44	(1.92)	0	0	0
Zimbabwe	8,960	8,710	(0.57)	8,876	8,626	(0.57)	84	110	5.39	24,850	47,654	0
NORTH AMERICA	**453,270**	**457,086**	**0.17**	**X**	**X**	**X**	**X**	**X**	**X**	**1,782,470**	**0**	**35,477**
Canada	243,698	244,571	0.07	X	X	X	X	X	X	212,189	0	0
United States	209,572	212,515	0.28	X	X	X	X	X	X	1,570,281	0	35,477
C. AMERICA & CARIBBEAN	**84,609**	**79,425**	**(1.26)**	**84,124**	**78,940**	**(1.27)**	**485**	**813**	**10.34**	**295,037**	**14,298**	**115,758**
Belize	1,995	1,962	(0.33)	1,993	1,960	(0.34)	2	3	4.46	95,800	0	0
Costa Rica	1,455	1,248	(3.07)	1,427	1,220	(3.13)	28	129	30.49	8,246	14,275	17,632
Cuba	1,960	1,842	(1.24)	1,715	1,597	(1.42)	245	470	13.03	0	0	0
Dominican Rep	1,714	1,582	(1.60)	1,707	1,575	(1.61)	7	20	21.00	0	0	0
El Salvador	124	105	(3.33)	120	101	(3.51)	4	8	13.86	0	0	0
Guatemala	4,253	3,841	(2.04)	4,225	3,813	(2.05)	28	68	17.63	100,026	0	0
Haiti	25	21	(3.49)	17	13	(5.25)	8	12	8.11	0	0	0
Honduras	4,626	4,115	(2.34)	4,623	4,112	(2.34)	3	40	51.81	19,876	0	0
Jamaica	254	175	(7.45)	239	160	(8.03)	15	6	(17.67)	0	0	0
Mexico	57,927	55,387	(0.90)	57,818	55,278	(0.90)	109	X	X	71,089	0	98,126
Nicaragua	6,314	5,560	(2.54)	6,300	5,546	(2.55)	14	23	9.93	0	0	0
Panama	3,118	2,800	(2.15)	3,112	2,794	(2.16)	6	16	19.87	0	23	0
Trinidad and Tobago	174	161	(1.55)	161	148	(1.65)	13	18	6.51	0	0	0
SOUTH AMERICA	**894,466**	**870,594**	**(0.54)**	**887,187**	**863,315**	**(0.55)**	**7,279**	**9,150**	**4.58**	**710,704**	**584,987**	**30,000**
Argentina	34,389	33,942	(0.26)	33,842	33,395	(0.27)	547	830	8.34	0	0	0
Bolivia	51,217	48,310	(1.17)	51,189	48,282	(1.17)	28	20	(6.73)	630,133	0	30,000
Brazil	563,911	551,139	(0.46)	559,011	546,239	(0.46)	4,900	4,805	(0.39)	80,571	584,987	0
Chile	8,038	7,892	(0.37)	7,023	6,877	(0.42)	1,015	1,747	10.86	0	0	0
Colombia	54,299	52,988	(0.49)	54,173	52,862	(0.49)	126	300	17.35	0	0	0
Ecuador	12,082	11,137	(1.63)	12,037	11,092	(1.64)	45	120	19.62	0	0	0
Guyana	18,620	18,577	(0.05)	18,612	18,569	(0.05)	8	12	8.11	0	0	0
Paraguay	13,160	11,527	(2.65)	13,151	11,518	(2.65)	9	18	13.86	0	0	0
Peru	68,646	67,562	(0.32)	68,462	67,378	(0.32)	184	349	12.80	0	0	0
Suriname	14,782	14,721	(0.08)	14,774	14,713	(0.08)	8	12	8.11	0	0	0
Uruguay	816	814	(0.05)	660	658	(0.05)	156	348	16.05	0	0	0
Venezuela	46,512	43,995	(1.11)	46,259	43,742	(1.12)	253	589	16.90	0	0	0
OCEANIA {g}	**91,143**	**90,689**	**(0.10)**	**42,501**	**41,746**	**(0.36)**	**X**	**1,250**	**X**	**9,222**	**2,305**	**38,000**
Australia	40,823	40,908	0.04	X	X	X	X	1,068	X	0	0	0
Fiji	853	835	(0.43)	775	757	(0.47)	78	103	5.56	0	0	0
New Zealand	7,667	7,884	0.56	X	X	X	X	X	X	0	2,305	0
Papua New Guinea	37,605	36,939	(0.36)	37,575	36,909	(0.36)	30	50	10.22	4,310	0	0
Solomon Islands	2,412	2,389	(0.19)	2,394	2,371	(0.19)	16	29	11.89	4,912	0	38,000
DEVELOPING	**2,017,529**	**1,952,577**	**(0.65)**	**1,939,744**	**1,871,165**	**(0.72)**	**67,484**	**68,350**	**0.25**	**1,149,316**	**721,943**	**193,758**
DEVELOPED	**1,493,067**	**1,501,774**	**0.12**	**X**	**X**	**X**	**X**	**X**	**X**	**4,472,313**	**280,287**	**10,713,835**

Notes: Negative numbers are shown in parentheses. a. With the exception of South Africa and Australia, forest areas in developed countries are not broken down into the subcategories of natural forests and plantations due to the difficulty in distinguishing the two in many countries. b. Includes areas with natural forests and plantations, areas with seminatural forests, and areas with mixed seminatural forests and plantations. c. World totals have been excluded because data for natural forests and plantations are not available for developed nations. d. Includes Taiwan, Province of China. e. Total forest figures for Belgium include Luxembourg. f. Plantation figures are for the Former People's Democratic Republic of Ethiopia, which included Eritrea. g. Plantation figures include data for Australia even though it is a developed nation.

Sources: Food and Agriculture Organization of the United Nations and Forest Stewardship Council

	Land Area (000 ha)	Original Forest as a Percent of Land Area {a}	Closed Forests as a Percent of Original Forest: Current Forests {b} 1996	Closed Forests as a Percent of Original Forest: Frontier Forests {c} 1996	Percent Frontier Forest {d} Threatened 1996	Forest Ecosystems (1990s): Tropical Forests Area (000 ha)	Tropical Forests Percent Protected	Nontropical Forests Area (000 ha)	Nontropical Forests Percent Protected	Sparse Trees and Parkland Area (000 ha)	Sparse Trees and Parkland Percent Protected	Number of Tree Species Threatened 1990s
WORLD	**13,048,407**	**47.7**	**53.4**	**21.7**	**39.5**	**1,407,649**	**11.7**	**1,823,787**	**6.0**	**541,616**	**5.5**	**5,904 e**
ASIA (EXCL. MIDDLE EAST)	**2,406,823**	**X**	**X**	**X**	**X**	**210,720**	**16.4**	**132,065**	**5.3**	**42,384**	**7.0**	**X**
Armenia	2,820	45.6	21.1	0.0	0.0	0	0.0	355	10.1	0	0.0	0
Azerbaijan	8,660	32.0	32.0	0.0	0.0	0	0.0	1,133	0.0	0	0.0	0
Bangladesh	13,017	100.0	7.9	3.8	100.0	862	3.7	0	0.0	0	0.0	12
Bhutan	4,700	67.3	61.8	24.0	100.0	966	22.9	1,129	37.6	0	0.0	4
Cambodia	17,652	100.0	65.1	10.3	100.0	11,516	25.6	0	0.0	0	0.0	30
China	932,641	51.8	21.6	1.8	92.8	109	13.0	82,710	3.9	26,715	4.7	190 f
Georgia	6,970	76.6	57.3	0.0	0.0	0	0.0	3,158	0.0	0	0.0	1
India	297,319	79.0	20.5	1.3	57.2	44,450	8.9	9,260	8.3	0	0.0	266
Indonesia	181,157	100.0	64.6	28.5	53.8	88,744	20.9	0	0.0	0	0.0	426
Japan	37,652	91.4	58.2	0.0	0.0	0	0.0	5,677	9.7	0	0.0	11
Kazakhstan	267,073	2.6	22.9	2.9	100.0	0	0.0	2,638	9.6	0	0.0	X
Korea, Dem People's Rep	12,041	97.3	38.7	0.0	0.0	0	0.0	3,967	1.1	506	0.0	3
Korea, Rep	9,873	88.5	16.5	0.0	0.0	0	0.0	1,426	3.4	0	0.0	0
Kyrgyzstan	19,180	8.4	14.0	0.0	0.0	0	0.0	785	0.2	0	0.0	1
Lao People's Dem Rep	23,080	99.9	30.0	2.1	100.0	3,639	23.0	849	5.0	0	0.0	20
Malaysia	32,855	99.5	63.8	14.5	48.5	13,007	11.7	0	0.0	0	0.0	737
Mongolia	156,650	22.5	49.6	8.2	0.0	0	0.0	2,636	25.2	14,697	11.2	0
Myanmar	65,755	100.0	40.6	0.0	0.0	20,661	0.8	9,574	2.0	0	0.0	44
Nepal	14,300	83.6	22.4	0.0	0.0	1,162	18.8	2,660	20.7	0	0.0	3
Pakistan	77,088	44.8	5.8	0.0	0.0	807	0.6	2,083	4.7	0	0.0	2
Philippines	29,817	95.3	6.0	0.0	0.0	2,402	5.2	0	0.0	0	0.0	209
Singapore	61	78.8	3.1	0.0	0.0	X	X	X	X	X	X	61
Sri Lanka	6,463	94.7	18.1	11.9	76.2	1,581	27.6	0	0.0	466	19.4	294
Tajikistan	14,060	50.7	4.2	0.0	0.0	X	X	X	X	X	X	1
Thailand	51,089	100.0	22.2	4.9	100.0	16,237	31.2	361	11.3	0	0.0	91
Turkmenistan	46,993	7.9	4.1	0.0	0.0	0	0.0	216	0.7	0	0.0	X
Uzbekistan	41,424	2.8	10.2	0.0	0.0	0	0.0	231	0.0	0	0.0	0
Viet Nam	32,549	99.7	17.2	1.9	100.0	4,218	10.3	723	8.8	0	0.0	132
EUROPE	**2,263,394**	**72.7**	**58.4**	**21.3**	**18.7**	**0**	**X**	**991,346**	**3.0**	**10,350**	**1.0**	**X**
Albania	2,740	92.5	37.3	0.0	0.0	0	0.0	1,066	1.2	0	0.0	0
Austria	8,273	95.1	52.8	0.0	0.0	0	0.0	3,593	20.9	0	0.0	X
Belarus	20,748	91.1	27.2	0.0	0.0	0	0.0	6,280	6.5	0	0.0	X
Belgium	3,023	95.2	21.0	0.0	0.0	0	0.0	687	7.8	0	0.0	X
Bosnia and Herzegovina	5,100	X	X	X	X	0	0.0	2,303	1.0	0	0.0	1
Bulgaria	11,055	92.3	31.7	0.0	0.0	0	0.0	3,787	5.9	0	0.0	0
Croatia	5,592	X	X	X	X	0	0.0	1,391	9.9	0	0.0	1
Czech Rep	7,728	X	X	X	X	0	0.0	2,481	27.5	0	0.0	X
Denmark	4,243	88.5	0.8	0.0	0.0	0	0.0	459	4.2	0	0.0	1
Estonia	4,227	96.0	29.4	0.0	0.0	0	0.0	1,524	6.7	0	0.0	X
Finland	30,459	100.0	82.3	1.1	100.0	0	0.0	25,309	5.8	0	0.0	X
France	55,010	95.3	16.5	0.0	0.0	0	0.0	10,831	13.7	0	0.0	1
Germany	34,927	92.6	26.3	0.0	0.0	0	0.0	10,401	24.8	0	0.0	8
Greece	12,890	84.3	17.0	0.0	0.0	0	0.0	4,423	1.7	0	0.0	2
Hungary	9,234	50.4	8.2	0.0	0.0	0	0.0	777	26.1	0	0.0	X
Iceland	10,025	43.7	0.0	0.0	0.0	X	X	X	X	X	X	X
Ireland	6,889	64.4	3.6	0.0	0.0	0	0.0	457	2.1	0	0.0	1
Italy	29,406	94.5	20.4	0.0	0.0	0	0.0	6,757	6.1	0	0.0	2
Latvia	6,205	100.0	19.8	0.0	0.0	0	0.0	1,624	10.1	0	0.0	X
Lithuania	6,480	99.3	16.0	0.0	0.0	0	0.0	1,509	10.3	0	0.0	X
Macedonia, FYR	2,543	X	X	X	X	0	0.0	1,091	9.9	0	0.0	X
Moldova, Rep	3,297	48.3	3.7	0.0	0.0	0	0.0	143	5.4	0	0.0	X
Netherlands	3,392	46.2	4.8	0.0	0.0	0	0.0	235	7.4	0	0.0	X
Norway	30,683	33.8	90.4	0.0	0.0	0	0.0	8,139	2.0	0	0.0	X
Poland	30,442	95.6	22.2	0.0	0.0	0	0.0	8,939	12.9	0	0.0	2
Portugal	9,150	91.2	9.4	0.0	0.0	0	0.0	2,661	5.6	0	0.0	13
Romania	23,034	75.2	41.5	0.0	0.0	0	0.0	8,137	2.5	0	0.0	2
Russian Federation	1,688,850	69.6	68.7	29.3	18.6	0	0.0	815,551	1.8	10,350	1.0	2
Slovakia	4,808	X	X	X	X	0	0.0	2,308	29.6	0	0.0	X
Slovenia	2,012	X	X	X	X	0	0.0	696	8.9	0	0.0	0
Spain	49,944	94.2	15.1	0.0	0.0	0	0.0	14,024	10.8	0	0.0	21
Sweden	41,162	93.3	86.0	2.9	100.0	0	0.0	29,364	1.6	0	0.0	1
Switzerland	3,955	85.5	44.8	0.0	0.0	0	0.0	1,309	12.6	0	0.0	X
Ukraine	57,935	45.7	20.4	0.0	0.0	0	0.0	7,046	1.6	0	0.0	2
United Kingdom	24,160	77.5	6.0	0.0	0.0	0	0.0	2,303	20.7	0	0.0	10
Yugoslavia	10,200	X	X	X	X	0	0.0	3,664	3.4	0	0.0	0
MIDDLE EAST & N. AFRICA	**1,253,551**	**X**	**X**	**X**	**X**	**134**	**0.0**	**17,950**	**3.3**	**1**	**76.9**	**X**
Afghanistan	65,209	48.1	6.5	0.0	0.0	0	0.0	2,076	0.0	0	0.0	1
Algeria	238,174	4.6	12.0	0.0	0.0	0	0.0	2,694	3.7	1	83.3	2
Egypt	99,545	0.6	0.0	0.0	0.0	134	0.0	4	0.0	0	0.0	2
Iran, Islamic Rep	162,200	43.1	3.3	0.0	0.0	0	0.0	2,348	12.0	0	0.0	1
Iraq	43,737	13.4	0.0	0.0	0.0	X	X	X	X	X	X	0
Israel	2,062	86.6	0.0	0.0	0.0	X	X	X	X	X	X	0
Jordan	8,893	13.1	0.0	0.0	0.0	X	X	X	X	X	X	0
Kuwait	1,782	0.0	0.0	0.0	0.0	X	X	X	X	X	X	0
Lebanon	1,023	73.7	0.7	0.0	0.0	0	0.0	36	0.0	0	0.0	0
Libyan Arab Jamahiriya	175,954	0.8	0.0	0.0	0.0	0	0.0	53	0.0	0	0.0	1
Morocco	44,630	21.6	7.3	0.0	0.0	0	0.0	1,862	2.6	0	0.0	5
Oman	21,246	0.0	0.0	0.0	0.0	X	X	X	X	X	X	7
Saudi Arabia	214,969	0.0	0.0	0.0	0.0	X	X	X	X	X	X	3
Syrian Arab Rep	18,378	19.2	0.0	0.0	0.0	0	0.0	47	0.0	0	0.0	0
Tunisia	15,536	18.1	4.7	0.0	0.0	0	0.0	300	2.2	0	0.0	0
Turkey	76,963	66.4	11.3	0.0	0.0	0	0.0	8,390	1.2	0	0.0	5
United Arab Emirates	8,360	0.0	0.0	0.0	0.0	X	X	X	X	X	X	X
Yemen	52,797	0.0	0.0	0.0	0.0	X	X	X	X	X	X	56

	Land Area (000 ha)	Original Forest as a Percent of Land Area {a}	Closed Forests as a Percent of Original Forest: Current Forests {b} 1996	Closed Forests as a Percent of Original Forest: Frontier Forests {c} 1996	Percent Frontier Forest {d} Threatened 1996	Forest Ecosystems (1990s): Tropical Forests Area (000 ha)	Tropical Forests Percent Protected	Nontropical Forests Area (000 ha)	Nontropical Forests Percent Protected	Sparse Trees and Parkland Area (000 ha)	Sparse Trees and Parkland Percent Protected	Number of Tree Species Threatened 1990s
SUB-SAHARAN AFRICA	**2,363,090**	**X**	**X**	**X**	**X**	**448,063**	**9.1**	**52**	**26.4**	**69,709**	**11.3**	**X**
Angola	124,670	19.8	15.3	0.0	0.0	37,564	2.6	0	0.0	0	0.0	X
Benin	11,062	15.5	3.5	0.0	0.0	1,516	18.2	0	0.0	585	2.3	12
Botswana	56,673	2.1	100.0	0.0	0.0	12,123	19.9	0	0.0	0	0.0	0
Burkina Faso	27,360	0.0	0.0	0.0	0.0	0	0.0	0	0.0	5,667	15.9	2
Burundi	2,568	46.3	3.5	0.0	0.0	219	18.2	0	0.0	139	2.6	1
Cameroon	46,540	80.4	42.4	7.9	97.4	20,009	6.0	0	0.0	2,416	21.8	104
Central African Rep	62,298	51.8	15.9	4.4	100.0	17,101	20.1	0	0.0	1,451	47.7	9
Chad	125,920	0.0	0.0	0.0	0.0	3,516	3.6	0	0.0	2,857	0.7	2
Congo	34,150	100.0	67.8	28.7	64.6	24,321	4.4	0	0.0	0	0.0	X
Congo, Dem Rep	226,705	82.5	60.4	15.6	70.4	135,071	6.6	0	0.0	172	40.0	58
Côte d'Ivoire	31,800	74.9	9.9	2.2	100.0	2,702	22.8	0	0.0	625	18.4	104
Equatorial Guinea	2,805	95.6	38.4	0.0	0.0	1,749	0.0	0	0.0	0	0.0	15
Eritrea	10,100	X	X	X	X	1	0.0	0	0.0	0	0.0	4
Ethiopia {g}	100,000	24.5	17.3	0.0	0.0	11,937	18.8	0	0.0	4,804	21.4	24
Gabon	25,767	100.0	90.4	32.4	100.0	21,481	3.6	0	0.0	0	0.0	70
Gambia	1,000	39.1	61.9	0.0	0.0	188	5.1	0	0.0	244	2.5	3
Ghana	22,754	65.9	8.6	0.0	0.0	1,694	7.1	0	0.0	336	16.7	117
Guinea	24,572	75.6	5.0	0.0	0.0	3,073	1.1	0	0.0	2,723	1.2	21
Guinea-Bissau	2,812	100.0	33.7	0.0	0.0	1,141	0.0	0	0.0	550	0.0	4
Kenya	56,914	16.8	18.5	0.0	0.0	3,423	8.3	0	0.0	2,754	2.8	125
Lesotho	3,035	2.4	0.0	0.0	0.0	89	8.7	0	0.0	0	0.0	X
Liberia	9,632	99.6	44.2	0.0	0.0	3,149	2.9	0	0.0	1	0.0	47
Madagascar	58,154	92.6	13.1	0.0	0.0	6,940	5.5	0	0.0	0	0.0	164
Malawi	9,408	12.2	0.0	0.0	0.0	3,830	8.5	0	0.0	0	0.0	23
Mali	122,019	0.0	0.0	0.0	0.0	6,132	2.3	0	0.0	336	0.4	6
Mauritania	102,522	0.0	0.0	0.0	0.0	X	X	X	X	X	X	X
Mozambique	78,409	33.2	13.6	0.0	0.0	20,863	7.5	0	0.0	14,414	6.6	43
Namibia	82,329	0.0	95.3	0.0	0.0	3,436	10.6	0	0.0	0	0.0	4
Niger	126,670	0.0	0.0	0.0	0.0	27	15.6	0	0.0	0	0.0	2
Nigeria	91,077	45.1	10.7	0.6	100.0	11,634	7.4	0	0.0	10,588	4.0	112
Rwanda	2,467	36.1	16.1	0.0	0.0	291	77.0	0	0.0	162	1.8	X
Senegal	19,253	14.2	16.0	0.0	0.0	2,076	7.0	0	0.0	8,816	13.1	7
Sierra Leone	7,162	100.0	9.7	0.0	0.0	260	20.3	0	0.0	104	0.0	44
Somalia	62,734	4.2	0.0	0.0	0.0	11,800	1.1	0	0.0	1,530	0.9	19
South Africa	122,104	12.8	0.2	0.0	0.0	10,333	5.2	52	26.4	0	0.0	37
Sudan	237,600	1.2	0.0	0.0	0.0	12,288	12.3	0	0.0	5,870	8.9	17
Tanzania, United Rep	88,359	22.4	9.1	0.0	0.0	14,356	15.8	0	0.0	583	3.3	317
Togo	5,439	32.9	7.0	0.0	0.0	224	2.6	0	0.0	91	9.1	9
Uganda	19,965	70.0	4.4	0.0	0.0	3,772	17.0	0	0.0	1,850	65.2	32
Zambia	74,339	7.1	70.1	0.0	0.0	21,989	31.9	0	0.0	39	14.1	11
Zimbabwe	38,685	7.0	67.3	0.0	0.0	15,397	12.2	0	0.0	0	0.0	18
NORTH AMERICA	**1,838,009**	**59.7**	**77.3**	**34.1**	**26.2**	**443**	**6.7**	**683,700**	**8.9**	**148,827**	**5.7**	**X**
Canada	922,097	65.8	91.2	56.5	20.9	0	0.0	404,313	7.9	143,573	5.6	1
United States	915,912	53.5	60.2	6.3	84.7	443	6.7	279,386	10.4	5,254	9.0	198
C. AMERICA & CARIBBEAN	**264,711**	**67.2**	**54.5**	**9.7**	**87.0**	**70,812**	**12.0**	**21,293**	**3.1**	**4**	**1.2**	**X**
Belize	2,280	91.8	95.7	35.5	66.1	1,440	43.6	0	0.0	0	0.0	28
Costa Rica	5,106	98.4	34.9	9.5	100.0	1,464	44.8	0	0.0	0	0.0	114
Cuba	10,982	90.4	28.8	0.0	0.0	1,761	15.3	0	0.0	0	0.0	169
Dominican Rep	4,838	97.7	25.1	0.0	0.0	1,171	16.9	0	0.0	0	0.0	32
El Salvador	2,072	99.4	9.9	0.0	0.0	111	4.5	0	0.0	0	0.0	28
Guatemala	10,843	98.7	46.2	2.2	100.0	3,862	31.9	0	0.0	0	0.0	83
Haiti	2,756	93.2	0.8	0.0	0.0	64	2.0	0	0.0	0	0.0	32
Honduras	11,189	99.5	51.6	16.4	100.0	5,273	18.3	0	0.0	0	0.0	116
Jamaica	1,083	96.7	35.6	0.0	0.0	399	20.5	0	0.0	0	0.0	229
Mexico	190,869	55.7	63.4	8.1	77.0	45,765	4.3	21,293	3.1	0	0.0	186
Nicaragua	12,140	100.0	44.3	21.6	100.0	5,322	24.7	0	0.0	0	0.0	42
Panama	7,443	97.2	62.0	34.8	100.0	3,744	30.9	0	0.0	0	0.0	208
Trinidad and Tobago	513	93.5	35.5	0.0	0.0	124	7.0	0	0.0	0	0.0	1
SOUTH AMERICA	**1,751,708**	**55.6**	**69.1**	**45.6**	**54.0**	**620,514**	**12.2**	**39,178**	**15.8**	**168,216**	**2.4**	**X**
Argentina	273,669	5.5	59.5	6.3	99.9	4,360	5.5	19,094	10.1	6,392	9.4	42
Bolivia	108,438	53.9	77.2	43.6	96.9	68,638	12.1	0	0.0	0	0.0	79
Brazil	845,651	64.0	66.4	42.2	47.8	301,273	6.9	2,613	6.9	137,494	1.3	351
Chile	74,880	39.7	40.6	54.5	76.0	0	0.0	14,526	27.1	263	0.6	40
Colombia	103,870	92.2	53.5	36.4	18.7	53,186	10.8	0	0.0	0	0.0	227
Ecuador	27,684	78.8	66.4	36.9	99.5	13,508	23.9	0	0.0	755	46.9	175
Guyana	19,685	98.9	97.4	81.8	41.1	17,844	1.3	0	0.0	0	0.0	22
Paraguay	39,730	21.9	44.5	0.0	0.0	9,290	2.6	2,848	4.9	16,253	2.7	13
Peru	128,000	74.4	86.6	56.7	99.6	75,636	5.1	0	0.0	2,660	3.0	281
Suriname	15,600	91.7	95.6	92.2	21.7	13,219	4.0	0	0.0	0	0.0	27
Uruguay	17,481	1.6	0.0	0.0	0.0	2	0.0	97	1.0	325	1.6	0
Venezuela	88,205	74.7	83.6	59.3	37.3	55,615	59.0	0	0.0	3,997	18.6	70
OCEANIA {h}	**848,870**	**16.9**	**64.9**	**22.3**	**76.3**	**53,560**	**9.1**	**27,088**	**18.7**	**102,126**	**6.1**	**X**
Australia	768,230	9.3	64.3	17.8	62.8	14,088	7.3	22,877	14.1	101,485	6.1	34
Fiji	1,827	84.1	49.9	0.0	0.0	641	1.0	0	0.0	0	0.0	67
New Zealand	26,799	84.7	29.2	8.9	100.0	0	0.0	4,212	43.8	0	0.0	21
Papua New Guinea	45,286	96.0	85.4	39.6	83.5	35,791	10.7	0	0.0	0	0.0	165
Solomon Islands	2,799	82.1	93.9	0.0	0.0	2,669	0.0	0	0.0	0	0.0	18
ASIA	**3,085,414**	**49.1**	**28.2**	**5.3**	**63.1**	**210,720**	**16.4**	**145,101**	**5.1**	**42,384**	**7.0**	**X**
AFRICA	**2,963,568**	**22.9**	**33.9**	**7.8**	**76.8**	**448,197**	**9.1**	**8,249**	**2.0**	**69,710**	**11.3**	**X**

Notes: a. Original forest refers to estimated forest cover about 8,000 years ago assuming current climatic conditions. b. Includes frontier and nonfrontier forests. These represent estimated closed forests within the last 10 years or so. Only closed moist forests are depicted for Africa and Asia. Woodlands and shrublands are excluded. c. Frontier forests are relatively undisturbed large, intact natural forest ecosystems. d. Threatened frontier forests are areas where ongoing or planned human activities are likely to result in significant loss of ecosystem integrity (i.e., declines or local extinctions of species, changes in age structure of forests, etc.). e. The world total comes from: S. Oldfield, C. Lusty and A. MacKinven (eds.), World Conservation Press, 1998 (see sources). f. China includes Hong Kong, SAR. g. Values for closed forests are for the Former People's Democratic Republic of Ethiopia, which included Eritrea. h. Includes only Australia, Papua New Guinea, and New Zealand.

Sources: World Conservation Monitoring Centre, World Resources Institute, and Food and Agriculture Organization of the United Nations

	Average Annual Roundwood Production						Average Annual Forest Products Production				Trade in Forest Products {a}		
	Total Roundwood		Wood Fuel		Industrial Roundwood		Wood-Based Panels		Paper and Paperboard				
	Cubic Meters (000) 1996-98	Percent Change Since 1986-88	Cubic Meters (000) 1996-98	Percent Change Since 1986-88	Cubic Meters (000) 1996-98	Percent Change Since 1986-88	Cubic Meters (000) 1996-98	Percent Change Since 1986-88	Metric Tons (000) 1996-98	Percent Change Since 1986-88	Import Value (million US$) 1996-98	Export Value (million US$) 1996-98	Percent of Total Exports 1997
WORLD {b}	**3,261,621**	**1.5**	**1,742,064**	**10.4**	**1,522,116**	**(7.1)**	**151,390**	**24**	**288,285**	**34**	**142,935.0**	**135,313.0**	**1.97**
ASIA (EXCL. MIDDLE EAST) {b,c}	**1,111,958**	**X**	**863,316**	**X**	**268,470**	**X**	**39,057**	**X**	**82,277**	**X**	**37,114.7**	**16,434.0**	**X**
Armenia {c}	X	X	X	X	19,828	X	X	X	X	X	1.2	0.1	0.02
Azerbaijan {c}	X	X	X	X	X	X	X	X	X	X	21.6	0.4	0.03
Bangladesh	32,505	16.5	31,894	18.3	610	(35.1)	9	(9)	69	(39)	93.3	5.0	0.10
Bhutan	1,655	16.0	1,610	27.2	45	(72.1)	13	1,850	0	0	2.0	0.1	0.08
Cambodia	7,852	37.4	6,812	32.3	1,040	83.4	51	2,103	0	0	5.6	130.3	14.53
China {d}	305,787	10.9	199,669	12.4	106,118	8.1	12,015	257	31,670	123	11,792.2	3,600.4	0.83
Georgia {c}	X	X	X	X	X	X	10	X	X	X	X	X	0.00
India	294,905	19.0	269,841	20.6	25,064	3.9	348	(43)	3,103	65	747.6	32.9	0.07
Indonesia	198,345	18.8	154,770	17.1	43,575	25.0	9,803	41	4,810	502	903.8	4,583.5	7.25
Japan	21,878	(30.1)	768	112.3	21,110	(31.8)	6,765	(27)	30,305	33	15,499.1	1,521.2	0.32
Kazakhstan {c}	X	X	315	X	X	X	X	X	X	X	51.5	0.7	0.01
Korea, Dem People's Rep	6,233	40.2	5,000	30.0	1,233	105.6	X	X	80	0	14.7	21.5	X
Korea, Rep	1,622	(69.1)	394	(89.9)	1,228	(8.3)	2,172	58	7,921	148	3,422.8	1,321.4	0.80
Kyrgyzstan {c}	X	X	X	X	X	X	X	X	X	X	9.8	0.4	0.06
Lao People's Dem Rep	4,522	41.5	3,803	31.7	719	133.3	122	1,503	0	0	1.8	59.4	14.23
Malaysia	37,081	(11.4)	7,410	26.8	29,671	(17.6)	6,315	351	715	640	929.4	3,615.3	3.89
Mongolia	631	(69.1)	186	(78.8)	445	(61.7)	2	(54)	0	0	2.5	16.5	3.28
Myanmar	21,995	11.5	18,749	12.5	3,246	6.0	8	(45)	30	127	11.2	197.4	13.72
Nepal	20,993	27.7	20,373	28.3	620	10.7	30	X	13	550	1.4	1.0	0.08
Pakistan	31,528	41.0	29,312	38.4	2,217	86.0	109	36	487	352	144.8	0.0	0.00
Philippines	41,676	13.8	38,233	25.1	3,443	(43.1)	486	(18)	613	107	636.8	65.6	0.16
Singapore	0	0.0	0	0.0	0	0.0	355	(27)	87	770	1,110.0	607.9	0.39
Sri Lanka	10,408	10.0	9,708	10.5	700	3.0	15	51	25	(5)	81.1	2.6	0.05
Tajikistan {c}	X	X	X	X	X	X	X	X	X	X	1.8	0.0	0.00
Thailand	35,960	6.0	33,117	12.7	2,843	(37.5)	421	105	2,225	373	1,480.3	597.6	0.83
Turkmenistan {c}	X	X	X	X	X	X	X	X	X	X	2.6	0.3	0.04
Uzbekistan {c}	X	X	X	X	X	X	X	X	X	X	20.2	0.1	0.00
Viet Nam	35,712	16.1	31,219	22.1	4,493	(13.3)	39	(3)	125	108	104.3	46.1	0.40
EUROPE {b,c}	**498,720**	**X**	**92,609**	**X**	**394,704**	**X**	**48,023**	**X**	**87,019**	**X**	**63,934.6**	**64,797.8**	**X**
Albania	409	(82.1)	346	(78.5)	64	(90.7)	16	35	44	211	13.1	3.8	1.70
Austria	14,989	6.3	3,465	34.0	117	1.7	2,079	43	3,826	59	2,131.3	4,060.5	4.76
Belarus {c}	17,012	X	809	X	16,203	X	386	X	131	X	52.2	74.0	0.89
Belgium {e}	4,102	11.2	533	(2.9)	3,568	13.6	2,853	36	1,432	43	3,546.3	2,570.1	1.39
Bosnia and Herzegovina {c}	X	X	X	X	40	X	34	X	X	X	15.3	42.9	X
Bulgaria	3,096	(19.0)	1,181	3.8	1,915	(28.6)	233	(58)	150	(68)	78.5	69.9	1.12
Croatia {c}	2,997	X	982	X	2,015	X	80	X	367	X	323.9	248.6	3.03
Czech Rep {c}	13,291	X	646	X	12,645	X	939	X	751	X	612.1	842.4	2.82
Denmark	2,180	(0.9)	485	7.9	1,695	(3.2)	425	29	376	17	1,891.5	442.3	0.69
Estonia {c}	5,118	X	1,163	X	3,955	X	261	X	44	X	91.6	306.8	8.50
Finland	50,532	18.9	4,056	36.0	46,476	17.6	1,637	18	11,765	46	750.4	10,577.9	21.93
France	41,443	1.8	9,800	0.0	31,643	2.4	4,488	70	8,947	54	6,120.8	4,707.6	1.29
Germany	38,091	(9.4)	2,602	(39.7)	35,489	(6.0)	11,522	37	15,658	38	11,163.0	9,784.2	1.66
Greece	1,816	(40.7)	1,257	(42.4)	559	(36.5)	321	(21)	574	104	702.3	70.7	0.48
Hungary	4,020	(40.1)	1,878	(36.1)	2,143	(43.2)	497	35	684	30	506.9	273.8	1.12
Iceland	0	0.0	0	0.0	0	0.0	0	0	0	0	64.2	1.4	0.05
Ireland	2,246	72.8	67	42.3	2,178	73.9	545	139	0	(100)	799.5	222.8	0.36
Italy	9,272	0.7	5,121	12.1	4,151	(10.5)	4,097	33	7,744	55	7,369.4	2,921.6	0.94
Latvia {c}	8,936	X	2,746	X	6,189	X	311	X	14	X	83.7	498.0	17.34
Lithuania {c}	5,189	X	1,183	X	4,006	X	245	X	31	X	99.7	186.3	3.57
Macedonia, FYR {c}	774	X	616	X	158	X	2	X	21	X	56.7	17.7	1.33
Moldova, Rep {c}	397	X	346	X	51	X	10	X	X	X	17.7	2.8	0.27
Netherlands	1,028	(13.4)	132	20.0	896	(16.8)	101	11	3,109	39	4,878.6	2,718.7	1.27
Norway	8,251	(21.0)	355	(61.4)	7,896	(17.1)	568	(15)	2,159	34	1,230.0	1,808.3	2.86
Poland	21,772	(6.7)	1,492	(58.8)	20,280	2.9	3,024	43	1,636	18	1,098.9	758.6	1.91
Portugal	8,978	(5.5)	550	10.0	8,428	(6.3)	1,215	31	1,081	71	1,162.6	1,420.4	4.39
Romania	12,476	(34.0)	3,174	(8.8)	9,302	(39.7)	390	(69)	317	(61)	147.6	315.0	3.20
Russian Federation {c}	115,693	X	36,670	X	79,023	X	3,286	X	3,368	X	630.0	2,947.7	2.88
Slovakia {c}	5,312	X	416	X	4,896	X	330	X	531	X	205.0	334.4	3.05
Slovenia {c}	2,111	X	482	X	1,628	X	437	X	459	X	355.3	485.3	4.64
Spain	15,631	5.4	3,198	73.8	12,433	(4.3)	2,970	48	3,977	22	3,713.3	1,568.1	1.06
Sweden	58,200	9.5	3,800	(13.6)	54,400	11.6	940	(29)	9,551	23	1,430.0	10,442.7	10.34
Switzerland	4,365	(5.8)	931	4.6	3,434	(8.3)	686	1	1,545	34	2,385.0	1,873.3	X
Ukraine {c}	10,176	X	1,832	X	8,344	X	289	X	270	X	264.2	118.1	0.58
United Kingdom	7,403	33.2	232	63.0	7,171	32.4	2,652	98	6,382	54	9,609.5	2,031.9	0.55
Yugoslavia {c}	1,320	X	50	X	1,270	X	132	X	78	X	166.4	45.0	X
MIDDLE EAST & N. AFRICA {b}	**44,094**	**14.8**	**25,620**	**4.9**	**18,474**	**32.1**	**2,978**	**92**	**2,321**	**52**	**5,376.8**	**228.4**	**X**
Afghanistan	7,928	39.8	6,210	46.9	1,717	19.1	1	0	0	0	1.3	0.2	X
Algeria	2,676	33.5	2,239	27.2	437	79.5	50	0	62	(48)	336.9	0.3	0.00
Egypt	2,776	23.6	2,647	23.6	129	23.3	81	27	263	70	948.9	8.2	0.05
Iran, Islamic Rep	6,881	7.5	1,997	0.0	4,884	10.8	444	117	180	100	344.4	1.3	0.01
Iraq	134	106.7	81	442.2	53	6.0	4	22	19	(33)	1.1	0.1	X
Israel	118	0.3	13	15.2	105	(1.3)	230	55	275	72	694.2	23.8	0.08
Jordan	11	33.3	7	66.7	4	0.0	0	0	32	164	125.9	16.3	0.42
Kuwait	0	0.0	0	0.0	0	0.0	0	0	0	0	94.0	0.4	0.00
Lebanon	400	17.8	393	20.9	7	(51.2)	46	0	42	4	172.1	5.1	0.33
Libyan Arab Jamahiriya	650	1.7	536	0.0	114	10.7	0	0	6	0	46.2	0.0	X
Morocco {f}	1,324	(7.4)	539	(31.2)	785	21.4	35	(75)	108	2	335.7	59.6	0.63
Oman	0	0.0	0	0.0	0	0.0	0	0	0	0	27.1	0.7	0.01
Saudi Arabia	0	0.0	0	0.0	0	0.0	0	0	0	0	631.6	6.9	0.01
Syrian Arab Rep	50	(21.6)	16	53.9	35	(36.1)	27	0	1	(91)	117.0	0.3	0.00
Tunisia	2,770	22.6	2,559	20.1	211	64.5	104	16	97	58	203.2	17.3	0.21
Turkey	18,376	7.0	8,383	(20.1)	9,994	49.6	1,956	150	1,236	68	918.0	80.8	0.16
United Arab Emirates	0	0.0	0	0.0	0	0.0	0	0	0	0	290.1	6.7	X
Yemen	0	0.0	0	0.0	0	0.0	0	0	0	0	42.9	0.2	0.01

	Average Annual Roundwood Production						Average Annual Forest Products Production				Trade in Forest Products {a}		
	Total Roundwood		Wood Fuel		Industrial Roundwood		Wood-Based Panels		Paper and Paperboard		Import	Export	
	Cubic Meters (000) 1996-98	Percent Change Since 1986-88	Cubic Meters (000) 1996-98	Percent Change Since 1986-88	Cubic Meters (000) 1996-98	Percent Change Since 1986-88	Cubic Meters (000) 1996-98	Percent Change Since 1986-88	Metric Tons (000) 1996-98	Percent Change Since 1986-88	Value (million US$) 1996-98	Value (million US$) 1996-98	Percent of Total Exports 1997
SUB-SAHARAN AFRICA {b}	**512,491**	**28.5**	**445,783**	**28.4**	**66,709**	**29.6**	**1,630**	**14**	**2,302**	**16**	**980.6**	**2,628.9**	**X**
Angola	6,272	37.3	5,220	38.5	1,052	31.5	11	(55)	0	0	4.6	1.2	0.02
Benin	5,839	31.8	5,507	31.5	332	35.9	0	0	0	0	2.1	1.8	0.34
Botswana	1,641	33.4	1,540	33.4	101	33.3	0	0	0	0	0.0	0.0	X
Burkina Faso	10,506	32.0	10,022	32.0	484	31.9	0	0	0	0	2.3	0.0	0.00
Burundi	1,669	36.7	1,487	26.7	182	286.9	0	0	0	0	2.9	0.1	0.06
Cameroon	15,191	29.1	11,869	31.9	3,323	20.2	124	42	2	(67)	22.5	426.6	17.46
Central African Rep	3,388	(1.4)	2,660	(12.3)	728	80.1	1	(67)	0	0	0.2	27.2	12.75
Chad	1,871	36.5	1,147	35.7	724	37.8	0	0	0	0	1.9	0.1	0.03
Congo	3,985	29.6	2,456	32.7	1,529	25.0	54	(5)	0	0	1.4	147.4	8.19
Congo, Dem Rep	48,372	40.9	44,814	42.0	3,557	27.9	21	(47)	3	13	6.0	63.3	4.38
Côte d'Ivoire	12,983	21.1	9,970	32.8	3,013	(6.4)	306	42	0	0	36.7	280.1	5.68
Equatorial Guinea	811	27.4	447	0.0	364	92.3	9	13	0	0	1.0	56.8	13.18
Eritrea {c}	X	X	2,110	X	X	X	0	0	0	0	9.3	0.0	0.00
Ethiopia {c}	48,990	X	46,522	X	2,468	X	13	X	9	X	15.9	0.1	0.01
Gabon	5,144	62.1	2,491	33.2	2,653	103.8	36	(80)	0	0	4.3	287.6	8.73
Gambia	813	28.7	700	17.6	113	207.9	0	0	0	0	0.8	0.2	0.08
Ghana	21,931	61.3	20,678	64.4	1,253	22.9	136	112	0	0	15.8	144.7	8.73
Guinea	8,643	137.5	7,977	154.5	666	32.1	0	0	0	0	5.6	8.7	1.17
Guinea-Bissau	586	4.5	422	0.0	164	18.0	0	0	0	0	0.2	3.1	5.55
Kenya	28,813	32.5	26,879	33.6	1,934	19.0	52	26	129	41	30.0	1.1	0.04
Lesotho	1,556	27.4	1,556	27.4	0	0.0	0	0	0	0	0.0	0.0	0.00
Liberia	2,936	(19.2)	2,700	0.0	236	(74.7)	45	644	0	0	1.0	11.9	X
Madagascar	9,254	19.5	8,878	28.0	376	(53.4)	5	20	4	(62)	9.3	6.0	0.77
Malawi	9,449	25.2	8,950	24.2	499	44.6	18	213	0	0	4.5	2.3	0.38
Mali	6,284	24.0	5,890	24.2	394	21.4	0	0	0	0	5.3	1.4	0.21
Mauritania	15	28.6	9	28.6	6	28.6	0	0	0	0	2.3	0.4	0.08
Mozambique	17,973	18.2	16,724	17.0	1,249	37.2	3	(64)	0	(100)	1.4	9.6	1.92
Namibia	X	X	X	X	X	X	X	X	X	X	X	X	0.00
Niger	6,260	38.8	5,873	38.8	386	38.8	0	0	0	0	1.8	0.1	0.04
Nigeria	95,993	17.6	87,001	18.0	8,992	14.3	38	(80)	59	(28)	79.8	29.8	0.19
Rwanda	3,020	(48.3)	3,000	(46.4)	20	(91.7)	0	(100)	0	0	1.9	0.1	0.05
Senegal	4,785	29.7	4,029	30.0	756	28.0	0	0	0	0	23.7	0.1	0.01
Sierra Leone	3,215	16.2	3,092	17.6	124	(11.7)	0	0	0	0	2.2	0.6	0.64
Somalia	7,616	7.7	7,513	7.7	102	5.1	0	(100)	0	0	0.2	0.1	X
South Africa {g}	32,906	46.1	14,467	36.5	18,439	54.6	653	64	1,988	19	481.6	1,025.0	2.89
Sudan	9,289	23.0	7,207	23.1	2,082	22.5	2	0	3	(69)	34.4	0.1	0.02
Tanzania, United Rep	38,193	35.0	35,947	35.4	2,246	27.9	4	(67)	25	32	6.0	4.8	0.40
Togo	1,156	41.2	872	33.2	284	73.0	0	0	0	0	3.4	2.2	0.35
Uganda	15,236	30.1	13,080	29.7	2,156	32.8	5	47	3	50	3.3	0.0	0.00
Zambia	8,051	31.2	7,219	28.7	832	57.8	18	(20)	3	(2)	5.0	0.6	0.05
Zimbabwe	8,192	8.3	6,260	2.2	1,932	34.4	68	148	74	(7)	29.0	22.1	0.72
NORTH AMERICA {b}	**679,734**	**(0.4)**	**79,960**	**(16.5)**	**599,774**	**2.3**	**50,958**	**27**	**97,608**	**17**	**27,786.9**	**43,315.3**	**X**
Canada	190,711	2.3	5,319	(19.5)	185,392	3.1	11,154	79	19,530	22	3,663.6	25,202.6	10.19
United States	489,023	(1.3)	74,641	(16.3)	414,382	1.9	39,804	17	78,078	16	24,115.6	18,112.6	1.93
C. AMERICA & CARIBBEAN {b}	**68,541**	**15.7**	**59,230**	**17.8**	**11,743**	**7.0**	**715**	**(31)**	**3,749**	**23**	**2,455.5**	**489.8**	**X**
Belize	188	5.5	126	0.0	62	19.0	X	X	0	0	3.6	2.6	0.81
Costa Rica	5,215	40.6	3,548	34.0	1,668	57.4	74	56	20	40	132.4	14.0	0.31
Cuba	2,756	(5.9)	2,145	(7.9)	611	2.2	149	12	57	(61)	23.1	0.1	X
Dominican Rep	562	0.0	556	0.0	6	0.0	406	563	57	497	161.9	0.6	0.01
El Salvador	4,881	(11.0)	4,386	(18.6)	496	426.7	0	0	56	234	65.1	5.8	0.21
Guatemala	13,026	24.6	12,794	23.8	232	99.5	43	506	31	106	101.0	11.6	0.36
Haiti	6,295	26.9	6,056	28.3	239	0.0	X	X	0	0	12.8	0.0	0.00
Honduras	6,929	24.9	6,209	34.3	720	(22.0)	17	92	96	100	76.7	35.7	1.63
Jamaica	343	(64.0)	300	(60.4)	43	(78.0)	0	(100)	0	(100)	73.4	0.3	0.01
Mexico	22,940	13.0	15,677	20.0	7,263	0.5	0	(100)	3,404	21	1,429.7	388.1	0.32
Nicaragua	4,105	34.1	3,864	40.0	241	(19.7)	5	(17)	0	0	11.8	21.4	2.48
Panama	1,080	15.0	985	20.6	95	(22.3)	21	320	28	17	83.3	7.0	0.08
Trinidad and Tobago	71	45.8	2,559	20.1	61	57.6	0	0	0	0	72.6	0.7	0.03
SOUTH AMERICA {b}	**296,757**	**4.8**	**167,017**	**(7.9)**	**129,740**	**27.4**	**5,373**	**40**	**9,720**	**29**	**3,364.1**	**4,791.5**	**X**
Argentina	11,428	19.4	3,741	24.7	7,687	16.9	619	67	1,098	10	822.5	269.0	0.94
Bolivia	2,129	49.7	1,324	26.5	806	114.2	21	106	2	33	29.2	66.0	4.84
Brazil	198,402	(6.0)	114,052	(18.7)	84,350	19.0	3,181	17	6,295	36	1,161.4	2,563.2	4.25
Chile	30,493	70.9	10,404	51.6	20,089	83.0	985	305	645	51	256.0	1,393.6	6.76
Colombia	18,622	14.6	16,402	30.6	2,221	(39.7)	118	(35)	703	46	365.8	65.5	0.41
Ecuador	11,081	89.5	5,296	76.3	5,785	103.4	0	0	90	159	209.4	98.8	1.65
Guyana	501	134.8	10	3.3	491	141.3	80	100	0	0	3.4	38.3	X
Paraguay	8,097	12.4	4,220	6.3	3,877	19.9	161	58	13	35	28.4	81.0	1.87
Peru	8,494	(0.1)	7,044	(4.3)	1,450	27.4	66	38	114	(44)	150.2	38.4	0.46
Suriname	193	7.3	1	(46.4)	192	7.9	8	(13)	0	0	1.4	5.6	X
Uruguay	5,058	56.7	3,588	40.0	1,470	120.9	6	(44)	88	40	84.1	77.1	1.81
Venezuela	2,140	34.1	876	26.0	1,264	40.3	127	(17)	672	2	249.5	92.5	0.37
OCEANIA {b}	**49,325**	**27.6**	**8,529**	**(0.3)**	**32,503**	**31.5**	**2,657**	**74**	**3,290**	**42**	**1,921.8**	**2,627.2**	**X**
Australia	22,528	14.8	2,700	0.0	11,524	0.1	1,440	48	2,427	48	1,560.4	765.1	0.91
Fiji	670	133.7	37	0.0	633	153.5	16	(2)	0	0	9.9	28.9	2.41
New Zealand	16,275	63.4	17	(66.7)	16,259	64.1	1,186	132	863	29	299.6	1,365.4	7.66
Papua New Guinea	8,772	8.1	5,533	0.0	3,239	25.3	15	(42)	0	0	8.8	339.3	13.78
Solomon Islands	872	83.5	138	6.7	734	112.1	0	0	0	0	0.3	122.8	51.39
AFRICA {b}	**522,687**	**28.3**	**454,303**	**28.1**	**68,384**	**29.8**	**1,899**	**7**	**2,838**	**17**	**2,851.4**	**2,714.3**	**X**
ASIA {b,c}	**1,145,856**	**X**	**880,416**	**X**	**285,268**	**X**	**41,765**	**X**	**84,062**	**X**	**40,620.7**	**16,577.0**	**X**
DEVELOPING {b}	**1,989,087**	**15.8**	**1,551,162**	**16.5**	**440,367**	**13.1**	**42,124**	**87**	**67,800**	**102**	**32,574.1**	**22,491.3**	**X**
DEVELOPED {b}	**1,272,533**	**(14.9)**	**190,901**	**(22.6)**	**1,081,749**	**(13.3)**	**109,266**	**9**	**220,485**	**21**	**110,361.0**	**112,821.7**	**X**

Notes: Negative numbers are shown in parentheses. a. Includes trade in industrial roundwood, wood fuel, sawnwood, wood-based panels, pulp, paper and paperboard, recovered paper, chips and particles, and wood residues. b. Regional and world totals calculated by the Food and Agriculture Organization of the United Nations. They include countries not listed in this table. c. Due to recent independence of countries in the former Soviet Union, Yugoslavia, Czechoslovakia, and Ethiopia, percentage changes from 1986-88 to 1996-98 cannot be shown for these countries or for the regional totals for Asia (excluding the Middle East), Europe, and Asia. d. Includes mainland China; Hong Kong, SAR; Taiwan, Province of China; and Macau. e. Data for Belgium includes Luxembourg. f. Data are for Morocco and Western Sahara. g. Data are for South Africa and Namibia.

Sources: Food and Agriculture Organization of the United Nations and World Bank

	Cattle {a}		Sheep and Goats {a}		Equines {a}		Buffaloes and Camels {a}		Grains Fed to Livestock as Percent of Total Grain Consumption {a}		Meat Production (000 metric tons) Beef and Veal		Meat Production (000 metric tons) Sheep and Goats	
	Annual Average (000) 1996-98	Percent Change Since 1986-88	Annual Average (000) 1996-98	Percent Change Since 1986-88	Annual Average (000) 1996-98	Percent Change Since 1986-88	Annual Average (000) 1996-98	Percent Change Since 1986-88	1988	1998	Average Annual 1996-98	Percent Change Since 1986-88	Average Annual 1996-98	Percent Change Since 1986-88
WORLD	**1,328,037**	**4.7**	**1,760,712**	**5.9**	**119,121**	**2.3**	**180,281**	**12.1**	**38**	**37**	**53,921**	**6**	**10,954**	**26**
ASIA (EXCL. MIDDLE EAST)	**424,516**	**X**	**652,805**	**X**	**37,216**	**X**	**158,580**	**X**	**X**	**X**	**8,785**	**138**	**4,357**	**116**
Armenia	504	X	582	X	14	X	0	X	52	25	31	X	6	X
Azerbaijan	1,768	X	5,429	X	54	X	322	X	47	17	48	X	26	X
Bangladesh	23,645	4.8	34,910	95.8	X	X	845	27.1	0	0	153	13	122	99
Bhutan	435	12.9	101	47.7	58	12.0	4	(18.9)	0	0	6	13	0	61
Cambodia	2,867	57.9	X	X	22	57.1	761	14.0	0	0	41	84	X	X
China {b}	95,493	35.2	256,879	53.2	24,166	(9.4)	21,706	3.0	18	26	3,775	506	2,038	185
Georgia	1,003	X	654	X	27	X	19	X	53	30	55	X	9	X
India	209,156	4.9	176,279	14.2	2,190	5.1	92,240	18.5	2	4	1,383	31	680	18
Indonesia	12,029	23.8	22,307	38.8	634	(7.2)	3,143	(5.6)	5	8	350	78	101	30
Japan	4,759	1.2	47	(36.0)	30	36.8	X	X	49	47	537	(5)	0	(12)
Kazakhstan	5,428	X	14,288	X	1,353	X	123	X	59	37	404	X	144	X
Korea, Dem People's Rep	558	(53.5)	1,183	19.9	40	(4.8)	X	X	0	0	20	(52)	5	17
Korea, Rep	3,318	37.6	629	254.3	7	128.0	X	X	39	47	321	62	3	323
Kyrgyzstan	849	X	3,836	X	323	X	49	X	69	25	92	X	47	X
Lao People's Dem Rep	1,175	66.8	149	81.1	26	(38.1)	1,176	15.7	0	0	13	156	0	99
Malaysia	723	13.7	575	31.1	4	(27.0)	150	(30.1)	37	41	18	59	1	0
Mongolia	3,469	40.4	23,122	31.5	2,773	37.8	358	(35.1)	0	0	87	19	114	(4)
Myanmar	10,306	3.8	1,626	9.9	130	(9.7)	2,300	4.7	0	0	99	4	8	10
Nepal	7,019	10.4	6,768	13.8	X	X	3,355	14.9	0	0	48	17	37	25
Pakistan	17,933	2.0	78,181	41.8	4,579	31.1	21,819	27.9	3	4	368	43	784	98
Philippines	2,263	29.1	6,440	44.4	210	6.8	2,938	0.8	19	23	131	90	78	262
Singapore	0	(30.4)	0	(77.3)	X	X	0	(85.0)	42	14	0	(84)	0	(84)
Sri Lanka	1,607	(10.3)	536	(4.3)	2	0.0	736	(24.8)	0	0	25	3	2	(14)
Tajikistan	1,090	X	2,335	X	78	X	45	X	36	10	25	X	11	X
Thailand	6,959	39.9	173	(3.2)	14	(25.3)	3,470	(42.0)	25	31	213	34	1	2
Turkmenistan	1,019	X	6,036	X	43	X	40	X	43	23	56	X	55	X
Uzbekistan	5,240	X	9,222	X	316	X	26	X	40	12	381	X	79	X
Viet Nam	3,896	31.5	514	22.7	122	(9.9)	2,950	7.7	0	0	83	32	5	45
EUROPE	**164,643**	**X**	**182,758**	**X**	**8,868**	**X**	**226**	**X**	**64**	**58**	**13,254**	**20**	**1,619**	**17**
Albania	786	18.8	3,143	29.6	196	11.0	0	(94.0)	0	0	32	92	16	66
Austria	2,265	(13.7)	432	51.5	73	64.9	X	X	70	69	208	(8)	7	60
Belarus	4,903	X	221	X	239	X	X	X	75	58	269	X	3	X
Belgium {c}	3,276	3.3	169	(4.4)	24	(0.9)	X	X	33	58	335	4	4	(40)
Bosnia and Herzegovina	278	X	276	X	50	X	1	X	X	X	13	X	3	X
Bulgaria	609	(63.7)	3,966	(59.7)	466	(4.2)	12	(54.6)	59	49	70	(44)	52	(35)
Croatia	452	X	537	X	23	X	X	X	X	X	25	X	2	X
Czech Rep	1,848	X	155	X	19	X	X	X	X	58	151	X	2	X
Denmark	2,032	(14.2)	151	63.9	33	5.1	X	X	80	78	172	(26)	2	65
Estonia	342	X	41	X	4	X	X	X	74	45	20	X	0	X
Finland	1,158	(22.5)	113	63.2	54	44.3	X	X	66	70	97	(21)	1	1
France	20,571	(7.3)	11,638	(11.0)	380	24.9	X	X	61	70	1,695	(11)	149	(17)
Germany	15,626	(25.8)	2,443	(39.9)	680	44.0	X	X	X	64	1,519	(26)	43	(1)
Greece	557	(21.5)	15,253	13.7	162	(49.9)	1	(24.1)	59	59	70	(17)	130	0
Hungary	903	(47.5)	1,004	(58.1)	75	(23.8)	X	X	73	76	64	(47)	1	(87)
Iceland	75	6.3	473	(24.8)	80	34.1	X	X	X	X	3	4	8	(35)
Ireland	6,947	21.8	5,550	47.6	65	(10.6)	X	X	63	67	563	16	84	76
Italy	7,198	(18.6)	12,215	(3.1)	369	(6.1)	161	56.7	49	53	1,150	(2)	76	8
Latvia	474	X	51	X	25	X	X	X	72	54	26	X	0	X
Lithuania	1,062	X	46	X	80	X	X	X	80	60	90	X	1	X
Macedonia, FYR	292	X	1,980	X	66	X	1	X	X	X	7	X	9	X
Moldova, Rep	619	X	1,304	X	64	X	X	X	69	55	34	X	3	X
Netherlands	4,405	(9.3)	1,724	65.5	97	51.8	X	X	42	49	560	6	16	44
Norway	1,014	6.9	2,547	8.0	23	40.8	X	X	73	67	89	16	27	4
Poland	7,133	(32.6)	498	(89.4)	562	(51.4)	X	X	65	64	425	(41)	3	(90)
Portugal	1,310	(1.1)	6,933	17.5	228	(18.3)	X	X	49	60	101	(6)	26	(0)
Romania	3,389	(49.0)	10,317	(42.8)	846	17.6	0	0.0	76	66	182	(7)	64	(24)
Russian Federation	35,732	X	23,186	X	2,276	X	34	X	64	41	2,424	X	203	X
Slovakia	875	X	447	X	12	X	X	X	X	58	62	X	2	X
Slovenia	476	X	37	X	8	X	X	X	X	X	50	X	0	X
Spain	5,759	16.4	26,094	16.5	410	(13.7)	X	X	71	76	596	33	243	9
Sweden	1,759	4.9	439	9.9	87	51.4	X	X	76	75	143	5	4	(27)
Switzerland	1,694	(9.2)	485	10.1	57	13.3	X	X	65	59	153	(8)	6	31
Ukraine	15,210	X	3,169	X	768	X	X	X	65	42	920	X	24	X
United Kingdom	11,688	(4.2)	42,853	10.0	173	4.4	X	X	49	49	700	(33)	372	23
Yugoslavia	1,906	X	2,846	X	90	X	17	X	X	X	236	X	31	X
MIDDLE EAST & N. AFRICA	**33,818**	**3.6**	**250,065**	**5.4**	**11,339**	**(6.3)**	**5,818**	**7.8**	**35**	**32**	**1,475**	**24**	**1,920**	**25**
Afghanistan	1,500	(0.7)	16,500	16.1	1,483	(14.3)	265	(4.2)	0	0	65	(2)	141	17
Algeria	1,244	(11.1)	20,069	8.1	345	(36.9)	139	13.0	34	25	101	38	183	70
Egypt	3,082	33.3	7,433	35.7	3,032	56.6	3,184	23.7	38	32	269	98	120	41
Iran, Islamic Rep	8,564	31.4	78,419	22.3	1,787	(26.5)	605	11.6	25	20	297	70	401	33
Iraq	1,274	(19.7)	7,900	(24.8)	436	(13.8)	71	(62.1)	31	12	43	(2)	26	(20)
Israel	379	15.2	420	(3.7)	11	0.0	5	(9.1)	57	60	42	26	6	12
Jordan	63	110.3	2,924	81.3	25	(0.3)	18	15.0	40	50	3	224	14	125
Kuwait	21	(17.9)	542	83.8	1	(66.3)	9	14.2	57	46	1	(70)	40	19
Lebanon	75	28.4	798	32.2	35	46.0	1	76.9	46	50	12	(20)	15	46
Libyan Arab Jamahiriya	150	(31.8)	6,850	23.1	69	(19.4)	102	(20.3)	35	26	19	(34)	45	7
Morocco	2,519	(17.6)	21,002	5.8	1,609	4.0	36	(5.1)	25	27	109	(5)	124	42
Oman	146	9.1	877	(5.1)	27	12.1	95	17.6	0	0	3	(3)	15	43
Saudi Arabia	220	(5.0)	12,352	23.1	100	(9.1)	422	8.8	71	72	17	(24)	102	29
Syrian Arab Rep	856	17.8	14,742	7.7	238	(6.8)	9	40.5	31	37	42	39	155	50
Tunisia	724	12.8	7,701	15.5	367	5.2	231	6.5	33	34	46	26	49	15
Turkey	11,620	(8.0)	41,180	(27.3)	1,261	(36.5)	230	(57.5)	34	34	347	(5)	374	(2)
United Arab Emirates	75	70.9	1,377	85.0	0	22.8	170	84.0	43	25	10	84	37	55
Yemen	1,215	8.6	8,081	22.0	503	(4.5)	180	18.3	X	0	43	21	42	17

	Cattle {a}		Sheep and Goats {a}		Equines {a}		Buffaloes and Camels {a}		Grains Fed to Livestock as Percent of Total Grain Consumption {a}		Meat Production (000 metric tons) Beef and Veal		Sheep and Goats	
	Annual Average (000) 1996-98	Percent Change Since 1986-88	Annual Average (000) 1996-98	Percent Change Since 1986-88	Annual Average (000) 1996-98	Percent Change Since 1986-88	Annual Average (000) 1996-98	Percent Change Since 1986-88	1988	1998	Average Annual 1996-98	Percent Change Since 1986-88	Average Annual 1996-98	Percent Change Since 1986-88
SUB-SAHARAN AFRICA	**206,563**	**21.1**	**365,718**	**25.0**	**15,610**	**15.2**	**13,850**	**4.9**	**2**	**2**	**2,969**	**8**	**1,339**	**33**
Angola	3,455	4.7	1,713	0.3	6	10.4	X	X	0	0	62	9	5	18
Benin	1,383	52.8	1,621	(7.9)	6	0.0	X	X	0	0	20	50	6	20
Botswana	2,383	2.1	2,090	20.1	270	57.1	X	X	0	0	41	8	6	26
Burkina Faso	4,492	21.0	13,914	32.4	496	21.4	14	17.4	1	4	49	37	34	50
Burundi	369	(16.9)	1,213	11.8	X	X	X	X	0	0	11	(5)	4	17
Cameroon	5,717	31.0	7,600	41.6	51	10.1	X	X	0	0	88	43	31	48
Central African Rep	2,926	30.2	2,406	91.4	X	X	X	X	0	0	54	58	8	81
Chad	5,298	32.6	6,940	60.0	505	22.4	662	31.8	0	0	39	12	26	47
Congo	73	4.9	401	19.4	0	(18.8)	X	X	0	0	2	(10)	1	17
Congo, Dem Rep	1,083	(21.9)	5,144	29.3	X	X	X	X	5	0	16	(37)	24	65
Côte d'Ivoire	1,303	39.9	2,380	25.8	X	X	X	X	4	3	40	42	10	15
Equatorial Guinea	5	6.7	44	4.1	X	X	X	X	X	X	0	7	0	13
Eritrea	1,320	X	2,930	X	X	X	69	X	X	X	11	X	10	X
Ethiopia {d}	29,900	6.8	38,667	(6.5)	8,580	11.0	1,027	1.0	0	0	270	12	144	(2)
Gabon	39	85.6	259	14.0	X	X	X	X	0	0	1	61	1	26
Gambia	339	10.8	415	11.8	53	(5.9)	X	X	0	0	3	5	1	22
Ghana	1,183	2.8	4,382	15.6	15	1.7	X	X	4	4	21	7	12	24
Guinea	2,291	100.0	1,439	110.5	5	43.8	X	X	0	0	15	81	4	77
Guinea-Bissau	507	44.3	575	38.3	7	34.2	X	X	0	0	4	46	2	39
Kenya	13,789	8.0	13,133	0.5	2	0.0	822	4.4	3	1	258	18	52	10
Lesotho	585	(1.2)	1,897	(28.6)	267	1.3	X	X	53	16	14	10	6	(4)
Liberia	36	(14.3)	430	(10.0)	X	X	X	X	0	0	1	(25)	1	(0)
Madagascar	10,329	1.1	2,093	3.7	1	(11.9)	X	X	0	0	145	5	10	5
Malawi	750	(23.1)	1,373	39.0	2	35.8	X	X	2	4	18	12	5	40
Mali	5,719	24.3	14,279	34.2	777	64.3	352	51.5	1	1	88	40	54	56
Mauritania	1,312	7.0	10,332	39.3	174	5.4	1,164	38.0	0	0	10	(34)	24	38
Mozambique	1,287	(4.7)	508	4.2	21	5.0	X	X	0	0	37	(2)	3	7
Namibia	2,079	7.6	4,011	(9.4)	132	8.1	X	X	X	X	50	38	11	(19)
Niger	2,079	39.0	10,070	43.3	532	16.1	390	(2.3)	0	0	36	42	37	22
Nigeria	19,300	43.6	38,500	22.4	1,204	15.3	18	0.2	2	2	290	11	240	68
Rwanda	492	(17.0)	1,237	(10.3)	0	0.0	X	X	11	6	13	(3)	4	(10)
Senegal	2,887	15.6	7,791	46.0	880	43.9	7	(43.3)	1	0	46	17	29	50
Sierra Leone	397	19.1	535	28.0	X	X	X	X	0	0	6	28	1	12
Somalia	5,433	13.8	26,533	(19.8)	40	(18.7)	6,200	(6.2)	2	7	59	37	61	(25)
South Africa	13,619	9.2	36,139	1.9	479	5.5	X	X	43	42	495	(20)	131	(15)
Sudan	33,119	67.8	76,000	131.1	725	7.6	3,063	12.9	0	0	230	4	247	165
Tanzania, United Rep	14,163	10.9	13,640	21.1	178	4.1	X	X	2	2	211	20	36	22
Togo	215	(10.5)	1,793	(24.4)	5	17.8	X	X	16	15	7	46	7	39
Uganda	5,345	20.0	5,525	34.8	18	6.0	X	X	0	0	90	47	25	41
Zambia	3,150	21.8	669	27.2	2	41.0	X	X	4	4	31	(5)	3	65
Zimbabwe	5,429	(7.0)	3,245	18.5	132	7.0	X	X	13	14	64	(19)	12	41
NORTH AMERICA	**114,949**	**1.6**	**10,318**	**(19.4)**	**6,596**	**13.9**	**0**	**0.0**	**63**	**65**	**12,836**	**7**	**128**	**(19)**
Canada	13,300	23.2	664	26.6	399	(14.9)	X	X	74	76	1,081	11	10	26
United States	101,649	(0.7)	9,655	(21.4)	6,197	16.5	X	X	66	66	11,755	7	118	(21)
C. AMERICA & CARIBBEAN	**46,649**	**(10.9)**	**19,356**	**(2.3)**	**16,063**	**2.0**	**5**	**(37.2)**	**19**	**29**	**1,885**	**2**	**83**	**10**
Belize	62	25.4	4	(12.8)	9	0.0	X	X	X	X	1	27	0	0
Costa Rica	1,547	(31.7)	4	0.4	127	1.0	X	X	35	50	87	(5)	0	2
Cuba	4,617	(7.7)	428	(12.7)	657	(13.0)	X	X	0	0	70	(51)	2	(23)
Dominican Rep	2,482	18.6	705	16.3	609	5.9	X	X	44	57	79	13	3	16
El Salvador	1,202	9.8	21	7.0	123	4.2	X	X	20	33	32	52	0	8
Guatemala	2,319	14.1	660	29.5	166	8.1	X	X	25	28	54	16	3	23
Haiti	1,272	6.0	1,580	29.7	780	7.8	X	X	2	2	29	(1)	5	30
Honduras	2,091	(17.4)	42	21.6	267	2.4	X	X	37	44	27	(33)	0	30
Jamaica	407	11.9	441	0.5	37	0.5	X	X	34	34	15	6	2	0
Mexico	26,890	(14.8)	15,038	(6.0)	12,770	2.3	X	X	27	45	1,350	7	66	10
Nicaragua	1,736	(8.3)	10	6.1	299	(2.9)	X	X	0	10	51	52	0	4
Panama	1,415	(0.4)	5	(15.2)	169	17.1	X	X	31	44	66	15	X	X
Trinidad and Tobago	37	(41.2)	71	2.3	5	0.0	5	(37.2)	43	33	1	(26)	0	(6)
SOUTH AMERICA	**300,552**	**16.3**	**109,840**	**(11.4)**	**23,006**	**9.8**	**1,701**	**57.0**	**46**	**50**	**10,214**	**23**	**352**	**4**
Argentina	54,367	4.7	20,638	(35.8)	3,565	10.6	X	X	35	55	2,685	(1)	64	(34)
Bolivia	6,248	19.4	9,723	16.0	1,034	3.1	X	X	33	40	147	20	20	9
Brazil	163,000	20.0	30,667	(0.3)	9,750	7.1	1,700	57.0	54	55	5,113	35	130	36
Chile	3,842	14.6	4,727	(23.0)	625	18.3	X	X	31	45	259	42	15	(17)
Colombia	27,431	14.6	3,388	(3.0)	3,750	17.7	X	X	19	31	694	17	15	19
Ecuador	5,195	33.8	2,209	48.0	943	32.5	X	X	25	35	156	64	7	67
Guyana	230	89.0	209	3.0	4	0.0	X	X	3	4	3	71	1	5
Paraguay	9,784	31.6	510	(5.6)	416	13.3	X	X	3	2	228	95	3	(1)
Peru	4,621	10.8	15,158	3.0	1,409	3.2	X	X	28	37	117	12	28	4
Suriname	92	21.6	16	14.9	0	(5.6)	1	24.8	0	0	2	39	0	21
Uruguay	10,561	7.1	18,593	(22.6)	498	7.8	X	X	19	20	433	40	60	12
Venezuela	15,173	20.3	3,999	103.0	1,012	0.5	X	X	60	40	376	26	8	6
OCEANIA	**36,313**	**16.5**	**168,187**	**(22.4)**	**401**	**(21.3)**	**0**	**X**	**50**	**62**	**2,497**	**22**	**1,135**	**(6)**
Australia	26,622	18.8	120,545	(19.7)	232	(30.5)	X	X	53	66	1,839	23	600	1
Fiji	360	38.5	223	19.2	44	5.7	X	X	0	0	9	(6)	1	35
New Zealand	8,904	9.8	47,345	(28.8)	85	(12.4)	X	X	41	46	638	20	534	(13)
Papua New Guinea	87	(18.8)	7	38.6	2	28.4	X	X	0	0	2	(14)	0	43
Solomon Islands	10	(35.8)	X	X	0	3.4	X	X	X	X	0	(35)	X	X
DEVELOPED	**350,777**	**(14.2)**	**439,954**	**(24.0)**	**18,509**	**(1.7)**	**855**	**1.9**	**X**	**X**	**30,733**	**(11)**	**3,395**	**(12)**
DEVELOPING	**977,228**	**13.7**	**1,319,094**	**22.0**	**100,590**	**3.1**	**179,325**	**12.2**	**X**	**X**	**23,183**	**40**	**7,538**	**57**

Notes: It is assumed that all countries with livestock have reported their numbers to the Food and Agriculture Organization of the United Nations. a. World and regional totals calculated by the Economic Research Service, U.S. Department of Agriculture. b. Includes Taiwan. c. Data for Belgium include Luxembourg. d. Data for Ethiopia before 1993 include Eritrea.

Sources: Food and Agriculture Organization of the United Nations and U.S. Department of Agriculture

	International Geosphere-Biosphere Programme (IGBP) Classification (000 km²)			Pilot Analysis of Global Ecosystems (PAGE) Classification				
						Carbon Stocks {e}		
	IGBP Land Area	Urban Area {a}	Agri-cultural Mosaic Area {b}	PAGE Area (000 km²) {c}	Population (000) {d}	Low Estimate (gigatons)	High Estimate (gigatons)	Protected Area (000 km²) {e}
GRASSLANDS	**53,544**	**1,010**	**7,172**	**52,554**	**792,711**	**405**	**806**	**3,989**
Asia (excl. Middle East)	9,033	141	1,281	8,892	249,495	46	102	586
Europe	7,072	116	189	6,966	20,491	118	188	248
Middle East & N. Africa	3,031	161	159	2,871	111,882	5	20	216
Sub-Saharan Africa	14,546	83	3,531	14,464	312,935	97	218	1,329
North America	6,816	238	518	6,583	6,032	63	110	791
C. America & Caribbean	1,130	82	24	1,048	30,533	6	13	56
South America	5,017	150	1,416	4,872	57,529	36	78	307
Oceania	6,898	40	54	6,859	3,814	33	78	457
FORESTS	**29,905**	**930**	**1,727**	**28,974**	**446,470**	**471**	**929**	**2,453**
Asia (excl. Middle East)	3,812	91	192	3,721	231,782	45	103	302
Europe	6,957	226	338	6,731	43,713	137	224	155
Middle East & N. Africa	100	10	44	90	6,724	1	1	1
Sub-Saharan Africa	2,672	13	162	2,659	53,823	29	81	155
North America	7,564	449	965	7,115	30,764	166	258	711
C. America & Caribbean	997	59	1	939	33,940	11	23	88
South America	6,928	67	26	6,861	39,860	74	218	957
Oceania	874	17	0	857	5,864	10	21	84
AGRICULTURE	**27,890**	**2,407**	**X**	**36,234**	**2,790,582**	**336**	**628**	**1,594**
Asia (excl. Middle East)	8,874	683	X	10,370	1,991,214	82	158	268
Europe	6,840	763	X	7,448	311,923	89	147	338
Middle East & N. Africa	1,025	136	X	1,230	99,662	6	14	11
Sub-Saharan Africa	2,141	38	X	5,837	204,901	45	99	476
North America	2,867	511	X	4,406	47,927	55	90	97
C. America & Caribbean	517	48	X	611	26,973	5	11	43
South America	4,991	216	X	5,642	105,083	49	101	317
Oceania	635	13	X	690	2,899	5	9	45
OTHER	**18,136**	**395**	**180**	**22,343**	**1,812,688**	**91**	**223**	**X**
ECOSYSTEM TOTALS {f}	**129,476**	**4,745**	**9,079**	**X**	**X**	**X**	**X**	**X**

Notes: a. Area defined by the Nightime Lights of the World database within each IGBP ecosystem category (i.e., grasslands, forests, other). b. Area classified as 30-40% cropland within each IGBP ecosystem category (i.e., grasslands, forests, other). c. Boundaries for each PAGE ecosystem category are defined independently resulting in an overlap of agriculture ecosystem area with grassland and forest ecosystem area. Area estimates for grasslands and forest exclude urban areas as defined by the Nighttime Lights of the World database; area estimates for agriculture include urban areas. d. Estimates for grassland and forest ecosystems are based on boundaries of PAGE ecosystem categories; estimates for agriculture ecosystems area are based on PAGE ecosystem area definition for agriculture minus the urban areas as defined by the Nighttime Lights of the World database. See Technical Notes for population estimates of total PAGE agriculture ecosystem area. e. Estimates are based on boundaries of PAGE ecosystem categories. f. Global totals cannot be calculated for PAGE categories because the agriculture ecosystem area overlaps with the grasslands and forest ecosystem areas.

Sources: Various

	Marine Fish Catch {a}		Freshwater Fish Catch {b}		Mollusc and Crustacean Catch {c}		Total Aquaculture Production		Aquaculture Production 1995-97 (metric tons)				
	Metric Tons (000) 1995-97	Percent Change Since 1985-87	Metric Tons (000) 1995-97	Percent Change Since 1985-87	Metric Tons (000) 1995-97	Percent Change Since 1985-87	Metric Tons (000) 1995-97	Percent Change Since 1985-87	Marine Fish	Diad-romous Fish	Fresh-water Fish	Molluscs & Crus-taceans	Aquatic Plants
WORLD	**73,218.7**	**7**	**6,384.5**	**19**	**11,404.2**	**39**	**33,684.7**	**166**	**653,112**	**1,695,983**	**14,590,897**	**9,679,120**	**7,065,544**
ASIA (EXCL. MIDDLE EAST) {d}	**28,352.1**	**X**	**3,333.2**	**X**	**7,392.4**	**X**	**30,577.6**	**X**	**558,786**	**649,822**	**13,895,613**	**8,507,320**	**6,966,058**
Armenia {d}	X	X	0.5	X	X	X	3.1	X	X	1,433	1,630	X	X
Azerbaijan {d}	X	X	8.6	X	X	X	0.5	X	0	0	502	X	X
Bangladesh	255.0	27	533.1	23	24.2	199	447.6	192	X	X	338,509	109,113	X
Bhutan	X	X	0.3	6	X	X	0.0	X	X	X	30	X	X
Cambodia	22.5	118	69.6	14	8.0	396	10.3	301	X	X	9,771	532	X
China	8,765.4	203	974.2	134	3,371.9	170	22,054.4	318	248,696	144,841	10,595,404	6,608,371	4,457,078
Georgia {d}	3.7	X	0.5	X	0.5	X	0.0	X	X	2	37	X	X
India	2,423.3	68	608.0	25	341.5	29	1,748.8	149	X	X	1,666,100	82,656	X
Indonesia	2,864.9	72	313.3	21	310.3	77	843.3	106	10,763	171,472	371,310	154,117	135,667
Japan	4,524.7	(53)	59.0	(23)	1,276.9	9	1,350.8	10	253,109	65,896	15,392	475,432	540,956
Kazakhstan {d}	X	X	44.3	X	X	X	1.1	X	X	0	1,133	X	X
Korea, Dem People's Rep	219.9	(86)	20.0	(73)	32.2	184	670.0	(25)	X	X	3,995	71,011	595,032
Korea, Rep	1,635.3	(15)	7.2	(79)	654.5	47	961.9	13	19,622	5,323	17,040	307,892	611,979
Kyrgyzstan {d}	X	X	0.2	X	X	X	0.2	X	X	10	153	X	X
Lao People's Dem Rep	X	X	25.9	20	X	X	14.3	185	X	X	14,267	X	X
Malaysia	927.6	46	4.0	34	180.0	30	115.2	131	3,182	7,094	18,342	86,586	X
Mongolia	X	X	0.2	(51)	X	X	X	X	X	X	X	X	X
Myanmar	616.3	19	159.5	11	22.0	215	76.6	1,305	X	X	76,558	12	X
Nepal	X	X	11.2	106	X	X	10.8	153	X	X	10,848	X	X
Pakistan	369.6	21	138.8	85	38.1	35	18.4	116	X	7	18,333	56	X
Philippines	1,503.1	20	49.3	(39)	137.1	50	952.7	87	875	154,145	90,793	101,329	605,587
Singapore	7.9	(51)	0.0	(100)	1.8	(37)	3.8	156	314	388	90	2,968	X
Sri Lanka	207.3	46	17.7	(43)	6.8	34	6.5	48	X	X	2,516	3,961	X
Tajikistan {d}	X	X	0.2	X	X	X	0.2	X	X	0	154	X	X
Thailand	2,324.5	29	208.9	130	398.2	(9)	561.9	288	1,348	4,064	227,561	328,960	X
Turkmenistan {d}	X	X	9.0	X	X	X	0.4	X	X	X	414	X	X
Uzbekistan {d}	X	X	2.7	X	X	X	7.6	X	X	0	7,564	X	X
Viet Nam	646.5	42	66.7	(49)	317.3	171	435.1	227	X	X	333,005	92,600	9,500
EUROPE {d}	**15,595.2**	**X**	**415.5**	**X**	**1,087.2**	**X**	**1,665.1**	**X**	**51,661**	**687,591**	**192,607**	**723,641**	**9,596**
Albania	1.1	(85)	0.2	(90)	0.2	(31)	0.3	(87)	0	1	39	213	X
Austria	X	X	0.4	(32)	X	X	3.0	(27)	X	2,201	760	3	X
Belarus {d}	X	X	0.3	X	X	X	4.7	X	X	3	4,696	X	X
Belgium	29.4	(24)	0.5	6	2.4	(8)	0.9	68	X	579	300	X	X
Bosnia and Herzegovina {d}	0.0	X	2.6	X	X	X	X	X	X	X	X	X	X
Bulgaria	4.3	(95)	1.3	(26)	3.8	(39)	4.9	(58)	X	658	4,143	125	X
Croatia {d}	14.2	X	0.4	X	1.7	X	3.5	X	341	334	2,518	282	X
Czech Rep {d}	X	X	3.6	X	X	X	18.1	X	X	778	17,368	X	X
Denmark	1,723.0	4	0.2	(42)	112.5	9	41.7	68	X	41,687	X	X	X
Estonia {d}	113.4	X	2.4	X	4.3	X	0.3	X	X	241	41	X	X
Finland	115.1	18	47.7	(18)	X	X	17.1	52	X	17,100	44	X	X
France	491.5	(11)	4.5	51	75.3	16	284.7	22	3,844	53,817	10,199	216,770	75
Germany	203.0	(34)	23.0	187	17.7	(34)	64.3	(3)	0	25,034	12,667	26,555	X
Greece	116.8	15	16.7	119	32.4	144	38.9	1,497	26,722	2,538	438	9,246	X
Hungary	X	X	13.0	(31)	X	X	8.9	(50)	X	6	8,918	X	X
Iceland	1,857.9	16	0.6	16	100.6	95	3.6	963	13	3,599	X	X	X
Ireland	306.3	46	3.8	2,779	27.0	102	33.0	128	X	15,234	X	17,723	X
Italy	234.8	(25)	9.8	(36)	124.1	(10)	219.3	99	10,883	53,200	4,100	146,122	5,000
Latvia {d}	128.9	X	0.5	X	2.9	X	0.4	X	X	3	414	X	X
Lithuania {d}	36.4	X	1.4	X	2.6	X	1.6	X	X	0	1,589	X	X
Macedonia, FYR {d}	X	X	0.1	X	X	X	1.1	X	X	525	547	X	X
Moldova, Rep {d}	X	X	0.0	X	X	X	1.2	X	X	X	1,230	X	X
Netherlands	398.2	10	2.9	(34)	32.5	(41)	94.0	(7)	21	2,429	1,142	90,415	X
Norway	2,626.8	42	0.6	43	46.0	(50)	321.8	584	648	320,813	X	358	X
Poland	333.5	(40)	23.0	132	16.8	(78)	27.2	38	X	6,090	21,073	X	X
Portugal	224.0	(35)	X	X	23.6	44	5.8	(34)	1,088	1,182	0	3,570	X
Romania	18.7	(90)	6.5	(68)	0.0	0	15.0	(66)	X	408	14,558	X	X
Russian Federation {d}	4,164.4	X	223.9	X	155.0	X	60.5	X	97	3,657	50,865	1,386	4,522
Slovakia {d}	X	X	1.6	X	X	X	1.3	X	X	715	560	X	X
Slovenia {d}	2.0	X	0.3	X	0.0	X	0.9	X	71	513	241	33	X
Spain	977.9	(9)	5.8	(11)	135.7	(9)	231.6	(14)	6,345	26,357	179	198,730	X
Sweden	371.6	75	1.9	(10)	3.7	31	7.5	52	X	5,917	0	1,599	X
Switzerland	X	X	1.8	(59)	X	X	1.2	338	X	1,122	35	X	X
Ukraine {d}	349.6	X	7.5	X	31.8	X	32.7	X	155	108	32,251	179	X
United Kingdom	751.2	(4)	1.9	(2)	134.4	49	111.2	353	X	100,733	X	10,329	X
Yugoslavia {d}	0.3	X	5.0	X	0.0	X	1.6	X	X	6	1,600	1	X
MIDDLE EAST & N. AFRICA	**2,049.2**	**23**	**375.5**	**112**	**176.6**	**64**	**167.0**	**85**	**35,704**	**23,607**	**105,069**	**2,634**	**0**
Afghanistan	X	X	1.3	48	X	X	X	X	X	X	X	X	X
Algeria	98.7	42	0.0	(100)	2.8	(48)	0.3	155	37	18	256	27	X
Egypt	82.3	99	215.1	67	7.9	155	70.4	58	19,799	X	50,569	X	X
Iran, Islamic Rep	231.2	90	99.6	1,279	12.7	63	29.6	40	X	1,991	27,366	274	X
Iraq	9.2	79	20.8	87	X	X	3.2	(33)	X	X	3,167	X	X
Israel	2.6	(78)	1.4	(12)	0.4	45	16.9	29	2,294	688	13,948	X	X
Jordan	0.0	0	0.4	75	X	X	0.2	206	X	X	184	X	X
Kuwait	6.2	(4)	0.0	0	2.1	(1)	0.1	3,240	111	X	X	X	X
Lebanon	3.8	153	0.0	X	0.1	500	0.3	9	X	317	X	X	X
Libyan Arab Jamahiriya	33.7	108	0.0	0	0.0	(100)	0.1	900	X	X	100	X	X
Morocco	662.8	39	1.7	30	90.7	117	2.1	1,278	972	155	767	229	X
Oman	120.1	10	0.0	0	5.4	117	0.0	(100)	X	X	X	X	X
Saudi Arabia	39.1	(3)	0.0	0	8.5	65	3.7	6,579	0	X	3,330	411	X
Syrian Arab Rep	2.3	95	3.5	90	0.1	110	4.3	77	X	X	4,336	X	X
Tunisia	70.3	(10)	0.7	X	13.7	(14)	1.5	539	886	138	380	111	X
Turkey	463.7	(13)	28.7	19	22.6	59	33.4	1,005	10,983	20,199	668	1,569	X
United Arab Emirates	109.0	38	0.0	0	0.1	36	0.0	(100)	0	X	0	0	X
Yemen	100.5	52	2.2	X	5.1	(10)	X	X	X	X	X	X	X

	Marine Fish Catch {a}		Freshwater Fish Catch {b}		Mollusc and Crustacean Catch {c}		Total Aquaculture Production		Aquaculture Production 1995-97 (metric tons)				
	Metric Tons (000) 1995-97	Percent Change Since 1985-87	Metric Tons (000) 1995-97	Percent Change Since 1985-87	Metric Tons (000) 1995-97	Percent Change Since 1985-87	Metric Tons (000) 1995-97	Percent Change Since 1985-87	Marine Fish	Diad-romous Fish	Fresh-water Fish	Molluscs & Crus-taceans	Aquatic Plants
SUB-SAHARAN AFRICA	**2,389.9**	**9**	**1,690.9**	**21**	**136.7**	**11**	**41.4**	**285**	**399**	**1,069**	**31,040**	**5,297**	**3,609**
Angola	73.0	15	6.0	(23)	0.6	71	X	X	X	X	X	X	X
Benin	8.6	2	25.0	(6)	0.1	X	X	X	X	X	X	X	X
Botswana	X	X	2.0	20	X	X	X	X	X	X	X	X	X
Burkina Faso	X	X	8.0	5	X	X	0.0	(16)	X	X	25	X	X
Burundi	X	X	14.8	26	X	X	0.0	46	X	X	23	X	X
Cameroon	63.8	24	23.0	15	0.5	(37)	0.1	(58)	X	X	55	X	X
Central African Rep	X	X	12.7	(2)	X	X	0.4	95	X	X	370	X	X
Chad	X	X	91.7	49	X	X	X	X	X	X	X	X	X
Congo	18.7	(1)	23.9	67	0.5	591	0.1	46	X	X	115	X	X
Congo, Dem Rep	3.9	93	157.4	3	X	X	0.8	41	X	X	750	X	X
Côte d'Ivoire	56.4	(28)	11.3	(54)	0.3	(23)	0.7	291	X	X	655	X	X
Equatorial Guinea	3.1	9	0.7	76	0.6	7	X	X	X	X	X	X	X
Eritrea {d}	2.6	X	0.0	X	0.0	X	X	X	X	X	X	X	X
Ethiopia {d}	0.0	X	8.5	X	X	X	0.0	X	X	X	46	X	X
Gabon	32.8	89	8.8	365	1.4	(26)	0.1	1,875	X	X	53	X	X
Gambia	25.7	211	2.5	(7)	0.7	(64)	0.0	X	X	X	3	X	X
Ghana	352.1	29	67.9	35	5.5	66	0.5	36	X	X	500	X	X
Guinea	76.6	154	3.2	26	8.0	X	0.0	14	X	X	3	X	X
Guinea-Bissau	5.4	94	0.3	188	1.3	25	X	X	X	X	X	X	X
Kenya	4.3	(20)	172.1	53	1.4	51	0.6	168	X	71	506	0	X
Lesotho	X	X	0.0	X	X	X	0.0	(43)	X	4	10	X	X
Liberia	4.2	(62)	4.0	0	0.1	(73)	0.0	(100)	X	X	0	X	X
Madagascar	71.0	101	30.0	(21)	12.7	49	6.3	2,767	X	0	4,153	2,146	X
Malawi	X	X	57.9	(22)	X	X	0.2	20	X	X	227	2	X
Mali	X	X	114.8	102	X	X	0.1	686	X	X	73	X	X
Mauritania	52.0	14	5.8	(3)	28.5	(42)	X	X	X	X	X	X	X
Mozambique	13.5	(47)	8.0	129	12.2	17	0.0	58	X	X	14	X	X
Namibia	281.6	1,466	1.3	51	0.6	(63)	0.0	X	X	X	5	43	0
Niger	X	X	4.7	112	X	X	0.0	103	X	X	20	X	X
Nigeria	224.3	46	106.0	14	20.8	286	17.4	174	X	24	17,014	X	X
Rwanda	X	X	3.1	134	X	X	0.0	(13)	X	X	47	X	X
Senegal	351.9	66	54.4	262	25.6	30	0.1	104	X	X	49	22	X
Sierra Leone	48.2	38	14.7	(9)	1.2	2	0.0	107	X	X	28	X	X
Somalia	14.8	(9)	0.3	(26)	0.9	92	X	X	X	X	X	X	X
South Africa	497.0	(50)	0.8	4	8.9	(33)	3.9	287	3	870	101	2,616	276
Sudan	4.5	382	40.8	64	X	X	1.0	2,595	X	X	1,000	24	X
Tanzania, United Rep	42.5	5	310.7	13	3.0	139	3.6	12,002	X	X	217	X	3,333
Togo	8.8	(24)	5.0	43	0.0	873	0.0	31	X	X	21	X	X
Uganda	X	X	207.3	11	X	X	0.2	479	X	X	205	X	X
Zambia	X	X	64.6	(2)	X	X	4.6	535	X	X	4,550	1	X
Zimbabwe	X	X	16.9	(8)	X	X	0.2	20	X	100	63	15	X
NORTH AMERICA	**4,535.1**	**(11)**	**66.3**	**(41)**	**1,344.5**	**3**	**488.8**	**32**	**3,714**	**96,096**	**235,542**	**153,415**	**0**
Canada	550.4	(57)	38.9	(15)	311.8	76	73.7	565	0	55,452	X	18,291	X
United States	3,984.8	5	27.4	(59)	1,032.7	(8)	415.0	15	3,714	40,644	235,542	135,125	X
C. AMERICA & CARIBBEAN	**1,337.2**	**(8)**	**127.5**	**12**	**348.7**	**33**	**100.9**	**165**	**666**	**1,409**	**54,463**	**44,362**	**1**
Belize	0.1	(73)	0.0	(100)	1.0	6	1.1	14,470	X	X	146	971	X
Costa Rica	17.5	61	2.0	568	4.2	(48)	6.9	4,203	X	139	4,015	2,760	X
Cuba	55.7	(69)	0.3	(27)	18.1	(35)	34.2	102	X	X	32,091	2,075	X
Dominican Rep	11.1	(23)	1.5	(16)	2.9	31	1.2	357	215	13	888	129	X
El Salvador	3.6	(30)	3.3	56	5.7	(49)	0.4	(43)	25	X	175	223	X
Guatemala	0.7	4	4.4	1,805	2.5	68	3.8	602	X	X	1,685	2,150	X
Haiti	3.7	(25)	0.5	67	1.5	97	X	X	X	X	X	X	X
Honduras	7.7	71	0.1	X	8.0	(37)	9.1	683	X	X	229	8,898	X
Jamaica	7.3	(18)	0.7	220	2.6	1,296	3.1	128	X	X	3,013	80	X
Mexico	1,037.6	2	113.8	5	270.5	60	32.1	139	416	1,258	11,905	18,495	X
Nicaragua	5.4	324	0.9	766	6.9	175	2.7	7,297	X	X	4	2,733	X
Panama	145.2	(15)	0.1	(40)	14.4	(23)	5.9	83	X	X	220	5,713	X
Trinidad and Tobago	13.5	176	0.0	0	0.7	(6)	0.0	X	X	X	5	7	X
SOUTH AMERICA	**17,755.0**	**49**	**361.0**	**8**	**656.2**	**80**	**530.1**	**588**	**161**	**207,309**	**76,210**	**160,582**	**85,875**
Argentina	922.1	122	11.9	35	315.0	736	1.3	347	X	1,299	7	25	X
Bolivia	X	X	5.1	58	X	X	0.5	2,973	X	377	84	X	X
Brazil	460.8	(28)	202.2	0	64.5	(32)	58.7	404	X	1,036	49,317	8,382	X
Chile	6,483.7	32	X	X	107.5	(13)	301.5	2,783	161	196,144	X	19,472	85,721
Colombia	107.4	526	22.4	(59)	6.1	(19)	36.8	2,874	0	7,209	22,823	6,745	X
Ecuador	565.6	(34)	0.3	(63)	15.1	8	116.8	143	X	X	587	116,218	X
Guyana	36.5	9	0.7	(11)	14.0	362	0.2	311	X	X	150	43	X
Paraguay	X	X	22.4	120	X	X	0.3	3,945	X	X	297	X	X
Peru	8,676.0	85	37.3	16	60.4	18	6.8	42	X	877	43	5,767	118
Suriname	12.7	278	0.1	(1)	0.2	(71)	0.0	X	X	X	1	0	X
Uruguay	114.9	(15)	1.2	45	12.8	488	0.0	1,367	X	7	6	1	X
Venezuela	371.7	70	57.4	171	56.4	90	7.2	1,198	X	360	2,896	3,929	36
OCEANIA	**741.5**	**78**	**14.5**	**(2)**	**141.9**	**52**	**99.4**	**276**	**2,016**	**15,097**	**186**	**81,694**	**405**
Australia	126.8	21	1.7	(12)	69.6	6	24.7	163	2,016	9,783	28	12,920	X
Fiji	24.2	0	0.0	24	5.2	194	0.3	70	0	X	146	107	25
New Zealand	458.8	141	1.1	150	60.0	209	72.8	353	X	5,250	X	67,564	X
Papua New Guinea	24.8	161	11.7	(6)	1.6	34	0.0	X	X	9	10	4	X
Solomon Islands	58.1	21	0.0	0	0.1	170	0.0	433	X	X	X	11	X
DEVELOPED	**25,743.9**	**(29)**	**611.7**	**(37)**	**3,848.0**	**(3)**	**3,636.0**	**21**	**312,797**	**867,620**	**469,203**	**1,435,587**	**550,828**
DEVELOPING	**47,011.4**	**47**	**5,772.8**	**32**	**7,436.2**	**79**	**30,034.3**	**211**	**340,310**	**814,380**	**14,121,527**	**8,243,359**	**6,514,716**

Notes: Negative values are shown in parentheses. a. Includes marine fish and diadromous fish caught in marine areas. b. Includes freshwater fish and diadromous fish caught in inland waters or low-salinity marine areas. c. Includes marine and freshwater molluscs and crustaceans. d. Due to the recent independence of countries in the former USSR, Yugoslavia, Czechoslovakia, and Ethiopia percentage changes from 1985-87 to 1995-97 cannot be shown for these countries or for the regional totals for Asia (excluding the Middle East) and Europe.

Source: Food and Agriculture Organization of the United Nations

	Trade in Fish and Fishery Products						Food Supply from Fish & Fishery Products					
	Fish and Fish Products (million US$)		Molluscs and Crustaceans (million US$)		Meals and Solubles (million US$)		Total (000 metric tons in live weight)	Per Capita {a} (kg per person)	Fish Protein as a Percent of All		No. of Fishers	No. of Decked Fishery Vessels {b}
	Exports 1996-98	Imports 1996-98	Exports 1996-98	Imports 1996-98	Exports 1996-98	Imports 1996-98	1997	1997	Animal Protein 1997	Protein Supply 1997	1996	1995
WORLD	**X**	**X**	**X**	**X**	**X**	**X**	**93,818**	**16.1**	**16.5**	**6.0**	**X**	**1,256,841**
ASIA (EXCL. MIDDLE EAST) {c}	**8,239.29**	**11,665.98**	**8,480.36**	**8,686.44**	**49.21**	**1,229.40**	**61,872**	**18.8**	**X**	**X**	**X**	**1,057,966**
Armenia	0.00	0.42	0.13	0.13	X	X	3	1.0	1.6	0.5	X	6
Azerbaijan	0.90	0.55	0.00	0.05	X	X	8	1.1	1.7	0.5	X	X
Bangladesh	32.37	3.52	261.12	0.48	X	0.20	1,295	10.6	48.3	6.6	1,444,960	61
Bhutan	X	X	X	X	X	X	0	0.2	1.5	0.3	X	X
Cambodia	21.85	3.90	4.77	0.11	0.01	0.04	90	8.6	28.3	5.2	78,345	X
China	1,889.68	445.51	924.46	158.59	1.77	505.08	31,474	25.7	23.9	8.2	12,076,192	432,674
Georgia	0.15	2.40	0.11	0.05	0.00	X	11	2.1	3.6	1.0	X	82
India	215.01	6.68	910.91	0.47	0.14	7.79	4,546	4.7	15.3	2.5	5,958,744 d	56,600
Indonesia	572.49	9.84	1,064.41	10.01	4.02	66.97	3,700	18.2	53.1	9.7	4,668,482	67,325
Japan	464.11	8,209.70	293.38	6,598.09	6.17	281.51	8,374	66.4	45.8	25.3	287,370	360,747
Kazakhstan	18.45	12.74	0.01	0.22	0.92	0.11	47	2.9	2.5	0.9	16,000 d	1,970
Korea, Dem People's Rep	6.47	1.00	47.14	0.98	0.00	0.45	405	17.6	55.7	7.7	X	2,900
Korea, Rep	893.43	553.69	468.38	286.59	13.86	27.16	2,320	50.7	43.3	17.7	343,869	76,801
Kyrgyzstan	X	2.04	X	0.01	X	X	3	0.6	0.5	0.2	154 d	X
Lao People's Dem Rep	0.01	1.16	0.07	0.05	X	0.02	43	8.5	29.7	4.8	X	X
Malaysia	110.43	226.00	210.41	68.87	3.59	10.01	1,178	56.2	34.5	19.8	100,666	17,965
Mongolia	0.10	0.11	0.00	0.00	X	X	0	0.1	0.1	0.0	X	X
Myanmar	10.62	0.54	95.18	0.04	0.28	0.01	769	17.5	45.4	6.0	X	140
Nepal	0.05	0.20	0.00	0.01	X	0.02	23	1.1	3.6	0.5	263,175	X
Pakistan	49.57	0.06	101.03	0.00	0.00	0.02	299	2.1	3.2	1.0	401,407	5,064
Philippines	219.07	69.41	219.79	3.93	0.07	49.10	2,161	30.3	42.8	19.9	990,872	3,220
Singapore	313.00	311.61	162.38	247.90	1.26	1.79	116	34.0	17.0	9.2	611	110
Sri Lanka	21.27	65.16	59.35	1.62	0.00	4.30	369	20.2	54.3	13.4	83,776	2,990
Tajikistan	0.05	0.12	0.01	0.06	X	X	1	0.1	0.2	0.0	X	X
Thailand	1,255.39	578.87	2,896.91	189.60	6.19	66.60	2,013	33.7	41.5	18.5	438,934 d	17,600
Turkmenistan	0.13	0.22	0.00	0.02	0.00	X	9	2.2	2.9	1.0	716	45
Uzbekistan	0.24	1.46	X	0.05	0.00	0.05	12	0.5	0.6	0.2	5,800	X
Viet Nam	93.08	3.68	493.01	1.20	0.76	2.36	1,329	17.4	39.4	8.8	3,030,000 e	140
EUROPE {c}	**13,738.82**	**14,573.29**	**3,543.49**	**6,290.59**	**714.47**	**952.64**	**13,655**	**18.5**	**10.3**	**5.7**	**X**	**105,324**
Albania	3.08	2.26	2.08	0.20	X	0.00	4	1.3	0.8	0.3	1,402	2
Austria	7.61	157.68	0.42	27.49	0.92	9.85	90	11.2	4.1	2.7	2,300	X
Belarus	4.25	50.39	3.66	3.95	X	3.68	11	1.0	0.6	0.3	X	X
Belgium	205.86	523.88	222.73	435.42	5.17	33.33	X	X	X	X	600	156
Bosnia and Herzegovina	X	2.67	X	0.25	X	0.26	5	1.3	2.6	0.6	X	X
Bulgaria	7.79	7.37	5.32	5.59	1.66	3.05	28	3.4	2.6	1.3	1,483 e	30
Croatia	45.57	19.74	5.28	9.69	0.01	10.61	19	4.3	4.4	1.9	50,227	305
Czech Rep	29.99	78.97	0.25	2.58	0.16	9.95	91	8.8	6.0	3.4	2,065	X
Denmark	1,956.15	1,153.44	516.69	374.03	220.95	66.29	124	23.5	10.1	6.2	4,792	4,285
Estonia	91.34	34.55	2.18	1.39	1.21	1.16	27	18.8	12.6	6.9	10,468	186
Finland	19.09	90.01	0.91	18.38	0.10	20.61	170	33.1	14.7	9.9	6,373	3,838
France	778.66	2,085.62	267.49	1,093.53	9.87	54.46	1,606	27.5	8.6	5.7	26,522	6,586
Germany	796.02	1,959.05	83.09	368.93	144.51	161.21	1,038	12.7	6.9	4.1	3,894	2,406
Greece	173.17	168.37	36.19	92.92	0.69	28.31	280	26.5	12.1	6.5	22,192	18,375
Hungary	6.87	22.52	0.33	0.97	1.06	25.84	43	4.2	2.6	1.4	4,200	X
Iceland	943.08	32.90	242.15	13.31	158.36	0.17	25	91.1 f	25.6	16.8	5,635	826
Ireland	287.01	72.91	94.57	22.09	12.86	16.27	56	15.3	6.3	3.9	7,500	1,353
Italy	241.18	1,653.86	124.70	932.40	8.05	56.15	1,272	22.2	10.8	5.8	40,236	16,000
Latvia	90.08	35.75	0.05	1.01	0.93	0.68	27	11.1	9.2	4.5	4,751 d	351
Lithuania	55.90	59.45	2.85	3.73	1.46	4.19	56	15.0	10.8	4.6	X	131
Macedonia, FYR	0.04	6.54	0.00	0.08	0.00	1.25	9	4.3	4.5	1.6	8,446	X
Moldova, Rep	0.52	8.90	0.00	0.00	X	0.00	4	0.9	1.1	0.4	X	X
Netherlands	939.43	730.94	441.08	331.20	21.38	55.18	236	15.1	7.4	5.0	3,810	1,008
Norway	3,218.80	334.99	159.19	75.19	70.17	82.95	223	50.7	24.8	15.3	28,129	8,664
Poland	219.25	259.31	21.59	13.88	17.23	1.43	445	11.5	11.0	5.5	9,178	445
Portugal	207.11	645.25	54.94	166.30	0.43	7.69	577	58.5	22.8	13.1	28,458	9,265
Romania	0.62	25.54	1.68	2.23	0.00	13.40	38	1.7	1.5	0.7	19,249	33
Russian Federation	1,068.02	340.49	332.55	16.08	1.80	5.55	3,240	21.9	15.3	7.7	X	3,584
Slovakia	2.52	37.25	0.02	0.37	0.02	6.50	28	5.2	3.3	1.6	X	X
Slovenia	5.36	18.70	0.35	7.94	0.00	2.49	13	6.5	3.3	1.9	175	11
Spain	993.62	1,729.12	469.96	1,459.71	11.69	50.45	1,618	40.9	18.5	11.3	83,731	15,243
Sweden	341.70	445.02	21.00	153.70	3.92	2.58	231	26.1	12.2	8.0	3,287 e	1,240
Switzerland	2.88	282.12	0.37	89.81	0.36	7.16	101	13.9	6.5	4.1	432	X
Ukraine	67.45	40.77	7.86	5.99	0.05	0.00	435	8.5	8.6	3.5	X	444
United Kingdom	922.40	1,421.98	421.70	555.85	19.38	173.50	1,240	21.1	10.0	5.7	19,044	9,562
Yugoslavia	0.73	21.04	0.26	1.05	0.06	34.81	31	2.9	1.5	0.8	1,429 d	5
MIDDLE EAST & N. AFRICA {c}	**552.54**	**537.78**	**628.28**	**46.47**	**5.43**	**115.76**	**2,473**	**6.5**	**X**	**X**	**X**	**21,990**
Afghanistan	X	X	X	X	X	X	1	0.1	0.2	0.0	X	X
Algeria	0.64	7.80	2.40	0.32	X	0.00	103	3.5	6.7	1.5	24,190 d	2,184
Egypt	1.15	109.58	1.70	2.17	X	7.88	648	10.0	18.4	3.3	237,496	X
Iran, Islamic Rep	44.96	7.85	21.19	0.11	X	44.49	310	4.8	7.7	1.9	108,398	900
Iraq	X	0.81	X	0.00	0.00	0.18	35	1.7	8.3	0.9	X	8
Israel	9.73	119.29	0.06	2.87	0.00	9.67	135	23.1	9.3	4.7	1,408 d	384
Jordan	1.17	28.39	0.03	0.62	0.00	0.60	13	2.9	4.0	1.0	504	X
Kuwait	0.23	9.18	7.18	7.95	X	0.07	20	11.3	5.0	2.6	2,994	917
Lebanon	X	25.49	X	1.73	0.00	2.64	20	6.5	7.4	2.1	9,825	5
Libyan Arab Jamahiriya	31.47	10.07	0.30	0.39	X	1.85	31	6.0	6.8	2.0	9,500 d	93
Morocco	275.50	5.88	438.86	1.86	5.39	3.23	210	7.8	15.1	2.8	109,062	3,052
Oman	62.27	3.31	5.43	0.72	X	0.23	55	24.1	21.5	9.8	25,575	390
Saudi Arabia	4.25	85.20	2.29	6.94	0.00	3.67	129	6.6	5.4	2.2	22,213	23
Syrian Arab Rep	0.11	29.40	0.00	0.34	X	4.77	15	1.0	1.4	0.3	X	5
Tunisia	19.95	12.45	77.29	0.42	0.00	0.07	84	9.1	13.7	3.1	60,741	17
Turkey	71.73	40.16	35.12	3.20	0.02	29.80	454	7.2	8.8	2.3	33,614 e	9,710
United Arab Emirates	16.92	13.92	16.18	5.65	0.01	1.10	66	28.8	13.7	7.2	13,411	4,050
Yemen	7.44	4.61	11.60	0.02	0.00	X	113	7.0	23.2	4.1	12,134 d	71

	Trade in Fish and Fishery Products						Food Supply from Fish & Fishery Products					
	Fish and Fish Products (million US$)		Molluscs and Crustaceans (million US$)		Meals and Solubles (million US$)		Total (000 metric tons in live weight)	Per Capita {a} (kg per person)	Fish Protein as a Percent of All		No. of Fishers	No. of Decked Fishery Vessels {b}
	Exports 1995-97	Imports 1995-97	Exports 1995-97	Imports 1995-97	Exports 1995-97	Imports 1995-97	1997	1997	Animal Protein 1997	Protein Supply 1997	1996	1995
SUB-SAHARAN AFRICA {c}	**1,255.82**	**799.93**	**503.30**	**48.76**	**33.11**	**46.79**	**4,122**	**6.9**	**X**	**X**	**X**	**3,637**
Angola	4.15	12.08	3.35	0.30	0.47	X	78	6.6	27.1	6.3	30,364 e	580
Benin	0.00	5.34	1.83	0.00	X	0.00	53	9.4	28.5	4.6	53,025	5
Botswana	0.12	5.03	0.02	0.33	0.02	0.12	9	5.7	6.4	2.3	X	X
Burkina Faso	0.00	3.06	0.02	0.01	X	0.00	14	1.2	5.0	0.6	8,000	X
Burundi	0.22	0.28	0.00	0.00	X	X	20	3.2	29.6	1.9	14,077 d	X
Cameroon	0.15	22.49	1.88	0.08	0.00	0.28	128	9.2	25.0	5.8	24,136 e	25
Central African Rep	X	0.36	X	0.01	X	0.00	13	3.9	8.4	2.6	5,400 e	X
Chad	X	0.62	0.00	0.00	0.00	X	48	6.8	19.3	2.9	350,000	X
Congo	0.92	27.55	6.99	0.03	X	0.01	68	25.3	48.8	18.5	2,129	26
Congo, Dem Rep	0.90	51.32	X	0.09	X	0.00	272	5.7	31.0	5.6	100,000	23
Côte d'Ivoire	218.31	182.92	6.96	2.46	0.13	0.05	157	11.1	36.9	6.7	19,707	63
Equatorial Guinea	1.95	1.71	0.09	0.01	X	X	10	22.6	61.9	33.2	6,731	5
Eritrea	0.15	0.01	0.01	0.01	X	X	1	0.3	1.2	0.1	14,500	X
Ethiopia	0.06	0.16	X	0.00	X	X	10	0.2	0.8	0.1	4,000	X
Gabon	0.83	9.70	6.63	0.10	X	0.01	51	44.6	35.0	17.5	X	39
Gambia	1.97	1.05	2.62	0.03	X	0.00	28	23.7	61.7	15.0	1,877	X
Ghana	81.40	16.97	2.22	0.15	X	0.63	420	22.5	63.2	14.9	230,749	500
Guinea	18.24	9.79	4.74	0.04	0.00	0.02	117	16.0	60.2	9.4	10,707	15
Guinea-Bissau	6.27	0.57	2.89	0.02	X	0.00	3	2.7	8.7	1.6	X	8
Kenya	42.07	8.88	5.17	0.02	0.01	0.46	135	4.8	9.7	2.9	43,488 d	32
Lesotho	X	X	X	0.00	X	X	X	X	X	X	X	X
Liberia	0.01	1.94	0.00	0.07	X	X	12	4.9	23.0	4.2	4,239	14
Madagascar	24.81	11.43	76.84	0.03	0.39	0.03	111	7.6	16.9	4.8	142,666	65
Malawi	0.30	0.26	0.08	0.02	0.00	0.36	57	5.7	37.7	3.1	42,271	57
Mali	0.42	1.00	X	0.00	X	X	99	9.5	16.1	4.2	70,000	X
Mauritania	23.26	1.16	104.59	0.02	X	0.00	35	14.2	13.0	5.1	6,700 d	126
Mozambique	1.67	7.11	73.54	0.01	X	0.16	36	2.0	19.2	1.8	X	291
Namibia	220.45	X	X	X	24.71	X	20	12.4	16.8	5.4	X	218
Niger	0.13	1.03	0.08	0.04	X	0.01	8	0.8	2.9	0.4	X	X
Nigeria	3.36	211.77	19.89	0.24	X	1.02	602	5.8	21.6	2.8	437,702	318
Rwanda	X	0.07	X	0.00	X	X	3	0.6	4.2	0.4	4,500	X
Senegal	221.85	15.94	73.36	0.20	3.46	0.02	318	36.3	47.4	16.1	60,417	180
Sierra Leone	16.28	3.18	9.90	0.01	X	X	59	13.4	63.0	11.6	16,722 d	27
Somalia	4.08	0.02	3.55	X	X	X	12	1.3	1.3	0.9	X	12
South Africa	150.75	48.92	68.00	25.31	2.34	39.40	304	7.8	9.2	3.1	X	600
Sudan	0.10	0.08	0.00	0.01	X	0.08	48	1.7	1.9	0.7	22,950	X
Tanzania, United Rep	53.63	1.95	8.94	0.00	0.17	0.15	323	10.3	33.6	7.0	62,593	30
Togo	1.95	24.46	4.98	0.22	X	0.00	74	17.3	50.2	8.3	17,060	3
Uganda	38.49	0.04	X	0.00	0.00	X	195	9.8	30.0	6.5	X	X
Zambia	0.33	0.68	0.06	0.00	0.00	0.21	71	8.2	22.8	4.6	23,833	235
Zimbabwe	1.84	16.56	0.01	1.20	0.00	0.88	37	3.3	10.0	1.9	X	X
NORTH AMERICA {c}	**3,139.97**	**4,047.00**	**1,814.10**	**4,927.18**	**66.15**	**77.48**	**6,313**	**20.9**	**X**	**X**	**X**	**45,480**
Canada	1,200.30	583.14	1,063.81	516.72	6.63	40.91	657	21.7	9.9	5.6	84,775 d	18,280
United States	1,939.68	3,463.86	750.29	4,410.46	59.52	36.57	5,657	20.8	6.8	4.3	X	27,200
C. AMERICA & CARIBBEAN {c}	**397.82**	**286.10**	**1,185.90**	**66.28**	**13.74**	**15.23**	**1,472**	**8.9**	**X**	**X**	**X**	**7,161**
Belize	0.26	0.23	15.23	0.52	X	0.00	2	7.2	6.8	2.9	X	12
Costa Rica	175.74	33.44	69.80	3.49	0.11	0.32	24	6.4	5.9	2.3	X	1,003
Cuba	1.88	20.45	105.69	0.43	1.09	0.00	140	12.6	16.4	6.1	X	1,250
Dominican Rep	0.16	44.61	1.02	0.48	X	0.01	70	8.6	10.6	5.1	X	X
El Salvador	2.94	5.08	34.35	0.34	0.00	0.11	14	2.3	3.9	1.0	X	80
Guatemala	1.03	3.43	22.35	0.76	0.00	1.53	13	1.2	2.6	0.6	25,975	85
Haiti	5.97	6.35	3.83	0.01	X	0.00	20	2.6	12.1	1.8	X	1
Honduras	3.66	2.47	58.19	5.31	0.03	1.63	21	3.5	5.5	1.8	X	280
Jamaica	4.54	34.59	13.44	1.34	X	0.03	29	11.7	12.6	6.1	8,204	5
Mexico	124.65	39.85	626.44	27.99	8.02	10.99	1,025	10.9	9.5	3.9	257,206	3,100
Nicaragua	14.92	2.44	54.88	0.32	0.03	0.06	6	1.3	3.2	0.8	13,608	280
Panama	44.76	12.08	119.77	1.74	4.45	0.49	36	13.3	11.8	6.3	9,374	695
Trinidad and Tobago	9.49	6.17	1.99	0.21	X	0.00	16	12.4	16.0	6.3	13,646	19
SOUTH AMERICA {c}	**2,201.34**	**688.25**	**1,782.32**	**48.23**	**1,290.72**	**37.09**	**3,307**	**10.0**	**7.3**	**3.6**	**X**	**13,106**
Argentina	521.24	63.32	397.30	18.36	6.49	0.06	349	9.8	4.7	2.9	12,320	800
Bolivia	0.09	3.09	X	0.27	X	0.02	13	1.7	1.9	0.8	7,754 d	X
Brazil	49.81	463.08	76.68	9.11	1.23	0.46	1,120	6.8	4.7	2.4	X	1,450
Chile	967.87	27.61	200.06	5.00	504.59	0.35	296	20.2	12.4	6.1	80,730	563
Colombia	113.64	80.42	96.58	7.63	0.03	22.45	204	5.1	5.8	2.6	200,410	167
Ecuador	282.75	8.67	789.49	1.62	29.45	5.91	91	7.6	9.0	4.1	162,870	515
Guyana	5.38	1.01	21.81	0.03	X	X	54	64.2	51.4	24.5	4,563 d	55
Paraguay	0.09	3.13	X	0.43	0.00	0.00	31	6.0	4.3	2.3	5,500 d	X
Peru	115.80	6.82	109.46	1.21	747.56	0.00	652	26.8	26.1	10.1	63,973	7,710
Suriname	9.86	3.30	3.07	0.41	X	0.01	9	22.0	26.7	10.9	X	22
Uruguay	97.83	9.21	14.87	1.67	0.49	0.70	26	8.0	5.2	3.3	3,404 d	958
Venezuela	34.58	17.06	61.63	2.24	0.89	7.10	458	20.1	20.4	10.1	39,621	866
OCEANIA {c}	**751.92**	**350.07**	**943.58**	**212.88**	**24.28**	**23.77**	**579**	**19.9**	**9.1**	**5.5**	**X**	**1,917**
Australia	151.27	271.66	667.48	190.01	4.52	13.06	338	18.4	6.5	4.1	15,800 e	246
Fiji	31.68	20.07	3.95	0.80	0.06	0.34	26	32.9	21.4	9.1	59,500 d	X
New Zealand	501.57	20.09	247.49	14.80	19.59	9.95	91	24.1	9.8	6.3	2,499	1,375
Papua New Guinea	3.76	17.48	12.88	0.29	0.05	0.32	63	13.9	28.0	9.0	16,000	35
Solomon Islands	45.79	0.08	1.04	0.03	0.06	0.00	14	34.5	73.4	24.3	51,250	130
DEVELOPED {c}	**18,176.14**	**27,309.92**	**6,634.26**	**18,049.45**	**814.17**	**1,383.87**	**29,304**	**22.5**	**12.3**	**6.9**	**X**	**516,259**
DEVELOPING {c}	**12,101.39**	**5,638.48**	**12,247.08**	**2,277.39**	**1,382.95**	**1,114.30**	**64,490**	**14.2**	**20.0**	**5.7**	**X**	**740,322**

Notes: a. Per capita values are expressed on a live-weight equivalent basis, which means that all parts of the fish, including bones, are taken into account when estimating consumption of fish and fishery products. b. Includes fishing vessels such as trawlers, long liners, etc., and nonfishing vessels such as motherships, fish carriers, etc. c. Regional totals for food and protein supply from fish and fishery products were calculated by the Food and Agriculture Organization of the United Nations. d. Data are for 1994. e. Data are for 1995. f. Per capita fish consumption in Iceland includes quantities of fish and fish products destined for the export market.

Source: Food and Agriculture Organization of the United Nations

	Coastal Length {a} (km)	Area of Continental Shelf (up to 200 m depth) (000 km²)	Territorial Sea (up to 12 nm) (000 km²)	Claimed Exclusive Economic Zone (000 km²)	Exclusive Fishing Zone (000 km²)	Population Within 100 km from the Coast (percent)	Coastal Biodiversity (1990s): Mangroves {b}: Area (km²)	Mangroves {b}: Area Protected (km²)	Mangroves {b}: Number of Species	Number of Seagrass Species	Number of *Scleractinia* Coral Genera {c}	International Legal Net Trade in Live Coral 1997 {d} (no. of pieces)
WORLD	**1,634,701**	**24,287.1**	**18,816.9 e**	**102,108.4**	**12,885.2**	**39.0**	**181,077**	**22,617**	**70**	**58 f**	**X**	**1,045,123**
ASIA (EXCL. MIDDLE EAST)	**288,459**	**5,515.4**	**5,730.9**	**11,844.2**	**249.5**	**X**	**75,173**	**10,705**	**51**	**27**	**79**	**(773,430)**
Armenia	0	0.0	X	X	X	0.0	X	X	X	X	X	X
Azerbaijan {g}	871	78.0	X	X	X	55.7	X	X	X	X	X	X
Bangladesh	3,306	59.6	40.3	39.9	X	54.8	5,767	367	21	X	X	X
Bhutan	0	0.0	X	X	X	0.0	X	X	X	X	X	X
Cambodia	1,127	34.6	19.9	X	X	23.8	851	310	5	1	X	X
China	30,017	810.4 h	348.1	X	X	24.0	366	0	23 i	5	36	X
Georgia	376	2.7	6.1	18.9	X	38.8	X	X	X	X	X	X
India	17,181	372.4	193.8	2,103.4	X	26.3	6,700	1,506	28	12	59	X
Indonesia	95,181	1,847.7	3,205.7	2,915.0	X	95.9	42,550	7,834	45	12	77	(787,045)
Japan	29,020	304.2	373.8	3,648.4	X	96.3	4	0	11	8	75	38,636
Kazakhstan {g}	4,528	139.1	X	X	X	3.6	X	X	X	X	X	X
Korea, Dem People's Rep	4,009	26.3	12.7	72.8	X	92.9	X	X	X	X	X	X
Korea, Rep	12,478	226.3	81.1	202.6	X	100.0	X	X	X	X	X	13,609
Kyrgyzstan	0	0.0	X	X	X	0.0	X	X	X	X	X	X
Lao People's Dem Rep	0	0.0	X	X	X	5.6	X	X	X	X	X	X
Malaysia	9,323	335.9	152.4	198.2	X	98.0	6,424	109	36	9	72	(130)
Mongolia	0	0.0	X	X	X	0.0	X	X	X	X	X	X
Myanmar	14,708	216.4	154.8	358.5	X	49.0	3,786	0	24	3	67	X
Nepal	0	0.0	X	X	X	0.0	X	X	X	X	X	X
Pakistan	2,599	43.7	31.4	201.5	X	9.1	1,683	290	4	X	X	X
Philippines	33,900	244.5	679.8	293.8	X	100.0	1,607	0	30	19	74	(3,785)
Singapore	268	0.7	0.7	X	0.7	100.0	6	X	31 j	11	66	X
Sri Lanka	2,825	19.2	30.5	500.8	X	100.0	89	8	23	7	45	X
Tajikistan	0	0.0	X	X	X	0.0	X	X	X	X	X	X
Thailand	7,066	185.4	75.9	176.5	X	38.7	2,641	256	35	14	68	(41,448)
Turkmenistan {g}	1,289	72.4	X	X	X	8.1	X	X	X	X	X	X
Uzbekistan	1,707	26.1	X	X	X	2.6	X	X	X	X	X	X
Viet Nam	11,409	352.4	158.6	237.8	X	82.8	2,525	16	29	9	1	(37)
EUROPE	**325,892**	**6,316.0**	**2,589.4**	**11,447.1**	**1,783.0**	**X**	**0**	**0**	**0**	**9**	**X**	**162,425**
Albania	649	6.1 h	6.2	X	6.2	97.1	X	X	X	X	X	X
Austria	0	0.0	X	X	X	2.2	X	X	X	X	X	1,081
Belarus	0	0.0	X	X	X	0.0	X	X	X	X	X	X
Belgium	76	3.6	1.5	X	2.1	83.0	X	X	X	X	X	1,122
Bosnia and Herzegovina	23	0.0	X	X	X	46.6	X	X	X	X	X	X
Bulgaria	457	10.9	6.5	25.7	X	29.2	X	X	X	1	X	X
Croatia	5,663	44.9 h	31.7	X	X	37.9	X	X	X	X	X	X
Czech Rep	0	0.0	X	X	X	0.0	X	X	X	X	X	520
Denmark {k}	5,316	102.4	24.8	80.4	X	100.0	X	X	X	X	X	2,101
Estonia	2,956	36.2	24.3	11.6	X	85.9	X	X	X	X	X	X
Finland	31,119	82.5 h	55.1	X	55.1	72.8	X	X	X	1	X	490
France	7,330	160.7	73.4	706.4	73.4	39.6	X	X	X	4	X	60,779
Germany	3,624	55.5	18.4	37.4	X	14.6	X	X	X	X	X	50,198
Greece	15,147	94.3 h	114.9	X	114.9	99.2	X	X	X	4	X	X
Hungary	0	0.0	X	X	X	0.0	X	X	X	X	X	132
Iceland	8,506	108.7	73.0	678.7	X	99.9	X	X	X	X	X	X
Ireland	6,437	151.9	39.4	X	358.9	99.9	X	X	X	X	X	X
Italy	9,226	110.8 h	155.6	X	155.6	79.1	X	X	X	3	X	13,475
Latvia	565	28.0	12.6	15.6	X	75.2	X	X	X	X	X	X
Lithuania	258	5.7	2.0	3.6	X	22.9	X	X	X	X	X	X
Macedonia, FYR	0	0.0	X	X	X	14.3	X	X	X	X	X	X
Moldova, Rep	0	0.0	X	X	X	9.1	X	X	X	X	X	X
Netherlands	1,914	64.0	13.2	X	50.3	93.4	X	X	X	1	X	16,294
Norway	53,199	218.5	111.2	1,095.1	X	95.4	X	X	X	2	X	470
Poland	1,032	30.0	10.6	19.4	X	13.5	X	X	X	1	X	X
Portugal	2,830	20.1	64.1	1,656.4	X	92.7	X	X	X	3	X	X
Romania	696	18.6	5.3	18.0	X	6.3	X	X	X	X	X	X
Russian Federation {g}	110,310	4,137.0	1,318.1	6,255.8	X	14.9	X	X	X	X	X	(18)
Slovakia	0	0.0	X	X	X	0.0	X	X	X	X	X	X
Slovenia	41	0.2	0.2	X	X	60.6	X	X	X	X	X	X
Spain	7,268	62.1	115.8	683.2	205.2	67.9	X	X	X	3	1	2,383
Sweden	26,384	153.8	85.3	73.2	X	87.7	X	X	X	2	X	1,577
Switzerland	0	0.0	X	X	X	0.0	X	X	X	X	X	3,016
Ukraine	4,953	78.0	53.9	86.4	X	20.9	X	X	X	7	X	X
United Kingdom	19,717	522.6	168.1	X	753.8	98.6	X	X	X	2	X	8,805
Yugoslavia	X	3.1 h	X	X	X	8.1	X	X	X	X	X	X
MIDDLE EAST & N. AFRICA	**47,282**	**786.5**	**649.7 e**	**2,016.0**	**196.0**	**X**	**1,492**	**0**	**3**	**13**	**63**	**2**
Afghanistan	0	0.0	X	X	X	0.0	X	X	X	X	X	X
Algeria	1,557	9.7	27.9	X	60.5	68.8	X	X	X	3	X	X
Egypt	5,898	50.1	57.0	185.3	X	53.1	861	0	2	9	57	X
Iran, Islamic Rep {g}	5,890	160.2	76.4	129.7	X	23.9	207	0	2	X	27	X
Iraq	105	1.0 h	0.7	X	X	5.7	X	X	X	X	X	X
Israel	205	3.2 h	4.1	X	X	96.6	X	X	X	4	51	2
Jordan	27	0.1	0.1	X	0.1	29.0	X	X	X	X	44	X
Kuwait	756	6.5 h	5.4	X	X	100.0	X	X	X	2	23	X
Lebanon	294	1.2	4.7	X	X	100.0	X	X	X	X	X	X
Libyan Arab Jamahiriya	2,025	63.6 h,l	38.1 e	222.4	20.9	78.7	X	X	X	1	X	X
Morocco	2,008	70.4	37.5	328.4	X	65.1	X	X	X	1	X	X
Oman	2,809	46.7	51.8	487.4	X	88.5	20	X	1	X	40	X
Saudi Arabia	7,572	95.6 h	82.0	X	X	30.2	292	X	3	5	54	X
Syrian Arab Rep	212	0.9 l	3.9 e	X	X	34.5	X	X	X	1	X	X
Tunisia	1,927	65.3 h	36.8	X	X	84.0	X	X	X	3	X	X
Turkey	8,140	53.3	81.0	176.6	81.0	57.5	X	X	X	3	X	X
United Arab Emirates	2,871	51.4	31.0	21.2	X	84.9	30	X	1	1	28	X
Yemen	3,149	65.3	82.4	465.0	X	63.5	81	X	2	8	51	X

	Coastal Length {a} (km)	Area of Continental Shelf (up to 200 m depth) (000 km²)	Territorial Sea (up to 12 nm) (000 km²)	Claimed Exclusive Economic Zone (000 km²)	Exclusive Fishing Zone (000 km²)	Population Within 100 km from the Coast (percent)	Coastal Biodiversity (1990s): Mangroves {b}: Area (km²)	Mangroves {b}: Area Protected (km²)	Mangroves {b}: Number of Species	Number of Seagrass Species	Number of *Scleractinia* Coral Genera {c}	International Legal Net Trade in Live Coral 1997 {d} (no. of pieces)
SUB-SAHARAN AFRICA	**63,124**	**987.0**	**871.9 e**	**7,866.1**	**3,111.1**	**X**	**36,512**	**548**	**17**	**15**	**68**	**(202)**
Angola	2,252	44.2 l	34.7 e	X	438.0	29.4	1,250	0	7	X	2	X
Benin	153	2.8 l	2.5 e	X	26.8	62.4	17	0	6	X	X	X
Botswana	0	0.0	X	X	X	0.0	X	X	X	X	X	X
Burkina Faso	0	0.0	X	X	X	0.0	X	X	X	X	X	X
Burundi	0	0.0	X	X	X	0.0	X	X	X	X	X	X
Cameroon	1,799	13.1 l	8.5 e	10.9	X	21.9	2,494	44	8	X	1	X
Central African Rep	0	0.0	X	X	X	0.0	X	X	X	X	X	X
Chad	0	0.0	X	X	X	0.0	X	X	X	X	X	X
Congo	205	7.4 l	3.5 e	X	41.4	24.5	120	143	2	X	X	X
Congo, Dem Rep	177	0.8	1.0	X	121.0	2.7	226	0	6	X	X	X
Côte d'Ivoire	797	8.6	12.3	157.4	X	39.7	644	0	4	X	1	X
Equatorial Guinea	603	8.6	12.9	291.4	X	72.3	277	0	2	X	2	X
Eritrea	3,446	47.5 h	39.2	X	X	73.5	581	0	3	X	56	X
Ethiopia	0	0.0	X	X	X	1.4	X	X	X	X	X	X
Gabon	2,019	36.8	19.6	180.7	X	62.8	2,500	44	7	X	2	X
Gambia	503	5.7	2.3	20.5	X	90.8	497	24	7	X	X	X
Ghana	758	18.1	11.9	216.9	X	42.5	100	0	6	X	1	X
Guinea	1,614	49.7	14.2	97.0	X	40.9	2,963	0	7	X	1	X
Guinea-Bissau	3,176	37.2	19.5	86.7	X	94.6	2,484	0	6	X	X	X
Kenya	1,586	8.5	12.4	104.1	X	7.6	530	0	8	9	54	X
Lesotho	0	0.0	X	X	X	0.0	X	X	X	X	X	X
Liberia	842	14.9 l	12.7 e	X	239.1	57.9	190	0	5	X	1	X
Madagascar	9,935	96.7	124.9	1,079.7	X	55.1	3,403	6	9	8	58	(155)
Malawi	0	0.0	X	X	X	0.0	X	X	X	X	X	X
Mali	0	0.0	X	X	X	0.0	X	X	X	X	X	X
Mauritania	1,268	28.4	19.5	141.3	X	39.6	1	X	3	1	X	X
Mozambique	6,942	73.3	70.9	493.7	X	59.0	925	211	10	8	49	X
Namibia	1,754	95.0	32.7	536.8	X	4.7	X	X	X	X	X	(30)
Niger	0	0.0	X	X	X	0.0	X	X	X	X	X	X
Nigeria	3,122	41.8	19.3 e	164.1	X	25.7	10,515	2	8	X	X	X
Rwanda	0	0.0	X	X	X	0.0	X	X	X	X	X	X
Senegal	1,327	21.0	11.5	147.2	X	83.2	1,853	45	7	X	2	X
Sierra Leone	1,677	23.2 l	11.2 e	X	155.9	54.7	1,838	14	6	1	1	X
Somalia	3,898	40.4	68.8	X	759.3	54.8	910	0	6	4	50	X
South Africa	3,751	160.9	74.7	X	1,450.6	38.9	11	0	6	3	8	(1)
Sudan	2,245	15.9	32.6	X	X	2.8	937	0	3	2	56	X
Tanzania, United Rep	3,461	17.9	36.6	204.3	X	21.1	1,155	14	10	7	57	X
Togo	53	0.6	1.0 e	10.8	X	44.6	26	0	2	X	X	(16)
Uganda	0	0.0	X	X	X	0.0	X	X	X	X	X	X
Zambia	0	0.0	X	X	X	0.0	X	X	X	X	X	X
Zimbabwe	0	0.0	X	X	X	0.0	X	X	X	X	X	X
NORTH AMERICA	**398,835**	**5,107.5**	**3,484.1**	**11,084.4**	**X**	**X**	**1,990**	**1,195**	**X**	**10**	**37**	**819,118**
Canada	265,523	2,877.6	2,687.7	3,006.2	X	23.9	X	X	X	2	X	11,430
United States	133,312	2,229.9	796.4	8,078.2	X	43.3	1,990	1,195	6	10	37	807,688
C. AMERICA & CARIBBEAN	**73,703**	**806.6**	**1,050.0 e**	**6,489.0**	**197.2**	**X**	**22,759**	**2,149**	**13**	**9**	**30**	**616**
Belize	1,996	8.7	18.5	12.8	X	100.0	719	29	5	X	24	X
Costa Rica	2,069	14.8	24.2	542.1	X	100.0	370	10	9	1	28	X
Cuba	14,519	51.0	122.8	222.2	X	100.0	7,848	538	5	4	25	X
Dominican Rep	1,612	5.9	14.0	246.5	X	100.0	325	362	6	4	25	49
El Salvador	756	17.7 l	6.6 e	X	87.5	98.8	268	0	6	X	X	X
Guatemala	445	13.0	7.7	104.5	X	61.2	161	27	5	X	X	(1)
Haiti	1,977	5.9	40.1	86.4	X	99.6	134	0	6	X	25	X
Honduras	1,878	58.8	36.5	201.2	X	65.5	1,458	974	5	1	25	X
Jamaica	895	5.6	16.0	234.8	X	100.0	106	9	5	3	25	X
Mexico	23,761	393.3	291.6	2,997.7	X	28.7	5,315	0	5	6	25	558
Nicaragua	1,915	68.6 l	31.6 e	X	94.9	71.6	1,718	140	9	1	25	X
Panama	5,637	44.2 l	57.8 e	274.6	X	100.0	1,814	38	12	3	24	X
Trinidad and Tobago	704	22.6	13.0	60.7	X	100.0	>70	3	7	2	25	X
SOUTH AMERICA	**144,567**	**2,203.0**	**1,030.0 e**	**9,358.8**	**1,814.1**	**X**	**24,084**	**9,397**	**12**	**2**	**32**	**(389)**
Argentina	8,397	798.5	142.5	925.4	X	45.1	X	X	X	X	X	116
Bolivia	0	0.0	X	X	X	0.0	X	X	X	X	X	X
Brazil	33,379	711.5	218.1	3,442.5	X	48.6	13,400	3,811	7	1	19	3,187
Chile	78,563	218.9	271.9	3,415.9	X	81.5	X	X	X	1	7	X
Colombia	5,874	16.2	44.0	706.1	X	29.9	3,659	817	11	X	X	8
Ecuador	4,597	31.5 l	107.3 e	X	957.0	60.5	2,469	337	7 m	X	10	(3,700)
Guyana	1,154	48.8	10.9	122.0	X	76.6	800	0	5	X	X	X
Paraguay	0	0.0	X	X	X	0.0	X	X	X	X	X	X
Peru	3,362	84.8 l	59.6 e	X	746.5	57.2	51	0	5	X	X	X
Suriname	620	56.9	9.0	119.1	X	87.0	1,150	391	4	X	X	X
Uruguay	1,096	68.8 l	22.5 e	110.5	110.5	78.5	X	X	X	X	X	X
Venezuela	6,762	123.6	136.0	385.7	X	73.1	2,500	4,041	7	4	25	X
OCEANIA	**137,772**	**2,565.0**	**2,830.4**	**30,155.0**	**X**	**X**	**18,788**	**3,614**	**46**	**22**	**76**	**(208,359)**
Australia	66,530	2,065.2	773.1	6,664.1	X	89.8	11,500	2,544	39	21	75	292
Fiji	4,637	19.5	162.2	1,055.0	X	99.9	385	0	9	5	56	(131,221)
New Zealand	17,209	247.8	176.6	3,887.4	X	100.0	287	0	1	1	16	248
Papua New Guinea	20,197	132.4	752.3	1,613.8	X	61.2	5,399	1,063	44	7	73	X
Solomon Islands	9,880	25.9	212.3	1,377.1	X	100.0	642	0	22	3	68	(49,192)
DEVELOPED	**850,213**	**14,523.2**	**7,482.0**	**36,750.3**	**3,233.6**	**X**	**13,792**	**3,739**	**X**	**X**	**X**	**1,020,720**
DEVELOPING	**629,421**	**9,763.8**	**10,754.3 e**	**53,510.2**	**4,117.3**	**X**	**167,007**	**23,868**	**X**	**X**	**X**	**(1,020,939)**

Notes: a. Figures should be interpreted as approximations because of the difficulty of measuring coastline length. Estimates may differ from other published sources. b. "X" signifies that either the country has no mangroves or data are not available. c. "Scleractinia" corals are reef-forming corals (i.e., true or stony corals). d. Refers to trade as reported by CITES. Figures show the balance of imports minus exports. Exports are shown as negative balances (in parentheses). World totals reflect total number of coral colonies traded. e. Excludes excessive territorial seas claims. For the world, the area of territorial seas in dispute is 2,867,050 square kilometers. f. Total is from J.E. Maragos et al. (eds.), 1995. g. No areas claimed in the Caspian Sea have been included. h. Includes continental shelf area of the potential exclusive economic zone even though the country may have not claimed it. i. Includes Taiwan. j. This figure is based on relatively old data, therefore it is likely that many of these species are now extinct in Singapore. k. Excludes Greenland. l. The breadth of the territorial sea is disputed. m. Number of mangrove species for Ecuador excludes the Galapagos, which has four species.

Source: Various

	Average Marine Fish Production {a} (000 metric tons)								Average Marine Production {a,d} (000 metric tons)				Status of the Fisheries in {e}	Fully Fished by:	Discards (as a % of overall catch {f})
	Demersal Fishes {b}				Pelagic Fishes {c}										
	1965-67	1975-77	1985-87	1995-97	1965-67	1975-77	1985-87	1995-97	1965-67	1975-77	1985-87	1995-97	1994		1988-92
WORLD	**12,391**	**18,507**	**20,856**	**19,499**	**25,293**	**27,358**	**37,835**	**42,917**	**47,480**	**60,102**	**79,061**	**93,643**	**X**	**1999**	**25**
ATLANTIC OCEAN {g}	**7,816**	**9,009**	**8,718**	**7,445**	**9,181**	**12,305**	**9,864**	**10,528**	**19,420**	**24,427**	**23,179**	**22,625**	**I-F**	**1983**	**25**
Northeast Atlantic	3,771	5,732	5,318	5,064	5,345	5,945	4,074	5,466	9,875	12,578	10,667	11,748	Ov	1983	19
Northwest Atlantic	2,664	1,385	1,353	330	640	1,160	677	666	3,941	3,351	2,964	2,031	Ov	1971	27
Eastern Central Atlantic	219	447	385	347	671	2,419	2,100	2,532	1,257	3,434	3,051	3,412	Ov	1984	10
Western Central Atlantic	129	166	142	182	629	757	1,123	866	1,200	1,502	2,137	1,832	Ov	1987	14
Southeast Atlantic	525	788	606	351	1,716	1,780	1,607	744	2,372	2,671	2,325	1,138	Ov	1978	27
Southwest Atlantic	508	490	914	1,172	179	245	283	255	775	891	2,035	2,464	I	1997	14
PACIFIC OCEAN {g}	**3,953**	**8,481**	**10,910**	**10,078**	**14,638**	**13,044**	**24,783**	**28,356**	**24,969**	**30,783**	**48,276**	**60,824**	**I-F**	**1999**	**24**
Northeast Pacific	979	1,590	2,470	2,158	168	130	84	88	1,362	1,961	2,803	2,512	Ov	1990	26
Northwest Pacific	2,395	5,814	7,002	5,584	3,548	5,213	9,345	7,587	10,375	17,613	25,142	30,327	I	1998	22
Eastern Central Pacific	40	68	66	96	366	946	1,312	921	568	1,151	1,678	1,632	Ov	1988	27
Western Central Pacific	339	612	763	1,124	942	2,076	3,051	4,467	2,680	4,659	6,489	9,275	I	2003	33
Southeast Pacific	153	229	305	642	9,538	4,566	10,822	15,160	9,826	5,022	11,546	16,248	I	2001	21
Southwest Pacific	45	167	304	475	76	113	170	132	159	377	616	830	Ov	1991	15
INDIAN OCEAN {g}	**423**	**754**	**871**	**1,554**	**953**	**1,270**	**1,945**	**2,966**	**2,147**	**3,512**	**5,215**	**7,968**	**I**	**X**	**26**
Eastern Indian	100	196	293	523	250	376	720	1,148	875	1,500	2,536	3,938	I	2037	30
Western Indian	323	557	578	1,031	703	894	1,225	1,817	1,272	2,012	2,679	4,030	I	2051	22
MEDITERRANEAN AND THE BLACK SEA	**193**	**180**	**282**	**337**	**518**	**724**	**1,207**	**984**	**933**	**1,216**	**1,887**	**1,827**	**F**	**X**	**25**
ANTARCTIC {g}	**5**	**83**	**73**	**11**	**0**	**0**	**1**	**0**	**5**	**136**	**414**	**113**	**Ov**	**1980**	**10**
ARCTIC	**0**	**0**	**0**	**0**	**X**	**X**	**X**	**X**	**0**	**0**	**0**	**0**	**X**	**X**	**X**

Notes: a. Production includes capture and aquaculture. b. Demersal or bottom-dwelling fish species include flounders, halibuts, soles, cods, hakes, haddocks, redfishes, basses, congers, sharks, rays, and chimeras. c. Pelagic fish are those that live in ocean surface waters or open seas. Pelagic fish species include jacks, mullets, sauries, herrings, sardines, anchovies, tunas, bonitos, billfishes, mackerels, snooks, and cutlassfishes. d. Marine production includes marine fish (except diadromous fish), marine crustaceans, and marine molluscs. e. Status assessed by FAO: Ov = overfished; F = fully fished; and I = catch is increasing. f. Refers to the percentage of the overall catch (discards plus landings) that consisted of nontarget or low-value species and undersized fish of targeted species. g. Ocean totals, except for the Antarctic, do not include the Antarctic regions. Antarctic portions of the Atlantic, Indian, and Pacific Oceans are included in the regional total for the Antarctic.

Source: Food and Agriculture Organization of the United Nations

	Index of Agricultural Production (1989-91 = 100) Total 1986-88	Total 1996-98	Per Capita 1986-88	Per Capita 1996-98	Average Cereal Crop Yields (kg per hectare) 1996-98	Percent Change Since 1986-88	Average Production of Cereals (000 metric tons) 1996-98	Percent Change Since 1986-88	Average Roots and Tubers Crop Yields (kg per hectare) 1996-98	Percent Change Since 1986-88	Average Production of Roots and Tubers (000 metric tons) 1996-98	Percent Change Since 1986-88	Average Meat Production (000 metric tons) 1996-98	Percent Change Since 1986-88
WORLD	**94**	**115**	**98**	**104**	**2,949**	**17**	**2,074,498**	**17**	**12,958**	**3.1**	**638,438**	**13**	**214,557**	**46**
ASIA (EXCL. MIDDLE EAST)	**X**	**X**	**X**	**X**	**X**	**X**	**932,868**	**X**	**X**	**X**	**259,954**	**X**	**77,147**	**X**
Armenia	X	76	X	74	1,772	X	335	X	12,529	X	403	X	45	X
Azerbaijan	X	57	X	53	1,647	X	1,017	X	9,328	X	249	X	92	X
Bangladesh	91	111	96	100	2,705	23	29,883	23	10,870	4.5	1,930	12	409	72
Bhutan	101	107	108	95	1,097	(8)	112	(16)	10,750	26.7	56	2	8	21
Cambodia	87	131	95	108	1,785	24	3,496	60	5,114	(9.8)	108	(25)	164	103
China {a}	88	149	92	139	4,837	23	448,904	27	17,308	14.6	170,478	22	53,747	191
Georgia	X	72	X	73	X	X	730	X	X	X	331	X	125	X
India	87	121	93	107	2,204	34	220,841	31	16,906	16.3	29,909	59	4,604	57
Indonesia	89	120	93	108	3,915	10	59,029	27	11,663	6.9	18,804	9	1,974	113
Japan	103	95	104	93	6,023	6	12,995	(12)	26,265	2.9	4,937	(17)	3,081	(20)
Kazakhstan	X	56	X	56	670	X	9,985	X	8,255	X	1,464	X	740	X
Korea, Dem People's Rep	X	X	103	63	2,251	(50)	3,142	(59)	11,213	(16.4)	684	(32)	146	(76)
Korea, Rep	95	123	98	115	6,589	12	7,748	(12)	21,346	(1.2)	997	(7)	1,614	158
Kyrgyzstan	X	93	X	91	X	X	1,609	X	X	X	655	X	187	X
Lao People's Dem Rep	83	122	92	99	2,642	26	1,670	32	7,055	(16.1)	197	(10)	71	103
Malaysia	86	118	93	100	X	X	2,019	14	X	X	488	(0)	989	160
Mongolia	95	89	104	77	687	(45)	218	(72)	7,636	(27.6)	55	(57)	243	6
Myanmar	110	139	116	123	2,947	4	17,938	26	9,667	0.0	317	13	382	26
Nepal	86	117	93	97	1,959	18	6,416	37	7,958	41.5	1,101	89	220	40
Pakistan	86	129	95	107	2,077	21	25,787	32	13,578	22.5	1,556	78	2,083	133
Philippines	93	121	99	104	2,336	24	14,563	10	6,728	1.0	2,767	2	1,848	182
Singapore	123	32	131	28	X	X	X	X	X	X	0	(94)	151	11
Sri Lanka	101	111	105	103	3,103	5	2,366	(3)	8,508	(8.4)	379	(41)	91	107
Tajikistan	X	57	X	50	1,682	X	510	X	10,268	X	118	X	36	X
Thailand	92	114	96	107	2,481	18	27,652	17	14,314	4.2	17,348	(10)	1,709	69
Turkmenistan	X	72	X	62	1,138	X	735	X	6,917	X	28	X	117	X
Uzbekistan	X	94	X	81	2,279	X	3,690	X	11,908	X	694	X	494	X
Viet Nam	88	143	94	125	3,760	38	29,330	76	7,009	(8.8)	3,894	(25)	1,530	89
EUROPE	**X**	**87 b**	**X**	**86 b**	**3,030 b**	**X**	**404,660**	**X**	**15,986 b**	**X**	**150,050**	**X**	**52,170**	**X**
Albania	101	125	106	119	2,584	(12)	592	(42)	10,949	71.0	131	52	59	X
Austria	100	101	101	95	5,659	11	4,746	(8)	29,014	3.4	703	(26)	886	15
Belarus	X	66	X	65	2,204	X	5,327	X	13,112	X	9,274	X	642	X
Belgium {c}	96	114	97	111	X	X	2,502	X	X	X	2,776	X	X	X
Bosnia and Herzegovina	X	39	X	45	2,382	X	479	X	6,744	X	241	X	34	X
Bulgaria	108	66	106	68	2,644	(31)	5,006	(36)	9,677	(5.9)	415	7	470	(58)
Croatia	X	60	X	60	4,733	X	3,050	X	10,063	X	650	X	107	X
Czech Rep	X	80	X	80	4,178	X	6,832	X	20,379	X	1,567	X	841	X
Denmark	92	103	93	101	6,130	23	9,351	21	39,017	10.0	1,569	41	1,904	40
Estonia	X	47	X	51	1,978	X	630	X	11,717	X	417	X	57	X
Finland	94	94	95	91	3,381	41	3,772	33	22,477	33.6	758	7	343	2
France	100	105	102	102	7,109	25	64,578	22	37,017	8.2	6,378	(11)	6,533	27
Germany	102	92	104	89	6,366	26	44,067	21	39,841	40.3	12,446	(35)	6,069	(27)
Greece	95	102	97	99	3,535	(5)	4,621	(15)	18,712	2.0	915	(10)	512	(1)
Hungary	102	76	101	79	4,275	(15)	12,317	(15)	19,410	9.5	1,153	(8)	1,068	(52)
Iceland	113	90	117	84	X	X	X	X	X	X	9	(37)	20	(15)
Ireland	97	107	96	105	6,721	12	1,984	(3)	27,384	20.9	562	(15)	981	46
Italy	101	101	101	100	4,920	27	20,486	13	22,997	22.5	2,113	(14)	4,061	11
Latvia	X	46	X	49	2,123	X	988	X	13,048	X	907	X	73	X
Lithuania	X	70	X	70	2,474	X	2,808	X	15,692	X	1,925	X	193	X
Macedonia, FYR	X	100	X	94	2,721	X	605	X	11,794	X	165	X	37	X
Moldova, Rep	X	55	X	54	2,860	X	2,658	X	X	X	412	X	120	X
Netherlands	99	99	101	94	7,445	12	1,489	24	43,664	2.4	7,919	13	2,864	23
Norway	97	101	98	98	4,041	19	1,346	17	25,195	6.3	453	9	260	31
Poland	99	87	100	85	2,937	(3)	25,911	3	18,741	(1.0)	24,647	(33)	2,904	1
Portugal	85	98	85	98	2,398	38	1,476	(9)	13,867	14.8	1,198	(22)	689	73
Romania	111	93	112	96	2,868	(7)	17,122	(8)	13,157	(1.2)	3,372	(22)	1,168	(39)
Russian Federation	X	63	X	63	1,295	X	67,064	X	10,668	X	35,664	X	4,953	X
Slovakia	X	78	X	77	4,159	X	3,515	X	16,289	X	564	X	386	X
Slovenia	X	100	X	99	5,357	X	542	X	22,081	X	498	X	181	X
Spain	95	109	96	108	3,156	22	21,281	5	21,382	23.2	3,483	(32)	4,355	63
Sweden	96	101	98	98	4,707	24	5,893	12	35,375	16.1	1,226	7	584	23
Switzerland	99	96	102	90	6,709	22	1,271	21	45,385	23.8	0	X	429	(17)
Ukraine	X	55	X	55	2,214	X	27,867	X	11,183	X	17,537	X	1,905	X
United Kingdom	98	100	99	99	6,880	21	23,585	5	41,105	10.4	6,952	4	3,658	14
Yugoslavia	X	102	X	100	3,797	X	8,894	X	8,896	X	1,012	X	1,046	X
MIDDLE EAST & N. AFRICA	**X**	**X**	**X**	**X**	**X**	**X**	**91,441**	**23**	**X**	**X**	**15,862**	**37**	**7,430**	**57**
Afghanistan	100	112	105	75	1,363	3	3,694	17	16,786	7.3	235	5	231	X
Algeria	88	127	95	108	980	35	2,934	60	13,835	59.5	1,071	23	511	X
Egypt	91	138	98	121	6,681	37	17,328	86	19,278	(11.4)	2,439	29	1,221	116
Iran, Islamic Rep	90	141	101	117	1,863	X	16,714	X	X	X	3,242	X	1,437	X
Iraq	101	98	110	84	835	(9)	2,497	13	15,813	3.0	403	176	111	(93)
Israel	99	107	105	86	1,776	(22)	144	(43)	40,704	8.0	305	41	323	86
Jordan	91	166	106	123	1,037	22	100	6	28,742	41.8	125	184	115	89
Kuwait	163	163	179	200	X	X	3	(6)	X	X	31	1,664	74	14
Lebanon	80	140	80	115	2,456	43	94	48	22,839	28.0	308	48	101	26
Libyan Arab Jamahiriya	88	129	98	102	820	26	185	(34)	7,323	(1.0)	205	71	170	49
Morocco	83	107	88	93	1,215	(6)	6,947	4	18,694	5.4	1,195	50	510	77
Oman	95	112	108	83	X	X	6	40	X	X	6	89	27	63
Saudi Arabia	77	85	88	69	3,880	(4)	2,273	(25)	14,482	(21.0)	250	649	574	75
Syrian Arab Rep	109	147	121	122	1,543	27	5,182	48	18,879	15.1	419	16	296	63
Tunisia	77	120	82	105	1,240	55	1,863	97	14,165	22.4	311	73	173	49
Turkey	97	111	103	99	2,196	1	30,758	3	24,556	13.4	5,122	21	1,206	12
United Arab Emirates	66	206	74	172	X	X	1	(19)	X	X	4	17	86	118
Yemen	95	122	106	87	974	12	714	(5)	12,712	(21.9)	192	24	144	11

	Index of Agricultural Production (1989-91 = 100) Total		Per Capita		Average Cereal Crop Yields (kg per hectare)	Percent Change Since	Average Production of Cereals (000 metric tons)	Percent Change Since	Average Roots and Tubers Crop Yields (kg per hectare)	Percent Change Since	Average Production of Roots and Tubers (000 metric tons)	Percent Change Since	Average Meat Production (000 metric tons)	Percent Change Since
	1986-88	1996-98	1986-88	1996-98	1996-98	1986-88	1996-98	1986-88	1996-98	1986-88	1996-98	1986-88	1996-98	1986-88
SUB-SAHARAN AFRICA	**91 b**	**122 b**	**99 b**	**100 b**	**997 b**	**2**	**86,990**	**25**	**7,694 b**	**11.3**	**132,744**	**54**	**7,568**	**55**
Angola	102	127	111	101	642	76	539	55	5,102	32.1	2,909	73	105	X
Benin	82	159	90	130	1,096	37	816	68	9,385	10.6	3,019	84	70	42
Botswana	89	98	99	82	258	5	41	(18)	6,167	14.5	10	38	67	45
Burkina Faso	95	129	103	106	729	6	2,167	16	5,664	(29.4)	65	(52)	122	60
Burundi	98	94	106	80	1,337	10	279	0	6,602	(4.6)	1,378	1	26	3
Cameroon	96	120	104	99	1,274	1	1,222	35	5,777	0.7	2,588	32	202	47
Central African Rep	95	128	103	110	1,002	(8)	145	(3)	3,724	0.9	1,002	29	85	83
Chad	88	140	93	116	647	4	1,140	71	4,262	(11.5)	611	1	74	35
Congo	99	111	108	91	707	(13)	4	(82)	7,138	0.5	859	10	24	22
Congo, Dem Rep	91	97	101	76	746	(7)	1,468	16	7,393	(6.7)	17,560	(0)	239	27
Côte d'Ivoire	91	125	101	102	1,102	28	1,775	62	5,752	(0.6)	4,987	24	143	34
Equatorial Guinea	101	94	108	79	X	X	X	X	X	X	85	17	0	20
Eritrea	X	101	X	85	451	X	131	X	2,828	X	110	X	28	X
Ethiopia {d}	94	121	103	97	1,206	9	8,683	56	3,704	2.9	2,058	11	622	153,747
Gabon	92	108	100	88	1,742	11	31	34	5,617	4.8	414	22	30	24
Gambia	106	86	120	67	1,006	(16)	104	6	3,000	0.0	6	0	7	29
Ghana	93	152	102	125	1,365	36	1,751	71	11,027	66.6	11,109	127	140	4
Guinea	85	135	94	103	1,297	32	931	55	6,277	(10.3)	989	99	27	95
Guinea-Bissau	92	117	98	101	1,410	(9)	180	16	7,171	1.5	77	10	17	36
Kenya	92	108	102	89	1,535	(11)	2,850	(10)	7,583	(6.7)	1,976	42	401	26
Lesotho	96	112	104	94	1,011	44	197	15	15,838	4.6	75	118	29	11
Liberia	X	X	137	86	1,262	0	158	(46)	6,840	(12.4)	327	(24)	18	4
Madagascar	95	107	106	85	1,961	7	2,679	14	6,417	0.2	3,333	8	264	23
Malawi	94	115	107	106	1,224	13	1,718	25	4,692	5.2	579	17	49	29
Mali	86	124	94	99	1,004	11	2,203	38	5,003	20.9	30	43	195	55
Mauritania	90	104	97	87	743	(13)	168	12	2,077	8.4	5	(6)	58	6
Mozambique	95	131	98	102	827	86	1,533	135	5,290	28.6	5,375	43	81	8
Namibia	92	123	100	103	315	(18)	105	34	8,570	(1.6)	240	17	71	33
Niger	91	126	100	100	339	(17)	2,263	20	7,541	2.9	268	11	118	57
Nigeria	78	137	85	111	1,207	(1)	22,107	41	10,310	10.7	53,717	157	1,187	89
Rwanda	99	78	104	91	1,188	(3)	210	(29)	6,314	(13.9)	1,386	(13)	29	1
Senegal	99	102	107	85	719	(7)	847	(9)	3,037	(28.6)	51	(35)	159	96
Sierra Leone	97	100	103	90	1,223	(8)	460	(16)	4,667	14.9	352	165	21	25
Somalia	X	X	111	85	410	(47)	269	(53)	9,988	(6.3)	56	21	160	15
South Africa	94	100	100	85	2,220	36	12,104	1	22,068	49.4	1,633	49	1,212	(3)
Sudan	107	156	115	134	602	16	5,458	54	2,707	(4.4)	166	(24)	588	71
Tanzania, United Rep	94	101	103	82	1,261	5	4,026	5	7,436	(9.3)	6,603	(16)	306	37
Togo	87	138	95	113	876	12	679	64	7,064	(9.1)	1,251	58	39	99
Uganda	85	109	92	88	1,248	(8)	1,660	35	4,781	(24.7)	4,397	(12)	220	83
Zambia	93	107	100	91	1,585	(21)	1,171	(23)	5,311	(3.8)	774	98	107	38
Zimbabwe	97	112	106	95	1,283	(7)	2,543	(1)	4,599	(7.6)	190	64	134	10
NORTH AMERICA	**93 b**	**117 b**	**96 b**	**109 b**	**4,744 b**	**24**	**394,005**	**25**	**35,487 b**	**16.0**	**26,397**	**31**	**38,405**	**44**
Canada	93	116	97	108	2,737	22	53,020	10	27,091	6.3	4,005	41	3,320	27
United States	94	117	96	110	5,352	22	340,985	28	37,570	18.7	22,392	30	35,085	46
C. AMERICA & CARIBBEAN	**X**	**X**	**X**	**X**	**2,465 b**	**15**	**34,246**	**22**	**8,063 b**	**19.9**	**4,234**	**15**	**5,526**	**41**
Belize	87	157	93	131	2,096	22	53	99	21,765	(0.5)	4	6	10	79
Costa Rica	85	124	92	105	3,162	35	219	(31)	23,673	86.9	227	158	178	63
Cuba	99	65	102	62	1,973	(23)	501	(17)	5,783	(1.7)	815	(9)	210	(52)
Dominican Rep	96	105	102	92	3,851	5	544	(2)	6,118	4.2	230	8	299	105
El Salvador	95	107	100	91	1,906	12	821	18	16,719	19.7	92	208	87	51
Guatemala	90	117	98	96	1,882	7	1,152	(17)	5,568	2.3	72	5	191	85
Haiti	106	93	113	81	969	(0)	442	(6)	3,796	0.1	734	(3)	74	17
Honduras	88	117	97	95	1,528	9	772	37	9,316	7.6	33	46	98	68
Jamaica	95	120	96	113	1,227	(15)	3	(39)	16,612	33.8	327	33	82	67
Mexico	99	119	105	105	2,653	18	28,839	27	21,290	52.5	1,441	41	3,911	44
Nicaragua	97	117	103	96	1,611	(9)	637	33	11,295	(4.4)	82	4	90	75
Panama	97	99	103	87	2,164	26	243	(18)	5,816	(24.4)	79	(8)	132	58
Trinidad and Tobago	93	102	96	97	3,628	41	15	58	10,322	10.6	12	38	29	5
SOUTH AMERICA	**93 b**	**122 b**	**98 b**	**109 b**	**2,740 b**	**34**	**96,761**	**20**	**12,334 b**	**5.5**	**45,027**	**5**	**21,435**	**80**
Argentina	101	122	105	112	3,260	31	33,639	41	22,822	19.5	3,084	6	3,643	6
Bolivia	85	133	91	112	1,592	21	1,180	39	5,668	(7.2)	1,123	(14)	359	89
Brazil	92	123	97	112	2,458	36	44,725	8	12,754	4.4	26,476	(1)	12,184	124
Chile	84	128	89	115	4,455	32	2,824	2	14,669	4.2	918	12	815	167
Colombia	85	104	90	92	2,798	9	3,378	(1)	12,848	15.0	4,830	29	1,325	59
Ecuador	88	145	95	125	1,746	14	1,858	44	7,650	16.6	678	34	420	129
Guyana	117	185	117	174	3,982	27	551	120	10,280	26.5	46	49	16	154
Paraguay	77	107	85	88	2,365	29	1,470	133	13,762	(11.8)	3,110	(12)	385	82
Peru	100	135	106	120	2,731	12	2,582	17	9,456	13.4	3,662	34	715	66
Suriname	106	88	110	81	3,746	(3)	221	(21)	16,392	73.4	7	142	6	(66)
Uruguay	92	124	94	119	3,370	55	2,090	95	11,627	101.0	194	4	574	63
Venezuela	97	114	105	97	2,952	46	2,212	(7)	11,343	35.0	886	37	991	34
OCEANIA	**96 b**	**118 b**	**101 b**	**107 b**	**2,032 b**	**29**	**33,494**	**46**	**12,116 b**	**10.8**	**3,214**	**11**	**4,786**	**25**
Australia	93	120	98	111	1,995	30	32,534	48	31,061	14.7	1,304	27	3,335	31
Fiji	91	98	93	100	2,094	(3)	19	(32)	11,141	107.0	86	87	24	57
New Zealand	104	117	106	108	5,637	26	930	(6)	42,676	44.2	480	69	1,334	11
Papua New Guinea	100	104	106	89	3,840	116	9	272	5,870	(18.6)	1,005	(18)	67	54
Solomon Islands	104	116	116	92	X	(100)	X	(100)	17,236	3.3	126	27	3	7
DEVELOPED	**98 b**	**98 b**	**100 b**	**94 b**	**3,372 b**	**16**	**875,983**	**6**	**17,436 b**	**1.7**	**189,047**	**(10)**	**101,697**	**38**
DEVELOPING	**90 b**	**129 b**	**96 b**	**114 b**	**2,701 b**	**20**	**1,198,482**	**26**	**11,688 b**	**7.8**	**448,585**	**26**	**112,771**	**117**

Notes: Negative values are shown in parantheses. Zero is either 0 or less than one-half the unit of measure. (0) indicates a value less than 0 and greater than negative one-half. a. Data for China include Taiwan. b. Regional total calculated by the Food and Agriculture Organization of the United Nations. c. Data for Belgium include Luxembourg. d. Data for Ethiopia before 1993 include Eritrea.

Source: Food and Agriculture Organization of the United Nations

	Cropland				Irrigated Land as a Percentage of Cropland		Average Annual Fertilizer Use (kilograms per hectare of cropland)		Pesticide Use (kilograms per hectare of cropland)	Tractors	
	Total Hectares (000)		Hectares Per 1000 People							Number	Percent Change Since
	1987	1997	1987	1997	1987	1997	1985-87	1995-97	1996	1997	1987
WORLD	**1,489,051**	**1,510,442**	**297**	**259**	**15**	**18**	**97**	**97**	**X**	**26,334,690**	**3**
ASIA (EXCL. MIDDLE EAST)	**X**	**482,752**	**X**	**147**	**X**	**34**	**X**	**139**	**X**	**5,757,829**	**X**
Armenia	X	559	X	157	X	52	X	14	2	17,500	X
Azerbaijan	X	1,935	X	253	X	75	X	14	X	32,917	X
Bangladesh	9,248	8,241	89	67	24	45	73	138	176 a	5,400	8
Bhutan	130	160	83	82	26	25	1	1	670 a	X	X
Cambodia	3,080	3,807	391	363	7	7	0	2	X	1,190	(1)
China {b}	128,786	135,365	117	109	35	38	150	265	X	703,117	(21)
Georgia	X	1,066	X	208	X	44	X	33	X	16,600	X
India	169,770	169,850	212	176	25	34	52	89	436 c	1,450,000	108
Indonesia	30,644	30,987	176	152	14	16	73	92	88 a	70,000	296
Japan	4,708	4,295	39	34	62 d	63 d	492	440	X	2,210,000	16
Kazakhstan	X	30,135	X	1,840	X	7	X	4	X	108,121	X
Korea, Dem People's Rep	1,980	2,000	101	87	67	73	397	63	X	75,000	7
Korea, Rep	2,143	1,924	51	42	63 d	60 d	448	693	13,829 a	131,358	561
Kyrgyzstan	X	1,425	X	308	X	75	X	22	1,860 c	19,000	X
Lao People's Dem Rep	850	852	224	169	14	19	1	5	57 a	890	9
Malaysia	5,920	7,605	358	362	6	4	125	158	5,982 c	43,300	155
Mongolia	1,336	1,320	658	520	5	6	16	2	X	7,000	(40)
Myanmar	10,060	10,151	259	231	11	15	16	19	16 a	8,036	(24)
Nepal	2,341	2,968	135	133	36	38	24	35	21 e	4,600	28
Pakistan	20,920	21,600	193	150	78	81	77	114	365 e	320,500	60
Philippines	9,020	9,520	158	133	17	16	44	82	X	11,500	59
Singapore	3	1	1	<1	0	0	1,325	3,247	X	65	12
Sri Lanka	1,895	1,888	115	103	28 d	32 d	102	111	6,261 a	6,672	(11)
Tajikistan	X	890	X	150	X	81	X	65	X	30,000	X
Thailand	20,490	20,445	387	342	20	25	28	75	1,116 f	149,500	270
Turkmenistan	X	1,695	X	400	X	106	X	82	6,744 a	50,000	X
Uzbekistan	X	4,850	X	209	X	88	X	109	X	170,000	X
Viet Nam	6,387	7,202	102	94	28	32	76	206	X	115,487	412
EUROPE	**X**	**311,205**	**X**	**427**	**X**	**8**	**X**	**89**	**X**	**11,197,796**	**X**
Albania	714	702	229	224	57	48	133	7	435 e	7,900	(26)
Austria	1,513	1,479	199	183	0	0	258	168	2,710 c	352,375	8
Belarus	X	6,319	X	610	X	2	X	157	X	96,300	X
Belgium {g}	681	785	69	78	3	4	1,531	1,475	X	106,667	(9)
Bosnia and Herzegovina	X	650	X	185	X	0	X	11	X	29,000	X
Bulgaria	4,131	4,511	465	538	30	18	220	43	966	25,000	(53)
Croatia	X	1,442	X	322	X	0	X	161	3,060 e	2,985	X
Czech Rep	X	3,331	X	323	X	1	X	107	1,169 e	86,000	X
Denmark	2,588	2,373	506	451	16	20	267	167	2,200 e	141,293	(14)
Estonia	X	1,143	X	790	X	0	X	29	105	50,607	X
Finland	2,292	2,129	464	414	3	3	265	147	410 e	194,750	(19)
France	19,459	19,468	349	333	6	9	303	262	X	1,312,000	(11)
Germany	12,390	12,060	159	147	4	4	413	250	2,085 a	1,215,700	(26)
Greece	3,942	3,915	392	370	30	35	178	133	X	236,100	16
Hungary	5,289	5,047	504	497	3 h	4 h	279	83	2,863 e	92,250	72
Iceland	7	6	28	22	0	0	3,491	3,429	X	10,519	(19)
Ireland	983	1,346	278	368	0	0	713	533	X	167,500	3
Italy	12,070	10,927	212	190	20	25	191	227	19,288 a	1,480,000	13
Latvia	X	1,830	X	744	X	1	X	30	208 e	56,938	X
Lithuania	X	3,006	X	811	X	0	X	41	312 a	77,871	X
Macedonia, FYR	X	658	X	331	X	8	X	67	7,718 a	54,000	X
Moldova, Rep	X	2,183	X	499	X	14	X	54	1,434 a	49,000	X
Netherlands	890	935	61	60	61	60	951	821	11,842 a	173,000	(5)
Norway	869	902	208	205	11	14	413	218	941 i	148,000	(2)
Poland	14,817	14,424	394	373	1	1	237	122	490 e	1,310,500	26
Portugal	3,163	2,900	320	294	20	22	90	84	2,584 c	150,000	23
Romania	10,686	9,900	465	439	31	31	150	41	1,617 c	163,016	(11)
Russian Federation	X	127,962	X	867	X	4	X	17	407 c	886,490	X
Slovakia	X	1,605	X	299	X	12	X	77	4,148 e	25,726	X
Slovenia	X	285	X	143	X	1	X	258	6,389 e	104,751	X
Spain	20,390	19,164	525	484	16	19	95	123	X	841,932	24
Sweden	2,890	2,799	343	316	4	4	164	113	509 a	165,000	(8)
Switzerland	412	444 j	62	61 j	6	6 j	450	258 j	4,576 e	112,000	4
Ukraine	X	34,081	X	667	X	7	X	26	2,001 c	349,000	X
United Kingdom	7,004	6,425	123	110	2	2	372	343	4,745 c	500,000	(4)
Yugoslavia	X	4,058	X	382	X	2	X	44	887 c	423,130	X
MIDDLE EAST & N. AFRICA	**95,025**	**102,320**	**320**	**270**	**24**	**27**	**58**	**59**	**X**	**1,593,873**	**34**
Afghanistan	8,054	8,054	570	385	35	35	10	1	X	840	2
Algeria	7,624	8,040	330	274	4	7	39	8	835 i	92,893	1
Egypt	2,547	3,300	49	51	100	100	362	343	1,293 a	90,000	72
Iran, Islamic Rep	16,530	19,400	322	300	42	37	57	55	1,881 c	235,000	X
Iraq	5,520	5,540	336	262	32	64	35	61	X	49,600	24
Israel	433	437	99	75	48	46	381	360	X	24,500	(15)
Jordan	365	390	84	64	17	19	16	52	2,495 a	4,773	(16)
Kuwait	4	7	2	4	50	71	1,886	(855) k	X	100	(9)
Lebanon	304	308	117	98	28	38	81	188	X	5,610	87
Libyan Arab Jamahiriya	2,130	2,115	525	406	17	22	36	36	X	34,000	24
Morocco	8,532	9,595	378	357	15	13	41	32	X	43,226	29
Oman	53	63	34	27	91	98	45	122	24,125 i	150	15
Saudi Arabia	2,980	3,830	212	197	47	42	152	98	X	9,500	116
Syrian Arab Rep	5,630	5,521	503	369	12	21	42	68	X	87,442	67
Tunisia	4,860	4,900	634	532	6	8	25	25	X	35,100	35
Turkey	27,927	29,162	530	460	12	14	58	66	1,145 a	874,995	38
United Arab Emirates	38	81	22	35	158	89	28	421	X	272	56
Yemen	1,479	1,555	143	95	21	31	10	9	X	5,800	(0)

	Cropland				Irrigated Land as a Percentage of Cropland		Average Annual Fertilizer Use (kilograms per hectare of cropland)		Pesticide Use (kilograms per hectare of cropland)	Tractors	
	Total Hectares (000)		Hectares Per 1000 People							Number	Percent Change Since
	1987	1997	1987	1997	1987	1997	1985-87	1995-97	1996	1997	1987
SUB-SAHARAN AFRICA	**159,495**	**171,384**	**349**	**288**	**4**	**4**	**13**	**12**	**X**	**261,984**	**(10)**
Angola	3,400	3,500	403	299	2	2	4	1	42 a	10,300	0
Benin	1,490	1,595	349	283	0	1	7	22	X	142	18
Botswana	410	346	354	225	0	0	2	9	40 c	6,000	90
Burkina Faso	3,140	3,440	377	313	1	1	5	10	1 e	1,993	564
Burundi	1,180	1,100	235	173	1	1	2	3	268	170	13
Cameroon	7,210	7,160	684	514	0	0	7	5	253 e	500	0
Central African Rep	2,005	2,020	733	591	0	0	1	0	12 e	65	8
Chad	3,205	3,256	599	460	0	1	2	3	223 c	170	3
Congo	179	185	88	68	1	1	17	20	216 f	700	1
Congo, Dem Rep	7,850	7,880	233	164	0	0	1	1	X	2,430	6
Côte d'Ivoire	5,400	7,350	511	523	1	1	7	16	X	3,800	12
Equatorial Guinea	230	230	690	547	0	0	0	0	X	100	0
Eritrea	X	393	X	114	X	7	X	11	X	440	X
Ethiopia	13,930 l	10,500	299 l	170	1 l	2	5 l	16	34	3,000 l	(23)
Gabon	452	495	530	435	1	1	5	1	X	1,500	7
Gambia	184	200	228	168	1	1	21	5	46	45	5
Ghana	4,000	4,550	290	244	0	0	3	4	2,333 e	3,570	(13)
Guinea	1,196	1,485	229	203	8	6	1	3	274	542	126
Guinea-Bissau	335	350	367	308	5	5	1	1	83 a	19	6
Kenya	4,490	4,520	211	159	1	1	25	27	X	14,400	20
Lesotho	326	325	203	161	1	1	12	19	X	2,000	18
Liberia	371	327	154	136	1	1	6	0	X	325	2
Madagascar	3,067	3,108	287	213	29	35	3	4	28 e	3,550	26
Malawi	1,610	1,710	199	170	1	2	25	31	X	1,420	4
Mali	2,076	4,650	250	446	3	2	8	8	136 c	2,550	59
Mauritania	345	502	185	204	14	10	4	8	X	380	15
Mozambique	3,090	3,180	225	172	3	3	2	2	X	5,750	0
Namibia	662	820	532	506	1	1	0	0	X a	3,150	9
Niger	3,592	5,000	511	512	1	1	1	2	X	180	14
Nigeria	31,482	30,738	393	296	1	1	10	5	X	30,000	50
Rwanda	1,142	1,150	172	193	0	0	1	0	260 c	90	5
Senegal	2,350	2,266	348	258	4	3	11	12	183	550	17
Sierra Leone	539	546	143	124	5	5	5	6	X	81	(84)
Somalia	1,035	1,061	147	120	17	19	3	0	X	1,845	(11)
South Africa	13,500	16,300	422	421	9	8	63	51	57 e	100,000	(38)
Sudan	12,900	16,900	573	610	15	12	5	4	106	10,500	7
Tanzania, United Rep	3,501	4,000	151	127	4	4	14	9	X	7,600	5
Togo	2,360	2,430	735	567	0	0	4	6	95	80	(18)
Uganda	6,705	6,810	436	340	0	0	0	0	17 e	4,700	18
Zambia	5,208	5,265	775	613	1	1	13	10	317	6,000	7
Zimbabwe	2,814	3,210	313	286	3	5	57	59	531	26,000	63
NORTH AMERICA	**233,766**	**224,700**	**855**	**744**	**8**	**10**	**102**	**134**	**X**	**5,511,335**	**(0)**
Canada	45,990	45,700	1,726	1,510	2	2	51	67	644 i	711,335	(4)
United States	187,776	179,000	761	659	10	12	114	151	1,599 a	4,800,000	0
C. AMERICA & CARIBBEAN	**39,467**	**43,408**	**287**	**263**	**17**	**19**	**78**	**58**	**X**	**295,669**	**7**
Belize	57	89	327	397	4	3	67	50	17,804 c	1,170	15
Costa Rica	526	505	188	135	22	25	181	322	18,726 e	7,000	11
Cuba	3,670	4,450	356	402	24 m	20 m	186	52	X	78,000	5
Dominican Rep	1,437	1,500	216	185	16	17	58	59	X	2,350	3
El Salvador	777	816	159	138	15	15	103	106	2,642	3,430	1
Guatemala	1,785	1,905	220	181	6	7	68	99	574 c	4,300	4
Haiti	903	910	140	116	8	10	3	9	23 e	136	(35)
Honduras	1,785	2,045	401	342	4	4	16	50	6,521 c	4,900	45
Jamaica	220	274	94	109	15	12	81	85	X	3,080	2
Mexico	25,500	27,300	325	290	20	24	71	54	X	172,000	6
Nicaragua	1,949	2,746	546	587	4	3	33	14	357 e	2,700	9
Panama	615	655	272	241	5	5	54	54	X	5,000	(4)
Trinidad and Tobago	120	122	100	96	18	18	53	203	11,827 a	2,700	3
SOUTH AMERICA	**106,536**	**116,186**	**381**	**351**	**7**	**9**	**53**	**66**	**X**	**1,291,323**	**14**
Argentina	27,200	27,200	872	763	6	6	6	28	1,266 e	280,000	14
Bolivia	2,205	2,100	359	270	5	4	3	5	1,514 a	5,700	14
Brazil	54,000	65,300	384	399	4	5	71	78	836 e	805,000	17
Chile	3,313	2,297	266	157	38	55	83	210	3,240 e	53,710	30
Colombia	5,371	4,430	163	111	9	24	83	125	6,134 c	21,000	(39)
Ecuador	2,836	3,001	297	251	11	8	23	48	1,696 e	8,900	9
Guyana	495	496	623	588	26	26	32	29	X	3,630	2
Paraguay	2,088	2,285	543	449	3	3	6	13	1,542 c	16,500	30
Peru	3,790	4,200	186	172	32	42	45	43	X	13,191	6
Suriname	68	67	173	163	84	90	177	93	4,877 f	1,330	8
Uruguay	1,304	1,307	428	400	8	11	45	89	1,316 a	33,000	(5)
Venezuela	3,860	3,490	214	153	4	6	156	86	1,403 c	49,000	7
OCEANIA	**52,148**	**57,766**	**2,065**	**1,972**	**4**	**5**	**32**	**53**	**X**	**401,025**	**(2)**
Australia	47,106	53,100	2,918	2,896	4	5	26	43	2,535 c	315,000	(2)
Fiji	210	285	295	362	0	1	93	66	2,333 c	7,000	17
New Zealand	3,844	3,280	1,171	872	7	9	108	216	2,215 e	76,000	(5)
Papua New Guinea	571	670	159	149	0	0	21	19	1,750 e	1,160	1
Solomon Islands	57	60	197	148	0	0	0	0	X	X	X
DEVELOPED	**487,736**	**476,872**	**397**	**369**	**13**	**14**	**188**	**136**	**X**	**19,878,769**	**(5)**
DEVELOPING	**812,966**	**853,994**	**215**	**189**	**20**	**24**	**64**	**96**	**X**	**6,448,965**	**36**

Notes: Negative values are shown in parentheses. Zero is either 0 or less than one-half the unit of measure. (0) indicates a value less than 0 and greater than negative one-half.
a. Data are from 1993. b. Data for China include Taiwan. c. Data are from 1992. d. Data refer to irrigated rice only. e. Data are from 1995. f. Data are from 1991. g. Data for Belgium include Luxembourg. h. Data exclude complementary farm plots and individual farms. i. Data are from 1994. j. Due to a change in statistical methods, data previous to 1992 should not be compared to data after 1992. k. The export of fertilizer stocks can cause the balance of trade approach to give negative values. l. Data for Ethiopia before 1993 include Eritrea. m. Irrigated land refers to land in the state sector only.

Sources: Food and Agriculture Organization of the United Nations and World Bank

	Average Per Capita Cereal Production (metric tons per 1000 people)		Variation in Domestic Cereal Production (average % variation from mean)		Net Cereal Imports and Food Aid as a Percent of Total Cereal Consumption {a}		Food Aid as a Percent of Total Imports		Average Daily Per Capita Calorie Supply (kilocalories)		Average Daily Per Capita Calories from Animal Products (kilocalories)		Percent of Children Underweight
	1986-88	1996-98	1979-88	1989-98	1985-87	1995-97	1985-87	1995-97	1987	1997	1987	1997	1990-97 {b}
WORLD	**355**	**356**	**5**	**3**	**1**	**0**	**6**	**2**	**2,667**	**2,782**	**408**	**441**	**29**
ASIA (EXCL. MIDDLE EAST)	**257**	**283**	**X**	**5**	**6**	**7**	**X**	**2**	**X**	**X**	**X**	**X**	**X**
Armenia	X	94	X	14 c	X	55	X	X	X	2,371	X	404	X
Azerbaijan	X	133	X	43 c	X	29	X	X	X	2,236	X	367	10
Bangladesh	235	244	5	3	13	8	45	27	2,062	2,086	60	69	56
Bhutan	85	58	10	4	12	32	14	7	X	X	X	X	38 d
Cambodia	276	334	21	17	4	3	24	47	1,868	2,048	133	163	52
China	321 e	361 e	9 e	5 e	2 e	3 e	3 e	1 e	2,608 e	2,897 e	255 e	510 e	16
Georgia	X	143	X	22 c	X	42	X	X	X	2,614	X	316	X
India	210	229	9	4	(0)	(1)	52	40	2,228	2,496	149	174	53
Indonesia	267	290	12	5	4	10	11	0	2,458	2,886	97	134	34
Japan	121	103	6	8	64	67	X	X	2,870	2,932	578	598	X
Kazakhstan	X	609	X	41 c	X	(52)	X	X	X	3,085	X	609	8
Korea, Dem People's Rep	388	136	8	38	2	35	0	42	2,509	1,837	252	150	X
Korea, Rep	210	169	6	6	47	62	0	0	3,110	3,155	328	502	X
Kyrgyzstan	X	348	X	16 c	X	9	X	X	X	2,447	X	543	X
Lao People's Dem Rep	333	332	12	8	1	3	41	44	2,102	2,108	107	136	40
Malaysia	107	96	7	4	54	66	0	0	2,616	2,977	449	570	19
Mongolia	390	86	32	43	(4)	28	0	10	2,034	1,917	841	845	10
Myanmar	367	408	6	10	(4)	(1)	0	11	2,697	2,862	119	119	43
Nepal	269	288	12	7	0	2	43	35	2,144	2,366	153	152	47
Pakistan	180	179	5	8	3	3	39	3	2,224	2,476	290	370	38
Philippines	231	204	8	5	11	16	22	1	2,244	2,366	243	364	28
Singapore	0	0	X	X	100	100	0	0	X	X	X	X	X
Sri Lanka	149	129	9	9	31	36	31	7	2,253	2,302	119	143	34
Tajikistan	X	86	X	35 c	X	50	X	X	X	2,001	X	143	X
Thailand	446	463	9	7	(45)	(19)	26	1	2,133	2,360	212	282	19
Turkmenistan	X	173	X	19	X	31	X	X	X	2,306	X	417	X
Uzbekistan	X	159	X	21	X	31	X	X	X	2,433	X	432	19
Viet Nam	267	384	11	13	2	(9)	11	9	2,193	2,484	173	226	41
EUROPE	**394**	**556**	**X**	**12**	**(5)**	**(3)**	**X**	**0**	**3,383**	**3,208 f**	**1,058**	**906 f**	**X**
Albania	329	189	5	20	6	32	0	9	2,556	2,961	410	682	X
Austria	677	586	7	7	(17)	(7)	X	X	3,419	3,536	1,243	1,256	X
Belarus	X	515	X	12 c	X	16	X	X	X	3,226	X	919	X
Belgium {g}	222	237	X	X	47	60	X	X	3,454	3,619	1,223	1,151	X
Bosnia and Herzegovina	X	137	X	40 c	X	30	X	X	X	2,266	X	228	X
Bulgaria	885	597	10	20	13	(1)	0	1	3,699	2,686	871	617	X
Croatia	X	680	X	8 c	X	(0)	X	X	X	2,445	X	459	1
Czech Rep	X	663	X	29 c	X	(5)	X	X	X	3,244	X	816	1
Denmark	1,512	1,779	7	8	(23)	(25)	X	X	3,211	3,407	1,191	1,259	X
Estonia	X	435	X	11 c	X	22	X	X	X	2,849	X	797	X
Finland	576	734	14	9	(16)	(10)	X	X	2,941	3,100	1,228	1,195	X
France	952	1,104	8	7	(98)	(86)	X	X	3,543	3,518	1,360	1,334	X
Germany	466	537	6	8	9	(16)	X	X	3,478	3,382	1,165	1,050	X
Greece	544	437	8	9	(4)	12	X	X	3,481	3,649	733	798	X
Hungary	1,380	1,213	6	17	(14)	(28)	0	0	3,768	3,313	1,419	1,046	2 d
Iceland	0	0	X	X	100	100	0	0	3,208	3,117	1,371	1,224	X
Ireland	579	543	5	7	7	10	X	X	3,623	3,565	1,293	1,126	X
Italy	319	357	3	5	20	20	X	X	3,512	3,507	906	902	X
Latvia	X	402	X	12 c	X	11	X	X	X	2,864	X	705	X
Lithuania	X	758	X	14 c	X	5	X	X	X	3,261	X	794	X
Macedonia, FYR	X	304	X	37 c	X	23	X	X	X	2,664	X	488	X
Moldova, Rep	X	607	X	22 c	X	(1)	X	X	X	2,567	X	388	X
Netherlands	82	95	5	6	76	76	X	X	3,076	3,284	1,009	1,135	X
Norway	276	306	8	9	27	31	X	X	3,304	3,357	1,220	1,124	X
Poland	670	670	11	8	8	8	0	0	3,441	3,366	1,115	884	X
Portugal	163	150	12	10	54	61	0	0	3,400	3,667	676	995	X
Romania	811	759	6	14	1	(6)	0	0	2,944	3,253	705	697	6
Russian Federation	X	454	X	21 c	X	4	X	X	X	2,904	X	726	3
Slovakia	X	654	X	29 c	X	(11)	X	X	X	2,984	X	786	X
Slovenia	X	272	X	11 c	X	48	X	X	X	3,101	X	841	X
Spain	524	537	21	15	8	24	X	X	3,150	3,310	780	860	X
Sweden	623	666	7	11	(28)	(17)	X	X	2,898	3,194	1,053	1,075	X
Switzerland	157	175	11	4	52	29	X	X	3,358	3,223	1,277	1,119	X
Ukraine	X	546	X	14 c	X	(4)	X	X	X	2,795	X	583	X
United Kingdom	394	403	8	5	(18)	(13)	X	X	3,215	3,276	1,128	1,024	X
Yugoslavia	X	837	X	12 c	X	(3)	X	X	X	3,031	X	1,041	2
MIDDLE EAST & N. AFRICA	**249**	**241**	**9**	**7**	**37**	**35**	**7**	**1**	**X**	**X**	**X**	**X**	**X**
Afghanistan	223	177	9	11	13	9	74	37	2,281	1,745	180	138	48
Algeria	80	100	26	45	64	68	0	0	2,757	2,853	315	265	13
Egypt	178	268	5	10	53	34	19	2	3,120	3,287	238	221	15
Iran, Islamic Rep	236	259	11	10	27	31	0	0	2,659	2,836	268	279	16
Iraq	134	118	19	16	61	37	1	6	3,418	2,619	264	95	23
Israel	57	25	29	27	89	94	0	0	3,075	3,278	652	589	X
Jordan	22	16	42	13	91	94	4	8	2,780	3,014	385	266	9 d
Kuwait	2	2	84	35	99	99	X	X	3,021	3,096	707	728	6
Lebanon	24	30	32	8	93	90	9	0	3,040	3,277	431	448	3
Libyan Arab Jamahiriya	69	35	14	20	83	91	0	0	3,308	3,289	464	341	5
Morocco	297	259	27	41	30	43	16	0	3,047	3,078	190	210	9
Oman	3	2	36	5	99	99	X	X	X	X	X	X	23
Saudi Arabia	214	116	70	29	74	72	X	X	2,488	2,783	446	415	X
Syrian Arab Rep	310	347	29	26	29	(9)	4	2	3,166	3,352	395	399	13
Tunisia	124	203	33	42	49	58	21	1	3,067	3,283	275	279	9
Turkey	566	485	7	6	2	5	0	0	3,496	3,525	419	397	10
United Arab Emirates	1	0	29	57	100	100	X	X	3,038	3,390	793	827	14
Yemen	73	44	22	13	64	73	11	2	2,126	2,051	166	144	39

	Average Per Capita Cereal Production (metric tons per 1000 people)		Variation in Domestic Cereal Production (average % variation from mean)		Net Cereal Imports and Food Aid as a Percent of Total Cereal Consumption {a}		Food Aid as a Percent of Total Imports		Average Daily Per Capita Calorie Supply (kilocalories)		Average Daily Per Capita Calories from Animal Products (kilocalories)		Percent of Children Underweight
	1986-88	1996-98	1979-88	1989-98	1985-87	1995-97	1985-87	1995-97	1987	1997	1987	1997	1990-97 {b}
SUB-SAHARAN AFRICA	**152**	**146**	**12**	**8**	**16**	**13**	**29**	**14**	**2,063 f**	**2,183 f**	**149 f**	**143 f**	**31 f**
Angola	41	46	8	26	49	59	31	29	1,864	1,903	251	138	42
Benin	113	145	17	15	16	14	11	13	1,961	2,487	105	104	29
Botswana	43	27	68	36	92	77	25	0	2,337	2,183	339	396	17
Burkina Faso	225	197	22	13	13	6	26	18	2,181	2,121	106	109	30
Burundi	56	44	10	7	10	9	14	6	2,023	1,685	56	44	37
Cameroon	86	88	10	15	24	15	4	1	2,178	2,111	146	125	14
Central African Rep	55	42	15	18	23	15	27	3	1,914	2,016	171	190	27
Chad	124	161	21	20	15	6	45	31	1,596	2,032	116	113	39
Congo	10	1	24	53	82	96	2	8	2,326	2,144	162	144	17 d
Congo, Dem Rep	37	31	12	7	31	17	20	6	2,132	1,755	64	47	34
Côte d'Ivoire	103	126	11	14	36	29	0	5	2,677	2,610	160	97	24
Equatorial Guinea	X	0	X	X	100	100	28	8	X	X	X	X	X
Eritrea	X	38	X	26	X	58	X	X	X	1,622	X	92	44
Ethiopia	127 h	150	10 h	20	25 h	11	57 h	50	1,677 h	1,858	107 h	101	48
Gabon	27	28	31	11	79	79	0	0	2,460	2,556	383	327	X
Gambia	122	88	17	5	44	51	23	4	2,498	2,350	121	126	26
Ghana	74	94	22	17	22	18	35	15	1,979	2,611	107	84	27
Guinea	115	127	8	15	23	27	29	2	2,060	2,232	63	62	X
Guinea-Bissau	170	158	16	6	24	28	33	5	2,378	2,430	172	165	23 d
Kenya	149	100	16	9	6	17	36	10	2,025	1,977	272	241	23
Lesotho	106	98	20	36	55	67	24	6	2,216	2,244	158	136	16
Liberia	123	64	6	46	35	73	26	28	2,344	2,044	98	65	X
Madagascar	220	183	3	3	12	5	39	19	2,292	2,022	239	194	40
Malawi	169	170	6	22	3	12	45	26	2,027	2,043	73	56	30
Mali	192	211	15	8	16	4	25	21	1,967	2,030	192	214	40
Mauritania	81	69	49	27	69	59	29	10	2,509	2,622	540	443	23
Mozambique	48	83	8	42	55	30	53	30	1,785	1,832	68	44	27
Namibia	63	65	8	32	56	56	0	4	2,199	2,183	273	272	26
Niger	268	232	12	9	12	4	24	36	2,033	2,097	118	113	43
Nigeria	196	213	28	7	10	6	0	0	2,103	2,735	84	103	36
Rwanda	45	35	5	22	13	71	33	56	2,042	2,057	57	60	27
Senegal	139	97	22	9	37	43	18	1	2,104	2,418	207	193	22
Sierra Leone	146	104	4	10	26	43	31	14	2,126	2,035	73	66	29
Somalia	82	31	24	38	43	28	43	11	1,995	1,566	635	737	X
South Africa	374	313	16	22	(8)	1	0	0	2,976	2,990	458	409	9
Sudan	157	197	35	32	32	7	51	16	2,208	2,395	373	473	34
Tanzania, United Rep	166	128	11	10	7	6	23	18	2,288	1,995	143	129	27
Togo	128	159	15	12	18	12	12	5	1,946	2,469	92	102	19
Uganda	80	83	10	8	3	2	46	51	2,113	2,085	128	138	26
Zambia	224	137	20	26	19	13	43	15	2,017	1,970	107	113	24
Zimbabwe	285	227	34	30	(17)	(19)	23	1	2,112	2,145	205	180	16
NORTH AMERICA	**1,156**	**1,305**	**12**	**9**	**(38)**	**(43)**	**X**	**X**	**3,398 f**	**3,641 f**	**1,921 f**	**1,837 f**	**X**
Canada	1,807	1,753	13	6	(86)	(72)	X	X	3,105	3,119	942	843	X
United States	1,086	1,255	14	10	(32)	(40)	X	X	3,430	3,699	979	995	1
C. AMERICA & CARIBBEAN	**205**	**208**	**9**	**6**	**24**	**30**	**9**	**1**	**2,798 f**	**2,794 f**	**436 f**	**428 f**	**X**
Belize	152	235	11	24	32	28	3	0	2,565	2,907	720	648	6
Costa Rica	114	58	8	11	44	75	42	0	2,717	2,649	425	469	2
Cuba	59	45	6	23	78	77	X	X	3,125	2,480	671	301	9
Dominican Rep	83	67	12	8	47	66	X	X	2,330	2,288	271	339	6
El Salvador	142	139	8	8	41	35	51	2	2,312	2,562	259	297	11
Guatemala	171	110	8	10	22	31	44	5	2,384	2,339	164	205	27
Haiti	73	56	5	8	31	47	X	X	1,848	1,869	123	99	28
Honduras	127	129	10	7	32	30	52	10	2,206	2,403	255	332	18
Jamaica	2	1	17	12	98	99	X	X	2,630	2,553	438	455	10
Mexico	290	306	11	7	14	22	0	0	3,022	3,097	493	520	14 d
Nicaragua	134	136	8	13	34	26	44	15	2,330	2,186	268	165	12
Panama	132	90	8	14	30	51	0	0	2,302	2,430	483	537	7
Trinidad and Tobago	8	11	34	19	97	95	X	X	2,975	2,661	544	382	7 d
SOUTH AMERICA	**288**	**293**	**7**	**9**	**(5)**	**1**	**6**	**1**	**2,654 f**	**2,800 f**	**487 f**	**561 f**	**X**
Argentina	764	943	15	17	(148)	(117)	X	X	3,096	3,093	1,004	886	X
Bolivia	138	152	16	12	41	24	42	33	2,153	2,174	356	406	16
Brazil	295	273	12	9	12	15	0	0	2,745	2,974	443	585	6
Chile	222	193	22	5	11	29	4	0	2,518	2,796	409	612	1
Colombia	103	84	5	7	22	47	3	0	2,341	2,597	342	428	8
Ecuador	135	156	27	12	24	15	8	2	2,430	2,679	349	471	17 d
Guyana	316	654	8	33	5	(42)	42	40	2,469	2,530	200	343	12
Paraguay	164	289	13	21	2	(18)	8	0	2,564	2,566	455	594	4
Peru	108	106	16	15	47	51	15	3	2,276	2,302	377	339	8
Suriname	711	536	7	7	(31)	(21)	0	26	2,435	2,665	313	332	X
Uruguay	352	640	9	20	(34)	(75)	0	0	2,613	2,816	962	944	5
Venezuela	132	97	21	8	51	48	X	X	2,602	2,321	433	361	5
OCEANIA	**910**	**1,144**	**16**	**18**	**(857)**	**(150)**	**0**	**0**	**2,980 f**	**3,054 f**	**960 f**	**902 f**	**X**
Australia	1,360	1,775	17	18	(1,735)	(194)	X	X	3,159	3,224	1,093	1,021	X
Fiji	39	24	18	17	77	87	0	0	2,626	2,865	496	589	8
New Zealand	301	247	16	9	(36)	24	X	X	3,201	3,395	1,291	1,303	X
Papua New Guinea	1	2	20	28	99	98	0	0	2,137	2,224	268	247	30 d
Solomon Islands	3	0	58	X	86	100	3	0	2,195	2,122	239	197	21 d
DEVELOPED	**675**	**677**	**6**	**4**	**(9)**	**(15)**	**0**	**0**	**3,324 f**	**3,240 f**	**940 f**	**854 f**	**X**
DEVELOPING	**252**	**265**	**8**	**5**	**8**	**9**	**11**	**3**	**2,453 f**	**2,650 f**	**234 f**	**322 f**	**31 f**

Notes: Zero is either 0 or less than one-half the unit of measure. (0) indicates a value less than 0 and greater than negative one-half. a. Negative values indicate net exports of grain. b. Data are for the most recent year within the time range given. c. Data are for the years 1992-98. d. Data refer to years or periods other than those specified in the column heading, differ from the standard definition, or refer to only part of that country. e. Data for China include Taiwan. f. Regional totals calculated by the data source. g. Data for Belgium include Luxembourg. h. Data for Ethiopia previous to 1993 include Eritrea.

Sources: Food and Agriculture Organization of the United Nations, United Nations Population Division, and United Nations Children's Fund

	Average Annual Internal Renewable Water Resources		Annual River Flows		Annual Withdrawals				Sectoral Withdrawals (percent)		
	Total (cubic km)	Per Capita (cubic meters) Year 2000	From Other Countries (cubic km)	To Other Countries (cubic km)	Year of Data	Total (cubic km)	Percentage of Water Resources	Per Capita (cubic meters)	Domestic	Industry	Agriculture
WORLD {a}	**42,650.00**	**7,044**	**X**	**X**	**1990**	**3,414.00**	**8**	**648**	**9**	**20**	**71**
ASIA (EXCL. MIDDLE EAST)	**X**	**X**	**X**	**X**	**X**	**X**	**X**	**X**	**X**	**X**	**X**
Armenia	9.07	2,577	1.5	3.2	1994	2.93	32	817	30	4	66
Azerbaijan	8.12	1,049	22.2	X	1995	16.53	204	2,186	5	25	70
Bangladesh	104.99	813	1,105.6	X	1990	14.64	14	134	12	2	86
Bhutan	95.00	44,728	X	X	1987	0.02	0	13	36	10	54
Cambodia	120.56	10,795	355.6	471.4	1987	0.52	0	66	5	1	94
China	2,812.40	2,201	17.2	719.0	1993	525.46	19	439	5	18	77
Georgia	58.13	11,702	5.2	14.4	1990	3.47	6	635	21	20	59
India	1,260.56	1,244	647.2	1,307.0	1990	500.00	40	588	5	3	92
Indonesia	2,838.00	13,380	X	X	1990	74.34	3	407	6	1	93
Japan	430.00	3,393	0.0	0.0	1992	91.40	21	735	19	17	64
Kazakhstan	75.42	4,649	34.2	51.7	1993	33.67	45	2,019	2	17	81
Korea, Dem People's Rep	67.00	2,787	10.1	4.9	1987	14.16	21	726	11	16	73
Korea, Rep	64.85	1,384	4.9	X	1994	23.67	36	531	26	11	63
Kyrgyzstan	46.45	9,884	X	25.9	1994	10.09	22	2,219	3	3	94
Lao People's Dem Rep	190.42	35,049	91.2	424.8	1987	0.99	1	260	8	10	82
Malaysia	580.00	26,074	X	X	1995	12.73	2	633	11	13	76
Mongolia	34.80	13,073	X	25.0	1993	0.43	1	182	20	27	53
Myanmar	880.59	19,306	128.2	X	1987	3.96	0	102	7	3	90
Nepal	198.20	8,282	12.0	133.2	1994	29.00	15	1,397	1	0	99
Pakistan	84.73	541	170.3	X	1991	155.60	184	1,269	2	2	97
Philippines	479.00	6,305	X	X	1995	55.42	12	811	8	4	88
Singapore	X	X	0.0	X	X	X	X	X	X	X	X
Sri Lanka	50.00	2,656	X	X	1990	9.77	20	573	2	2	96
Tajikistan	66.30	10,714	13.3	63.6	1994	11.87	18	2,095	4	4	92
Thailand	210.00	3,420	199.9	X	1990	33.13	16	596	5	4	91
Turkmenistan	1.36	305	44.1	26.4	1994	23.78	1,748	5,947	1	1	98
Uzbekistan	16.34	672	X	21.8	1994	58.05	355	2,626	4	2	94
Viet Nam	366.51	4,591	524.7	X	1990	54.33	15	815	4	10	86
EUROPE {a}	**2,900.00**	**3,981**	**X**	**X**	**1990**	**476.10**	**16**	**660**	**14**	**45**	**41**
Albania	26.92	8,646	15.7	X	1995	1.40	5	441	29	0	71
Austria	55.00	6,699	29.0	X	1995	2.23	4	278	31	60	9
Belarus	37.20	3,634	20.8	58.0	1990	2.73	7	266	22	43	35
Belgium	12.00	1,181	4.0	X	1980	9.03	75	917	X	X	X
Bosnia and Herzegovina	35.50	8,938	2.0	37.0	X	X	X	X	X	X	X
Bulgaria	18.00	2,188	X	X	X	X	X	X	X	X	X
Croatia	37.70	8,429	33.7	39.3	1996	0.76	2	170	50	50	X
Czech Rep	15.00	1,464	1.0	X	1995	2.52	17	244	39	57	1
Denmark	6.00	1,134	X	X	1995	0.89	15	170	53	9	16
Estonia	12.71	9,105	0.1	X	1995	0.16	1	106	56	39	5
Finland	107.00	20,673	3.0	X	1995	2.44	2	477	17	82	0
France	180.00	3,047	11.0	20.5	1995	40.64	23	700	15	73	12
Germany	107.00	1,301	71.0	X	1990	46.27	43	583	14	86	0
Greece	54.00	5,073	15.0	3.0	1990	7.03	13	688	16	3	81
Hungary	6.00	598	114.0	X	1995	6.26	104	612	14	70	5
Iceland	170.00	605,049	X	X	1995	0.16	0	611	50	6	0
Ireland	49.00	13,136	3.0	X	1995	1.18	2	326	40	21	15
Italy	160.68	2,804	6.8	0.0	1993	57.54	36	1,005	18	37	45
Latvia	16.74	7,104	18.7	0.7	1994	0.29	2	111	55	32	13
Lithuania	15.56	4,239	9.3	24.7	1995	0.25	2	68	81	16	3
Macedonia, FYR	6.00	2,965	1.0	7.0	X	X	X	X	X	X	X
Moldova, Rep	1.00	228	10.7	11.7	1992	2.96	296	677	9	65	26
Netherlands	11.00	697	80.0	X	1990	7.81	71	522	16	68	0
Norway	382.00	85,560	11.0	X	1985	2.03	1	488	27	68	3
Poland	55.00	1,419	8.0	X	1995	12.07	22	313	20	67	3
Portugal	37.00	3,747	35.0	0.0	1990	7.29	20	739	8	40	53
Romania	37.00	1,657	X	X	X	X	X	X	X	X	X
Russian Federation	4,313.70	29,358	184.5	X	1994	77.10	2	520	19	62	20
Slovakia	13.00	2,413	70.0	X	1995	1.41	11	263	39	50	8
Slovenia	18.50	9,317	0.0	17.3	1994	0.50	3	250	50	50	X
Spain	111.80	2,821	0.3	29.0	1997	35.52	32	897	13	18	68
Sweden	178.00	19,977	0.0	X	1995	2.73	2	310	35	30	4
Switzerland	40.00	5,416	13.0	X	1995	2.60	6	363	42	58	0
Ukraine	53.10	1,052	86.5	5.8	1992	25.99	49	501	18	52	30
United Kingdom	145.00	2,465	2.0	X	1995	9.34	6	160	65	8	2
Yugoslavia	44.00	4,135	144.0	182.0	X	X	X	X	X	X	X
MIDDLE EAST & N. AFRICA	**X**	**X**	**X**	**X**	**X**	**X**	**X**	**X**	**X**	**X**	**X**
Afghanistan	55.00	2,421	10.0	X	1987	26.11	47	1,846	1	0	99
Algeria	13.90	442	0.4	0.4	1990	4.50	32	180	25	15	60
Egypt	2.30	34	56.0	0.0	1993	55.10	2,396	920	6	8	86
Iran, Islamic Rep	128.50	1,898	X	55.9	1993	70.03	55	1,165	6	2	92
Iraq	35.20	1,523	X	0.0	1990	42.80	122	2,368	3	5	92
Israel	0.75	121	0.3	0.0	1997	1.71	228	292	29	7	64
Jordan	0.68	102	X	X	1993	0.98	145	187	22	3	75
Kuwait	0.02	10	0.0	0.0	1994	0.54	2,690	307	37	2	60
Lebanon	4.80	1,463	0.0	0.7	1994	1.29	27	444	28	4	68
Libyan Arab Jamahiriya	0.80	143	0.0	0.0	1995	3.89	486	783	9	4	87
Morocco	30.00	1,058	0.0	0.2	1991	11.05	37	454	5	3	92
Oman	0.99	388	X	X	1991	1.20	122	646	5	2	94
Saudi Arabia	2.40	111	0.0	X	1992	17.00	708	1,002	9	1	90
Syrian Arab Rep	7.00	434	37.7	32.0	1993	14.41	206	1,069	4	2	94
Tunisia	3.52	367	0.6	0.0	1996	2.83	80	312	13	2	86
Turkey	196.00	2,943	7.6	60.4	1997	35.50	18	560	16	11	73
United Arab Emirates	0.15	61	0.0	X	1995	2.11	1,405	954	24	9	67
Yemen	4.10	226	X	X	1990	2.93	72	253	7	1	92

	Average Annual Internal Renewable Water Resources		Annual River Flows		Annual Withdrawals				Sectoral Withdrawals (percent)		
	Total (cubic km)	Per Capita (cubic meters) Year 2000	From Other Countries (cubic km)	To Other Countries (cubic km)	Year of Data	Total (cubic km)	Percentage of Water Resources	Per Capita (cubic meters)	Domestic	Industry	Agriculture
SUB-SAHARAN AFRICA	**X**	**X**	**X**	**X**	**X**	**X**	**X**	**X**	**X**	**X**	**X**
Angola	184.00	14,288	X	X	1987	0.48	0	57	14	10	76
Benin	10.30	1,689	15.5	X	1994	0.15	1	28	23	10	67
Botswana	2.90	1,788	11.8	X	1992	0.11	4	81	32	20	48
Burkina Faso	17.50	1,466	X	X	1992	0.38	2	39	19	0	81
Burundi	3.60	538	X	X	1987	0.10	3	20	36	0	64
Cameroon	268.00	17,766	0.0	0.0	1987	0.40	0	38	46	19	35
Central African Rep	141.00	39,001	X	X	1987	0.07	0	26	21	6	73
Chad	15.00	1,961	28.0	X	1990	0.19	1	33	16	2	82 b
Congo	221.90	75,387	610.0	X	1987	0.04	0	20	62	27	11
Congo, Dem Rep	935.00	18,101	84.0	X	1994	0.36	0	8	61	16	23
Côte d'Ivoire	76.70	5,187	1.0	X	1987	0.70	1	66	22	11	67
Equatorial Guinea	30.00	66,275	0.0	X	1987	0.01	0	30	81	13	6
Eritrea	2.80	727	6.0	X	X	X	X	X	X	X	X
Ethiopia	110.00	1,758	0.0	X	1987	2.20	2	50	11	3	86
Gabon	164.00	133,754	0.0	X	1987	0.06	0	70	72	22	6
Gambia	3.00	2,298	5.0	X	1990	0.03	1	33	7	2	91 c
Ghana	30.30	1,499	22.9	X	1970	0.30	1	35	35	13	52
Guinea	226.00	30,416	0.0	X	1987	0.74	0	141	10	3	87
Guinea-Bissau	16.00	13,189	11.0	X	1991	0.02	0	17	60	4	36
Kenya	20.20	672	10.0	X	1990	2.05	10	87	20	4	76
Lesotho	5.23	2,430	0.0	X	1987	0.05	1	31	22	22	56
Liberia	200.00	63,412	32.0	X	1987	0.13	0	54	27	13	60
Madagascar	337.00	21,139	0.0	X	1990	19.70	6	1,694	1	0	99 d
Malawi	17.54	1,605	1.1	X	1994	0.94	5	98	10	3	86
Mali	60.00	5,341	40.0	X	1987	1.36	2	164	2	1	97
Mauritania	0.40	150	11.0	X	1990	16.30	4,075	8,046	6	2	92 e
Mozambique	100.00	5,081	116.0	X	1992	0.61	1	40	9	2	89
Namibia	6.20	3,592	39.3	X	1990	0.25	4	185	29	3	68
Niger	3.50	326	29.0	X	1990	0.50	14	65	16	2	82 f
Nigeria	221.00	1,982	59.0	X	1990	4.00	2	46	31	15	54 b
Rwanda	6.30	815	X	X	1993	0.77	12	134	5	1	94
Senegal	26.40	2,784	13.0	X	1990	1.50	6	205	5	3	92 b
Sierra Leone	160.00	32,960	0.0	X	1987	0.37	0	98	7	4	89
Somalia	6.00	594	9.7	X	1987	0.81	14	115	3	0	97
South Africa	44.80	1,110	5.2	X	1990	13.30	30	391	17	11	72
Sudan	35.00	1,187	119.0	X	1995	17.80	51	669	5	1	94
Tanzania, United Rep	80.00	2,387	9.0	X	1994	1.17	1	40	9	2	89
Togo	11.50	2,484	0.5	X	1987	0.09	1	28	62	13	25
Uganda	39.00	1,791	27.0	X	1970	0.20	1	20	32	8	60
Zambia	80.20	8,747	35.8	X	1994	1.71	2	214	16	7	77
Zimbabwe	14.10	1,208	5.9	X	1987	1.22	9	136	14	7	79
NORTH AMERICA {a} {g}	**7,770.00**	**25,105**	**X**	**X**	**1990**	**617.10**	**8**	**2,189**	**11**	**42**	**47**
Canada	2,740.00	87,971	52.0	X	1990	45.10	2	1,623	11	80	9
United States	2,460.00	8,838	18.0	0.0	1995	447.71	18	1,677	8	65	27
C. AMERICA & CARIBBEAN	**X**	**X**	**X**	**X**	**X**	**X**	**X**	**X**	**X**	**X**	**X**
Belize	16.00	66,470	X	X	1993	0.10	1	469	12	88	0
Costa Rica	112.40	27,936	X	X	1997	5.77	5	1,540	13	7	80
Cuba	38.00	3,393	X	X	1995	5.21	14	475	49	0	51
Dominican Rep	21.00	2,472	X	X	1994	8.34	40	1,085	11	0	89
El Salvador	17.70	2,820	X	X	1992	0.73	4	137	34	20	46
Guatemala	134.40	11,805	X	X	1992	1.16	1	126	9	17	74
Haiti	12.11	1,473	X	X	1991	0.98	8	139	5	1	94
Honduras	96.10	14,818	X	X	1992	1.52	2	293	4	5	91
Jamaica	9.40	3,640	X	X	1993	0.90	10	371	15	7	77
Mexico	409.00	4,136	49.0	X	1998	77.81	19	812	17	5	78
Nicaragua	190.20	37,484	X	X	1998	1.29	1	267	14	2	84
Panama	147.40	51,616	X	X	1990	1.64	1	685	28	2	70
Trinidad and Tobago	X	X	X	X	X	X	X	X	X	X	X
SOUTH AMERICA {a}	**12,030.00**	**34,791**	**X**	**X**	**1990**	**140.70**	**1**	**477**	**20**	**11**	**69**
Argentina	360.00	9,721	X	X	1995	28.58	8	822	16	9	75
Bolivia	316.00	37,941	X	X	1990	1.38	0	210	32	20	48 b
Brazil	5,418.00	31,849	X	X	1992	54.87	1	359	21	18	61
Chile	928.00	61,007	X	X	1990	21.40	2	1,634	5	11	84 b
Colombia	2,133.00	50,400	X	X	1996	8.94	0	228	59	4	37
Ecuador	442.00	34,952	X	X	1997	16.99	4	1,423	12	6	82
Guyana	241.00	279,799	X	X	1992	1.46	1	1,811	1	0	98
Paraguay	94.00	17,102	X	X	1987	0.43	0	112	15	7	78
Peru	1,746.00	68,039	X	X	1992	18.97	1	849	7	7	86
Suriname	200.00	479,467	X	X	1990	0.49	0	1,220	6	5	89 b
Uruguay	59.00	17,680	X	X	1990	4.20	7	1,352	6	3	91 h
Venezuela	846.00	35,002	X	X	1970	4.10	0	382	44	10	46
OCEANIA {a}	**2,400.00**	**78,886**	**X**	**X**	**1990**	**24.30**	**1**	**919**	**57**	**14**	**30**
Australia	352.00	18,638	0.0	X	1995	15.06	4	839	12	6	70
Fiji	X	X	X	X	X	X	X	X	0	0	0
New Zealand	327.00	84,673	X	X	1995	2.00	1	545	9	13	55
Papua New Guinea	801.00	166,644	X	X	1987	0.10	0	28	29	22	49
Solomon Islands	X	X	X	X	X	X	X	X	X	X	X
AFRICA {a}	**4,040.00**	**5,152**	**X**	**X**	**1990**	**148.80**	**4**	**242**	**9**	**6**	**85**
ASIA {a}	**13,508.00**	**3,949**	**X**	**X**	**1990**	**2,007.00**	**15**	**675**	**7**	**9**	**84**

Notes: Zero is either 0 or less than one-half the unit of measure. Total withdrawals may exceed 100 percent due to groundwater drawdowns, withdrawals from river inflows, and the operation of desalinization plants. a. World and regional totals are reported by I.A. Shiklomanov. b. Sectoral withdrawal estimates are for 1987. c. Sectoral withdrawal estimates are for 1982. d. Sectoral withdrawal estimates are for 1984. e. Sectoral withdrawal estimates are for 1985. f. Sectoral withdrawal estimates are for 1988. g. Data include Central America and the Caribbean. h. Sectoral withdrawal estimates are for 1965.

Sources: Various

	Average Annual Groundwater Recharge		Annual Groundwater Withdrawals							Desalinated Water Production (million cubic meters)
	Total (cubic km)	Per Capita (cubic meters)			Percentage	Per Capita	Sectoral Share (percentage)			
	Years Vary	Year 2000	Year	Total (cubic km)	of Annual Recharge	(cubic meters)	Domestic	Industry	Agriculture	1990
WORLD {a}	**X**	**X**	**1990**	**760.0**	**X**	**125.5**	**24**	**72**	**5**	**X**
ASIA (EXCL. MIDDLE EAST)	**X**	**X**	**X**	**X**	**X**	**X**	**X**	**X**	**X**	**X**
Armenia	4.2	1,193	X	X	X	X	X	X	X	X
Azerbaijan	6.5	842	X	X	X	X	X	X	X	X
Bangladesh	21.0	163	1990	10.7	50.9	82.7	13	1	86 b	X
Bhutan	X	X	X	X	X	X	X	X	X	X
Cambodia	17.6 c	1,576	X	X	X	X	X	X	X	X
China	828.8 c	649	1988	52.9	6.4	41.4	X	46 d	54 b	X
Georgia	17.2	3,469	X	X	X	X	X	X	X	X
India	418.5 c	413	1990	190.0	45.4	187.4	9	2	89	X
Indonesia	455.0 c	2,145	X	X	X	X	X	X	X	X
Japan	185.0 c	1,460	1995	13.6	7.3	107.1	29	41	30 b	40.0
Kazakhstan	35.9	2,211	1993	2.4	6.7	147.9	21	71	8 e	1,328.0 f
Korea, Dem People's Rep	21.0 c	874	X	X	X	X	X	X	X	X
Korea, Rep	13.3 c	284	1995	2.5	18.6	52.9	X	83 d	17 b	X
Kyrgyzstan	13.6	2,894	1994	0.6	4.4	127.7	X	X	X	X
Lao People's Dem Rep	38.0	6,994	X	X	X	X	X	X	X	X
Malaysia	64.0 c	2,877	1995	0.4	0.6	17.2	62	33	5	X
Mongolia	6.1	2,291	1993	0.4	5.8	132.5	X	X	X	X
Myanmar	156.0 c	3,420	X	X	X	X	X	X	X	X
Nepal	X	X	X	X	X	X	X	X	X	X
Pakistan	55.0	351	1991	55.0	100.0	351.5	X	11 d	89 b	X
Philippines	180.0 c	2,369	1980	4.0	2.2	52.7	50	50	0	X
Singapore	X	X	X	X	X	X	X	X	X	X
Sri Lanka	7.8	414	X	X	X	X	X	X	X	X
Tajikistan	6.0	970	1994	2.3	37.7	365.2	X	X	X	X
Thailand	41.9 c	682	1980	0.7	1.7	11.4	60	26	14	X
Turkmenistan	3.4	753	1994	0.4	11.9	89.9	53	9	38	X
Uzbekistan	19.7	809	1994	7.4	37.6	303.9	33	11	57	X
Viet Nam	48.0 c	601	1990	0.8	1.7	9.9	X	X	X	X
EUROPE	**X**	**X**	**X**	**X**	**X**	**X**	**X**	**X**	**X**	**X**
Albania	7.0 c	2,248	1989	0.6	9.0	202.3	48	0	52	X
Austria	22.3 c	2,716	1995	1.4	6.2	168.1	52	43	5 b	X
Belarus	18.0	1,758	1989	1.2	6.6	115.5	56	14	30 b	X
Belgium	0.9 g	89	1980	0.8	86.4	76.6	68	27	5	X
Bosnia and Herzegovina	X	X	X	X	X	X	X	X	X	X
Bulgaria	3.1 g	377	1988	5.0	161.3	607.9	X	X	X	X
Croatia	11.0	2,459	X	X	X	X	X	X	X	X
Czech Rep	X	X	1995	0.5	X	48.4	X	X	X	X
Denmark	4.3 h	812	1995	0.9	20.6	167.6	40	22	38 b	X
Estonia	4.0	2,865	X	X	X	X	X	X	X	X
Finland	2.2 g	425	1995	0.2	11.1	47.1	65	11	24 b	X
France	100.0 c	1,693	1994	6.0	6.0	101.5	56	27	17	X
Germany	45.7 c	556	1990	7.1	15.5	86.3	49	48	4 b	X
Greece	10.8	235	1990	2.0	80.0	187.9	37	5	58	X
Hungary	6.8 c	678	1995	1.0	14.5	98.4	35	48	18 b	X
Iceland	24.0 c	85,419	1995	0.2	0.6	533.9	X	X	X	X
Ireland	3.5 g	928	1995	0.2	6.5	60.3	35	37	29 b	X
Italy	43.0	750	1992	13.9	46.3	242.6	39	4	58	X
Latvia	2.2	934	X	X	X	X	X	X	X	X
Lithuania	1.2	327	1995	0.2	17.1	55.9	X	X	X	X
Macedonia, FYR	X	X	X	X	X	X	X	X	X	X
Moldova, Rep	0.4	91	X	X	X	X	X	X	X	X
Netherlands	4.5 h	285	1990	1.0	23.3	66.5	32	45	23 b	X
Norway	96.0 c	21,502	1985	0.4	0.4	90.7	27	73	0	X
Poland	36.0	929	1995	2.0	5.5	51.3	70	30	0 b	X
Portugal	5.1 g	516	1995	3.5	68.6	354.4	39	23	39 b	X
Romania	8.3 h	372	1993	3.6	43.7	162.6	61	38	1 b	X
Russian Federation	788.0	5,363	1988	12.6	1.6	85.4	X	X	X	X
Slovakia	X	X	1995	0.6	X	112.3	X	X	X	X
Slovenia	X	X	1994	0.2	X	88.6	X	X	X	X
Spain	28.9	522	1995	5.4	26.1	136.5	18	2 j	80 j	X
Sweden	20.0 g	2,245	1995	0.6	3.2	71.9	92	8	0 b	X
Switzerland	2.7	366	1995	0.9	33.4	122.1	64	36	0 b	X
Ukraine	20.0	396	1989	4.0	20.1	79.6	30	18	52 b	X
United Kingdom	9.8	167	1995	2.5	25.2	42.0	51	47	2 b	X
Yugoslavia	3.0	282	X	X	X	X	X	X	X	X
MIDDLE EAST & N. AFRICA	**X**	**X**	**X**	**X**	**X**	**X**	**X**	**X**	**X**	**X**
Afghanistan	29.0 c	1,276	X	X	X	X	X	X	X	X
Algeria	1.7 g	54	1989	2.9	167.6	90.6	46	5	49	64.0
Egypt	1.3 g	19	1992	2.7	407.7	77.4	58	0	42	25.0
Iran, Islamic Rep	42.0	620	1980	29.0	69.0	428.3	X	X	X	2.9 k
Iraq	13.0	562	1985	0.2	1.5	8.7	56	44	0	X
Israel	0.5	80	1996	1.2	234.0	188.2	18	2	80	20.0
Jordan	0.6 g	87	1993	0.5	91.4	79.5	30	4	66 l	2.0 f
Kuwait	X	X	1994	0.3	X	126.8	0	0	100 l	231.0 f
Lebanon	4.8	1,463	1991	0.4	12.5	121.9	13	9	8	X
Libyan Arab Jamahiriya	0.7	116	1995	3.7	561.5	651.2	9	4	87	70.0 o
Morocco	9.0	317	1998	2.7	29.8	94.5	16	0 k	84 k	3.4 j
Oman	1.0	376	1985	0.4	41.9	157.4	X	X	X	34.0 m
Saudi Arabia	2.2 c	102	1990	14.5	660.2	672.2	10	0	90	714.0 m
Syrian Arab Rep	6.6	409	1993	1.8	27.3	111.6	13	4	83	X
Tunisia	4.2	433	1995	1.6	134.4	169.6	10	4	86	8.3
Turkey	20.0 h	300	1995	7.6	38.0	114.1	31	9	60 b	0.5
United Arab Emirates	0.1	49	1995	1.6	1345.8	661.5	X	20 d	80	385.0 m
Yemen	1.5	84	1985	1.4	88.5	74.5	X	X	X	10.0 n

	Average Annual Groundwater Recharge		Annual Groundwater Withdrawals							Desalinated Water Production (million cubic meters)
	Total (cubic km)	Per Capita (cubic meters)			Percentage	Per Capita	Sectoral Share (percentage)			
	Years Vary	Year 2000	Year	Total (cubic km)	of Annual Recharge	(cubic meters)	Domestic	Industry	Agriculture	1990
SUB-SAHARAN AFRICA	**X**	**X**	**X**	**X**	**X**	**X**	**X**	**X**	**X**	**X**
Angola	72.0 c	5,591	X	X	X	X	X	X	X	0.1
Benin	1.8 g	295	X	X	X	X	X	X	X	X
Botswana	1.7 g	1,048	X	X	X	X	X	X	X	X
Burkina Faso	9.5 g	796	X	X	X	X	X	X	X	X
Burundi	2.1 c	314	X	X	X	X	X	X	X	X
Cameroon	100.0 c	6,629	X	X	X	X	X	X	X	X
Central African Rep	56.0 c	15,490	X	X	X	X	X	X	X	X
Chad	11.5 c	1,503	1990	0.1	0.8	11.8	29	0	71	X
Congo	198.0 c	67,268	X	X	X	X	X	X	X	0.2
Congo, Dem Rep	421.0 c	8,150	X	X	X	X	X	X	X	X
Côte d'Ivoire	37.7 g	2,550	X	X	X	X	X	X	X	X
Equatorial Guinea	10.0 c	22,092	X	X	X	X	X	X	X	X
Eritrea	X	X	X	X	X	X	X	X	X	X
Ethiopia	44.0 c	703	X	X	X	X	X	X	X	X
Gabon	62.0 c	50,566	1989	0.0	0.0	0.4	100	0	0	X
Gambia	0.5 c	383	X	X	X	X	X	X	X	X
Ghana	26.3 g	1,301	X	X	X	X	X	X	X	X
Guinea	38.0 c	5,114	X	X	X	X	X	X	X	X
Guinea-Bissau	14.0 c	11,541	X	X	X	X	X	X	X	X
Kenya	3.0 g	100	X	X	X	X	X	X	X	X
Lesotho	0.5 c	232	X	X	X	X	X	X	X	X
Liberia	60.0 c	19,023	X	X	X	X	X	X	X	X
Madagascar	55.0 c	3,450	1984	4.8	8.7	298.6	100	0	0	X
Malawi	1.4 c	128	X	X	X	X	X	X	X	X
Mali	20.0 g	1,780	1989	0.1	0.5	8.9	7	0	93	X
Mauritania	0.3 g	112	1985	0.9	293.3	329.6	X	X	X	1.7
Mozambique	17.0 c	864	X	X	X	X	X	X	X	0.1
Namibia	2.1 c	1,217	X	X	X	X	X	X	X	3.0
Niger	2.5 g	233	1988	0.1	5.2	12.1	58	4	39	X
Nigeria	87.0 c	780	X	X	X	X	X	X	X	3.0
Rwanda	3.6 c	466	X	X	X	X	X	X	X	X
Senegal	7.6 c	802	1985	0.3	3.3	26.4	25	0	75	0.1
Sierra Leone	50.0 c	10,300	X	X	X	X	X	X	X	X
Somalia	3.3 c	327	1985	0.3	9.1	29.7	X	X	X	0.1
South Africa	4.8	119	1980	1.8	37.3	44.3	11	6	84	17.5
Sudan	7.0	237	1985	0.3	4.0	9.5	X	X	X	0.4
Tanzania, United Rep	30.0 c	895	X	X	X	X	X	X	X	X
Togo	5.7 g	1,231	X	X	X	X	X	X	X	X
Uganda	29.0 c	1,332	X	X	X	X	X	X	X	X
Zambia	47.1	5,137	X	X	X	X	X	X	X	X
Zimbabwe	5.0 c	428	X	X	X	X	X	X	X	X
NORTH AMERICA	**X**	**X**	**X**	**X**	**X**	**X**	**X**	**X**	**X**	**X**
Canada	370.0	11,879	1990	1.0	0.3	33.3	43	14	43 b	X
United States	660.0	2,371	1995	109.8	16.2	384.5	24	10	66	X
C. AMERICA & CARIBBEAN	**X**	**X**	**X**	**X**	**X**	**X**	**X**	**X**	**X**	**X**
Belize	X	X	X	X	X	X	X	X	X	X
Costa Rica	21.0 c	5,219	X	X	X	X	X	X	X	X
Cuba	8.0 c	714	1975	3.8	47.5	339.3	X	X	X	X
Dominican Rep	3.0 c	353	X	X	X	X	X	X	X	X
El Salvador	X	X	X	X	X	X	X	X	X	X
Guatemala	31.0 c	2,723	X	X	X	X	X	X	X	X
Haiti	2.5 h	304	X	X	X	X	X	X	X	X
Honduras	39.0 c	6,013	X	X	X	X	X	X	X	X
Jamaica	X	X	X	X	X	X	X	X	X	X
Mexico	139.0 c	1,406	1995	25.1	18.1	253.8	13	23	64 b	X
Nicaragua	59.0 c	11,627	X	X	X	X	X	X	X	X
Panama	42.0 c	14,708	X	X	X	X	X	X	X	X
Trinidad and Tobago	X	X	X	X	X	X	X	X	X	X
SOUTH AMERICA	**X**	**X**	**X**	**X**	**X**	**X**	**X**	**X**	**X**	**X**
Argentina	128.0 c	3,456	1975	4.7	3.7	126.9	11	19	70	X
Bolivia	130.0 c	15,609	X	X	X	X	X	X	X	X
Brazil	1,874.0 c	11,016	1987	8.0	0.4	47.0	38	25	38	X
Chile	140.0 c	9,204	X	X	X	X	X	X	X	X
Colombia	510.0 c	12,051	X	X	X	X	X	X	X	X
Ecuador	134.0 c	10,596	X	X	X	X	X	X	X	X
Guyana	103.0 c	119,582	X	X	X	X	X	X	X	X
Paraguay	41.0 c	7,459	X	X	X	X	X	X	X	X
Peru	303.0 c	11,807	1973	2.0	0.7	77.9	25	15	60	X
Suriname	80.0 c	191,787	X	X	X	X	X	X	X	X
Uruguay	23.0 c	6,892	X	X	X	X	X	X	X	X
Venezuela	227.0 c	9,392	X	X	X	X	X	X	X	X
OCEANIA	**X**	**X**	**X**	**X**	**X**	**X**	**X**	**X**	**X**	**X**
Australia	72.0 g	3,812	1985	2.2	3.1	118.6	X	23 d	77 b	X
Fiji	X	X	X	X	X	X	X	X	X	X
New Zealand	198.0 c	51,270	X	X	X	X	X	X	X	X
Papua New Guinea	X	X	X	X	X	X	X	X	X	X
Solomon Islands	X	X	X	X	X	X	X	X	X	X

Notes: Zero is either 0 or less than one-half the unit of measure. a. World totals reported by I.A. Shiklomanov. b. Sectoral data predate other withdrawal data. c. Sum of all groundwater flows (as a constituent of surface water flows). d. Domestic and industrial withdrawals combined. e. Both withdrawal and sectoral data estimated from a FAO bar graph (Food and Agriculture Organization of the United Nations [FAO] Water Report). f. Data are from 1993. g. Sum of all acquifer recharge flows. h. Sum of the total groundwater flow that is exploitable. i. Sum of groundwater flows collected by water courses and those that discharge directly to the sea. j. Data are from 1992. k. Data are from 1991. l. Sectoral data are estimated from a FAO bar graph (FAO Water Report). m. Data are from 1995. n. Data are from 1989. o. Data are from 1994.

Sources: Various

Major Watersheds	Modeled Watershed Area {a} (km²)	Countries within the Watershed (number)	Average Population Density (per km²)	Percent of Watershed that is: Cropland	Forest	Grassland	Built-up Area {b}	Irrigated Area	Arid Area	Wetlands	Ramsar Sites {c} (number)	Water Available Per Person (m³/person/year)	Large Dams in Progress (number)	Degree of River Fragmentation {d}
ASIA														
Amu Darya	534,739	5	39	22.4	0.1	57.3	3.7	7.5	72.0	0.0	0	3,211	2	High
Amur	1,929,955	3 e	34	18.4	53.8	8.8	2.6	0.8	15.1	4.4	7	4,917	4	Medium
Brahmaputra	651,335	4 e	178	29.4	18.5	44.7	2.4	3.7	0.0	20.7	1	X	3	X
Chao Phrya	178,785	1	122	44.7	35.4	11.1	9.2	12.5	0.0	8.4	0	1,237	0	High
Ganges	1,016,124	4 e	398	72.4	4.2	13.4	6.3	22.7	26.0	17.7	4	X	5	X
Godavari	319,810	1	203	64.0	6.8	22.5	6.7	11.7	42.9	1.2	0	1,602	0	X
Hong	170,888	2	191	36.3	43.2	15.5	2.1	3.9	0.0	5.4	0	3,083	3	X
Huang He	944,970	1	157	29.5	1.5	60.0	5.9	7.2	37.5	1.1	0	361	7	High
Indigirka	274,818	1	<1	0.0	2.4	69.7	0.1	0.0	0.0	3.0	0	973,515	0	Low
Indus	1,081,718	4	163	30.0	0.4	46.4	4.6	24.1	62.6	4.2	10	830	3	X
Irrawaddy	413,710	3	78	30.5	56.2	9.7	1.9	3.4	0.0	6.3	0	18,614	0	X
Kizil	122,277	1	55	38.0	1.6	52.0	6.4	4.1	84.9	3.4	2	1,171	9	X
Kolyma	679,934	1	<1	0.0	0.7	45.3	0.3	0.0	0.0	1.0	0	722,456	1	Medium
Krishna	226,037	1	263	66.4	2.8	22.7	8.8	16.2	41.3	16.2	0	786	2	X
Kura-Araks	205,037	5	75	54.0	7.1	30.6	6.3	10.7	25.4	0.9	2	1,121	4	High
Lake Balkhash	512,015	2	11	23.2	4.0	61.1	1.5	1.9	91.6	4.7	0	439	0	X
Lena	2,306,743	1	1	1.7	64.7	11.4	0.4	0.0	0.7	0.6	0	161,359	0	Medium
Mahanadi	145,816	1	198	59.5	8.1	26.7	4.9	17.4	0.0	0.2	0	2,171	1	X
Mekong	805,604	6	71	37.8	41.5	17.2	2.1	2.9	0.0	8.7	0	8,934	3	Medium
Narmada	96,271	1	177	76.5	0.8	15.8	6.1	24.0	25.8	0.8	0	2,159	2	X
Ob	2,972,493	4	10	36.9	33.9	16.0	3.0	0.5	42.5	11.2	4	14,937	0	Medium
Salween	271,914	3	22	5.5	43.4	48.3	0.5	0.4	0.0	9.5	0	23,796	1	X
Syr Darya	782,617	4	28	22.2	2.4	67.4	3.2	5.4	88.5	2.0	1	1,171	4	High
Tapti	74,627	1	239	78.3	0.2	14.7	7.6	13.3	63.7	0.8	0	1,107	1	X
Tarim	1,152,448	2	7	2.3	0.0	35.3	0.3	0.6	61.4	16.3	0	754	0	High
Tigris & Euphrates	765,742	4	57	25.4	1.2	47.7	6.2	9.1	90.9	2.9	0	2,189	19	X
Xun Jiang	409,480	2	194	66.5	9.6	6.1	5.3	5.2	0.0	1.3	0	3,169	10	X
Yalu Jiang	48,331	2	102	41.6	51.2	2.2	2.9	4.4	0.0	1.0	0	3,628	0	High
Yangtze	1,722,193	1	212	47.6	6.3	28.2	3.0	7.1	0.0	3.0	2	2,265	38	Medium
Yenisey	2,554,388	2	3	12.8	39.7	32.4	1.3	0.0	10.9	2.7	1	79,083	1	High
EUROPE														
Dalalven	30,410	1	10	2.3	71.6	3.4	13.5	0.0	0.0	19.1	1	18,476	0	High
Danube	795,656	13 e	102	66.9	18.2	3.2	10.7	5.2	2.6	1.4	47	2,519	11	High
Dnieper	533,966	3	62	86.5	2.2	1.3	8.8	1.8	3.4	5.9	0	1,552	0	High
Dniester	68,627	3	106	82.8	5.2	1.9	9.3	3.8	5.8	1.1	0	1,621	0	X
Don	458,694	2	46	83.1	1.4	5.4	8.8	3.2	33.1	0.5	1	1,422	0	High
Duero	98,258	2	43	67.4	0.6	21.5	9.9	5.6	9.7	0.4	1	4,093	2	High
Ebro	82,587	1 e	34	58.2	5.1	22.1	13.7	10.0	39.8	0.9	4	8,235	5	High
Elbe	148,919	4	170	75.7	2.0	3.2	18.2	1.3	0.0	1.7	11	1,195	0	High
Garonne	53,540	3	61	75.9	6.4	2.1	15.4	4.0	0.0	0.1	0	5,504	0	High
Glåma	41,795	1	27	1.3	46.0	25.0	15.4	0.7	0.0	1.8	3	17,907	0	High
Guadalquivir	52,664	1	73	52.7	0.5	27.2	18.9	10.4	34.6	3.2	3	2,645	1	X
Kemijoki	52,456	3	3	0.3	77.8	11.6	6.2	0.0	0.0	2.9	1	132,939	0	High
Loire	115,282	1	65	83.7	1.6	0.1	14.3	0.7	0.0	0.9	3	3,386	0	Medium
North Dvina	357,075	1	6	11.7	83.4	0.4	3.0	0.0	0.0	1.5	0	48,450	0	Medium
Oder	124,164	3	121	82.3	0.7	2.7	13.3	0.4	0.0	0.3	4	1,271	0	Medium
Pechora	289,532	1	2	0.2	49.5	45.1	1.2	0.0	0.0	5.0	0	215,057	0	Low
Po	76,997	2	214	50.2	13.0	12.7	20.0	16.2	0.0	1.8	9	2,731	1	X
Rhine-Maas	198,735	8	310	64.7	6.8	1.4	25.7	3.3	0.0	1.0	20	1,396	0	Medium
Rhône	100,543	2	100	62.0	11.0	5.3	20.1	4.1	0.0	1.0	3	5,401	0	High
Seine	78,919	1 e	199	79.0	1.6	0.0	19.2	1.4	0.0	0.1	1	965	0	Medium
Tagus	78,467	2	118	46.8	0.1	34.8	16.6	5.2	31.4	1.6	2	2,248	0	High
Ural	244,334	2	14	59.3	2.3	33.4	4.2	0.9	99.6	0.2	0	2,003	0	X
Vistula	180,156	4 e	139	83.2	1.8	2.2	11.7	0.2	0.0	3.2	2	1,367	0	Medium
Volga	1,410,951	2 e	42	60.2	22.5	7.3	8.2	0.4	19.6	1.1	2	4,260	0	High
West Dvina	79,389	3	29	84.8	7.2	0.0	5.5	0.2	0.0	3.4	0	6,626	0	High
Weser	45,138	1	203	78.7	1.8	1.0	18.3	1.7	0.0	0.1	4	1,567	0	High

Major Watersheds	Modeled Watershed Area {a} (km²)	Countries within the Watershed (number)	Average Population Density (per km²)	Percent of Watershed that is: Crop-land	Forest	Grass-land	Built-up Area {b}	Irrigated Area	Arid Area	Wetlands	Ramsar Sites {c} (number)	Water Available Per Person (m³/ person/ year)	Large Dams in Progress (number)	Degree of River Fragmen-tation {d}
AFRICA														
Congo	3,730,881	9	15	7.2	44.0	45.4	0.2	0.0	0.0	9.0	3	22,752	0	Medium
Cuanza	149,688	1	24	2.8	16.2	79.6	0.3	0.0	5.8	2.1	0	17,126	0	Medium
Cunene	109,832	2	10	2.6	3.3	90.9	0.1	0.1	15.8	2.9	0	13,216	0	X
Jubba	497,626	3	12	6.6	2.7	87.9	0.2	0.1	71.5	3.5	2	1,076	0	X
Lake Chad {f}	2,497,738	8	12	3.1	0.2	45.2	0.2	0.0	82.8	8.2	1	7,922	0	Low
Lake Turkana	209,096	4	60	20.8	11.9	50.2	0.1	0.3	33.0	5.9	0	4,450	0	X
Limpopo	421,123	4	32	26.3	0.7	67.7	4.5	0.9	47.3	2.8	1	716	0	High
Mangoky	58,851	1	18	4.5	3.3	90.8	0.1	2.3	39.1	0.2	0	19,059	0	Low
Mania	56,118	1	25	2.5	5.7	89.8	0.2	2.6	0.1	0.9	0	25,913	0	Low
Niger	2,261,741	10	32	4.4	0.9	68.6	0.5	0.1	65.4	4.1	6	4,076	1	High
Nile	3,254,853	10	44	10.7	2.0	53.0	1.0	1.4	67.4	6.1	2	2,207	0	High
Ogooué	223,946	4	2	0.8	75.1	21.7	0.5	0.0	0.0	6.2	1	289,401	0	X
Okavango {g}	721,258	4	2	5.5	1.7	91.1	0.2	0.0	75.8	4.1	1	X	0	X
Orange	941,351	4	11	6.0	0.2	85.0	2.2	0.5	77.0	0.8	1	1,050	1	High
Oued Draa	114,544	3	10	0.3	0.2	12.0	0.5	3.2	95.3	0.2	0	2	1	X
Rufiji	204,780	1	21	19.7	2.1	77.4	0.2	0.1	0.0	7.8	0	6,466	0	Low
Senegal	419,575	4	10	4.8	0.1	68.2	0.1	0.0	82.0	3.6	4	5,775	0	High
Shaballe	336,604	2	30	7.1	1.2	87.9	0.1	0.5	80.5	1.8	0	X	0	X
Volta	407,093	6	42	10.4	0.7	85.6	0.5	0.1	59.9	4.6	3	2,054	0	High
Zambezi	1,332,412	8	18	19.9	4.0	72.0	0.7	0.1	8.8	7.6	1	X	0	High
NORTH & CENTRAL AMERICA														
Alabama-Tombigbee	138,139	1	31	9.1	73.0	0.2	17.4	0.0	0.0	4.0	0	15,832	0	High
Balsas	117,095	1	85	4.1	37.6	46.6	11.5	3.1	12.4	0.0	0	1,650	0	High
Brazos	137,098	1	18	25.0	1.9	58.8	13.8	5.6	80.2	14.8	0	1,288	0	X
Colorado	703,148	2	10	0.9	17.0	74.9	6.9	2.0	89.1	2.5	0	2,105	1	High
Columbia	657,501	2	9	6.4	50.0	35.5	7.3	3.6	48.7	6.3	1	39,474	0	High
Fraser	248,016	2	5	0.4	85.7	6.2	4.1	0.0	2.5	1.7	0	60,950	0	Medium
Hudson	41,906	1	133	0.3	76.3	0.0	22.8	0.0	0.0	15.0	0	3,335	0	Medium
Mackenzie	1,706,388	1	<1	2.6	66.0	14.7	1.9	0.0	0.0	48.9	3	408,243	0	Medium
Mississippi	3,202,185	2	22	35.8	22.2	28.5	12.6	3.1	35.5	20.0	6	8,973	0	High
Nelson-Saskatchewan	1,093,141	2	5	47.4	31.9	6.1	7.1	0.5	21.5	86.8	5	15,167	0	High
Rio Grande	607,965	2	18	5.2	7.5	80.9	6.0	2.6	96.0	2.1	1	621	0	X
Rio Grande de Santiago	136,694	1	111	4.2	36.3	45.0	13.9	9.0	24.9	0.0	0	655	0	High
San Pedro & Usumacinta	78,720	3	28	30.5	58.7	7.6	2.6	0.2	2.8	0.0	1	30,120	0	Medium
Sacramento	78,757	1	32	6.0	48.6	33.3	11.5	11.5	26.5	3.0	0	3,474	0	High
St. Lawrence {h}	1,049,636	2	43	16.4	43.5	0.1	14.5	0.2	0.0	47.2	7	9,095	0	High
Susquehanna	78,672	1	57	7.0	73.3	0.0	19.2	0.0	0.0	4.0	0	9,029	0	Medium
Thelon	239,245	1	<1	0.0	5.6	21.1	0.0	0.0	0.0	11.0	0	14,641,336	0	Low
Yaqui	79,162	2	8	1.9	61.5	33.0	3.0	2.1	99.9	0.0	0	173	0	X
Yukon	847,620	2	<1	0.0	64.0	27.6	0.4	0.0	0.0	27.8	1	1,249,832	0	Low
SOUTH AMERICA														
Amazon	6,145,186	7	4	14.1	73.4	10.2	0.6	0.1	4.0	8.3	3	273,767	0	Medium
Chubut	182,622	2	1	0.6	24.8	67.7	0.6	0.0	61.4	0.0	0	171,362	0	X
Lakes Titicaca & Salar de Uyuni	193,090	3	7	0.6	0.1	89.4	0.9	0.4	65.4	0.0	5	15,980	0	X
Magdalena	263,773	1	78	35.8	37.2	14.8	10.3	2.4	7.2	0.2	0	10,191	3	Medium
Orinoco	953,675	2	17	7.6	50.5	37.8	2.6	0.2	8.5	15.3	0	90,482	3	Medium
Paraná	2,582,704	4	27	43.3	18.1	33.0	4.2	0.5	9.9	10.9	7	8,025	4	High
Parnaíba	322,887	1	10	44.8	5.8	47.4	1.8	0.1	41.7	18.8	0	7,729	0	Medium
Rio Colorado	403,005	2	6	9.7	1.1	71.2	2.0	1.3	71.0	2.0	1	3,196	0	X
São Francisco	617,814	1	19	60.2	3.1	31.8	2.8	0.3	32.0	9.7	0	8,261	0	High
Tocantins	764,213	1	5	61.5	9.9	26.2	1.3	0.0	0.0	19.1	1	103,383	1	Medium
Uruguay	297,211	3	18	42.7	8.2	44.4	3.5	0.3	0.0	3.9	0	32,731	1	High
OCEANIA														
Belyando	146,219	1	1	2.2	3.6	93.5	0.6	1.3	42.6	0.3	1	239,338	0	X
Dawson	152,375	1	1	3.1	20.8	73.2	2.6	1.8	28.3	0.3	0	39,587	0	X
Fly	78,855	2	4	2.4	75.8	18.3	0.3	0.0	0.0	41.7	0	555,800	0	X
Kapuas	88,781	1	20	33.2	64.8	0.2	0.3	0.0	0.0	1.7	1	105,159	0	X
Mahakam	98,194	1	22	17.2	79.4	0.5	1.3	0.0	0.0	7.7	0	135,955	0	X
Murray-Darling	1,050,116	1	2	28.4	8.0	62.1	1.2	1.6	67.1	3.4	9	11,549	0	X
Sepik	80,321	2	10	6.6	76.3	15.1	0.1	0.0	0.0	33.8	0	143,175	0	X

Notes: Percentages presented in this table do not add up to 100 because different sources were used to estimate land cover and land use within watersheds, land cover types overlap, and not all land cover types were accounted for. Zero is either 0 or less than one-half the unit of measure. a. Watershed area was digitally derived from elevation data using a geographic information system; thus, area may differ from other published sources. b. Based on stable nighttime lights data. These figures overestimate the actual area lit. c. Sites designated as "wetlands of international importance" under the Convention on Wetlands. d. Indicates the level of modification of a river due to dams, reservoirs, interbasin transfers, and irrigation consumption. e. Countries that have <1 percent area in the watershed are excluded. f. Watershed includes intermittent tributaries in northern Chad, Niger, and Algeria. g. Watershed includes intermittent tributaries in Botswana (northern Kalahari Desert). h. Basin excludes the tidal area of the St. Lawrence River.

Sources: Various

	Carbon Dioxide (CO2) Emissions (000 metric tons)								Per Capita CO2 Emissions (kilograms)	CO2 Emitted Per Million Int$ (PPP) of Gross Domestic Product {a} (metric tons)		Signatory to the Kyoto Protocol
	Solid Fuels 1996	Liquid Fuels 1996	Gaseous Fuels 1996	Gas Flaring 1996	Cement Manufact-uring 1996	Total 1990	Total 1996	Total Contribution Since 1950	1996	1990	1996	1999 (Yes/No)
WORLD	**9,013,440**	**9,497,088**	**4,382,144**	**245,488**	**740,128**	**22,361,392**	**23,881,952**	**718,514,064**	**4,157**	**X**	**X**	**X**
ASIA (EXCL. MIDDLE EAST)	**4,251,273**	**2,201,364**	**555,697**	**10,936**	**423,960**	**5,194,233**	**7,452,030**	**145,131,386**	**2,296**	**X**	**X**	**X**
Armenia	15	1,528	2,015	0	141	X	3,697	130,452 b	1,037	X	425	N
Azerbaijan	15	18,166	11,736	0	100	X	30,019	1,637,062 b	3,945	X	2,621	N
Bangladesh	0	8,537	14,279	0	142	15,359	22,959	269,033	190	189	187	N
Bhutan	62	121	0	0	80	128	260	2,015	137	X	X	N
Cambodia	0	498	0	0	0	451	498	14,759	49	56	37	N
China	2,597,109	483,740	38,520	0	244,169	2,401,741	3,363,541	57,580,606	2,729	1,523	949	Y
Georgia	509	645	1,762	0	50	X	2,968	503,077 b	572	X	308	X c,d
India	677,411	229,213	50,230	2,549	37,981	675,261	997,385	15,516,498	1,050	710	652	N
Indonesia	42,026	108,290	74,643	7,641	12,458	165,210	245,056	3,538,050	1,223	447	365	Y
Japan	341,239	651,020	128,321	0	47,086	1,070,665	1,167,666	31,157,964	9,284	463	389	Y
Kazakhstan	122,968	31,913	17,715	0	1,246	X	173,846	10,119,364 b	10,577	X	3,134	Y
Korea, Dem People's Rep	234,741	11,113	0	0	8,471	244,634	254,326	4,954,149	11,249	X	X	N
Korea, Rep	125,459	229,436	24,596	0	28,570	241,179	408,060	4,988,074	8,999	X	X	Y
Kyrgyzstan	1,905	1,957	2,023	0	249	X	6,134	401,943 b	1,335	X	X	N
Lao People's Dem Rep	4	330	0	0	5	231	337	9,413	69	X	X	N
Malaysia	6,698	64,354	36,252	6	6,147	55,279	119,069	1,185,652 e	5,794	637	725	Y
Mongolia	7,159	1,667	0	0	53	9,981	8,882	219,349	3,560	2,751	2,368	N
Myanmar	169	3,851	3,026	15	252	4,148	7,310	177,209	168	X	X	N
Nepal	132	1,312	0	0	171	630	1,612	21,160	74	42	69	N
Pakistan	9,358	49,904	30,638	0	4,435	67,872	94,333	1,188,653	674	516	468	N
Philippines	6,046	51,215	0	0	5,980	44,305	63,241	1,228,649	905	246	256	Y
Singapore	117	64,076	0	0	1,644	41,920	65,835	981,736	19,505	958	801	N
Sri Lanka	4	6,625	0	0	451	3,855	7,079	155,068	391	137	163	N
Tajikistan	267	3,422	2,132	0	25	X	5,844	700,550 b	1,001	X	898	N
Thailand	45,932	117,658	24,329	0	17,441	95,740	205,360	1,899,256	3,471	433	501	Y
Turkmenistan	264	13,015	20,738	0	225	X	34,244	1,219,985 b	8,240	X	3,623	Y
Uzbekistan	3,158	19,427	69,902	0	2,492	X	94,978	3,772,482 b	4,157	X	1,656	Y
Viet Nam	14,777	16,821	15	X	2,840	22,464	37,644	766,842 f	501	X	318	Y
EUROPE	**2,131,682**	**2,076,872**	**1,768,290**	**50,808**	**128,310**	**X**	**6,124,896**	**264,991,558**	**8,414**	**X**	**X**	**X**
Albania	147	1,649	44	0	100	8,380	1,942	164,854	616	1,225	255	N
Austria	12,835	29,037	15,436	0	1,993	57,393	59,302	2,130,213	7,364	422	334	Y
Belarus	5,382	30,686	24,945	0	731	X	61,742	3,241,636 b	5,949	X	1,380	N
Belgium	31,888	43,118	27,040	0	3,986	97,437	106,032	4,923,287	10,489	538	467	Y
Bosnia and Herzegovina	942	1,583	484	0	100	X	3,111	X	909	X	X	N
Bulgaria	28,748	14,770	10,710	0	1,046	75,339	55,271	2,537,228	6,543	1,920	1,539	Y
Croatia	458	11,205	4,961	0	918	X	17,543	X	3,909	X	770	Y
Czech Rep	88,431	18,518	17,257	0	2,497	X	126,701	6,499,129 b	12,282	X	X	Y
Denmark	23,512	23,336	8,035	372	1,310	50,724	56,561	2,244,068	10,792	560	467	Y
Estonia	12,241	2,627	1,323	0	199	X	16,389	742,658 b	11,180	X	2,378	Y
Finland	28,488	23,424	6,789	0	472	51,072	59,174	1,658,982	11,544	605	600	Y
France	62,185	214,868	74,801	0	9,966	353,184	361,820	16,826,436	6,211	348	286	Y
Germany	350,956	322,066	167,350	919	19,932	X	861,223	42,689,711 g	10,514	X	499	Y
Greece	35,215	39,311	110	3	5,980	72,199	80,615	1,712,436	7,655	714	617	Y
Hungary	17,067	17,617	23,402	0	1,383	64,120	59,470	2,868,938	5,834	955	848	N
Iceland	242	1,913	0	0	40	2,019	2,195	70,547	8,099	431	370	N
Ireland	12,025	16,067	6,068	0	747	29,847	34,907	959,653	9,606	731	505	Y
Italy	42,821	238,043	105,421	0	16,942	398,852	403,231	12,319,486	7,029	427	349	Y
Latvia	967	6,379	1,799	0	134	X	9,281	454,055 b	3,714	X	1,015	Y
Lithuania	967	8,116	4,466	0	299	X	13,850	733,527 b	3,728	X	932	Y
Macedonia, FYR	8,028	4,411	0	0	274	X	12,714	X	6,438	X	2,006	N
Moldova, Rep	2,466	2,876	6,723	0	35	X	12,099	742,909 b	2,765	X	X	N
Netherlands	35,021	32,796	85,302	399	1,644	138,891	155,163	5,192,482	9,984	578	484	Y
Norway	3,807	20,819	7,357	34,232	797	47,669	67,015	1,787,369	15,327	670	640	Y
Poland	280,761	47,251	21,852	0	6,916	347,585	356,782	14,369,915	9,229	1,947	1,506	Y
Portugal	12,879	30,920	0	0	4,136	42,327	47,932	996,652	4,862	405	351	Y
Romania	40,330	35,537	40,000	0	3,417	155,071	119,282	5,528,382	5,270	1,610	1,142	Y
Russian Federation	484,546	338,865	732,393	9,859	13,853	X	1,579,514	68,412,659 b	10,681	X	X	Y
Slovakia	19,489	7,878	11,036	0	1,246	X	39,644	1,967,818 b	7,389	X	X	Y
Slovenia	3,257	7,786	1,502	0	498	X	13,040	X	6,537	X	575	Y
Spain	69,843	130,328	19,767	12	12,536	211,710	232,484	6,183,432	5,872	442	382	Y
Sweden	11,564	39,688	1,667	0	1,219	48,537	54,139	2,797,548	6,130	336	311	Y
Switzerland	550	36,046	5,437	0	2,193	42,689	44,224	1,525,906	6,144	281	250	Y
Ukraine	184,112	51,197	159,494	0	2,492	X	397,291	21,708,318 b	7,751	X	3,441	Y
United Kingdom	169,427	207,983	168,778	5,013	5,780	563,281	556,983	27,167,794	9,532	606	472	Y
Yugoslavia	23,831	6,126	5,141	0	1,099	X	36,197	3,299,234 h	3,413	X	X	N
MIDDLE EAST & N. AFRICA	**104,241**	**765,021**	**399,574**	**72,170**	**67,379**	**1,057,086**	**1,408,390**	**24,846,874**	**3,792**	**X**	**X**	**X**
Afghanistan	11	780	308	11	58	2,620	1,176	68,000	58	X	X	N
Algeria	3,188	22,651	50,567	14,400	3,488	80,443	94,297	1,725,913	3,283	755	728	N
Egypt	3,419	57,778	27,707	0	8,969	75,434	97,873	1,856,890	1,541	X	X	Y
Iran, Islamic Rep	3,873	153,936	77,549	23,082	8,222	212,354	266,662	5,056,708	4,201	X	X	N
Iraq	0	84,023	6,221	96	1,046	49,262	91,387	1,493,703	4,435	X	X	N
Israel	21,068	28,847	26	0	2,392	34,628	52,329	907,258	9,145	557	503	Y
Jordan	0	11,989	0	0	1,744	10,182	13,733	203,696	2,313	1,142	908	N
Kuwait	0	23,145	17,463	986	997	42,206	42,590	981,102	25,257	2,614	X	N
Lebanon	542	11,879	0	0	1,744	9,094	14,165	263,654	4,594	855	597	N
Libyan Arab Jamahiriya	15	25,348	9,966	3,482	1,769	37,772	40,579	855,830	7,978	X	X	N
Morocco	6,225	18,426	40	0	3,189	23,486	27,879	558,859	1,055	345	301	N
Oman	0	4,752	9,065	726	598	11,538	15,143	228,051	6,791	831	X	N
Saudi Arabia	0	156,343	79,351	23,948	8,191	177,096	267,831	4,107,234	14,225	1,168	1,338	N
Syrian Arab Rep	11	32,782	5,016	4,437	2,043	35,845	44,290	738,717	3,040	X	X	N
Tunisia	238	9,167	4,250	254	2,276	13,260	16,184	318,035	1,782	428	364	N
Turkey	65,600	80,901	15,649	0	16,195	143,819	178,342	3,212,251	2,861	554	474	N
United Arab Emirates	0	21,460	56,645	750	2,990	58,433	81,843	1,202,675	36,220	1,534	X	N
Yemen	0	16,488	0	0	518	9,603	17,008	247,111 i	1,085	X	X	N

	Carbon Dioxide (CO2) Emissions (000 metric tons)								Per Capita Carbon Emissions (kilograms)	CO2 Emitted Per Million Int$ (PPP) of Gross Domestic Product {a} (metric tons)		Signatory to the Kyoto Protocol 1999
	Solid Fuels 1996	Liquid Fuels 1996	Gaseous Fuels 1996	Gas Flaring 1996	Cement Manufact-uring 1996	Total 1990	Total 1996	Total Contribution Since 1950	1996	1990	1996	(Yes/No)
SUB-SAHARAN AFRICA	**268,395**	**170,270**	**14,070**	**54,136**	**12,685**	**466,991**	**519,548**	**12,928,091**	**894**	**X**	**X**	**X**
Angola	0	2,151	344	2,465	149	4,650	5,108	145,552	450	291	288	N
Benin	0	465	0	0	189	564	656	14,762	120	122	94	N
Botswana	6,383	12,579	4	9	691	2,415	19,665	51,505	13,035	337	1,778	N
Burkina Faso	0	967	0	0	0	1,008	967	14,682	90	146	96	N
Burundi	18	202	0	0	0	194	220	3,975	35	46	54	N
Cameroon	4	3,213	0	0	299	1,488	3,517	86,507	260	65	139	N
Central African Rep	0	234	0	0	0	198	234	5,866	70	X	X	N
Chad	0	99	0	0	0	143	99	5,027	14	35	15	N
Congo	0	4,741	7	167	50	2,037	4,961	40,161	1,884	X	X	N
Congo, Dem Rep	854	1,436	0	0	5	4,096	2,294	142,475	49	X	X	N
Côte d'Ivoire	0	12,817	0	0	249	9,907	13,066	185,109	946	X	X	N
Equatorial Guinea	0	143	0	0	0	117	143	2,862	349	X	X	N
Eritrea	X	X	X	X	X	X	X	X	X	X	X	N
Ethiopia	0	3,048	0	0	320	2,964	3,367	66,509	59	144	117	N
Gabon	0	2,041	1,587	0	65	6,112	3,690	135,432	3,333	1,019	437	N
Gambia	0	216	0	0	0	191	216	4,481	188	X	X	N
Ghana	7	3,342	0	0	698	3,539	4,045	109,271	223	187	142	N
Guinea	0	1,092	0	0	0	1,011	1,092	32,445	150	126	88	N
Guinea-Bissau	0	231	0	0	0	209	231	4,477	208	X	X	N
Kenya	267	5,760	0	0	747	5,822	6,775	180,020	243	228	203	N
Lesotho	X	X	X	X	X	X	X	X	X	X	X	N
Liberia	0	326	0	0	0	465	326	33,189	148	X	X	N
Madagascar	44	1,121	0	0	30	945	1,198	37,860	84	87	94	N
Malawi	44	616	0	0	70	601	733	18,184	75	127	105	N
Mali	0	462	0	0	10	421	473	12,025	46	81	66	Y
Mauritania	15	2,876	0	0	60	2,634	2,950	40,648	1,232	943	720	N
Mozambique	158	821	0	0	15	997	997	89,149	56	132	91	N
Namibia	X	X	X	X	X	X	X	X	X	X	X	N
Niger	458	649	0	0	15	1,048	1,121	20,863	119	162	139	Y
Nigeria	150	21,596	8,596	51,494	1,495	88,665	83,330	1,944,327	822	1,196	802	N
Rwanda	0	484	0	2	5	528	491	9,167	90	96	105	N
Senegal	0	2,770	0	0	294	2,895	3,063	76,142	358	263	211	N
Sierra Leone	0	447	0	0	0	333	447	19,844	104	128	182	N
Somalia	0	0	0	0	15	18	15	13,747	2	X	X	N
South Africa	227,707	56,821	3,532	0	4,684	291,108	292,746	8,541,575	7,678	1,236	986	N
Sudan	0	3,283	0	0	189	3,459	3,473	137,001	128	148	83	N
Tanzania, United Rep	15	2,030	0	0	399	2,272	2,444	61,694	80	166	146	N
Togo	0	575	0	0	174	689	751	17,023	180	140	121	N
Uganda	0	986	0	0	50	846	1,033	32,837	53	67	46	N
Zambia	773	1,499	0	0	174	2,444	2,444	111,151	291	344	277	Y
Zimbabwe	13,623	4,217	0	0	573	16,646	18,412	369,232	1,667	818	705	N
NORTH AMERICA	**2,012,492**	**2,279,862**	**1,354,255**	**18,726**	**45,005**	**5,233,610**	**5,710,344**	**200,969,374**	**19,074**	**X**	**X**	**X**
Canada	94,927	148,293	156,434	4,190	5,506	409,628	409,353	14,855,347	13,669	790	622	Y
United States	1,917,565	2,131,569	1,197,820	14,535	39,499	4,823,982	5,300,991	186,114,027	19,674	858	706	Y
C. AMERICA & CARIBBEAN	**20,196**	**367,202**	**81,885**	**6,560**	**16,832**	**411,973**	**499,033**	**11,518,037**	**3,078**	**X**	**X**	**X**
Belize	0	355	0	0	0	311	355	7,478	1,624	469	368	N
Costa Rica	0	4,192	0	0	493	2,917	4,683	85,756	1,282	188	209	Y
Cuba	509	29,854	84	0	724	31,818	31,170	1,018,401	2,829	X	X	Y
Dominican Rep	341	11,805	0	0	747	9,435	12,890	237,779	1,619	X	X	N
El Salvador	0	3,572	0	0	472	2,616	4,045	82,535	699	254	245	Y d
Guatemala	0	6,210	22	0	543	5,086	6,775	138,961	661	181	163	Y d
Haiti	0	1,048	0	0	25	993	1,070	25,238	139	107	114	N
Honduras	0	3,550	0	0	479	2,590	4,027	72,591	692	268	315	Y
Jamaica	169	9,603	0	0	277	7,958	10,050	234,650	4,029	1,069	1,113	X c,d
Mexico	17,910	248,104	67,165	3,549	11,376	294,974	348,106	7,760,957	3,754	525	470	Y
Nicaragua	0	2,686	0	0	174	2,601	2,862	70,015	629	455	322	Y
Panama	150	6,240	174	0	174	3,129	6,679	129,933	2,495	259	357	Y d
Trinidad and Tobago	0	4,536	14,385	3,010	307	16,924	22,237	489,957	17,508	2,521	2,570	Y d
SOUTH AMERICA	**73,514**	**460,568**	**143,413**	**24,600**	**33,789**	**572,181**	**735,885**	**17,875,788**	**2,260**	**X**	**X**	**X**
Argentina	3,019	63,853	54,502	5,929	2,550	109,729	129,852	3,853,154	3,687	519	381	Y
Bolivia	0	4,639	3,239	1,759	465	5,500	10,102	159,637	1,330	377	468	Y
Brazil	44,642	197,585	10,622	3,284	17,240	202,612	273,371	5,706,610	1,692	273	266	Y
Chile	11,439	31,573	3,635	321	1,811	36,263	48,779	1,076,150	3,383	383	278	Y
Colombia	11,871	38,901	9,658	719	4,159	55,850	65,307	1,589,993	1,662	313	246	N
Ecuador	0	20,918	1,271	960	1,334	16,569	24,487	421,888	2,093	408	428	Y
Guyana	0	953	0	0	0	1,132	953	51,915	1,138	848	379	N
Paraguay	0	3,386	0	0	309	2,261	3,697	53,960	746	163	204	Y d
Peru	1,447	22,416	352	43	1,917	22,175	26,176	771,034	1,093	357	244	Y
Suriname	0	2,074	0	0	25	1,810	2,099	63,647	5,115	X	X	N
Uruguay	4	5,298	0	0	341	3,909	5,643	217,601	1,741	200	196	Y
Venezuela	1,092	68,055	60,134	11,584	3,638	113,569	144,501	3,897,477	6,477	798	765	N
OCEANIA	**198,424**	**92,047**	**48,061**	**0**	**3,833**	**295,524**	**342,360**	**8,582,495**	**11,842**	**X**	**X**	**X**
Australia	192,774	72,723	37,897	0	3,239	266,010	306,633	7,622,743	16,902	1,011	820	N
Fiji	59	660	0	0	46	813	762	23,058	981	326	230	Y d
New Zealand	5,141	14,110	10,003	0	498	23,596	29,752	791,083	7,997	508	463	Y
Papua New Guinea	4	2,246	161	0	0	2,429	2,407	56,202	547	X	X	Y
Solomon Islands	0	161	0	0	0	161	161	3,488	412	263	171	Y

Notes: Negative values are shown in parantheses. Zero is either 0 or less than one-half the unit of measure. a. Gross domestic product figures are calculated using purchasing power parity (PPP) and are in current international dollars. b. Data are estimated for the countries of the former Soviet Union and former Czechoslovakia before 1992. (See technical notes for further explanation.) c. Country went straight to ratification.d. Country has ratified the Kyoto Protocol. e. Includes Singapore until 1957. f. Data before 1970 refer to the Democratic Republic of Viet Nam and South Viet Nam. g. Data before 1991refer to the Democratic Republic of Germany and the Federal Republic of Germany. h. Data include Bosnia and Herzegovina, Croatia, Macedonia (FYR), Slovenia, Yugoslavia, and the former Socialist Federal Republic of Yugoslavia. i. Data before 1991 refer to the People's Democratic Republic of Yemen and the Republic of Yemen.

Source: Carbon Dioxide Information Analysis Center

	Sulfur Dioxide (000 metric tons)			Nitrogen Oxides (000 metric tons)			Carbon Monoxide (000 metric tons)			Volatile Organic Compounds (000 metric tons)		
	1980	1990	1996	1980	1990	1996	1980	1990	1996	1980	1990	1996
ASIA (EXCL. MIDDLE EAST)	**X**	**X**	**X**	**X**	**X**	**X**	**X**	**X**	**X**	**X**	**X**	**X**
Armenia	X	72	2	X	46	11	X	304	125	X	81	18
EUROPE	**X**	**X**	**X**	**X**	**X**	**X**	**X**	**X**	**X**	**X**	**X**	**X**
Austria	400	91	52	231	194	163	1,690	1,287	1,021	355	351	261
Belarus	740	637	246	234	285	173	X	1,722	1,242	549	533	328
Belgium	828	322	240	442	343	334	X	1,631	1,434	X	358	324
Bosnia and Herzegovina	X	480	X	X	X	X	X	X	X	X	X	X
Bulgaria	2,050	2,020	1,420	X	376	259	X	891	613	X	217	147
Croatia	150	180	58	X	87	67	X	651	375	X	105	79
Czech Rep	2,257	1,876	946	937	742	432	894	1,055	886	X	435	284
Denmark	450	182	186	282	282	288	681	794	597	203	178	136
Finland	584	260	105	295	300	267	660	556	430	X	209	173
France	3,338	1,298	1,031	1,823	1,585	1,641	9,216	10,736	8,850	X	2,404	2,570
Germany {a}	7,514	5,313	1,543	3,334	2,693	1,887	14,046	11,165	6,717	3,224	3,195	1,877
Greece	400	509	543	X	343	374	X	1,338	1,334	X	373	409
Hungary	1,633	1,010	673	273	238	196	1,019	767	727	215	205	150
Iceland	18	24	X	X	20	X	X	26	X	X	6	X
Ireland	222	178	147	73	115	121	X	429	307	X	197	103
Italy	3,757	1,651	X	1,638	1,938	X	7,588	8,003	X	2,179	2,213	X
Latvia	X	119	59	X	93	35	X	388	176	X	152	41
Lithuania	311	222	93	152	158	65	541	521	312	108	111	87
Macedonia, FYR	X	X	X	X	X	X	X	X	X	X	X	X
Moldova, Rep	308	231	X	58	39	X	136	193	X	X	11	X
Netherlands	490	202	135	583	580	501	X	1,143	903	X	502	362
Norway	140	53	34	189	222	223	896	858	720	176	301	369
Poland	4,100	3,210	2,368	1,229	1,280	1,154	X	7,406	4,837	1,036	831	766
Portugal	266	362	X	X	348	X	X	1,020	X	X	640	X
Romania	1,055	1,311	X	523	546	X	3,245	3,186	X	673	616	X
Russian Federation	7,161	4,460	2,685	1,734	3,600	2,467	13,520	13,174	9,312	2,843	3,566	2,576
Slovakia	780	543	227	X	225	130	X	487	346	X	149	105
Slovenia	234	194	110	51	62	70	68	79	95	X	42	X
Spain	3,319	2,266	X	950	1,177	X	X	4,752	X	X	1,134	X
Sweden	491	119	83	404	338	302	X	1,210	1,082	X	526	446
Switzerland	116	43	30	170	166	130	1,280	707	485	323	292	203
Ukraine	3,849	2,782	1,293	1,145	1,097	467	X	8,141	2,567	X	1,369	718
United Kingdom	4,862	3,731	2,017	2,511	2,686	2,029	7,642	7,111	5,000	2,309	2,552	2,046
Yugoslavia	406	508	434	47	66	57	X	X	X	X	X	X
MIDDLE EAST & N. AFRICA	**X**	**X**	**X**	**X**	**X**	**X**	**X**	**X**	**X**	**X**	**X**	**X**
Turkey	860	X	X	275	497	X	X	X	X	X	X	X
NORTH AMERICA	**X**	**X**	**X**	**X**	**X**	**X**	**X**	**X**	**X**	**X**	**X**	**X**
Canada	4,643	3,236	2,722	1,959	2,104	2,011	10,273	10,596	X	2,099	2,880	X
United States	23,501	20,989	17,339	22,501	21,584	21,222	105,872	87,576	80,579	23,596	19,037	17,315

Note: a. The figures for 1980 are the combined figures from the German Democratic Republic and the Federal Republic of Germany.

Source: Co-Operative Programme for Monitoring and Evaluation of the Long-Range Transmission of Air Pollutants in Europe

Years	Carbon Dioxide (CO_2) ppm	Methane (CH_4) ppb	Nitrous Oxide (N_2O) ppb	Carbon Tetra-Chloride (CCl_4) ppt	Methyl Chloro-form (CH_3CCl_3) ppt	CFC-11 (CCl_3F) ppt	CFC-12 (CCl_2F_2) ppt	CFC-113 ($C_2Cl_3F_3$) ppt	Total Gaseous Chlorine ppt
Preindustrial (1860)	**286-288** a	**848** b	**285** c	**0**	**0**	**0**	**0**	**0**	**0**
1965	320	X	X	X	X	X	X	X	X
1966	321	X	X	X	X	X	X	X	X
1967	322	X	X	X	X	X	X	X	X
1968	323	X	X	X	X	X	X	X	X
1969	324	X	X	X	X	X	X	X	X
1970	326	X	X	X	X	X	X	X	X
1971	326	X	X	X	X	X	X	X	X
1972	327	X	X	X	X	X	X	X	X
1973	330	X	X	X	X	X	X	X	X
1974	330	X	X	X	X	X	X	X	X
1975	331	X	X	X	X	X	X	X	X
1976	332	X	X	X	X	X	X	X	X
1977	334	X	X	X	X	X	X	X	X
1978	335	X	298 d	88 d	57 d	139 d	257 d	X	1,453 e, f
1979	337	X	298	87	62	147	272	X	1,522 f
1980	339	X	299	89	69	158	293	X	1,624 f
1981	340	X	299	90	75	166	305	X	1,692 f
1982	341	X	301	92	81	175	325	26 g	1,865 h
1983	343	X	302	93	85	182	341	28	1,939
1984	344	X	303	94	88	190	355	31	2,016
1985	346	X	304	96	92	200	376	36	2,121
1986	347	1,600 i	305	97	96	210	394	40	2,216
1987	349	1,610	305	99	98	221	413	48	2,322
1988	351	1,619	306	100	103	231	433	53	2,425
1989	353	1,641	306	100	107	240	452	59	2,524
1990	354	1,645	306	101	110	249	470	66	2,620
1991	355	1,657	307	101	113	254	484	71	2,685
1992	356	1,673	308	101	116	259	496	77	2,751
1993	357	1,671	308	101	112	260	503	80	2,764
1994	359	1,674	309	100	106	261	512	81	2,769
1995	361	1,681	309	99	97	261	518	82	2,753
1996	363	1,684	310	98	85	261	523	82	2,725
1997	364	1,690	311	97	73	260	528	83	2,693
1998	367	1,693 j	311 j	96 j	64 j	259 j	530 j	82 j	2,664 j

Notes: All estimates are by volume; ppb = parts per billion, ppm = parts per million, and ppt = parts per trillion. a. Historical CO_2 record from the Siple Station and Law Dome ice core records. b. Historical CH_4 record from the Law Dome (East Side, "DE08" site), Antarctica, ice core. c. Historical N_2O record from the Law Dome (Summit, "BHD" site) ice core. d. Annual average only includes the months of July through December. e. Annual average only includes the months of July through December for carbon tetrachloride, methyl chloroform, CFC-11, and CFC-12. f. Annual average only includes carbon tetrachloride, methyl chloroform, CFC-11, and CFC-12. g. Annual average only includes the months of June through December. h. Annual average includes carbon tetrachloride, methyl chloroform, CFC-11, CFC-12, and CFC-113. The CFC-113 figure only includes data from June through December. i. Annual average only includes the months of May through December. j. Annual average only includes the months of January through September.

Source: Carbon Dioxide Information Analysis Center

	Energy Production								Total Electricity Generated	
	From All Sources			From Non-Renewable Energy Sources {a}						
	(1000 metric toe) {b} 1997	Percent Change Since 1987	Per Capita (kg oil equivalent) 1997	Solid Fuels (1000 metric toe) {b} 1997	Liquid Fuels (1000 metric toe) {b} 1997	Gaseous Fuels (1000 metric toe) {b} 1997	Nuclear Fuels (1000 metric toe) {b} 1997	Other Sources (1000 metric toe) {b} 1997	(1000 metric toe) {b} 1997	Percent Change Since 1987
WORLD	**9,664,837**	**17**	**1,660**	**2,274,208**	**3,530,469**	**1,916,758**	**624,162**	**X**	**1,199,591**	**32**
ASIA (EXCL. MIDDLE EAST)	**2,321,274**	**28**	**705**	**938,270**	**389,545**	**248,231**	**119,586**	**X**	**313,682**	**80**
Armenia	537	(57)	151	0	0	0	417	0	518	(60)
Azerbaijan	14,027	(42)	1,836	0	9,066	4,829	0	0	1,445	(27)
Bangladesh	21,894	27	179	0	7	5,794	0	0	1,020	112
Bhutan	X	X	X	X	X	X	X	X	X	X
Cambodia	X	X	X	X	X	X	X	X	X	X
China	1,097,160	36	882	686,410	160,741	21,134	3,757	0	97,565	128
Georgia	694	(59)	136	2	135	2	0	0	617	(51)
India	404,503	35	419	147,233	37,586	17,815	2,632	0	39,853	112
Indonesia	221,549	52	1,089	33,888	77,291	63,043	0	0	6,436	236
Japan	106,895	48	848	2,356	795	2,009	83,148	83	88,536	44
Kazakhstan	64,784	(30)	3,957	32,009	25,568	6,574	0	0	4,472	(41)
Korea, Dem People's Rep	20,461	(38)	890	17,564	0	0	0	X	2,923	(32)
Korea, Rep	24,037	10	526	2,030	0	0	20,089	0	20,987	230
Kyrgyzstan	1,408	(36)	305	320	85	34	0	0	1,084	35
Lao People's Dem Rep	X	X	X	X	X	X	X	X	X	X
Malaysia	73,979	78	3,526	153	38,039	33,096	0	0	4,977	241
Mongolia	X	X	X	X	X	X	X	X	X	X
Myanmar	12,249	14	279	9	400	1,428	0	0	362	81
Nepal	6,559	25	294	0	0	0	0	0	105	120
Pakistan	42,048	39	292	1,413	2,918	13,434	90	0	5,085	106
Philippines	16,616	14	233	446	16	5	0	0	3,424	76
Singapore	61	X	18	0	0	0	0	0	2,313	126
Sri Lanka	4,345	9	238	0	0	0	0	0	442	90
Tajikistan	1,253	(38)	211	4	26	34	0	0	1,204	(12)
Thailand	46,166	124	773	6,785	5,707	12,406	0	0	8,020	225
Turkmenistan	18,739	(76)	4,427	0	4,723	14,016	0	0	808	(29)
Uzbekistan	49,054	33	2,113	1,037	7,798	39,722	0	0	3,961	(16)
Viet Nam	43,525	108	570	6,550	9,875	3,311	0	0	1,647	216
EUROPE	**2,221,880**	**(10)**	**3,051**	**416,434**	**648,575**	**720,694**	**300,479**	**X**	**352,135**	**1**
Albania	912	(69)	291	16	357	15	0	0	482	27
Austria	8,007	0	989	294	1,018	1,215	0	0	4,777	12
Belarus	3,275	0	316	553	1,831	204	0	0	2,241	(30)
Belgium	12,933	(8)	1,277	219	0	0	12,355	219	6,713	25
Bosnia and Herzegovina	626	143	178	348	0	0	0	0	189	(84)
Bulgaria	9,981	(1)	1,189	4,895	28	28	4,633	0	3,574	(4)
Croatia	4,011	122	894	29	1,800	1,402	0	0	833	3
Czech Rep	30,989	(30)	3,008	26,955	439	163	3,256	550	5,523	3
Denmark	20,260	152	3,855	14	11,587	6,956	0	14	3,809	50
Estonia	3,421	X	2,364	2,795	367	0	0	367	793	(49)
Finland	15,039	41	2,925	2,627	64	0	5,445	20	5,949	30
France	127,520	33	2,181	4,432	2,198	2,126	103,065	323	42,904	33
Germany	139,678	(30)	1,702	70,224	3,461	16,074	44,406	56	47,127	3
Greece	9,645	20	913	7,709	476	45	0	0	3,723	44
Hungary	12,747	(24)	1,255	3,299	1,992	3,355	3,640	0	3,044	19
Iceland	1,481	20	5,414	0	0	0	0	0	480	33
Ireland	2,871	(5)	785	740	0	1,906	0	0	1,694	56
Italy	29,311	23	511	12	6,175	15,775	0	0	21,200	24
Latvia	1,636	343	665	78	0	0	0	0	387	(24)
Lithuania	3,970	62	1,072	18	213	0	3,196	0	1,237	(37)
Macedonia, FYR	1,700	70	856	1,423	0	0	0	0	578	88
Moldova, Rep	98	362	22	0	0	0	0	0	453	(69)
Netherlands	65,298	5	4,182	0	3,040	60,574	628	0	7,453	27
Norway	212,653	145	48,378	259	160,788	41,007	0	0	9,501	6
Poland	99,474	(21)	2,571	92,033	366	3,203	0	1,461	12,119	(2)
Portugal	2,317	19	235	0	0	0	0	0	2,934	70
Romania	31,013	(41)	1,375	6,229	6,579	11,905	1,407	0	4,915	(23)
Russian Federation	927,261	(25)	6,280	101,118	305,520	461,124	28,774	80	71,644	(20)
Slovakia	4,688	(14)	873	1,146	65	225	2,814	0	2,092	5
Slovenia	2,870	0	1,438	1,052	1	9	1,308	0	1,132	5
Spain	31,316	12	791	9,796	378	163	14,411	42	15,975	40
Sweden	33,067	12	3,734	252	0	0	18,224	0	12,850	2
Switzerland	10,993	12	1,516	0	0	0	6,640	0	5,299	5
Ukraine	81,175	(45)	1,590	39,590	4,588	15,204	20,701	0	15,293	(37)
United Kingdom	268,119	11	4,580	29,611	134,242	77,457	25,577	867	29,572	14
Yugoslavia	11,481	543	1,080	8,668	1,000	557	0	0	3,467	230
MIDDLE EAST & N. AFRICA	**1,497,312**	**55**	**3,952**	**13,938**	**1,241,839**	**223,169**	**X**	**X**	**50,805**	**84**
Afghanistan	X	X	X	X	X	X	X	X	X	X
Algeria	125,576	37	4,272	0	61,670	63,381	0	0	1,865	70
Egypt	57,997	9	896	0	44,640	11,070	0	0	4,958	52
Iran, Islamic Rep	224,935	68	3,480	567	184,550	38,398	0	0	8,238	107
Iraq	62,088	(45)	2,931	0	58,311	3,701	0	0	2,542	31
Israel	601	66	103	42	26	13	0	0	3,018	101
Jordan	193	698	31	0	0	189	0	0	539	80
Kuwait	116,087	52	67,022	0	111,633	4,455	0	0	2,330	50
Lebanon	207	31	66	0	0	0	0	0	732	85
Libyan Arab Jamahiriya	78,942	43	15,151	0	73,067	5,750	0	0	1,563	17
Morocco	836	3	31	211	242	23	0	231	1,129	64
Oman	51,620	64	22,391	0	46,921	4,699	0	0	629	116
Saudi Arabia	487,095	96	25,006	0	449,511	37,584	0	0	8,927	105
Syrian Arab Rep	32,794	154	2,194	0	29,860	2,066	0	0	1,544	125
Tunisia	6,655	2	723	0	4,013	1,466	0	0	686	75
Turkey	27,556	11	435	13,118	3,525	208	0	0	8,883	133
United Arab Emirates	153,555	68	66,553	0	123,003	30,552	0	0	1,769	51
Yemen	19,105	X	1,173	0	19,028	0	0	0	203	102

	Energy Production								Total Electricity Generated	
	From All Sources			From Non-Renewable Energy Sources {a}						
	(1000 metric toe) {b} 1997	Percent Change Since 1987	Per Capita (kg oil equivalent) 1997	Solid Fuels (1000 metric toe) {b} 1997	Liquid Fuels (1000 metric toe) {b} 1997	Gaseous Fuels (1000 metric toe) {b} 1997	Nuclear Fuels (1000 metric toe) {b} 1997	Other Sources (1000 metric toe) {b} 1997	(1000 metric toe) {b} 1997	Percent Change Since 1987
SUB-SAHARAN AFRICA	**X**	**X**	**X**	**X**	**X**	**X**	**X**	**X**	**X**	**X**
Angola	41,430	89	3,537	0	35,700	465	0	0	95	37
Benin	1,897	8	337	0	67	0	0	0	4	92
Botswana	X	X	X	X	X	X	X	X	X	X
Burkina Faso	X	X	X	X	X	X	X	X	X	X
Burundi	X	X	X	X	X	X	X	X	X	X
Cameroon	11,250	(7)	808	0	6,357	0	0	0	269	25
Central African Rep	X	X	X	X	X	X	X	X	X	X
Chad	X	X	X	X	X	X	X	X	X	X
Congo	13,540	89	4,998	0	12,638	3	0	0	37	53
Congo, Dem Rep	14,364	29	299	56	1,315	0	0	0	517	11
Côte d'Ivoire	4,908	31	349	0	803	0	0	0	276	55
Equatorial Guinea	X	X	X	X	X	X	X	X	X	X
Eritrea	X	X	X	X	X	X	X	X	X	X
Ethiopia	16,316	27	280	0	0	0	0	0	115	50
Gabon	19,786	129	17,403	0	18,794	68	0	0	87	13
Gambia	X	X	X	X	X	X	X	X	X	X
Ghana	5,843	48	313	0	361	0	0	0	529	26
Guinea	X	X	X	X	X	X	X	X	X	X
Guinea-Bissau	X	X	X	X	X	X	X	X	X	X
Kenya	11,651	20	410	0	0	0	0	0	364	60
Lesotho	X	X	X	X	X	X	X	X	X	X
Liberia	X	X	X	X	X	X	X	X	X	X
Madagascar	X	X	X	X	X	X	X	X	X	X
Malawi	X	X	X	X	X	X	X	X	X	X
Mali	X	X	X	X	X	X	X	X	X	X
Mauritania	X	X	X	X	X	X	X	X	X	X
Mozambique	6,994	(1)	379	0	0	0	0	0	86	199
Namibia	X	X	X	X	X	X	X	X	X	X
Niger	X	X	X	X	X	X	X	X	X	X
Nigeria	191,034	55	1,839	86	117,249	4,429	0	0	1,305	35
Rwanda	X	X	X	X	X	X	X	X	X	X
Senegal	1,654	32	189	0	0	20	0	0	108	52
Sierra Leone	X	X	X	X	X	X	X	X	X	X
Somalia	X	X	X	X	X	X	X	X	X	X
South Africa	142,139	27	3,667	124,678	401	1,543	3,296	0	17,866	38
Sudan	9,881	18	356	0	259	0	0	0	169	43
Tanzania, United Rep	13,529	23	431	3	0	0	0	0	166	52
Togo	X	X	X	X	X	X	X	X	X	X
Uganda	X	X	X	X	X	X	X	X	X	X
Zambia	5,556	16	647	106	0	0	0	0	689	(8)
Zimbabwe	8,152	9	727	2,712	0	0	0	0	628	15
NORTH AMERICA	**2,046,512**	**11**	**6,776**	**604,959**	**516,778**	**579,512**	**195,194**	**0**	**365,125**	**32**
Canada	362,701	44	11,986	43,032	120,156	137,397	21,536	0	49,453	16
United States	1,683,811	6	6,196	561,928	396,622	442,115	173,658	0	315,672	35
C. AMERICA & CARIBBEAN	**259,862**	**18**	**1,575**	**4,970**	**181,633**	**35,102**	**2,725**	**X**	**20,028**	**64**
Belize	X	X	X	X	X	X	X	X	X	X
Costa Rica	1,157	14	309	0	0	0	0	0	482	79
Cuba	7,255	48	655	0	1,685	18	0	0	1,211	4
Dominican Rep	1,423	(0)	176	0	0	0	0	0	631	57
El Salvador	2,649	34	448	0	0	0	0	0	313	90
Guatemala	4,433	47	421	0	1,080	0	0	0	421	148
Haiti	1,298	4	166	0	0	0	0	0	54	27
Honduras	2,003	28	335	0	0	0	0	0	283	81
Jamaica	595	133	237	0	0	0	0	0	538	237
Mexico	223,132	16	2,367	4,970	172,149	28,255	2,725	0	15,054	68
Nicaragua	1,529	23	327	0	0	0	0	0	164	47
Panama	808	37	297	0	0	0	0	0	357	47
Trinidad and Tobago	13,579	17	10,635	0	6,719	6,828	0	0	429	43
SOUTH AMERICA	**529,055**	**58**	**1,600**	**26,396**	**320,514**	**73,038**	**2,883**	**X**	**53,576**	**55**
Argentina	80,134	90	2,246	148	44,801	28,059	2,057	0	6,187	46
Bolivia	5,953	33	766	0	1,876	3,003	0	0	295	99
Brazil	120,236	23	734	2,162	45,988	5,288	826	0	26,428	51
Chile	8,168	18	558	709	520	1,619	0	0	2,923	117
Colombia	67,524	70	1,686	19,934	34,164	5,339	0	0	3,966	47
Ecuador	22,792	118	1,909	0	21,072	0	0	0	825	78
Guyana	X	X	X	X	X	X	X	X	X	X
Paraguay	6,960	90	1,368	0	0	0	0	0	4,353	172
Peru	12,225	(14)	502	29	5,946	768	0	0	1,544	28
Suriname	X	X	X	X	X	X	X	X	X	X
Uruguay	1,086	(14)	332	0	0	0	0	0	615	(6)
Venezuela	203,979	77	8,955	3,414	166,146	28,962	0	0	6,438	38
OCEANIA	**213,325**	**38**	**7,282**	**140,939**	**31,034**	**30,254**	**0**	**X**	**18,861**	**36**
Australia	199,167	37	10,864	138,967	27,885	25,550	0	0	15,700	39
Fiji	X	X	X	X	X	X	X	X	X	X
New Zealand	14,158	44	3,764	1,972	3,149	4,704	0	0	3,161	25
Papua New Guinea	X	X	X	X	X	X	X	X	X	X
Solomon Islands	X	X	X	X	X	X	X	X	X	X
DEVELOPED	**4,881,847**	**(0)**	**3,774**	**1,322,778**	**1,245,011**	**1,399,236**	**582,534**	**X**	**859,650**	**17**
DEVELOPING	**4,712,931**	**44**	**1,042**	**950,711**	**2,277,537**	**517,294**	**41,628**	**X**	**337,357**	**99**

Notes: Negative values are shown in parentheses. Zero is either 0 or less than one-half the unit measure. (0) indicates a value less than 0 and greater than negative one-half.
a. See ERC.4 for data on renewable energy. b. Tons of oil equivalent. See technical notes for more information on toe.

Source: International Energy Agency

	Energy Consumption							Final	Trade in Energy	
	From all Sources			From Non-Renewable Energy Sources {a}						
	(1000 metric toe) {b} 1997	Percent Change Since 1987	Per Capita (kg oil equivalent) 1997	Solid Fuels (1000 metric toe) {b} 1997	Liquid Fuels (1000 metric toe) {b} 1997	Gaseous Fuels (1000 metric toe) {b} 1997	Nuclear Fuels (1000 metric toe) {b} 1997	Consumption of Electricity (1000 toe) {b} 1997	Imports (1000 metric toe) {b} 1997	Exports (1000 metric toe) {b} 1997
WORLD	**9,521,506**	**16**	**1,635**	**2,254,969**	**3,409,241**	**1,911,171**	**624,162**	**986,677**	**3,454,233**	**3,419,104**
ASIA (EXCL. MIDDLE EAST)	**2,958,844**	**36**	**899**	**1,028,240**	**926,399**	**258,379**	**119,586**	**254,119**	**1,076,015**	**392,653**
Armenia	1,804	(77)	508	2	155	1,110	417	371	1,267	0
Azerbaijan	11,987	(63)	1,568	3	6,951	4,829	0	1,099	461	2,502
Bangladesh	24,327	28	198	0	2,440	5,794	0	813	2,440	0
Bhutan	X	X	X	X	X	X	X	X	X	X
Cambodia	X	X	X	X	X	X	X	X	X	X
China	1,098,931	42	883	658,083	193,761	18,825	3,757	72,534	66,531	53,769
Georgia	2,295	(63)	448	2	955	767	0	533	1,841	176
India	461,032	45	477	153,279	87,937	17,815	2,632	30,018	59,162	2,376
Indonesia	138,779	71	682	9,545	50,503	31,506	0	5,674	23,282	106,647
Japan	514,898	37	4,085	86,532	271,600	54,948	83,148	78,524	427,020	10,557
Kazakhstan	38,418	(65)	2,346	21,888	9,119	6,201	0	3,515	4,673	31,038
Korea, Dem People's Rep	23,323	(39)	1,015	18,959	1,466	0	0	468	X	301
Korea, Rep	176,351	166	3,856	32,799	108,245	13,326	20,089	19,172	190,773	31,247
Kyrgyzstan	2,793	45	605	959	500	512	0	547	2,218	833
Lao People's Dem Rep	X	X	X	X	X	X	X	X	X	X
Malaysia	48,473	161	2,310	1,581	25,026	19,191	0	4,383	13,718	39,025
Mongolia	X	X	X	X	X	X	X	X	X	X
Myanmar	13,009	23	296	9	1,307	1,281	0	215	953	7
Nepal	7,160	30	321	71	531	0	0	75	610	22
Pakistan	56,818	51	394	2,045	17,056	13,434	90	3,673	15,110	241
Philippines	38,251	55	536	2,648	19,449	5	0	2,734	22,269	1,344
Singapore	26,878	202	7,843	0	25,541	1,275	0	2,120	90,301	47,725
Sri Lanka	7,159	35	392	1	2,813	0	0	362	3,288	203
Tajikistan	3,384	61	571	48	1,228	851	0	1,127	2,512	381
Thailand	79,963	157	1,339	8,631	37,595	12,406	0	7,089	40,546	5,830
Turkmenistan	12,181	(84)	2,878	0	3,659	8,750	0	374	845	7,403
Uzbekistan	42,553	1	1,833	984	6,953	34,039	0	3,349	3,361	10,346
Viet Nam	39,306	68	515	4,435	7,770	3,311	0	1,316	7,730	11,949
EUROPE	**2,553,858**	**(20)**	**3,507**	**499,633**	**851,600**	**762,970**	**300,479**	**283,538**	**1,341,347**	**961,926**
Albania	1,048	(63)	335	16	476	15	0	242	135	0
Austria	27,761	14	3,428	3,624	12,087	6,537	0	4,201	21,501	2,239
Belarus	25,142	(40)	2,429	1,011	9,013	13,775	0	2,302	25,705	3,371
Belgium	57,125	21	5,641	8,482	24,167	11,261	12,355	6,177	71,574	22,701
Bosnia and Herzegovina	1,750	134	497	348	896	211	0	150	1,140	16
Bulgaria	20,616	(33)	2,457	7,492	4,713	3,699	4,633	2,290	12,916	1,972
Croatia	7,650	178	1,706	258	4,027	2,246	0	948	5,621	1,884
Czech Rep	40,576	(18)	3,939	21,111	7,920	7,667	3,256	4,269	19,105	9,347
Denmark	21,107	4	4,016	6,579	9,570	3,861	0	2,739	18,975	15,267
Estonia	5,556	(1,713)	3,839	3,168	1,262	618	0	435	2,691	925
Finland	33,075	10	6,433	6,854	10,288	2,906	5,445	6,051	23,294	4,362
France	247,534	19	4,233	14,631	88,035	31,330	103,065	30,545	150,764	27,667
Germany	347,272	(5)	4,232	86,264	139,311	71,928	44,406	39,709	233,959	23,769
Greece	25,556	36	2,418	8,450	15,322	171	0	3,153	23,462	3,984
Hungary	25,311	(18)	2,492	4,354	6,975	9,696	3,640	2,480	15,730	2,609
Iceland	2,330	22	8,516	56	793	0	0	408	882	0
Ireland	12,491	31	3,415	2,955	6,541	2,771	0	1,435	11,068	1,353
Italy	163,315	14	2,846	11,302	93,533	47,472	0	21,346	157,047	21,677
Latvia	4,460	505	1,812	160	1,744	1,065	0	373	2,883	242
Lithuania	8,806	(40)	2,377	139	3,229	2,001	3,196	579	9,118	4,178
Macedonia, FYR	2,841	184	1,430	1,496	1,064	0	0	454	1,147	5
Moldova, Rep	4,436	(54)	1,014	151	899	3,127	0	451	4,440	2
Netherlands	74,910	15	4,798	9,228	27,590	35,323	628	7,698	119,534	96,357
Norway	24,226	12	5,511	1,029	8,345	3,921	0	8,900	6,888	194,412
Poland	105,155	(20)	2,718	71,277	19,323	9,415	0	8,148	29,990	21,824
Portugal	20,400	58	2,068	3,529	14,218	87	0	2,742	21,261	2,441
Romania	44,135	(36)	1,957	9,169	12,727	15,934	1,407	3,305	17,325	3,113
Russian Federation	591,982	(52)	4,009	97,216	127,422	309,719	28,774	50,428	20,163	353,530
Slovakia	17,216	(23)	3,204	4,694	3,284	5,634	2,814	1,964	15,001	2,457
Slovenia	6,380	19	3,197	1,269	2,705	715	1,308	846	3,850	292
Spain	107,328	43	2,709	18,154	57,113	11,305	14,411	13,184	89,804	8,765
Sweden	51,934	6	5,864	2,498	16,135	719	18,224	10,687	31,362	10,992
Switzerland	26,218	12	3,616	110	13,394	2,294	6,640	4,197	17,903	2,875
Ukraine	150,059	(24)	2,939	43,532	18,431	66,314	20,701	10,677	73,701	4,818
United Kingdom	227,977	9	3,894	40,010	82,489	76,378	25,577	26,596	76,963	112,399
Yugoslavia	15,842	319	1,491	8,706	3,650	2,230	0	2,872	X	0
MIDDLE EAST & N. AFRICA	**525,927**	**58**	**1,388**	**30,732**	**301,815**	**174,602**	**0**	**41,487**	**129,890**	**1,075,597**
Afghanistan	X	X	X	X	X	X	X	X	X	X
Algeria	26,497	23	901	272	8,780	16,919	0	1,426	763	99,278
Egypt	39,581	41	611	644	25,598	11,070	0	4,170	2,677	11,838
Iran, Islamic Rep	108,289	83	1,676	927	67,626	38,316	0	6,094	12,079	128,389
Iraq	27,091	33	1,279	0	23,314	3,701	0	2,542	0	34,997
Israel	17,591	70	3,002	5,478	11,669	13	0	2,544	18,942	2,146
Jordan	4,795	52	783	0	4,602	189	0	456	4,602	0
Kuwait	16,165	29	9,332	0	11,706	4,455	0	2,005	5	99,742
Lebanon	5,244	76	1,669	132	4,852	0	0	688	5,038	0
Libyan Arab Jamahiriya	15,090	33	2,896	0	10,424	4,541	0	1,563	29	63,795
Morocco	9,275	61	345	2,019	6,631	23	0	994	8,714	320
Oman	6,775	265	2,939	0	2,482	4,294	0	507	340	45,030
Saudi Arabia	98,449	60	5,054	0	60,861	37,584	0	7,049	4	386,792
Syrian Arab Rep	14,642	46	979	1	11,706	2,066	0	995	737	18,890
Tunisia	6,805	43	739	69	3,470	2,087	0	562	4,236	3,685
Turkey	71,273	54	1,124	21,176	30,863	8,339	0	6,853	44,674	1,333
United Arab Emirates	30,874	62	13,381	0	7,149	23,708	0	1,769	12,575	124,913
Yemen	3,355	30	206	0	3,278	0	0	129	1	13,785

	Energy Consumption							Final	Trade in Energy	
	From all Sources			From Non-Renewable Energy Sources {a}						
	(1000 metric toe) {b} 1997	Percent Change Since 1987	Per Capita (kg oil equivalent) 1997	Solid Fuels (1000 metric toe) {b} 1997	Liquid Fuels (1000 metric toe) {b} 1997	Gaseous Fuels (1000 metric toe) {b} 1997	Nuclear Fuels (1000 metric toe) {b} 1997	Consumption of Electricity (1000 toe) {b} 1997	Imports (1000 metric toe) {b} 1997	Exports (1000 metric toe) {b} 1997
SUB-SAHARAN AFRICA	**X**	**X**	**X**	**X**	**X**	**X**	**X**	**X**	**X**	**X**
Angola	6,848	37	585	0	1,118	465	0	64	65	33,997
Benin	2,182	36	388	0	331	0	0	22	340	70
Botswana	X	X	X	X	X	X	X	X	X	X
Burkina Faso	X	X	X	X	X	X	X	X	X	X
Burundi	X	X	X	X	X	X	X	X	X	X
Cameroon	5,756	24	413	0	863	0	0	216	15	5,454
Central African Rep	X	X	X	X	X	X	X	X	X	X
Chad	X	X	X	X	X	X	X	X	X	X
Congo	1,242	(12)	459	0	331	3	0	46	24	12,312
Congo, Dem Rep	14,539	36	303	241	1,317	0	0	484	1,271	1,063
Côte d'Ivoire	5,597	39	398	0	1,489	0	0	222	2,073	1,218
Equatorial Guinea	X	X	X	X	X	X	X	X	X	X
Eritrea	X	X	X	X	X	X	X	X	X	X
Ethiopia	17,131	25	294	0	815	0	0	110	962	79
Gabon	1,635	10	1,438	0	643	68	0	75	218	18,116
Gambia	X	X	X	X	X	X	X	X	X	X
Ghana	6,896	44	370	2	1,447	0	0	427	1,316	236
Guinea	X	X	X	X	X	X	X	X	X	X
Guinea-Bissau	X	X	X	X	X	X	X	X	X	X
Kenya	14,138	19	497	57	2,418	0	0	312	2,863	170
Lesotho	X	X	X	X	X	X	X	X	X	X
Liberia	X	X	X	X	X	X	X	X	X	X
Madagascar	X	X	X	X	X	X	X	X	X	X
Malawi	X	X	X	X	X	X	X	X	X	X
Mali	X	X	X	X	X	X	X	X	X	X
Mauritania	X	X	X	X	X	X	X	X	X	X
Mozambique	7,664	1	416	12	641	0	0	67	777	43
Namibia	X	X	X	X	X	X	X	X	X	X
Niger	X	X	X	X	X	X	X	X	X	X
Nigeria	88,652	36	853	86	14,867	4,429	0	850	43	102,342
Rwanda	X	X	X	X	X	X	X	X	X	X
Senegal	2,770	27	316	0	1,116	20	0	81	1,192	65
Sierra Leone	X	X	X	X	X	X	X	X	X	X
Somalia	X	X	X	X	X	X	X	X	X	X
South Africa	107,220	15	2,766	80,492	10,459	1,543	3,296	13,292	12,525	43,715
Sudan	11,480	21	414	0	1,858	0	0	116	1,601	0
Tanzania, United Rep	14,258	21	454	3	725	0	0	146	781	29
Togo	X	X	X	X	X	X	X	X	X	X
Uganda	X	X	X	X	X	X	X	X	X	X
Zambia	5,987	13	697	102	564	0	0	457	606	175
Zimbabwe	9,926	17	885	2,636	1,504	0	0	906	1,893	88
NORTH AMERICA	**2,400,174**	**16**	**7,947**	**540,674**	**935,332**	**578,709**	**195,194**	**313,067**	**673,759**	**285,889**
Canada	237,983	17	7,864	27,376	80,826	70,738	21,536	40,854	59,098	180,771
United States	2,162,191	16	7,956	513,299	854,506	507,971	173,658	272,213	614,662	105,118
C. AMERICA & CARIBBEAN	**198,317**	**26**	**1,202**	**6,801**	**117,549**	**35,678**	**2,725**	**15,838**	**58,677**	**115,951**
Belize	X	X	X	X	X	X	X	X	X	X
Costa Rica	2,663	48	710	0	1,517	0	0	403	1,631	133
Cuba	14,273	(5)	1,290	47	8,656	18	0	993	6,977	0
Dominican Rep	5,453	25	673	85	3,945	0	0	432	4,021	0
El Salvador	4,095	49	693	1	1,438	0	0	274	1,450	65
Guatemala	5,633	47	535	0	2,279	0	0	365	2,328	1,015
Haiti	1,779	16	227	0	480	0	0	27	480	0
Honduras	3,182	44	532	1	1,178	0	0	212	1,150	0
Jamaica	3,963	104	1,575	41	3,326	0	0	477	3,352	67
Mexico	141,520	24	1,501	6,592	88,214	28,832	2,725	11,831	15,103	95,567
Nicaragua	2,573	26	550	0	1,030	0	0	115	991	16
Panama	2,328	44	855	35	1,490	0	0	269	3,092	564
Trinidad and Tobago	8,196	77	6,419	0	1,336	6,828	0	370	2,140	7,003
SOUTH AMERICA	**379,732**	**35**	**1,148**	**20,679**	**179,005**	**72,463**	**2,883**	**42,975**	**91,009**	**236,800**
Argentina	61,710	36	1,730	856	24,425	28,857	2,057	5,012	4,766	22,951
Bolivia	4,254	67	547	0	2,095	1,085	0	261	244	1,918
Brazil	172,030	31	1,051	12,315	85,673	5,288	826	24,540	59,393	3,282
Chile	23,012	121	1,573	4,240	11,288	2,164	0	2,529	15,030	150
Colombia	30,481	29	761	2,704	14,332	5,339	0	3,047	1,427	38,000
Ecuador	8,513	44	713	0	6,794	0	0	627	924	14,890
Guyana	X	X	X	X	X	X	X	X	X	X
Paraguay	4,191	49	824	0	1,162	0	0	332	1,118	3,948
Peru	15,127	20	621	305	8,572	768	0	1,272	5,721	3,453
Suriname	X	X	X	X	X	X	X	X	X	X
Uruguay	2,883	30	883	1	1,809	0	0	480	2,189	93
Venezuela	57,530	32	2,526	258	22,855	28,962	0	4,873	196	148,115
OCEANIA	**118,305**	**31**	**4,038**	**43,703**	**41,892**	**21,613**	**0**	**15,950**	**30,122**	**122,289**
Australia	101,626	30	5,543	42,346	35,607	16,908	0	13,239	24,483	119,382
Fiji	X	X	X	X	X	X	X	X	X	X
New Zealand	16,679	39	4,434	1,357	6,285	4,705	0	2,711	5,638	2,906
Papua New Guinea	X	X	X	X	X	X	X	X	X	X
Solomon Islands	X	X	X	X	X	X	X	X	X	X
DEVELOPED	**5,827,461**	**(4)**	**4,505**	**1,280,399**	**2,152,070**	**1,476,854**	**582,534**	**717,830**	**2,520,891**	**1,479,201**
DEVELOPING	**3,631,617**	**49**	**803**	**973,694**	**1,244,028**	**434,089**	**41,628**	**267,035**	**908,491**	**1,931,077**

Notes: Negative values are shown in parentheses. Zero is either 0 or less than one-half the unit of measure. (0) indicates a value less than 0 and greater than negative one-half.
a. See ERC.4 for data on renewable energy. b. Tons of oil equivalent. See the technical notes for more information on toe.

Source: International Energy Agency

Energy Consumption by Economic Sector (% of total consumption)

	Industry				Transportation								Commercial and Public Services			
	All Industries		Iron and Steel		Total		Air		Road		Agriculture		Commercial and Public Services		Residential	
	1987	1997	1987	1997	1987	1997	1987	1997	1987	1997	1987	1997	1987	1997	1987	1997
WORLD	**38.2**	**32.2**	**2.7**	**4.1**	**24.7**	**24.8**	**3.1**	**3.1**	**18.8**	**19.6**	**3.8**	**2.8**	**8.3**	**7.4**	**19.1**	**27.1**
ASIA (EXCL. MIDDLE EAST)	**48.6**	**37.4**	**7.7**	**5.6**	**16.4**	**15.9**	**1.5**	**1.8**	**11.9**	**12.3**	**4.0**	**3.0**	**5.1**	**5.4**	**21.2**	**33.6**
Armenia	X	27.1	X	0.0	X	7.2	X	2.7	X	2.8	X	4.7	X	3.7	X	34.9
Azerbaijan	X	39.0	X	0.0	X	15.3	X	6.7	X	7.4	X	11.5	X	1.1	X	24.6
Bangladesh	X	36.2	X	0.0	X	5.1	X	0.5	X	3.3	X	2.1	X	0.5	X	5.1
Bhutan	X	X	X	X	X	X	X	X	X	X	X	X	X	X	X	X
Cambodia	X	X	X	X	X	X	X	X	X	X	X	X	X	X	X	X
China	X	42.5	X	8.1	X	8.8	X	0.5	X	5.7	X	3.4	X	4.4	X	36.8
Georgia	X	20.6	X	3.5	X	22.3	X	2.2	X	19.1	X	3.4	X	12.6	X	17.6
India	X	29.7	X	3.4	X	11.7	X	0.7	X	10.1	X	2.4	X	0.7	X	54.2
Indonesia	11.7	22.8	0.8	2.5	12.9	18.9	1.1	1.7	11.6	15.4	1.1	1.8	0.5	1.1	67.9	54.1
Japan	42.8	39.8	8.1	6.9	24.2	27.0	2.3	3.1	19.4	22.1	3.5	3.2	11.0	12.2	14.1	14.1
Kazakhstan	X	X	X	X	X	X	X	X	X	X	X	X	X	X	X	X
Korea, Dem People's Rep	81.2	83.9	0.0	0.0	8.2	6.1	0.0	0.0	8.2	6.1	X	X	0.0	0.0	0.9	0.7
Korea, Rep	37.9	44.9	6.4	7.1	18.8	23.2	1.5	2.3	13.8	17.3	2.4	2.8	8.5	16.6	28.3	8.5
Kyrgyzstan	82.0	35.3	0.0	0.0	0.5	8.2	0.0	2.8	0.0	5.0	8.4	8.7	0.0	0.0	2.9	5.6
Lao People's Dem Rep	X	X	X	X	X	X	X	X	X	X	X	X	X	X	X	X
Malaysia	38.3	40.4	0.0	0.5	35.9	33.4	4.0	5.5	31.9	27.9	X	1.9	3.7	4.3	17.8	13.4
Mongolia	X	X	X	X	X	X	X	X	X	X	X	X	X	X	X	X
Myanmar	X	11.0	X	0.0	X	7.2	X	0.5	X	6.6	X	0.1	X	0.4	X	81.1
Nepal	2.1	4.2	0.0	0.0	1.9	2.8	0.4	0.4	1.5	2.4	0.1	0.2	0.9	1.2	94.9	91.4
Pakistan	X	26.1	X	0.2	X	16.6	X	1.3	X	14.8	X	1.9	X	2.3	X	52.4
Philippines	19.6	39.3	1.7	2.3	12.2	19.7	2.6	4.5	8.3	13.0	5.1	8.6	2.5	3.7	48.9	23.6
Singapore	34.8	36.7	0.6	0.8	46.5	47.0	24.3	28.2	22.2	18.6	0.0	0.0	6.1	6.6	4.7	4.6
Sri Lanka	13.5	20.0	0.0	0.0	16.2	20.6	2.4	4.5	13.3	15.4	0.0	0.2	1.8	3.3	67.7	55.3
Tajikistan	57.2	19.6	0.0	0.0	0.6	37.8	0.0	0.2	0.0	37.3	19.4	13.0	0.0	0.0	6.1	5.8
Thailand	26.8	33.1	1.0	1.2	33.4	38.9	5.9	5.2	25.6	33.3	6.7	2.8	2.9	4.6	29.0	19.1
Turkmenistan	X	X	X	X	X	X	X	X	X	X	X	X	X	X	X	X
Uzbekistan	X	18.5	X	0.0	X	11.5	X	0.9	X	6.0	X	7.5	X	8.9	X	42.5
Viet Nam	X	19.0	X	0.0	X	13.6	X	0.8	X	12.3	X	1.2	X	0.0	X	64.9
EUROPE	**37.9**	**32.9**	**4.7**	**4.9**	**20.9**	**23.0**	**2.1**	**2.8**	**16.5**	**17.3**	**3.1**	**4.0**	**8.0**	**7.3**	**23.1**	**27.1**
Albania	31.8	13.5	1.3	0.0	14.0	41.2	0.0	0.0	14.0	41.2	X	0.6	0.0	2.8	17.3	31.1
Austria	27.1	23.9	6.3	5.6	24.2	29.0	1.2	2.4	21.5	24.8	2.8	3.9	3.7	3.4	32.3	29.8
Belarus	X	36.9	X	0.8	X	10.8	X	0.0	X	7.8	X	5.8	X	1.3	X	35.4
Belgium	36.2	36.9	8.7	7.6	21.6	23.3	1.9	3.4	18.6	18.4	1.4	2.6	9.4	9.4	28.2	24.7
Bosnia and Herzegovina	X	7.8	X	3.1	X	48.3	X	5.1	X	30.5	X	0.0	X	0.0	X	20.8
Bulgaria	51.6	58.9	3.9	9.1	9.4	10.4	2.0	1.9	6.7	7.5	1.3	3.3	2.3	1.6	13.2	21.6
Croatia	X	33.4	X	1.4	X	24.3	X	1.5	X	20.8	X	3.5	X	7.8	X	27.8
Czech Rep	53.1	48.5	8.7	10.4	6.5	14.5	0.6	0.6	5.0	12.6	4.2	2.2	5.8	7.5	25.0	22.3
Denmark	18.4	19.3	1.0	0.7	31.3	31.0	5.2	5.4	21.4	23.6	4.7	6.5	9.9	11.7	31.5	28.7
Estonia	X	30.1	X	0.0	X	18.7	X	0.8	X	16.1	X	2.6	X	8.1	X	38.6
Finland	42.3	45.0	4.1	4.6	18.0	18.2	1.4	2.0	15.7	15.3	4.0	3.0	3.6	6.0	24.5	21.9
France	30.6	28.1	3.8	3.3	27.2	29.7	2.3	3.3	23.5	25.1	2.2	2.0	13.1	13.1	24.1	23.2
Germany	33.7	29.6	5.0	4.1	20.7	26.6	1.9	2.7	17.7	22.8	1.2	1.1	12.1	10.4	28.2	28.3
Greece	28.6	24.6	1.3	0.6	36.9	38.3	8.4	6.9	25.8	27.9	7.1	5.9	3.5	6.0	19.5	22.9
Hungary	39.3	28.3	7.8	4.0	13.5	16.4	0.7	1.1	10.9	14.2	5.9	4.1	16.7	17.4	17.4	30.1
Iceland	25.5	27.4	6.0	5.5	16.3	15.9	4.0	5.6	10.9	9.6	15.4	17.8	3.2	4.1	36.4	29.3
Ireland	34.9	26.4	0.8	0.5	23.9	31.2	4.4	4.8	18.9	25.1	2.7	2.7	8.4	14.0	28.1	23.6
Italy	34.0	32.7	5.2	4.2	29.2	32.4	1.9	2.5	25.8	28.7	2.4	2.6	2.5	3.7	28.1	25.4
Latvia	X	26.3	X	3.9	X	21.4	X	0.9	X	16.1	X	3.0	X	10.3	X	38.8
Lithuania	X	28.5	X	0.1	X	24.5	X	0.6	X	21.8	X	3.5	X	11.0	X	29.2
Macedonia, FYR	X	31.0	X	10.6	X	30.6	X	1.8	X	26.7	X	4.4	X	4.1	X	27.4
Moldova, Rep	X	20.3	X	0.0	X	15.7	X	0.7	X	11.5	X	9.2	X	20.6	X	28.4
Netherlands	38.7	32.9	2.0	3.2	18.4	23.8	2.9	5.4	14.2	16.9	5.8	7.1	2.8	4.8	21.2	18.3
Norway	41.9	37.4	7.1	6.0	21.7	24.3	3.0	3.4	14.0	16.3	1.5	3.8	10.6	11.0	20.1	19.6
Poland	43.9	36.9	9.4	6.4	11.0	14.7	0.4	0.7	8.3	13.0	3.7	7.2	8.7	7.4	22.6	31.5
Portugal	45.1	38.3	1.7	1.2	29.7	33.9	5.0	3.9	23.4	29.1	4.0	3.4	5.4	6.5	12.6	12.5
Romania	55.2	44.0	9.8	10.3	4.2	14.4	0.0	0.5	3.8	11.1	2.2	3.1	0.7	0.0	20.4	32.4
Russian Federation	X	34.9	X	7.7	X	13.5	X	1.9	X	4.7	X	8.1	X	2.0	X	32.8
Slovakia	58.9	48.2	2.0	11.1	7.4	10.1	0.0	0.3	6.7	9.1	5.3	3.2	10.0	14.0	8.8	19.0
Slovenia	X	28.4	X	4.0	X	33.7	X	0.4	X	32.7	X	0.0	X	14.9	X	23.0
Spain	38.5	32.8	5.9	4.0	34.5	38.1	4.4	5.0	26.0	29.9	5.3	2.9	6.2	6.7	11.6	14.1
Sweden	37.4	38.2	3.5	3.1	21.5	22.0	2.1	2.5	17.7	18.4	2.0	1.5	13.0	13.5	23.6	22.5
Switzerland	19.7	18.6	0.0	0.0	29.3	33.3	5.4	7.2	22.7	25.0	1.0	1.3	16.8	16.5	28.9	27.9
Ukraine	X	42.4	X	4.2	X	7.6	X	0.8	X	3.9	X	4.2	X	10.8	X	30.5
United Kingdom	28.0	25.6	3.4	2.9	28.4	32.2	4.3	5.7	22.6	25.0	0.9	0.8	9.4	9.1	27.5	26.2
Yugoslavia	X	33.8	X	0.3	X	12.2	X	1.1	X	11.0	X	0.2	X	0.3	X	16.2
MIDDLE EAST & N. AFRICA	**31.7**	**30.4**	**1.1**	**1.0**	**31.2**	**25.7**	**3.7**	**3.1**	**27.0**	**22.1**	**2.7**	**2.5**	**2.7**	**3.5**	**17.5**	**21.2**
Afghanistan	X	X	X	X	X	X	X	X	X	X	X	X	X	X	X	X
Algeria	21.2	20.6	4.3	2.3	39.1	18.7	3.3	2.2	34.9	12.2	0.3	0.0	1.5	0.0	23.0	29.2
Egypt	49.6	47.8	0.7	0.5	19.8	20.6	2.9	3.6	16.9	16.9	0.5	0.6	0.0	0.0	23.2	22.9
Iran, Islamic Rep	37.7	31.4	0.1	0.1	24.6	23.2	1.2	0.8	23.5	22.4	7.3	5.5	7.0	7.6	12.9	28.0
Iraq	28.2	28.6	0.0	0.0	47.1	44.4	4.8	2.2	42.3	42.3	0.0	0.0	0.0	0.0	16.1	11.2
Israel	29.9	26.5	0.8	0.7	36.8	32.2	8.4	7.4	28.0	24.1	1.0	1.1	4.7	4.6	14.8	16.2
Jordan	22.8	19.3	0.1	0.0	51.6	41.3	8.5	10.1	43.2	31.2	1.6	2.9	1.4	2.2	15.1	18.1
Kuwait	53.5	27.0	0.0	0.0	27.9	38.3	4.4	7.0	23.5	31.2	0.0	0.0	0.2	0.0	12.3	33.3
Lebanon	X	27.9	X	0.0	X	37.6	X	2.8	X	34.7	X	0.0	X	2.8	X	23.9
Libyan Arab Jamahiriya	32.5	33.8	0.0	0.0	36.8	38.0	6.1	3.4	30.7	34.6	0.0	0.0	0.0	0.0	7.2	9.5
Morocco	31.4	23.7	0.1	0.0	28.0	11.8	5.2	3.7	19.7	5.7	4.9	0.7	6.5	2.4	25.7	24.7
Oman	16.8	37.9	0.0	0.0	42.7	29.1	10.2	5.3	32.5	23.8	0.0	0.0	6.5	5.2	12.7	20.9
Saudi Arabia	20.6	19.0	0.0	0.0	32.3	23.6	5.0	4.8	27.2	18.8	0.0	0.3	3.1	3.9	6.9	9.2
Syrian Arab Rep	30.3	19.4	0.0	0.0	38.2	12.6	3.6	2.1	34.6	10.6	3.3	0.0	0.0	0.0	9.5	10.8
Tunisia	31.7	22.7	1.2	0.6	33.1	28.0	5.8	5.7	21.9	21.9	5.5	6.2	7.5	7.7	22.0	32.9
Turkey	27.7	33.7	3.9	3.8	22.9	22.8	0.8	2.9	20.3	18.9	4.9	5.3	1.3	2.8	40.0	31.4
United Arab Emirates	X	29.1	X	0.0	X	41.7	X	2.6	X	39.1	X	0.0	X	0.0	X	6.0
Yemen	15.0	6.1	0.0	0.0	62.5	63.4	8.3	3.7	54.2	59.7	0.0	0.0	0.4	0.0	13.8	22.6

	Energy Consumption by Economic Sector (% of total consumption)															
	Industry				Transportation								Commercial and			
	All Industries		Iron and Steel		Total		Air		Road		Agriculture		Public Services		Residential	
	1987	1997	1987	1997	1987	1997	1987	1997	1987	1997	1987	1997	1987	1997	1987	1997
SUB-SAHARAN AFRICA	**X**	**X**	**X**	**X**	**X**	**X**	**X**	**X**	**X**	**X**	**X**	**X**	**X**	**X**	**X**	**X**
Angola	X	11.6	X	0.0	X	12.3	X	6.1	X	6.2	X	X	X	0.0	X	75.6
Benin	X	17.1	X	0.0	X	12.6	X	2.4	X	10.1	X	X	X	0.0	X	70.3
Botswana	X	X	X	X	X	X	X	X	X	X	X	X	X	X	X	X
Burkina Faso	X	X	X	X	X	X	X	X	X	X	X	X	X	X	X	X
Burundi	X	X	X	X	X	X	X	X	X	X	X	X	X	X	X	X
Cameroon	X	17.0	X	0.0	X	11.6	X	1.0	X	10.6	X	X	X	0.5	X	69.7
Central African Rep	X	X	X	X	X	X	X	X	X	X	X	X	X	X	X	X
Chad	X	X	X	X	X	X	X	X	X	X	X	X	X	X	X	X
Congo	X	14.2	X	0.0	X	18.4	X	6.3	X	12.1	X	X	X	0.0	X	65.7
Congo, Dem Rep	X	21.8	X	0.1	X	5.4	X	1.2	X	4.2	X	X	X	0.0	X	70.9
Côte d'Ivoire	X	7.0	X	0.0	X	16.0	X	3.1	X	12.5	X	1.2	X	12.1	X	62.0
Equatorial Guinea	X	X	X	X	X	X	X	X	X	X	X	X	X	X	X	X
Eritrea	X	X	X	X	X	X	X	X	X	X	X	X	X	X	X	X
Ethiopia	X	X	X	X	X	X	X	X	X	X	X	X	X	X	X	X
Gabon	X	23.2	X	0.0	X	19.6	X	6.7	X	11.0	X	X	X	0.9	X	51.5
Gambia	X	X	X	X	X	X	X	X	X	X	X	X	X	X	X	X
Ghana	16.0	14.1	0.0	0.0	11.0	13.2	1.4	1.2	9.0	11.4	0.7	1.0	0.5	0.7	69.8	69.9
Guinea	X	X	X	X	X	X	X	X	X	X	X	X	X	X	X	X
Guinea-Bissau	X	X	X	X	X	X	X	X	X	X	X	X	X	X	X	X
Kenya	X	11.4	X	0.0	X	12.6	X	4.5	X	7.7	X	6.5	X	0.7	X	67.7
Lesotho	X	X	X	X	X	X	X	X	X	X	X	X	X	X	X	X
Liberia	X	X	X	X	X	X	X	X	X	X	X	X	X	X	X	X
Madagascar	X	X	X	X	X	X	X	X	X	X	X	X	X	X	X	X
Malawi	X	X	X	X	X	X	X	X	X	X	X	X	X	X	X	X
Mali	X	X	X	X	X	X	X	X	X	X	X	X	X	X	X	X
Mauritania	X	X	X	X	X	X	X	X	X	X	X	X	X	X	X	X
Mozambique	X	7.8	X	0.0	X	1.5	X	0.4	X	1.1	X	0.6	X	3.4	X	85.0
Namibia	X	X	X	X	X	X	X	X	X	X	X	X	X	X	X	X
Niger	X	X	X	X	X	X	X	X	X	X	X	X	X	X	X	X
Nigeria	X	10.9	X	0.1	X	6.7	X	0.7	X	5.9	X	X	X	0.3	X	79.7
Rwanda	X	X	X	X	X	X	X	X	X	X	X	X	X	X	X	X
Senegal	X	17.0	X	0.0	X	24.0	X	10.0	X	12.6	X	2.3	X	0.6	X	55.3
Sierra Leone	X	X	X	X	X	X	X	X	X	X	X	X	X	X	X	X
Somalia	X	X	X	X	X	X	X	X	X	X	X	X	X	X	X	X
South Africa	48.4	36.6	17.6	7.8	21.0	23.5	1.3	2.5	17.9	19.2	2.6	3.4	4.1	5.2	19.3	21.3
Sudan	X	7.4	X	0.0	X	17.8	X	1.0	X	16.9	X	0.1	X	2.0	X	70.9
Tanzania, United Rep	X	11.7	X	0.0	X	1.7	X	0.3	X	1.4	X	3.2	X	0.3	X	79.1
Togo	X	X	X	X	X	X	X	X	X	X	X	X	X	X	X	X
Uganda	X	X	X	X	X	X	X	X	X	X	X	X	X	X	X	X
Zambia	X	23.6	X	0.1	X	5.1	X	0.9	X	4.2	X	0.7	X	2.0	X	66.6
Zimbabwe	X	12.2	X	2.0	X	10.0	X	1.5	X	8.3	X	9.2	X	3.6	X	61.8
NORTH AMERICA	**30.1**	**26.3**	**1.4**	**1.9**	**35.1**	**38.1**	**5.0**	**5.2**	**27.9**	**30.3**	**1.3**	**1.2**	**11.6**	**12.6**	**16.0**	**17.3**
Canada	37.9	36.3	3.2	2.5	26.9	27.8	2.7	2.7	20.8	20.2	1.8	2.3	12.7	13.5	16.9	16.6
United States	29.2	25.0	1.1	1.8	36.0	39.5	5.2	5.6	28.7	31.7	1.2	1.1	11.4	12.5	15.9	17.3
C. AMERICA & CARIBBEAN	**41.1**	**39.4**	**4.2**	**4.8**	**30.0**	**31.8**	**2.0**	**2.5**	**26.3**	**28.3**	**2.9**	**2.4**	**2.8**	**3.4**	**20.6**	**21.3**
Belize	X	X	X	X	X	X	X	X	X	X	X	X	X	X	X	X
Costa Rica	24.4	24.0	0.0	0.0	29.0	49.1	1.7	5.3	10.7	42.5	X	3.8	0.4	7.6	45.3	14.4
Cuba	49.7	70.9	2.0	0.4	27.7	10.5	2.1	2.9	21.8	7.5	6.2	3.1	2.5	2.1	10.0	7.5
Dominican Rep	28.8	20.5	0.0	0.0	36.3	29.9	3.6	1.7	24.4	14.0	0.6	1.0	0.0	0.0	34.4	46.3
El Salvador	22.0	20.5	0.0	0.0	19.0	24.0	1.1	1.6	17.9	22.4	0.1	X	2.3	3.0	56.2	51.6
Guatemala	10.6	13.9	0.0	0.0	14.8	21.2	1.3	1.0	13.6	20.1	0.9	0.7	2.1	4.0	70.4	58.8
Haiti	14.7	11.7	0.0	0.0	10.7	15.9	1.7	1.5	4.2	6.8	X	X	1.3	4.4	73.3	67.3
Honduras	23.8	27.3	0.0	0.0	15.1	17.4	1.3	0.8	13.8	16.5	0.1	0.2	4.3	4.9	56.7	50.3
Jamaica	43.3	26.1	0.0	0.0	33.6	41.7	10.5	12.1	23.1	21.0	3.1	11.9	0.2	6.4	13.5	13.2
Mexico	43.0	38.6	5.4	6.4	32.0	36.4	1.9	2.5	29.0	33.3	3.0	2.7	3.1	3.6	16.1	17.6
Nicaragua	19.1	11.4	0.0	0.0	21.0	23.2	2.4	1.6	18.0	20.1	2.6	1.0	2.3	6.1	52.8	56.7
Panama	26.9	20.5	0.0	0.0	29.3	35.3	0.1	0.2	29.2	34.7	X	X	1.7	2.4	41.6	39.3
Trinidad and Tobago	X	85.6	X	7.4	X	10.9	X	1.3	X	9.6	X	X	X	1.1	X	2.1
SOUTH AMERICA	**37.8**	**37.3**	**5.7**	**5.2**	**29.5**	**32.7**	**2.3**	**2.4**	**26.1**	**29.0**	**4.9**	**4.8**	**3.2**	**5.0**	**20.9**	**16.8**
Argentina	31.6	29.2	0.4	0.5	32.1	33.8	2.3	3.1	29.6	30.4	5.8	6.7	2.2	6.6	23.8	19.8
Bolivia	12.9	28.7	0.0	0.0	38.0	34.9	4.6	5.0	32.8	28.6	0.7	1.6	0.0	1.6	47.9	32.7
Brazil	42.4	41.4	8.3	7.1	27.3	32.3	2.2	2.0	23.8	28.7	5.9	5.2	3.7	4.8	17.5	13.5
Chile	36.8	41.8	3.5	2.1	29.2	30.0	2.7	3.2	24.5	24.3	X	1.3	1.8	0.8	30.8	25.4
Colombia	28.3	29.6	1.4	1.6	29.6	33.0	2.5	2.7	26.8	29.4	5.9	6.7	3.6	4.2	26.6	18.1
Ecuador	21.6	18.5	0.0	0.0	39.6	42.4	3.3	3.3	26.5	30.6	5.6	5.4	2.4	6.6	24.9	24.8
Guyana	X	X	X	X	X	X	X	X	X	X	X	X	X	X	X	X
Paraguay	30.0	38.0	0.0	0.0	19.0	25.5	1.3	0.1	17.2	24.9	X	X	0.9	1.7	49.6	34.5
Peru	20.8	22.5	0.5	4.4	25.8	27.2	2.8	3.8	21.6	23.0	7.5	6.2	2.2	3.0	42.4	40.0
Suriname	X	X	X	X	X	X	X	X	X	X	X	X	X	X	X	X
Uruguay	28.0	21.0	0.0	0.0	23.9	33.3	0.8	0.3	21.8	33.0	6.6	8.1	5.7	7.5	32.1	26.8
Venezuela	46.1	44.2	10.3	9.8	35.4	34.7	2.0	2.3	33.1	32.4	0.1	1.7	3.1	7.6	9.4	7.8
OCEANIA	**35.5**	**35.5**	**4.5**	**3.9**	**39.1**	**39.0**	**4.6**	**6.3**	**30.9**	**29.2**	**2.6**	**2.3**	**6.0**	**6.7**	**13.2**	**12.4**
Australia	35.1	34.3	4.1	3.6	39.8	39.5	4.3	6.2	32.4	31.2	2.5	2.3	5.6	6.7	13.3	12.8
Fiji	X	X	X	X	X	X	X	X	X	X	X	X	X	X	X	X
New Zealand	37.6	41.7	6.9	5.9	35.0	36.4	6.2	6.9	22.0	18.4	3.1	2.5	8.3	6.7	12.4	10.7
Papua New Guinea	X	X	X	X	X	X	X	X	X	X	X	X	X	X	X	X
Solomon Islands	X	X	X	X	X	X	X	X	X	X	X	X	X	X	X	X
DEVELOPED	**34.8**	**30.7**	**3.6**	**3.8**	**27.9**	**29.8**	**3.5**	**3.9**	**22.2**	**23.3**	**2.3**	**2.8**	**9.7**	**9.8**	**18.6**	**21.4**
DEVELOPING	**44.4**	**34.9**	**6.1**	**4.5**	**21.2**	**17.8**	**1.9**	**1.8**	**17.0**	**14.5**	**3.8**	**2.9**	**3.2**	**3.9**	**22.2**	**34.4**

Notes: Zero is either 0 or less than one-half the unit of measure.

Source: International Energy Agency

	Renewable Energy Production: All Renewable Sources (1000 toe) {a} 1997	Renewable Energy Production: All Renewable Sources Percent Change Since 1987	Renewable Energy Consumption: All Renewable Sources (1000 toe) {a} 1997	All Renewable Sources % of Total Consumption from All Sources 1997	Fuels and Waste (1000 toe) {a} 1997	Fuels and Waste Percent Change Since 1987	Hydroelectric (1000 toe) {a} 1997	Hydroelectric Percent Change Since 1987	Geothermal (1000 toe) {a} 1997	Solar (1000 toe) {a} 1997	Wind (1000 toe) {a} 1997
WORLD	**1,323,051**	**19**	**1,321,093**	**14**	**1,061,775**	**16**	**220,656**	**26**	**36,659**	**1,015**	**989**
ASIA (EXCL. MIDDLE EAST)	**623,784**	**18**	**625,537**	**21**	**569,937**	**15**	**43,653**	**46**	**11,926**	**16**	**5**
Armenia	120	(4)	120	7	1	X	119	(5)	0	0	0
Azerbaijan	132	92	135	1	4	X	131	90	0	0	0
Bangladesh	16,093	14	16,093	66	16,031	14	62	39	0	0	0
Bhutan	X	X	X	X	X	X	X	X	X	X	X
Cambodia	X	X	X	X	X	X	X	X	X	X	X
China	225,117	12	225,117	20	208,262	8	16,855	96	0	0	0
Georgia	555	(28)	555	24	35	X	520	(32)	0	0	0
India	199,237	15	199,237	43	192,801	14	6,431	58	0	0	5
Indonesia	47,327	19	47,226	34	44,493	15	515	16	2,217	0	0
Japan	18,669	38	18,669	4	7,461	27	7,723	20	3,485	0	0
Kazakhstan	633	28	633	2	74	X	559	13	0	0	0
Korea, Dem People's Rep	1,019	10	2,898	12	1,019	10	1,878	(25)	0	0	0
Korea, Rep	1,917	317	1,892	1	1,634	X	242	(47)	0	16	0
Kyrgyzstan	969	135	969	35	4	X	965	134	0	0	0
Lao People's Dem Rep	0	X	0	X	X	X	X	X	X	X	X
Malaysia	2,690	14	2,673	6	2,390	22	283	(33)	0	0	0
Mongolia	0	X	0	X	X	X	X	X	X	X	X
Myanmar	10,412	18	10,412	80	10,270	18	142	62	0	0	0
Nepal	6,559	25	6,559	92	6,464	24	95	102	0	0	0
Pakistan	24,192	31	24,192	43	22,398	30	1,794	37	0	0	0
Philippines	16,149	17	16,149	42	9,403	(0)	522	16	6,224	0	0
Singapore	X	X	X	0	0	X	0	X	0	0	0
Sri Lanka	4,345	9	4,345	61	4,049	7	296	58	0	0	0
Tajikistan	1,189	(5)	1,189	35	0	X	1,189	(5)	0	0	0
Thailand	21,268	69	21,275	27	20,656	69	619	77	0	0	0
Turkmenistan	X	X	X	X	X	X	X	X	X	X	X
Uzbekistan	497	(15)	497	1	0	X	497	(15)	0	0	0
Viet Nam	23,790	34	23,790	61	22,398	27	1,392	1,069	0	0	0
EUROPE	**139,258**	**56**	**139,247**	**5**	**76,996**	**154**	**57,645**	**1**	**3,691**	**236**	**630**
Albania	523	(21)	523	50	60	(84)	464	61	0	0	0
Austria	5,479	7	5,578	20	2,483	13	3,095	(0)	0	0	0
Belarus	687	X	687	3	685	X	2	17	0	0	0
Belgium	578	147	578	1	549	178	26	(29)	1	1	1
Bosnia and Herzegovina	277	X	277	16	155	X	122	(53)	0	0	0
Bulgaria	396	(8)	385	2	240	14	146	(33)	0	0	0
Croatia	779	X	779	10	324	X	456	(9)	0	0	0
Czech Rep	686	X	686	2	540	X	146	(31)	0	0	0
Denmark	1,714	67	1,714	8	1,538	52	2	(31)	1	7	166
Estonia	625	X	592	11	592	X	0	X	0	0	0
Finland	6,923	39	6,923	21	5,869	48	1,053	(11)	0	0	1
France	16,022	X	16,098	7	10,564	X	5,351	(10)	116	16	2
Germany	5,569	(4)	5,569	2	3,816	(5)	1,493	(15)	0	3	258
Greece	1,415	92	1,415	6	952	90	334	40	4	123	3
Hungary	461	(5)	461	2	443	(5)	19	28	0	0	0
Iceland	1,481	20	1,481	64	1	X	448	32	1,032	0	0
Ireland	225	X	225	2	162	X	58	(0)	0	0	4
Italy	7,290	19	7,610	5	1,562	54	3,578	5	2,452	8	10
Latvia	1,557	X	1,335	30	1,081	X	254	(20)	0	0	0
Lithuania	543	X	544	6	519	X	25	(18)	0	0	0
Macedonia, FYR	277	X	280	10	190	X	77	X	13	0	0
Moldova, Rep	100	X	94	2	61	X	33	55	0	0	0
Netherlands	1,038	X	1,038	1	979	X	8	X	0	11	41
Norway	10,597	8	10,600	44	1,160	26	9,439	6	0	0	1
Poland	5,333	86	5,328	5	5,160	89	169	18	0	0	0
Portugal	2,317	26	2,317	11	1,126	7	1,127	43	45	16	3
Romania	4,893	X	4,880	11	3,374	X	1,506	56	0	0	0
Russian Federation	30,806	X	30,546	5	17,023	X	13,499	(3)	24	0	0
Slovakia	438	44	439	3	83	(40)	356	114	0	0	0
Slovenia	500	64	528	8	263	X	266	(13)	0	0	0
Spain	6,609	133	6,609	6	3,539	X	2,976	26	3	28	64
Sweden	14,271	22	14,271	27	8,313	51	5,936	(4)	0	4	18
Switzerland	4,353	15	4,361	17	1,414	72	2,928	(2)	0	19	0
Ukraine	1,093	32	1,093	1	246	X	848	3	0	0	0
United Kingdom	2,097	X	2,097	1	1,685	X	355	(0)	0	0	57
Yugoslavia	1,256	20	1,256	8	210	X	1,046	(0)	0	0	0
MIDDLE EAST & N. AFRICA	**18,597**	**19**	**18,613**	**4**	**11,562**	**(0)**	**6,275**	**72**	**179**	**597**	**0**
Afghanistan	0	X	0	X	X	X	X	X	X	X	X
Algeria	525	16	525	2	519	27	6	(85)	0	0	0
Egypt	2,286	30	2,268	6	1,238	22	1,031	38	0	0	0
Iran, Islamic Rep	1,420	4	1,420	1	786	21	634	(12)	0	0	0
Iraq	76	(69)	76	0	26	21	50	(78)	0	0	0
Israel	519	66	522	3	6	126	6	X	0	511	0
Jordan	4	48	5	0	3	42	2	16	0	0	0
Kuwait	X	X	4	0	4	(37)	0	X	0	0	0
Lebanon	207	31	208	4	125	16	77	48	0	6	0
Libyan Arab Jamahiriya	125	0	125	1	125	0	0	X	0	0	0
Morocco	592	56	592	6	415	35	177	150	0	0	0
Oman	X	X	X	0	0	X	0	X	0	0	0
Saudi Arabia	X	X	4	0	4	(54)	0	X	0	0	0
Syrian Arab Rep	868	366	868	6	5	56	863	373	0	0	0
Tunisia	1,176	19	1,176	17	1,172	20	4	(61)	0	0	0
Turkey	10,705	12	10,705	15	7,022	(11)	3,424	114	179	80	0
United Arab Emirates	X	X	17	0	17	X	0	X	0	0	0
Yemen	77	6	77	2	77	6	0	X	0	0	0

	Renewable Energy Production: All Renewable Sources		Renewable Energy Consumption: All Renewable Sources		Fuels and Waste		Hydroelectric		Geothermal	Solar	Wind
	(1000 toe) {a} 1997	Percent Change Since 1987	(1000 toe) {a} 1997	% of Total Consumption from All Sources 1997	(1000 toe) {a} 1997	Percent Change Since 1987	(1000 toe) {a} 1997	Percent Change Since 1987	(1000 toe) {a} 1997	(1000 toe) {a} 1997	(1000 toe) {a} 1997
SUB-SAHARAN AFRICA	**X**	**X**	**X**	**X**	**X**	**X**	**X**	**X**	**X**	**X**	**X**
Angola	5,264	28	5,264	77	5,178	28	87	44	0	0	0
Benin	1,830	26	1,830	84	1,830	26	0	X	0	0	0
Botswana	X	X	X	X	X	X	X	X	X	X	X
Burkina Faso	X	X	X	X	X	X	X	X	X	X	X
Burundi	X	X	X	X	X	X	X	X	X	X	X
Cameroon	4,893	31	4,893	85	4,627	31	266	26	0	0	0
Central African Rep	X	X	X	X	X	X	X	X	X	X	X
Chad	X	X	X	X	X	X	X	X	X	X	X
Congo	899	28	899	72	862	27	37	55	0	0	0
Congo, Dem Rep	12,993	37	12,993	89	12,488	38	506	12	0	0	0
Côte d'Ivoire	4,105	40	4,105	73	3,931	38	174	131	0	0	0
Equatorial Guinea	X	X	X	X	X	X	X	X	X	X	X
Eritrea	X	X	X	X	X	X	X	X	X	X	X
Ethiopia	16,316	27	16,316	95	16,126	27	101	62	89	0	0
Gabon	924	24	924	57	860	26	64	6	0	0	0
Gambia	X	X	X	X	X	X	X	X	X	X	X
Ghana	5,482	39	5,482	79	4,953	40	529	31	0	0	0
Guinea	X	X	X	X	X	X	X	X	X	X	X
Guinea-Bissau	X	X	X	X	X	X	X	X	X	X	X
Kenya	11,651	20	11,651	82	11,013	19	300	83	338	0	0
Lesotho	X	X	X	X	X	X	X	X	X	X	X
Liberia	X	X	X	X	X	X	X	X	X	X	X
Madagascar	X	X	X	X	X	X	X	X	X	X	X
Malawi	X	X	X	X	X	X	X	X	X	X	X
Mali	X	X	X	X	X	X	X	X	X	X	X
Mauritania	X	X	X	X	X	X	X	X	X	X	X
Mozambique	6,993	(1)	6,993	91	6,925	(1)	68	339	0	0	0
Namibia	X	X	X	X	X	X	X	X	X	X	X
Niger	X	X	X	X	X	X	X	X	X	X	X
Nigeria	69,270	32	69,270	78	68,789	32	481	70	0	0	0
Rwanda	X	X	X	X	X	X	X	X	X	X	X
Senegal	1,634	31	1,634	59	1,634	31	0	X	0	0	0
Sierra Leone	X	X	X	X	X	X	X	X	X	X	X
Somalia	X	X	X	X	X	X	X	X	X	X	X
South Africa	12,221	32	11,999	11	11,819	31	180	29	0	0	0
Sudan	9,622	15	9,622	84	9,532	15	90	16	0	0	0
Tanzania, United Rep	13,526	23	13,526	95	13,402	23	125	26	0	0	0
Togo	0	X	0	X	X	X	X	X	X	X	X
Uganda	0	X	0	X	X	X	X	X	X	X	X
Zambia	5,450	20	5,450	91	4,765	26	685	(8)	0	0	0
Zimbabwe	5,441	21	5,441	55	5,288	20	153	60	0	0	0
NORTH AMERICA	**150,068**	**8**	**150,027**	**6**	**78,240**	**(5)**	**58,590**	**23**	**12,820**	**77**	**297**
Canada	40,581	12	40,581	17	10,377	16	30,196	11	0	0	5
United States	109,487	6	109,446	5	67,863	(7)	28,395	39	12,820	77	292
C. AMERICA & CARIBBEAN	**35,432**	**21**	**35,432**	**18**	**25,641**	**16**	**3,815**	**46**	**5,968**	**0**	**7**
Belize	0	X	0	X	X	X	X	X	X	X	X
Costa Rica	1,157	14	1,157	43	269	(64)	413	60	468	0	7
Cuba	5,552	39	5,552	39	5,544	39	8	112	0	0	0
Dominican Rep	1,423	(0)	1,423	26	1,308	2	115	(21)	0	0	0
El Salvador	2,649	34	2,649	65	1,926	43	103	30	620	0	0
Guatemala	3,353	18	3,353	60	3,030	13	323	113	0	0	0
Haiti	1,298	4	1,298	73	1,281	5	17	(38)	0	0	0
Honduras	2,003	28	2,003	63	1,724	22	279	85	0	0	0
Jamaica	595	133	595	15	585	138	10	(3)	0	0	0
Mexico	15,033	15	15,033	11	8,059	4	2,273	44	4,701	0	0
Nicaragua	1,529	23	1,529	59	1,314	32	35	(17)	180	0	0
Panama	808	37	808	35	570	38	238	36	0	0	0
Trinidad and Tobago	32	(37)	32	0	32	(37)	0	X	0	0	0
SOUTH AMERICA	**106,225**	**16**	**104,699**	**28**	**62,052**	**(3)**	**42,600**	**57**	**0**	**0**	**48**
Argentina	5,069	34	5,069	8	2,647	39	2,422	29	0	0	0
Bolivia	1,074	44	1,074	25	876	34	198	117	0	0	0
Brazil	65,972	7	64,448	37	40,448	(11)	24,000	50	0	0	0
Chile	5,320	53	5,320	23	3,691	53	1,629	54	0	0	0
Colombia	8,088	11	8,088	27	5,267	0	2,821	41	0	0	0
Ecuador	1,720	21	1,720	20	1,137	10	583	48	0	0	0
Guyana	X	X	X	X	X	X	X	X	X	X	X
Paraguay	6,960	90	6,957	166	2,619	27	4,339	172	0	0	0
Peru	5,482	15	5,482	36	4,297	12	1,136	21	0	0	48
Suriname	X	X	X	X	X	X	X	X	X	X	X
Uruguay	1,086	(14)	1,086	38	529	(17)	558	(11)	0	0	0
Venezuela	5,456	71	5,456	9	541	(0)	4,915	85	0	0	0
OCEANIA	**11,098**	**32**	**11,098**	**9**	**5,924**	**41**	**3,437**	**12**	**1,646**	**89**	**1**
Australia	6,765	39	6,765	7	5,237	45	1,439	23	0	89	1
Fiji	X	X	X	X	X	X	X	X	X	X	X
New Zealand	4,332	23	4,332	26	687	14	1,998	5	1,646	0	1
Papua New Guinea	X	X	X	X	X	X	X	X	X	X	X
Solomon Islands	X	X	X	X	X	X	X	X	X	X	X
DEVELOPED	**335,929**	**27**	**335,661**	**6**	**180,564**	**37**	**131,561**	**12**	**21,643**	**913**	**929**
DEVELOPING	**924,052**	**19**	**924,291**	**25**	**821,322**	**15**	**87,792**	**58**	**15,016**	**102**	**60**

Notes: Negative values are shown in parentheses. Zero is either 0 or less than one-half the unit of measure. (0) indicates a value less than 0 and greater than negative one-half.
a. Tons of oil equivalent. See the technical notes for more information on toe.

Source: International Energy Agency

	Passenger Cars (per 1000 people)		Annual Motor Gasoline Consumption (liters per person)		Annual Meat Consumption (kg per person)		Annual Paper Consumption (kg per person)		Annual Coffee Consumption (kg per person)
	1990	1996	1987	1997	1988	1998	1988	1998	1997
WORLD	**77 a**	**84 a**	**186**	**182**	**33**	**37**	**44.6**	**49.2**	**X**
ASIA (EXCL. MIDDLE EAST)	**16**	**23**	**30**	**50**	**15**	**25**	**18.1**	**26.0**	**X**
Armenia	1	0	261	8	0	22	X	X	X
Azerbaijan	36	36	X	86	0	16	X	X	X
Bangladesh	0	0	1	2	3	3	0.9	1.3	X
Bhutan	X	X	X	X	X	X	X	X	X
Cambodia	0	5	X	X	14	15	0.0	0.4	X
China {b}	1	3	20	35	23	47	14.6	29.8	0.0
Georgia	89	79	X	123	X	X	X	X	X
India	2	4	5	7	4	4	2.7	3.7	0.1
Indonesia	7	12	27	49	7	9	5.0	17.1	0.6
Japan	283	373	308	422	38	42	204.0	238.6	2.9
Kazakhstan	50	63	X	155	0	41	X	X	X
Korea, Dem People's Rep	X	X	X	X	15	7	4.2	3.6	X
Korea, Rep	48	151	39	245	21	38	83.0	113.7	1.4
Kyrgyzstan	44	32	X	33	0	38	X	X	X
Lao People's Dem Rep	6	3	X	X	10	14	X	X	X
Malaysia	101	139	176	305	33	52	30.7	69.1	0.4
Mongolia	5	12	X	X	94	90	X	X	X
Myanmar	X	1	6	8	9	9	0.7	0.8	X
Nepal	X	X	1	2	10	10	0.3	0.6	X
Pakistan	4	5	11	10	11	14	3.3	3.2	X
Philippines	7	10	28	47	16	27	8.7	12.4	0.7
Singapore	101	120	205	268	75	76	129.2	167.7	X
Sri Lanka	6	6	11	14	3	5	7.7	6.9	0.1
Tajikistan	0	0	X	224	X	X	0.0	0.2 c	X
Thailand	14	28	48	120	21	26	13.7	22.6	0.4
Turkmenistan	X	X	X	140	X	X	X	X	X
Uzbekistan	X	X	X	81	0	29	X	X	X
Viet Nam	X	X	9	28	15	21	0.9	3.3	0.2
EUROPE	**224**	**289**	**232**	**303**	**58**	**72**	**80.0**	**111.5**	**X**
Albania	2	20	47	76	16	25	8.7	17.7	X
Austria	387	458	438	345	104	106	133.0	243.3	8.1
Belarus	59	101	X	159	0	63	X	18.3	X
Belgium	385	424	383	334	X	X	X	X	X
Bosnia and Herzegovina	101	43	X	11	X	X	X	X	X
Bulgaria	146	204	183	95	75	63	63.9	27.2	2.0
Croatia	X	X	X	202	0	22	X	107.4	4.8
Czech Rep	228	325	137	238	0	83	X	78.5	2.8
Denmark	320	331	398	502	103	127	247.9	245.2	9.0
Estonia	154	277	X	281	0	59	X	34.5	X
Finland	386	379	469	488	63	69	302.1	419.9	11.0
France	405	437	443	334	98	100	141.0	180.9	5.7
Germany	386	500	482	491	100	87	176.6	205.2	7.1
Greece	171	223	265	383	70	82	50.4	97.1	4.3
Hungary	188	239	191	178	110	79	68.9	107.7	3.5
Iceland	470	463	638	658	70	70	119.1	107.6 c	X
Ireland	227	272	314	428	101	111	95.4	104.3	1.6
Italy	476	533	288	434	82	88	112.4	167.1	5.1
Latvia	106	153	X	203	0	37	X	36.3	X
Lithuania	132	212	X	237	0	48	X	23.1	1.7
Macedonia, FYR	X	X	X	144	0	30	X	29.6	X
Moldova, Rep	48	39	268	74	0	19	0.0	3.3 c	X
Netherlands	368	370	310	353	81	106	187.8	246.8	9.2
Norway	380	379	559	505	51	60	146.6	173.8	9.2
Poland	138	209	104	170	75	72	40.5	53.8	3.4
Portugal	162	269	141	263	53	85	70.1	106.1	3.9
Romania	56	107	106	92	59	56	32.3	14.8	1.7
Russian Federation	X	107	X	201	0	45	X	14.9	0.4
Slovakia	163	198	116	132	0	77	X	116.2	2.6
Slovenia	289	365	339	610	0	88	X	141.5	4.8
Spain	309	376	235	302	89	111	100.2	156.2	4.6
Sweden	426	413	649	621	59	71	242.2	212.8	8.5
Switzerland	449	462	668	703	82	71	X	X	6.0
Ukraine	63	93	X	84	0	30	X	7.9	X
United Kingdom	341	360	519	507	72	76	159.2	201.0	2.5
Yugoslavia	133	150	X	84	0	100	X	19.6	1.7
MIDDLE EAST & N. AFRICA	**X**	**43**	**127**	**149**	**21**	**22**	**9.9**	**15.5**	**X**
Afghanistan	X	1	X	X	X	X	X	X	X
Algeria	X	25	113	79	17	18	13.0	7.9	3.1
Egypt	21	23	54	43	17	21	7.9	12.7	0.2
Iran, Islamic Rep	25	30	123	171	19	24	5.0	7.3	X
Iraq	1	36	204	185	27	6	6.6	1.0	X
Israel	174	208	381	433	60	65	85.0	107.2	4.3
Jordan	X	50	96	116	30	21	11.5	18.4	X
Kuwait	X	317	803	1,309	61	82	X	X	X
Lebanon	300	299	349	556	45	40	37.1	46.8	3.5
Libyan Arab Jamahiriya	X	159	322	454	35	33	5.9	2.8	X
Morocco	28	38	21	19	17	19	7.7	10.2	0.8
Oman	83	97	343	392	37	36	7.0	5.5 c	X
Saudi Arabia	98	90	587	660	44	44	X	X	0.7
Syrian Arab Rep	10	10	112	101	19	20	6.8	9.2	0.9
Tunisia	23	30	38	49	17	20	15.3	20.9	1.1
Turkey	34	55	63	96	20	19	14.5	31.9	0.2
United Arab Emirates	97	82	652	677	69	100	X	X	X
Yemen	14	15	61	86	16	10	0.1	2.3 c	X

	Passenger Cars (per 1000 people)		Annual Motor Gasoline Consumption (liters per person)		Annual Meat Consumption (kg per person)		Annual Paper Consumption (kg per person)		Annual Coffee Consumption (kg per person)
	1990	1996	1987	1997	1988	1998	1988	1998	1997
SUB-SAHARAN AFRICA	**11**	**14**	**31**	**31**	**12**	**13**	**5.3**	**4.8**	**X**
Angola	15	18	12	11	16	11	2.5	0.2	0.2
Benin	2	7	13	30	13	15	X	X	<0.1
Botswana	10	15	X	X	23	32	X	X	X
Burkina Faso	2	4	X	X	11	11	X	X	X
Burundi	X	X	X	X	X	X	X	X	<0.1
Cameroon	6	7	38	22	16	15	4.5	2.7	0.4
Central African Rep	1	0	X	X	21	24	X	X	0.2
Chad	1	2	X	X	11	10	X	X	X
Congo	12	14	40	26	16	17	X	X	0.1
Congo, Dem Rep	13	17	7	7	7	5	0.3	0.1	0.3
Côte d'Ivoire	16	21	24	14	15	11	X	X	0.2
Equatorial Guinea	3	4	X	X	X	X	X	X	0.0
Eritrea	1	2	X	X	X	X	X	X	X
Ethiopia	1	1	4	3	0	10	0.2	0.4	1.6
Gabon	19	22	45	48	53	49	X	X	0.1
Gambia	7	8	X	X	9	7	0.4	0.1 c	X
Ghana	X	5	21	26	10	8	X	X	0.0
Guinea	2	2	X	X	X	X	X	X	0.4
Guinea-Bissau	4	6	X	X	15	15	X	X	X
Kenya	10	10	26	18	16	13	5.6	6.2	0.1
Lesotho	3	6	X	X	20	16	X	X	X
Liberia	7	3	X	X	9	8	X	X	0.1
Madagascar	4	5	X	X	21	18	0.6	0.9	0.7
Malawi	2	3	X	X	5	5	X	X	0.0
Mali	2	3	X	X	18	19	X	X	X
Mauritania	6	8	X	X	30	23	X	X	X
Mozambique	3	0	3	3	6	5	1.1	0.1	X
Namibia	40	47	X	X	29	36	X	X	X
Niger	5	4	X	X	13	12	X	X	X
Nigeria	X	7	51	42	9	12	2.4	1.9	0.0
Rwanda	1	2	X	X	X	X	X	X	0.0
Senegal	8	10	17	12	15	18	X	X	X
Sierra Leone	7	4	X	X	5	5	X	X	0.1
Somalia	1	0	X	X	22	18	X	X	X
South Africa	97	100	219	267	41	34	51.5	53.5	0.4
Sudan	8	10	8	11	17	21	1.8	0.9	X
Tanzania, United Rep	1	1	5	4	11	10	1.9	1.3	0.0
Togo	15	19	X	X	12	11	X	X	0.0
Uganda	1	2	X	X	9	11	0.5	0.5	0.2
Zambia	8	17	21	17	13	12	1.5	1.6	0.0
Zimbabwe	X	29	24	52	12	11	9.0	8.7	0.0
NORTH AMERICA	**563**	**484**	**1,618**	**1,637**	**113**	**120**	**300.3**	**295.0**	**X**
Canada	468	441	1,222	1,178	99	99	224.2	316.7	4.5
United States	573	489	1,660	1,688	114	122	308.5	292.6	4.0
C. AMERICA & CARIBBEAN	**X**	**69**	**182**	**204**	**32**	**39**	**31.1**	**34.7**	**X**
Belize	36	42	X	X	52	52	X	X	X
Costa Rica	55	81	80	159	37	42	21.1	72.3	4.1
Cuba	X	20	137	50	37	24	30.0	7.3	1.1
Dominican Rep	21	28	125	97	28	37	13.9	32.4	2.4
El Salvador	17	29	42	69	14	17	10.4	18.6	1.9
Guatemala	X	10	43	70	15	20	7.4	18.9	1.7
Haiti	X	4	10	14	10	10	X	X	2.5
Honduras	X	X	37	48	14	17	27.3	34.2	1.6
Jamaica	X	41	113	216	40	55	23.3	16.8	0.5
Mexico	82	93	260	303	39	51	42.7	46.2	0.6
Nicaragua	10	16	45	33	12	15	X	X	1.6
Panama	60	76	117	167	39	48	26.8	27.2	1.8
Trinidad and Tobago	X	94	448	322	29	28	25.0	40.8 c	0.7
SOUTH AMERICA	**X**	**32**	**122**	**148**	**48**	**61**	**26.5**	**34.9**	**X**
Argentina	134	127	202	161	93	98	30.9	48.0	1.2
Bolivia	25	29	82	78	40	47	1.6	12.9	0.3
Brazil	X	X	47	97	46	69	27.3	39.8	4.2
Chile	52	71	117	204	34	62	28.0	53.2	0.6
Colombia	X	19	168	191	31	34	20.2	24.6	2.4
Ecuador	31	40	137	146	24	32	21.5	34.7	1.5
Guyana	X	X	X	X	10	25	X	X	X
Paraguay	X	14	38	44	64	67	7.0	11.5	0.2
Peru	X	59	86	60	28	32	17.0	9.8	0.5
Suriname	90	122	X	X	X	X	X	X	X
Uruguay	122	150	81	131	103	110	20.3	38.9	X
Venezuela	X	68	567	539	47	44	51.5	27.4	2.2
OCEANIA	**344**	**364**	**710**	**692**	**98**	**93**	**126.9**	**133.7**	**X**
Australia	450	485	970	950	114	110	161.1	176.1	2.6
Fiji	X	37	X	X	35	45	X	X	0.3
New Zealand	445	451	695	755	150	137	179.3	176.5	2.6
Papua New Guinea	X	7	X	X	26	25	X	X	0.0
Solomon Islands	X	X	X	X	X	X	X	X	X
DEVELOPED	**296**	**326**	**548**	**626**	**66**	**77**	**137.9**	**160.2**	**X**
DEVELOPING	**9**	**15**	**39**	**55**	**18**	**26**	**11.0**	**17.5**	**X**

Notes: Zero is either 0 or less than one-half the unit of measure. a. World total calculated by the World Resources Institute. b. Data for China do not include Hong Kong or Taiwan. c. Production assumed to be 0; net imports therefore equal consumption. See the technical notes for more detail.

Sources: Food and Agriculture Organization of the United Nations, International Coffee Organization, International Energy Agency, World Bank

	Population (thousands)			Average Annual Population Change (percent)		Percentage of Population in Specific Age Groups 2000 {a}			Total Fertility Rate (average number of children per woman)	
	1950	2000 {a}	2025 {a}	1975-80	1995-00 {a}	<15	15-65	>65	1975-80	1995-00 {a}
WORLD	**2,521,495**	**6,055,049**	**7,823,703**	**1.7**	**1.3**	**30**	**63**	**7**	**3.9**	**2.7**
ASIA (EXCL. MIDDLE EAST)	**1,333,740**	**3,420,234**	**4,307,665**	**X**	**X**	**48**	**47**	**5**	**X**	**X**
Armenia	1,354	3,520	3,946	1.8	(0.3)	25	67	9	2.5	1.7
Azerbaijan	2,896	7,734	9,403	1.6	0.4	29	64	7	3.6	2.0
Bangladesh	41,783	129,155	178,751	2.8	1.7	35	62	3	6.7	3.1
Bhutan	734	2,124	3,904	2.3	2.8	43	53	4	5.9	5.5
Cambodia	4,346	11,168	16,526	(1.8)	2.2	41	56	3	4.1	4.6
China	554,760	1,277,558	1,480,412	1.5	0.9	58	37	5	3.3	1.8
Georgia	3,527	4,968	5,178	0.7	(1.1)	22	65	13	2.4	1.9
India	357,561	1,013,662	1,330,449	2.1	1.6	33	62	5	4.8	3.1
Indonesia	79,538	212,107	273,442	2.1	1.4	31	65	5	4.7	2.6
Japan	83,625	126,714	121,150	0.9	0.2	55	34	12	1.8	1.4
Kazakhstan	6,703	16,223	17,698	1.1	(0.3)	28	65	7	3.1	2.3
Korea, Dem People's Rep	9,488	24,039	29,388	1.6	1.6	60	36	4	3.3	2.1
Korea, Rep	20,357	46,844	52,533	1.6	0.8	57	37	5	2.9	1.7
Kyrgyzstan	1,740	4,699	6,096	1.9	0.6	35	59	6	4.1	3.2
Lao People's Dem Rep	1,755	5,433	9,653	1.2	2.6	44	53	3	6.7	5.8
Malaysia	6,110	22,244	30,968	2.3	2.0	34	62	4	4.2	3.2
Mongolia	761	2,662	3,709	2.8	1.6	61	36	3	6.6	2.6
Myanmar	17,832	45,611	58,120	2.1	1.2	28	67	5	5.3	2.4
Nepal	7,862	23,930	38,010	2.5	2.4	41	55	4	6.2	4.5
Pakistan	39,513	156,483	263,000	2.6	2.8	42	55	3	7.0	5.0
Philippines	20,988	75,967	108,251	2.3	2.1	37	60	4	5.0	3.6
Singapore	1,022	3,567	4,168	1.3	1.4	22	71	7	1.9	1.7
Sri Lanka	7,678	18,827	23,547	1.7	1.0	26	67	7	3.8	2.1
Tajikistan	1,532	6,188	8,857	2.8	1.5	40	55	5	5.9	4.2
Thailand	20,010	61,399	72,717	2.4	0.9	25	69	6	4.3	1.7
Turkmenistan	1,211	4,459	6,287	2.5	1.8	38	58	4	5.3	3.6
Uzbekistan	6,314	24,318	33,355	2.6	1.6	37	58	5	5.1	3.4
Viet Nam	29,954	79,832	108,037	2.2	1.6	33	61	5	5.6	2.6
EUROPE	**547,053**	**728,416**	**701,734**	**0.5 b**	**0.0 b**	**17**	**68**	**15**	**2.0 b**	**1.4 b**
Albania	1,230	3,113	3,820	1.9	(0.4)	29	64	6	4.2	2.5
Austria	6,935	8,211	8,186	(0.1)	0.5	17	68	15	1.6	1.4
Belarus	7,745	10,236	9,496	0.6	(0.3)	19	68	14	2.1	1.4
Belgium	8,639	10,161	9,918	0.1	0.1	17	66	17	1.7	1.6
Bosnia and Herzegovina	2,661	3,972	4,324	0.9	3.0	19	71	10	2.2	1.4
Bulgaria	7,251	8,225	7,023	0.3	(0.7)	16	68	16	2.2	1.2
Croatia	3,850	4,473	4,193	0.5	(0.1)	17	68	15	2.0	1.6
Czech Rep	8,925	10,244	9,512	0.6	(0.2)	17	70	14	2.3	1.2
Denmark	4,271	5,293	5,238	0.2	0.3	18	67	15	1.7	1.7
Estonia	1,101	1,396	1,131	0.6	(1.2)	17	69	14	2.1	1.3
Finland	4,009	5,176	5,254	0.3	0.3	18	67	15	1.6	1.7
France	41,829	59,080	61,662	0.4	0.4	19	65	16	1.9	1.7
Germany	68,376	82,220	80,238	(0.1)	0.1	16	68	16	1.5	1.3
Greece	7,566	10,645	9,863	1.3	0.3	15	67	18	2.3	1.3
Hungary	9,338	10,036	8,900	0.3	(0.4)	17	68	15	2.1	1.4
Iceland	143	281	328	0.9	0.9	23	65	12	2.3	2.1
Ireland	2,969	3,730	4,404	1.4	0.7	21	67	11	3.5	1.9
Italy	47,104	57,298	51,270	0.4	(0.0)	14	68	18	1.9	1.2
Latvia	1,949	2,357	1,936	0.4	(1.5)	18	68	14	2.0	1.3
Lithuania	2,567	3,670	3,399	0.7	(0.3)	19	67	13	2.1	1.4
Macedonia, FYR	1,230	2,024	2,258	1.4	0.6	23	67	10	2.7	2.1
Moldova, Rep	2,341	4,380	4,547	0.9	0.0	23	67	10	2.4	1.8
Netherlands	10,114	15,786	15,782	0.7	0.4	18	68	14	1.6	1.5
Norway	3,265	4,465	4,817	0.4	0.5	20	65	15	1.8	1.9
Poland	24,824	38,765	39,069	0.9	0.1	19	69	12	2.3	1.5
Portugal	8,405	9,875	9,348	1.4	0.0	16	68	16	2.4	1.4
Romania	16,311	22,327	19,945	0.9	(0.4)	18	69	13	2.6	1.2
Russian Federation	102,192	146,934	137,933	0.6	(0.2)	18	69	13	1.9	1.3
Slovakia	3,463	5,387	5,393	1.0	0.1	20	69	11	2.5	1.4
Slovenia	1,473	1,986	1,818	1.0	(0.0)	16	70	14	2.2	1.3
Spain	28,009	39,630	36,658	1.1	0.0	15	68	17	2.6	1.2
Sweden	7,014	8,910	9,097	0.3	0.2	18	64	17	1.6	1.6
Switzerland	4,694	7,386	7,587	(0.1)	0.7	17	68	15	1.5	1.5
Ukraine	36,906	50,456	45,688	0.4	(0.4)	18	68	14	2.0	1.4
United Kingdom	50,616	58,830	59,961	0.0	0.2	19	65	16	1.7	1.7
Yugoslavia	7,131	10,640	10,844	0.9	0.1	20	67	13	2.4	1.8
MIDDLE EAST & N. AFRICA	**111,706**	**403,793**	**613,894**	**X**	**X**	**36**	**60**	**4**	**X**	**X**
Afghanistan	8,958	22,720	44,934	0.9	2.9	44	53	3	7.2	6.9
Algeria	8,753	31,471	46,611	3.1	2.3	37	60	4	7.2	3.8
Egypt	21,834	68,470	95,615	2.4	1.9	35	61	4	5.3	3.4
Iran, Islamic Rep	16,913	67,702	94,463	3.3	1.7	36	59	4	6.5	2.8
Iraq	5,158	23,115	41,014	3.3	2.8	41	56	3	6.6	5.3
Israel	1,258	6,217	8,277	2.3	2.2	28	62	10	3.4	2.7
Jordan	1,237	6,669	12,063	2.3	3.0	42	55	3	7.4	4.9
Kuwait	152	1,972	2,974	6.2	3.1	34	64	2	5.9	2.9
Lebanon	1,443	3,282	4,400	(0.7)	1.7	33	61	6	4.3	2.7
Libyan Arab Jamahiriya	1,029	5,605	8,647	4.4	2.4	38	59	3	7.4	3.8
Morocco	8,953	28,351	38,670	2.3	1.8	33	63	4	5.9	3.1
Oman	456	2,542	5,352	5.0	3.3	44	53	3	7.2	5.9
Saudi Arabia	3,201	21,607	39,965	5.6	3.4	41	57	3	7.3	5.8
Syrian Arab Rep	3,495	16,125	26,292	3.1	2.5	41	56	3	7.4	4.0
Tunisia	3,530	9,586	12,843	2.6	1.4	30	64	6	5.7	2.6
Turkey	20,809	66,591	87,869	2.1	1.7	28	66	6	4.5	2.5
United Arab Emirates	70	2,441	3,284	14.0	2.0	28	69	3	5.7	3.4
Yemen	4,316	18,112	38,985	3.2	3.7	48	49	2	7.6	7.6

	Population (thousands)			Average Annual Population Change (percent)		Percentage of Population in Specific Age Groups 2000 {a}			Total Fertility Rate (average number of children per woman)	
	1950	2000 {a}	2025 {a}	1975-80	1995-00 {a}	<15	15-65	>65	1975-80	1995-00 {a}
SUB-SAHARAN AFRICA	**176,816**	**640,663**	**1,095,448**	**X**	**X**	**44**	**53**	**3**	**X**	**X**
Angola	4,131	12,878	25,107	2.7	3.2	48	50	3	6.8	6.8
Benin	2,046	6,097	11,109	2.5	2.7	46	51	3	7.1	5.8
Botswana	389	1,622	2,242	3.5	1.9	42	55	2	6.4	4.4
Burkina Faso	3,654	11,937	23,321	2.5	2.7	47	50	3	7.8	6.6
Burundi	2,456	6,695	11,569	2.3	1.7	46	51	3	6.8	6.3
Cameroon	4,466	15,085	26,484	2.8	2.7	43	53	4	6.5	5.3
Central African Rep	1,314	3,615	5,704	2.3	1.9	43	54	4	5.9	4.9
Chad	2,658	7,651	13,908	2.1	2.6	46	51	3	6.6	6.1
Congo	808	2,943	5,689	2.9	2.8	46	50	3	6.3	6.1
Congo, Dem Rep	12,184	51,654	104,788	3.0	2.6	48	49	3	6.5	6.4
Côte d'Ivoire	2,776	14,786	23,345	3.9	1.8	43	54	3	7.4	5.1
Equatorial Guinea	226	453	795	(0.7)	2.5	43	53	4	5.7	5.6
Eritrea	1,140	3,850	6,681	2.6	3.8	44	53	3	6.4	5.7
Ethiopia	18,434	62,565	115,382	2.4	2.4	46	51	3	6.8	6.3
Gabon	469	1,226	1,981	3.1	2.6	40	54	6	4.4	5.4
Gambia	294	1,305	2,151	3.1	3.2	0	95	5	6.5	5.2
Ghana	4,900	20,212	36,876	1.9	2.7	43	54	3	6.5	5.2
Guinea	2,550	7,430	12,497	1.5	0.8	44	53	3	7.0	5.5
Guinea-Bissau	505	1,213	1,946	4.7	2.2	43	53	4	5.6	5.8
Kenya	6,265	30,080	41,756	3.8	2.0	43	54	3	8.1	4.5
Lesotho	734	2,153	3,506	2.5	2.2	40	56	4	5.7	4.8
Liberia	824	3,154	6,618	3.1	8.2	42	55	3	6.8	6.3
Madagascar	4,230	15,942	28,964	2.5	3.0	44	53	3	6.6	5.4
Malawi	2,881	10,925	19,958	3.3	2.4	47	50	3	7.6	6.8
Mali	3,520	11,234	21,295	2.1	2.4	46	50	4	7.1	6.6
Mauritania	825	2,670	4,766	2.5	2.7	43	53	3	6.5	5.5
Mozambique	6,198	19,680	30,612	2.8	2.5	45	52	3	6.5	6.3
Namibia	511	1,726	2,338	2.7	2.2	42	55	4	6.0	4.9
Niger	2,400	10,730	21,495	3.2	3.2	48	49	2	8.1	6.8
Nigeria	30,703	111,506	183,041	2.8	2.4	43	54	3	6.9	5.2
Rwanda	2,120	7,733	12,427	3.3	7.7	45	52	2	8.5	6.2
Senegal	2,500	9,481	16,743	2.8	2.6	45	53	2	7.0	5.6
Sierra Leone	1,944	4,854	8,085	2.0	3.0	44	53	3	6.5	6.1
Somalia	2,264	10,097	21,211	7.0	4.2	48	50	2	7.3	7.3
South Africa	13,683	40,377	46,015	2.2	1.5	35	61	4	4.5	3.3
Sudan	9,190	29,490	46,264	3.1	2.1	39	57	3	6.7	4.6
Tanzania, United Rep	7,886	33,517	57,918	3.1	2.3	45	52	3	6.8	5.5
Togo	1,329	4,629	8,482	2.7	2.6	46	51	3	6.6	6.1
Uganda	4,762	21,778	44,435	3.2	2.8	50	48	2	6.9	7.1
Zambia	2,440	9,169	15,616	3.4	2.2	47	51	2	7.2	5.6
Zimbabwe	2,730	11,669	15,092	3.0	1.4	41	56	3	6.6	3.8
NORTH AMERICA	**171,550**	**309,504**	**363,469**	**0.9 b**	**0.8 b**	**72**	**23**	**4**	**1.8 b**	**1.9 b**
Canada	13,737	31,147	37,896	1.2	1.0	19	68	13	1.8	1.6
United States	157,813	278,357	325,573	0.9	0.8	21	66	13	1.8	2.0
C. AMERICA & CARIBBEAN	**53,958**	**173,292**	**235,662**	**X**	**X**	**34**	**61**	**5**	**X**	**X**
Belize	69	241	370	1.7	2.4	40	56	4	6.2	3.7
Costa Rica	862	4,023	5,929	3.0	2.5	32	63	5	3.9	2.8
Cuba	5,850	11,201	11,798	0.9	0.4	21	69	10	2.1	1.6
Dominican Rep	2,353	8,495	11,164	2.4	1.6	33	62	4	4.7	2.8
El Salvador	1,951	6,276	9,062	2.1	2.0	36	59	5	5.6	3.2
Guatemala	2,969	11,385	19,816	2.5	2.6	44	53	4	6.4	4.9
Haiti	3,261	8,222	11,988	2.1	1.7	41	56	4	6.0	4.4
Honduras	1,380	6,485	10,656	3.4	2.7	42	55	3	6.6	4.3
Jamaica	1,403	2,583	3,245	1.2	0.9	31	62	7	4.0	2.5
Mexico	27,737	98,881	130,196	2.7	1.6	33	62	5	5.3	2.8
Nicaragua	1,134	5,074	8,696	3.1	2.7	43	54	3	6.4	4.4
Panama	860	2,856	3,779	2.5	1.6	31	63	6	4.1	2.6
Trinidad and Tobago	636	1,295	1,493	1.3	0.5	25	68	7	3.4	1.7
SOUTH AMERICA	**112,992**	**345,779**	**460,864**	**2.3 b**	**1.5 b**	**30**	**64**	**6**	**4.3 b**	**2.6 b**
Argentina	17,150	37,032	47,160	1.5	1.3	28	63	10	3.4	2.6
Bolivia	2,714	8,329	13,131	2.4	2.3	40	56	4	5.8	4.4
Brazil	53,975	170,115	217,930	2.4	1.3	29	66	5	4.3	2.3
Chile	6,082	15,211	19,548	1.5	1.4	28	64	7	3.0	2.4
Colombia	12,568	42,321	59,758	2.3	1.9	33	63	5	4.3	2.8
Ecuador	3,387	12,646	17,796	2.8	2.0	34	61	5	5.4	3.1
Guyana	423	861	1,045	0.7	0.7	30	66	4	3.9	2.3
Paraguay	1,488	5,496	9,355	3.2	2.6	40	57	3	5.2	4.2
Peru	7,632	25,662	35,518	2.7	1.7	33	62	5	5.4	3.0
Suriname	215	417	525	(0.5)	0.4	30	64	6	4.2	2.2
Uruguay	2,239	3,337	3,907	0.6	0.7	25	62	13	2.9	2.4
Venezuela	5,094	24,170	34,775	3.4	2.0	34	62	4	4.5	3.0
OCEANIA	**12,605**	**30,424**	**39,686**	**1.1 b**	**1.3 b**	**25**	**65**	**10**	**2.8 b**	**2.4 b**
Australia	8,219	18,886	23,098	0.9	1.0	21	67	12	2.1	1.8
Fiji	289	817	1,104	1.9	1.2	31	64	5	4.0	2.7
New Zealand	1,908	3,862	4,695	0.2	1.0	23	66	12	2.2	2.0
Papua New Guinea	1,613	4,807	7,460	2.5	2.2	39	58	3	5.9	4.6
Solomon Islands	90	444	817	3.5	3.1	43	54	3	7.1	4.9
DEVELOPED	**852,572**	**1,306,083**	**1,359,258**	**0.6 b**	**0.3 b**	**31**	**57**	**12**	**1.9 b**	**1.6 b**
DEVELOPING	**1,667,848**	**4,746,022**	**6,459,163**	**2.1 b**	**1.6 b**	**45**	**50**	**5**	**4.7 b**	**3.0 b**

Notes: Negative values are shown in parentheses. Zero is either 0 or less than one-half the unit of measure. (0) indicates a value less than 0 and greater than negative one-half. a. Data include projections based on 1990 base year population data. See the technical notes for more information. b. Regional totals were calculated by the data source.

Source: United Nations Population Division

	Infant Mortality Rate {a} (per 1000 live births) 1995-00	Under-5 Mortality Rate (per 1000 live births) 1997	Life Expectancy at Birth (years): Female 1975-80	Female 1995-00	Male 1975-80	Male 1995-00	Births Attended by Trained Personnel (percent) 1995-97 {b}	Adults and Children Infected with HIV/AIDS (number) 1997	Percent of Adults Infected with HIV/AIDS {c} 1997	Number of Children Orphaned by AIDS Since Beginning of Epidemic
WORLD	**57**	**87**	**61**	**68**	**58**	**63**	**60**	**30,600,000**	**0.97**	**8,200,000**
ASIA (EXCL. MIDDLE EAST)	**X**	**X**	**X**	**X**	**X**	**X**	**X**	**6,220,900**	**X**	**202,170**
Armenia	26	30	75	74	69	67	96	<100	0.01	X
Azerbaijan	36	45	72	74	64	66	99	<100	<0.005	X
Bangladesh	79	109	46	58	47	58	8	21,000	0.03	810
Bhutan	63	121	47	62	45	60	15	<100 d, e	<0.005 e	X
Cambodia	103	167	33	55	30	51	31	130,000	2.40	7,300
China	41	47	66	72	64	68	89	400,000	0.06	720
Georgia	20	29	74	77	67	69	X	<100	<0.005	X
India	72	108	52	63	53	62	34	4,100,000	0.82	120,000
Indonesia	48	68	54	67	52	63	54	52,000	0.05	1,000
Japan	4	6	78	83	73	77	100 f	6,800	0.01	<100
Kazakhstan	35	44	70	72	60	63	100	2,500	0.03	X
Korea, Dem People's Rep	22	30	69	75	62	69	100 f	X	X	X
Korea, Rep	10	6	68	76	61	69	98	3,100	0.01	<100
Kyrgyzstan	40	48	68	72	60	63	98	<100	<0.005	X
Lao People's Dem Rep	93	122	45	55	42	52	X	1,100	0.04	150
Malaysia	11	11	67	74	64	70	99	68,000	0.62	1,500
Mongolia	51	150	58	67	55	64	100	<100	0.01	X
Myanmar	79	114	53	62	50	59	56	440,000	1.79	14,000
Nepal	83	104	45	57	47	58	9	26,000	0.24	750
Pakistan	74	136	54	65	53	63	18	64,000	0.09	5,000
Philippines	36	41	61	70	58	67	64	24,000	0.06	480
Singapore	5	4	73	79	69	75	100 f	3,100	0.15	<100
Sri Lanka	18	19	69	75	65	71	94	6,900	0.07	450
Tajikistan	57	76	67	70	62	64	79	<100	<0.005	X
Thailand	29	38	63	72	59	66	71 f	780,000	2.23	48,000
Turkmenistan	55	78	65	69	58	62	96	<100	0.01	X
Uzbekistan	44	60	68	71	62	64	98	<100	<0.005	X
Viet Nam	38	43	58	70	54	65	85	88,000	0.22	1,900
EUROPE	**12 g**	**X**	**75 g**	**77 g**	**67 g**	**69 g**	**X**	**662,650 g**	**X**	**X**
Albania	30	40	71	76	67	70	99 f	<100	0.01	X
Austria	6	5	76	80	69	74	100	7,500	0.18	X
Belarus	23	18	76	74	66	62	100 f	9,000	0.17	X
Belgium	7	7	76	81	69	74	100 f	7,500	0.14	X
Bosnia and Herzegovina	15	16	72	76	67	71	97	750 d	0.04	X
Bulgaria	17	19	74	75	69	68	100 f	300 d	0.01	X
Croatia	10	9	74	77	67	69	X	300 d	0.01	X
Czech Rep	6	7	74	77	67	70	X	2,000	0.04	X
Denmark	7	6	77	78	71	73	100 f	3,100	0.12	X
Estonia	19	14	74	75	65	63	X	<100	0.01	X
Finland	6	4	77	81	68	73	100	500	0.02	X
France	6	5	78	82	70	74	99	110,000	0.37	X
Germany	5	5	76	80	69	74	99	35,000	0.08	X
Greece	8	8	76	81	72	76	97 f	7,500	0.14	X
Hungary	10	11	73	75	66	67	99 f	2,000	0.04	X
Iceland	5	5	79	81	73	77	100 f	200	0.14	X
Ireland	7	7	75	79	70	74	X	1,700	0.09	X
Italy	7	6	77	81	70	75	X	90,000	0.31	X
Latvia	18	20	74	74	64	62	X	<100	0.01	X
Lithuania	21	15	75	76	66	64	X	<100	0.01	X
Macedonia, FYR	23	23	71	75	68	71	95	<100	0.01	X
Moldova, Rep	29	31	69	72	62	64	X	2,500	0.11	X
Netherlands	6	6	79	81	72	75	100 f	14,000	0.17	X
Norway	5	4	79	81	72	75	100 f	1,300	0.06	X
Poland	15	11	75	77	67	68	99 f	12,000	0.06	X
Portugal	9	8	74	79	67	72	90 f	35,000	0.69	X
Romania	23	26	72	74	67	66	100 f	5,000	0.01	X
Russian Federation	18	25	73	73	62	61	99	40,000	0.05	X
Slovakia	11	11	74	77	67	69	X	<100	<0.005	X
Slovenia	7	6	75	78	67	71	X	<100	0.01	X
Spain	7	5	77	82	71	75	96 f	120,000	0.57	X
Sweden	5	4	78	81	72	76	100 f	3,000	0.07	X
Switzerland	6	5	79	82	72	75	99 f	12,000	0.32	X
Ukraine	19	24	74	74	64	64	100	110,000	0.43	X
United Kingdom	7	7	76	80	70	75	100 f	25,000	0.09	X
Yugoslavia	18	21	73	75	68	70	93	5,000 d	0.10	X
MIDDLE EAST & N. AFRICA	**X**	**X**	**X**	**X**	**X**	**X**	**X**	**210,000**	**0.13**	**14,000**
Afghanistan	152	257	40	46	40	45	9 f	<100 d, e	<0.005 f	X
Algeria	44	39	59	70	57	68	77	11,000 d	0.07	X
Egypt	51	73	55	68	53	65	56	8,100 d	0.03	750
Iran, Islamic Rep	35	35	59	70	58	69	86	1,000 d	<0.005 f	X
Iraq	95	122	62	64	61	61	54 f	300 d	<0.005 f	X
Israel	8	6	75	80	71	76	99 f	2,100 d	0.07	X
Jordan	26	24	63	72	59	69	97	660 d	0.02	X
Kuwait	12	13	72	78	68	74	99 f	1,100 d	0.12	X
Lebanon	29	37	67	72	63	68	98	1,500 d	0.09	X
Libyan Arab Jamahiriya	28	25	59	72	56	68	76 f	1,400 d	0.05	X
Morocco	51	72	58	69	54	65	43	5,000 d	0.03	X
Oman	25	18	56	73	54	69	93	1,200 d	0.11	X
Saudi Arabia	23	28	60	73	58	70	90	1,100 d	0.01	X
Syrian Arab Rep	33	33	62	71	58	67	67	800 d	0.01	X
Tunisia	30	33	61	71	60	68	81	2,200 d	0.04	X
Turkey	45	45	63	72	58	67	76	2,000 d	0.01	X
United Arab Emirates	16	10	69	76	65	74	96 f	2,400 d	0.18	X
Yemen	80	100	44	58	44	57	43	900 d	0.01	X

	Infant Mortality Rate {a} (per 1000 live births) 1995-00	Under-5 Mortality Rate (per 1000 live births) 1997	Life Expectancy at Birth (years) Female 1975-80	Female 1995-00	Male 1975-80	Male 1995-00	Births Attended by Trained Personnel (percent) 1995-97 {b}	Adults and Children Infected with HIV/AIDS (number) 1997	Percent of Adults Infected with HIV/AIDS {c} 1997	Number of Children Orphaned by AIDS Since Beginning of Epidemic
SUB-SAHARAN AFRICA	**X**	**170 g**	**X**	**X**	**X**	**X**	**37 g**	**21,014,160**	**7.41**	**7,771,690**
Angola	125	292	42	48	38	45	15 f	110,000	2.12	19,000
Benin	88	167	49	55	45	52	60	54,000	2.06	11,000
Botswana	59	49	58	48	55	46	78 f	190,000	25.10	28,000
Burkina Faso	99	169	44	45	42	44	42	370,000	7.17	200,000
Burundi	119	176	48	44	44	41	19 f	260,000	8.30	160,000
Cameroon	74	99	50	56	47	53	64	320,000	4.89	74,000
Central African Rep	98	173	47	47	42	43	46	180,000	10.77	65,000
Chad	112	198	43	49	39	46	15	87,000	2.72	55,000
Congo	90	108	51	51	46	46	X	100,000	7.78	64,000
Congo, Dem Rep	90	207	50	52	46	49	X	950,000	4.35	410,000
Côte d'Ivoire	87	150	50	47	46	46	45	700,000	10.06	320,000
Equatorial Guinea	108	172	44	52	40	48	58 f	2,400	1.21	1,600
Eritrea	91	116	47	52	44	49	21	49,000 d	3.17	X
Ethiopia	116	175	44	44	40	42	14 f	2,600,000	9.31	840,000
Gabon	87	145	49	54	45	51	80 f	23,000	4.25	4,800
Gambia	122	87	41	49	37	45	44	13,000	2.24	8,400
Ghana	66	107	54	62	50	58	41	210,000	2.38	130,000
Guinea	124	201	39	47	38	46	31	74,000	2.09	18,000
Guinea-Bissau	130	220	39	46	36	43	27 f	12,000	2.25	990
Kenya	66	87	55	53	51	51	45	1,600,000	11.64	440,000
Lesotho	93	137	54	57	50	55	50	85,000	8.35	9,500
Liberia	116	235	51	49	48	46	58 f	44,000	3.65	21,000
Madagascar	83	158	51	59	48	56	47	8,600	0.12	1,300
Malawi	138	215	44	40	42	39	55	710,000	14.92	360,000
Mali	118	239	46	55	44	52	25	89,000	1.67	33,000
Mauritania	92	183	47	55	44	52	40	6,100	0.52	1,400
Mozambique	114	208	45	47	42	44	44	1,200,000	14.17	170,000
Namibia	65	75	53	53	50	52	68	150,000	19.94	7,800
Niger	115	320	42	50	39	47	15	65,000	1.45	20,000
Nigeria	81	187	47	52	43	49	31	2,300,000	4.12	410,000
Rwanda	124	170	47	42	43	39	26	370,000	12.75	120,000
Senegal	63	124	46	54	42	51	47	75,000	1.77	49,000
Sierra Leone	170	316	37	39	34	36	25 f	68,000	3.17	47,000
Somalia	122	211	44	49	40	45	2 f	11,000 d	0.25	X
South Africa	59	65	59	58	52	52	82	2,900,000	12.91	200,000
Sudan	71	115	48	56	45	54	69	140,000 d	0.99	X
Tanzania, United Rep	82	143	51	49	47	47	38	1,400,000	9.42	730,000
Togo	84	125	50	50	46	48	54 f	170,000	8.52	110,000
Uganda	107	137	49	40	45	39	38	930,000	9.51	1,700,000
Zambia	82	202	51	41	48	40	47	770,000	19.07	470,000
Zimbabwe	69	80	56	45	52	44	69	1,500,000	25.84	450,000
NORTH AMERICA	**7 g**	**X**	**77 g**	**80 g**	**70 g**	**74 g**	**X**	**864,000**	**0.55 g**	**71,000**
Canada	6	7	78	82	71	76	99 f	44,000	0.33	1,000
United States	7	8	77	80	69	73	99 f	820,000	0.76	70,000
C. AMERICA & CARIBBEAN	**X**	**X**	**X**	**X**	**X**	**X**	**X**	**599,000**	**X**	**76,330**
Belize	29	43	71	76	69	73	79	2,100	1.89	340
Costa Rica	12	14	73	79	69	74	98	10,000	0.55	940
Cuba	9	8	75	78	71	74	99	1,400	0.02	160
Dominican Rep	34	53	64	73	60	69	96	83,000	1.89	3,700
El Salvador	32	36	62	73	52	67	87	18,000	0.58	2,200
Guatemala	46	55	58	67	54	61	35	27,000	0.52	3,600
Haiti	68	132	52	56	49	51	21	190,000	5.17	40,000
Honduras	35	45	60	72	56	68	61	43,000	1.46	6,100
Jamaica	22	11	72	77	68	73	91	14,000	0.99	180
Mexico	31	35	69	75	62	70	91	180,000	0.35	16,000
Nicaragua	43	57	60	71	55	66	61	4,100	0.19	120
Panama	21	20	71	76	67	72	86	9,000	0.61	1,000
Trinidad and Tobago	15	17	71	76	66	72	98 f	6,800	0.94	760
SOUTH AMERICA	**37 g**	**X**	**65 g**	**72 g**	**60 g**	**65 g**	**X**	**983,800**	**X**	**9,600**
Argentina	22	24	72	77	65	70	97	120,000	0.69	2,400
Bolivia	66	96	52	63	48	60	47	2,600	0.07	150
Brazil	42	44	64	71	60	63	92	580,000	0.63	X
Chile	13	13	71	78	64	72	100	16,000	0.20	530
Colombia	30	30	66	74	62	67	85	72,000	0.36	1,500
Ecuador	46	39	63	72	60	67	64	18,000	0.28	1,100
Guyana	58	82	63	68	58	61	95	10,000	2.13	660
Paraguay	39	33	69	72	64	67	61	3,200	0.13	340
Peru	45	56	60	71	57	66	56	72,000	0.56	990
Suriname	29	30	68	73	63	68	91 f	2,800	1.17	390
Uruguay	18	21	73	78	66	70	96 f	5,200	0.33	340
Venezuela	21	25	71	76	65	70	69 f	82,000	0.69	1,200
OCEANIA	**24 g**	**X**	**71 g**	**76 g**	**66 g**	**71 g**	**X**	**17,060**	**X**	**1,420**
Australia	6	6	77	81	70	75	100	11,000	0.14	X
Fiji	20	24	69	75	66	71	96 f	260	0.06	<100
New Zealand	7	7	76	80	69	74	99 f	1,300	0.07	120
Papua New Guinea	61	112	50	59	50	57	53	4,500	0.19	1,300
Solomon Islands	23	28	67	74	63	70	87 f	X	X	X
DEVELOPED	**9 g**	**7 g**	**76 g**	**79 g**	**68 g**	**71 g**	**99 g**	**4,450,350**	**X**	**271,120**
DEVELOPING	**63 g**	**96 g**	**58 g**	**65 g**	**56 g**	**62 g**	**55 g**	**25,954,780**	**X**	**7,861,840**

Notes: a. Under 1 year of age. b. Data are for the most recent year within the range specified. c. Values are reproduced as reported by the data source, including values < 0.5%. d. Data refer only to adults (ages 15-49). e. Data are from 1994. f. Data refer to years or periods other than those specified in the column heading, differ from the standard definition, or refer to only part of that country. g. Regional totals were calculated by the source.

Sources: United Nations Population Division, United Nations Children's Fund, World Heath Organization, Joint United Nations Programme on HIV/AIDS

	Population with Access to Safe Water (percent)			Population with Access to Sanitation (percent)			Net Primary School Enrollment (percent)		Net Secondary School Enrollment (percent)		Gross Tertiary School Enrollment		Adult Literacy Rate (percent)			
	Rural 1990-97 {a}	Urban 1990-97 {a}	Total 1990-97 {a}	Rural 1990-97 {a}	Urban 1990-97 {a}	Total 1990-97 {a}	Female 1996-97 {a}	Male 1996-97 {a}	Female 1996-97 {a}	Male 1996-97 {a}	Total (percent) 1996-97	Females (percent of total) 1996-97	Female 1985	Female 1995	Male 1985	Male 1995
WORLD	**62**	**90**	**72**	**25**	**78**	**44**	**X**	**X**	**X**	**X**	**X**	**X**	**X**	**X**	**X**	**X**
ASIA (EXCL. MIDDLE EAST)	**X**	**X**	**X**	**X**	**X**	**X**	**X**	**X**	**X**	**X**	**X**	**X**	**X**	**X**	**X**	**X**
Armenia	X	X	X	X	X	X	X	X	X	X	12	56	96	97	99	99
Azerbaijan	X	X	X	X	X	X	X	X	X	X	18	50	X	X	X	X
Bangladesh	95	99	95	38	83	43	70	80	16	27	6	X	22	29	45	51
Bhutan	54	75	58	66	90	70	X	X	X	X	0	X	X	X	X	X
Cambodia	25	X	30	9	X	19	100	100	31	47	1	19	13	20	48	57
China	56	97	67	7	74	24	100	100	65	74	6	X	65	75	85	91
Georgia	X	X	X	X	X	X	89	89	75	76	41	51	X	X	X	X
India	79	85	81	14	70	29	71	83	48	71	7	36	34	44	61	67
Indonesia	66	91	75	49	77	59	99	100	53	59	11	X	70	81	86	91
Japan	X	X	X	X	X	X	100	100	100	100	43	X	X	X	X	X
Kazakhstan	84	99	93	98	100	99	X	X	X	X	32	X	X	X	X	X
Korea, Dem People's Rep	X	X	81	X	X	X	X	X	X	X	X	X	X	X	X	X
Korea, Rep	76	100	93	100	100	100	100	100	100	100	100	68	37	96	98	99
Kyrgyzstan	X	X	71	X	X	94	99	100	79	77	12	X	X	X	X	X
Lao People's Dem Rep	X	X	44	X	X	18	69	77	53	74	3	30	18	30	50	62
Malaysia	66	96	78	94	94	94	100	100	69	60	11	X	73	82	86	91
Mongolia	3	73	40	74	99	86	88	83	64	48	19	68	38	51	63	72
Myanmar	50	78	60	36	56	43	99	100	53	55	6	X	73	80	87	89
Nepal	68	93	71	14	28	16	63	93	40	68	5	X	13	22	46	57
Pakistan	73	89	79	39	93	56	X	X	X	X	4	X	20	29	48	58
Philippines	80	93	84	63	89	75	100	100	79	77	35	X	91	95	93	95
Singapore	X	100 b	100 b	X	X	X	91	92	75	76	39	X	82	88	94	96
Sri Lanka	52	88	57	62	68	63	100	100	79	73	5	X	84	88	93	94
Tajikistan	49	82	60	X	46	X	X	X	X	X	20	X	97	99	99	100
Thailand	73	88	81	94	97	96	89	87	47	48	21	X	88	93	95	97
Turkmenistan	X	X	74	X	X	91	X	X	X	X	20	X	X	X	X	X
Uzbekistan	88	99	90	99	100	100	X	X	X	X	36	X	76	83	88	93
Viet Nam	42	47	43	15	43	21	100	100	54	56	7	X	86	91	94	95
EUROPE	**X**	**X**	**X**	**X**	**X**	**X**	**X**	**X**	**X**	**X**	**X**	**X**	**X**	**X**	**X**	**X**
Albania	X	X	X	X	X	X	X	X	X	X	11	57	66	76	85	91
Austria	X	X	X	X	X	X	100	100	97	98	48	49	X	X	X	X
Belarus	X	X	X	X	X	X	X	X	X	X	44	55	99	99	100	100
Belgium	X	X	X	X	X	X	100	100	100	100	57	X	X	X	X	X
Bosnia and Herzegovina	X	X	X	X	X	X	X	X	X	X	X	X	X	X	X	X
Bulgaria	X	X	X	X	X	X	99	97	75	80	41	61	96	98	98	99
Croatia	X	X	X	X	X	X	100	100	73	72	28	51	94	97	99	99
Czech Rep	X	X	X	X	X	X	100	100	100	100	24	48	X	X	X	X
Denmark	X	X	X	X	X	X	100	100	95	94	X	X	X	X	X	X
Estonia	X	X	X	X	X	X	100	100	87	85	45	53	X	X	X	X
Finland	X	X	X	X	X	X	100	100	96	95	74	53	X	X	X	X
France	X	X	X	X	X	X	100	100	99	99	51	55	X	X	X	X
Germany	X	X	X	X	X	X	100	100	95	96	47	46	X	X	X	X
Greece	X	X	X	X	X	X	100	100	93	90	47	48	91	96	98	98
Hungary	X	X	X	X	X	X	97	98	98	96	25	X	99	99	99	99
Iceland	X	X	X	X	X	X	100	100	88	87	36	58	X	X	X	X
Ireland	X	X	X	X	X	X	100	100	100	100	41	52	X	X	X	X
Italy	X	X	X	X	X	X	100	100	96	94	47	54	97	98	98	99
Latvia	X	X	X	X	X	X	100	100	81	81	33	60	100	100	100	100
Lithuania	X	X	X	X	X	X	X	X	X	X	31	59	99	99	100	100
Macedonia, FYR	X	X	X	X	X	X	95	96	55	57	20	54	X	X	X	X
Moldova, Rep	18	98	55	8	90	50	X	X	X	X	27	55	95	98	99	100
Netherlands	X	X	X	X	X	X	100	100	100	100	47	48	X	X	X	X
Norway	X	X	X	X	X	X	100	100	98	97	62	56	X	X	X	X
Poland	X	X	X	X	X	X	99	100	89	85	24	X	99	100	100	100
Portugal	X	X	X	X	X	X	100	100	91	88	38	X	83	89	90	94
Romania	X	X	X	X	X	X	100	100	76	75	23	53	95	97	99	99
Russian Federation	X	X	X	X	X	X	100	100	91	85	41	X	99	99	100	100
Slovakia	X	X	X	X	X	X	X	X	X	X	22	50	X	X	X	X
Slovenia	X	X	X	X	X	X	94	95	X	X	36	56	100	100	100	100
Spain	X	X	X	X	X	X	100	100	93	91	53	53	94	97	98	98
Sweden	X	X	X	X	X	X	100	100	100	100	50	56	X	X	X	X
Switzerland	X	X	X	X	X	X	100	100	80	87	34	X	X	X	X	X
Ukraine	X	X	X	X	X	X	X	X	X	X	42	X	99	99	100	100
United Kingdom	X	X	X	X	X	X	100	100	93	91	52	52	X	X	X	X
Yugoslavia	57	98	76	49	92	69	X	X	X	X	22	54	X	X	X	X
MIDDLE EAST & N. AFRICA	**X**	**X**	**X**	**X**	**X**	**X**	**X**	**X**	**X**	**X**	**X**	**X**	**X**	**X**	**X**	**X**
Afghanistan	5	39	12	1	38	8	33	66	14	30	2	X	6	19	36	50
Algeria	64	91	78	80	99	91	93	99	64	73	13	X	38	54	65	77
Egypt	79	97	87	79	98	88	91	100	70	80	23	X	32	42	59	66
Iran, Islamic Rep	82	98	90	74	86	81	89	91	76	86	18	36	52	67	71	82
Iraq	48	96	81	31	93	75	70	80	34	52	11	X	31	43	55	64
Israel	X	X	X	X	X	X	X	X	X	X	44	X	90	94	96	98
Jordan	X	X	98	X	X	77	X	X	X	X	19	47	68	83	88	94
Kuwait	X	X	X	X	100 b	X	64	66	63	63	19	62	72	79	80	83
Lebanon	88	96	94	8	81	63	X	X	X	X	27	X	71	79	87	92
Libyan Arab Jamahiriya	97	97	97	94	99	98	100	100	100	100	20	X	48	65	81	90
Morocco	34	98	65	24	94	58	67	86	32	43	11	41	23	34	51	60
Oman	X	X	85	57	90	78	67	69	65	68	8	45	34	58	65	78
Saudi Arabia	74 b	100 b	95 b	30 b	100 b	86 b	58	62	53	64	16	47	47	64	77	83
Syrian Arab Rep	77	95	86	31	96	67	91	99	39	45	15	X	45	58	80	87
Tunisia	95	100	98	52	96	80	100	100	72	76	14	45	44	58	69	79
Turkey	25	66	49	56	95	80	98	100	49	68	21	35	65	75	88	93
United Arab Emirates	X	X	97	X	X	92	81	83	80	76	12	X	69	77	70	73
Yemen	55	88	61	17	47	24	X	X	X	X	4	13	10	23	51	66

	Population with Access to Safe Water (percent) Rural 1990-97 {a}	Urban 1990-97 {a}	Total 1990-97 {a}	Population with Access to Sanitation (percent) Rural 1990-97 {a}	Urban 1990-97 {a}	Total 1990-97 {a}	Net Primary School Enrollment (percent) Female 1996-97 {a}	Male 1996-97 {a}	Net Secondary School Enrollment (percent) Female 1996-97 {a}	Male 1996-97 {a}	Gross Tertiary School Enrollment Total (percent) 1996-97	Females (percent of total) 1996-97	Adult Literacy Rate (percent) Female 1985	Female 1995	Male 1985	Male 1995
SUB-SAHARAN AFRICA	**39 c**	**75 c**	**50 c**	**34 c**	**66 c**	**44 c**	**X**	**X**	**X**	**X**	**X**	**X**	**X**	**X**	**X**	**X**
Angola	22	46	31	27	62	40	34	35	28	34	1	X	X	X	X	X
Benin	71	46	56	8	57	27	50	85	18	38	3	19	15	23	38	54
Botswana	88	100	90	41	91	55	83	78	91	86	6	47	68	78	64	73
Burkina Faso	37	66	42	33	41	37	25	39	9	16	1	23	7	13	24	32
Burundi	49	92	52	50	60	51	33	38	14	20	1	X	25	38	49	55
Cameroon	43	57	50	36	64	50	59	64	35	45	4	X	50	67	70	80
Central African Rep	21	55	38	16	38	27	38	55	13	26	1	X	19	32	45	58
Chad	17	48	24	7	73	21	35	61	10	26	1	X	17	31	34	49
Congo	7	53	34	X	X	69	76	81	74	94	8	X	54	72	75	86
Congo, Dem Rep	26	89	42	6	53	18	48	69	29	46	2	X	31	47	59	71
Côte d'Ivoire	32	56	42	17	71	39	50	66	24	45	5	X	21	36	41	53
Equatorial Guinea	100	88	95	48	61	54	80	79	65	72	2	X	59	72	85	91
Eritrea	8	60	22	0	48	13	28	31	34	41	1	13	26	38	57	66
Ethiopia	19	91	25	7	97	19	27	44	18	32	1	20	18	31	34	42
Gabon	30	80	67	X	72	X	X	X	X	X	8	X	X	X	X	X
Gambia	65	80	69	23	83	37	58	74	25	42	2	X	17	28	29	42
Ghana	52	88	65	44	62	55	X	X	X	X	1	X	44	60	68	79
Guinea	36	69	46	19	54	31	33	58	7	22	1	11	X	X	X	X
Guinea-Bissau	67	32	43	X	X	46	39	66	16	32	X	X	11	17	44	57
Kenya	49	67	53	81	69	77	67	63	57	65	2	X	57	74	79	88
Lesotho	57	91	62	35	56	38	74	63	80	66	2	54	88	93	64	71
Liberia	13	79	46	4	56	30	X	X	X	X	3	X	25	34	55	67
Madagascar	12	68	26	30	68	40	X	X	X	X	2	45	48	58	65	72
Malawi	40	95	47	1	18	3	100	97	54	91	1	X	36	44	68	73
Mali	55	87	66	3	12	6	31	45	13	23	1	20	17	31	30	46
Mauritania	59	88	74	19	44	32	X	X	X	X	4	X	25	31	46	52
Mozambique	X	X	63	X	X	54	34	45	17	28	1	24	16	27	47	58
Namibia	71	100	83	20	93	62	94	89	84	77	9	X	71	80	76	82
Niger	44	76	48	5	79	17	19	30	7	12	1	X	4	7	17	22
Nigeria	40	58	49	32	50	41	X	X	X	X	4	X	35	53	57	70
Rwanda	79	X	X	85	X	X	X	X	X	X	1	X	44	57	63	72
Senegal	44	90	63	15	71	39	54	65	16	24	3	X	17	26	37	45
Sierra Leone	21	58	34	8	17	11	X	X	X	X	2	X	X	X	X	X
Somalia	X	X	26	X	X	X	X	X	X	X	2	X	X	X	X	X
South Africa	70	99	87	80	92	87	100	100	97	93	X	X	79	84	81	85
Sudan	X	X	73	X	X	51	X	X	X	X	4	X	29	43	59	68
Tanzania, United Rep	58	92	66	83	98	86	49	48	X	X	1	20	48	64	75	83
Togo	41	82	55	22	76	41	70	94	40	77	4	17	27	38	62	73
Uganda	41	77	46	55	75	57	X	X	X	X	2	33	41	54	68	76
Zambia	10	84	38	57	94	71	72	73	35	49	3	X	56	69	77	84
Zimbabwe	69	99	79	32	96	52	92	94	56	62	7	36	73	83	85	92
NORTH AMERICA	**X**	**X**	**X**	**X**	**X**	**X**	**X**	**X**	**X**	**X**	**X**	**X**	**X**	**X**	**X**	**X**
Canada	X	X	X	X	X	X	100	100	94	96	90	X	X	X	X	X
United States	X	X	X	X	X	X	100	100	96	97	81	X	X	X	X	X
C. AMERICA & CARIBBEAN	**X**	**X**	**X**	**X**	**X**	**X**	**X**	**X**	**X**	**X**	**X**	**X**	**X**	**X**	**X**	**X**
Belize	69	100	83	87	23	57	100	100	63	65	1	X	87	92	89	93
Costa Rica	92	100	96	70	95	84	89	89	X	X	33	X	94	95	94	95
Cuba	85	96	93	51	71	66	100	100	73	67	12	60	95	96	95	97
Dominican Rep	X	80	65	83	76	78	94	89	82	75	23	57	78	83	79	83
El Salvador	40	84	66	80	98	90	89	89	37	36	18	50	67	75	75	81
Guatemala	78	76	77	74	95	83	70	77	32	38	8	X	52	60	68	75
Haiti	28	50	37	17	49	25	X	X	X	X	1	X	35	46	41	50
Honduras	62	X	76	57	X	74	89	86	38	34	11	X	66	74	68	73
Jamaica	X	X	86	80	100	89	96	96	72	68	8	X	85	90	77	82
Mexico	X	X	85	X	X	72	100	100	64	68	16	48	84	89	90	93
Nicaragua	32	88	62	35	34	35	80	77	53	49	12	53	65	69	63	66
Panama	X	X	93	X	X	83	90	90	72	71	32	X	88	91	89	92
Trinidad and Tobago	91	99	97	98	99	79	100	100	72	71	8	X	88	92	94	95
SOUTH AMERICA	**X**	**X**	**X**	**X**	**X**	**X**	**X**	**X**	**X**	**X**	**X**	**X**	**X**	**X**	**X**	**X**
Argentina	29	77	71	37	73	68	100	100	80	74	42	X	95	97	96	97
Bolivia	32	86	63	37	74	58	95	100	37	43	24	X	68	78	86	91
Brazil	25	88	76	30	80	70	94	100	67	65	15	X	79	85	81	85
Chile	41	99	91	X	90	X	89	92	87	83	31	46	93	95	94	96
Colombia	56	97	85	56	97	85	89	89	78	75	17	52	88	91	88	91
Ecuador	49	80	68	49	95	76	100	100	51	51	26	X	84	89	89	93
Guyana	85	96	91	85	90	88	93	93	76	73	11	51	96	98	98	99
Paraguay	X	X	60	14	65	41	97	96	60	62	10	55	87	92	92	94
Peru	33	84	67	37	89	72	93	94	81	87	26	X	78	84	91	94
Suriname	X	X	X	43	X	X	X	X	X	X	14	X	X	X	X	X
Uruguay	X	95	X	X	98	X	95	94	89	79	30	X	97	98	96	97
Venezuela	75	80	79	30	64	58	84	81	54	44	25	X	87	91	90	93
OCEANIA	**X**	**X**	**X**	**X**	**X**	**X**	**X**	**X**	**X**	**X**	**X**	**X**	**X**	**X**	**X**	**X**
Australia	X	X	X	X	X	X	100	100	96	96	80	51	X	X	X	X
Fiji	X	X	77	85	100	92	100	100	84	84	13	X	84	90	91	94
New Zealand	82	100	97	X	X	X	100	100	94	92	63	57	X	X	X	X
Papua New Guinea	23	78	32	80	93	83	X	X	X	X	3	X	47	55	65	71
Solomon Islands	62	80	X	9	60	X	X	X	X	X	X	X	X	X	X	X
DEVELOPED	**X**	**X**	**X**	**X**	**X**	**X**	**X**	**X**	**X**	**X**	**X**	**X**	**X**	**X**	**X**	**X**
DEVELOPING	**62 c**	**89 c**	**71 c**	**25 c**	**78 c**	**44 c**	**X**	**X**	**X**	**X**	**X**	**X**	**X**	**X**	**X**	**X**

Notes: Zero is either 0 or less than one-half the unit of measure. a. Data are for the most recent year available within the range specified. b. Data refer to years or periods other than those specified in the column heading, differ from the standard definition, or refer to only part of a country. c. Total calculated by source.

Sources: United Nations Children's Fund and the United Nations Educational, Scientific and Cultural Organizaiton

Data Table EI.1

Basic Economic Indicators

Gross Domestic Product and Trade Values

	Gross Domestic Product (GDP): Purchasing Power Parity (PPP) (Current Intl$) Total Value (million Int$) 1997	Purchasing Power Parity (PPP) Per Capita (Int$) 1997	Exchange Rate Based (GDP) (1995 US$) Total (million US$) 1997	Exchange Rate Based Per Capita (US$) 1997	Average Annual GDP Growth Rate (percent) {a} 1978-87	Average Annual GDP Growth Rate (percent) {a} 1988-97	Distribution (percent) {b} Agriculture 1997	Distribution Industry 1997	Distribution Services 1997	Exports of Goods and Services (million US$) 1997	Imports of Goods and Services (million US$) 1997
WORLD	X	X	30,643,610	5,262	2.5	2.2	4 b	32 b	61 b	6,869,405	6,760,876
ASIA (EXCL. MIDDLE EAST)	X	X	X	X	X	X	X	X	X	X	X
Armenia	8,939	2,518	3,149	887	X	X	41	36	23	330	952
Azerbaijan	11,771	1,540	3,651	478	X	(11.6)	22	18	60	1,154	1,900
Bangladesh	129,666	1,057	41,816	341	4.2	4.1	24	27	49	5,096	7,677
Bhutan	X	X	X	X	X	X	38	38	25	121	162
Cambodia	13,494	1,288	3,175	303	X	4.2	51	15	34	896	1,252
China	3,837,794	3,085	834,973	671	8.3	8.2	19	49	32	207,251	166,754
Georgia	10,650	2,080	3,716	726	2.6	(13.7)	32	23	45	622	1,192
India	1,609,656	1,666	377,811	391	3.7	5.0	25	30	45	44,102	59,236
Indonesia	699,784	3,441	228,639	1,124	5.4	6.7	16	43	41	63,238	62,830
Japan	3,035,467	24,084	5,384,154	42,719	3.2	2.2	2 b	38 b	60 b	478,542	431,094
Kazakhstan	56,223	3,434	20,365	1,244	X	(5.7)	12	27	61	7,611	8,279
Korea, Dem People's Rep	X	X	X	X	X	X	X	X	X	X	X
Korea, Rep	624,914	13,665	515,527	11,273	6.5	6.4	6	43	51	164,922	171,297
Kyrgyzstan	10,423	2,256	3,913	847	X	(4.7)	45	23	33	676	817
Lao People's Dem Rep	6,317	1,255	2,007	399	X	6.4	52	21	26	417	715
Malaysia	176,444	8,409	102,258	4,873	4.7	7.5	12	47	41	92,897	91,522
Mongolia	3,318	1,308	1,014	400	X	(0.7)	37	24	40	504	535
Myanmar	X	X	X	X	X	X	59	10	31	1,439	2,415
Nepal	24,270	1,088	4,813	216	3.0	4.3	41	22	36	1,295	1,855
Pakistan	199,960	1,388	63,835	443	5.8	3.7	25	25	50	9,956	14,677
Philippines	258,882	3,624	82,555	1,156	1.2	3.0	19	32	49	40,365	50,477
Singapore	88,317	25,772	98,071	28,619	5.9	7.3	0	35	65	156,252	144,168
Sri Lanka	46,205	2,528	14,376	787	4.2	4.4	22	26	52	5,514	6,569
Tajikistan	6,505 b	1,115 b	1,601 b	274 b	X	X	X	X	X	804	801
Thailand	405,587	6,790	176,648	2,957	5.1	6.8	11	40	49	72,415	72,437
Turkmenistan	9,453 b	2,275 b	4,197 b	1,010 b	X	X	X	X	X	759	1,004
Uzbekistan	57,347 b	2,510 b	24,026	1,035	X	(0.8)	31	27	42	3,980	4,417
Viet Nam	125,332	1,641	24,081	315	X	6.9	26	31	43	11,485	13,465
EUROPE	X	X	X	X	X	X	X	X	X	X	X
Albania	7,046	2,250	2,460	785	X	(1.3)	63	18	19	222	809
Austria	178,176	22,000	244,751	30,220	1.9	2.4	1 b	30 b	68 b	85,323	87,821
Belarus	49,758	4,807	20,976	2,026	X	(2.5)	14	44	42	8,306	9,103
Belgium	231,822	22,892	285,386	28,182	1.4	1.7	1 b	27 b	72 b	185,404 c	174,267 c
Bosnia and Herzegovina	X	X	X	X	X	X	X	X	X	X	X
Bulgaria	33,303	3,968	10,960	1,306	X	(4.4)	23	26	50	6,251	5,685
Croatia	22,797 b	5,080 b	18,859 b	4,202 b	X	X	12 d	25 d	62 d	8,198	11,402
Czech Rep	108,298	10,514	53,385	5,183	X	(0.6)	6 e	40 e	54 e	29,869	32,713
Denmark	125,197	23,819	193,417	36,799	2.0	1.9	4 d	27 d	69 d	63,680	57,971
Estonia	7,647	5,284	5,547	3,833	X	(2.8)	7	28	65	3,609	4,142
Finland	103,572	20,145	138,237	26,888	3.2	1.2	4 b	34 b	62 b	48,228	37,976
France	1,290,861	22,077	1,594,793	27,275	1.7	1.6	2 b	26 b	72 b	365,342	319,781
Germany	1,744,929	21,265	2,486,982	30,308	X	X	1 b	X	44 b	590,985	558,836
Greece	131,946	12,485	117,760 b	11,182 b	1.3	X	11 d	18 d	71 d	14,863	25,601
Hungary	73,095	7,197	47,343	4,662	2.1	(0.9)	6	34	60	24,514	25,067
Iceland	5,932 b	21,889 b	7,346 b	27,105 b	3.6	X	11 d	26 d	63 d	2,687	2,650
Ireland	75,805	20,725	77,112	21,083	2.3	6.0	6 d	X	62 d	61,447	51,711
Italy	1,167,362	20,345	1,111,646	19,374	2.2	1.3	3 b	31 b	66 b	310,550	261,885
Latvia	9,705	3,944	6,883	2,797	3.1	(5.6)	7	31	62	2,871	3,348
Lithuania	15,653	4,225	7,624	2,058	X	(3.2)	13	32	55	5,224	6,237
Macedonia, FYR	6,407	3,225	2,124	1,069	X	X	12	27	61	1,330	1,862
Moldova, Rep	6,470	1,478	2,889	660	X	(9.5)	31	35	34	1,023	1,430
Netherlands	329,522	21,104	424,445	27,183	1.4	2.6	3 d	27 d	70 d	213,503	191,173
Norway	107,688	24,499	159,952	36,389	3.0	3.0	2 b	32 b	66 b	63,213	52,286
Poland	252,121	6,516	134,724	3,482	X	1.1	6 d	39 d	55 d	39,717	46,367
Portugal	141,897	14,386	112,328	11,388	2.5	2.5	4 d	32 d	64 d	32,339	40,684
Romania	97,102	4,306	31,874	1,414	2.0	(2.6)	20	45	36	9,853	12,448
Russian Federation	644,202	4,363	338,180	2,290	X	X	8	37	55	102,196	90,065
Slovakia	42,575	7,925	19,746	3,675	X	(0.4)	5	33	62	10,959	12,367
Slovenia	23,444	11,749	20,062	10,055	X	X	5	39	57	10,449	10,631
Spain	626,346	15,812	593,347	14,979	1.8	2.1	3 b	32 d	25 b	148,357	142,478
Sweden	175,145	19,777	237,042	26,766	1.9	0.9	2 d	30 d	68 d	100,989	84,779
Switzerland	178,919	24,677	312,634	43,120	1.6	0.9	X	X	X	121,737 a	109,063 a
Ukraine	111,126	2,176	74,762	1,464	X	(9.1)	12	40	48	20,355	21,891
United Kingdom	1,223,547	20,900	1,172,327	20,025	2.0	1.5	2 b	31 b	67 b	366,505	371,843
Yugoslavia	X	X	X	X	X	X	X	X	X	X	X
MIDDLE EAST & N. AFRICA	X	X	X	X	X	X	X	X	X	X	X
Afghanistan	X	X	X	X	X	X	X	X	X	X	X
Algeria	130,735	4,448	43,381	1,476	3.2	0.9	11	49	39	14,890	10,280
Egypt	184,009	2,843	65,469	1,011	5.3	3.7	18	32	51	16,171	18,296
Iran, Islamic Rep	X	X	X	X	(0.6)	X	25 d	34 d	40 d	18,978	17,446
Iraq	X	X	X	X	X	X	X	X	X	X	X
Israel	105,912	18,073	92,603	15,802	3.9	4.6	X	X	X	30,320	38,810
Jordan	15,305	2,499	6,757	1,103	6.1	2.5	3	25	71	3,893	5,652
Kuwait	28,684 d	16,978 d	26,558 d	15,750 d	(1.8)	X	0 d	54 d	46 d	16,044	12,998
Lebanon	24,609	7,830	X	X	X	X	12	27	61	1,557	8,053
Libyan Arab Jamahiriya	X	X	X	X	X	X	X	X	X	X	X
Morocco	90,313	3,359	36,232	1,347	3.5	2.0	15	33	51	9,510	10,627
Oman	21,307 d	9,887 d	12,102 d	5,427 d	7.9	X	X	X	X	7,649	5,815
Saudi Arabia	203,149	10,429	129,884	6,668	0.5	2.5	6	45	49	64,168	51,632
Syrian Arab Rep	48,392	3,237	17,596	1,177	2.6	4.0	X	X	X	5,661	5,092
Tunisia	48,857	5,305	20,302	2,204	3.9	4.1	13	29	58	8,081	8,644
Turkey	404,498	6,380	194,681	3,071	3.9	3.9	15	28	57	52,004	56,536
United Arab Emirates	42,385 d	19,181 d	X	X	X	X	2 e	57 e	40 e	X	X
Yemen	12,951	795	4,063	249	X	X	18	49	34	2,522	3,005

	Gross Domestic Product (GDP)									Exports of Goods and Services (million US$) 1997	Imports of Goods and Services (million US$) 1997
	Purchasing Power Parity (PPP) (Current Intl$)		Exchange Rate Based (GDP) (1995 US$)		Average Annual GDP Growth Rate (percent) {a}		Distribution (percent) {b}				
	Total Value (million Int$) 1997	Per Capita (Int$) 1997	Total (million US$) 1997	Per Capita (US$) 1997	1978-87	1988-97	Agriculture 1997	Industry 1997	Services 1997		
SUB-SAHARAN AFRICA	X	X	X	X	X	X	X	X	X	X	X
Angola	16,704	1,426	6,075	519	X	(0.1)	9	62	29	5,286	4,681
Benin	7,377	1,311	2,241	398	3.6	3.5	38	14	48	530	696
Botswana	11,796	7,657	5,243	3,403	8.8	5.3	3	48	49	2,381 a	1,770 a
Burkina Faso	10,543	958	2,626	239	3.1	2.7	35	27	38	332	723
Burundi	4,040	635	918	144	3.7	(1.5)	53	17	30	96	139
Cameroon	26,406	1,896	8,790	631	5.3	(0.8)	41	21	38	2,443	2,041
Central African Rep	4,546	1,329	1,182	346	0.2	0.7	54	18	28	213	237
Chad	6,918	976	1,588	224	(1.3)	3.6	39	15	46	271	587
Congo	4,397	1,623	2,206	815	6.7	1.0	10	57	33	1,800	1,368
Congo, Dem Rep	40,882	852	5,923	123	1.9	(5.2)	58	17	25	1,445	1,385
Côte d'Ivoire	26,134	1,858	11,206	797	(0.5)	2.3	27	21	51	4,927	3,694
Equatorial Guinea	X	X	373	887	X	11.8	23	67	10	431	694
Eritrea	3,097	902	661	193	X	X	9	30	61	201	583
Ethiopia {f}	30,194	519	6,750	116	X	2.7	55	7	38	1,017	1,683
Gabon	8,704	7,655	5,358	4,712	(0.4)	3.6	7	55	37	3,295	2,165
Gambia	1,731	1,455	412	346	2.7	2.7	30	15	55	229	282
Ghana	29,494	1,581	7,042	377	0.6	3.8	36	26	39	1,657	2,641
Guinea	13,016	1,777	4,026	550	X	3.9	23	35	42	741	834
Guinea-Bissau	X	X	279	246	1.7	3.7	54	11	35	56	107
Kenya	33,917	1,192	9,625	338	3.8	2.3	29	16	56	2,994	3,771
Lesotho	3,751	1,861	1,037	514	1.4	6.4	11	42	47	309	1,214
Liberia	X	X	X	X	X	X	X	X	X	X	X
Madagascar	13,109	897	3,343	229	0.5	1.1	32	14	55	772	1,063
Malawi	7,278	723	1,703	169	1.6	3.5	36	18	46	615	1,153
Mali	7,637	732	2,735	262	1.6	3.5	49	17	34	644	897
Mauritania	4,265	1,733	1,168	475	2.0	3.0	25	29	46	463	387
Mozambique	12,324	668	2,332	126	X	4.2	31	24	45	499	1,007
Namibia	8,137	5,017	3,494	2,154	X	2.9	11	33	56	1,726	1,908
Niger	8,292	849	2,009	206	(0.3)	1.0	38	18	44	301	442
Nigeria	107,959	1,039	31,080	299	(0.3)	3.7	33	47	20	15,994	14,213
Rwanda	5,172	868	1,647	276	3.8	(2.3)	37	26	36	152	488
Senegal	15,230	1,736	4,998	570	2.7	2.0	18	22	59	1,475	1,716
Sierra Leone	1,954	442	726	164	0.8	(3.7)	50	21	29	92	146
Somalia	X	X	X	X	X	X	X	X	X	X	X
South Africa	299,577	7,729	140,257	3,619	2.0	1.1	5	39	57	35,440	34,625
Sudan	43,389	1,565	8,089	292	0.4	5.1	X	X	X	613	1,614
Tanzania, United Rep {g}	18,091	576	5,013	160	X	2.8	47	21 b	31 b	1,200	1,962
Togo	6,463	1,509	1,496	349	0.9	1.6	42	21	37	632	724
Uganda	23,622	1,181	6,622	331	X	6.0	44	17	39	825	1,651
Zambia	9,087	1,058	3,855	449	0.6	0.7	16	31	52	1,276	1,474
Zimbabwe	26,931	2,401	7,902	705	4.1	2.5	19	25	56	3,045	3,873
NORTH AMERICA	X	X	X	X	X	X	X	X	X	X	X
Canada	680,923	22,502	612,021	20,225	2.6	1.7	3 e	30 e	67 e	247,438	236,225
United States	7,765,237	28,573	7,786,572	28,651	2.0	2.5	2 b	27 b	71 b	937,434	1,043,477
C. AMERICA & CARIBBEAN	X	X	X	X	X	X	X	X	X	X	X
Belize	986	4,396	616	2,749	3.8	4.7	23	28	49	317	362
Costa Rica	23,049	6,149	9,256	2,470	1.7	3.4	15	23	62	4,478	4,667
Cuba	X	X	X	X	X	X	X	X	X	X	X
Dominican Rep	39,097	4,828	13,843	1,710	3.5	3.4	12	32	55	6,420	7,124
El Salvador	17,085	2,890	10,058	1,702	(2.8)	4.1	13	28	60	2,706	3,885
Guatemala	43,091	4,096	15,734	1,496	0.6	3.5	24	20	56	3,187	4,193
Haiti	9,485	1,213	2,738	350	0.8	(1.2)	30	20	50	218	809
Honduras	13,274	2,219	4,273	714	2.0	2.8	20	28	52	2,191	2,511
Jamaica	8,776	3,488	4,011	1,594	0.2	1.3	8	35	57	3,177	3,984
Mexico	790,026	8,380	321,957	3,415	2.0	2.8	5	26	69	121,829	122,424
Nicaragua	8,877 b	1,950 b	2,371 b	521 b	(2.7)	X	34 b	22 b	44 b	863	1,609
Panama	18,713 b	6,991 b	8,106 b	3,028 b	2.4	X	8 b	18 b	73 b	8,276	8,581
Trinidad and Tobago	8,936	6,999	5,670	4,441	1.6	1.4	2	46	52	2,698	2,857
SOUTH AMERICA	X	X	X	X	X	X	X	X	X	X	X
Argentina	367,487	10,302	319,501	8,957	1.1	3.2	7	33	61	28,494	34,758
Bolivia	22,399	2,881	7,284	937	X	3.6	16	33	51	1,362	2,049
Brazil	1,059,971	6,475	746,778	4,562	3.2	2.0	8	35	57	60,256	79,817
Chile	186,067	12,723	68,220	4,665	3.2	6.9	7	31	61	20,608	22,218
Colombia	272,847	6,814	84,695	2,115	3.1	3.5	11	20	69	15,861	18,784
Ecuador	58,979	4,941	18,913	1,584	1.7	2.5	12	35	53	6,000	5,787
Guyana	2,721	3,226	X	X	(2.3)	X	36 d	37 d	27 d	629 b	699
Paraguay	20,246	3,979	9,415	1,850	3.7	2.9	23	22	55	4,343	4,960
Peru	114,080	4,682	64,906	2,664	2.5	2.2	7	36	57	8,356	10,842
Suriname	X	X	X	X	(4.2)	X	26 e	26 e	48 e	519 b	454 b
Uruguay	30,046	9,201	19,956	6,111	0.9	3.0	8	27	64	4,256	4,450
Venezuela	201,843	8,862	80,843	3,549	0.1	1.9	4	41	55	25,120	18,282
OCEANIA	X	X	X	X	X	X	X	X	X	X	X
Australia	374,536	20,430	382,114	20,843	2.8	2.4	3 b	26 b	71 b	83,705	81,877
Fiji	3,249	4,132	1,983	2,522	1.0	2.9	18	26	56	1,202	1,242
New Zealand	65,488	17,412	63,317	16,834	2.0	2.0	7 d	26 d	67 d	17,835	18,434
Papua New Guinea	X	X	4,901	1,089	1.4	3.0	28	36	36	2,462	2,656
Solomon Islands	932	2,306	328	810	6.7	3.7	X	X	X	239	292
LOW INCOME {h}	**X**	**X**	**737,334**	**X**	**3.1**	**3.4**	**28**	**28**	**43**	**140,087**	**180,867**
MIDDLE INCOME {h}	**X**	**X**	**5,053,564**	**X**	**3.0**	**2.3**	**11**	**37**	**52**	**1,382,960**	**1,393,256**
HIGH INCOME {h}	**X**	**X**	**24,848,389**	**X**	**2.4**	**2.2**	**2 b**	**31 b**	**63 b**	**5,346,334**	**5,186,801**

Notes: Negative values are shown in parantheses. Zero is either 0 or less than one-half the unit of measure. a. Figures were calculated using GDP in constant 1995 U.S. dollars. b. Data are from 1996. c. Data for Belgium include Luxembourg. d. Data are from 1995. e. Data are from 1993. f. Data for Ethiopia before 1993 include Eritrea. g. Data refer to mainland only. h. As defined by the World Bank (World Development Indicators 1999).

Sources: United Nations Population Division and World Bank

	Average Annual Official Development Assistance (ODA) and Official Aid (OA) (million US$) {a}		ODA and OA as a Percentage of Gross Domestic Product (GDP) {b}	ODA and OA Per Capita (US$) {a}	Direct Foreign Investment (million US$)		Total External Debt (million US$)		Debt Service as a Percentage of Total Exports	International Tourism Receipts (million US$)	
	1985-87	1995-97	1995-97	1995-97	1985-87	1995-97	1985-87	1995-97	1995-97	1985-87	1995-97
WORLD	**41,778**	**63,183**	**X**	**11**	**88,182**	**339,354**	**X**	**X**	**X**	**140,488**	**427,057**
ASIA (EXCL. MIDDLE EAST)	**X**	**X**	**X**	**X**	**X**	**X**	**X**	**X**	**X**	**14,830**	**65,888**
Armenia	X	229	2.7	64	X	28	X	529	7	X	8
Azerbaijan	X	136	1.2	18	X	509	X	420	3	X	154
Bangladesh	1,470	1,182	1.0	10	2	51	8,433	15,819	12	16	33
Bhutan	35	69	X	36	0	0	23	88	6	1	6
Cambodia	166	465	3.5	45	X	216	X	2,088	1	X	121
China	1,140	2,751	0.1	2	1,949	40,088	25,252	131,201	9	1,548	10,336
Georgia	X	259	2.7	50	X	33	X	1,331	4	X	X
India	1,881	1,787	0.1	2	145	2,640	48,199	94,076	24	1,223	2,908
Indonesia	812	1,123	0.2	6	318	5,073	44,042	129,837	32	651	5,968
Japan	(5,591)	(11,095)	(0.4)	(88)	675	1,147	X	X	X	1,566	3,875
Kazakhstan	X	108	0.2	7	X	1,141	X	3,717	5	X	X
Korea, Dem People's Rep	63	86	X	4	0	4 c	X	X	X	X	X
Korea, Rep	(5)	(82)	X	(2)	436	2,315	52,318	130,048	9	1,544	5,406
Kyrgyzstan	X	253	X	55	X	64	X	767	9	X	4
Lao People's Dem Rep	156	331	X	68	0	94	883	2,249	6	X	52
Malaysia	262	(193)	(0.1)	(9)	535	4,772	21,663	40,415	8	651	3,895
Mongolia	674	223	6.3	89	X	7	X	585	10	X	21
Myanmar	373	91	X	2	(1)	98	3,768	5,343	13	14	34
Nepal	300	418	1.8	19	1	17	782	2,409	7	57	118
Pakistan	744	765	0.4	5	122	783	15,072	29,879	29	181	126
Philippines	704	831	0.3	12	149	1,406	28,209	41,136	13	1,010	2,662
Singapore	26	11	0.0	3	1,864	7,820	X	X	X	1,838	8,111
Sri Lanka	500	467	1.1	26	38	202	4,125	7,957	7	82	203
Tajikistan	X	93	X	16	X	17	X	745	2	X	X
Thailand	465	775	0.2	13	259	2,716	18,823	89,040	13	1,513	8,343
Turkmenistan	X	21	X	5	X	64	X	975	23 d	X	7 d
Uzbekistan	X	101	X	4	X	152	X	2,304	9	X	X
Viet Nam	1,696	933	0.8	12	X	1,567	X	24,437	6	28 e	87
EUROPE	**X**	**X**	**X**	**X**	**X**	**X**	**X**	**X**	**X**	**72,752**	**203,888**
Albania	X	194	2.7	62	X	69	X	674	4	X	9
Austria	(216)	(617)	(0.4)	(77)	256	2,951	X	X	X	6,967	13,659
Belarus	X	113	0.2	11	X	98	X	1,308	2	X	40
Belgium	(558)	(904)	(0.4)	(89)	X	X	X	X	X	2,305	5,870
Bosnia and Herzegovina	X	884	X	256	X	X	X	X	X	X	13
Bulgaria	X	168	0.5	20	0	232	5,994	10,072	17	398	417
Croatia	X	77	X	17	X	341	X	5,168	7	X	1,843
Czech Rep	X	129	X	12	0	1,763	4,347	19,166	10	X	3,550
Denmark	(665)	(1,677)	(1.4)	(320)	120	2,568	X	X	X	1,768	3,419
Estonia	X	62	0.9	42	X	206	X	450	1	X	433
Finland	(319)	(392)	(0.4)	(76)	242	1,430	X	X	X	655	1,616
France	(4,142)	(7,400)	(0.6)	(127)	3,664	22,917	X	X	X	9,845	28,067
Germany	(3,722)	(6,994)	(0.4)	(85)	1,110	3,469	X	X	X	6,240	17,338
Greece	21	X	X	0	534	1,032	X	X	X	1,843	3,886
Hungary	X	39	0.1	4	0	2,860	16,817	27,641	37	375	2,180
Iceland	X	X	X	0	12	66	X	X	X	62	159
Ireland	(51)	(173)	(0.2)	(48)	71	2,264	X	X	X	670	2,980
Italy	(2,039)	(1,768)	(0.2)	(31)	1,692	4,034	X	X	X	10,262	29,247
Latvia	X	75	0.8	30	X	361	X	480	3	X	128
Lithuania	X	125	0.8	34	X	193	X	1,177	3	X	276
Macedonia, FYR	X	111	1.8	56	X	12	X	1,546	5	X	X
Moldova, Rep	X	56	X	13	X	36	X	855	8	X	41
Netherlands	(1,657)	(3,140)	(1.0)	(202)	2,554	9,328	X	X	X	2,192	6,205
Norway	(754)	(1,287)	(1.2)	(294)	259	3,048	X	X	X	1,023	2,429
Poland	X	1,860	0.8	48	14	4,355	37,517	42,542	8	146	7,900
Portugal	(24)	(242)	(0.2)	(25)	326	1,035	X	X	X	1,605	4,288
Romania	X	250	0.2	11	0	632	6,857	8,764	13	X	556
Russian Federation	X	1,193	X	8	0	3,579	32,447	123,752	7	X	5,952
Slovakia	X	104	X	19	0	210	1,289	7,786	12	X	609
Slovenia	X	77	0.3	39	X	227	X	X	X	X	1,196
Spain	(201)	(1,278)	(0.2)	(32)	3,330	6,070	X	X	X	11,656	26,553
Sweden	(1,102)	(1,811)	(1.0)	(205)	705	10,100	X	X	X	1,592	3,633
Switzerland	(424)	(1,007)	(0.6)	(140)	1,903	3,558 c	X	X	X	4,239	8,739
Ukraine	X	296	0.3	6	X	470	X	9,597	6	X	199
United Kingdom	(1,713)	(3,278)	(0.3)	(56)	9,925	28,061	X	X	X	8,503	19,473
Yugoslavia	X	88	X	8	X	0	22,074 f	14,128	X	X	42
MIDDLE EAST & N. AFRICA	**X**	**X**	**X**	**X**	**X**	**X**	**X**	**X**	**X**	**9,207**	**21,428**
Afghanistan	260	241	X	12	0	0 c	X	X	X	1	1
Algeria	148	290	0.2	10	3	5	21,771	32,386	30	98	24
Egypt	1,677	2,061	X	32	1,114	708	40,018	31,471	11	1,091	3,245
Iran, Islamic Rep	2	187	X	3	(153)	31	6,009	16,801	30	27	227
Iraq	42	336	X	16	5	0 c	X	X	X	109	13
Israel	1,722	1,248	1.2	218	156	2,115	X	X	X	1,138	2,893
Jordan	561	506	3.4	85	29	17	5,080	8,138	12	544	721
Kuwait	4	4	X	3	0 g	124	X	X	X	88	135
Lebanon	86	221	0.9	72	6	88	895	3,999	9	X	808
Libyan Arab Jamahiriya	8	9	X	X	(59)	95 c	X	X	X	2	6
Morocco	521	537	0.6	20	27	600	18,011	21,219	29	765	1,248
Oman	59	47	X	21	112	68	2,711	3,399	8	49	100
Saudi Arabia	27	22	0.0	1	234	(1,503) c	X	X	X	2,326	1,313
Syrian Arab Rep	815	262	X	18	36	90	13,093	21,201	6	228	1,265
Tunisia	217	131	0.3	14	88	273	5,881	11,234	16	570	1,464
Turkey	286	183	0.0	3	113	804	33,297	82,269	23	1,473	5,973
United Arab Emirates	51	7	X	3	X	X	X	X	X	X	X
Yemen	457	267	X	17	3	(139)	3,935	5,478	3	64	48

	Average Annual Official Development Assistance (ODA) and Official Aid (OA) (million US$) {a}		ODA and OA as a Percentage of Gross Domestic Product (GDP) {b}	ODA and OA Per Capita (US$) {a}	Direct Foreign Investment (million US$)		Total External Debt (million US$)		Debt Service as a Percentage of Total Exports	International Tourism Receipts (million US$)	
	1985-87	1995-97	1995-97	1995-97	1985-87	1995-97	1985-87	1995-97	1995-97	1985-87	1995-97
SUB-SAHARAN AFRICA	**X**	**X**	**X**	**X**	**X**	**X**	**X**	**X**	**X**	**1,676**	**5,338**
Angola	133	466	2.3	41	210	300	4,035	10,739	14	X	9
Benin	125	267	3.8	49	0	2	1,012	1,611	8	30	28
Botswana	120	99	0.9	66	79	87	438	626	4 c	38	174
Burkina Faso	256	428	4.3	40	1	0	659	1,286	11	6	32
Burundi	178	204	4.9	33	1	1	598	1,117	37	1	1
Cameroon	192	453	1.8	33	116	29	4,003	9,394	21	39	38
Central African Rep	142	144	X	43	8	5	474	921	7	5	5
Chad	183	257	3.9	37	30	15	275	975	10	6	10
Congo	99	274	X	104	26	7	3,625	5,439	14	6	5
Congo, Dem Rep	469	177	X	4	7	1	7,373	12,799	2	13	4
Côte d'Ivoire	178	875	X	63	62	248	11,562	18,010	26	49	78
Equatorial Guinea	34	30	X	73	0	174	162	286	2	X	2
Eritrea	X	145	5.0	44	X	0	X	52	0	X	67
Ethiopia	784	793	2.8	14	(1)	6	6,234 h	10,155	24	10	30
Gabon	77	104	1.2	94	72	(93)	1,923	4,318	13	9	7
Gambia	86	43	X	37	1	10	281	437	13	19	22
Ghana	320	602	2.1	33	5	119	2,779	5,992	27	28	249
Guinea	174	365	2.9	50	7	9	1,767	3,334	20	X	4
Guinea-Bissau	82	143	X	128	1	1	390	918	41	X	X
Kenya	482	602	1.8	22	31	22	4,841	6,922	26	302	440
Lesotho	96	105	3.2	53	4	27	211	669	6	8	19
Liberia	90	143	X	64	2	18	1,461	2,091	X	X	X
Madagascar	290	502	3.9	35	6	11	3,073	4,191	15	7	64
Malawi	195	429	6.3	43	0	1	1,182	2,253	19	7	7
Mali	377	502	7.0	49	(4)	17	1,749	2,970	14	31	19
Mauritania	236	252	6.2	105	4	5	1,740	2,405	23	9	11
Mozambique	571	996	8.9	56	3	36	3,496	5,833	26	X	X
Namibia	13	182	2.3	115	0	153	X	X	X	X	214
Niger	337	292	3.6	31	17	1	1,411	1,567	18	7	17
Nigeria	54	202	0.2	2	416	1,404	23,292	31,318	12	86	75
Rwanda	212	659	14.1	118	17	1	474	1,061	18	7	1
Senegal	512	560	3.9	66	(9)	36	3,275	3,725	16	110	147
Sierra Leone	79	178	7.9	41	(44)	3	870	1,169	43	10	9
Somalia	486	129	X	15	22	0	1,816	2,628	X	X	X
South Africa	X	418	0.1	11	X	1,159	X	25,543	11	476	1,962
Sudan	1,000	219	0.5	8	(1)	0	9,945	16,967	9	32	8
Tanzania, United Rep	698	913	5.4	30	2	143	6,506	7,345	17	20	313
Togo	137	161	2.6	39	10	0	1,078	1,427	7	31	13
Uganda	226	787	3.6	40	(1)	141	1,522	3,652	21	3	94
Zambia	403	1,089	12.5	130	51	65	5,655	6,933	76	7	57
Zimbabwe	261	398	1.5	36	(7)	58	2,631	5,006	22	29	208
NORTH AMERICA	**X**	**X**	**X**	**X**	**X**	**X**	**X**	**X**	**X**	**24,467**	**78,050**
Canada	(1,737)	(1,969)	(0.3)	(66)	4,107	7,722	X	X	X	3,897	8,597
United States	(9,361)	(7,874)	(0.1)	(29)	37,955	76,241	X	X	X	20,570	69,453
C. AMERICA & CARIBBEAN	**X**	**X**	**X**	**X**	**X**	**X**	**X**	**X**	**X**	**8,633**	**20,145**
Belize	23	18	1.9	84	5	18	126	311	11	33	85
Costa Rica	237	11	0.0	3	70	293	4,566	3,612	14	129	700
Cuba	750	67	X	6	0	11 c	X	X	X	137	1,182
Dominican Rep	152	103	X	13	58	401	3,704	4,339	6	453	1,841
El Salvador	376	310	1.9	54	18	15	1,900	2,935	9	42	51
Guatemala	153	246	0.6	24	94	81	2,768	3,838	11	82	295
Haiti	181	480	5.1	62	5	5	757	920	26	72	81
Honduras	272	363	2.8	62	32	87	3,002	4,600	26	26	105
Jamaica	171	79	0.9	32	13	218	4,349	4,052	17	506	1,097
Mexico	184	262	0.0	3	1,753	10,396	102,400	158,106	32	3,130	6,902
Nicaragua	279	681	X	150	0	113	6,803	7,323	32	8	61
Panama	54	90	X	34	18	482	5,083	6,227	10	200	342
Trinidad and Tobago	20	25	0.3	20	7	320	1,724	2,380	17	91	96
SOUTH AMERICA	**X**	**X**	**X**	**X**	**X**	**X**	**X**	**X**	**X**	**4,304**	**12,494**
Argentina	75	241	0.1	7	491	5,506	53,951	107,439	46	1,119	4,649
Bolivia	280	768	3.6	101	19	483	5,405	5,243	31	37	162
Brazil	197	421	0.0	3	964	11,904	110,816	177,392	45	1,507	2,389
Chile	18	167	0.1	12	153	4,373	21,006	28,135	26	163	962
Colombia	68	252	0.1	6	672	3,858	15,544	28,575	30	306	908
Ecuador	171	227	0.4	19	69	498	9,504	14,469	27	157	275
Guyana	28	168	6.7	201	1	82	1,623	1,782	17	20	41
Paraguay	66	124	0.7	25	2	218	2,142	2,151	5	125	874
Peru	295	443	0.4	18	18	2,419	15,082	30,229	27	202	581
Suriname	16	88	X	215	(31)	(6) c	X	X	X	9	18
Uruguay	16	64	0.2	20	29	151	4,041	5,958	18	234	696
Venezuela	15	40	0.0	2	35	2,752	34,748	35,577	23	425	940
OCEANIA	**X**	**X**	**X**	**X**	**X**	**X**	**X**	**X**	**X**	**2,525**	**13,799**
Australia	(709)	(1,110)	(0.3)	(61)	4,221	9,183	X	X	X	1,489	8,503
Fiji	38	45	1.4	58	14	30	451	227	4	144	297
New Zealand	(72)	(133)	(0.2)	(36)	1,254	2,016	X	X	X	657	2,368
Papua New Guinea	278	369	X	84	89	255	2,124	2,380	16	16	67
Solomon Islands	36	44	4.7	113	5	20	80	146	3	5	11
LOW INCOME {i}	**19,150**	**25,007**	**X**	**13**	**1,304**	**9,070**	**216,402**	**400,398**	**19**	**2,607**	**6,143**
MIDDLE INCOME {i}	**18,356**	**30,189**	**X**	**11**	**9,521**	**120,361**	**889,428**	**1,708,726**	**16**	**25,547**	**107,526**
HIGH INCOME {i}	**2,607**	**2,257**	**X**	**3**	**77,357**	**209,923**	**52,318** j	**130,048** j	**X**	**112,335**	**313,388**

Notes: Zero is either 0 or less than one-half the unit of measure. a. ODA and OA flows to recipients are shown as positive numbers; flows from donors are shown as negative numbers (in parentheses). b. GDP figures are calculated using purchasing power parity. c. Data are from 1995-96. d. Data are from 1996-97. e. Data are from 1986-87. f. Data before 1993 refer to the former Socialist Federal Republic of Yugoslavia. g. Data are from 1985-86. h. Data for Ethiopia before 1993 include Eritrea. i. As defined by the World Bank (World Development Indicators 1999). j. Data represent only the Republic of Korea.

Sources: Organisation for Economic Co-Operation and Development and World Bank

Basic Economic Indicators

	Income Distribution							Population in Poverty					
			Percentage of Income in Each Quintile of Population					International Poverty Line (<1 Int$/day) {b}		National Poverty (percent)			
	Survey Year	Gini Coefficient {a}	Lowest 0-20	20-40	40-60	60-80	Highest 80-100	Year(s)	(percent)	Year(s)	National	Rural	Urban
WORLD	**X**	**X**	**X**	**X**	**X**	**X**	**X**	**X**	**X**	**X**	**X**	**X**	**X**
ASIA (EXCL. MIDDLE EAST)	**X**	**X**	**X**	**X**	**X**	**X**	**X**	**X**	**X**	**X**	**X**	**X**	**X**
Armenia	X	X	X	X	X	X	X	X	X	X	X	X	X
Azerbaijan	X	X	X	X	X	X	X	X	X	1995	68.1	X	X
Bangladesh	1992 c	28	9.4	13.5	17.2	22.0	37.9	X	X	1995-96	35.6	39.8	14.3
Bhutan	X	X	X	X	X	X	X	X	X	X	X	X	X
Cambodia	X	X	X	X	X	X	X	X	X	1997	36.1	40.1	21.1
China	1995 d	42	5.5	9.8	14.9	22.3	47.5	1995	22.2	1996	6.0	7.9	<2.0 e
Georgia	X	X	X	X	X	X	X	X	X	X	X	X	X
India	1994 c	30	9.2	13.0	16.8	21.7	39.3	1994	47.0	1994	35.0	36.7	30.5
Indonesia	1996 d	37	8.0	11.3	15.1	20.8	44.9	1996	7.7	1990	15.1	14.3	16.8
Japan	X	X	X	X	X	X	X	1992	0.2 f	1992	6.9	X	X f
Kazakhstan	1993 d	33	7.5	12.3	16.9	22.9	40.4	X	X	1996	34.6	39.0	3.0
Korea, Dem People's Rep	X	X	X	X	X	X	X	X	X	X	X	X	X
Korea, Rep	X	X	X	X	X	X	X	X	X	X	X	X	X
Kyrgyzstan	1993 d	35	6.7	11.5	16.4	23.1	42.3	1993	18.9	1993	40.0	48.1	28.7
Lao People's Dem Rep	1992 c	30	9.6	12.9	16.3	21.0	40.2	X	X	1993	46.1	53.0	24.0
Malaysia	1989 d	48	4.6	8.3	13.0	20.4	53.7	1995	4.3	1989	15.5	X	X
Mongolia	1995 c	33	7.3	12.2	16.6	23.0	40.9	X	X	1995	36.3	33.1	38.5
Myanmar	X	X	X	X	X	X	X	X	X	X	X	X	X
Nepal	1995-96 c	37	7.6	11.5	15.1	21.0	44.8	1995	50.3	1995-96	42.0	44.0	23.0
Pakistan	1996 c	31	9.4	13.0	16.0	20.3	41.2	1991	11.6	1991	34.0	36.9	28.0
Philippines	1994 c	43	5.9	9.6	13.9	21.1	49.6	1994	26.9	1997	37.5	51.2	22.5
Singapore	X	X	X	X	X	X	X	X	X	X	X	X	X
Sri Lanka	1990 c	30	8.9	13.1	16.9	21.7	39.3	1990	4.0	1991	22.4	24.4	18.3
Tajikistan	X	X	X	X	X	X	X	X	X	X	X	X	X
Thailand	1992 c	46	5.6	8.7	13.0	20.0	52.7	1992	<2.0 e	1992	13.1	15.5	10.2
Turkmenistan	1993 d	36	6.7	11.4	16.3	22.8	42.8	1993	4.9	X	X	X	X
Uzbekistan	X	X	X	X	X	X	X	X	X	X	X	X	X
Viet Nam	1993 c	36	7.8	11.4	15.4	21.4	44.0	X	X	1993	50.9	57.2	25.9
EUROPE	**X**	**X**	**X**	**X**	**X**	**X**	**X**	**X**	**X**	**X**	**X**	**X**	**X**
Albania	X	X	X	X	X	X	X	X	X	1996	X	X	19.6
Austria	1987 d	23	10.4	14.8	18.5	22.9	33.3	X	X	X	X	X	X
Belarus	1995 d	29	8.5	13.5	17.7	23.1	37.2	1993	<2.0 e	1995	22.5	X	X
Belgium	1992 d	25	9.5	14.6	18.4	23.0	34.5	1992	0.9 f	1992	2.6	X	X e
Bosnia and Herzegovina	X	X	X	X	X	X	X	X	X	X	X	X	X
Bulgaria	1992 d	31	8.3	13.0	17.0	22.3	39.3	1992	2.6	X	X	X	X
Croatia	X	X	X	X	X	X	X	X	X	X	X	X	X
Czech Rep	1993 d	27	10.5	13.9	16.9	21.3	37.4	1993	3.1	X	X	X	X
Denmark	1992 d	25	9.6	14.9	18.3	22.7	34.5	1992	0.9 f	1992	4.1	X	X e
Estonia	1995 d	35	6.2	12.0	17.0	23.1	41.8	1993	6.0	1994	8.9	14.7	6.8
Finland	1991 d	26	10.0	14.2	17.6	22.3	35.8	1991	0.1 f	1991	2.8	X	X e
France	1989 d	33	7.2	12.7	17.1	22.8	40.1	1984	0.8 f	1984	4.3	X	X e
Germany	1989 d	28	9.0	13.5	17.5	22.9	37.1	1989	0.7 f	1989	5.2	X	X e
Greece	X	X	X	X	X	X	X	X	X	X	X	X	X
Hungary	1993 d	28	9.7	13.9	16.9	21.4	38.1	1993	0.7	1993	25.3	X	X
Iceland	X	X	X	X	X	X	X	X	X	X	X	X	X
Ireland	1987 d	36	6.7	11.6	16.4	22.4	42.9	1987	1.4 f	1987	4.4	X	X e
Italy	1991 d	31	7.6	12.9	17.3	23.2	38.9	X	X	1991	2.7	X	X e
Latvia	1995 d	29	8.3	13.8	18.0	22.9	37.0	1993	<2.0 e	X	X	X	X
Lithuania	1993 d	34	8.1	12.3	16.2	21.3	42.1	1993	<2.0 e	X	X	X	X
Macedonia, FYR	X	X	X	X	X	X	X	X	X	X	X	X	X
Moldova, Rep	1992 d	34	6.9	11.9	16.7	23.1	41.5	1992	6.8	X	X	X	X
Netherlands	1991 d	32	8.0	13.0	16.7	22.5	39.9	1991	1.9 f	1991	4.3	X	X e
Norway	1991 d	25	10.0	14.3	17.9	22.4	35.3	1991	0.4 f	1991	2.4	X	X e
Poland	1992 c	27	9.3	13.8	17.7	22.6	36.6	1993	6.8	1993	23.8	X	X
Portugal	X	X	X	X	X	X	X	X	X	X	X	X	X
Romania	1994 d	28	8.9	13.6	17.6	22.6	37.3	1992	17.7	1994	21.5	28.0	15.6
Russian Federation	1996 c	48	4.2	8.8	13.6	20.7	52.8	1993	<2.0 e	1994	30.9	X	X
Slovakia	1992 d	20	11.9	15.8	18.8	22.2	31.4	1992	12.8	X	X	X	X
Slovenia	1993 d	29	9.3	13.3	16.9	21.9	38.6	1993	<2.0 e	X	X	X	X
Spain	1990 d	33	7.5	12.6	17.0	22.6	40.3	1990	0.5 f	1990	5.5	X	X e
Sweden	1992 d	25	9.6	14.5	18.1	23.2	34.5	1992	0.3 f	1992	4.2	X	X e
Switzerland	1982 d	36	7.4	11.6	15.6	21.9	43.5	X	X	X	X	X	X
Ukraine	1995 d	47	4.3	9.0	13.8	20.8	52.2	1992	<2.0 e	1995	31.7	X	X
United Kingdom	1986 d	33	7.1	12.8	17.2	23.1	39.8	1991	0.5 f	1991	6.7	X	X e
Yugoslavia	X	X	X	X	X	X	X	X	X	X	X	X	X
MIDDLE EAST & N. AFRICA	**X**	**X**	**X**	**X**	**X**	**X**	**X**	**X**	**X**	**X**	**X**	**X**	**X**
Afghanistan	X	X	X	X	X	X	X	X	X	X	X	X	X
Algeria	1995 c	35	7.0	11.6	16.1	22.7	42.6	1995	<2.0 e	1995	22.6	30.3	14.7
Egypt	1991 c	32	8.7	12.5	16.3	21.4	41.1	1990-91	7.6	X	X	X	X
Iran, Islamic Rep	X	X	X	X	X	X	X	X	X	X	X	X	X
Iraq	X	X	X	X	X	X	X	X	X	X	X	X	X
Israel	1992 d	36	6.9	11.4	16.3	22.9	42.5	X	X	X	X	X	X
Jordan	1991 c	43	5.9	9.8	13.9	20.3	50.1	1992	2.5	1991	15.0	X	X
Kuwait	X	X	X	X	X	X	X	X	X	X	X	X	X
Lebanon	X	X	X	X	X	X	X	X	X	X	X	X	X
Libyan Arab Jamahiriya	X	X	X	X	X	X	X	X	X	X	X	X	X
Morocco	1990-91 c	39	6.6	10.5	15.0	21.7	46.3	1990-91	<2.0 e	1990-91	13.1	18.0	7.6
Oman	X	X	X	X	X	X	X	X	X	X	X	X	X
Saudi Arabia	X	X	X	X	X	X	X	X	X	X	X	X	X
Syrian Arab Rep	X	X	X	X	X	X	X	X	X	X	X	X	X
Tunisia	1990 c	40	5.9	10.4	15.3	22.1	46.3	1990	3.9	1990	14.1	21.6	8.9
Turkey	X	X	X	X	X	X	X	X	X	X	X	X	X
United Arab Emirates	X	X	X	X	X	X	X	X	X	X	X	X	X
Yemen	1992 c	40	6.1	10.9	15.3	21.6	46.1	X	X	1992	19.1	19.2	18.6

	Income Distribution								Population in Poverty					
				Percentage of Income in Each Quintile of Population					International Poverty Line (<1 Int$/day) {b}		National Poverty (percent)			
	Survey Year		Gini Coefficient {a}	Lowest 0-20	20-40	40-60	60-80	Highest 80-100	Year(s)	(percent)	Year(s)	National	Rural	Urban
SUB-SAHARAN AFRICA	**X**		**X**	**X**	**X**	**X**	**X**	**X**	**X**	**X**	**X**	**X**	**X**	**X**
Angola	X		X	X	X	X	X	X	X	X	X	X	X	X
Benin	X		X	X	X	X	X	X	X	X	1995	33.0	X	X
Botswana	X		X	X	X	X	X	X	1985-86	33.0	X	X	X	X
Burkina Faso	1994	c	48	5.5	8.7	12.0	18.7	55.0	X	X	X	X	X	X
Burundi	X		X	X	X	X	X	X	X	X	1990	36.2	X	X
Cameroon	X		X	X	X	X	X	X	X	X	1984	40.0	32.4	44.4
Central African Rep	X		X	X	X	X	X	X	X	X	X	X	X	X
Chad	X		X	X	X	X	X	X	X	X	1995-96	64.0	67.0	63.0
Congo	X		X	X	X	X	X	X	X	X	X	X	X	X
Congo, Dem Rep	X		X	X	X	X	X	X	X	X	X	X	X	X
Côte d'Ivoire	1988	c	37	6.8	11.2	15.8	22.2	44.1	1988	17.7	X	X	X	X
Equatorial Guinea	X		X	X	X	X	X	X	X	X	X	X	X	X
Eritrea	X		X	X	X	X	X	X	X	X	X	X	X	X
Ethiopia	1995	c	40	7.1	10.9	14.5	19.8	47.7	1981-82	46.0	X	X	X	X
Gabon	X		X	X	X	X	X	X	X	X	X	X	X	X
Gambia	1992	c	48	4.4	9.0	13.5	20.4	52.8	X	X	1992	64.0	X	X
Ghana	1997	c	33	8.4	12.2	15.8	21.9	41.7	X	X	1992	31.4	34.3	26.7
Guinea	1994	c	40	6.4	10.4	14.8	21.2	47.2	1991	26.3	X	X	X	X
Guinea-Bissau	1991	c	56	2.1	6.5	12.0	20.6	58.9	1991	88.2	1991	48.8	60.9	24.1
Kenya	1994	c	45	5.0	9.7	14.2	20.9	50.2	1992	50.2	1992	46.4	46.4	29.3
Lesotho	1986-87	c	56	2.8	6.5	11.2	19.4	60.1	1986-87	48.8	1993	49.2	53.9	27.8
Liberia	X		X	X	X	X	X	X	X	X	X	X	X	X
Madagascar	1993	c	46	5.1	9.4	13.3	20.1	52.1	1993	72.3	X	X	X	X
Malawi	X		X	X	X	X	X	X	X	X	1990-91	54.0	X	X
Mali	1994	c	51	4.6	8.0	11.9	19.3	56.2	X	X	X	X	X	X
Mauritania	1995	c	39	6.2	10.8	15.4	22.0	45.6	1988	31.4	1990	57.0	X	X
Mozambique	X		X	X	X	X	X	X	X	X	X	X	X	X
Namibia	X		X	X	X	X	X	X	X	X	X	X	X	X
Niger	1995	c	51	2.6	7.1	13.9	23.1	53.3	1992	61.5	1989-93	63.0	66.0	52.0
Nigeria	1992-93	c	45	4.0	8.9	14.4	23.4	49.4	1992-93	31.1	1992-93	34.1	36.4	30.4
Rwanda	1983-85	c	29	9.7	13.2	16.5	21.6	39.1	1983-85	45.7	1993	51.2	X	X
Senegal	1991	c	54	3.1	7.4	12.1	19.5	57.9	1991-92	54.0	1991	33.4	40.4	16.4
Sierra Leone	1989	c	63	1.1	2.0	9.8	23.7	63.4	X	X	1989	68.0	76.0	53.0
Somalia	X		X	X	X	X	X	X	X	X	X	X	X	X
South Africa	1993-94	c	59	2.9	5.5	9.2	17.7	64.8	1993	23.7	X	X	X	X
Sudan	X		X	X	X	X	X	X	X	X	X	X	X	X
Tanzania, United Rep	1993	c	38	6.8	11.0	15.1	21.6	45.5	X	X	1991	51.1	X	X
Togo	X		X	X	X	X	X	X	X	X	1987-89	32.3	X	X
Uganda	1992-93	c	39	6.6	10.9	15.2	21.3	46.1	1989-90	69.3	1993	55.0	X	X
Zambia	1996	c	50	4.2	8.2	12.8	20.1	54.8	1993	84.6	1993	86.0	X	X
Zimbabwe	1990	c	57	4.0	6.3	10.0	17.4	62.3	1990-91	41.0	1990-91	25.5	X	X
NORTH AMERICA	**X**		**X**	**X**	**X**	**X**	**X**	**X**	**X**	**X**	**X**	**X**	**X**	**X**
Canada	1994	d	32	7.5	12.9	17.2	23.0	39.3	1991	0.3 f	1991	7.0	X	X e
United States	1994	d	40	4.8	10.5	16.0	23.5	45.2	1994	1.4 f	1994	13.2	X	X e
C. AMERICA & CARIBBEAN	**X**		**X**	**X**	**X**	**X**	**X**	**X**	**X**	**X**	**X**	**X**	**X**	**X**
Belize	X		X	X	X	X	X	X	X	X	X	X	X	X
Costa Rica	1996	d	47	4.0	8.8	13.7	21.7	51.8	1989	18.9	X	X	X	X
Cuba	X		X	X	X	X	X	X	X	X	X	X	X	X
Dominican Rep	1989	d	51	4.2	7.9	12.5	19.7	55.7	1989	19.9	1992	20.6	29.8	10.9
El Salvador	1995	d	50	3.7	8.3	13.1	20.5	54.4	X	X	1992	48.3	55.7	43.1
Guatemala	1989	d	60	2.1	5.8	10.5	18.6	63.0	1989	53.3	X	X	X	X
Haiti	X		X	X	X	X	X	X	X	X	1987	65.0	X	X
Honduras	1996	d	54	3.4	7.1	11.7	19.7	58.0	1992	46.9	1992	50.0	46.0	56.0
Jamaica	1991	c	41	5.8	10.2	14.9	21.6	47.5	1993	4.3	1992	34.2	X	X
Mexico	1995	d	54	3.6	7.2	11.8	19.2	58.2	1992	14.9	1988	10.1	X	X
Nicaragua	1993	c	50	4.2	8.0	12.6	20.0	55.2	1993	43.8	1993	50.3	76.1	31.9
Panama	1995	d	57	2.3	6.2	11.3	19.8	60.4	1989	25.6	X	X	X	X
Trinidad and Tobago	X		X	X	X	X	X	X	X	X	1992	21.0	X	X
SOUTH AMERICA	**X**		**X**	**X**	**X**	**X**	**X**	**X**	**X**	**X**	**X**	**X**	**X**	**X**
Argentina	X		X	X	X	X	X	X	X	X	1991	25.5	X	X
Bolivia	1990	d	42	5.6	9.7	14.5	22.0	48.2	X	X	X	X	X	X
Brazil	1995	d	60	2.5	5.7	9.9	17.7	64.2	1995	23.6	1990	17.4	32.6	13.1
Chile	1994	d	57	3.5	6.6	10.9	18.1	61.0	1992	15.0	1994	20.5	X	X
Colombia	1995	d	57	3.1	6.8	10.9	17.6	61.5	1991	7.4	1992	17.7	31.2	8.0
Ecuador	1994	c	47	5.4	8.9	13.2	19.9	52.6	1994	30.4	1994	35.0	47.0	25.0
Guyana	1993	c	40	6.3	10.7	15.0	21.2	46.9	X	X	X	X	X	X
Paraguay	1995	d	59	2.3	5.9	10.7	18.7	62.4	X	X	1991	21.8	28.5	19.7
Peru	1996	d	46	4.4	9.1	14.1	21.3	51.2	X	X	1997	49.0	64.7	40.4
Suriname	X		X	X	X	X	X	X	X	X	X	X	X	X
Uruguay	X		X	X	X	X	X	X	X	X	X	X	X	X
Venezuela	1995	d	47	4.3	8.8	13.8	21.3	51.8	1991	11.8	1989	31.3	X	X
OCEANIA	**X**		**X**	**X**	**X**	**X**	**X**	**X**	**X**	**X**	**X**	**X**	**X**	**X**
Australia	1989	d	34	7.0	12.2	16.6	23.3	40.9	1989	0.8 f	1989	7.0	X	X e
Fiji	X		X	X	X	X	X	X	X	X	X	X	X	X
New Zealand	X		X	X	X	X	X	X	X	X	X	X	X	X
Papua New Guinea	1996	c	51	4.5	7.9	11.9	19.2	56.5	X	X	X	X	X	X
Solomon Islands	X		X	X	X	X	X	X	X	X	X	X	X	X

Notes: a. Gini coefficients measure the inequality in income distribution within the population (0 = perfectly equal, 100 = perfectly unequal). b. Estimated using GDP based on purchasing power parity. c. Refers to expenditure shares by percentiles of population. Rankings are based on per capita expenditure. d. Refers to income shares by percentiles of population. Rankings are based on per capita income. e. As reported by the World Bank. f. Luxembourg Income Study estimates (national poverty is defined here as 40 percent of the national median income).

Sources: Luxembourg Income Study and World Bank

	Population (000) 2000	Land Area (000 ha) 1998	Natural Forest Extent (000 ha) 1995	Natural Forest Annual Average Percent Change 1990-95	Crop-land Area (000 ha) 1997	*Scleractinia* Coral Genera {a} (number)	Claimed Exclusive Economic Zone (000 km²) 2000	Marine Fish Catch (metric tons) Average Total Catch 1995-97	Marine Fish Catch (metric tons) Percent Change Since 1985-87	Food Supply from Fish & Fishery Products Per Capita {b} (kg per person) 1997	Food Supply from Fish & Fishery Products Fish Protein as a % of All Protein 1997	Under-5 Mortality Rate (per 000 live births) 1997	Gross Domestic Product (PPP) (million Int$) 1997
SMALL NATIONS/ISLANDS													
Antigua and Barbuda	68	44	9	(0.4)	8	25	103	377	(82)	24	9	21	X
Aruba	103	19	X	X	2	X	X	162	(77)	9	X	X	X
Bahamas	307	1,001	158	(2.6)	10	24	369	1,510	(15)	20	7	21	X
Bahrain	617	69	0	0.0	5	29	X	6,861	16	14	4	22	9,665
Barbados	270	43	0	0.0	145	26	183	3,162	(20)	32	11	12	X
Bermuda	65	5	0	0.0	X	16	447	435	(41)	40	13	X	X
Brunei Darussalam	328	527	434	(0.7)	7	64	6	5,204	67	21	7	10	X
Cape Verde	428	403	6	(1.3)	41	6	742	9,231	19	25	11	73	1,201
Comoros	694	223	9	(4.9)	118	47	162	12,882	149	20	16	93	791
Cyprus	786	924	X	X	145	X	X	2,122	(8)	20	5	9	X
Djibouti	638	2,318	22	(0.4)	X	55	2	350	(12)	3	2	156	X
Dominica	71	75	46	0.2	15	25	25	843	42	27	8	20	319
Grenada	94	34	4	2.1	11	25	20	1,350	(39)	20	10	29	451
Kiribati	83	73	0	0.0	37	40	3,388	19,387	(21)	74	33	75	X
Luxembourg	431	259	X	X	X	X	X	X	X	X	X	7	12,524
Maldives	286	30	X	X	3	63	871	105,563	78	165 c	54	74	945
Malta	389	32	X	X	11	X	X	825	(42)	30	7	10	4,942
Marshall Islands	64	18	X	X	3	61	1,877	397	86	9	X	92	X
Mauritius	1,158	203	3	1.4	106	56	1,275	13,968	0	21	9	23	10,689
Micronesia, Fed States	543	70	X	X	X	68	2,906	9,199	163	40	20	24	X
Qatar	599	1,100	0	0.0	17	27	X	4,593	103	10	3	20	X
St. Lucia	154	61	5	(1.9)	17	25	11	1,222	46	18	7	29	837
St. Vincent & Grenadines	114	39	11	(0.4)	11	25	32	1,338	125	15	7	21	464
Samoa	180	283	127	(1.2)	122	55	110	3,636	13	63	23	52	619
Sao Tome & Principe	147	96	56	0.0	41	5	143	2,988	(6)	21	16	78	X
Seychelles	77	45	3	(1.9)	7	62	1,289	4,246	4	66	26	18	X
Swaziland	1,008	1,720	74	0.0	180	X	X	X	X	0	0	94	3,208
Tonga	99	72	0	0.0	48	45	845	2,563	(9)	32	14	23	X
Vanuatu	190	1,219	893	(0.8)	120	65	530	1,636	(23)	23	10	50	617
OVERSEAS TERRITORIES													
French Guiana	181	8,815	7,990	(0.0)	13	X	132	3,611	69	28	9	X	X
French Polynesia	235	366	0	0.0	28	44	4,553	10,148	344	64	20	14 d	X
Guadeloupe	456	169	80	(1.7)	25	25	116	9,243	15	28	11	11 d	X
Martinique	395	106	38	(1.1)	22	25	11	4,413	16	26	10	9 d	X
Netherlands Antilles	217	80	0	0.0	8	25	X	1,027	(3)	19	7	16 d	X
New Caledonia	214	1,828	689	(0.1)	13	65	1,348	1,833	11	24	8	16 d	X
Puerto Rico	3,869	887	272	(0.9)	78	25	188	2,128	89	1	1	14 d	X
Réunion	699	250	82	0.1	38	54	310	4,933	278	10	4	10 d	X
Virgin Islands, U.S.	93	34	0	0.0	X	25	X	793	6	10	44	X	X

Notes: Negative values are shown in parantheses. Zero is either 0 or less than one-half the unit of measure. a. "Scleractinia" corals are reef-forming corals (i.e., true or stony corals). b. Per capita values are expressed on a live-weight-equivalent basis, which means that all parts of the fish, including bones, are taken into account when estimating consumption of fish and fishery products.c. This value is an overestimate. Fish consumption in the Maldives is estimated based on the resident and the present population (including tourists). d. Data are from the United Nations Population Division and refer to the period 1995-2000.

Sources: Various

Regional Groupings of Countries as Listed by Various Organizations

World Resources Institute - Regions
Food and Agricultural Organization of the United Nations - Developing, Developed
World Bank - High, Medium, Low Income
UNICEF - Industrialized, Developing and Least Developed, Other Regions
United Nations Population Division - Regions, Developing, Developed

World Resources Institute - Regions

ASIA (excluding the Middle East, does not include former U.S.S.R.)
Armenia
Azerbaijan
Bangladesh
Bhutan
Brunei Darussalam
Cambodia
China
Georgia
Hong Kong, P.R. China, SAR
India
Indonesia
Japan
Kazakhstan
Korea, Dem People's Rep
Korea, Rep
Kyrgyzstan
Lao People's Dem Rep
Macau
Malaysia
Maldives
Mongolia
Myanmar
Nepal
Pakistan
Philippines
Singapore
Sri Lanka
Taiwan, Province of China
Tajikistan
Thailand
Turkmenistan
Uzbekistan
Vietnam

EUROPE (does not include former U.S.S.R.)
Albania
Austria
Belarus
Belgium
Belgium/Luxembourg
Bosnia and Herzegovina
Bulgaria
Croatia
Czech Rep
Czechoslovakia (former)
Denmark
Estonia
Finland
France
German Dem Rep (former)
Germany
Germany, Fed Rep (former)
Greece
Hungary
Iceland
Ireland
Italy
Latvia
Lithuania
Luxembourg
Macedonia, FYR
Malta
Moldova, Rep
Netherlands
Norway
Poland
Portugal
Romania
Russian Federation
Slovakia
Slovenia
Spain
Sweden
Switzerland
Ukraine
United Kingdom
Yugoslavia
Yugoslavia (former)

MIDDLE EAST AND NORTH AFRICA
Afghanistan
Algeria
Bahrain
Cyprus
Egypt
Iran, Islamic Rep
Iraq
Israel
Jordan
Kuwait
Lebanon
Libyan Arab Jamahiriya
Morocco
Oman
Qatar
Saudi Arabia
Syrian Arab Rep
Tunisia
Turkey
United Arab Emirates
Yemen

SUB-SAHARAN AFRICA
Angola
Benin
Botswana
Burkina Faso
Burundi
Cameroon
Cape Verde
Central African Rep
Chad
Comoros
Congo
Congo, Dem Rep
Côte d'Ivoire
Djibouti
Equatorial Guinea
Eritrea
Ethiopia
Gabon
Gambia
Ghana
Guinea
Guinea-Bissau
Kenya
Lesotho
Liberia
Madagascar
Malawi
Mali
Mauritania
Mauritius
Mozambique
Namibia
Niger
Nigeria
Reunion
Rwanda
Sao Tome and Principe
Senegal
Seychelles
Sierra Leone
Somalia
South Africa
Sudan
Swaziland
Tanzania
Togo
Uganda
Zambia
Zimbabwe

NORTH AMERICA
Canada
United States

CENTRAL AMERICA AND CARIBBEAN
Antigua and Barbuda
Aruba
Bahamas
Barbados
Belize
Bermuda
Costa Rica
Cuba
Dominica
Dominican Rep
El Salvador
Grenada
Guadeloupe
Guatemala
Haiti
Honduras
Jamaica
Martinique
Mexico
Netherlands Antilles
Nicaragua
Panama
Puerto Rico
St. Lucia
St. Vincent and Grenadines
Trinidad and Tobago
Virgin Islands, U.S.

SOUTH AMERICA
Argentina
Bolivia
Brazil
Chile
Colombia
Ecuador
French Guiana
Guyana
Paraguay
Peru
Suriname
Uruguay
Venezuela

OCEANIA
Australia
Fiji
French Polynesia
Kiribati
Marshall Islands
Micronesia, Fed States
New Caledonia
New Zealand
Papua New Guinea
Samoa
Solomon Islands
Tonga
Vanuatu

Food and Agricultural Organization of the United Nations - Developing and Developed Countries

Developing

Afghanistan
Algeria
Angola
Antigua and Barbuda
Argentina
Aruba
Bahamas
Bahrain
Bangladesh
Barbados
Belize
Benin
Bermuda
Bhutan
Bolivia
Botswana
Brazil
Brunei Darussalam
Burkina Faso
Burundi
Cambodia
Cameroon
Cape Verde
Central African Rep
Chad
Chile
China
Colombia
Comoros
Congo
Congo, Dem Rep
Costa Rica
Côte d'Ivoire
Cuba
Cyprus
Djibouti
Dominica
Dominican Rep
Ecuador
Egypt
El Salvador
Equatorial Guinea
Eritrea
Ethiopia
Fiji
French Guiana
French Polynesia
Gabon
Gambia
Ghana
Grenada
Guadeloupe
Guatemala
Guinea
Guinea-Bissau
Guyana
Haiti
Honduras
Hong Kong, China SAR
India
Indonesia
Iran, Islamic Rep
Iraq
Jamaica
Jordan
Kenya
Kiribati
Korea, Dem People's Rep
Korea, Rep
Kuwait
Lao People's Dem Rep
Lebanon
Lesotho
Liberia
Libyan Arab Jamahiriya
Macau
Madagascar
Malawi
Malaysia
Maldives
Mali
Marshall Islands
Martinique
Mauritania
Mauritius
Mexico
Micronesia, Fed States
Mongolia
Morocco
Mozambique
Myanmar
Namibia
Nepal
Netherlands Antilles
New Caledonia
Nicaragua
Niger
Nigeria
Oman
Pakistan
Panama
Papua New Guinea
Paraguay
Peru
Philippines
Puerto Rico
Qatar
Réunion
Rwanda
Samoa
Sao Tome & Principe
Saudi Arabia
Senegal
Seychelles
Sierra Leone
Singapore
Solomon Islands
Somalia
Sri Lanka
St. Lucia
St. Vincent & Grenadines
Sudan
Suriname
Swaziland
Syrian Arab Rep
Taiwan, Province of China
Tanzania, United Rep
Thailand
Togo
Tonga
Trinidad and Tobago
Tunisia
Turkey
Uganda
United Arab Emirates
Uruguay
Vanuatu
Venezuela
Viet Nam
Virgin Islands, U.S.
Yemen
Zambia
Zimbabwe

Developed (includes former U.S.S.R.)

Albania
Armenia
Australia
Austria
Azerbaijan
Belarus
Belgium
Belgium/Luxembourg
Bosnia and Herzegovina
Bulgaria
Canada
Croatia
Czech Rep
Czechoslovakia (former)
Denmark
Estonia
Finland
France
Georgia
German Dem Rep (former)
Germany
Germany, Fed Rep (former)
Greece
Hungary
Iceland
Ireland
Israel
Italy
Japan
Kazakhstan
Kyrgyzstan
Latvia
Lithuania
Luxembourg
Macedonia, FYR
Malta
Moldova, Rep
Netherlands
New Zealand
Norway
Poland
Portugal
Romania
Russian Federation
Slovakia
Slovenia
South Africa
Spain
Sweden
Switzerland
Tajikistan
Turkmenistan
Ukraine
United Kingdom
United States
U.S.S.R. (former)
Uzbekistan
Yugoslavia
Yugoslavia (former)

The World Bank - High, Middle, and Low Income Countries

High Income

Andorra
Aruba
Australia
Austria
Bahamas, The
Belgium
Bermuda
Brunei
Canada
Cayman Islands
Channel Islands
Cyprus
Denmark
Faeroe Islands
Finland
France
French Guiana
French Polynesia
Germany
Greece
Greenland
Guam
Hong Kong, China
Iceland
Ireland
Israel
Italy
Japan
Korea, Rep.
Kuwait
Liechtenstein
Luxembourg
Macao
Martinique
Monaco
Netherlands
Netherlands Antilles
New Caledonia
New Zealand
Northern Mariana
Islands
Norway
Portugal
Qatar
Reunion
Singapore
Slovenia
Spain
Sweden
Switzerland
United Arab Emirates
United Kingdom
United States
Virgin Islands (U.S.)

Middle Income

Algeria
American Samoa
Antigua and Barbuda
Argentina
Bahrain
Barbados
Belarus
Belize
Bolivia
Botswana
Brazil
Bulgaria
Cape Verde
Chile
China
Colombia
Costa Rica
Croatia
Cuba
Czech Republic
Djibouti
Dominica
Dominican Republic
Ecuador
Egypt, Arab Rep.
El Salvador
Equatorial Guinea
Estonia
Fiji
Gabon
Georgia
Grenada
Guadeloupe
Guatemala
Guyana
Hungary
Indonesia
Iran, Islamic Rep.
Iraq
Isle of Man
Jamaica
Jordan
Kazakhstan
Kiribati
Korea, Dem. Rep.
Latvia
Lebanon
Libya
Lithuania
Macedonia, FYR
Malaysia
Maldives
Malta
Marshall Islands
Mauritius
Mayotte
Mexico
Micronesia, Fed. Sts.
Morocco
Namibia
Oman
Palau
Panama
Papua New Guinea
Paraguay
Peru
Philippines
Poland
Puerto Rico
Romania
Russian Federation
Samoa
Saudi Arabia
Seychelles
Slovak Republic
Solomon Islands
South Africa
Sri Lanka
St. Kitts and Nevis
St. Lucia
St. Vincent and the
Grenadines
Suriname
Swaziland
Syrian Arab Republic
Thailand
Tonga
Trinidad and Tobago
Tunisia
Turkey
Ukraine
Uruguay
Uzbekistan
Vanuatu
Venezuela
West Bank and Gaza
Yugoslavia, FYR

Low Income

Afghanistan
Albania
Angola
Armenia
Azerbaijan
Bangladesh
Benin
Bhutan
Bosnia and
Herzegovina
Burkina Faso
Burundi
Cambodia
Cameroon
Central African
Republic
Chad
Comoros
Congo, Dem. Rep.
Congo, Rep.
Côte d'Ivoire
Eritrea
Ethiopia
Gambia, The
Ghana
Guinea
Guinea-Bissau
Haiti
Honduras
India
Kenya
Kyrgyz Republic
Lao PDR
Lesotho
Liberia
Madagascar
Malawi
Mali
Mauritania
Moldova
Mongolia
Mozambique
Myanmar
Nepal
Nicaragua
Niger
Nigeria
Pakistan
Rwanda
Sao Tome and Principe
Senegal
Sierra Leone
Somalia
Sudan
Tajikistan
Tanzania
Togo
Turkmenistan
Uganda
Vietnam
Yemen, Rep.
Zambia
Zimbabwe

UNICEF

Industrialized Countries

Andorra
Australia
Austria
Belgium
Canada
Denmark
Finland
France
Germany
Greece
Holy See
Iceland
Ireland
Israel
Italy
Japan
Liechtenstain
Luxembourg
Malta
Monaco
Netherlands
New Zealand
Norway
Portugal
San Marino
Slovenia
Spain
Sweden
Switzerland
United Kingdom
United States

Developing Countries

Afghanistan
Algeria
Angola
Antigua and Barbuda
Argentina
Armenia
Azerbaijan
Bahamas
Bahrain
Bangladesh
Barbados
Belize
Benin
Bhutan
Bolivia
Botswana
Brazil
Brunei Darussalam
Burkina Faso
Burundi
Cambodia
Cameroon
Cape Verde
Central African Rep
Chad
Chile
China
Colombia
Comoros
Congo
Congo, Dem Rep
Cook Islands
Costa Rica
Côte d'Ivoire
Cuba
Cyprus
Djibouti
Dominica
Dominican Rep
Ecuador
Egypt
El Salvador
Equatorial Guinea
Eritrea
Ethiopia
Fiji
Gabon
Gambia
Georgia
Ghana
Grenada
Guatemala
Guinea
Guinea-Bissau
Guyana
Haiti
Honduras
India
Indonesia
Iran
Iraq
Israel
Jamaica
Jordan
Kazakhstan
Kenya
Kiribati
Korea, Dem People's Rep
Korea, Rep of
Kuwait
Kyrgyzstan
Lao People's Dem Rep
Lebanon
Lesotho
Liberia
Libya
Madagascar
Malawi
Malaysia
Maldives
Mali
Marshall Islands
Mauritania
Mauritius
Mexico
Micronesia, Fed States of
Mongolia
Morocco
Mozambique
Myanmar
Namibia
Nauru
Nepal
Nicaragua
Niger
Nigeria
Niue
Oman
Pakistan
Palau
Panama
Papua New Guinea
Paraguay
Peru
Philippines
Qatar
Rwanda
Saint Kitts and Nevis
Saint Lucia
Saint Vincent/ Grenadines
Samoa
Sao Tome & Principe
Saudi Arabia
Senegal
Seychelles
Sierra Leone
Singapore
Solomon Islands
Somalia
South Africa
Sri Lanka
Sudan
Suriname
Swaziland
Syria
Tajikistan
Tanzania
Thailand
Togo
Tonga
Trinidad and Tobago
Tunisia
Turkey
Turkmenistan
Tuvalu
Uganda
United Arab Emirates
Uruguay
Uzbekistan
Vanuatu
Venezuela
Viet Nam
Yemen
Zambia
Zimbabwe

Least Developed Countries

Afghanistan
Angola
Bangladesh
Benin
Bhutan
Burkina Faso
Burundi
Cambodia
Cape Verde
Central African Republic
Chad
Comoros
Congo, Dem Rep
Djibouti
Equatorial Guinea
Eritrea
Ethiopia
Gambia
Guinea
Guinea-Bissau
Haiti
Kiribati
Lao People's Dem Rep
Lesotho
Liberia
Madagascar
Malawi
Maldives
Mali
Mauritania
Mozambique
Myanmar
Nepal
Niger
Rwanda
Samoa
Sao Tome and Principe
Sierra Leone
Solomon Islands
Somalia
Sudan
Tanzania
Togo
Tuvalu
Uganda
Vanuatu
Yemen
Zambia

UNICEF REGIONS

Sub-Saharan Africa
Angola
Benin
Botswana
Burkina Faso
Burundi
Cameroon
Cape Verde
Central African Rep
Chad
Comoros
Congo
Congo, Dem Rep
Côte d'Ivoire
Equatorial Guinea
Eritrea
Ethiopia
Gabon
Gambia
Ghana
Guinea
Guinea-Bissau
Kenya
Lesotho
Liberia
Madagascar
Malawi
Mali
Mauritania
Mauritius
Mozambique
Namibia
Niger
Nigeria
Rwanda
Sao Tome & Principe
Senegal
Seychelles
Sierra Leone
Somalia
South Africa
Swaziland
Tanzania
Togo
Uganda
Zambia
Zimbabwe

Middle East and North Africa
Algeria
Bahrain
Cyprus
Djibouti
Egypt
Iran
Iraq
Jordan
Kuwait
Lebanon
Libya
Morocco
Oman
Qatar
Saudi Arabia
Sudan
Syria
Tunisia
United Arab Emirates
Yemen

South Asia
Afghanistan
Bangladesh
Bhutan
India
Maldives
Nepal
Pakistan
Sri Lanka

East Asia and Pacific
Brunei Darussalam
Cambodia
China
Cook Islands
Fiji
Indonesia
Kiribati
Korea, Dem People's Rep
Korea, Rep of
Lao People's Dem Rep
Malaysia
Marshall Islands
Micronesia, Fed States of
Mongolia
Myanmar
Nauru
Niue
Palau
Papua New Guinea
Philippines
Samoa
Singapore
Solomon Islands
Thailand
Tonga
Tuvalu
Vanuatu
Viet Nam

Latin America and Caribbean
Antigua and Barbuda
Argentina
Bahamas
Barbados
Belize
Bolivia
Brazil
Chile
Colombia
Costa Rica
Cuba
Dominica
Dominican Rep
Ecuador
El Salvador
Grenada
Guatemala
Guyana
Haiti
Honduras
Jamaica
Mexico
Nicaragua
Panama
Paraguay
Peru
Saint Kitts and Nevis
Saint Lucia
Saint Vincent/Grenadines
Suriname
Trinidad and Tobago
Uruguay
Venezuela

Central and Eastern Europe (CEE), the Commonwealth of Independent States (CIS), and the Baltic States
Albania
Armenia
Azerbaijan
Belarus
Bosnia and Herzegovina
Bulgaria
Croatia
Czech Rep
Estonia
Georgia
Hungary
Kazakhstan
Kyrgyzstan
Latvia
Lithuania
Moldova, Rep of
Poland
Romania
Russian Federation
Slovakia
Tajikistan
TFYR Macedonia
Turkey
Turkmenistan
Ukraine
Uzbekistan
Yugoslavia

UN POPULATION DIVISION

More developed regions comprise all regions of Europe and Northern America, Australia/New Zealand and Japan.

Less developed regions comprise all regions of Africa, Asia (excluding Japan), Latin America and the Caribbean, and the regions of Melanesia, Micronesia and Polynesia.

Least developed countries as defined by the United Nations General Assembly, as of 1998, include 48 countries, of which 33 are in Africa, 9 in Asia, 1 in Latin America and the Caribbean and 5 in Oceania: Afghanistan, Angola, Bangladesh, Benin, Bhutan, Burkina Faso, Burundi, Cambodia, Cape Verde, Central African Republic, Chad, Comoros, Dem. Republic of the Congo, Djibouti, Equatorial Guinea, Eritrea, Ethiopia, Gambia, Guinea, Guinea-Bissau, Haiti, Kiribati, Lao People's Dem. Republic, Lesotho, Liberia, Madagascar, Malawi, Maldives, Mali, Mauritania, Mozambique, Myanmar, Nepal, Niger, Rwanda, Samoa, Sao Tome and Principe, Sierra Leone, Solomon Islands, Somalia, Sudan, Togo, Tuvalu, Uganda, United Rep. of Tanzania, Vanuatu, Yemen, Zambia. They are included in the less developed regions.

Data Table BI.1
National and International Protection of Natural Areas

Sources: National protection systems: Protected Areas Database of the World Conservation Monitoring Centre (WCMC), unpublished data (WCMC, Cambridge, U.K., May/August 1999). Biosphere reserves: United Nations Educational, Scientific, and Cultural Organization (UNESCO), Man and the Biosphere Programme, *List of Biosphere Reserves*, available online at: http://www.unesco.org/mab/wnbr.htm.

World heritage sites: UNESCO, World Heritage Committee, *List of World Heritage Sites*, available online at: http://www.unesco.org/whc/nwhc/pages/sites/main.htm. Wetlands of international importance: Ramsar Convention Bureau, *List of Wetlands of International Importance*, available online at: http://ramsar.org/sitelist.pdf.

Protected areas (IUCN management categories I–V) combine natural areas of at least 1,000 hectares in five of six World Conservation Union (IUCN) management categories. IUCN defines a 'protected area' as: "an area of land and/or sea especially dedicated to the protection and maintenance of biological diversity, and of natural and associated cultural resources, and managed through legal or other effective means." Definitions of IUCN management categories I–V follow:

Category Ia. Strict nature reserve: A protected area managed mainly for science; an area of land and/or sea possessing some outstanding or representative ecosystems, geological or physiological features and/or species, available primarily for scientific research and/or environmental monitoring.

Category Ib. Wilderness area: A protected area managed mainly for wilderness protection; a large area of unmodified or slightly modified land and/or sea retaining its natural character and influence, without permanent or significant habitation, which is protected and managed so as to preserve its natural condition.

Category II. National park: A protected area managed mainly for ecosystem protection and recreation; a natural area of land and/or sea designated to: a) protect the ecological integrity of one or more ecosystems for present and future generations; b) exclude exploitation or occupation inimical to the purposes of designation of the area; and c) provide a foundation for spiritual, scientific, educational, recreational, and visitor opportunities, all of which must be environmentally and culturally compatible.

Category III. Natural monument: A protected area managed mainly for conservation of specific natural features; an area containing one or more specific natural or natural/cultural features that is of outstanding or unique value because of its inherent rarity, representative or aesthetic qualities, or cultural significance.

Category IV. Habitat/species management area: A protected area managed mainly for conservation through management intervention; an area of land and/or sea subject to active intervention for management purposes so as to ensure the maintenance of habitats and/or to meet the requirements of specific species.

Category V. Protected landscape/seascape: A protected area managed mainly for landscape/seascape conservation and recreation; an area of land, with coast and sea as appropriate, where the interaction of people and nature over time has produced an area of distinct character with significant aesthetic, ecological, and/or cultural value, and often with high biological diversity. Safeguarding the integrity of this traditional interaction is vital to the protection, maintenance, and evolution of such an area.

IUCN has an additional management category, category VI, which includes areas "managed mainly for the sustainable use of natural ecosystems." These areas contain predominantly unmodified natural systems, managed to ensure longterm protection and maintenance of biological diversity, while also providing a sustainable flow of natural products and services to meet community needs. Areas classified under this category are not included in this table.

The world total excludes protected areas in Antarctica and Greenland. Greenland has two protected areas with a combined area of 98,250,000 hectares, and Antarctica has 55 areas, which occupy 302,800 hectares.

The extent of protected areas in some countries may include overlaps between different designations (e.g., in Venezuela some areas within different IUCN categories overlap, and, therefore, the area may be counted twice) or may include marine components (e.g., Australia, Ecuador), which may lead to some unexpectedly high figures for *percent of land area* protected.

Protected areas at least 100,000 hectares and *at least 1 million hectares in size* refer to all IUCN categories I–V protected areas that fall within these two size classifications. The *number* of sites does not account for agglomerations of protected areas that together might exceed 100,000 or 1 million hectares.

Marine protected areas include protected areas in any of the IUCN categories (I–VI) that are marine or have a marine component. IUCN defines a 'marine protected area' as: "any area of intertidal or subtidal terrain, together with its overlying water and associated flora and fauna, historical and cultural features, which has been reserved by law or other effective means to protect part or all of the enclosed environment." Marine protected areas (MPAs) include areas that are fully marine as well as areas that have only a small area of intertidal land. Many of these MPAs have large terrestrial areas. The extent of the marine portion of most protected areas is rarely documented. The degree of protection varies from one country to another, and it may bear little relationship to the legal status of any site. The categories of *marine* and *littoral* are not exclusive. One protected area can be both littoral and marine, therefore adding numbers under these two different categories may produce a higher number than the total for MPAs. Littoral is defined as any site that is known to incorporate at least some intertidal area. Such sites can also include marine and/or terrestrial elements. All sites with mangrove and saltmarsh communities are recorded as littoral. Marine is defined as any site that is known to incorporate at least some subtidal area permanently submerged under the ocean. Such sites can also include littoral and terrestrial elements. All sites with coral reefs and seagrasses are recorded as marine.

Protected areas that are part of global agreements usually include sites that are listed under national protection systems. *Biosphere reserves* are representative of terrestrial and coastal/marine environments that have been internationally recognized under UNESCO's Man and the Biosphere Programme. They have been selected for their value to conservation and are intended to foster the scientific knowledge and skills necessary for improving the balance between people and nature, and for promoting sustainable development. Each reserve must contain a diverse, natural ecosystem large enough to be an effective conservation unit. Each biosphere reserve consists of a minimally disturbed core area for conservation and research, a buffer zone where traditional land uses, experimental ecosystem research, and ecosystem rehabilitation may be permitted, and a transition area that may expand over time when cooperation increases. Only the core area requires legal protection. Biosphere reserves are nominated by national governments and remain under the sovereign jurisdiction of the state where they are located. Reserves must meet a minimal set of criteria and adhere to a minimal set of conditions, set out in the Statutory Framework, before being admitted into the World Network of Biosphere Reserves. Several countries share biosphere reserves. These sites are counted only once in regional and world totals. The transition area, according to the Statutory Framework, does not have to be defined, therefore the area of the biosphere reserves presented in this table may not correspond exactly to the actual territory concerned.

World heritage sites represent areas of "outstanding universal value" for their natural features, their cultural value, or for both natural and cultural values. The table includes only natural and mixed natural/cultural sites. Any party to the World Heritage Convention (adopted in 1972) may nominate sites. Natural sites contain "examples of a major stage of the Earth's evolutionary history; outstanding physical, biological, and geological processes; a unique or superlative natural phenomenon, formation, or feature; a habitat for a threatened species; or areas of value on scientific or aesthetic grounds, or from the point of view of conservation." Cultural sites contain "monuments, groups of buildings or sites of historical, aesthetic, archaeological, scientific, ethnological, or anthropological value." Parties to the Convention commit to the conservation of the listed sites. Several countries share world heritage sites. These sites, referred to as international heritage sites, are counted only once in continental and world totals, although they are tallied in each country that shares that site.

The Convention on Wetlands of International Importance Especially as Waterfowl Habitat (Ramsar, Iran, 1971) is an intergovernmental treaty for the conservation and sustainable use of wetlands. When a country becomes a party to the Convention, it agrees to designate at least one wetland for inclusion in the *List of Wetlands of International Importance* (the "Ramsar list") and to promote its conservation. Ramsar sites are wetlands with "international significance in terms of ecology, botany, zoology, limnology, or hydrology." Of the sites designated in 1976 by the former Soviet Union as Ramsar sites, three are now in the Russian Federation, four are in Ukraine, and one is in Estonia. Of the remaining five sites, which are listed by Ramsar as former Soviet Union sites, one is in Azerbaijan, two are in Kazakhstan, one is in Kyrgyzstan, and one is in Turkmenistan. The Ramsar Bureau is still waiting for confirmation from the members of the Commonwealth of Independent states of their status as parties to the Convention; therefore, in Data Table BI.1 these sites are included in Europe's total. Sites in Monaco and Liechtenstein are also included in Europe's total.

Some internationally protected areas simultaneously encompass areas protected under other systems (e.g., national parks or nature reserves) and other internationally recognized sites. For example, there are 54 biosphere reserves that are wholly or partially world heritage sites; 52 sites are inscribed as both Ramsar sites and biosphere reserves, and 20 sites are inscribed in both the Ramsar and the World Heritage lists. Listings of these common sites are available online at: http://www.unesco.org/mab/BR-WH.htm, http://ramsar.org/world_heritage.htm, and http://ramsar.org/mab_sites .htm.

Data Table BI.2
Globally Threatened Species: Mammals, Birds, and Reptiles

Sources: Total and endemic species of mammals, birds, and reptiles: World Conservation Monitoring Centre (WCMC) Species Database, unpublished data (WCMC, Cambridge, U.K., December 1999). Threatened species of mammals, birds, and reptiles: World Conservation Union (IUCN), *1996 IUCN Red List of Threatened Animals* (IUCN, Gland, Switzerland, 1996). Number of species per 10,000 square kilometers: calculated by the World Resources Institute using land area figures from the Food and Agriculture Organization of the United Nations (FAO), *FAOSTAT On-line Statistical Service* (FAO, Rome, 1999), available online at: http://www.fao.org. World total for number of known mammal species comes from D.E. Wilson, and D.M. Reeder (eds.), *Mammal Species of the World: A Taxonomic and Geographic Reference* (2nd Edition) (Smithsonian Institution Press, Washington, D.C., 1993). World total for number of known bird species comes from C.G. Sibley and B. L. Monroe, *Distribution and Taxonomy of Birds of the World* (Yale University Press, New Haven, 1993) and from C.G. Sibley and B. L. Monroe, *A Supplement to Distribution and Taxonomy of Birds of the World* (Yale University Press, New Haven, 1993). World total for number of known reptile species was estimated by WCMC based on the latest taxonomic references for each reptilian group. Currently there is no standard species checklist for reptiles.

The *total number of known species* may include introductions in some instances. Data on mammals exclude marine mammals except where otherwise indicated. Total bird species listed includes only birds that breed in that country, not those that migrate or winter there.

The number of *endemic species* refers to those species known to be found only within the country listed.

Figures are not necessarily comparable among countries because taxonomic concepts and the extent of knowledge vary from one country to another. For this reason, country totals of species and endemics may be underestimates. In general, numbers of mammals and birds are fairly well known, while reptiles have not been as well inventoried.

The number of *threatened species* listed for all countries includes full species that are Critically Endangered, Endangered, or Vulnerable, but excludes introduced species, species whose status is insufficiently known (categorized by IUCN as "data deficient" [DD]), those known to be extinct, and those for which status has not been assessed (categorized by IUCN as "not evaluated" [NE]). Threatened species data for animals presented in Data Tables BI.2 and BI.3 reflect estimates presented in the *1996 IUCN Red List of Threatened Animals.* BirdLife International compiles and maintains the bird species section of the IUCN *Red List*. Threatened species data for birds are from *Birds to Watch 2* (Collar et al. BirdLife International, Cambridge, U.K., 1994). The number of threatened species of birds are listed for countries included within their breeding or wintering ranges. Threatened marine turtles are excluded from country totals.

The *1996 IUCN Red List of Threatened Animals* was the first time that all *known* mammal and bird species had been assessed. This was also the first time that the new IUCN threat categories and criteria were applied to assess the risk of extinction of species. A brief description of the new system follows. For more detailed information, please refer to the original source.

Definitions for IUCN threat categories:

Critically Endangered: "When a taxon is facing an extremely high risk of extinction in the wild in the immediate future as defined by any of the criteria A–E" (see below).

Endangered: "When a taxon is not Critically Endangered but is facing a very high risk of extinction in the wild in the near future as defined by any of the criteria A–E."

Vulnerable: "When a taxon is not Critically Endangered or Endangered but is facing a high risk of extinction in the wild in the medium-term future as defined by any of the criteria A–E."

For each threat category there are five criteria (A–E, see below) used to assess species status. This process provides a more rigorous approach from that used prior to revisions of the IUCN criteria. Species need to meet only one of the five criteria to be listed under that particular threat category. The five criteria are:

A–Declining population rate,
B–Small population and decline or fluctuation,
C–Small population size and decline rate,
D–Very small population/very restricted distribution, and
E–Quantitative analysis indicating the probability of extinction in the wild (e.g., Population Viability Analysis).

In addition, there are subcriteria that provide further information on the reasons to list a species, potential causes of threat, etc.

Number of species per 10,000 km² provides a relative estimate for comparing numbers of species among countries of differing size. Because the relationship between area and species number is nonlinear (i.e., as the area sampled increases, the number of new species located decreases), a species-area curve has been used to standardize these species numbers. The curve predicts how many species a country would have, given its current number of species, if it was a uniform 10,000 square kilometers in size. This number is calculated using the formula: $S = cA^z$, where S = the number of species, A = area, and c and z are constants. The slope of the species-area curve is determined by the constant z, which is approximately 0.33 for large areas containing many habitats. This constant is based on data from previous studies of species-area relationships. In reality, the constant z would differ among regions and countries, because of differences in species' range size (which tend to be smaller in the tropics) and differences in varieties of habitats present. A tropical country with a broad variety of habitats would be expected to have a steeper species-area curve than a temperate, homogenous country because one would predict a greater number of species per unit area. Species-area curves also are steeper for islands than for mainland countries. At present, there are insufficient regional data to estimate separate slopes for each country.

Data Table BI.3
Globally Threatened Species: Amphibians, Freshwater Fish, and Plants

Sources: Total and endemic species data: World Conservation Monitoring Centre (WCMC) Species Database, unpublished data (WCMC, Cambridge, U.K., December 1999). Threatened species of amphibians and freshwater fish: World Conservation Union (IUCN), *1996 IUCN Red List of Threatened Animals* (IUCN, Gland, Switzerland, 1996). Threatened species of plants and world total for plants: WCMC Species Database, unpublished data (WCMC, Cambridge, U.K., December 1999). These figures are generated from the dataset underlying the data published in K.S. Walter and H.J. Gillett (eds.), *1997 IUCN Red List of Threatened Plants* (IUCN, Gland, Switzerland and Cambridge, U.K., 1998) compiled by WCMC. Number of species per 10,000 square kilometers: calculated by the World Resources Institute using land area figures from the Food and Agriculture Organization of the United Nations (FAO), *FAOSTAT On-line Statistical Service* (FAO, Rome, 1999), available online at: http://www.fao.org. World total for number of known amphibian species comes from D.R. Frost (ed.), *Amphibian Species of the World: A Taxonomic and Geographical Reference* (Allan Press, Inc. and the Association of Systematic Collections, Lawrence, Kansas, 1985), and from W.E. Duellman, *Amphibian Species of the World: Additions and Corrections* (University of Kansas Museum of Natural History, Special Pub. No. 21, Lawrence, 1993). World total for number of known fish species comes from W.N. Eschmeyer, C.J. Ferraris, M. Dang Hoang, and D. Long, *Catalog of Fishes, Vols. 1–3* (California Academy of Sciences, San Francisco, 1998) available online at: http://www.calacademy.org/research /ichthyology/ catalog/index.html.

The *total number of known species* may include introductions in some instances. *Higher plants* includes estimates for flowering plants, conifers and cycads, and ferns and fern-allies.

The number of *endemic species* refers to those species known to be found only within the country listed. The world total for the number of known freshwater fish species also includes marine species. Of this total, around 40–45 percent are estimated to be freshwater species. Most marine fish are excluded from country totals.

Figures are not necessarily comparable among countries because taxonomic concepts and the extent of knowledge vary from one country to another. For this reason, country totals of species and endemics may be underestimates.

The number of *threatened amphibian* and *freshwater fish species* listed for all countries follow the *1996 IUCN Red List of Threatened Animals* classification. For details, please refer to the Sources and Technical Notes for Data Table BI.2. Country totals for the threatened freshwater fish category may include some marine species. Even though marine species are poorly represented in the IUCN *Red List*, more than 100 marine fish species were assessed for threat; the remaining 14,000 species of known marine fish were not assessed. Similarly, for many regions of the world, most freshwater fish species have not been assessed. The IUCN *Red List* was the first time that the *Red List* criteria were applied to marine fish. These figures should, therefore, be taken as a preliminary assessment that needs further evaluation.

The number of *threatened species of higher plants* listed includes ferns and fern allies, conifers and cycads, and flowering plants that have been classified as threatened by IUCN following the pre-1994 categories, which have since been revised (see below). Threat categories for plant species, therefore, vary from those used to classify animals. The number of threatened species listed includes full species that are categorized by IUCN as being endangered, vulnerable, rare, or indeterminate. It excludes introduced species, species whose status is insufficiently known, those known to be extinct, those that are suspected to have recently become extinct, and those whose status has not been assessed. Threatened species data in this table reflect a subset of those estimates presented in the *1997 IUCN Red List of Threatened Plants*. This subset omits subspecies and includes only species that are recorded as occurring in one country. "Country" is used in a general sense; for example, figures are provided for overseas dependencies (e.g., Aruba). This subset of 25,971 records includes 83 percent of the 31,195 records published in the *1997 IUCN Red List*. It was the first published list of vascular plants recorded as globally threatened.

A brief description of the pre-1994 threat categories follows. For more detailed information, please refer to the original source.

IUCN classified threatened species of plants as "all full species categorized at the global level as Endangered, Vulnerable, Rare, or Indeterminate." The definitions for these categories follow:

Endangered: "taxa in danger of extinction and whose survival is unlikely if the causal factors continue operating." Included in this category are taxa that "may be extinct but have definitely been seen in the wild in the past 50 years."

Vulnerable: "taxa believed to move likely into the 'Endangered' category in the near future if the causal factors continue operating."

Rare: "taxa with small world populations that are not at present 'Endangered' or 'Vulnerable', but are at risk."

Indeterminate: "taxa that are known to be 'Endangered,' 'Vulnerable,' or 'Rare,' but where there is not enough information to say which of the three categories is appropriate."

The *number of species per 10,000 km²* provides a relative estimate for comparing numbers of species among countries of differing size. For details, refer to the Sources and Technical notes for Data Table BI.2.

Data Table BI.4
Trade in Wildlife and Wildlife Products Reported by CITES

Source: Trade data: Convention on International Trade in Endangered Species of Wild Fauna and Flora (CITES) annual report data, World Conservation Monitoring Centre (WCMC) CITES Trade Database, (WCMC, Cambridge, U.K., December 1999).

The international trade in wildlife and wildlife products, worth billions of dollars annually, causes serious declines in the numbers of many species of animals and plants. The degree to which species are being affected by the overexploitation for trade aroused such concern for the survival of species that CITES was drawn up in 1973 to protect wildlife against such overexploitation and to prevent international trade from threatening species with extinction. The Convention entered into force on July 1, 1975. It is one of the world's largest conservation agreements in existence, with 151 member states as of February 2000.

CITES entered into force is the year the Convention entered into force in that particular country. Members of CITES agree to ban the commercial international trade in an agreed-upon list of endangered species and to monitor trade in species that might become endangered. Species are listed in appendixes to CITES on the basis of their degree of rarity and the threat posed by trade. International trade in species listed in the appendixes or products derived from the species requires export, import, and reexport permits or certificates.

A brief description of the appendixes and permit regulation follows.

1. The most endangered species are listed in Appendix I. This appendix includes "all species threatened with extinction which are or may be affected by trade." International trade in these species is subject to particularly strict regulations and must be authorized by the corresponding National CITES Management and Scientific Authorities. Trade permits for these species are only granted under exceptional circumstances.
2. Other species at serious risk are included in Appendix II. This appendix includes "all species which although not necessarily currently threatened with extinction may become so unless trade is subject to strict regulation." Any international trade in Appendix II species requires export and import permits as well as reexport certificates, granted in accordance with conditions set forth in the Convention. Permits are granted by the National CITES Management and Scientific Authorities.
3. The species listed in Appendix III are "all species which any Party identifies as being subject to regulation within its jurisdiction for the purpose of preventing or restricting exploitation. The cooperation of other Parties, is therefore, needed." Any international trade in Appendix III species requires an export permit, a certificate of origin, and sometimes a reexport certificate. Permits and certificates are granted by the National CITES Management and Scientific Authorities in accordance with conditions set forth in the Convention.

Additional information on conditions and trade regulations can be found online at: http://www.cites.org/CITES/eng/text.shtml#II.

Parties to the Convention are required to submit annual reports, including trade records, to the CITES Secretariat. These trade records are compiled by WCMC in the CITES Trade Database. *CITES reporting requirements met as of 1997* indicates the number of reports submitted by a country to the CITES Secretariat as a percentage of number of reports expected since it became a party to the Convention. Figures presented are for reports submitted for years up to 1997.

Countries that have ratified CITES after 1997, therefore, are marked with "X" under reporting requirements met.

Net trade in 1997 is the balance of imports minus exports. Exports are shown as a negative balance in parentheses. Figures are for trade reported in 1997. Data on net exports and net imports as reported by CITES correspond to legal international trade and are based on data from permits issued, not on actual items traded. Figures may be overestimates if not all permits are used that year. Some permits issued in one year are used at a later date; therefore, numbers of exports and imports may not match exactly for any given year. World totals show the total number of imports, since calculating the balance of trade for the world would have canceled most figures.

Species traded within national borders and illegal trade in wildlife and wildlife products are not reflected in these figures. Illegal trade in wildlife products is estimated to be in the billions of dollars annually. In addition, data on mortality of individuals during capture or collection, transit, or quarantine are also not reflected in these numbers.

Number of live primates includes all species of monkeys, apes, and prosimians listed under CITES that were traded live in 1997.

Number of live parrots includes all *Psittaciformes* species listed under CITES that were traded live in 1997.

Number of live tortoises includes all *Testudines* species listed under CITES that were traded live in 1997.

Number of live lizards includes individuals from the *Sauria* and *Rhynchocephalia* species listed under CITES that were traded live in 1997.

Number of live snakes includes individuals from all *Serpentes* species listed under CITES that were traded live in 1997.

Number of wild orchids includes individuals from all *Orchidaceae* species listed under CITES that were collected from the wild and traded in 1997.

Number of cat skins includes whole skins of all *Felidae* species listed under CITES that were traded in 1997.

Number of crocodile skins includes whole skins of all *Crocodylia* species listed under CITES that were traded in 1997.

Number of lizard skins includes whole skins of all *Sauria* and *Rhynchocephalia* species listed under CITES that were traded in 1997.

Number of snake skins includes whole skins of all Serpentes species listed under CITES that were traded in 1997.

Data Table FG.1
Forest Cover and Change, and Certified Forest Area

Sources: Data on natural forest extent for 1990: Food and Agriculture Organization of the United Nations (FAO), Forest Resources Division, unpublished data (FAO, Rome, 1997). Data for total forest and plantation extent for 1990: *Forest Resources Assessment 1990: Global Synthesis* (FAO, Rome, 1995). Data for total and natural forest extent for 1995: *State of the World's Forests 1999* (FAO, Rome, 1999). Data for extent of plantations in 1995: D. Pandey, *Tropical Forest Plantation Areas 1995* (Report to FAO project GCP/INT/628/UK, Draft, November 1997). Average annual percentage change was calculated by the World Resources Institute. Certified forests data are from *Forest Stewardship Council (FSC) On-line Data Service*, available online at: http://www.fscoax.org (FSC, Oaxaca, Mexico, February 2000).

Total forest consists of all forest area for temperate developed countries, and the sum of natural forest and plantation area categories for tropical and temperate developing countries. For developed countries, forest cover data was obtained from official sources in response to a questionnaire from the United Nations Economic Commission for Europe and the FAO liaison office in Geneva (UN-ECE/FAO). The countries surveyed include countries in Europe, North America, and the former Soviet Union, as well as Japan, Australia, and New Zealand. Forest areas in developed countries are defined as land where tree crowns cover 20 percent of the area, including: open forests, forest roads and fire breaks; temporarily cleared areas, young stands expected to achieve at least 20 percent crown cover upon maturity, etc. Forest areas in developed countries are not broken down into the subcategories of natural forests and plantations due to the difficulty in distinguishing the two in many countries.

FAO defines a *natural forest* in tropical and temperate developing countries as a forest composed primarily of indigenous (native) tree species. Natural forests include closed forest, where trees cover a high proportion of the ground and where grass does not form a continuous layer on the forest floor (e.g., broadleaved forests, coniferous forests, and bamboo forests), and open forest, which FAO defines as mixed forest/grasslands with at least 10 percent tree cover and a continuous grass layer on the forest floor. Natural forests in tropical and temperate developing countries encompass all stands except plantations and include stands that have been degraded to some degree by agriculture, fire, logging, and other factors. For all regions, trees are distinguished from shrubs on the basis of height. A mature tree has a single well-defined stem and is taller than 7 meters. A mature shrub is usually less than 7 meters tall.

Average annual percent change reflects the increase or decrease in forest cover between 1990 and 1995. It is shown as a percentage of the exponential growth rate. If negative (in parentheses), these figures reflect net deforestation, which is defined as the clearing of forest lands for all forms of agricultural uses (shifting cultivation, permanent agriculture, and ranching) and for other land uses such as settlements, other infrastructure, and mining. In tropical countries, this entails clearing that reduces tree crown cover to less than 10 percent. It should be noted that deforestation, as defined here, does not reflect changes within the forest stand or site such as selective logging (unless the forest cover is permanently reduced to less than 10 percent). Such changes are termed "forest degradation" and they can substantially affect forests, forest soil, wildlife and its habitat, and the global carbon cycle. Thus, the effects from the reported deforestation figures may be less than the effects from the total deforestation that includes all types of forest alterations. Positive change figures reflect net afforestation within a country or region.

FAO's forest assessments produce consistent estimates on forest status for common reference years (1990 and 1995) and forest area change for the periods between these years. The estimates are made using a model to adjust baseline forest inventory data from each country to the common reference years. This model correlates the share of forest cover for each subnational unit to population density and growth, initial forest extent, and ecological zone. Existing forest inventory data at national and subnational levels are reviewed, adjusted to a common set of classifications and concepts, and combined in a database. To accomplish this, FAO uses a geographic information system to integrate statistical and map data. The reliability of these modeled estimates hinges partly on the quality of the primary data sources feeding into the model. The variation in quality, comprehensiveness, and timeliness of the forest information is tremendous, and acute information deficits as regards forest resources can easily be observed, particularly in developing countries. On average, the forest cover figures for most developing countries for the 1990 assessment were close to 10 years old. For the latest update (1995), forest inventory data additional to that available for the 1990 assessment was used for Bolivia, Brazil, Cambodia, Côte d'Ivoire, Guinea-Bissau, Mexico, Papua New Guinea, the Philippines, and Sierra Leone.

Although the deforestation model allowed standardization of country data to a common baseline, a number of additional factors may have contributed to discrepancies in forest area and change estimates for specific countries. Potential forest cover estimates for dry forests and the related adjustment function are of unknown reliability; in addition, for some countries, socioeconomic factors such as livestock projects in Central America and resettlement schemes in Indonesia may have played a larger role in deforestation. FAO acknowledged these shortcomings implicitly and noted that country estimates are not intended to replace the original country information, which remain unique sources of reference.

Because of these shortcomings, readers are encouraged to refer to the original sources and the latest country inventories that use satellite data or extensive ground data for estimates of forest cover and deforestation.

FAO is currently finalizing the *Global Forest Resource Assessment 2000*, which is expected to be published in the year 2000. This assessment is expected to contain country profiles on its forest resources for every country as well as a new global overview and maps with forest cover and change. Readers are encouraged to check this new report, which will be available online at: http://www.fao.org.

Plantations refers to forest stands established artificially by afforestation and reforestation for industrial and nonindustrial usage. Reforestation does not include regeneration of old tree crops (through either natural regeneration or forest management), although some countries may report regeneration as reforestation. Many trees are also planted for nonindustrial uses, such as village wood lots. Nonindustrial plantations include those established for fuelwood production, soil protection, or other purposes. They do not include plantations of agroforestry crops, such as rubber and oil palm. Reforestation data often exclude this component. The data presented here reflect plantation survival rates as estimated by FAO. Plantation figures correspond to reported areas, that is, areas reported to be present either by government, industry, or some other outside source. Reported areas are usually "accumulated planted areas" and may be different from the actual plantation areas that exist on the ground. Plantation areas in developed countries, with a couple of exceptions, are not presented in this table because of the difficulty in distinguishing between natural forests and plantations in many countries.

The data for *plantations extent 1995* cover the areas of forest plantations of all species in 90 tropical and subtropical countries. Tropical countries are defined as those countries having more than 50 percent of their land area between the Tropic of Cancer and Tropic of Capricorn, whereas subtropical countries are those bordering the tropics.

The Forest Stewardship Council (FSC) is an international nonprofit organization founded in 1993 to support "environmentally appropriate, socially beneficial, and economically viable management of the world's forests." It is the only independent third-party certification body for forests and forest products. Membership in FSC consists of representatives from social and environmental groups, timber trade and forestry professionals, indigenous peoples' organizations, community forestry groups, and forest product certification organizations. FSC accredits certification organizations or bodies, which then carry out the actual assessments of forest management.

FSC is introducing an international labeling scheme for forest products that will provide a credible guarantee that the product comes from a well-managed forest. According to FSC, "all forest products carrying their logo have been independently certified as coming from forests that meet the internationally recognized FSC Principles and Criteria of Forest Stewardship."

Certification can be granted only when the following FSC principles and criteria are met:

1. Forest management operations shall respect all applicable national laws and relevant international agreements signed by that country.
2. Long-term tenure and use rights to the land and forest resources are clearly defined, documented, and legally established.
3. The legal and customary rights of indigenous peoples to own, use, and manage their lands, territories, and resources are recognized and respected.
4. Forest management operations shall maintain or enhance the long-term social and economic well-being of forest workers and local communities.
5. Forest management operations shall encourage the efficient use of the forest's multiple products and services to ensure economic viability and a wide range of environmental and social benefits.
6. Forest management operations shall conserve biological diversity and its associated values as well as maintain the ecological functions and the integrity of the forest.
7. Forest management operations need to have a documented management plan with clearly stated goals and objectives.
8. Forest condition, yields, and management activities, as well as social and environmental impacts, must be monitored.
9. Management activities in high-conservation-value forests shall maintain or enhance the attributes that define such forests.
10. In addition to the principles stated above, plantations need to meet Principle 10 and its criteria, by which: Plantations should complement the management of, reduce pressures on, and promote the restoration and conservation of natural forests.

Full certification of forest products involves an assessment of forest management at the site to ensure that management practices meet the criteria of resource sustainability and environmental service provision; it also involves a chain-of-custody certification, whereby the channels through which forest products travel, from certified forest to processor and distributors to the final consumer, are verified. Many companies that have certified forests also possess certified mills and are distributors of certified wood as well.

Certified *natural forests* are defined as areas where many of the principal characteristics and key elements of native ecosystems such as complexity, structure, and diversity are present, as defined by FSC-approved national and regional standards of forest management.

Certified *plantations* are areas lacking most of the principal characteristics and key elements of native ecosystems as defined by FSC-approved national and regional standards of forest stewardship, which result from the human activities of planting, sowing, or intensive silvicultural treatments. According to FSC, certified plantations should "decrease the pressures on natural forests; have diversity in composition in species and age classes; include, where possible, native over exotic species; serve to improve soil function, fertility and structure; and have some proportion of their area managed for the restoration of natural forest cover."

Certified forests labeled as *mixed forests* include areas with natural forests and plantations, areas with seminatural forests, and areas with mixed seminatural forests and plantations.

FSC is currently revising their classification into three forest categories: natural forests, plantations, and high-conservation-value forests. For more information, please refer to the original source.

Data Table FG.2
Forest Ecosystems and Threatened Tree Species

Sources: Land area: Food and Agriculture Organization of the United Nations (FAO), *FAOSTAT On-line Statistical Service,* available online at: http://www.fao.org. Closed forest data: D. Bryant, D. Nielsen, and L. Tangley, *The Last Frontier Forests: Ecosystems and Economies on the Edge* (World Resources Institute, Washington, D.C., 1997). Forest ecosystem data: S. Iremonger, C. Ravilious, and T. Quinton, "A statistical analysis of global forest conservation," in S. Iremonger, C. Ravilious, and T. Quinton (eds.), *A Global Overview of Forest Conservation CD-ROM* (World Conservation Monitoring Centre [WCMC] and Centre for International Forestry Research [CIFOR], Cambridge, U.K., 1997). Threatened tree species data: Tree Conservation Database, WCMC. This database is used to generate *The World List of Threatened Trees*, compiled by WCMC, S. Oldfield, C. Lusty, and A. MacKinven (eds.), World Conservation Press, 1998, available online at: http://www.wcmc.org.uk/trees/Background/country_stats.htm. *The World List of Threatened Trees* is a collaboration between individual scientists worldwide, expert opinions derived from three regional workshops, and many organizations–notably the WCMC and the Species Survival Commission of IUCN-The World Conservation Union (IUCN).

Original forest as a percent of land area refers to the estimate of the percentage of land that would have been covered by closed forest about 8,000 years ago assuming current climatic conditions, before large-scale disturbance by human society. This map of estimated forest cover was developed by WCMC based on numerous global and regional biogeographic maps. It overestimates where forests were in the northern boreal regions, particularly in Russia, because it considers tundra-forest transition zones as forest. The map of original forest cover is not a direct measure of forest cover; it depicts where forests might be expected to occur today in absence of humans, based on climate, topography, and other variables. For further information, please refer to: C. Billington, V. Kapos, M. Edwards, S. Blyth, and S. Iremonger, *Estimated Original Forest Cover Map–A First Attempt* (WCMC, Cambridge, U.K., 1996).

Current forests refers to estimated closed forest cover within the last 10 years or so (this varies by country). Only closed moist forests are depicted for Africa and Asia. Woodlands and shrublands are not included. Current forest cover figures were produced from *The World Forest Map* (WCMC, Cambridge, U.K., 1996). This map was developed by WCMC in collaboration with the World Wildlife Fund and CIFOR for the WRI forest frontiers assessment; it is based on country and regional maps derived from national and international sources.

Frontier forests are large, relatively intact forest ecosystems. They represent undisturbed forest areas that are large enough to maintain all of their biodiversity, including viable populations of wide-ranging species associated with each forest type. A frontier forest must meet the following criteria:

- It is primarily forested.
- It is large enough to support viable populations of all species associated with that forest type, even in the face of natural disasters (e.g., hurricanes, etc.).
- Its structure and composition are determined mainly by natural events (e.g., fires, etc.), and it remains relatively unmanaged by humans, although limited human disturbance by traditional activities is acceptable.
- In forests where patches of trees of different ages occur naturally, the landscape shows this type of heterogeneity.
- It is dominated by indigenous tree species.
- It is home to most, if not all, other plants and animals that typically live in this type of forest.

Percent frontier forest threatened refers to those frontier forests where ongoing or planned human activities such as logging, mining, and other large-scale disturbances will eventually degrade the ecosystem through such effects as species decline or extinction, and drastic changes in the forest's age structure. If continued, these effects would result in the violation of one of the above mentioned criteria.

The data on extent and area protected by forest type under the *forest ecosystems* categories are for 1996. WCMC carried out this analysis by overlaying forest and protected areas coverages in a geographic information system (GIS). The forest GIS coverages were created by WCMC using many national and regional data such as land cover, forest, or vegetation maps. The legends of these maps were harmonized into 15 tropical and 11 nontropical forest types, defined specifically for this study. In general, WCMC assumed that the land cover categories shown in the source maps were correct and translated the legends directly to the 25 type classes for the world without attempting to assess the accuracy of the source data.

The WCMC publication does not include shrub-dominated lands but does include areas with sparse tree cover. *Sparse trees and parkland* are natural forests where the tree canopy cover is between 10 and 30 percent, such as in the savannah and steppe regions of the world. The forest type categories were split between "tropical" and "nontropical." *Tropical forests* included all forests located between the Tropic of Cancer and Tropic of Capricorn. All other forests were put into the *nontropical* categories. Montane forests within the tropics that were classified in the source maps as "temperate" were registered in the "tropical forests" categories in this study. Most of the plantation forests in the world were not recorded as such in the WCMC study and, therefore, are included in the categories describing natural forests. For example, all forests in Europe are classified as "natural" forests.

Percent protected includes forest areas within the protected areas that are listed by IUCN as being within their management categories I–VI. Please refer to the original source or to the Sources and Technical Notes of Data Table BI.1 for a description of these management categories.

The *number of tree species threatened* listed for all countries includes full species that are categorized by IUCN as being Critically Endangered, Endangered, or Vulnerable. These categories correspond to the new IUCN threat definitions. For a description of these threat categories, please refer to the Sources and Technical Notes of Data Table BI.2. These figures present the results of the first survey of the conservation status of tree species worldwide, published in *The World List of Threatened Trees*. Figures are not necessarily comparable among countries because taxonomic concepts and the extent of knowledge vary. The world's tree flora is estimated to be around 100,000 species.

Some taxonomic groups of trees were not evaluated, such as tree ferns in the families *Cyatheaceae* and *Dicksoniaceae*, tree species in the cycad families *Cycadaceae* and *Zamiaceae*, and arborescent members of the *Cactaceae* family.

In addition to the threatened tree species listed in this data table, *The World List of Threatened Trees* has several appendixes that provide a list of:

1. 141 globally threatened Australian tree species,
2. 202 globally threatened Japanese tree species, and
3. 1,022 tree species classified as "threatened" under the *1997 IUCN Red List of Threatened Plants* that were not evaluated using the new IUCN categories. For a description of the pre-1994 IUCN categories, please refer to the Sources and Technical Notes of Data Table BI.3.

Data Table FG.3
Wood Production and Trade

Sources: Wood production and wood import and export values: Food and Agriculture Organization of the United Nations (FAO), *FAOSTAT On-line Statistical Service* (FAO, Rome, 1999), available online at: http://www.fao.org. Total exports of goods and services data: World Bank, *World Development Indicators 1999*, on CD-ROM (Development Data Group, World Bank, Washington, D.C., 1999).

Total roundwood production refers to all wood in the rough, whether destined for industrial or fuelwood uses. All wood felled or harvested from forests and trees outside the forest, with or without bark, round, split, roughly squared, or in other forms, such as roots and stumps is included. All production figures are national totals averaged over a 3-year period in thousands of cubic meters.

Wood fuel production covers all rough wood used for cooking, heating, and power production. Wood used in charcoal production, pit kilns, and portable ovens is included. FAO data include only wood from direct sources such as natural forests, plantations, and other wooded land, such as homesteads and roadsides. FAO data do not currently include wood fuel from indirect sources such as industrial by-products derived from primary and secondary wood industries, recovered sources (i.e., wood waste from construction sites demolition, packaging, etc.), and black liquor (derived from by-products of the pulp industry). Statistics on wood fuel production are often lacking for many countries. FAO wood fuel estimates are partly based on household consumption surveys dating from the 1960s and per capita consumption estimates from the 1980s. Estimates are updated in line with population growth, therefore caution is recommended when using these figures.

Industrial roundwood production comprises all roundwood products other than fuelwood and charcoal. It includes sawlogs or veneer logs, posts, pitprops, pulpwood, and other roundwood industrial products.

Wood-based panels includes the following commodities: veneer sheets, plywood, particleboard, and compressed or noncompressed fiberboard. Starting from 1995, compressed fiberboard has been disaggregated into hardboard and medium-density fiberboard, and noncompressed fiberboard has been relabeled "insulating board."

Wood production data are national totals averaged over a 3-year period in thousand cubic meters.

Paper and paperboard production includes newsprint, printing and writing paper, packaging paper, household and sanitary paper, and other paper and paperboard. Recovered paper is excluded. Figures are national totals averaged over a 3-year period in thousand metric tons.

Trade in forest products includes trade in industrial roundwood, wood fuel, sawnwood, wood-based panels, pulp, paper and paperboard, recovered paper, chips and particles, and wood residues. Import and export figures are national totals averaged over a 3-year period in millions of U.S. dollars. Imports are usually on a cost, insurance, and freight basis (i.e., insurance and freight costs are added in). Exports are generally on a free-on-board basis (i.e., not including insurance or freight costs). "In-transit" shipments are excluded wherever possible.

All wood production and trade data refer to both coniferous and nonconiferous species. FAO compiles forest products data from responses to annual questionnaires sent to national governments. Data from other sources, such as national statistical yearbooks, are also used. In some cases, FAO prepares its own estimates. FAO continually revises its data using new information; the latest figures are subject to revision.

Percent of total exports in 1997 is the export value of forest products as a percentage of all exports (i.e., all goods and other market services). The value of merchandise, freight, insurance, travel, and other nonfactor services is included. Factor and property income (formerly called factor services), such as investment income, interest, and labor income, is excluded.

For more information, please refer to the original source.

Data Table FG.4
Livestock Populations, Grain Consumed as Feed, and Meat Production

Sources: Livestock and meat production data: Food and Agriculture Organization of the United Nations (FAO), *FAOSTAT On-line Statistical Service* (FAO, Rome, 1999), available online at: http://www.fao.org. Feed data: Economic Research Service, U. S. Department of Agriculture (USDA), *Production, Supply, and Distribution View* (USDA, Washington, D.C., 1999). Data available online at: http://usda.mannlib.cornell.edu/data-sets/international/93002/.

Data on livestock include all animals in the country, regardless of the place or purpose of their breeding. Data are collected annually by FAO. The figures for an indicated year reflect data reported by countries for any day between October of the previous year and September of the year indicated. Estimates are made by FAO for countries that either do not report data or only partially report data.

Cattle figures include the common ox (*Bos taurus*), zebu, humped ox (*Bos indicus*), Asiatic ox (subgenus *Bibos*), and Tibetan yak (*Poephagus grunniens*). *Sheep* (*Ovis spp.*) figures include Uriel, Argali, Bighorn, Karakul, and Astrakhan. *Goat* (*Capra spp.*) figures include Hircus, Ibex, Nubiana, Pyrenaica, Tibetana, Kashmir, and Angora. *Equine* figures include horses (*Equus caballus*), asses (*Equus asinus*), and hybrids between horses and asses. *Buffalo* figures include Indian, Asiatic, pigmy, water buffalo (*Bubalus bubalus*, *B. arnee*, and *B. depressicornis*), African buffalo (*Syncerus spp.*), American bison (*Bison bison*), European bison (*B. bonasus*), and hybrids between bison and cattle. Wild bisons and buffaloes are excluded. *Camel* figures include the Bactrian camel (*Camelus bactrianus*) and Arabian camel (*C. dromedarius*).

Grains fed to livestock as percent of total grain consumption is calculated by dividing the total feed grain consumed by the total domestic grain consumed. Grains include wheat, rice (milled weight), corn, barley, sorghum, millet, rye, oats, and mixed grains. Grain consumption is the total domestic use during the local marketing year of the individual country. It is the sum of feed, food, seed, and industrial uses.

Meat production of *beef and veal* are from bovine animals, whether salted, in brine, dried, or smoked. *Meat production* of *sheep and goats* includes lambs and kids. These figures refer to fresh, chilled, and frozen meat, bone in or boneless.

FAO defines "meat" as the flesh of animals used for food. Meat production data normally include bone, but exclude meat that is unfit for human consumption. As reported by individual countries, these data may refer either to commercial production (meat entering marketing channels), inspected production (from animals slaughtered under sanitary inspection), or total production (the total of the above mentioned categories plus slaughter for personal consumption). Production figures relate to animals slaughtered within national boundaries regardless of their origin, whether indigenous or foreign.

For more information, please refer to the original source.

Data Table FG.5
PAGE Ecosystems: Area, Population, Carbon Stocks, and Protected Areas

Sources: Landcover: Global Land Cover Characteristics Database Version 1.2 (T.R. Loveland, B.C. Reed, J.F. Brown, D.O. Ohlen, Z. Zhu, L. Yang, J. Merchant. 2000. "Development of Global Land Cover Characteristics Database and IGBP DISCover from 1 km AVHRR data." *International Journal of Remote Sensing* 21 [6]:1303–1330). Available online at: http://edcdaac.usgs.gov/glcc/glcc.html. U.S. Geological Survey (USGS) Earth Resources Observation Systems (EROS) Data Center (USGS/EDC) 1-km Land Cover Characterization Database, Revisions for Latin America (USGS/EDC, Sioux Falls, SD, 1999). Country boundaries: Environmental Systems Research Institute (ESRI), *Digital Chart of the World CD-ROM* (ESRI, Redlands, CA, 1993). Urban area: National Oceanographic and Atmospheric Administration–National Geophysical Data Center (NOAA/NGDC) Stable Lights and Radiance Calibrated Lights of the World CD-ROM (NOAA/NGDC, Boulder, CO. View Nighttime Lights of the World database online at: http://julius.ngdc.noaa.gov:8080/production/html/BIOMASS/night.html. Population: Center for International Earth Science Information Network (CIESIN); Columbia University; International Food Policy Research Institute (IFPRI); and World Resources Institute (WRI) *Gridded Population of the World, Version 2 alpha*. (CIESIN, Columbia University, Palisades, NY, 2000). Available online at: http://sedac.ciesin.org/plue/gpw. Carbon stocks in vegetation: Based on a 1999 unpublished map by USGS/EDC that applied carbon density numbers from a previous study (J.S. Olson, J.A. Watts, and L.J. Allison, "Carbon in Live Vegetation of Major World Ecosystems" Report ORNL-5862 [Oak Ridge, Tennessee, Oak Ridge National Laboratory, 1983]) to a more recent global vegetation map (Global Land cover characteristics Database Versions 1.2 [Loveland, et al. 2000]). Carbon stocks in soils: N.H. Batjes. 1996. "Total Carbon and Nitrogen in the Soils of the World." *European Journal of Soil Science* 47:151–163 and N.H. Batjes and E.M. Bridges. 1994. "Potential Emissions of Radiatively Active Gases from Soil to Atmosphere with Special Reference to Methane: Development of a Global Database WISE" *Journal of Geophysical Research* 99 (D8):16479–16489. World Inventory of Soil Emission Potentials (WISE) data online at: http://www.isric.nl/WISE.html. Protected areas: Database of the World Conservation Monitoring Centre (WCMC), unpublished data (WCMC, Cambridge, U.K., May/August 1999).

This table provides summary statistics for ecosystems used in Chapter 2 of *World Resources 2000–2001*. All area estimates use a common global land cover characteristics database, which was initiated by the IGBP and produced by the USGS and the University of Nebraska-Lincoln. The land cover regions in the database based on interpretation of advanced very-high-resolution radiometer (AVHRR) satellite imagery consolidated into monthly global composites for the period April 1992 to March 1993.

The global land cover characteristics database identifies between approximately 130 and 260 seasonal land cover regions (SLCRs) per continent (e.g., 167 for South America and 205 for North America). Each SLCR represents an area with similar land cover associations and distinctive patterns of biomass production such as the onset, peak, and duration of greenness. As part of a broader global land cover characterization process, seven different thematic global maps, each with a separate land cover legend, were produced by aggregating these detailed SLCR map units. These seven maps have corresponding customized legends to make them useful for specific global applications such as environmental modeling, land management, and monitoring. Pilot Analysis of Global Ecosystems (PAGE) researchers worked directly from the SLCR units to delineate the extent of agricultural ecosystems. They used two of the seven global legends to delineate forest and grassland boundaries and to calculate carbon storage: 1) the IGBP legend, developed to assist global change investigations, and 2) the Olson legend, produced for carbon cycle studies.

IBGP land area is based on the IGBP legend, with the exception of grasslands. The 17 separate IGBP land cover classes were aggregated to four broad categories: grasslands, forests, agriculture, and other. *Grasslands* consist of areas classified as open shrubland, closed shrubland, woody savanna, savanna, and grassland under the IGBP legend. It also includes an additional land cover type, tundra, which is not explicitly defined in the IGBP legend. This class is defined here by the Olson legend and typically overlaps with the IGBP shrubland, barren, and snow/ice classes. *Forests* include the IGBP categories for evergreen needleleaf forest, evergreen broadleaf forest, deciduous needleleaf forest, deciduous broadleaf forest, and mixed forest. *Agriculture* aggregates the IGBP categories for cropland and cropland/

natural vegetation mosaic. *Other* includes the IGBP categories for wetlands, snow/ice, urban areas, and barren land.

While the IGBP legend includes a specific category that depicts urban areas, summarized under the category "other" in the table, the data used to delineate urban areas (*Digital Chart of the World*) are out-of-date and underestimate urban extent. Thus, PAGE researchers used a different data set to delineate *urban area*. It is based on the Nighttime Lights of the World database, a 1-kilometer resolution map derived from nighttime imagery from the Defense Meteorological Satellite Program (DMSP) Operational Linescan System (OLS) of the United States. The data set contains the location of stable lights, including frequently observed light sources such as fires and lightning. The extent of "lit" area may be slightly overestimated due to the sensor's resolution and factors such as reflection from water and other surface features. It is a good indicator of the spatial distribution of settlements and infrastructure, but should not be interpreted as a measure of population density. The mean settlement size required to produce enough light to be detected is much greater in developing countries than in industrialized countries because of differences in energy consumption. The urban area estimates result from an overlay of IGBP land area, as defined in the first column of this table, with the Nighttime Lights of the World database. Area estimates are summarized by grasslands, forests, agriculture, and other, and are used to define PAGE extent.

Agricultural mosaic area represents the area of the 30–40 percent agriculture class within the IGBP grassland and forest extent as defined by the PAGE area for agriculture (see definition below). It provides an indication of the overlap between IGBP forest and grassland area and the PAGE area for agriculture.

Boundaries for *PAGE area* are defined independently using different legends for grassland, forest, and agriculture.

PAGE area for grassland and forest uses the IGBP and Olson legends. The extent for PAGE grassland and forest were calculated by subtracting a map of urban area, as defined in the second column of this table, from a map of IGBP grassland and forest extent, as defined in the first column of this table.

PAGE area for agriculture is based directly on the SLCR map units. The classification system used in the original SLCR map units allows for opportunities to improve the data interpretation for agricultural purposes, and improve upon the thematic maps generally presented at a global level that use the IGBP and Olson legends. These global land cover interpretations do not explicitly recognize all occurrences of agriculture occupying a less than dominant share (60 percent) of a SLCR class.

In consultation with the USGS/EDC, the potential agricultural content of all 961 SLCRs defined globally was reassessed. For example, an area interpreted as containing more than 60 percent forest and classified as "deciduous broad-leaf forest" cover using the IGBP classification scheme might, upon closer inspection of its naming convention, contain an agricultural subcomponent (i.e., its detailed classification might describe it as "deciduous broad-leaf forest with cropland"). The reassessment aimed to identify all such occurrences of agriculture, even when they occurred as minor cover components, although this was limited by the SLCR naming convention which did not identify an agricultural component if it occupied less than 30 percent of the SLCR region.

This reassessment resulted in a global map with three primary agricultural cover categories showing agricultural area intensity at 30–40, 40–60, and greater than 60 percent agriculture. PAGE area for agriculture thus adds up all 1-kilometer by 1-kilometer cells of the land cover characteristics database for which agricultural area intensity is more than 30 percent. Relative to the IGBP classification, the PAGE classification expands the geographic extent of agriculture by including areas where agriculture is not the dominant land cover. In addition to the IGBP agricultural area, it includes about 6 percent of the IGBP forest area and 13 percent of the IGBP grassland area as shown in the second column of this table. Furthermore, PAGE area for agriculture includes urban areas (with at least 30 percent agriculture) because no explicit urban class was assigned in the SLCR map units. This makes the PAGE agriculture area not directly comparable to the PAGE grassland and forest extent, which subtracted urban areas.

The reinterpretation to obtain agricultural extent, however, does not address some weaknesses in the original land cover characteristics database such as regional variations in the reliability of the satellite data interpretation, reflecting differences in the structure of land cover and in the availability of reliable groundtruthing data. Specific agricultural land-cover types whose interpretation is considered to be problematic include irrigated areas, permanently cropped areas (especially tree crops in forest margins), and extensive pastureland. Similarly, the adjustments made to obtain a better estimate for urban areas do not overcome the intrinsic limitations of the original land cover characteristics database, namely the coarse resolution and lack of auxiliary data to fine tune the satellite image interpretation for selected regions (e.g., in Africa). Therefore, all area estimates are most useful in depicting the relative size of broad ecosystem categories rather than providing an absolute estimate of the world's surface covered by trees, grasses, and other vegetation types.

Population data came from an inventory of national censuses, which were compiled by administrative units. These data were standardized to 1995 and translated into a global grid consisting of 4.6-kilometer by 4.6-kilometer cells, each cell accurately reflecting population counts in the respective administrative units intersecting with this cell. The map of PAGE extent was scaled to match the resolution of the population data, i.e., a generalization from 1-kilometer by 1-kilometer to 4.6-kilometer by 4.6-kilometer cells. To calculate population by ecosystem category, population was assigned to the majority ecosystem type. For example, population in cells that had 51 percent forest and 49 percent grassland were allocated to the forest category.

All population estimates in the table are based on the boundaries of the PAGE extent. Population for the PAGE area for agriculture, however, excludes urban areas, as defined by the Nightime Lights of the World database, to be comparable with grassland and forest population estimates. The population estimates for total PAGE agriculture area, which includes urban areas, are as follows (in thousands): Agriculture Total - 4,002,386; Asia (excluding the Middle East) - 2,608,216; Europe - 616,663; Middle East and North Africa - 148,576; Sub-Saharan Africa - 232,742; North America - 179,263; Central America and the Caribbean - 40,527; South America - 171,869; and Oceania - 4,530. Because of overlapping areas between PAGE categories, population estimates in this table for PAGE cannot be added to produce a global total.

Carbon stocks represent total carbon amounts stored in both soil and vegetation. The estimates are based on two maps that use two global data sets, one of them modified and updated in line with best estimates of current vegetation cover.

The map of carbon stored in above- and below-ground live vegetation (1-kilometer by 1-kilometer resolution) is based on estimates developed by Olson et al. (1983). PAGE researchers worked in concert with USGS/EDC to apply Olson's estimates to the IGBP legend of land cover characteristics. While Olson's estimates of carbon storage in vegetation have, in some cases, been superseded by more recent studies at the national and regional level, they still represent the most commonly used global map of biomass carbon densities and provide the only consistent set of carbon storage estimates for vegetation types at this scale.

The map of carbon stored in soils (0.5 degree resolution) is based on the Global Data Set of Derived Soil Properties, compiled by the International Soil Reference and Information Centre (ISRIC) for the World Inventory of Soil Emission Potentials (WISE). The ISRIC-WISE data set also provides low and high estimates of carbon storage in different soil types based on soils with a median stone content and on so-called "stone-free" soils. The ISRIC estimates used here are those based, for the most part, on soil samples taken to a depth of 100 centimeters, the customary depth of measurement. One meter is the depth of soil believed to be most directly involved in interactions with the atmosphere and most sensitive to land use and environmental changes.

To calculate carbon stocks for this table, the two maps were first combined to produce a global map at a 0.5 degree resolution (requiring a generalization of the map showing carbon storage in vegetation) and then intersected with the PAGE area. The low and high estimates are based on combining low and high estimates of Olson et al. (above- and below-ground vegetation) and the ISRIC-WISE database (soil), respectively. The total carbon stored in terrestrial ecosystems ranges from 1,213 to 2,433 gigatons of carbon. (This number cannot be directly calculated from the table because of overlapping ecosystem categories.) While these results are consistent with other studies, the wide gap between the low and high estimates illustrates the major uncertainties in the underlying data.

Protected area represents the total area of each PAGE classification that falls within a protected area as designated by IUCN. A global map containing parks larger than 1,000 hectares and falling under IUCN management categories I-VI was produced for the analysis. For protected areas represented by points only, circular buffers corresponding to the size of the protected area were generated. This global map of protected areas was then intersected with the PAGE map and area estimates were summarized for each ecosystem category for all protected areas that had more than 50 percent in grassland, forest, and agriculture, respectively.

Data Table CMI.1
Marine and Freshwater Catches and Aquaculture Production

Sources: Marine catch, freshwater catch, and aquaculture production data: Food and Agriculture Organization of the United Nations (FAO), *Global Capture Dataset 1984–1997*, and *Aquaculture Quantities Dataset 1984–1997*, Fishery Statistics Databases, downloadable with *Fishstat-Plus* software at: http://www.fao.org/WAICENT/FAOINFO/FISHERY /statist/FISOFT/FISHPLUS.HTM.

Marine and *freshwater fish catch* data refer to marine and freshwater fish caught or trapped for commercial, industrial, and subsistence use (catches from recreational activities are included where available). Statistics for mariculture, aquaculture, and other kinds of fish farming are *not* included in the country totals. Marine fish includes demersal, pelagic, and diadromous fish caught in marine areas (i.e., sturgeons, paddlefish, river eels, salmons, trouts, smelt, shads, and miscellaneous diadromous fish). For a listing of groups of species under the demersal and pelagic categories, please refer to the Sources and Technical Notes of Data Table CMI.4. Freshwater fish includes fish caught in inland waters (i.e., carps, barbels, and other cyprinids; tilapias and other cichlids; and miscellaneous and freshwater fish) and diadromous fish caught in inland waters.

Mollusc and crustacean catch data refer to marine and freshwater molluscs and crustaceans caught or trapped for commercial, industrial, and subsistence use. Statistics for mariculture, aquaculture, and other kinds of shellfish farming are *not* included in the country totals. For a listing of species groupings in these categories, please refer to the Sources and Technical Notes of Data Table CMI.4.

Catch figures are the national totals averaged over a 3-year period; they include fish caught by a country's fleet anywhere in the world. Catches of freshwater species caught in low-salinity seas are included in the statistics of the appropriate marine area.

Data are represented as nominal catches, which are the landings converted to a live-weight basis, that is, the weight when caught. Fish catch does not include discards. Landings for some countries are identical to catches. Catch data are provided annually to the FAO Fisheries Department by national fishery offices and regional fishery commissions. Some countries' data are provisional for the latest year. If no data are submitted, FAO uses the previous year's figures or makes estimates based on other information. For details on data quality, please refer to the Sources and Technical Notes to Data Table CM.4.

Aquaculture is defined by FAO as "the farming of aquatic organisms, including fish, molluscs, crustaceans, and aquatic plants. Farming implies some form of intervention in the rearing process to enhance production, such as regular stocking, feeding, and protection from predators, etc. [It] also implies ownership of the stock being cultivated...." Aquatic organisms that are exploitable by the public as a common property resource are included in the harvest of fisheries.

FAO's global collection of aquaculture statistics by questionnaire to national fishery offices was begun in 1984. FAO's aquaculture database has 337 "species items" that are grouped into six categories. *Total aquaculture production* includes marine, freshwater, and diadromous fish, molluscs, crustaceans, and aquatic plants cultivated in marine, inland, or brackish environments. *Marine fish* include a variety of species groups such as mullets, seabasses, groupers, snappers, tunas, mackerels, etc. *Diadromous fish* include sturgeons, river eels, salmons, trouts, etc. *Freshwater fish* include carps, barbels and other cyprinids, tilapias and other cichlids, perches, catfish, etc. *Molluscs* include oysters, mussels, scallops, clams, abalones, cephalopods, and freshwater molluscs. *Crustaceans* include, among others, freshwater crustaceans, crabs, lobsters, shrimps, and prawns. *Aquatic plants* include brown, red, and green seaweeds, and miscellaneous aquatic plants. For a detailed listing of species, please refer to the original source.

Data Table CMI.2
Trade in Fish and Fishery Products, Fish Consumption, and Fishers and Fleet Information

Sources: Trade in fish and fishery products data: Food and Agriculture Organization of the United Nations (FAO), *Trade and Production Dataset 1976–1997*, Fishery Statistics Databases, downloadable with *Fishstat-Plus* software at: http:// www.fao.org/WAICENT/FAOINFO/FISHERY/statist/ FISOFT/FISHPLUS.HTM.

Food supply from fish and fishery products: E. Laureti, *Fish and Fishery Products: World Apparent Consumption Statistics Based on Food Balance Sheets (1961–1997)*, FAO Fisheries Circular No. 821, Rev. 5 (FAO, Rome, 1999). Data on fishers and fleets: FAO, Fishery Information, Data, and Statistics Unit (FIDI), December 1999.

Trade in *fish and fish products* include fish that are live, fresh, chilled, frozen, dried, salted, smoked, or canned, and other derived products and preparations. Trade in *molluscs and crustaceans* includes molluscs and crustaceans that are live, fresh, chilled, frozen, dried, salted, smoked, or canned, and other derived products and preparations. Trade in *meals and solubles* includes all meals and solubles from fish and other aquatic animals. Figures are the national totals averaged over a 3-year period in millions of U.S. dollars. *Exports* are generally on a free-on-board basis (i.e., not including insurance or freight costs). *Imports* are usually on a cost, insurance, and freight basis (i.e., insurance and freight costs added in).

Regional totals are calculated by adding up imports or exports of each country included in that region. Therefore, the regional totals should not be taken as a net trade for that region, since there may also be trade occurring within a region. To collate national data, FAO uses its International Standard Statistical Classification of Fishery Commodities. The commodities categories cover products derived from fish, crustaceans, molluscs, and other aquatic animals and residues caught for commercial, industrial, or subsistence uses. Commodities produced by aquaculture and other kinds of fish farming are also included.

Total food supply from fish and fishery products is defined as the quantity of both freshwater and marine fish, seafood, and derived products available for human consumption. Data are presented in thousands of metric tons in live weight (i.e., the weight of the fish and shellfish at the time of harvest or capture). Food supply figures were calculated by taking a country's fish production plus imports of fish and fishery products, minus exports, minus the amount of fishery production destined for nonfood uses (i.e., reduction to meal, etc.), and plus or minus variations in stocks, wherever possible.

Per capita food supply from fish and fishery products is defined as the estimate of the total supply available for human consumption, divided by the *de facto* population (i.e., those persons living within a country's borders or region). Data are expressed in live-weight equivalent.

Data on food supply from fish and fishery products represent apparent consumption on a live-weight basis, which means that the amounts of fish and fishery products consumed include all parts of the fish, including bones. The amount of fish and seafood actually consumed may be lower than the figures provided, depending on how much is lost during storage, preparation, and cooking, and how much is discarded.

Fish protein as a percent of all animal protein and *fish protein as a percent of all protein supply* are defined as the quantity of protein from both freshwater and marine fish, seafood, and derived products available for human consumption as a percentage of all *animal* protein and all protein available.

FAO compiles statistics on apparent consumption of fish and fishery products for 220 countries, based on Supply/Utilization Accounts (SUAs) maintained in FAOSTAT, the statistical component of the FAO World Agriculture Information Centre. An SUA contains an estimate of the supplies from different sources, which it matches against an estimate of how these supplies are used. A food balance sheet is derived from each SUA series for each primary and processed commodity. Currently FAO uses SUAs for eight groups of primary fisheries commodities and nine groups of processed products derived from fisheries commodities. Both official data from national fishery statistical offices and unofficial data from various sources are used to prepare the SUAs. For further information, please refer to the original source.

Number of fishers includes the number of people employed in commercial and subsistence fishing (both personnel on fishing vessels and on shore), operating in freshwater, brackish, and marine areas, and in aquaculture production activities. Data on the number of fishers are collected by FAO through annual questionnaires submitted to the national reporting offices of the member countries. When possible, other national or regional published sources are also used to estimate these figures. The numbers presented in this table are gross estimates. Many countries do not submit data on fishers or submit incomplete information; therefore, the quality of these data is poor. FAO recognizes that these statistics are incomplete and may not accurately reflect the current level of employment in the fishing sector. For further information on the variables and collection methodologies, please refer to the original source or to *Number of Fishers 1970–1996*, FAO Fisheries Circular No. 929, Rev. 1 (FAO, Rome, 1999).

FAO defines "fishery vessel[s]" as "mobile floating objects of any kind and size, operating in freshwater, brackish and marine areas, and used for catching, harvesting, searching, transporting, landing, preserving and/or processing fish, shellfish and other aquatic animals, residues and plants." Decked vessels are those that have a fixed structural deck covering the entire hull above the deepest operating waterline.

Number of decked fishery vessels includes trawlers, purse seiners, gill netters, long liners, trap setters, other seiners and liners, multipurpose vessels, dredgers, and other fishing vessels, as well as nonfishing vessels such as motherships, fish carriers, and fishery research vessels. Data on undecked vessels are being collected by FAO, but are not yet available. Recent international agreements such as the FAO Code of Conduct for Responsible Fisheries of 1995 may improve the collection and reporting of fishery fleet statistics.

Fleet data are collected by the FAO through questionnaires submitted to the national reporting offices of the member countries. Other national or regional published sources are also used to estimate fleet size. The flag of the vessel is used to assign its nationality, however, in many cases vessels are flagged in one country, while the ownership, landings, and trade resides with another nation. This approach is referred to as a "flag of convenience," and fishers or corporations use this approach to facilitate registration of a vessel (i.e., some countries have fewer registration restrictions), to gain access to fish in different Exclusive Economic Zones, or to avoid having to follow set fishing quotas in their own nation, among other reasons.

FAO recognizes that these fleet statistics are incomplete and may not accurately reflect the current world fishing capacity. These data may include vessels that are no longer in operation. The quality of the estimates varies because many countries lack the resources to adequately monitor and report on fleet size. For further information, please refer to the original source or to *Fishery Fleet Statistics, 1970, 1975, 1980, 1985, 1989–95*, Bulletin of Fishery Statistics No. 35 (FAO, Rome, 1998).

Data Table CMI.3
Coastal Statistics, Coastal Biodiversity, and Trade in Coral

Sources: Coastal length, area of continental shelf, territorial sea, claimed exclusive economic zone, and exclusive fishing zone: Figures were calculated by L. Pruett and J. Cimino, unpublished data, *Global Maritime Boundaries Database* (Veridian-MRJ Technology Solutions, Fairfax, Virginia, January 2000). Percentage of the population within 100 kilometers from the coast: Center for International Earth Science Information Network (CIESIN), World Resources Institute, and International Food Policy Research Institute, *Gridded Population of the World, Version 2 alpha* (Columbia University, Palisades, New York, 2000) available online at: http://sedac.ciesin.org/plue/gwp. Mangrove extent and mangrove species number: M. Spalding, F. Blasco, and C. Field (eds.), *World Mangrove Atlas* (International Society for Mangrove Ecosystems, Okinawa, Japan, 1997). Area of mangrove forests protected: S. Iremonger, C. Ravilious, and T. Quinton, "A Statistical Analysis of Global Forest Conservation," in S. Iremonger, C. Ravilious, and T. Quinton (eds.), *A Global Overview of Forest Conservation CD-ROM* (World Conservation Monitoring Centre [WCMC] and Centre for International Forestry Research, Cambridge, U.K., 1997). Species of seagrasses and coral genera: unpublished data (WCMC, Cambridge, U.K., August 1999). World total for number of seagrass species: J.E. Maragos et al. (eds.) *Marine and Coastal Biodiversity in the Tropical Island Pacific Region, Volume 1* (East-West Center, Honolulu, Hawaii, 1995). Trade in coral: Convention on International Trade in Endangered Species of Wild Fauna and Flora (CITES) annual report data, WCMC CITES Trade Database (WCMC, Cambridge, U.K., December 1999).

Coastal length was derived from the World Vector Shoreline database at 1:250,000 on the surface of a spheroid using proprietary Veridian-MRJ Technology Solutions software and a geographic information system (GIS). Because a higher resolution database of global shorelines does not exist, these figures should be interpreted as approximations and used with caution. The measurement of coastal length is scale dependent. Maps of individual islands, for example, frequently show great detail, whereas regional maps summarize complex coastlines into a few simple lines. Shoreline lengths are also affected by inclusion or exclusion of coastal features such as bays, lagoons, and river mouths. The only way to derive globally comparable statistics on coastline length is to use a single source with a constant scale. This is what has been attempted with the data presented in this table; however, highly complex coastlines will appear longer at higher resolutions. These estimates may differ from other published sources. In general, the coastline length of islands that are part of a country, but are not overseas territories, are included in the coastline estimate for that country (i.e., the Canary Islands are included in Spain). Coastline length for overseas territories and dependencies are not added to the total coastline of a country (i.e., Guam's coastline is not included in the coastal length for the United States). Disputed areas are not included in country or regional totals unless otherwise noted.

The United Nations Convention on the Law of the Sea (UNCLOS) is an international agreement that sets conditions and limits on the use and exploitation of the oceans. This convention also sets the rules on how the maritime jurisdictional boundaries of the different member states are set. UNCLOS was opened for signature on December 10, 1982 in Montego Bay, Jamaica, and it entered into force on November 16, 1994. As of January 2000, 132 countries have ratified UNCLOS.

Under UNCLOS, coastal states can claim sovereign rights over the national *area of* juridical continental shelf (seabed and subsoil) for exploration and exploitation. The claim on the continental shelf can extend beyond the territorial sea to a distance of more than 200 nautical miles from the territorial sea baseline and may include the physiographic continental shelf (to a depth of 200 meters), the continental slope, and the continental rise, and could potentially extend onto the abyssal plain. Areas of continental shelf that are disputed by overlapping claims by one or more nations have been excluded from this table. Areas of "cooperative joint development" between two or more nations have also been excluded. The data for the continental shelf presented in this table are for the physiographic continental shelf.

Territorial sea is defined under UNCLOS as the zone up to 12 nautical miles from the baseline or low-water line along the coast. The coastal state's sovereignty extends to the territorial sea, including its seabed, subsoil, and the air space above it. Foreign vessels are allowed "innocent passage" through those waters. Even though the established maximum limit for a territorial sea is 12 nautical miles, some countries claim larger areas. Disputed territorial seas due to overlapping claims are not included in this table unless specified with a footnote. Portions of the territorial sea claims in excess of UNCLOS guidelines are not included in the table.

Under UNCLOS, coastal states can claim sovereign rights over a 200-nautical mile exclusive economic zone (EEZ). This area is termed *claimed exclusive economic zone.* UNCLOS allows for sovereign rights over the EEZ in terms of exploration, exploitation, conservation, and management of all natural resources in the seabed, its subsoil, and overlaying waters. UNCLOS allows other states to navigate and fly over the EEZ, as well as to lay submarine cables and pipelines. The inner limit of the EEZ starts at the outer boundary of the territorial sea (i.e., 12 nautical miles from the low-water line along the coast). For cases in which a country's low-water lines are within 400 nautical miles of each other, the EEZ boundaries are generally established by treaty, though there are many cases in which these boundaries are in dispute. Some states have not ratified UNCLOS, and many have not yet claimed their EEZ. Where a country claims a territorial sea in excess of 12 nautical miles, the inner limit of the EEZ is calculated from the point at which the 12 nautical mile line would exist.

UNCLOS states that "land-locked and geographically disadvantaged States have the right to participate on an equitable basis in [the] exploitation of an appropriate part of the surplus of the living resources of the EEZ's of coastal States of the same region or sub-region."

The *exclusive fishing zone* refers to an area beyond the outer limit of the territorial sea (12 nautical miles from the coast) in which the coastal state has the right to fish, subject to any concessions that may be granted to foreign fishermen. Some countries have made no claim beyond the territorial sea, while others have claimed an exclusive fishing zone instead of the more encompassing EEZ.

Given the uncertainties surrounding much of the delimitation of the territorial seas, EEZ, and exclusive fishing zones, these figures should be used with caution.

Population within 100 km from the coast refers to estimates of the percentage of the population living within the coastal area based on 1995 population figures. These estimates were calculated using a dataset that provides information on the spatial distribution of the world's human population on a 2.5-minute grid. Populations are distributed according to administrative districts, which vary in scale, level, and size from country to country. A 100-kilometer coastal buffer was used in the GIS to calculate the number of people in the coastal zone for each country. The percentage of the population in the coastal zone was calculated from 1995 United Nations Population Division totals for each country.

Mangrove trees and shrubs are found along estuarine riverbanks and coastlines in tropical and subtropical countries. Their main characteristic is that they can tolerate salt and brackish water environments. Spalding et al. (1997) is the first compilation of data on *mangrove area* globally. Original data were compiled by WCMC from a variety of maps and other published sources including governments, mapping agencies, nongovernmental organizations, scientists, and international agencies. Data were incorporated into a GIS and area estimates were calculated from these GIS maps. Original published estimates of mangrove extent are presented when they are considered more accurate than GIS calculations. The year and quality of these data vary from country to country, therefore figures are not strictly comparable between countries. This is the first attempt to calculate mangrove extent globally, thus figures should be used with caution.

Area protected includes mangrove forest areas within the World Conservation Union-IUCN protected area Management Categories I-VI. For a description of these categories, please refer to the original source or to the Sources and Technical Notes of Data Table BI.1. WCMC carried out this analysis by overlaying forest and protected area coverages in a GIS. WCMC created the forest GIS coverages by using national and regional data such as land cover, forest, and vegetation maps. In general, WCMC assumed that the land cover categories shown in the source maps were correct and translated the legends directly to the forest type classes for the world without attempting to assess the accuracy of the source data. Documentation for the source data is given in full by Iremonger et al. (1997). *Number of species* includes mangrove species known to exist in each country.

Seagrasses are marine angiosperms that live in seawater. They are not true grasses, but have a grass-like appearance. They grow in soft substrates like sandy soils and form large underwater meadows in coastal regions of the world. All known seagrass species belong to only two families, and even though the total number of species worldwide is low–only 58–they play a key role in the functioning of the ecosystem by providing habitat, breeding, and feeding grounds for many species of fish and shellfish. Seagrasses are found both in tropical and temperate seas. *Number of seagrass species* includes seagrass species known to exist in each country.

Coral reefs are home to more than a quarter of all known marine fish species. In general, coral reefs are found in shallow waters, between the Tropic of Capricorn and the Tropic of Cancer. Most reef-forming corals belong to the family *Scle-*

ractinia, which are also called true or stony corals. They may be solitary or colonial and have a heavy external calcareous skeleton. Colonial species are restricted to shallow, clear waters in tropical seas, while solitary individuals can be found in deep waters and high latitudes. *Number of coral genera* includes only reef-building, colonial *Scleractinia* genera that are known to exist in each country.

The year and quality of the data on number of species and genera vary from country to country, therefore figures are not strictly comparable between countries.

International legal net trade in live coral includes pieces of coral species listed under CITES that were traded in 1997. Figures are the balance of imports minus exports. Exports are shown as negative balances in parentheses. CITES monitors the trade in more than 2,000 species of coral. The typical size of live coral pieces in trade is 10 by 6 centimeters in cross section, 6 centimeters in height, and weighing about 200 grams. For more information on trade in coral, please refer to E.P. Green and F. Shirley, *The Global Trade in Coral,* WCMC Biodiversity Series No. 9 (WCMC-World Conservation Press, Cambridge, U.K., 1999).

Data on net exports and net imports as reported by CITES correspond to legal international trade and are based on data from permits issued, not on actual items traded. Figures may be overestimates if not all permits are used that year. Some permits issued in one year are used at a later date, therefore numbers of exports and imports may not match up exactly for any given year. World totals show the total number of imports, as calculating the balance of trade for the world would have canceled most figures.

Species traded within national borders and illegal trade in wildlife and wildlife products are not reflected in these figures. Illegal trade in wildlife products is estimated to be in the billions of dollars annually. In addition, data on mortality of individuals during capture or collection, transit, or quarantine are also not reflected in these numbers. For more information on CITES, please refer to the Sources and Technical Notes of Data Table BI.4.

Data Table CMI.4
Marine Fisheries, Yield, and State of Exploitation

Sources: Marine fishery production data: Food and Agriculture Organization of the United Nations (FAO), *Global Production Dataset 1950–1997,* Fishery Statistics Database, downloadable with *Fishstat-Plus* software at: http://www.fao.org/WAICENT/FAOINFO/FISHERY/statist/FISOFT/FISHPLUS.HTM.

Fisheries status: R. J .R. Grainger and S. M. Garcia, *Chronicles of Marine Fishery Landings (1950–1994): Trends Analysis and Fisheries Potential,* FAO Fisheries Technical Paper No. 359 (FAO, Rome, 1996). Discards: D.L. Alverson, M.H. Freeberg, J.G. Pope, and S.A. Murawski, *A Global Assessment of Fisheries Bycatch and Discards,* FAO Fisheries Technical Paper No. 339 (FAO, Rome, 1994) and FAO, *Report of the Technical Consultation on Reduction of Wastage in Fisheries,* Tokyo, Japan, October 28–November 1, 1996, FAO Fisheries Report No. 547, p. 27 and Supplement, p.388 (FAO, Rome, 1996).

FAO divides the world into 27 major fishing areas (19 marine and 8 inland) and organizes annual *production* data by 1,167 "species items"–species groups separated at the family, genus, or species level. Sometimes items are classified at a higher taxonomic level than the family (e.g., marine fish not elsewhere included are classified as *Osteichthyes*). Data Table CMI.4 only shows data for the marine fishing areas. The Antarctic total includes Antarctic portions of the Atlantic, Indian, and Pacific Oceans.

Average marine fish production data refer to marine fish caught or cultivated for commercial, industrial, and subsistence use (catches from recreational activities are included where available). Statistics for mariculture, aquaculture, and other kinds of fish farming are included. FAO species groupings are as follows: *demersal fish*: flounders, halibuts, soles, etc.; cods, hakes, haddocks, etc.; redfish, basses, congers, etc.; and sharks, rays, chimeras, etc; *pelagic fish*: jacks, mullets, sauries, etc.; herrings, sardines, anchovies, etc.; tunas, bonitos, billfish, etc.; and mackerels, snooks, cutlassfish, etc.

Average marine production includes demersal, pelagic, and miscellaneous marine fish (except diadromous fish); marine crustaceans (sea-spiders, crabs, etc.; lobsters, spiny-rock lobsters, etc.; squat lobsters; shrimps, prawns, etc.; krill, planktonic crustaceans, etc.; and miscellaneous marine crustaceans); and marine molluscs (squids, cuttlefish, octopuses, etc.; abalones, winkles, conchs, etc.; oysters; mussels; scallops, pectens, etc.; clams, cockles, arkshells, etc.; and miscellaneous marine molluscs). The years shown are 3-year averages.

Marine capture figures are average landings and do not include discards (see below). Data are represented as nominal catches, which are the landings converted to a live-weight basis, that is, the weight when caught. Catches of freshwater species caught in low-salinity seas are included in the statistics of the appropriate marine area.

Fish production data are provided annually to the FAO Fisheries Department by national fishery offices and regional fishery commissions. Some countries' data are provisional for the latest year. If no data are submitted, FAO uses the previous year's figures or makes estimates based on other information. The quality of the production estimates varies because many countries lack the resources to adequately monitor landings within their borders. In addition, fishers sometimes underreport their catches because they have not kept within harvest limits established to manage the fishery. In some cases, catch statistics are inflated to increase the importance of the fishing industry to the national economy.

Years are calendar years except for Antarctic fisheries data, which are for split years (July 1–June 30). Data for Antarctic fisheries are given for the calendar year in which the split year ends.

Status of the fisheries: fully fished (F), overfished (Ov), or increasing (I) provide a measure of the degree to which fish stocks within FAO's marine fishing areas were exploited as of 1994. FAO used landings data collected between 1950 and 1994 for the top 200 species-/fishing-area combinations to estimate status. These 200 species-area combinations represent 77 percent of the world's marine production. Exploitation levels were determined by comparing the average landings from 1990 to 1994 to the estimated maximum potential landings for each fishing area. FAO estimated the potential landings for each area based on a generalized fishery development model. This model shows that a fishery has four phases: undeveloped (the fishery is not being exploited), developing (the fishery starts to develop), mature (fishery yields are at their peak and their rate of increase is zero), and senescent (yields are decreasing). Based on this model, FAO estimated the year in which the mature phase of the development had been or would be reached (i.e., the rate of increase is zero), the year the resource would be *fully fished*. For more detailed information on the analysis, please refer to the original source.

Discards as a percentage of overall catch refers to the percentage of overall catch (discards plus landings) during the 1988–92 period that consisted of nontarget or low-value species and undersized fish of targeted species. Globally, marine fisheries bycatch (i.e., discarded catch plus incidental catch) has been estimated to account for some 20 million metric tons per year or 25 percent of global marine fish catch. In certain fisheries bycatch can outweigh the catch of target species. For example, in the shrimp capture fishery discards may outweigh the target species catch by a ratio of 5:1. For more information on bycatch and discards, please refer to the original sources or to the following publication: *The State of World Fisheries and Aquaculture 1998*, (FAO, Rome, 1999).

Data Table AF.1
Food and Agricultural Production

Source: Food and Agriculture Organization of the United Nations (FAO), *FAOSTAT On-line Statistical Service* (FAO, Rome, 1999).

Total and *per capita index of agricultural production* indicate how a given year's agricultural production compare to the average of the production in the base years 1989–91. The index portrays disposable output (i.e., feed and seed are deducted from total production) and is based on the sums of price-weighted quantities of agricultural and food production.

For a given year in a given country, the index is calculated by 1) multiplying the disposable output of each commodity (in terms of weight or volume) by the relevant 1989–91 average international commodity price; 2) summing the values of all commodities included in the topic of interest; and 3) dividing this sum of the values by the average sum for the years 1989–91 and multiplying by 100 to obtain the index number. The process is repeated for each year.

The regional and world index values for a given year are calculated by totaling the disposable outputs of all relevant countries for each agricultural commodity. Each of these aggregates is multiplied by a respective 1989–91 average international producer price and is then summed to give a total agricultural output value for that region or for the world in terms of average 1989–91 prices. This method reduces distortion caused by the use of international exchange rates.

The multiplication of disposable outputs with the 1989–91 unit value eliminates inflationary or deflationary distortions. In addition, however, the base period's relative prices among the individual commodities are also preserved. Especially in economies with high inflation, price patterns among agricultural commodities can change dramatically over time.

The agricultural production index includes all crop and livestock products originating in each country.

Average cereal crop yields includes yields of cereal crops for feed and seed. Cereals comprise all cereals harvested for dry grain, exclusive of crops cut for hay or harvested green. *Average production of cereals* includes cereal production for feed and seed.

Average roots and tuber crop yields covers all root crops grown principally for human consumption such as potatoes, cassava, yucca, taro, and yams; root crops grown principally for feed are excluded. *Average production of roots and tubers* includes production of the these crops.

Average meat production refers to production of meats excluding fish. All data shown relate to total meat production, from both commercial and farm slaughter. Data are given in terms of dressed carcass weight, excluding offal and slaughter fats.

Further information on these data and FAO are available online at: http://www.fao.org.

Data Table AF.2
Agricultural Land and Inputs

Sources: Food and Agriculture Organization of the United Nations (FAO), *FAOSTAT On-line Statistical Service* (FAO, Rome, 1999). Pesticide use: FAO, as reported by the World Bank, *World Development Indicators 1999,* on CD-ROM (Development Data Group, World Bank, Washington, D.C., 1999).

Cropland and *cropland per 1000 people* refer to land under temporary and permanent crops, temporary meadows, market and kitchen gardens, and temporarily fallow land. Permanent cropland is land under crops that does not need to be replanted after each harvest, such as cocoa, coffee, fruit trees, rubber, and vines. Human population data used to calculate *cropland per 1000 people* are for 1987 and 1997.

Irrigated land as a percentage of cropland refers to areas purposely provided with water, including land flooded by river water for crop production or pasture improvement, whether or not this area is irrigated several times or only once during the year.

Annual fertilizer use refers to nutrients in terms of nitrogen (N), phosphate (P_2O_5), and potash (K_2O). Fertilizer use is calculated using a trade balance approach. As nations sometimes increase or decrease their stocks of fertilizer in a given year, actual use may be larger or smaller than the figure given. If the sale of fertilizer stocks is paticularly large, there is the potential for a negative fertilizer use value. Most fertilizer use data are reported yearly for the period July 1–June 30. For information on which countries report their data in ways that differ from the July 1–June 30 year, please refer to the FAO website noted below.

Pesticide use refers to per hectare use or sale to the agricultural sector of substances that reduce or eliminate unwanted plants or animals, especially insects. They include major groups of pesticides such as insecticides, mineral oils, herbicides, plant growth regulators, bacteria and seed treatments, and other active ingredients.

Tractors generally refer to wheeled and crawler tractors used in agriculture. Garden tractors are excluded.

Further information on these data and FAO are available online at: http://www.fao.org.

Table AF.3
Food Security

Sources: Cereal production, food aid and trade, and calorie-supply data: Food and Agriculture Organization of the United Nations (FAO), *FAOSTAT On-line Statistical Service* (FAO, Rome, 1999). Population data: United Nations (U.N.) Population Division, *World Population Prospects, 1950–2050 (The 1998 Revision)*, on diskette, (U.N., New York, 1998). Underweight children data: United Nations Children's Fund (UNICEF), *State of the World's Children 1999* (UNICEF, New York, 1997), including information from the World Health Organization, Demographic and Health Surveys, and Multiple Indicator Cluster Surveys.

This table addresses national food security (rather than household food security within nations). The variables were chosen based on FAO's definition of food security and causes of food insecurity. FAO defines food security as access by all people at all times to the food needed for a healthy and active life. FAO considers the failure to achieve food security to be a result of: 1) low productivity in agriculture (please refer to Data Table AF.1 for information on yields), 2) high seasonal and year-to-year variability in food supplies, and 3) lack of economic opportunity to overcome the above problems (please refer to the economics data tables EI.1-EI.3).

Average per capita cereal production is a rough indicator of whether a country produces enough food to feed its own population. The indicator includes cereal production for feed and seed. Cereals comprise all cereals harvested for dry grain, exclusive of crops cut for hay or harvested green. The population figures used are the midyear estimates published by the U.N. Population Division.

Variation in domestic cereal production is an indicator of whether cereal production is stable enough to ensure a predictable food supply. The variation is the average annual deviation, in percent, from the mean production for the time span given.

Net cereal imports and food aid as a percent of total cereal consumption indicates whether countries are able to produce sufficient grain for domestic consumption. *Food aid as a percent of total imports* shows whether countries are able to purchase imported grain, and are, therefore, more food secure, or whether they must depend on aid. Both indicators are based on trade data. The figures reflect net imports; exports were subtracted from imports. Consumption was calculated by adding imports and received food aid to production, and then subtracting exports and donated food aid. Countries that export more grain than they import appear as negative figures.

Food aid refers to the donation or highly concessional sale of food commodities. Cereals include wheat, rice, coarse grains, bulgur wheat, wheat flour, and the cereal component of blended foods. Food aid data are reported by donor countries and international organizations. Donors that have provided food aid at irregular intervals or in very small quantities are not listed individually.

Average daily per capita calorie supply and *average daily per capita calories from animal products* indicate whether food production and imports provide enough calories to maintain human health. These indicators are estimates of total per capita calorie supplies from all food and from animal products, respectively. These values were calculated by applying appropriate factors for all primary and processed products. Per capita supplies are derived from the total food calorie supplies available for human consumption by dividing the quantity of calories by the total population actually partaking of the food supplies. Nationals living abroad are excluded, but foreigners living in the country are included. Adjustments are made wherever possible for part-time presence or absence, such as temporary migrants, tourists, and refugees. Per capita supply figures represent only the average supply available for the population as a whole and do not necessarily indicate what is actually consumed by individuals. Even if they are taken as approximations of per capita consumption, it is important to bear in mind that there could be considerable variation in consumption between individuals. In almost all cases, the population figures used are the midyear estimates published by the U.N. Population Division. Calorie supplies are reported in kilocalories (1 calorie = 4.19 kilojoules).

Percent of children underweight indicates whether a country has achieved sufficient food security to keep its population healthy. Percent of children underweight refers to children under 5 whose weight-for-age is below -2 standard deviations (for moderate underweight) or below -3 standard deviations (for severe underweight) from the median weight-for-age of the reference population.

Data Table FW.1
Freshwater Resources and Withdrawals

Sources: Water resources and withdrawal data: J. Margat, *Les eaux souterraines dans le monde* (Bureau de recherches géologiques et minières [BRGM], Département eau, Orléans, France, December 1990); J. Margat and D. Vallée, *Water Resources and Uses in the Mediterranean Countries* (Blue Plan, Sophia Antipolis, 1999); I.A. Shiklomanov, *Comprehensive Assessment of the Freshwater Resources of the World* (Stockholm Environment Institute, Stockholm, 1997); Organisation for Economic Co-Operation and Development (OECD), *OECD Environmental Data Compendium 1997* (OECD, Paris, 1997); Food and Agriculture Organization of the United Nations (FAO), *Irrigation in Africa in Figures, Water Reports No. 7* (FAO, Rome, 1995); FAO, *Irrigation Potential in Africa, Land and Water Bulletin No. 4* (FAO, Rome, 1997); FAO, *Irrigation in the Near East Region in Figures, Water Reports No. 9* (FAO, Rome, 1997); FAO, *Irrigation in the Former Soviet Union in Figures, Water Reports No. 15* (FAO, Rome, 1997); FAO, *Irrigation in Asia in Figures, Water Reports No. 18* (FAO, Rome, 1999); and FAO, *Irrigation in Latin America in Figures, Water Reports* (FAO, Rome, in preparation). Withdrawal data for Israel: Statistical Abstract of Israel 1999, available online at: http://www .cbs.gov.il /engindex.htm. Population: United Nations (U.N.) Population Division, *World Population Prospects, 1950–2050 (The 1998 Revision)*, on diskette (U.N., New York, 1999).

In general, data are compiled from published documents (including national, U.N., and professional literature) and from estimates, when necessary, of resources and consumption from models using other data, such as area under irrigated agriculture, livestock populations, and precipitation.

Average annual internal renewable water resources refers to the average annual flow of rivers and recharge of groundwater generated from endogenous precipitation. Caution should be used when comparing different countries because these estimates are based on differing sources and dates. These annual averages also disguise large seasonal, interannual, and long-term variations.

When data for *annual river flows from* and *to other countries* are not shown, the internal renewable water resources figure *may* include these flows. When such data are shown, they are not included in a country's total internal renewable water resources.

Per capita annual internal renewable water resources data were calculated using 2000 population estimates.

Actual annual renewable water resources available for use is usually less than the sum of internal renewable resources and river flows. This is due to the fact that not all resources can be mobilized for use and that part of the flow coming from upstream countries or leaving for downstream countries might be reserved to those countries by treaty or other agreement. For example, actual water resources for Sudan include the flow of the Nile, less the amount that it is committed by treaty to deliver to Egypt at Aswan.

Annual withdrawals as a *percentage of water resources* refers to *total* water withdrawals, not counting evaporative losses from storage basins, as a percentage of internal renewable water resources. Water withdrawals also include water from nonrenewable groundwater sources, river flows from other countries, and desalinization plants in countries where that source is a significant part of all water withdrawals.

Per capita annual withdrawals were calculated using national population data for the year of data shown.

Sectoral withdrawals are classified as *domestic* (drinking water, homes, commercial establishments, public services [e.g., hospitals], and municipal use), *industry* (some countries include water withdrawn to cool thermoelectric plants, while others do not; these can be significant amounts of total water withdrawals), and *agriculture* (irrigation and livestock).

Totals may not add due to rounding.

Data Table FW. 2
Groundwater and Desalinization

Sources: Groundwater resources and withdrawal data: J. Margat, *Les eaux souterraines sans le monde* (Bureau de recherches géologiques et minières [BRGM], Département eau, Orléans, France, December 1990); J. Margat and D. Vallée, *Water Resources and Uses in the Mediterranean Countries* (Blue Plan, Sophia Antipolis, 1999); I.A. Shiklomanov, *Comprehensive Assessment of the Freshwater Resources of the World* (Stockholm Environment Institute, Stockholm, 1997); Organisation for Economic Co-Operation and Development (OECD), *OECD Environmental Data Compendium 1997* (OECD, Paris, 1997); and Economic Commission for Europe, *The Environment in Europe and North America* (United Nations, New York, 1992).

Groundwater resources and desalinization activities: J. Margat, *Lex Eaux Souterraines Dans Le Bassin Mediterraneen. Ressources et Utlisations Plan Bleu*, Doc. BRGM 282 (Ed. BRGM, Orléans, France, 1998); Food and Agriculture Organization of the United Nations (FAO), *Irrigation in*

Africa in Figures, Water Reports No. 7 (FAO, Rome, 1995); FAO, *Irrigation in the Near East Region in Figures, Water Reports No. 9* (FAO, Rome, 1997); FAO, *Irrigation in the Former Soviet Union in Figures, Water Report No. 15* (FAO, Rome, 1997); FAO, *Irrigation in Asia in Figures, Water Reports No. 18* (FAO, Rome, 1999); and FAO, *Irrigation in Latin America in Figures, Water Reports* (FAO, Rome, in preparation). Population data: United Nations (U.N.) Population Division, *World Population Prospects, 1950–2050 (The 1998 Revision),* on diskette (U.N., New York, 1999).

Average annual groundwater recharge is the amount of water that is estimated to annually infiltrate soils, including water from rivers and streams that lose it to underlying strata. In general, this figure would represent the maximum amount of water that could be withdrawn annually without ultimately depleting the groundwater resource. These data are estimated in a variety of ways and caution should be used in comparing values for different countries.

Per capita recharge is the amount of water that annually infiltrates soils on a per person basis, using 2000 population estimates from the U.N. Population Division.

Annual total groundwater withdrawals refers to abstractions from all groundwater sources–even nonrenewable sources. The *percentage of annual recharge* refers to *total* groundwater withdrawals.

Per capita annual withdrawals were calculated using national population data for the year of data shown.

Sectoral share of withdrawals of groundwater is classified as *domestic* (drinking water, homes, commercial establishments, public services, and municipal use), *industry* (including water withdrawn to cool thermoelectric plants), and *agriculture* (irrigation and livestock).

Desalinated water production refers to the removal of salt from saline waters–usually seawater–using a variety of techniques including reverse osmosis. Most desalinated water is used for domestic purposes.

Totals may not add due to rounding.

Data Table FW.3
Major Watersheds of the World

Sources: Watershed area: Center for Remote Sensing and Spatial Analysis (CRSSA) of Cook College, Rutgers University and U.S. Army Corps of Engineers Construction Engineering Research Laboratory, "Major Watershed Basins of the World," World Resources Institute (WRI), ed. *GlobalARC GIS CD-ROM Database* (CRSSA, New Brunswick, New Jersey, 1996). Number of countries within the watershed: Environmental Systems Research Institute (ESRI) software (ESRI, Redlands, California, 1995). Mean population density: Center for International Earth Science Information Network (CIESIN), WRI, and International Food Policy Research Institute, *Gridded Population of the World, Version 2 alpha* (Columbia University, Palisades, New York, 2000) available online at: http://sedac.ciesin.org/plue/ gwp. Cropland, forest, and grassland data: U.S. Geological Survey (USGS) and University of Nebraska-Joint Research Center for the European Commission, *Global Land Cover Characterization Database, Version 1.2,* distributed by USGS Earth Resources Observation System Data Center (USGS, 1997), available online at: http://edcdaac.usgs.gov/ glcc/globe_ int.html. Developed area: National Oceanic and Atmospheric Administration-National Geophysical Data Center (NOAA-NGDC), *Nighttime Lights of the World Database* (NOAA-NGDC, Boulder, Colorado, 1997). Irrigated area: Center for Environmental Systems Research, University of Kassel, *Global Map of Irrigated Areas* (University of Kassel, Kassel, Germany, 1999). Arid area: United Nations Environment Programme (UNEP), *World Atlas of Desertification Global Aridity Zone Map* (UNEP, Nairobi, 1992). Wetlands data: World Conservation Monitoring Centre (WCMC), *Biodiversity Map Library Database* (WCMC, Cambridge, U.K., 1996). Ramsar sites: Ramsar Convention Bureau, *List of Wetlands of International Importance* (Ramsar Convention Bureau, Gland, Switzerland, 1997). Water availability: B.M. Fekete, C.J. Vörösmarty, and W. Grabs, *Global Composite Runoff Fields Based on Observed River Discharge and Simulated Water Balance, Version 1.0* (University of New Hampshire, Durham, and Global Runoff Data Centre, Koblenz, Germany, 1999). Dams under construction: "1998 World Atlas and Industry Guide," *International Journal on Hydropower and Dams* (Aqua-Media International, Surrey, U.K., 1998). Degree of fragmentation: unpublished data, Landscape Ecology Group, Umeå University, (Umeå, Sweden, 2000) and M. Dynesius and C. Nilsson, "Fragmentation and Flow Regulation of River Systems in the Northern Third of the World," *Science* 266:753–762, 1994.

Major watersheds included in the table represent major river systems in the world and smaller river systems of regional significance. The basins in this table account for approximately 56 percent of the world's land area.

Most of the data in this table were obtained through geographic information system analysis of multiple datasets. The base data layer used for geographic definition of the watersheds was a 5-minute resolution dataset (1/20th of a degree of latitude/longitude) of major basins. There are some limitations associated with the scale of these base data: watershed boundaries are coarse, and some smaller basins and small tributaries are not identified. Basins were edited by WRI to capture some features such as deltas. Sub-basins were then aggregated to include all tributaries of the major river systems. Summary statistics for each watershed were digitally extracted by overlaying the basin map onto other existing digital datasets.

Modeled watershed area was estimated to a resolution of 1 square kilometer. These values only reflect horizontal extent (slopes are not accounted for) and may underestimate total land surface in the drainage area. Intermittent tributaries are included in most cases; for example, the northern part of the Kalahari Desert in Botswana within the Okavango basin is included, as well as many of the intermittent tributaries within the Lake Chad watershed. The tidal area of the St. Lawrence River is excluded. Water surface of rivers and lakes (e.g., the Great Lakes in the St. Lawrence River watershed) are included in the drainage area.

Countries within the watershed were identified using updated 1995 country boundaries from ESRI. Countries included in each basin are listed below in descending order as to their share of the basin (i.e., countries with more land within a basin are listed first). The countries listed may differ from other published sources due to the coarseness of the watershed boundaries. *Amu Darya:* Afghanistan, Uzbekistan, Tajikistan, Turkmenistan, and Kyrgyzstan. *Amur:* Russia, China, and Mongolia. *Brahmaputra:* China, India, Bangladesh, and Bhutan. *Chao Phrya:* Thailand. *Ganges:* India, Nepal, China, and Bangladesh. *Godavari:* India. *Hong:* Viet Nam and China. *Huang He:* China. *Indigirka:* Russia. *Indus:* Pakistan, India, Afghanistan, and China. *Irrawaddy:* Myanmar, China, and India. *Kizil:* Turkey. *Kolyma:* Russia. *Krishna:* India. *Kura-Araks:* Azerbaijan, Iran, Georgia, Armenia, and Turkey. *Lake Balkhash:* Kazakhstan and China. *Lena:* Russia. *Mahanadi:* India. *Mekong:* Laos, Thailand, China, Cambodia, Viet Nam, and Myanmar. *Narmada:* India. *Ob:* Russia, Kazakhstan, China, and Mongolia. *Salween:* China, Myanmar, and Thailand. *Syr Darya:* Kazakhstan, Kyrgyzstan, Uzbekistan, and Tajiskistan. *Tapti:* India. *Tarim:* China and Kyrgyzstan. *Tigris and Euphrates:* Iraq, Turkey, Iran, and Syria. *Xun Jiang:* China and Viet Nam. *Yalu Jiang:* China and Dem. People's Rep. Korea. *Yangtze:* China. *Yenisey:* Russia and Mongolia. *Dalalven:* Sweden. *Danube:* Romania, Hungary, Yugoslavia, Austria, Germany, Slovakia, Bosnia and Herzegovina, Bulgaria, Croatia, Ukraine, Czech Republic, Slovenia, and Moldova, and with less than 1 percent of the basin area: Switzerland, Italy, Poland, and Albania. *Dnieper:* Ukraine, Belarus, and Russia. *Dniester:* Ukraine, Moldova, and Poland. *Don:* Russia and Ukraine. *Duero:* Spain and Portugal. *Ebro:* Spain. *Elbe:* Germany, Czech Republic, Austria, and Poland. *Garonne:* France, Spain, and Andorra. *Gläma:* Norway. *Guadalquivir:* Spain. *Kemijoki:* Finland, Norway, and Russia. *Loire:* France. *North Dvina:* Russia. *Oder:* Poland, Czech Republic, and Germany. *Pechora:* Russia. *Po:* Italy and Switzerland. *Rhine-Maas:* Germany, France, Switzerland, Belgium, Netherlands, Luxembourg, Austria, and Liechtenstein. *Rhône:* France and Switzerland. *Seine:* France. *Tagus:* Spain and Portugal. *Ural:* Kazakhstan and Russia. *Vistula:* Poland, Ukraine, Belarus, and Slovakia. *Volga:* Russia and Kazakhstan. *West Dvina:* Belarus, Russia, and Latvia. *Weser:* Germany. *Congo:* Dem. Rep. Congo, Central African Republic, Angola, Rep. Congo, Tanzania, Zambia, Cameroon, Burundi, and Rwanda. *Cuanza:* Angola. *Cunene:* Angola and Namibia. *Jubba:* Kenya, Ethiopia, Somalia. *Lake Chad:* Chad, Niger, Central African Republic, Nigeria, Algeria, Sudan, Cameroon, and Libya. *Lake Turkana:* Ethiopia, Kenya, Sudan, and Uganda. *Limpopo:* South Africa, Botswana, Mozambique, and Zimbabwe. *Mangoky:* Madagascar. *Mania:* Madagascar. *Niger:* Mali, Nigeria, Niger, Algeria, Guinea, Cameroon, Burkina Faso, Benin, Côte d'Ivoire, and Chad. *Nile:* Sudan, Ethiopia, Egypt, Uganda, Tanzania, Kenya, Democratic Republic of Congo, Rwanda, Burundi, and Eritrea. *Ogooué:* Gabon, Rep. Congo, Cameroon, and Equatorial Guinea. *Okavango:* Botswana, Namibia, Angola, and Zimbabwe. *Orange:* South Africa, Namibia, Botswana, and Lesotho. *Oued Draa:* Morocco, Algeria, and Western Sahara (occupied by Morocco). *Rufiji:* Tanzania. *Senegal:* Mali, Mauritania, Senegal, and Guinea. *Shaballe:* Ethiopia and Somalia. *Volta:* Burkina Faso, Ghana, Mali, Togo, Côte d'Ivoire, and Benin. *Zambezi:* Zambia, Angola, Zimbabwe, Mozambique, Malawi, Botswana, Tanzania, and Namibia. *Alabama-Tombigbee:* United States. *Balsas:* Mexico. *Brazos:* United States. *Colorado:* United States and Mexico. *Columbia:* United States and Canada. *Fraser:* Canada and United States. *Hudson:* United States. *Mackenzie:* Canada. *Mississippi:* United States and Canada. *Nelson-Saskatchewan:* Canada and United States. *Rio Grande:* United States and Mexico. *Rio Grande de Santiago:* Mexico. *San Pedro* and *Usumacinta:* Mexico, Guatemala, and Belize. *Sacramento:* United States. *St. Lawrence:* Canada and United States. *Susquehanna:* United States. *Thelon:* Canada. *Yaqui:* Mexico and United States. *Yukon:* United States and Canada. *Amazon:* Brazil, Peru, Bolivia, Colombia, Ecuador, Venezuela, and Guyana. *Chubut:* Argentina and Chile. *Lakes Titicaca* and *Salar de Uyuni:* Bolivia, Peru and Chile. *Magdalena:* Colombia. *Orinoco:* Venezuela and Colombia. *Paraná:* Brazil, Argentina, Paraguay, and Bolivia. *Parnaíba:* Brazil. *Rio Colorado:* Argentina and Chile. *São Francisco:* Brazil. *Tocantins:* Brazil. *Uruguay:* Brazil, Uruguay, and Argentina. *Belyando:* Australia. *Dawson:* Australia. *Fly:* Papua New Guinea and Indonesia. *Kapuas:* Indonesia. *Mahakam:* Indonesia. *Murray-Darling:* Australia. *Sepik:* Papua New Guinea and Indonesia.

Average population density was extracted from a 2.5-minute resolution population map. Basins were overlaid on population data, and the population density was calculated for each basin. Data are presented as the number of people per square kilometer.

The USGS Global Land Cover Characterization database with the International Geosphere Biosphere Programme (IGBP) classification was used to identify the extent of different land cover types within each basin. The land cover database is derived from 1-kilometer resolution satellite data spanning April 1992 through March 1993. *Percent cropland* indicates the percentage of the basin defined as cropland or a crop/natural vegetation mosaic. *Percent forest* indicates the percentage of the basin defined as evergreen needleleaf forest, evergreen broadleaf forest, deciduous needleleaf forest, deciduous broadleaf forest, or mixed forest. *Percent grassland* includes IGBP classes defined as open shrublands, closed shrublands, woody savannas, savannas, and grasslands. *Percent built-up area* was estimated from a 1-kilometer by 1-kilometer resolution map derived from nighttime imagery from the Defense Meteorological Satellite Program Operational Linescan System of the United States. The dataset contains the locations of stable lights, including frequently observed light sources such as gas flares at oil drilling sites. Time-series analysis is used to exclude transient light sources such as fires and lightning. The extent of "lit" area may be slightly overestimated due to the sensor's resolution and factors such as reflection from water and other surface features. It is a good indicator of the spatial distribution of settlements and infrastructure, but should not be interpreted as a measure of population density. (The mean settlement size required to produce enough light to be detected is much greater in developing countries than in industrialized countries because of differences in energy consumption.) The *Nighttime Lights of the World* data are more highly correlated with measures of economic activity and energy consumption and are, therefore, considered a measure of relative development within the watershed. The *percent built-up area* was calculated by dividing the area within a watershed indicated as "lit" by the total area of the watershed.

Percent irrigated area indicates the percentage of the basin that has irrigated agriculture. This percentage was calculated by overlaying the boundaries of the major watersheds on an irrigated area map developed by the University of Kassel. The map is a 0.5° by 0.5° grid depicting the percentage of the area equipped for irrigation in 1995. The map was derived by combining information from large-scale irrigation maps, and national, subnational, and drainage basin level data on irrigated area. It was assumed that the irrigated areas are evenly distributed across each grid cell.

Percent arid area indicates the percentage of the basin that falls in an area defined as semiarid, arid, or hyperarid on the *World Atlas of Desertification Global Aridity Zone Map*. This map is based on an aridity index derived from the ratio of mean annual precipitation to the mean annual potential evapotranspiration.

Percent wetlands was calculated by dividing the sum of the areas of wetlands within the basin by the total watershed area. It includes all areas designated as wetlands in the *Biodiversity Map Library Database*.

Ramsar sites are sites designated as "wetlands of international importance" according to the terms of the Convention on Wetlands (Ramsar, Iran, 1971). Spatial accuracy of the coordinates varies from one site to another. For more information on Ramsar sites, please refer to the Sources and Technical Notes of Data Table BI.1.

Water available per person indicates the amount of total runoff available per person in each river basin. These estimates are based on a global runoff distribution database developed by the University of New Hampshire and the Global Runoff Data Centre. *Water availability per person* was calculated by dividing the total runoff available in a basin by the total number of people in that basin. Estimates are in cubic meters per person per year. The runoff distribution database has a spatial resolution of 0.5° and was calculated based on basin boundaries defined by the University of New Hampshire. The population database used was a 2.5-minute resolution population map from CIESIN and WRI for the years 1990 and 1995 (CIESIN and WRI, 1999).

Large dams in progress includes dams at least 60 meters high that were under construction in 1998. The approximate location of the dams was referenced based on continental-scale maps. Because of the lack of detailed dam location (i.e., geographical coordinates) and the coarseness of the watershed boundaries, some dams that fall near the watershed boundaries may be included in the adjacent watershed by referencing error.

Degree of river fragmentation indicates the level of modification of a river system due to dams, reservoirs, interbasin transfers, and irrigation consumption. Irrigation consumption refers to the water that is evaporated as a result of irrigation, but excludes the amount of water returned to the river after irrigation. River systems were classified into three levels of fragmentation: high, medium, and low. These categories are based on the number of dams in the main river channel and tributaries, the level of flow regulation, and the length of the main-channel segment without dams in relation to the entire length of the river. Generally, rivers with low fragmentation do not have dams in the main channel, and, if present, dams on tributaries do not change the river's discharge by more than 2 percent. Rivers with high fragmentation may have more than three-quarters of their main channels dammed or may have dams that substantially change the annual discharge. For more detailed information on the analysis, please refer to the original source.

Given the scale in the delineation of watershed boundaries and other datasets used in the production of this table, these figures should be used with caution.

Data Table AC.1
Emissions from Fossil Fuel Burning and Cement Manufacturing

Source: All fossil fuel carbon dioxide emissions data: Carbon Dioxide Information Analysis Center (CDIAC), *Global, Regional, and National Annual CO_2 Emissions from Fossil-Fuel Burning, Hydraulic Cement Production, and Gas Flaring: 1751–1996*, ORNL/CDIAC-25, NDP-030 (Environmental Sciences Division, Oak Ridge National Laboratory, Oak Ridge, Tennessee, March 1999). Data are available online at: http://cdiac.esd.ornl.gov/ftp/ndp030/. Gross domestic product at purchasing power parity in current international dollars: World Bank, *World Development Indicators 1999*, on CD-ROM (Development Data Group, World Bank, Washington, D.C., 1999). Population figures for per capita calculations: United Nations (U.N.) Population Division, *Annual Populations 1950–2050 (The 1998 Revision)*, on diskette (U.N., New York, 1999). Kyoto Protocol data: U.N. Framework Convention on Climate Change (an Internet-accessible numerical database) available online at: http://www .unfccc.de/.

Carbon dioxide (CO_2) emissions are often calculated and reported in terms of their content of elemental carbon. For this table, their values were converted to the actual mass of CO_2 by multiplying the carbon mass by 3.664 (the ratio of the mass of CO_2 to that of carbon).

These data from CDIAC represent a complete harmonized global dataset of CO_2 emissions. However, individual country estimates, based on more detailed information and a country-specific methodology, could differ. Guidelines were developed to assist in the preparation of national greenhouse gasses inventories. The Intergovernmental Panel on Climate Change (IPCC) accepted these guidelines at its Twelfth Session in Mexico City on September 11-13, 1996. The guidelines were published in *Revised 1996 IPCC Guidelines for National Greenhouse Gas Inventories* (IPCC, Cambridge, England, 1997). Such data are currently available for an increasing number of countries, but long time series are rare. Methods used by CDIAC have the advantage of calculating CO_2 emissions from a single common dataset available for all countries.

Solid fuels, *liquid fuels*, and *gaseous fuels* are primarily, but not exclusively, coals, petroleum products, and natural gas, respectively. *Gas flaring* is the practice of burning off gas released in the process of petroleum extraction, a practice that is declining. During *cement manufacturing*, cement is calcined to produce calcium oxide. In the process, 0.498 metric ton of CO_2 is released for each ton of cement produced. *Total emissions* for *1990* and *1996* consist of the sum of the CO_2 produced during the consumption of solid, liquid, and gaseous fuels, and from gas flaring and the manufacture of cement. However, these estimates do not include bunker fuels used in international transportation due to the difficulty of apportioning these fuels among the countries benefiting from that transport. For more information, please see the *World Resources 2000–01 Database CD-ROM* or the original source for data on emissions from bunker fuels.

CDIAC annually calculates emissions of CO_2 from the burning of fossil fuels and the manufacture of cement for most of the countries of the world. CDIAC calculates emissions from data on the net apparent consumption of fossil fuels (based on the World Energy Data Set maintained by the U.N. Statistical Division) and from data on world cement manufacturing (based on the Cement Manufacturing Data Set maintained by the U.S. Geological Survey). Emissions are calculated using conversion factors based on global average fuel chemistry and oxidation rates.

Total contribution since 1950 represents total carbon dioxide emitted from 1950 to 1996, excluding bunker fuels. For the independent republics of the former Soviet Union, CO_2 emissions from 1950 to 1991 are estimates based on each country's post-1991 share of all emissions from the entire former Soviet Union. Total 1992 CO_2 emissions for the former Soviet Union were 3,289,909 tons; the share of that total was then calculated for each of the former Soviet republics. For example, Kazakhstan's emissions in 1992 were calculated to be 8.90 percent of the total carbon emitted from the former Soviet Union. The same approach was used for the other former Soviet republics. Therefore, the total contributions for the former Soviet republics from 1950 to 1991 should be taken only as rough approximations. An equivalent method was used to calculate historical estimates for the former Czechoslovakia.

Per capita carbon emissions are calculated using 1996 CO_2 emissions and the 1996 population estimates from the U.N. Population Division (medium-case scenario).

The carbon dioxide intensity of a country's economic output is expressed as *CO_2 emitted per million international dollars (PPP) of gross domestic product in metric tons.* Gross domestic product (GDP) measures the final output of goods and services produced by the domestic economy. The international dollar values, which are different from U.S. dollar values, are obtained using special conversion factors designed to equalize the purchasing powers of different currencies. This conversion factor, the purchasing power parity (PPP), is defined as the number of units of a country's currency that are required to buy the same amounts of goods and services in the domestic market as $1 would buy in the United States. The computation involves deriving implicit quantities from national account expenditure data and specially collected price data, and then revaluing the implicit quantities in each country at a single set of average prices. Because the same international price averages are used for every country, cross-country comparisons reflect differences in quantities of goods and services free of price-level differences. This procedure is designed to bring cross-country comparisons in line with cross-time real-value comparisons that are based on constant price series. PPP estimates tend to lower per capita GDPs in industrialized countries and raise per capita GDPs in developing countries.

Information concerning whether a country is a *signatory* and whether they have *ratified* the Kyoto Protocol is current through October 1999. The Kyoto Protocol attempts to place legally binding limits on greenhouse gas emissions (carbon dioxide [CO_2], methane [CH_4], nitrous oxide [N_2O], hydrofluorocarbons [HFCs], perfluorocarbons [PFCs], and sulfur hexafluoride [SF_6]) from developed countries. By signing the treaty, a state recognizes the authentic text, intends to complete the procedures for becoming legally bound by it, and is committed not to act against the treaty's objectives before ratification. Ratification (or its alternatives of acceptance, approval, or accession) binds the state to observe the treaty. Depending on a country's system of governance, signing the treaty may be simply an executive decision while ratification may require legislative approval. Quantitative obligations by developed countries will be based on the base year of 1990. For further information about the Kyoto Protocol, please refer to the information available online at: http://www.unfccc.de /resource/convkp.html.

Data Table AC.2
Common Anthropogenic Pollutants

Source: Emissions data for anthropogenic pollutants: Co-Operative Programme for Monitoring and Evaluation of the Long-Range Transmission of Air Pollutants in Europe (EMEP), *1998 Major Review of Strategies and Policies for Air Pollution Abatement*, Report No. ECE/EB.AIR/65 (forthcoming). Data available online at: http://www.emep.int /emis_ tables/tab1.html.

Emissions of sulfur in the form of sulfur oxides and nitrogen in the form of its various oxides together contribute to acid rain and adversely affect agriculture, forests, aquatic habitats, and the weathering of building materials. Sulfate and nitrate aerosols impair visibility. These data on anthropogenic sources should be used with caution because different methods and procedures may have been used in each country. Therefore, the best comparative data may be within-country time trends.

Sulfur dioxide (SO_2) is created by natural as well as anthropogenic activities. High concentrations of SO_2 have important adverse health effects, and there is particular concern about its effects on the health of young children, the elderly, and people with respiratory illnesses (e.g., asthma). SO_2 in the presence of moisture contributes to acid precipitation as sulfuric acid.

Anthropogenic *nitrogen oxides* (NO_x) come mainly from industrial sources and contribute to the creation of photochemical smog and the production of tropospheric ozone–an important greenhouse gas. All oxides of nitrogen also contribute to acid precipitation, in the form of nitric acid.

This data table combines data from EMEP, Economic Commission for Europe (ECE), and Organisation for Economic Co-Operation and Development (OECD) to compile as complete a picture as possible of sulfur and nitrogen emissions. EMEP is an activity of the 1979 Convention on Long-Range Transboundary Air Pollution. Data on sulfur and nitrogen emissions are submitted to EMEP and ECE by parties as a part of their commitments under the protocols to the LRTAP Convention. In the event that official data are missing, EMEP interpolates between years for which official data exist. In the event that this is not possible, EMEP will use its own estimates or emissions estimates from other sources.

OECD polls its members on emissions with questionnaires that are completed by the relevant national statistical service or designee. OECD does not have any independent estimation capability.

EMEP and ECE report emissions in terms of the elemental content of sulfur, whereas OECD reports its emissions in terms of tons of oxides of sulfur. EMEP and ECE emissions estimates were converted to their weight in SO_2. EMEP and OECD report nitrogen emissions in terms of nitrogen dioxide. For further information, please consult the original sources.

This data table also reports OECD data for carbon monoxide and combines both EMEP and OECD data to describe the emissions of volatile organic compounds. Differences in definition can limit the comparability of these estimates.

Carbon monoxide (CO) is formed both naturally and from industrial processes, including the incomplete combustion of fossil and other carbon-bearing fuels. Automobile emissions are the most important source of CO, especially in urban environments. CO interferes with oxygen uptake in the blood, producing chronic anoxia leading to illness or, in the case of massive and acute poisoning, even death. CO also scavenges hydroxyl radicals that would otherwise contribute to the removal of methane, a potent greenhouse gas, from the atmosphere.

In the presence of sunlight, *volatile organic compounds* (VOCs) are, along with oxides of nitrogen, responsible for photochemical smog. Anthropogenic emissions of VOCs arise in part from the incomplete combustion of fuels or the evaporation of fuels, lubricants, and solvents, as well as from the incomplete burning of biomass.

Data Table AC.3
Atmospheric Concentration of Greenhouse and Ozone-Depleting Gases

Source: Carbon dioxide: Charles D. Keeling and T.P. Whorf, Carbon Dioxide Information Analysis Center (CDIAC), *Atmospheric CO_2 Concentrations–Mauna Loa Observatory, Hawaii, 1958–1998 (revised July 1999)*, ORNL/CDIAC-25, NDP-001/R9 (Environmental Sciences Division, Oak Ridge National Laboratory, Oak Ridge, Tennessee, July 1999). Data are accessible online at: http://cdiac.esd.ornl.gov/ftp /ndp001/maunaloa.co2. Preindustrial concentration for carbon dioxide: Neftel *et al.*, 1994, *Historical Carbon Dioxide Record from the Siple Station Ice Core* (Physics Institute, University of Bern, Switzerland) and data from the Law Dome ice core record. More details on the Law Dome ice core record may be found online at: http://cdiac.esd.ornl.gov/trends /co2/lawdome.html. Preindustrial concentration for methane: Etheridge et al., 1994, *Concentrations of CH_4 from the Law Dome (East Side, "DE08" Site) Ice Core* (Commonwealth Scientific and Industrial Research Organisation Aspendale, Victoria, Australia). Preindustrial concentration for nitrous oxide: Etheridge et al., 1988, *Concentrations of N_2O from the Law Dome (Summit, "BHD" Site) Ice Core* (Commonwealth Scientific and Industrial Research Organisation Aspendale, Victoria, Australia). Other greenhouse and ozone-depleting gases data: Atmospheric Lifetime Experiment (ALE) / Global Atmospheric Gases Experiment (GAGE) / Advanced GAGE (AGAGE) Network (updated and revised May 1999), CDIAC, Environmental Sciences Division, Oak Ridge National Laboratory, DB-1001 (an Internet-accessible numerical database), available online at: http://cdiac .ESD.ORNL.GOV/ndps/alegage.html (ALE/ GAGE/AGAGE Monthly Readings at Cape Grim, Tasmania), originally R.G. Prinn, R.F. Weiss, F.N. Alyea et al., "Atmospheric CFC-11 (CCl_3F), CFC-12 (CCl_2F_2), and N_2O from the ALE-GAGE Network," in T.A. Boden, D.P. Kaiser, R.J. Sepanski et al., eds., *Trends '93: A Compendium of Data on Global Change* (ORNL/CDIAC-65, CDIAC, Oak Ridge, Tennessee, 1994), pp. 396–420.

The greenhouse and ozone-depleting gases listed destroy atmospheric ozone, contribute to the greenhouse effect, or both. *Carbon dioxide* (CO_2) is emitted to the atmosphere by natural and anthropogenic processes. Due to the large concentration of CO_2 in the atmosphere compared to concentrations of other greenhouse gases, it is second only to water vapor in its potential effect on global warming. For further details, see the Source, and Technical Notes for Data Table EI.1.

Atmospheric CO_2 concentrations are monitored at many sites worldwide; the data presented here are from Mauna Loa, Hawaii (19∞ 32' North latitude, 155∞ 35' West longitude). Trends at Mauna Loa reflect global trends, although CO_2 concentrations differ significantly among monitoring sites at any given time. For example, the average annual concentration at the South Pole in 1997 was 2.3 parts per million lower than at Mauna Loa. The preindustrial concentration of CO_2 was estimated from an Antarctic ice core.

Annual means disguise large daily and seasonal variations in CO_2 concentrations. During the summer, photosynthetic plants store larger amounts of carbon than in the winter, resulting in seasonal variation in measurements. Some annual mean figures were derived from interpolated data. Data are revised to correct for drift in instrument calibration, hardware changes, and perturbations to "background" conditions. Details concerning data collection, revisions, and analysis are contained in C.D. Keeling, et al., "Measurement of the Concentration of Carbon Dioxide at Mauna Loa Observatory, Hawaii," in W.C. Clark, ed., *Carbon Dioxide Review: 1982* (Oxford University Press, New York, 1982).

Data for all other gases are from values monitored at Cape Grim, Tasmania (45∞ 41' South latitude, 144∞ 41' East longitude) under the ALE, GAGE, and Advanced GAGE. As with CO_2, gas concentrations at any given time vary among monitoring sites; the data reported here reflect global trends. Cape Grim generally receives unpolluted air from the Southeast and is the ALE/GAGE/AGAGE station with the longest, most complete dataset. Air samples were collected 4 times daily for ALE and 12 times daily for GAGE/AGAGE. The annual values shown here are averages of monthly values calculated by CDIAC. Missing values were interpolated. Preindustrial concentrations for nitrous oxide and methane were estimated from an Antarctic ice core.

Methane (CH_4) is emitted through the release of naturally occurring methane gas and as one of the products of anaerobic respiration. Emission sources associated with human activities include waste management (landfills), livestock management (enteric fermentation in ruminants), anaerobic respiration in the soils associated with wet rice agriculture, and combustion of fossil fuels and biomass (wood fuel and cleared forests). Sources of anaerobic respiration include the soils of moist forests, wetlands, bogs, tundra, and lakes. CH_4 acts to increase ozone in the troposphere and lower stratosphere. On a molecule-for-molecule basis, methane is 21 times more powerful than CO_2 at trapping heat in the atmosphere.

Nitrous oxide (N_2O) is emitted by aerobic decomposition of organic matter in oceans and soils, by bacteria, by combustion of fossil fuels and biomass (wood fuel and cleared forests), by the use of nitrogenous fertilizers, and through other processes. N_2O is an important depletor of stratospheric ozone and is 310 times more powerful than CO_2 at trapping heat in the atmosphere.

Carbon tetrachloride (CCl_4) is an intermediate product in the production of CFC-11 and CFC-12. It is also used in other chemical and pharmaceutical applications and for grain fumigation. Compared with other gases, CCl_4 makes a small contribution to the greenhouse effect and to stratospheric ozone depletion.

Methyl chloroform (CH_3CCl_3) is used primarily as an industrial degreasing agent and as a solvent for paints and adhesives. Its contribution to the greenhouse effect and to stratospheric ozone depletion also is small.

CFC-11 (CCl_3F), *CFC-12* (CCl_2F_2), and *CFC-113* ($C_2Cl_3F_3$) are potent depletors of stratospheric ozone. In addition, their cumulative effect on global warming may equal one-fourth that of CO_2. CFCs are used as solvents and in many applications including refrigeration, air conditioning, foam blowing, and cleaning of electronics components.

Total gaseous chlorine is calculated by multiplying the number of chlorine atoms in each of the chlorine-containing gases (carbon tetrachloride, methyl chloroform, and CFCs) by the concentration of that gas. Chlorine and bromine act as catalysts in the destruction of ozone. Chlorine is not consumed in the reaction and thus can react with and destroy ozone molecules many times over.

Data Table ERC.1
Energy Production by Source

Source: International Energy Agency (IEA), *Energy Balances of Organisation for Economic Co-operation and Development (OECD) Countries, 1960–1997*, on diskette (OECD, Paris, 1999), and *Energy Balances of Non-OECD Countries, 1971–1997*, on diskette (OECD, Paris, 1999).

We present all energy data in a common unit of 1,000 metric tons of oil equivalent (toe) to facilitate comparisons of energy sourcing, consumption, substitution, and conservation. A toe is defined as 41.868 gigajoules.

Energy production from all sources is the amount of energy from all sources produced by each country in the year specified. In addition to solid, liquid, and gaseous fuels and nuclear electricity, the total also includes hydropower, geothermal, solar, wind, tidal, wave, combustible renewables and waste, and indigenous heat production from heat pumps. *Per capita* shows the amount produced per person for that country.

Energy production from *solid fuels* is the energy produced from all types of primary coal (i.e., hard coal or lignite). Peat is also included in this category.

Energy production from *liquid fuels* is the energy produced from liquid fuels such as crude oil or natural gas liquids.

Energy production from *gaseous fuels* is the amount of energy produced from natural gas.

Energy production from *nuclear fuels* shows the primary heat equivalent of the electricity produced by nuclear power plants. Heat-to-electricity conversion efficiency is assumed to be 33 percent.

Total electricity generated is the toe equivalent of the electrical energy produced by thermal, nuclear, geothermal, hydropower (excluding pumped storage production–see notes to Data Table ERC.4), and other power plants. Electricity generated is not a primary energy source and should not be mistakenly added to the energy production from primary sources presented in this table. These data were converted from gigawatt-hours to toe using a conversion rate of 1 Gwh = 86 toe.

Data Table ERC.2
Energy Consumption by Source

Source: International Energy Agency (IEA), *Energy Balances of Organization for Economic Cooperation and Development (OECD) Countries, 1960–1997*, on diskette (OECD, Paris, 1999), and *Energy Balances of Non-OECD Countries, 1971–1997*, on diskette (OECD, Paris, 1999).

We present all energy data in a common unit of 1,000 metric tons of oil equivalent (toe) to facilitate comparisons of energy sourcing, consumption, substitution, and conservation. A toe is defined as 41.868 gigajoules.

Energy consumption from all sources is the amount of energy from all sources used by each country in the year specified. In addition to solid, liquid, and gaseous fuels and nuclear electricity, the total also includes hydropower, geothermal, solar, combustible renewables and waste, and indigenous heat production from heat pumps. *Per capita* shows the amount produced per person for that country.

It is important to note that the World Resources Institute (WRI) includes losses through transportation, friction, heat loss, and other inefficiencies as energy consumption. In these energy tables, total consumption equals IEA's total primary energy supply (TPES). TPES is indigenous production plus imports, minus exports, plus stock changes, minus international marine bunkers. (Note that IEA counts production after the removal of impurities such as sulfur.)

Stock change is the difference between stocks held by producers, transformation industries, importers, and large consumers the first day of the year and the last day of the year on national territory.

International marine bunker fuel is energy delivered to sea-going ships of all flags, including warships; ships traveling in inland and coastal waters are excluded.

Energy consumption from *solid fuels* is the total energy produced from all types of coal. Peat is also included in this category.

Energy consumption from *liquid fuels* is the energy consumed from liquid fuels such as crude oil or natural gas liquids.

Energy consumption from *gaseous fuels* is the amount of energy consumed from natural gas.

Energy consumption from *nuclear fuels* shows the primary heat equivalent of the electricity consumed from nuclear power plants. Heat-to-electricity conversion efficiency is assumed to be 33 percent.

Final consumption of electricity is the toe equivalent of the electrical energy consumed by the end user. This variable differs from the others in the table in that it does not present TPES but rather total final consumption, which does not include the energy lost to inefficiencies and transport losses.

Trade in energy imports and *exports* comprise the total amount of energy having crossed national territorial boundaries of the country, whether or not customs clearance has taken place. Coal imports and exports comprise the amount of fuels obtained from or supplied to other countries, whether or not there is an economic or customs union between the relevant countries. Coal in transit is not included. Oil and gas imported or exported under processing agreements (i.e., refining on account) are included. Quantities of oil in transit are excluded. Crude oil, natural gas liquids, and natural gas are reported as coming from the country of origin; refinery feedstocks and oil products are reported as coming from the country of last consignment. Re-exports of oil imported for processing within bonded areas are shown as exports of products from the processing country to the final destination. Electricity is considered as imported or exported when it has crossed the national boundaries of the country.

To account for the differences in quality between types of coal and other energy sources, the IEA has applied specific factors (to correctly convert that specific fuel to toe) supplied by national administrations for the main categories of energy sources and flows or uses (i.e., production, imports, exports, industry).

Energy statistics are expressed in terms of net calorific value and therefore may be slightly lower than values presented by other statistical compendia. The difference between the net and the gross calorific value for each fuel is the latent heat of vaporization of the water produced during combustion of the fuel. For oil and coal, net calorific value is 5 percent less than gross; for most forms of natural and manufactured gas the difference is 9-10 percent.

Data Table ERC.3
Energy Consumption by Economic Sector

Source: International Energy Agency (IEA), *Energy Balances of Organisation for Economic Co-operation and Development (OECD) Countries, 1960–1997,* on diskette (OECD, Paris, 1999) and *Energy Balances of Non-OECD Countries, 1971–1997,* on diskette (OECD, Paris, 1999).

Data for OECD and Economic Commission for Europe countries were compiled from information provided in questionnaires filled out by each country. Data from other large- and medium-sized energy consumers come mostly from individual country information. Data for the remaining countries were gathered from a variety of international organizations.

The World Resources Institute calculated the amount of total final energy consumption (TFC) used by each sector as a percentage of TFC for each country. TFC does not include distribution losses and fuels that are transformed to another form (e.g., fuels used for electricity generation, crude oil transformed into oil products in refineries, etc.).

Industry sector includes energy consumption by the iron and steel industry, chemical industry, nonferrous metals basic industries, nonmetallic mineral products (glass, ceramic, cement, etc.), transport equipment, machinery, mining and quarrying, food and tobacco, paper, pulp and print, wood and wood products, construction, textile and leather, and any nonspecified industry. *Iron and steel* consumption is the energy consumed by the iron and steel industry as a percentage of the total energy consumed by the country.

Transportation sector includes all fuel for air, road, and water transport except fuel used for international marine bunkers and for ocean, coastal, and inland fishing. *Air* transportation includes both international and domestic civil aviation. *Road* transportation includes all human and cargo transport along national road networks.

Agriculture sector refers to all agricultural and forestry activity, including ocean, coastal, and inland fishing.

Commercial and public services refer to service sectors such as stores, repair shops, and restaurants.

Residential sector includes household energy use.

IEA reports that it can be difficult to distinguish accurately between the agriculture, commercial, and public services sectors and that a figure for "total energy use" is more accurate than totals for the individual sectors.

Data Table ERC.4
Energy from Renewable Sources

Source: International Energy Agency (IEA), *Energy Balances of Organisation for Economic Co-operation and Development (OECD) Countries, 1960–1997,* on diskette (OECD, Paris, 1999), and *Energy Balances of Non-OECD Countries, 1971–1997,* on diskette (OECD, Paris, 1999).

Data for OECD and Economic Commission for Europe countries were compiled from information provided in questionnaires filled out by each country. Data from other large- and medium-sized energy consumers come mostly from individual country information. Data for the remaining countries were gathered from a variety of international organizations.

Renewable energy production and *renewable energy consumption* from *all renewable sources* show the total energy produced and consumed, respectively, from renewable energy sources. The totals include hydroelectric power, wind, solar, wave and tidal, geothermal, and combustible renewables and waste. Consumption in this table is equal to total primary energy supply (TPES), as in Data Table ERC.2. Please see the notes to that data table for more information on TPES.

Renewable sources as a *percent of total consumption from all sources* is the percentage of each country's total energy consumption supplied from renewables and waste.

Fuels and waste comprise solid biomass and animal products, gas/liquids from biomass, industrial waste, and municipal waste. Biomass is any plant matter used directly as fuel or converted into a fuel. This includes wood, wood or crop waste, ethanol, animal wastes, and sulfur lyes (lignin in black liquor from paper production). These data are often estimated on small sample surveys or other incomplete information.

Hydroelectric refers to the energy content of the electricity produced in hydroelectric power plants excluding output from pumped storage plants (in which electricity is produced from pumping water uphill during times of low demand for use in times of high demand).

Energy from *geothermal* and *solar* sources is counted for both heat and electricity. Geothermal electricity production efficiency is assumed to be 10 percent of the total energy contained in the heat used to drive the electrical generators.

Wind refers to the use of wind energy for electrical production.

Data Table ERC.5
Resource Consumption

Sources: Passenger cars: International Road Federation, *World Road Statistics* as reported by the World Bank, *World Development Indicators 1999,* on CD-ROM (Development Data Group, World Bank, Washington, D.C., 1999). Motor gasoline: International Energy Agency (IEA), *Energy Balances of Organisation for Economic Co-operation and Development (OECD) Countries, 1960–1997,* on diskette (OECD, Paris, 1999), and *Energy Balances of Non-OECD Countries, 1971–1997,* on diskette (OECD, Paris, 1999). Meat and Paper: Food and Agriculture Organization of the United Nations (FAO), *FAOSTAT On-line Statistical Service* (FAO, Rome, 1999). Coffee: International Coffee Organization (ICO), various sources (ICO, London, 1999). Population data for World Resources Institute (WRI) calculations: United Nations (U.N.) Population Division, *Annual Populations 1950–2050 (The 1998 Revision),* on diskette (U.N., New York, 1998).

In 1999, WRI published the report *Critical Consumption Trends and Implications: Degrading Earth's Ecosystems* by Emily Matthews and Allen Hammond. As essential inputs to both subsistence economies and advanced technological societies, natural resources are discussed as the basis of all human activity. The increase in consumption of these resources is driven by population growth, rising wealth, technological change, and urbanization. *Critical Consumption* examines consumption trends and associated effects on natural ecosystems. Among other topics, the report includes a discussion of wood fiber and food such as cereals, meat, and fish. *World Resources 1998–99* also examines trends in resource consumption and resulting environmental impacts. Specifically, *World Resources 1998–99* covers trends in consumption of paper, coffee, meat, vehicles, and other topics. The report can be found online at: http://www.wri.org/trends/index.html.

Passenger cars refer to the number of individual four-wheel vehicles per 1,000 people. These numbers exclude buses, freight vehicles, and two-wheelers such as mopeds and motorcycles.

Motor gasoline consumption depicts the per capita final consumption of fuel meant for use in internal combustion engines such as those in passenger cars. IEA reports the energy in metric tons of oil equivalent (toe). A conversion of 1 toe to 1,246 liters of motor gasoline was used to convert to liters of motor gasoline.

Meat consumption refers to the per capita total meat consumption. Total meat includes meat from animals slaughtered in countries, irrespective of their origin, and comprises horsemeat, poultry, and meat from all other domestic or wild animals such as camels, rabbits, reindeer, and game animals. Meat consumption was calculated using a trade balance approach (total production plus imports, minus exports).

Paper consumption refers to the per capita consumption of newsprint, printing and writing paper, construction paper and paperboard, household and sanitary paper, special thin paper, and wrapping and packaging paper. Paper consumption was calculated using a trade balance approach (total production plus imports, minus exports). For some countries for which the FAO has no production data, production was assumed to be negligible and was assigned a value of 0 instead of reporting X for these countries. These countries, marked with a footnote, have <500,000 hectares of forest land and have imported <10,000 metric tons of recovered paper.

Coffee consumption refers to the per capita consumption of all coffee and is given in kilograms of raw coffee beans per capita.

Data Table HD.1
Demographic Indicators

Source: Population data: United Nations (U.N.) Population Division, *Annual Populations 1950–2050 (The 1998 Revision)*, on diskette (U.N., New York, 1998). Population change data: U.N. Population Division, *World Populations 1950–2050 (The 1998 Revision)*, on diskette (U.N., New York, 1998). Percentage of population in specific age groups and fertility rates: U.N. Population Division, *Demographic Indicators, 1950–2050 (The 1998 Revision)*, on diskette (U.N., New York, 1998).

Population refers to the midyear population. Most data are estimates based on population censuses and surveys. All projections are for the medium-case scenario. (See the following discussion.) The *average annual population change* takes into account the effects of international migration.

Many of the population data in the Population and Human Development Data tables are estimated using demographic models based on several kinds of demographic parameters: a country's population size, age and sex distribution, fertility and mortality rates by age and sex groups, growth rates of urban and rural populations, and the levels of internal and international migration.

Information collected through recent population censuses and surveys is used to calculate or estimate these parameters, but accuracy varies. The U.N. Population Division compiles and evaluates census and survey results from all countries. These data are adjusted for overenumeration and underenumeration of certain age and sex groups (e.g., infants, female children, and young males), misreporting of age and sex distributions, and changes in definitions, when necessary. These adjustments incorporate data from civil registrations, population surveys, earlier censuses, and, when necessary, population models based on information from socioeconomically similar countries. (Because the figures have been adjusted, they are not strictly comparable to the official statistics compiled by the U.N. Statistical Office and published in the *Demographic Yearbook*.)

After the figures for population size and age/sex composition have been adjusted, these data are scaled to 1990. Similar estimates are made for each 5-year period between 1950 and 1990. Historical data are used when deemed accurate, also with adjustments and scaling. However, accurate historical data do not exist for many developing countries. In such cases, the U.N. Population Division uses available information and demographic models to estimate the main demographic parameters. Projections are based on estimates of the 1990 base-year population. Age- and sex-specific mortality rates are applied to the base-year population to determine the number of survivors at the end of each 5-year period. Births are projected by applying age-specific fertility rates to the projected female population. Births are distributed by an assumed sex ratio, and the appropriate age- and sex-specific survival rates are applied. Future migration rates are also estimated on an age- and sex-specific basis.

Assumptions about future mortality, fertility, and migration rates are made on a country-by-country basis and, when possible, are based on historical trends. Four scenarios of population growth (high, medium, low, and constant) are created by using different assumptions about these rates. For example, the medium-case scenario assumes medium levels of fertility, an assumption that may vary among countries. Refer to the original source for further details. Although projections may be of questionable quality, U.N. demographic models are based on surveys and censuses with well-understood qualities, which makes these data fairly reliable.

The *percentage of population in specific age groups* shows a country's age structure: the percentage of the population <15, 15–65, and >65 years old. It is useful for inferring dependency, needs for education and employment, potential fertility, and other age-related factors.

The *total fertility rate* is an estimate of the number of children an average woman would have if current age-specific fertility rates remained constant during her reproductive years.

Data Table HD.2
Trends in Mortality, Life Expectancy, and AIDS

Sources: Infant mortality data: United Nations (U.N.) *Demographic Indicators 1950–2050 Medium Fertility Variant (The 1998 Revision)* (U.N. Population Division, New York, 1998). Under-5 mortality rate: United Nations Children's Fund (UNICEF) *State of the World's Children 1999* (UNICEF, New York, 1999), including information from Multiple Indicator Cluster Surveys; U.N. Population Division; Demographic and Health Surveys; World Health Organization; and World Bank. Life expectancy at birth: U.N. Population Division, *Demographic Indicators, 1950–2050 (The 1998 Revision)*, on diskette (U.N., New York, 1998). AIDS data: Joint U.N. Programme on HIV/AIDS *Data Annex to the*

UNAIDS Report on the Global HIV/AIDS Epidemic (Joint U.N. Programme on HIV/AIDS, New York, June 1998).

The *infant mortality rate* is the probability of dying by age 1, multiplied by 1,000.

The *under-5 mortality rate* is the probability of dying by age 5, multiplied by 1,000. UNICEF provides this cohort measure, which is derived from *Child Mortality Since the 1960s: A Database for Developing Countries* (U.N., New York, 1992) and from infant mortality estimates provided by the U.N. Population Division. The mix is the result of a move from modeled estimates to estimates based on a periodically updated child mortality database. Nonetheless, this variable should not be compared to the U.N. Population Division's infant mortality rate, which is derived from population models where otherwise not available.

Life expectancy at birth is the average number of years that a newborn baby is expected to live if the age-specific mortality rates effective at the year of birth apply throughout his or her lifetime.

Births attended by trained personnel is the percentage of births attended by physicians, nurses, midwives, or primary health care workers trained in midwifery skills.

Adults and children infected with HIV/AIDS is the number adults and children infected with HIV, whether or not they have developed AIDS symptoms. *Percent of adults infected with HIV/AIDS* is the percentage of the population aged 15–49 infected with HIV, whether or not they have developed AIDS symptoms. *Number of children orphaned by AIDS since beginning of epidemic* is the estimated number of children under 15 who have lost their mother or both parents to AIDS since the beginning of the epidemic. UNAIDS has not specified dates for the beginning of the epidemic.

Data Table HD.3
Safe Water, Sanitation, School Enrollment, and Literacy

Sources: Access to safe drinking water and access to adequate sanitation services: United Nations Children's Fund (UNICEF), *The State of the World's Children 1999* (UNICEF, New York, 1999), including information from Multiple Indicator Cluster Surveys; United Nations (U.N.) Population Division; Demographic and Health Surveys; World Health Organization (WHO); and World Bank. Education data: United Nations Educational, Scientific and Cultural Organization (UNESCO), *World Education Indicators 1998*, on CD-ROM (UNESCO, Paris, 1999). Literacy data: UNESCO, *World Education Indicators 1998*, on CD-ROM (UNESCO, Paris, 1999).

WHO collected data on drinking water and sanitation from national governments in 1980, 1983, 1988, and 1990 using questionnaires completed by public health officials, WHO experts, and resident representatives of the U.N. Development Programme. In 1990, the Joint Monitoring Programme was established by WHO and UNICEF to build national capacity in measuring all aspects of water and sanitation services. The most recent round of data collection from this program presented figures for access to clean water and sanitation for many countries as of 1994.

Definitions of safe drinking water and appropriate access to sanitation and health services vary depending upon location and condition of local resources, thus, comparisons can be misleading. In addition, urban and rural populations were defined by each national government and might not be strictly comparable. The official definitions of access to safe drinking water, adequate sanitation, and health services are listed below, but countries are at liberty to adapt these definitions to reflect internal conditions.

Population with access to safe drinking water is the percentage of the population with reasonable access to an adequate amount of safe drinking water (including treated surface water and untreated water from protected springs, boreholes, and sanitary wells). WHO defines reasonable access to safe drinking water in *urban* areas as access to piped water or a public standpipe within 200 meters of a dwelling or housing unit. In *rural* areas, reasonable access implies that a family member need not spend a "disproportionate" part of the day fetching water.

Population with access to sanitation is the percentage of the population with at least adequate excreta disposal facilities that can effectively prevent human, animal, and insect contact with excreta. *Urban* areas with access to sanitation services are defined as urban populations served by connections to public sewers or household systems such as pit privies, pour-flush latrines, septic tanks, communal toilets, and other such facilities. *Rural* populations with access to sanitation services are defined as those with adequate disposal such as pit privies and pour-flush latrines.

UNESCO defines *net school enrollment* as the enrollment of the age group (as defined by the national education system) that corresponds to a given level of education, expressed as a percentage of the total population of that age group. *Gross school enrollment* is defined as the total enrollment, regardless of age, expressed as a percentage of the population that officially corresponds to the level of education. *Net primary school enrollment* is level 1 of the International Standard Classification of Education (ISCED), and its principal function is to provide the basic elements of education, such as those provided by elementary and primary schools. Intercountry comparisons should be made cautiously because regulations for this level are extremely flexible. *Net secondary school enrollment* includes levels 2 and 3 of the ISCED and is based upon at least 4 years of previous instruction at the first level. It is provided at teacher-training, middle, secondary, or high schools as well as vocational or technical schools. *Gross tertiary school enrollment* refers to levels 5 and 6 of the ISCED classification.

Adult literacy rate refers to the percentage of people aged 15 years and over who can read and write. UNESCO recommends defining a person as illiterate "who cannot both read and write with understanding a simple statement about his or her everyday life." This concept is widely accepted, but its interpretation and application vary from country to country when collecting data during demographic censuses and surveys. For the majority of countries, the literacy data shown in this publication are UNESCO estimates and projections that have been derived based on national population census data supplied by the countries to the U.N. Statistics Division or drawn from national publications.

Data Table EI.1
Gross Domestic Product and Trade Values

Sources: Gross domestic product (GDP), annual growth rates of GDP, distribution of GDP, exports of goods and services, and imports of goods and services data: World Bank, *World Development Indicators 1999,* on CD-ROM (Development Data Group, World Bank, Washington, D.C., 1999). Population data: United Nations (U.N.) Population Division, *Annual Populations 1950–2050 (The 1998 Revision)*, on diskette (U.N., New York, 1999).

Gross domestic product (GDP) using *purchasing power parity (PPP)* are estimates based on the purchasing power of currencies rather than on current exchange rates. GDP in PPP terms is derived by applying the ratio of GDP to gross national product (GNP) in local currency to the World Bank's estimates of GNP in PPP terms. The estimates are a blend of extrapolated and regression-based numbers, using the results of the International Comparison Programme (ICP). GNP data are available on the *World Resources 2000–01 Database CD-ROM.*

The ICP benchmark studies are essentially multilateral pricing exercises. The intercountry price comparisons have been reported in seven phases: 1970, 1973, 1975, 1980, 1985, 1990, and 1993. PPP studies recast traditional national accounts through special price collections and the disaggregation of GDP by expenditure components. National statistical offices report ICP details.

The international dollar values, which are different from the U.S. dollar values of GNP or GDP, are obtained using special conversion factors designed to equalize the purchasing powers of different currencies. This conversion factor, the PPP, is defined as the number of units of a country's currency required to buy the same amounts of goods and services in the domestic market as $1 would buy in the United States. The computation involves deriving implicit quantities from national accounts expenditure data and specially collected price data and then revaluing the implicit quantities in each country at a single set of average prices. Because the same international price averages are used for every country, cross-country comparisons reflect differences in quantities of goods and services free of price-level differences. This procedure is designed to bring cross-country comparisons in line with cross-time real-value comparisons that are based on constant price series. PPP estimates tend to lower per capita GDPs in industrialized countries and raise per capita GDPs in developing countries.

Gross domestic product (GDP) based on exchange rates are in 1995 U.S. dollars (based on 1995 exchange rates), and are the sum of GDP at factor cost (value added in the agriculture, industry, and services sectors) and indirect taxes, less subsidies. World Bank GDP estimates are in accord with the U.N. System of National Accounts.

GDP per capita estimates are calculated using population estimates from 1997 or from the year for which the most recent GDP data are available, as noted in the data table.

The *average annual growth rates* of GDP are least-squares estimates of the real growth of output over each decade. Growth rates are computed from constant 1995 U.S. dollar price data to exclude the effects of inflation.

Percent from *agriculture* (GDP) includes agricultural and livestock production, agricultural services, logging, forestry, fishing, and hunting. Percent from *industry* (GDP) comprises mining and quarrying; manufacturing; construction; and electricity, gas, and water. Percent from *services* includes wholesale and retail trade; transport, storage, and communications; banking, insurance, and real estate; public administration and defense; ownership of dwellings; and others. The *distribution* of GDP does not always add up to 100 percent due to rounding.

Exports and *imports of goods and services* represent the value of all goods and other marketed services provided to the world. Included is the value of merchandise, freight, insurance, travel, and other nonfactor services. Factor and property income (formerly called factor services), such as investment income, interest, and labor income, is excluded. Figures are in current U.S. dollars. *Agriculture* includes agricultural raw materials such as hides and skins, crude rubber, cork and wood, pulp and waste paper, textile fibers, and crude animal and vegetable materials. Other categories measured include food, fuel, metals, and manufacturing.

Although considerable effort has been made to standardize economic data according to the U.N. System of National Accounts, care should be taken when interpreting the indicators presented in this data table. Intercountry and intertemporal comparisons using economic data involve complicated technical problems that are not easily resolved; therefore, readers are urged to read these data as characterizing major differences between economies rather than as precise, quantitative measurements.

Data Table EI.2
International Financial Flows and Investment

Sources: Official development assistance (ODA) and official aid (OA): Organisation for Economic Co-Operation and Development (OECD), *1998 Development Co-Operation* (OECD, Paris, 1999), *Geographical Distribution of Financial Flows to Aid Recipients, 1993–97* (OECD, Paris, 1999), and *World Development Indicators 1999,* on CD-ROM (Development Data Group, World Bank, Washington, D.C., 1999). ODA and OA as a percentage of gross domestic product (GDP): calculated using data from World Bank, *World Development Indicators 1999,* on CD-ROM (Development Data Group, World Bank, Washington, D.C., 1999). Population figures: United Nations (U.N.) Population Division, *World Population Prospects, 1950–2050 (The 1998 Revision)*, on diskette (U.N., New York, 1999). Direct foreign investment, total external debt, debt service as a percentage of total exports, and international tourism receipts: World Bank, *World Development Indicators 1999,* on CD-ROM (Development Data Group, World Bank, Washington, D.C., 1999).

Average annual official development assistance (ODA) and official aid (OA) data are in current U.S. dollars and include the net amount of disbursed grants and concessional loans given or received by a country less repayments of concessional loans. Grants include gifts of money, goods, or services for which no repayment is required. A concessional loan has a grant element of 25 percent or more. The grant element is the amount by which the face value of the loan exceeds its present market value because of below-market interest rates, favorable maturity schedules, or repayment grace periods. Nonconcessional loans are not a component of ODA. ODA and OA contributions are shown as negative numbers (in parentheses); receipts are shown as positive numbers. Data for some developing countries (e.g., Republic of Korea) are shown as negative numbers because of net repayments of concessional loans. Data for donor countries include contributions made directly to developing countries and through multilateral institutions.

ODA and OA sources include the development assistance agencies of OECD and members of Organization of Petroleum Exporting Countries as well as other countries. Grants and concessional loans to and from multilateral development agencies are also included in contributions and receipts. OECD gathers ODA and OA data through questionnaires and reports from countries and multilateral agencies. Only limited data are available on ODA and OA flows among developing countries. These data are included when known.

GDP data used to calculate *ODA and OA as a percentage of gross domestic product (GDP)* are estimates calculated according to purchasing power parity. For an explanation of purchasing power parity, please refer to the Sources and Technical Notes for Data Table EI.1.

ODA and OA per capita estimates are calculated using the average 1995–97 ODA and OA estimates in current international dollars and U.N. Population Division population data for 1995–97.

Direct foreign investment is the net inflow of capital to acquire a lasting management interest (10 percent or more of the voting stock) in a country other than that of the investor. It includes equity capital, reinvestment of earnings, other long-term capital, and short-term capital as shown in the balance of payments between countries.

The World Bank operates the Debtor Reporting System (DRS), which compiles reports supplied by the Bank's member countries. Countries submit detailed reports on the annual status, transactions, and terms of the long-term external debt of public agencies and of publicly guaranteed private debt. Additional data are drawn from the World Bank, the International Monetary Fund (IMF), regional development banks, government lending agencies, and the Creditor Reporting System (CRS). The CRS is operated by OECD to compile reports from the members of its Development Assistance Committee. For further information on international debt, refer to *Global Development Finance 1999, Volumes 1 and 2* (World Bank, Washington, D.C., 1999).

Total external debt (current U.S. dollars) includes long-term debt outstanding, short-term debt, use of IMF credit, and private nonguaranteed debt outstanding. A long-term debt is an obligation with maturity of at least one year that is owed to nonresidents and is repayable in foreign currency, goods, or services. Long-term debt is divided into long-term public debt and long-term publicly guaranteed private debt. A short-term debt is a public or publicly guaranteed private debt that has a maturity of one year or less. This class of debt is especially difficult for countries to monitor. Only a few countries supply these data through the DRS; the World Bank supplements these data with creditor-country reports, information from international clearinghouse banks, and other sources to derive rough estimates of short-term debt.

Use of IMF credit refers to all drawings on the Fund's General Resources Account. Use of IMF credit is converted to dollars by applying the average special drawing right exchange rate in effect for the year being calculated.

A private debt is an external obligation of a private debtor that is not guaranteed by a public entity. Data for this class of debt are less extensive than those for public debt; many countries do not report these data through the DRS. These data are included in the total when available.

Debt service as a percentage of total exports (in foreign currencies, goods, and services) comprises interest payments and principal repayments made on the disbursed long-term public debt as well as private, nonguaranteed debt; IMF debt repurchases; IMF charges; and interest payments on short-term debt as a percent of exports.

Debt data are reported to the World Bank in the units of currency in which they are payable. The World Bank converts these data to U.S. dollars using the IMF par values, central rates, or the current market rates, where appropriate. Debt service data are converted to U.S. dollars at the average exchange rate for the given year. Comparability of data among countries and years is limited by variations in methods, definitions, and comprehensiveness of data collection and reporting.

External debt data pertain to only those countries within the DRS, which focuses on low- and middle-income economies. Many economies are not represented within the system, and the estimates that are presented may not be comprehensive due to different reporting frameworks.

These data do not account for the term structure and the concessionality mix of debt, which can lead to a misrepresentation of a country's underlying solvency.

Exports of goods and services represent the value of all goods and other marketed services provided to the world. Included are the value of merchandise, freight, insurance, travel, and other nonfactor services. Factor and property income (formerly called factor services), such as investment income, interest, and labor income, is excluded.

International tourism receipts include all expenditures (i.e., payments for goods and services) by international inbound visitors including money spent on national carriers for international transport. These receipts should include any payments made for goods and services in the destination country. Figures are in current U.S. dollars. The World Bank also reports expenditures of international outbound visitors in other countries.

Data Table EI.3
Distribution of Income and Poverty

Sources: Gini coefficient, income distribution, and poverty: World Bank, *World Development Indicators 1999*, on CD-ROM (World Bank, Washington, D.C., 1999). Poverty estimates for selected developed countries: Timothy M. Smeeding, *Financial Poverty in Developed Countries: The Evidence from the Luxembourg Income Study (Final Report to the United Nations Development Programme)*, Luxembourg Income Study, Working Paper No. 155 (Syracuse, New York and Walferdange, Luxembourg, 1997).

The World Bank uses a variety of data sources, including reports of governments, international organizations, and household surveys, when necessary, to compile estimates of the distribution of income within countries.

Survey year indicates the year data were collected for income distribution.

The *Gini coefficient* measures the extent to which the actual distribution of income differs from a perfectly equal distribution. A coefficient of zero would reflect perfect equality; a coefficient of 100 would represent perfect inequality. Graphically, the index is the area between a Lorenz curve (the cumulative percentage of total income against the cumulative percentage of recipients–starting with the poorest individual or household) and a hypothetical line of absolute equality (expressed as a percentage of the area under the line).

Percentage of income in each quintile of population is the share of total income that accrues to each of the different quintiles of the population ranked according to income (starting with the lowest 20 percent and moving up the income ladder to the highest 20 percent).

Personal or household income or consumption data come from national household surveys. Rankings are based on per capita income or consumption. Comparisons among figures for each quintile are difficult because of differences in household size and in extent of income sharing among household members. If original data from the household survey were available, income (or consumption) shares were calculated. If these data were unavailable, then shares were estimated from the best available grouped data.

Year indicates the year data were collected for the international poverty line.

The percentage of people falling below the *international poverty line* was calculated by the World Bank for most of the countries from primary household survey data obtained from government statistical agencies and World Bank country departments. Data for poverty in selected developed countries are the product of the Luxembourg Income Study (LIS).

The international poverty line is the percentage of people living on less than $1 a day (at 1985 international prices) adjusted for purchasing power parity. Purchasing power parity is defined as the number of units of a country's currency required to buy the same amounts of goods and services in the domestic market as $1 would buy in the United States. The computation involves deriving implicit quantities from national accounts expenditure data and specially collected price data and then revaluing the implicit quantities in each country at a single set of average prices. Because the same international price averages are used for every country, cross-country comparisons reflect differences in quantities of goods and services free of price-level differences. This procedure is designed to bring cross-country comparisons in line with cross-time real-value comparisons that are based on constant price series.

Year indicates the year data were collected for the national poverty line.

The *national* poverty rate is calculated using World Bank poverty assessments based on household surveys and reflects the population living below the national urban poverty line. Data for poverty in selected developed countries are the product of the LIS. The *rural* poverty rate refers to the rural population living below the national rural poverty line.

The *urban* poverty rate is the percentage of the population living under the national urban poverty line. Definitions of poverty vary among countries and consistent comparisons can be difficult. For further information, please refer to the original sources.

The definition of a *national poverty* line chosen here (40 percent of the median income) is just one of several alternative definitions offered by the LIS. Forty percent of the median income provides a close approximation to the percentage of poor reported by the United States (14.5 percent in 1994–using a more complex algorithm, U.S. Department of Commerce, *Statistical Abstract of the United States 1996* [U.S. Department of Commerce, Washington, D.C., 1996], p. 472). With the exception of the United States and the United Kingdom, these developed countries do not commonly report poverty estimates.

Data Table SCI.1
Small Nations and Islands

Sources: Population data: United Nations (U.N.) Population Division, *Annual Populations 1950–2050 (The 1998 Revision)*, on diskette (U.N., New York, 1998). Land area and cropland data: Food and Agriculture Organization of the United Nations (FAO), *FAOSTAT On-line Statistical Service* (FAO, Rome, 1999). Natural forest data: FAO, *State of the World's Forests 1999* (FAO, Rome, 1999). Number of coral genera: unpublished data (World Conservation Monitoring Centre, Cambridge, U.K., August 1999). Claimed exclusive economic zone: Figures calculated by L. Pruett and J. Cimino, unpublished data, *Global Maritime Boundaries Database* (Veridian-MRJ Technology Solutions, Fairfax, Virginia, January 2000). Marine fish catch: FAO, *Global Capture Dataset 1984–1997* and *Aquaculture Quantities Dataset 1984–1997* fishery statistics databases, downloadable with *Fishstat-Plus* software at: http://www.fao.org/WAICENT/FAOINFO/FISHERY/statist/FISOFT/FISHPLUS.HTM.

Food supply from fish and fishery products: E. Laureti, *Fish and Fishery Products: World Apparent Consumption Statistics Based on Food Balance Sheets (1961–1997)*, FAO Fisheries Circular No. 821, Rev. 5 (FAO, Rome, 1999). Under-5 mortality rate: United Nations Children Fund (UNICEF), *State of the World's Children 1999* (UNICEF, New York, 1999). Gross domestic product: World Bank, *World Development Indicators 1999*, on CD-ROM (Development Data Group, World Bank, Washington, D.C., 1999).

Population refers to the midyear population. Most data are estimates based on population censuses and surveys. All projections are for the medium-case scenario. (See the Source and Technical Notes of Data Table HD.1 for further discussion regarding this variable.)

Land area refers to the total area of a country excluding area under inland seas, lakes, and national claims to the continental shelf. FAO compiles these data through responses to questionnaires from national governments.

FAO defines *natural forest* in tropical and temperate developing countries as a forest composed primarily of indigenous (native) tree species. Natural forests include closed forest, where trees cover a high proportion of the ground and where grass does not form a continuous layer on the forest floor (e.g., broadleaved forests, coniferous forests, and bamboo forests), and open forest, which FAO defines as mixed forest/grasslands with at least 10 percent tree cover and a continuous grass layer on the forest floor. Natural forests in tropical and temperate developing countries encompass all stands except plantations and include stands that have been degraded to some degree by agriculture, fire, logging, and other factors. For all regions, trees are distinguished from shrubs on the basis of height. A mature tree has a single well-defined stem and is taller than 7 meters. A mature shrub is usually less than 7 meters tall. *Annual average percent change* reflects the increase or decrease in forest cover between 1990 and 1995. It is shown as a percentage of the exponential growth rate. If negative (in parentheses), these figures reflect net deforestation, which is defined as the clearing of forest lands for all forms of agricultural uses (shifting cultivation, permanent agriculture, and ranching) and for other land uses such as settlements, other infrastructure, and mining. In tropical countries, this entails clearing that reduces tree crown cover to less than 10 percent. It should be noted that deforestation, as defined here, does not reflect changes within the forest stand or site such as selective logging (unless the forest cover is permanently reduced to less than 10 percent). Such changes are termed "forest degradation," and they can substantially affect forests, forest soil, wildlife and its habitat, and the global carbon cycle. Thus, the effects from the reported deforestation figures may be less than the effects from the total deforestation that includes all types of forest alterations. Positive change figures reflect net afforestation within a country or region. For further discussion of how FAO derives these values, see the Sources and Technical Notes of Data Table FG.1.

Cropland area refers to land under temporary and permanent crops, temporary meadows, market and kitchen gardens, and temporarily fallow land. Permanent cropland is land under crops that does not need to be replanted after each harvest, such as cocoa, coffee, fruit trees, rubber, and vines.

Scleractinia coral genera include only reef-building, colonial *Scleractinia* genera that are known to exist in each country. Coral reefs are home to more than a quarter of all known marine fish species. In general, coral reefs are found in shallow waters, between the Tropic of Capricorn and the Tropic of Cancer. Most reef-forming corals belong to the family *Scleractinia*, which are also called true or stony corals. They may be solitary or colonial and have a heavy external calcareous skeleton. Colonial species are restricted to shallow, clear waters in tropical seas, while solitary individuals can be found in deep waters and high latitudes.

Claimed exclusive economic zone (EEZ) is the coastal area claimed by a country within its 200-nautical mile EEZ. The United Nations Convention on the Law of the Sea (UNCLOS) is an international agreement that sets conditions and limits on the use and exploitation of the oceans. This convention sets the rules on the maritime jurisdictional boundaries of the different member states. UNCLOS was opened for signature on December 10, 1982 in Montego Bay, Jamaica, and entered into force on November 16, 1994. As of January 2000, 132 countries had ratified UNCLOS. Under UNCLOS, coastal states can claim sovereign rights in a 200-nautical mile EEZ. This allows for sovereign rights over the EEZ in terms of exploration, exploitation, conservation, and management of all natural resources in the seabed, its subsoil, and overlaying waters. UNCLOS allows other states to navigate and fly over the EEZ, as well as to lay submarine cables and pipelines. The inner limit of the EEZ starts at the outer boundary of the territorial sea (i.e., 12 nautical miles from the low-water line along the coast). Some states have not ratified UNCLOS and many have not yet claimed their EEZ. For further information on EEZs, please refer to the Sources and Technical Notes of Data Table CMI.3.

Marine fish catch data refer to marine fish caught or trapped for commercial, industrial, and subsistence use (catches from recreational activities are included where available). Statistics for mariculture, aquaculture, and other kinds of fish farming are *not* included in the country totals. Marine fish includes demersal, pelagic, and diadromous fish caught in marine areas (i.e., sturgeons, paddlefishes, river eels, salmons, trouts, smelt, shads, and miscellaneous diadromous fish). For a listing of groups of species under the demersal and pelagic categories, please refer to the Sources and Technical Notes of Data Table CMI.4. Catch figures are the national totals averaged over a 3-year period; they include fish caught by a country's fleet anywhere in the world. For further information, please refer to the Sources and Technical Notes of Data Table CMI.1 and CMI.4.

Food supply from fish and fishery products is defined as the quantity of both freshwater and marine fish, seafood, and derived products available for human consumption. *Per capita* food supply from fish and fishery products is defined as the estimate of the total supply available for human consumption, divided by the *de facto* population (i.e., those persons living within a country's borders or region), unless otherwise noted. Data are expressed in live-weight equivalents. Data on food supply from fish and fishery products represent apparent consumption on a live-weight basis, which means that the amounts of fish and fishery products consumed include all parts of the fish, including bones. The amount of fish and seafood actually consumed may be lower than the figures provided, depending on how much is lost during storage, preparation, and cooking, and how much is discarded.

Fish protein as a percent of all protein is defined as the quantity of protein from freshwater and marine fish, seafood, and derived products available for human consumption as a percentage of all protein available.

The *under-5 mortality rate* is the probability of a child dying by age 5, multiplied by 1,000. Please see the Sources and Technical Notes of Data Table HD.2 for further information on how UNICEF derives this cohort measure.

Estimates of *gross domestic product* (GDP) using purchasing power parity (PPP) are based on the purchasing power of currencies rather than on current exchange rates. GDP in PPP terms is derived by applying the ratio of GDP to gross national product (GNP) in local currency to the World Bank's estimates of GNP in PPP terms. The estimates are a blend of extrapolated and regression-based numbers, using the results of the International Comparison Programme. GNP data are available on the *World Resources 2000–01 Database CD-ROM*. The Sources and Technical Notes of Data Table EI.1 contain more discussion on the derivation of GDP in PPP terms.

World Resources 2000-2001

SOURCES

Acronyms

Acknowledgments

Notes and References

Index

Acronyms

AAAS — American Association for the Advancement of Science
ACIAR — Australian Centre for International Agricultural Research
AGIDS — Amsterdam Research Institute for Global Issues and Development Studies
BGS — British Geological Survey
BP/RAC — Blue Plan for the Mediterranean/Regional Activity Centre
CANARI — Caribbean Natural Resources Institute
CARPE — Central African Regional Program for the Environment
CDIAC — Carbon Dioxide Information Analysis Center
CGIAR — Consultative Group on International Agricultural Research
CI — Conservation International
CIAT — International Center for Tropical Agriculture
CIESIN — Center for International Earth Science Information Network
CIMMYT — International Maize and Wheat Improvement Center, Mexico
CONABIO — National Commission for the Knowledge and Use of Biodiversity
COP-5 — Conference of the Parties to the Convention on Biological Diversity
CORAL — Coral Reef Alliance
CRSSA — Center for Remote Sensing and Spatial Analysis
C&SF Project — Central and South Florida Project
CSIR — Council for Scientific and Industrial Research, South Africa
CSRC — Complex Systems Research Center
DOE — United States Department of Energy
ECE — European Commission for Europe
EEA — European Environment Agency
EFI — European Forest Institute
EMEP — Co-Operative Programme for Monitoring and Evaluation of the Long-Range Transmission of Air Pollutants in Europe
EIA — Energy Information Administration
ESA — Ecological Society of America
ESRI — Environmental Systems Research Institute
EVRI — Environmental Valuation Reference Inventory
FAO — Food and Agriculture Organization of the United Nations
FAOSTAT — FAO Statistical Databases
FSC — Forest Stewardship Council
GAIM — Global Analysis, Integration and Modelling Task Force, International Geosphere-Biosphere Program
GCSSF — Governor's Commission for a Sustainable South Florida
GEF — Global Environment Facility
GESAMP — Joint Group of Experts on the Scientific Aspects of Marine Pollution
GLASOD — Global Assessment of Soil Degradation
GOOS — Global Ocean Observing System
GUO — Global Urban Observatory
IAI — Inter-American Institute for Global Change Research
ICLARM — International Center for Living Aquatic Resources Management
ICO — International Coffee Organization
ICOLD — International Commission on Large Dams
ICSU — International Council for Science
IEA — International Energy Agency
IFAD — International Fund for Agricultural Development
IFDC — International Fertilizer Development Center
IFPRI — International Food Policy Research Institute
IGBP — International Geosphere-Biosphere Programme
IIASA — International Institute for Applied Systems Analysis
IJHD — International Journal on Hydropower and Dams
IMERCSA — Musokotwane Environment Resource Centre for Southern Africa
IMF — International Monetary Fund
IPCC — Intergovernmental Panel on Climate Change
IRF — International Road Federation
IRN — International Rivers Network
ISRIC — International Soil Reference and Information Centre
ITTO — International Tropical Timber Organization
IUCN — IUCN-The World Conservation Union
MRC — Mekong River Commission
NASA — National Aeronautics and Space Administration
NOAA — National Oceanic and Atmospheric Administration
NOAA/NGDC — National Geophysical Data Center
NOAA/NOS — National Ocean Service
NRC — National Research Council

NRDC	Natural Resources Defense Council
ODI	Overseas Development Institute
OECD	Organisation for Economic Co-Operation and Development
ORNL	Oak Ridge National Laboratory
OVI	Ocean Voice International
PAGE	Pilot Analysis of Global Ecosystems
PRB	Population Reference Bureau
RFF	Resources for the Future
SARDC	Southern African Research and Documentation Centre
SFERTF	South Florida Ecosystem Restoration Task Force
SFWMD	South Florida Water Management District
Sida	Swedish International Development Cooperation Agency
TNC	The Nature Conservancy
UC Berkeley	University of California at Berkeley
UC Davis	University of California, Davis
UCSD	University of California, San Diego
UNCHS	United Nations Centre for Human Settlements (Habitat)
UNDP	United Nations Development Programme
UN-ECE	United Nations Economic Commission for Europe
UNEP	United Nations Environment Programme
UNESCO	United Nations Educational, Scientific and Cultural Organization
UNFIP	United Nations Fund for International Partnerships
UNFPA	United Nations Population Fund
UNICEF	United Nations Children's Fund
UNPD	United Nations Population Division
UNSTAT	United Nations Statistical Division
USACE	United States Army Corps of Engineers
USAID	United States Agency for International Development
USDA	United States Department of Agriculture
USDA/NASS	National Agricultural Statistics Service
USDA/NRCS	Natural Resources Conservation Service
USGS	United States Geological Survey
USGS/EDC	Earth Resources Observation Systems (EROS) Data Center
U.S. EPA	United States Environmental Protection Agency
USOTA	United States Office of Technology Assessment
UT Austin	University of Texas at Austin
WBCSD	World Business Council for Sustainable Development
WCFSD	World Commission on Forests and Sustainable Development
WCMC	World Conservation Monitoring Centre
WHO	World Health Organization
WMO	World Meteorological Organization
WRI	World Resources Institute
WTO	World Trade Organization
WWF	World Wildlife Fund

Abbreviations for Units of Measure

AVHRR	advanced very high resolution radiometer
Bha	billion hectares
cm	centimeter
GtC	billion tons or gigatons of carbon
km	kilometer
l	liter
m	meter
mi	mile
MtC	metric tonne of carbon
Mha	million hectares
ha	hectare
MW	megawatt
MMTCE	million metric tons of carbon equivalents
ppm	parts per million
ppb	parts per billion

Acknowledgments

World Resources 2000–2001 is the result of a unique partnership among the United Nations Environment Programme (UNEP), the United Nations Development Programme (UNDP), the World Bank, and the World Resources Institute (WRI). It is the only instance where UN agencies, a multilateral financial institution, and an NGO work together in a true partnership to determine the content, conclusions, and recommendations of a major report.

INSTITUTIONS

For this millennial edition, we give special acknowledgment to the generous support of the United Nations Foundation in improving the presentation and dissemination of the report and to the Netherlands Ministry of Foreign Affairs for increasing international collaboration on the report. We are also grateful to the following institutions for supporting the Pilot Analysis of Global Ecosystems and the international effort to establish a Millennium Ecosystem Assessment, and for contributing data, reviews, and encouragement to the whole project.

Aqua-Media International, U.K.
Australian Centre for International Agricultural Research
AVINA Foundation
BirdLife International
Blue Plan for the Mediterranean
Carbon Dioxide Information Analysis Center, Oak Ridge National Laboratory
Caribbean Association for Sustainable Tourism
Caribbean Tourism Organization
Center for International Earth Science Information Network
Center for Remote Sensing and Spatial Analysis
Consultative Group on International Agricultural Research
Co-Operative Programme for Monitoring and Evaluation of the Long-Range Transmission of Air Pollutants in Europe
COWI Consulting Engineers and Planners AS, Denmark
David and Lucile Packard Foundation
Declining Amphibian Populations Task Force
DHI Water and Environment, Denmark
Earth Resources Observation Systems Data Center, United States Geological Survey
Environmental Systems Research Institute
European Commission for Europe
European Environment Agency
European Forest Institute
Food and Agriculture Organization of the United Nations
Forest Stewardship Council
Global Environment Facility
Global Runoff Data Center, Germany
International Center for Tropical Agriculture
International Coffee Organization
International Energy Agency
International Fertilizer Development Center
International Food Policy Research Institute
International Institute for Applied Systems Analysis
International Livestock Research Institute
International Monetary Fund
International Potato Center
International Road Federation
International Soil Reference and Information Centre
International Tanker Owners Pollution Federation
Island Resources Foundation
IUCN-The World Conservation Union
Japan Oceanographic Data Center
Man and the Biosphere Program
National Agricultural Statistics Service
National Oceanic and Atmospheric Administration, National Geophysical Data Center and National Ocean Service
The Nature Conservancy
Netherlands Ministry of Foreign Affairs
Ocean Voice International
Ohio Environmental Protection Agency
Oregon Department of Fish and Wildlife
Organisation for Economic Co-Operation and Development
Patuxent Wildlife Research Laboratory
Ramsar Convention Bureau
Safari Club International
State Hydrological Institute, Russia
Swedish International Development Cooperation Agency
Umeå University, Sweden
United Nations Children's Fund
United Nations Economic Commission for Europe
United Nations Educational, Scientific and Cultural Organization
United Nations Fund for International Partnerships
United Nations Population Division
United Nations Statistical Division
United States Agency for International Development, Global Bureau
United States Army Corps of Engineers, Construction Engineering Research Labs
United States Department of Agriculture, Forest Service, National Agricultural Statistics Service, and National Resources Conservation Service
United States Fish and Wildlife Service, National Wetlands Inventory
United States Geological Survey
University of East Anglia, U.K.
University of Kassel, Center for Environmental Systems Research, Germany
University of Maryland, Geography Department

University of Nebraska-Joint Research Center for the European Commission
University of New Hampshire, Complex Systems Research Center
University of Rhode Island, Coastal Resources Center
Veridian-MRJ Technology Solutions
Washington Department of Fish and Wildlife
World Conservation Monitoring Centre
World Travel and Tourism Council
World Wildlife Fund-US
Yale School of Forestry and Environmental Studies

MILLENNIUM ASSESSMENT STEERING COMMITTEE
Special thanks are due to the members of the Millennium Assessment Steering Committee, who generously gave their time, insights, and expert review comments throughout the period of the Pilot Analysis of Global Ecosystems.

Edward Ayensu, Ghana
Mark Collins, WCMC
Angela Cropper, Trinidad and Tobago
Andrew Dearing, WBCSD
Michael Zammit Cutajar (invited), Framework Convention on Climate Change
Louise Fresco, FAO
Madhav Gadgil, Indian Institute of Science, Bangalore
Habiba Gitay, Australian National University
Gisbert Glaser, UNESCO
Zuzana Guziova, Ministry of the Environment, Slovak Republic
Calestous Juma, Harvard University
John Krebs, National Environment Research Council, U.K.
Jonathan Lash, WRI
Roberto Lenton, UNDP
Jane Lubchenco, Oregon State University
Jeffrey McNeely, IUCN-The World Conservation Union
Harold Mooney, ICSU
Ndegwa Ndiangui, UN Convention to Combat Desertification
Prabhu L. Pingali, CIMMYT
Per Pinstrup-Andersen, IFPRI
Mario Ramos, GEF
Peter Raven, Missouri Botanical Garden
Walter V. Reid, Secretariat
Cristian Samper, Instituto Alexander Von Humboldt, Colombia
José Sarukhán, CONABIO
Peter Schei, Directorate for Nature Management, Norway
Klaus Töpfer, UNEP
José Galízia Tundisi, International Institute of Ecology, Brazil
Robert Watson, World Bank
Xu Guanhua, Ministry of Science and Technology, P.R. of China
A.H. Zakri, Universiti Kebangsaan Malaysia

PUBLISHING SUPPORT AND ASSISTANCE
We also want to acknowledge publishing support and assistance from The Magazine Group in Washington, D.C., whose staff designed and typeset *World Resources 2000–2001;* Transcontinental Printing & Graphics, Inc., the printer of the hardcover and paperback English editions; Elsevier Science Ltd. in Oxford, U.K., the publisher of the English hardcover edition; Editions Eska in Paris, France, the publisher of the French edition; Ecoespaña Editorial in Madrid, Spain, the publisher of the Spanish edition; Nikkei Business Publications, Inc. in Tokyo, the publisher of the Japanese edition; Al-Ahram Center for Translation and Publishing in Cairo, Egypt, the publisher of the Arabic edition; and the State Environmental Protection Administration in Beijing, P.R. China, the publisher of the Chinese edition.

INDIVIDUALS
Many individuals contributed to the development of this report by providing expert advice, data, or careful review of manuscripts. While final responsibility for the contents rests with the *World Resources* staff, the contributions of these colleagues are reflected throughout the report.

Special thanks to Dan Claasen of UNEP, Robert Watson of the World Bank, and Roberto Lenton of UNDP, who coordinated access to pertinent experts at their organizations:

UNEP
Sheila Aggarwal-Khan, Nancy Bennet, Marion Cheatle, Gerry Cunningham, Til Darnhofer, Salif Diop, Sheila Edwards, Hiremagalur Gopalan, Sheila Heileman, Dave MacDevette, Timo Maukonen, Ricardo Sanchez, Surendra Shrestha, Ashbindu Singh, Anna Stabrawa, Bai-Mass Taal, Dik Tromp, Isabelle Vanderbeck, and Jinhua Zhang.

UNDP
Susan Becker, Karen Jorgensen, Kristen Lewis, Charles McNeil, Laura Mourino-Casas, and Ralph Schmidt.

WORLD BANK
Isabelle Alegre, J. Gonzalo Castro, John Dixon, Kirk Hamilton, Saeed Ordoubadi, Stefano Pagiola, and Gunars Platais.

PART I RETHINKING THE LINK

CHAPTER 1 LINKING PEOPLE AND ECOSYSTEMS

Main text
Editor: Gregory Mock (WRI). Contributing writers: John Dixon (World Bank), Kirk Hamilton (World Bank), Stefano Pagiola (World Bank), Christine Mlot (consultant), and Gregory Mock (WRI).

Box 1.1 History of Use and Abuse
Editor: Janet Overton (WRI). Writer: Lori Han (WRI). Reviewers: John McNeill (Georgetown University) and Walter V. Reid (consultant).

Box 1.2 Linking Ecosystems and People
Editors/writers: Gregory Mock (WRI), Christine Mlot (consultant), and Janet Overton (WRI).

Box 1.3 Water Filtration and Purification
Editor: Wendy Vanasselt (WRI). Contributing writers: Christine Mlot (consultant) and Wendy Vanasselt (WRI). Reviewer: Katherine C. Ewel (USDA Forest Service, Institute of Pacific Islands Forestry).

Box 1.4 Pollination
Editor/writer: Wendy Vanasselt (WRI). Reviewers: Eric H. Erickson (Carl Hayden Bee Research Center), David Inouye (Rocky Mountain Biological Lab), and Rainer Krell (FAO).

Box 1.5 Biological Diversity
Editor/writer: Wendy Vanasselt (WRI). Reviewer: Nels Johnson (WRI).

Box 1.6 Carbon Storage
Editors: Janet Overton (WRI) and Carol Rosen (WRI). Contributing writers: Christine Mlot (consultant), Wendy Vanasselt (WRI), Greg Mock (WRI), and Robert Livernash (consultant). Reviewer: Chas Feinstein (World Bank).

Box 1.7 Linking People and Ecosystems: Human-Induced Pressures
Editor/writer: Carol Rosen (WRI).

Box 1.8 Invasive Species
Editor/writer: Wendy Vanasselt (WRI). Reviewer: Nels Johnson (WRI).

Box 1.9 Trade-Offs: Lake Victoria's Ecosystem Balance Sheet
Editor: Gregory Mock (WRI). Writer: Carmen Revenga (WRI). Reviewer: Les Kaufman (Boston University).

Box 1.10 Domesticating the World: Conversion of Natural Ecosystems
Editor/writer: Gregory Mock (WRI). Reviewer: Norbert Henninger (WRI). Map: Siobhan Murray (WRI).

Box 1.11 How Much Do We Consume?
Editor/writer: Gregory Mock (WRI). Reviewer: Emily Matthews (WRI).

Box 1.12 Pollution and Ecosystems
Editor: Wendy Vanasselt (WRI). Contributing writers: Wendy Vanasselt (WRI), Greg Mock (WRI), and Robert Livernash (consultant).

Box 1.13 The Human Population
Editor/writer: Janet Overton (WRI).

Box 1.14 Valuing the Invaluable
Editor: Wendy Vanasselt (WRI). Contributing writer: Christine Mlot (consultant). Reviewers: John Dixon (World Bank), Stefano Pagiola (World Bank), and David Simpson (RFF).

Box 1.15 Ecotourism and Conservation: Are They Compatible?
Editor/writer: Wendy Vanasselt (WRI). Reviewers: Katrina Brandon (Organization for Tropical Studies) and James N. Sweeting (CI).

Box 1.16 Uprooting Communal Tenure in Indonesian Forests
Editor: Janet Overton (WRI). Contributing writer: Richard Payne (consultant). Reviewer: Owen J. Lynch (NRDC).

Box 1.17 Rural Poverty and Adaptation
Editor: Wendy Vanasselt (WRI). Contributing writers: Wendy Vanasselt (WRI) and Sara Scherr (University of Maryland). Reviewers: Simon Batterbury (London School of Economics) and Tim Forsyth (Kennedy School of Government, Harvard University).

CHAPTER 2 TAKING STOCK OF ECOSYSTEMS

Main text
Editor: Gregory Mock (WRI). Contributing writer: Walter V. Reid (consultant).

Pilot Analysis of Global Ecosystems (PAGE)
Project manager: Norbert Henninger (WRI).

The PAGE authors would like to express their gratitude to the many individuals who contributed data and advice, attended expert workshops in October 1998 or February 1999, and reviewed successive drafts of this report.

Agroecosystems
PAGE authors: Stanley Wood (IFPRI), Kate Sebastian (IFPRI), and Sara Scherr (University of Maryland).

Contributors: Joseph Alcamo (University of Kassel, Germany), Carlos Baanante (IFDC), K. Balasubramanian (JRD Tata Ecotechnology Centre), Mary-Jane Banks (IFPRI), Niels Batjes (ISRIC), Christine Bergmark (USAID), Ruchi Bhandari (WRI), Jesslyn Brown (USGS/EDC), Sally Bunning (FAO), Emily Chalmers (consultant), Connie Chan-Kang (IFPRI), Linda Collette (FAO), Uwe Deichmann (World Bank), Andrew Farrow (CIAT), Jean-Marc Faurès (FAO), Günther Fischer (IIASA), Kathleen Flaherty (IFPRI), Louise Fresco (FAO), Robert Friedmann (The H. John Heinz III Center for Science, Economics and the Environment), Arthur Getz (WRI), Luis Gomez (consultant), Richard Harwood (Michigan State University), Peter Hazell (IFPRI), Gerhard Heilig (IIASA), Julio Henao (IFDC), Norbert Henninger (WRI), Robert Hijmans (International Potato Center), Anthony C. Janetos (WRI), Peter Jones (CIAT), Sjef Kauffman (ISRIC), Parviz Koohafkan (FAO), Emily Matthews (WRI), Siobhan Murray (WRI), Freddy Nachtergaele (FAO), Robin O'Malley (The H. John Heinz III Center for Science, Economics and the Environment), Peter Oram (IFPRI), Phillip Pardey (IFPRI), Stephen Prince (University of Maryland), Armando Rabufetti (IAI), Claudia Ringler (IFPRI), Mark Rosegrant (IFPRI), Melinda Smale (IFPRI), Lori Ann Thrupp (U.S. EPA), Thomas Walker (International Potato Center), Manuel Winograd (CIAT), Hans Wolter (FAO), and Liangzhi You (IFPRI).

Coastal Ecosystems

PAGE authors: Lauretta Burke (WRI), Yumiko Kura (WRI), Ken Kassem (WRI), Mark Spalding (WCMC), Carmen Revenga (WRI), and Don McAllister (OVI).

Contributors: Tundi Agardy (CI), Salvatore Arico (UNESCO), Jaime Baquero (OVI), Barbara Best (USAID), Simon Blyth (WCMC), Suzanne Bricker (NOAA), John Caddy (FAO), Robert Cambell (OVI), Joe Cimino (Veridian-MRJ Technology Solutions), Steve Colwell (CORAL), Lucy Conway (WCMC), Neil Cox (WCMC), Ned Cyr (GOOS), Charlotte De Fontaubert (IUCN), Uwe Deichmann (World Bank), Robert Diaz (Virginia Institute of Marine Science), Charles Ehler(NOAA/NOS), Paul Epstein (Harvard Medical School), Jonathan Garber (U.S. EPA), Luca Garibaldi (FAO), Richard Grainger (FAO), Ed Green (WCMC), Brian Groombridge (WCMC), Ingrid Guch (NOAA), Chantal Hagen (WCMC), Lynne Hale (Coastal Resources Center, University of Rhode Island), Maria Haws (Coastal Resources Center, University of Rhode Island), Jim Hendee (NOAA), Joanna Hugues (WCMC), David James (FAO), John McManus (ICLARM), Tom O'Connor (NOAA), Paul Orlando (NOAA), Hal Palmer (Veridian-MRJ Technology Solutions), Bruce Potter (Island Resources Foundation), Lorin Pruett (Veridian-MRJ Technology Solutions), Corinna Ravillious (WCMC), Shawn Reifsteck (CORAL), Kelly Robinson (Caribbean Association for Sustainable Tourism), Pam Rubinoff (Coastal Resources Center, University of Rhode Island), Charles Sheppard (University of Warwick, U.K.), Ben Sherman (Univeristy of New Hampshire), Mercedes Silva (Caribbean Tourism Organization), Gary Spiller (OVI), Al Strong (NOAA), Matt Stutz (Duke University), James Tobey (Coastal Resources Center, University of Rhode Island), and Sylvia Tognetti (University of Maryland).

Forest Ecosystems

PAGE authors: Emily Matthews (WRI), Siobhan Murray (WRI), Richard Payne (consultant), and Mark Rohweder (WRI).

Contributors: Mark Ashton (Yale University), Jim Ball (FAO), Daniel Binkley (Colorado State University), Richard Birdsey (USDA Forest Service), Chris Brown (FAO), Sandra Brown (Winrock International), Dirk Bryant (WRI), Virginia Dale (ORNL), Robert Davis (FAO), Ruth de Fries (University of Maryland), Eric Dinerstein (WWF-US), John Dixon (ORNL), Robert Dixon (DOE), Nigel Dudley (Equilibrium, U.K.), Curt Flather (USDA Forest Service), Jeffrey Fox (East-West Center), Robert Friedman (The H. John Heinz III Center for Science, Economics and the Environment), Alan Grainger (Leeds University, U.K.), David Hall (Kings College London), John Hart, Richard Haynes (USDA Forest Service), Derek Holmes (World Bank), Richard Houghton (Woods Hole Research Center), Bill Jackson (Pacific Northwest Research Station, USDA Forest Service), Anthony C. Janetos (WRI), Nels Johnson (WRI), Valerie Kapos (WCMC), Tony King (ORNL), Lars Laestadius (WRI), Jonathan Loh (WWF International), Tim Moermond (University of Wisconsin, Madison), John Morrison (WWF-US), Gordon Orians (University of Washington), N.H. Ravindranath (ASTRA and Centre for Ecological Sciences, India), Kent Redford (Wildlife Conservation Society), Barry Rock (University of New Hampshire), Mark Sagoff (University of Maryland), Dan Simberloff (University of Tennessee), Jorge Soberon (University of Kansas), Robert Socolow (Princeton University), Miguel Trossero (FAO), Compton Tucker (University of Maryland), Emma Underwood (WWF-US), and Karen Waddell (USDA Forest Service).

Freshwater Systems

PAGE authors: Carmen Revenga (WRI), Jake Brunner (WRI), Norbert Henninger (WRI), Ken Kassem (WRI), and Richard Payne (consultant).

Contributors: Robin Abell (WWF-US), Devin Bartley (FAO), Amy Benson (USGS), Kajsa Berggren (Umeå University), Ger Bergkamp, (IUCN), Stephen J. Brady (USDA/NRCS), Jesslyn Brown (USGS/EDC), Morley Brownstein (Health Canada), Cynthia Carey (University of Colorado), John Cooper (Environment Canada), Thomas E. Dahl (National Wetlands Inventory, U.S. Fish and Wildlife Service), Nick Davidson (Ramsar Convention Bureau), Jean-Marc Faurès (FAO), Balázs Fekete (University of New Hampshire), Andy Fraser (Environment Canada), Stephen Foster (BGS), Scott Frazier (Wetlands International), Brij Gopal (Jawaharlal Nehru University, India), Wolfgang Grabs (Global Runoff Data Centre, Germany), Pia Hansson (Umeå University), Jippe Hoogeveen (FAO), Colette Jacono (USGS), Anthony C. Janetos (WRI), Jim Kapetsky (FAO), James Karr (University of Washington), Les Kaufman (Boston University), Yumiko Kura (WRI), Kim Martz (USGS), Don McAllister (OVI), Gregory Mock (WRI), Peter Moyle (UC Davis), Tom Neill (Oregon Department of Fish and Wildlife), Christer Nilsson (Umeå University), Kim W. Olesen (DHI Water & Environment, Denmark), Francisco Olivera (UT Austin), Sandra Postel (Global Water Policy Project), Edward T. Rankin (Ohio EPA), Corinna Ravilious (WCMC), Ilze Reiss (Environment Canada), Hans H. Riber (COWI Consulting Engineers and Planners AS, Denmark), Steve Rothert (IRN), Robert Rusin, NASA Goddard Space Flight Center, Dork Sahagian (IGBP/GAIM, University of New Hampshire), John R. Sauer (USGS), Teresa Scott (Washington Department of Fish & Wildlife), Igor Shiklomanov (State Hydrological Institute, Russia), Robert Slater (Environment Canada), Charles Spooner (U.S. EPA), Bruce Stein (TNC), Melanie J. Stiassny (American Museum of Natural History), Magnus Svedmark (Umeå University), Greg Thompson (Environment Canada), Kirsten Thompson (WRI), Niels Thyssen (EEA), Dan Tunstall (WRI), Joshua Viers (UC Davis), Zipangani M. Vokhiwa (Ministry of Research and Environmental Affairs, Malawi), Charles Vörösmarty (University of New Hampshire), David Wilcove (Environmental Defense), and Shaojun Xiong (Umeå University).

Grassland Ecosystems

PAGE authors: Robin White (WRI), Siobhan Murray (WRI), and Mark Rohweder (WRI).

Substantial contributions: Stephen Prince (University of Maryland, Geography Department) and Kirsten Thompson (WRI).

Contributors: Roy H. Behnke (ODI), Daniel Binkley (Colorado State University), Jesslyn Brown (USGS/EDC), Virginia Dale (ORNL), Andre DeGeorges (Safari Club International), Eric Dinerstein (WWF-US), James E. Ellis (Colorado State University), Hari Eswaran (USDA/NRCS), Louise Fresco (FAO), Robert Friedman (The H. John Heinz III Center for Science, Economics and the Environment), Ruth de Fries (University of Maryland), Peter Gilruth (UNDP), Scott Goetz (University of Maryland), Paul Goriup (Nature Conservation Bureau, U.K.), David Hall (Kings College, London), Allen Hammond (WRI), Richard Houghton, Woods Hole Research Center, JoAnn House Kings College, London), Anthony C. Janetos (WRI), John Kartesz (University of North Carolina, Chapel Hill), Tony King (ORNL), Kheryn Klubnikin (IUCN-Washington), Wayne Ostlie (TNC), Leslie Roberts (AAAS), Eric Rodenburg (USGS), Osvaldo Sala (Cátedra de Ecología Facultad de Agronomía, Argentina), Cristian Samper (Instituto Alexander von Humboldt, Colombia), David Sneath (University of Cambridge), Alison Stattersfield (Birdlife International), Bruce Stein (TNC), Thomas R. Vale (University of Wisconsin-Madison), and Keith L. White (University of Wisconsin-Green Bay).

Boxes

Editors: George Faraday (consultant), Deborah Farmer (consultant), and Carol Rosen (WRI).

Appendix
Mountain Ecosystems
Editor: Wendy Vanasselt (WRI). Contributing writers: Emily Matthews (WRI), Janet Overton (WRI), and Wendy Vanasselt (WRI). Reviewers: Thomas Kohler (University of Berne, Switzerland) and Martin Price (Environmental Change Institute, University of Oxford).

Polar Ecosystems
Editor: Wendy Vanasselt (WRI). Contributing writers: Lori Han (WRI), Steve Nadis (consultant), and Wendy Vanasselt (WRI). Reviewer: Lars Kullerund (GRID-Arendal).

Urban Ecosystems
Editors/writers: Wendy Vanasselt (WRI) and Gregory Mock (WRI). Contributors: Jeff Beattie (American Forests), Richard Haeuber (Ecological Society of America), Jay Moor (Global Urban Observatory), Dave Nowak (USDA Forest Service), Daniel Smith (American Forests), and Mark Walbridge (George Mason University).

CHAPTER 3 LIVING IN ECOSYSTEMS

Agroecosystems
Regaining the High Ground: Reviving the Hillsides of Machakos, Kenya
Editor: Wendy Vanasselt (WRI). Contributing writers: Laurie Conly (consultant) and Joel Bourne (consultant). Reviewers: Paul Kimeu (Machakos soil and water conservation officer), George N. Mbate (USAID) John Murton (British Embassy), and Mary Tiffen (Drylands Research, U.K.).

Cuba's Agricultural Revolution: A Return to Oxen and Organics
Editor: Wendy Vanasselt (WRI). Contributing writer: Joel Bourne (consultant). Reviewers: Miguel A. Altieri (UC Berkeley), J. Paul Mueller (North Carolina State University), and Peter Rosset (Institute for Food and Development Policy/Food First).

Coastal Ecosystems
Replumbing the Everglades: Wetlands Restoration in South Florida
Editors: Deborah Farmer (consultant) and Gregory Mock (WRI). Writer: Gregory Mock (WRI). Reviewers: Thomas Armentano (Everglades National Park), Nicholas G. Aumen (consultant), Steven Davis (SFWMD), Dale Galwick (SFWMD), Richard Harvey (U.S. EPA), Ronald Jones (Florida International University), and Charles Lee (Audubon of Florida). Additional contributions: Kevin Burger (SFERTF), Angela Chong (SFWMD), Bonnie Kranzer (GCSSF), Nancy Lin (SFWMD), Patrick Lynch (SFWMD), Terry Rice (Southeast Environmental Research Program), Kathryn Ronan (SFWMD), and Terrance Salt (SFERTF). Maps: Kirsten Thompson (WRI).

Bolinao Rallies Around its Reefs
Editor: Wendy Vanasselt (WRI). Contributing writers: Steve Nadis (consultant), Janet Overton (WRI), and Wendy Vanasselt (WRI). Reviewers: Tony LaVina (WRI) and Liana Talaue-McManus (University of the Philippines).

Managing Mankòtè Mangrove
Editor/writer: Wendy Vanasselt (WRI). Reviewers: Lauretta Burke (WRI) and Allan Smith (CANARI).

Forest Ecosystems
Up From the Roots: Regenerating Dhani Forest Through Community Action
Editors: Gregory Mock (WRI) and Wendy Vanasselt (WRI). Contributing writers: Prateep Nayak (Vasundhara, India), Neera M. Singh (Vasundhara, India), Greg Mock (WRI), Silanjan Bhattacharyya (Vivekananda College, India), Madhav Gadgil (Centre for Ecological Sciences, Indian Institute of Science), and Tapan Mishra (Raja Narendralal Khan Women's College, India). Reviewers: Madhav Gadgil and Anirban Ganguly (Centre for Ecological Sciences, Indian Institute of Science). Additional contributions: M.D. Subash Chandran (Dr. A.V. Baliga College of Arts and Science, India), Kalipada Chatterjee (Development Alternatives, India), Neeraj Negi (Seva Mandir, India), Usha Sekhar (Centre for Science and Environment, India), and Mamta Vardhan (Seva Mandir, India). Maps: Kirsten Thompson (WRI).

Freshwater Systems
Working for Water, Working for Human Welfare in South Africa
Editor/writer: Wendy Vanasselt (WRI). Reviewers: Mark Botha (Botanical Society, South Africa), Caroline Gelderblom (CSIR), Andrew Malk (WRI), Christo Marais (National Working for Water Programme), and Brian van Wilgen (CSIR). Contributing writer to box on South Africa's Water Law: Gwen Parker (WRI), reviewed by Geert Creemers (consultant) and Saliem Fakir (IUCN). Map: Siobhan Murray (WRI).

Managing the Mekong River: Will a Regional Approach Work?
Editor: Wendy Vanasselt (WRI). Contributing writers: Nathan Badenoch (WRI), Jake Brunner (WRI), and Greg Mock (WRI). Reviewers: John Dore (WRI/REPSI) and Glenn S. Morgan (World Bank). Map: Kirsten Thompson (WRI).

New York City's Watershed Protection Plan
Editor/writer: Wendy Vanasselt (WRI). Reviewers: Jeffrey Gratz (U.S. EPA), Mark Izeman (NRDC), Robin Marx (NRDC), Donald Reed (WRI), and Geoffrey Ryan (Department of Environmental Protection, New York City).

Grassland Ecosystems
Sustaining the Steppe: The Future of Mongolia's Grasslands
Editor: Wendy Vanasselt (WRI). Contributing writer and reviewer: David Sneath (University of Cambridge). Maps: Siobhan Murray and Kirsten Thompson (WRI).

Special thanks to Lori Han (WRI) and Amy Wagener (WRI) for graphics assistance throughout Chapter 3.

CHAPTER 4 ADOPTING AN ECOSYSTEM APPROACH

Editors/writers: Carol Rosen (WRI), Gregory Mock (WRI), and Wendy Vanasselt (WRI). Contributing writer: Walter V. Reid (consultant). Reviewers: Matthew Arnold (WRI), Gerard Cunningham (UNEP), Dave MacDevette (UNEP), Sheila Heileman (UNEP), Norbert Henninger (WRI), Anthony C. Janetos (WRI), Valerie Thompson (WRI), Dik Tromp (UNEP), and Dan Tunstall (WRI).

PART II DATA TABLES

Project manager: Robin White (WRI)
Copyeditor: Michael Edington (consultant)

Biodiversity and Protected Areas
Research and data compilation: Carmen Revenga (WRI). Reviewers and contributors: Antonia Agama (Man and the Biosphere, Spain), Javier Beltran (WCMC), John Caldwell (WCMC), Neil Cox (WCMC),

Harriet Gillet (WCMC), Rosanna Karam (UNESCO), Dwight Peck (Ramsar Convention Bureau), Mechtild Rossler (UNESCO), Mark Spalding (WCMC), and Katarina Vestin (UNESCO).

Forests and Grasslands
Research and data compilation: Carmen Revenga (WRI), Mark Rohweder (WRI), and Robin White (WRI). Reviewers and contributors: Tamara Finkler (FSC), Sue Irmonger (WCMC), Emily Matthews (WRI), D. Pandey (FAO), Corinna Ravilious (WCMC), Dan Tunstall (WRI), and Adrian Whiteman (FAO).

Coastal, Marine, and Inland Waters
Research and data compilation: Carmen Revenga (WRI). Reviewers and contributors: J. Cimino (Veridian-MRJ Technology Solutions), Adele Crispoldi (FAO), Rachel Donnelly (WCMC), Luca Garibaldi (FAO), David James (FAO), Ken Kassem (WRI), Yumiko Kura (WRI), Edmondo Laureti (FAO), Lorin Pruett (Veridian-MRJ Technology Solutions), Eric Rodenburg (USGS), Mark Spalding (WCMC), and Dan Tunstall (WRI).

Agriculture and Food
Research and data compilation: Christian Ottke (WRI). Reviewers and contributors: Alan Brewster (Yale), Mark Cohen (for P. Pinstrup-Anderson; IFPRI), Eric Rodenburg (USGS), Orio Tampieri (FAO), and Dan Tunstall (WRI).

Freshwater
Research and data compilation: Carmen Revenga (WRI) and Mark Rohweder (WRI). Reviewers and contributors: Aline Comeau (BP/RAC), Jean-Marc Faurès (FAO), Ken Kassem (WRI), Yumiko Kura (WRI), Jean Margat (BP/RAC), Eric Rodenburg (USGS), and Alexander Safian (Israel).

Atmosphere and Climate
Research and data compilation: Mark Rohweder (WRI). Reviewers and contributors: Kevin Baumert (WRI), Ruchi Bhandari (WRI), Tom Boden (CDIAC), Alan Brewster (Yale School of Forestry and Environmental Studies), Nancy Kete (WRI), Eric Rodenburg (USGS), Vigdis Vestrang (ECE), and Dan Tunstall (WRI).

Energy and Resource Use
Research and data compilation: Christian Ottke (WRI). Reviewers and contributors: Jonathan Loh (WWF), Jim MacKenzie (WRI), Emily Matthews (WRI), and Karen Treanton (IEA).

Population and Human Development
Research and data compilation: Christian Ottke (WRI). Reviewers and contributors: Alan Brewster (Yale School of Forestry and Environmental Studies), Vittoria Cavicchioni (UNESCO), Shiu-Kee Chu (UNESCO), Norbert Henninger (WRI), Anthony C. Janetos (WRI), Robert Johnston (UNSTAT), Laura Mourino-Casas (UNDP), Dan Tunstall (WRI), and Tessa Wardlaw (UNICEF).

Economic Indicators
Research and data compilation: Mark Rohweder (WRI). Reviewers and contributors: Duncan Austin (WRI), Gwen Parker (WRI), Dan Tunstall (WRI), Alan Brewster (Yale School of Forestry and Environmental Studies), Saeed Ordoubadi (World Bank), and Eric Rodenburg (USGS).

Small Nations and Islands
Research and data compilation: Christian Ottke (WRI), Carmen Revenga (WRI), and Mark Rohweder (WRI).

The *World Resources* staff also wishes to extend thanks to the following individuals for their various contributions:

Martha Ainsworth (World Bank), Patricia Ardila (consultant), Katya Balasubramaian (consultant), John Barnes; Beth Behrendt (WRI), Hyacinth Billings (WRI), Lynn Brown (World Bank), Mauricio Castro Salazar (Banco Centroamericano de Integracion Economica, Honduras), Elsa Chang (WRI), Munyaradzi Chenje (SARDC/ IMERCSA), Richard Cincotta (Population Action International), Diana Cornelius (PRB), Robert Crooks (World Bank), Angela Cropper, Maria Camila Diaz (Fundacion Pro-Sierra Nevada de Santa Marta), Laura Lee Dooley (WRI), Steven Erie (UCSD), Elizabeth Frankenberg (RAND), Jacob Gayle (UNAIDS), Julie Harlan (WRI), Gary Harrison (Chickaloon Village Traditional Council, Alaska), Beth Harvey (WRI), Carl Haub (PRB), Brian Hirsch (Earth Energy Systems, Ltd.), C.S. Holling (University of Florida), Susan Hunter, Andrei Iatsenia (World Bank), Lisa Jorgenson (consultant), Robert Kaplan (Inter-American Development Bank), Miwako Kurosaka (WRI), Judith Lancaster (Desert Research Institute), Gideon N. Louw, Magda Lovei (World Bank), Pilar Lozano (consultant), Kenton Miller (WRI), Becky Milton (WRI), Marta Miranda (WRI), Bill Pease (Environmental Defense), William Platt (Louisiana State University), Fred Powledge (consultant), Marc Reisner (Vidler Water Co.), Arsenio M. Rodriguez (World Bank), Maria Patricia Sanchez (consultant), Bernhard Schwartlander (UNAIDS), Mary Seely (Desert Research Foundation of Namibia), Grant Singleton (CSIRO Wildlife and Ecology), Henning Steinfeld (FAO), M.S. Swaminathan (M.S. Swaminathan Research Foundation), Charlotte M. Taylor (Missouri Botanical Garden), Jonathan Timberlake (Foundation for Africa), Helen Todd (Cashpor), Michael Totten (CI), John Williamson, Jacob Yaron (World Bank), and Hania Zlotnik (UN Population Division). We add special thanks to Judy Gibson (production manager), Brenda Waugh (typesetter), and Marisha Tapera (proofreader) of The Magazine Group.

Notes

1. *Extent and Growth.* To determine the extent of agroecosystems, the International Geosphere-Biosphere Programme (IGBP) defined agroecosystems on the basis of remote sensing imagery and defined agricultural regions as areas where more than 40 percent of the land is used for cropland or highly managed pasture. Using this definition, agroecosystems account for 21 percent of total land area (USGS EDC 1998). However, this excludes significant areas where there is overlap with forest and grassland ecosystems, since, in fact, land use is often fragmented spatially. Where agriculture mixes with other land uses–forests or grasslands–a mosaic of land cover is formed.

For the PAGE study, satellite data were reinterpreted to incorporate mosaic areas that have a 30 percent or more intensity of cropland or managed pasture. Using this approach, approximately 6 percent of areas classified as forest and 14 percent classified as grasslands by IGBP fall within the global extent of agroecosystems as defined by PAGE. Thus the percentage of agricultural land area totals 28 percent (Wood et al. [PAGE] 2000).

2. *Economic Importance.* The total value of agricultural production output was calculated by weighting 134 primary crop and 23 primary livestock commodity quantities by their respective average international agricultural prices (calculated by the Gary-Khamis method) during 1989–91.

3. *Soil Degradation.* It is difficult to reconcile these results with observed growth in food production in Asia, even allowing for past increases in fertilizer application rates. But this apparent incompatibility highlights the basic challenge of using existing data sets in making credible assessments of the state and changing capacity of ecosystems.

4. *Deforestation and Forest Loss.* See for example: Holmes, Derek (2000, draft of 25 February), Deforestation in Indonesia: A Review of the Situation in Sumatra, Kalimantan, and Sulawesi. (Draft report in preparation for the World Bank, based on mapping carried out by the Indonesian Ministry of Forestry and Estate Crops; data are subject to final revision, but are not expected to change significantly.)

5. *Supply and Demand.* PAGE researchers used a slightly lower estimate of global runoff than previous analyses and discounted the use of fossil water sources, since such use is unsustainable in the long term.

References

Chapter 1 text

Batterbury, S. and T. Forsyth. 1999. Fighting back: Human adaptations in marginal environments. *Environment* 41(6):6–11, 25–30.

Bryant, D., L. Burke, J. McManus and M. Spalding. 1998. *Reefs at Risk: A Map-Based Indicator of Threats to the World's Coral Reefs.* Washington, D.C.: World Resources Institute.

Bryant, D., D. Nielsen and L. Tangley. 1997. *The Last Frontier Forests.* Washington D.C.: World Resources Institute.

Da Rosa, C. and J. Lyon. 1997. *Golden Dreams, Poisoned Streams: How Reckless Mining Pollutes America's Waters, and How We Can Stop It.* Washington, D.C.: Mineral Policy Center.

Daily, G., ed. 1997. *Nature's Services: Societal Dependence on Natural Ecosystems.* Washington, D.C.: Island Press.

de Moor, A. and P. Calamai. 1997. *Subsidizing Unsustainable Development: Undermining the Earth with Public Funds.* San Jose, Costa Rica: The Earth Council.

Ecological Society of America (ESA). 1997a. Ecosystem Services: Benefits Supplied to Human Societies by Natural Ecosystems. *Issues in Ecology* 2(Spring).

Ecological Society of America (ESA). 1997b. Human Alteration of the Global Nitrogen Cycle: Causes and Consequences. *Issues in Ecology* 1(February).

Environmental Investigation Agency (EIA) and Telapak. 1999. *The Final Cut: Illegal Logging in Indonesia's Orangutan Parks.* London: EIA.

Food and Agriculture Organization (FAO). 1999. *The State of World Fisheries and Aquaculture, 1998.* Rome: FAO.

Gadgil, M. and R. Guha. 1992. *This Fissured Land: An Ecological History of India.* Delhi: Oxford University Press.

Garcia, S. M. and I. De Leiva Moreno. In Press. Trends in world fisheries and their resources: 1974–1999. In *The State of Fisheries and Aquaculture 2000.* FAO, ed. Rome: FAO.

Global Environment Facility (GEF). 1998. *Valuing the Global Environment: Actions and Investments for a 21st Century.* Washington, D.C.: GEF.

Harris, D. 1996. *The Last Stand: The War between Wall Street and Main Street over California's Ancient Redwoods.* San Francisco: Sierra Club Books.

Houghton, J., L. Filho, D. Griggs and K. Maskell. 1997. Stabilization of Atmospheric Greenhouse Gases: Physical, Biological, and Socio-economic Implications. Intergovernmental Panel on Climate Change (IPCC) Technical Paper. Geneva: World Meteorological Organization/U.N. Environment Programme.

Hughes, J., G. Daily and P. Ehrlich. 1997. Population diversity: Its extent and extinction. *Science* 278:689–692.

Kellert, S. R. and E. O. Wilson, eds. 1993. *The Biophilia Hypothesis.* Washington, D.C.: Island Press.

Koskela, J., C. Li and O. Luukkanen. 1999. Protective forest systems in China: Current status, problems, and perspectives. *Ambio* 28(4):341–345.

Living on Earth (LOE). 1996. Transcript of 29 March 1996 radio interview of David Harris by Steve Curwood, host of Living on Earth radio broadcast. Online at: http://www.loe.org/archives/960329.htm.

Milich, L. 1999. Resource mismanagement versus sustainable livelihoods: The collapse of the Newfoundland cod fishery. *Society and Natural Resources* 12:625–642.

National Oceanic and Atmospheric Administration (NOAA). 1998. Flooding in China Summer 1998. Online at: http://www.ncdc.noaa.gov/ol/reports/chinaflooding/chinaflooding.html. (20 November 1998).

Panayotou, T., Fellow, Harvard Institute for International Development, Harvard University. 1999. Personal Communication. E-mail. 12 October.

Parrotta, J. and J. Turnbull. 1997. Catalyzing Native Forest Regeneration on Degraded Forest Lands. *Forest Ecology and Management* 99 (Special Issues 1, 2):1–290.

Postel, S. 1999. *Pillar of Sand.* New York: Norton.

Prodanov, K., K. Midhailov, G. Dashkalov, C. Maxim, A. Chashchin, A. Arkhipov, V. Shlyakhov and E. Ozdamar. 1997. Environmental Management of Fish Resources in the Black Sea and Their Ratio-

nal Exploitation. General Fisheries Council for the Mediterranean, Studies and Reviews, No. 68. Rome: FAO.
Scherr, S. 1999. Background Paper on Poverty and Environment. Washington, D.C.: World Resources Institute.
Travis, J. 1993. Invader threatens Black, Azov Seas. *Science* 262:262–263.
U.N. Development Programme (UNDP). 1998. *Human Development Report 1998*. New York: UNDP.
U.N. Development Programme (UNDP). 2000. The Challenge of Poverty. Online at: http://www.undp.org/uncdf/pubs/cdf30/poverty.htm.
U.N. Population Division (UNPD). 1997. *World Urbanization Prospects: The 1996 Revision, Annex Tables*. New York: U.N. Population Division.
U.N. Population Division (UNPD). 1998. *World Population Prospects: The 1998 Revision*. 1. New York: U.N. Population Division.
Vitousek, P. M., H. A. Mooney, J. Lubchenco and J. M. Mellilo. 1997. Human domination of Earth's ecosystems. *Science* 277:494–499.
Watson, R., J. Dixon, S. Hamburg, A. Janetos and R. Moss. 1998. Protecting Our Planet, Securing Our Future. Washington, D.C.: U.N. Environment Programme, U.S. Aeronautics and Space Administration, The World Bank.
Wood, S., K. Sebastian and S. Scherr. 2000. Pilot Analysis of Global Ecosystems: Agroecosystems Technical Report. Washington, D.C.: World Resources Institute and International Food Policy Research Institute.
World Bank. 1997. Five Years after Rio: Innovations in Environmental Policy. Environmentally Sustainable Development Studies and Monographs Series, No. 18. Washington, D.C.: The World Bank.
World Bank. 1999a. China Yangtze Flood Emergency Rehabilitation Project. Online at: http://www.worldbank.org/pics/pid/cn63123.txt. (13 July 1999).
World Bank. 1999b. *World Development Indicators 1999*. Washington, D.C.: The World Bank.
World Commission on Forests and Sustainable Development (WCFSD). 1999. *Our Forests, Our Future*. Winnipeg, Manitoba, Canada: WCFSD.
World Meteorological Organization (WMO). 1997. *Comprehensive Assessment of the Freshwater Resources of the World*. Geneva: WMO.
World Resources Institute in collaboration with the United Nations Environment Programme, the United Nations Development Programme and the World Bank. 1998. *World Resources 1998–99*. New York: Oxford University Press.

Box 1.1 History of Use and Abuse

Clark, K. L. 1996. A Montreal Protocol for POPs? World Wide Fund for Nature International. Online at: http://www.chem.unep.ch/pops/indxhtms/manwg8.html.
Coe, M. 1999. *The Maya*. Sixth Edition. London: Thames and Hudson.
Crosby, A. W. 1986. *Ecological Imperialism: The Biological Expansion of Europe, 900–1900*. Cambridge: Cambridge University Press.
Gadgil, M. and R. Guha. 1992. *This Fissured Land: An Ecological History of India*. Delhi: Oxford University Press.
Hillel, D. J. 1991. *Out of the Earth: Civilization and the Life of the Soil*. New York: The Free Press.
Hughes, J. D. 1994. *Pans' Travail: Environmental Problems of the Ancient Greeks and Romans*. Baltimore: Johns Hopkins University Press.
McNeill, J. R. 2000. *Something New Under the Sun: An Environmental History of the Twentieth Century World*. New York: W. W. Norton.
Parker, L. 2000. Stratospheric Ozone Depletion: Implementation Issues. National Council for Science and the Environment. Online at: http://www.cnie.org/nle/strat-5.html. (11 January 2000).
Simmons, I. G. 1989. *Changing the Face of the Earth: Culture, Environment, and History*. Oxford: Basil Blackwell, Inc.
Vermeer, E. B. 1998. Population and ecology on the frontier. In *Sediments of Time: Environment and Society in Chinese History*. M. Elvin and L. Ts'ui-jung, eds. Cambridge: Cambridge University Press.

Box 1.3 Water Filtration and Purification

Beverage Industry. 1999. State of the industry report: Bottled water proves it's a big fish. *Beverage Industry* 90(7):38–40.
Bhatia, R. and M. Falkenmark. 1993. Water Resource Policies and the Urban Poor: Innovative Approaches and Policy Imperatives. Washington, D.C.: UNDP-World Bank Water and Sanitation Program.
Lerner, S. and W. Poole. 1999. The Economic Benefits of Parks and Open Space: How Land Conservation Helps Communities Grow Smart and Protect the Bottom Line. San Francisco: The Trust for Public Land.
Marinelli, J. 1990. After the next flush: The next generation. *Garbage* Jan/Feb:24–35.
Neander, J. N.D. City of Arcata Wastewater Treatment Facility and Marsh and Wildlife Sanctuary. Online at: http://www.epa.gov/owowwtr/coastal/cookbook/page90.html.
Reid, W. (forthcoming). Capturing the value of ecosystem services to protect biodiversity. In *Managing Human Dominated Ecosystems*. Washington, D.C.: Island Press.
Revenga, C., J. Brunner, N. Henninger, K. Kassem and R. Payne. 2000. Pilot Analysis of Global Ecosystems: Freshwater Ecosystems Technical Report. Washington, D.C.: World Resources Institute.
Trust for Public Land (TPL). 1997. Protecting the Source: Land Conservation and the Future of America's Drinking Water. San Francisco: The Trust for Public Land.
U.N. Environment Program (UNEP). 1999. *Global Environmental Outlook 2000*. London: Earthscan Publications Ltd.
UNICEF. 2000. State of the World's Children. Online at: http://www.unicef.org/sowc00/stat5.htm.
World Health Organization (WHO). 1996. Water and Sanitation Fact Sheet. Online at: http://www.who.org/inf-fs/en/fact112.html. (November).
World Resources Institute in collaboration with the United Nations Environment Programme, The United .Nations. Development Programme and The World Bank 1996. *World Resources 1996–97*. New York: Oxford University Press.

Box 1.4 Pollination

Buchmann, S. L. and G. P. Nabhan. 1997. *The Forgotten Pollinators*. Washington, D.C.: Island Press.
Food and Agriculture Organization (FAO). 2000. FAOSTAT Database. Online at: http://apps.fao.org/lim500/nph-wrap.pl?Production.Crops.Primary&Domain=SUA&servlet=1. (March 2000).
Kearns, C. A., D. W. Inouye and N. M. Waser. 1998. Endangered mutualisms: The conservation of plant-pollinator interactions. *Annual Review of Ecological Systems* 29:83–112.
Kenmore, P. and R. Krell. 1998. Global perspective and pollination in agriculture and agroecosystem management. Paper presented at the International Workshop on the Conservation and Sustainable Use of Pollinators in Agriculture, with Emphasis on Bees (FAO), Sao Paulo, Brazil. October 7–9.
Nabhan, G. P. and S. L. Buchmann. 1997. Services provided by pollinators. Pp: 133–150 in *Nature's Services: Societal Dependence on Natural Ecosystems*. G. C. Daily, ed. Washington, D.C.: Island Press.
Southwick, E. E. and L. Southwick Jr. 1992. Estimating the economic value of honey bees (*Hymenoptera: Apidae*) as agricultural pollinators in the United States. *Journal of Economic Entomology* 85(3):621–633.

Box 1.5 Biological Diversity

Goudie, A. 2000. *The Human Impact on the Natural Environment*. Cambridge, MA: MIT Press.
Grifo, F., D. Newman, A. S. Fairfield, B. Bhattacharya and J. T. Grupenhoff. 1997. The origins of prescription drugs. Pp: 131–163 in *Biodiversity and Human Health*. F. Grifo and J. Rosenthal, eds. Washington, D.C.: Island Press.
Hughes, J. B., G. C. Daily and P. R. Ehrlich. 1997. Population diversity: Its extent and extinction. *Science* 278:689–691.

References

Reid, W. V. and K. R. Miller. 1989. *Keeping Options Alive: The Scientific Basis for Conserving Biodiversity.* Washington, D.C.: World Resources Institute.

ten Kate, K. and S. A. Laird. 1999. *The Commercial Use of Biodiversity: Access to Genetic Resources and Benefit-sharing.* London: Earthscan Publications Ltd.

Thrupp, L. A. 1998. *Cultivating Diversity: Agrobiodiversity and Food Security.* Washington D.C.: World Resources Institute.

U.N. Environment Program (UNEP). 1995. *Global Biodiversity Assessment.* Cambridge, UK: Cambridge University Press.

World Conservation Monitoring Centre (WCMC). 1992. *Global Biodiversity: Status of the Earth's Living Resources.* London: Chapman and Hall.

World Conservation Monitoring Centre (WCMC). 1998. *1997 IUCN Red List of Threatened Plants.* K. S. Walter and H. J. Gillet eds. Cambridge, UK: The World Conservation Union (IUCN).

Box 1.6 Carbon Storage

Brown, P. 1998. *Climate, Biodiversity and Forests: Issues and Opportunities Emerging from the Kyoto Protocol.* Washington, D.C.: World Resources Institute.

Ciaias, P. 1999. Restless carbon pools. *Nature* 398:111–112.

Intergovernmental Panel on Climate Change (IPCC) (R. Watson, I. N., B. Bolin, N. Ravindranath, D. Verardo, and D. Dokken, eds.). 2000. *Land Use, Land-Use Change, and Forestry.* Cambridge, UK: Cambridge University Press.

Intergovernmental Panel on Climate Change (IPCC) (R. Watson, M. C. Z., R. H. Moss eds.). 1996. *Climate Change 1995: Impacts, Adaptations and Mitigation of Climate Change: Scientific-Technical Analyses.* New York: Cambridge University Press.

Matthews, E., R. Payne, M. Rohweder and S. Murray. 2000. Pilot Analysis of Global Ecosystems: Forest Ecosystems. Washington, D.C.: World Resources Institute.

Box 1.8 Invasive Species

Bright, C. 1998. *Life Out of Bounds: Bioinvasion in a Borderless World.* New York: W.W. Norton & Company.

Marine Conservation Biology Institute (MCBI). 1998. Scientists call on Secretary Babbitt to Keep Noxious Seaweed out of U.S. Waters. Online at: http://mcbi.org/caulerpa/caulerpa.html.

Office of Technology Assessment (OTA). 1993. Harmful Non-Indigenous Species in the United States. OTA-F-565. Washington, D.C.: U.S. Government Printing Office. September.

Ruesink, J. L., I. M. Parker, M. J. Groom and P. M. Kareiva. 1995. Reducing the risks of nonindigenous species introductions: Guilty until proven innocent. *BioScience* 45(7):465–477.

Travis, J. 1993. Invader threatens Black, Azov Seas. *Science* 262(26 November):1366–1367.

U.S. Geological Survey (USGS). 1999. West Nile Virus may be New Deadly Strain, USGS tells Congress. News Release, 14 December 1999. Online at: http://www.usgs.gov/public/press/public_affairs/press_release/pr1128.html.

Vitousek, P. M., C. M. D'Antonio, L. L. Loope and R. Westbrooks. 1996. Biological invasions as global environmental change. *American Scientist* 84:468–478.

Vitousek, P. M., C. M. D'Antonio, L. L. Loope and R. Westbrooks. 1997. Introduced species: A significant component of human-caused global change. *New Zealand Journal of Ecology* 21(1):1–16.

Box 1.9 Trade-Offs: Lake Victoria's Ecosystem Balance Sheet

Achieng, A. P. 1990. The impact of the introduction of Nile perch, *Lates niloticus* (L.) on the fisheries of Lake Victoria. *Journal of Fish Biology* 37(Supplement A):17–23.

Food and Agriculture Organization of the United Nations (FAO). 1999. FISHSTAT. Version 2.19 by Yury Shatz. Rome: FAO.

Kaufman, L. 1992. Catastrophic change in species-rich freshwater ecosystems: The lessons from Lake Victoria. *BioScience* 42(11):846–858.

Kaufman, L., Boston University Marine Program. 2000. Personal Communication. Interview. 7 February.

Witte, F., T. Goldschmidt, J. Wanink, M. van Oijen, K. Goudswaard, E. Witte-Mass and N. Bouton. 1992. The destruction of an endemic species flock: Quantitative data on the decline of the haplochromine cichlids of Lake Victoria. *Environmental Biology of Fishes* 34:1–28.

Box 1.10 Domesticating the World: Conversion of Natural Ecosystems

Loveland, T. R., B. C. Reed, J. F. Brown, D. O. Ohlen, Z. Zhu, L. Yang and J. Merchant. 2000. Development of a Global Land Cover Characteristics Database and IGBP DISCover from 1 km AVHRR data. *International Journal of Remote Sensing* 21(6):1303–1330. Online at: http://edcdaac.usgs.gov/glcc/glcc.html.

National Oceanic and Atmospheric Administration-National Geophysical Data Center (NOAA-NGDC). 1998. Stable lights and radiance calibrated lights of the world CD-ROM. NOAA-NGDC: Boulder, CO. Online at: http://julius.ngdc.noaa.gov:8080/production/html/BIOMASS/night.html. (December 1998).

Population Reference Bureau (PRB). 1998. *United States Population Data Sheet.* Washington, D.C.: PRB.

U.N. Environment Program (UNEP). 1999. *Global Environmental Outlook 2000.* London: Earthscan Publications Ltd.

Vitousek, P. M., H. A. Mooney, J. Lubchenco and J. M. Mellilo. 1997. Human domination of Earth's ecosystems. *Science* 277:494–499.

Walker, B. H., W. L. Steffen and J. Langridge. 1999. Interactive and integrated effects of global change on terrestrial ecosystems. Pp: 329–374 in *The Terrestrial Biosphere and Global Change: Implications for Natural and Managed Ecosystems.* B. Walker, W. Steffen, J. Canadell and J. Ingram, eds. Cambridge: Cambridge University Press.

World Wildlife Fund, US (WWF-US). 1999. Ecoregions database. Unpublished Database. Washington, D.C.: WWF-US.

Box 1.11 How Much Do We Consume?

Brown, C. 1999. Global Forest Products Outlook Study: Thematic Study on Plantations. Working Paper No. GFPOS/WP/03 (Draft). Rome: FAO.

Food and Agriculture Organization of the United Nations (FAO). 2000. FAOSTAT databases. Online at: http://apps.fao.org/.

Food and Agriculture Organization of the United Nations (FAO). 1999. *The State of World Fisheries and Aquaculture, 1998.* Rome: FAO.

Laureti, E. 1999. Fish and Fishery Products: World Apparent Consumption Statistics Based on Food Balance Sheets. FAO Fisheries Circular No. 821, Revision 5. Rome: FAO.

Matthews, E. and A. Hammond. 1999. *Critical Consumption Trends and Implications: Degrading the Earth's Ecosystems.* Washington, D.C.: World Resources Institute.

Pinstrup-Andersen, P., R. Pandya-Lorch and M. Rosegrant. 1999. *World Rood Prospects: Critical Issues for the Early Twenty-First Century.* Washington, D.C.: International Food Policy Research Institute.

U.N. Development Programme (UNDP). 1998. *Human Development Report 1998.* New York: UNDP.

U.S. Department of Agriculture (USDA). 2000. Production, Supply and Distribution Database. Online at: http://usda.mannlib.cornell.edu/data-sets/international/93002/PSDFAQ.TXT.

World Bank. 1999. *World Development Indicators 1999*. Washington, D.C.: The World Bank.

Box 1.12 The Human Population

U.N. Population Division (UNPD). 1998a. *World Population Estimates and Projections: The 1998 Revision*. Online at: http://www.popin.org/pop1998/6.htm.
U.N. Population Division (UNPD). 1998b. *World Population Prospects: The 1998 Revision*. 1. New York: U.N. Population Division.
U.N. Environment Program (UNEP). 1999. *Global Environmental Outlook 2000*. London: Earthscan Publications Ltd.
U.N. Population Fund (UNFPA). 1999. Population Change and People's Choices. Chapter 2 in *The State of the World's Population 1999*. New York: UNFPA. Online at: www.unfpa.org/swp/1999/chapter2c.htm.
Wood, S., K. Sebastian and S. Scherr. 2000. Pilot Analysis of Global Ecosystems: Agroecosystems Technical Report. Washington, D.C.: World Resources Institute and International Food Policy Research Institute.
World Health Organization (WHO). 1997. Health and Environment in Sustainable Development: Five Years after the Earth Summit. Geneva: WHO.

Box 1.13 Pollution and Ecosystems

Aspelin, S. L. and A. H. Grube. 1999. Pesticides Industry Sales and Usage: 1996 and 1997 Market Estimates. 733-R-99-001. Washington, D.C.: Environmental Protection Agency. November.
D'Esposito, S. and J. Feiler. 2000. Lessons from the disasters on the Danube: Is modern mining safe? *Mineral Policy Center Newsletter* Spring:1, 4–5, 17.
Etkin, D. S. 1998. International oil spill statistics: 1997. Arlington, MA: Cutter Information Corp.
European Environment Agency (EEA). 1999. *Environment in the European Union at the Turn of the Century*. Environmental Assessment Report No. 2. Copenhagen: EEA
Matthews, E. and A. Hammond. 1999. *Critical Consumption Trends and Implications: Degrading the Earth's Ecosystems*. Washington, D.C.: World Resources Institute.
National Oceanic and Atmospheric Administration (NOAA). 2000. Hypoxia in the Gulf of Mexico: Progress Towards the Completion of an Integrated Assessment. Online at: http://www.nox.noaa.gov/products/pubs_hypoxia.html#Topic2.
Rabalais, N. and D. Scavia. 1999. Origin, impact and implications of the "Dead Zone" in the Gulf of Mexico. Presented at the US Global Change Program Seminar Series. 19 July.
Rabalais, N. N. 1998. Oxygen Depletion in Coastal Waters. NOAA's State of the Coast Report. NOAA: Silver Spring, MD. Online at: http://state-of-coast.noaa.gov/bulletins/html/hyp_09/case.html.

Box 1.14 Valuing the Invaluable

Anderson, T. 1996. Enviro-Capitalists: Why and how to preserve their habitat. Pp: 189–221 in *Economics of Biodiversity Loss IUCN Workshop*. Gland, Switzerland. Online at: http://economics.iucn.org. (April).
Honey, M. 1999. *Ecotourism and Sustainable Development: Who Owns Paradise?* Washington, D.C.: Island Press.
Ryan, G., New York City Department of Environmental Protection, Bureau of Water Supply and Wastewater Collection. 1998. Personal Communication. Interview. 2 December.
Sweeting, J. E. N., A. G. Bruner and A. B. Rosenfeld. 1999. The Green Host Effect: An Integrated Approach to Sustainable Tourism and Resort Development. Conservation International Policy Paper. Washington, D.C.: Conservation International.
World Tourism Organization (WTO). 1997. *Tourism 2000: Building a Sustainable Future for Asia-Pacific*. Final report from Asia Pacific Ministers' Conference on Tourism and Environment. Madrid: WTO.

Box 1.15 Ecotourism and Conservation: Are They Compatible?

Ecotourism Society. 1998. Ecotourism Statistical Fact Sheet. Online at: http://www.ecotourism.org/textfiles/stats.text.
Epler, B. 1997. *An Economic and Social Analysis of Tourism in the Galapagos Islands*. Providence: University of Rhode Island, Coastal Resource Center.
Gossling, S. 1999. Ecotourism: A means to safeguard biodiversity and ecosystem function? *Ecological Economics* 29(2):303–320.
Honey, M. 1999. *Ecotourism and Sustainable Development: Who Owns Paradise?* Washington, D.C.: Island Press.
Lindberg, K. and R. H. Huber Jr. 1993. Economic Issues in Ecotourism Management. Pp: 82–115 in *Ecotourism: A Guide for Planners and Managers*. K. Lindberg and D. E. Hawkins, eds. North Bennington, VT: The Ecotourism Society.
Sweeting, J. E. N., A. G. Bruner and A. B. Rosenfeld. 1999. The Green Host Effect: An Integrated Approach to Sustainable Tourism and Resort Development. Conservation International Policy Paper. Washington, D.C.: Conservation International.
Wells, M. 1997. *Economic Perspectives on Nature Tourism, Conservation and Development*. Washington D.C.: The World Bank.
World Bank. 1999. *World Development Indicators 1999*. Washington, D.C.: The World Bank.

Box 1.16 Uprooting Communal Tenure in Indonesian Forests

Barber, C. V. 1997. *Environmental Scarcities, State Capacity, and Civil Violence: The Case of Indonesia*. Cambridge: American Academy of Arts and Sciences.
Bromley, D. W. and M. M. Cernea. 1989. The Management of Common Property Natural Resources: Some Conceptual and Operational Fallacies. World Bank Discussion Paper No. 57. Washington, D.C.: The World Bank.
Campbell, J., Program Officer, Ford Foundation. 1998. Personal Communication. E-mail. December.
Fox, J. and K. Atok. 1997. Forest-dweller demographics in West Kalimantan, Indonesia. *Environmental Conservation* 24(1):31–37.
Lynch, O. J. 2000. Personal Communication. E-mail. 14 February.
Lynch, O. J. and J. Alcorn. 1994. Tenurial rights and community based conservation. Pp: 373–392 in *Natural Connections: Perspectives in Community-Based Management*. D. Western and R. M. Wright, eds. Washington, D.C.: Island Press.
Lynch, O. J. and K. Talbott. 1995. *Balancing Acts: Community-Based Forest Management and National Law in Asia and the Pacific*. Washington, D.C.: World Resources Institute.
Michon, G. and H. de Foresta. 1995. The Indonesian agro-forest model: Forest resource management and biodiversity conservation. Pp: 90–106 in *Conserving Biodiversity Outside Protected Areas: The Role of Traditional Agro-Ecosystems*. P. Halladay and D. A. Gilmour, eds. Gland: IUCN.
Padoch, C. and M. Pinedo-Vasquez. 1996. Smallholder forest management: Looking beyond non-timber forest products. Pp: 103–117 in *Current Issues in Non-Timber Forest Products Research*. M. Ruiz Pérez and J. E. M. Arnold, eds. Bogor: Center for International Forestry Research.
Peluso, N. L. 1995. Whose woods are these? Counter-mapping forest territories in Kalimantan, Indonesia. *Antipode* 27(4):383–406.
Poffenberger, M., P. Walpole, E. D'Silva, K. Lawrence and A. Khare. 1997. Linking Government with Community Resource Management. Research Network Report Number 9. Surajkund, India: Asia Forest Network. 2–6 December.
Sirait, M., S. Prasodjo, N. Podger, A. Flavelle and J. Fox. 1994. Mapping customary land in East Kalimantan, Indonesia: A tool for forest management. *Ambio* 23(7):411–417.
Zerner, C. 1992. Indigenous Forest-Dwelling Communities in Indonesia's Outer Islands: Livelihood, Rights, and Environmental Management Institutions in the Era of Industrial Forest Exploitation. Unpublished report. Washington, D.C.: The World Bank.

References

Box 1.17 Rural Poverty and Adaptation

Agarwal, A. and S. Narain. 1999. Community and Household Water Management: The Key to Environmental Regeneration and Poverty Alleviation. Paper presented to EU-UNDP. Online at: http://www.undp.seed/pei/publication/water.pdf. (February 1999).

Batterbury, S. and T. Forsyth. 1999. Fighting back: Human adaptations in marginal environments. *Environment* 41(6):6–11, 25–30.

Consultative Group on International Agricultural Research (CGIAR). 1997. Report of the Study on CGIAR Research Priorities for Marginal Lands. TAC Working Document. Technical Advisory Committee Secretariat, FAO. Online at: http://www.fao.org/wairdocs/tac/x5784e02.htm. (3 March).

Fairhead, J. and M. Leach. 1996. *Misreading the African Landscape: Society and Ecology in a Forest-Savanna Mosaic*. Cambridge, UK: Cambridge University Press.

Forsyth, T. and M. Leach. 1998. Poverty and Environment: Priorities for Research and Policy: An Overview Study. Prepared for the UNDP and European Commission. Sussex, UK: Institute of Development Studies. August.

Hazell, P. and J. L. Garrett. 1996. Reducing Poverty and Protecting the Environment: The Overlooked Potential of Less-favored Lands. 2020 Brief 39. Online at: http://www.cimmyt.org/ifpri/2020/briefs/2br39.htm. (October).

International Fund for Agricultural Development (IFAD). 2000. *Poverty Incidence in West and Central Africa (Draft)*. Rome: IFAD.

Jazairy, I., M. Alamgir and T. Panuccio. 1992. *The State of the World Rural Poverty: An Inquiry into Its Causes and Consequences*. New York: New York University Press.

Lynch, O. J. and K. Talbott. 1995. *Balancing Acts: Community-Based Forest Management and National Law in Asia and the Pacific*. Washington, D.C.: World Resources Institute.

Sillitoe, P. 1996. *A Place Against Time: Land and Environment in the Papua New Guinea Highlands*. Amsterdam: Harwood Academic Press.

Sillitoe, P. 1998. It's all in the mound: Fertility management under stationary shifting cultivation in the Papua New Guinea highlands. *Mountain Research and Development* 18(2):123.

UN Centre for Human Settlements (Habitat)(UNCHS). 1996. *An Urbanizing World: Global Report on Human Settlements*. Oxford, UK: Oxford University Press.

World Bank. 1999. *World Development Report 1998/99*. New York: Oxford University Press.

Chapter 2

Introduction

Batjes, N. H. and E. M. Bridges. 1994. Potential emissions of radiatively active gases from soil to atmosphere with special reference to methane: Development of a global database WISE. *Journal of Geophysical Research* 99(D8):16, 479–489.

Batjes, N. H. 1996. Total Carbon and Nitrogen in Soils of the World. *European Journal of Soil Science* 47:151–163.

Intergovernmental Panel on Climate Change (IPCC) (R. Watson, I. N., B. Bolin, N. Ravindranath, D. Verardo, and D. Dokken, eds.). 2000. *Land Use, Land-Use Change, and Forestry*. Cambridge, UK: Cambridge University Press.

Matthews, E., R. Payne, M. Rohweder and S. Murray. 2000. Pilot Analysis of Global Ecosystems: Forest Ecosystems. Washington, D.C.: World Resources Institute.

Postel, S. L., G. C. Daily and P. R. Ehrlich. 1996. Human appropriations of renewable fresh water. *Science* 271:785–788.

Seckler, D., U. Amarasinghe, D. Molden, R. de Silva and R. Barker. 1998. World Water Demand and Supply, 1990 to 2025: Scenarios and Issues. Research Report 19. Colombo, Sri Lanka: International Water Management Institute (IWMI).

U.S. Geological Survey (USGS) Earth Resources Observation Systems (EROS) Data Center (USGS/EDC). 1999b. 1999 Unpublished map that applied carbon density numbers from a previous study (Olson, J., J. A. Watts and L. J. Allison. 1983. Carbon in Live Vegetation of Major World Ecosystems. Report ORNL-5862. Oak Ridge, TN: Oak Ridge National Laboratory) to a more recent global vegetation map (Loveland, T. R., B. C. Reed, J. F. Brown, D. O. Ohlen, Z. Zhu, L. Yang and J. Merchant. 2000. Development of a Global Land Cover Characteristics Database and IGBP DISCover from 1 km AVHRR data. *International Journal of Remote Sensing* 21(6):1303–1330. Online at: http://edcdaac.usgs.gov/glcc/glcc.html.).

Vitousek, P. M., J. Aber, R. W. Howarth, G. E. Likens, P. A. Matson, D. W. Schindler, W. H. Schlesinger and G. D. Tilman. 1997. Human alteration of the global nitrogen cycle: Causes and consequences. *Issues in Ecology* 1(Spring):1–15.

World Meteorological Organization (WMO). 1997. *Comprehensive Assessment of the Freshwater Resources of the World*. Geneva: WMO.

Agroecosystems

Bathrick, D. 1998. Fostering Global Well-Being: A New Paradigm to Revitalize Agricultural and Rural Development. 2020 Vision Food, Agriculture, and the Environment Discussion Paper No. 26. Washington, D.C.: IFPRI.

Bøjö, J. 1996. The costs of land degradation in SubSaharan Africa. *Ecological Economics* 16:161–173.

Cheema, G. S., F. Hartvelt, J. Rabinovitch, R. Work, J. Smit, A. Ratta and J. Nasr. 1996. Urban Agriculture: Food, Jobs and Sustainable Cities. New York: UN Development Programme.

Conway, G. 1997. *The Doubly Green Revolution: Food for all in the 21st Century*. Ithaca: Cornell University Press.

Delgado, C., M. Rosegrant, H. Steinfeld, S. Ehui and C. Courbois. 1999. Livestock to 2020: The Next Food Revolution. 2020 Vision Food, Agriculture and the Environment Discussion Paper No. 28. Washington, D.C.: IFPRI.

Döll, P. and S. Siebert. 1999. A Digital Global Map of Irrigated Areas. Kassel, Germany: Centre for Environmental Systems Research, University of Kassel.

Food and Agriculture Organization of the United Nations (FAO). 1997. Computer Printout of FAOSTAT's International Commodity Prices 1989–91. Personal Communication via Technical Advisory Committee, CGIAR. Rome: FAO.

Food and Agriculture Organization of the United Nations (FAO). 1998. *The State of the World's Plant Genetic Resources for Food and Agriculture*. Rome: FAO.

Food and Agriculture Organization of the United Nations (FAO). 1999a. Urban and Peri-Urban Agriculture. Report to the FAO Committee on Agriculture (COAG). Online at: http://www.fao.org/unfao/bodies/COAG/COAG15/X0076e.htm.

Food and Agriculture Organization of the United Nations (FAO). 1999b. The State of Food Insecurity in the World 1999. Online at: www.fao.org/NEWS/1999/img/SOFI99-E.pdf.

Food and Agriculture Organization of the United Nations (FAO). 2000. Statistical Databases. Online at: http://apps.fao.org. (5 April (Crops primary); 20 April (land use, fertilizer, irrigation); 1 June (food balance sheets); 15 June (population)).

Food and Agriculture Organization of the United Nations Statistical Databases (FAOSTAT). 1999. Online at: http://apps.fao.org.

Gleick, P. H. 1998. *The World's Water 1998–1999*. Washington, D.C.: Island Press.

Golkany, I. M. 1999. Meeting global food needs: The environmental tradeoffs between increasing land conversion and land productivity. *Technology* 6:107-130.
Henao, J. 1999. Assessment of Plant Nutrient Fluxes and Gross Balances in Soils of Agricultural Lands in Latin America. Report prepared as part of the Pilot Assessment of Global Ecosystems (PAGE). International Fertilizer Development Centre (IFDC).
Houghton, R. A., J. L. Hackler and K. T. Lawrence. 1999. The U.S. carbon budget: Contributions from land-use change. *Science* 285:574-578.
Intergovernmental Panel on Climate Change (IPCC) (R. Watson, I. N., B. Bolin, N. Ravindranath, D. Verardo, and D. Dokken, eds.). 2000. *Land Use, Land-Use Change, and Forestry*. Cambridge, UK: Cambridge University Press.
Lal, R. 1995. Erosion-crop productivity relationships for soil of Africa. *Soil Science Society of America Journal* 59(3):661-667.
Loveland, T. R., B. C. Reed, J. F. Brown, D. O. Ohlen, Z. Zhu, L. Yang and J. Merchant. 2000. Development of a Global Land Cover Characteristics Database and IGBP DISCover from 1 km AVHRR data. *International Journal of Remote Sensing* 21(6):1303-1330. Online at: http://edcdaac.usgs.gov/glcc/glcc.html.
Mantel, S. and V. W. P. van Engelen. 1997. The Impact of Land Degradation on Food Productivity: Case Studies of Uruguay, Argentina and Kenya. Wageningen: ISRIC.
McIntire, J. 1994. A review of the soil conservation sector in Mexico. In *Economic and Institutional Analyses of Soil Conservation Projects in Central America and the Caribbean*. E. Lutz, S. Pagiola and C. Reiche, eds. A CATIE-World Bank Project. World Bank Environment Paper 8. Washington, D.C.: The World Bank.
Morris, M. L. and P. W. Heisey. 1998. Achieving desirable levels of crop diversity in farmers' fields: Factors affecting the production and use of commercial seed. Pp: 217-238 in *Farmers, Gene Banks, and Crop Breeding: Economic Analyses of Diversity in Wheat, Maize and Rice*. M. Smale, ed. Boston: Kluwer Academic Publishers.
Nelson, M. and M. Maredia. 1999. Environmental Impacts of the CGIAR: An Initial Assessment. Impact Assessment and Evaluation Group Document ICW/99/08/d. Washington, D.C.: Consultative Group for International Agricultural Research.
Oldeman, L. R. 1998. Soil Degradation: A Threat to Food Security? 98/01. Wageningen: ISRIC.
Oldeman, L. R., R. T. A. Hakkeling and W. G. Sombroek. 1991. World Map of the Status of Human-Induced Soil Degradation: An Explanatory Note. Global Assessment of Soil Degradation (GLASOD), International Soil Reference Information Centre (ISRIC), and United Nations Environment Program (UNEP).
Pinstrup-Andersen, P., R. Pandya-Lorch and M. Rosegrant. 1999. World Food Prospects: Critical Issues for the Early Twenty-First Century. 2020 Food Policy Report. Washington, D.C.: IFPRI.
Postel, S. 1999. *Pillar of Sand*. New York: W. W. Norton & Co.
Rosegrant, M. and C. Ringler. 1999. Impact on Food Security and Rural Development of Reallocating Water from Agriculture. Environment and Production Technology Division Discussion Paper No. 47. Washington, D.C.: IFPRI.
Scherr, S. J. 1999. Soil Degradation: A Threat to Developing-Country Food Security. 2020 Vision brief No. 58. Washington, D.C.: IFPRI.
Seckler, D., U. Amarasinghe, D. Molden, R. de Silva and R. Barker. 1998. World Water Demand and Supply, 1990 to 2025: Scenarios and Issues. Research Report 19. Colombo, Sri Lanka: International Water Management Institute (IWMI).
Shiklomanov, I. A. 1993. World fresh water resources. Pp: 13-24 in *Water in Crisis*. P. Gleick, ed. New York: Oxford University Press.
Shiklomanov, I. A. 1997. Comprehensive Assessment of the Freshwater Resources of the World: Assessment of Water Resources and Water Availability in the World. Stockholm, Sweden: WMO and Stockholm Environment Institute.
Smaling, E. M. A., S. M. Nandwa and B. H. Janssen. 1997. Soil fertility in Africa is at stake. Pp: 47-62 in *Replenishing Soil Fertility in Africa*. R. Buresh, P. A. Sanchez and F. Calhoun, eds. Soil Science Society of America Special publication Number 41. Madison: American Society of Agronomy.
Sombroek, W. G. and R. Gommes. 1996. The climate change-agriculture conundrum. In *Global Climate Change and Agricultural Production*. F. Bazzaz and W. Sombroek, eds. West Sussex: Wiley.
Thrupp, L. A. 1998. *Cultivating Diversity: Agrobiodiversity and Food Security*. Washington D.C.: World Resources Institute.
U.S. Department of Agriculture, National Agriculture Statistical Service (USDA-NASS). 1999. Historical Track Records for Commodities. Online at: http://www.usda.gov/nass/pubs/histdata.htm.
U.S. Geological Survey (USGS) Earth Resources Observation Systems (EROS) Data Center (USGS/EDC). 1999a. 1 km Global Land Cover Characterization database, Revisions for Latin America. Sioux Falls, SD: USGS/EDC.
Van Lynden, G. W. J. and L. R. Oldeman. 1997. The Assessment of the Status of Human-Induced Soil Degradation in South and Southeast Asia (ASSOD). Wageningen: ISRIC, FAO, and UNEP.
Wood, S., K. Sebastian and S. Scherr. 2000. Pilot Analysis of Global Ecosystems: Agroecosystems Technical Report. Washington, D.C.: World Resources Institute and International Food Policy Research Institute.
World Bank. 1999. *World Development Indicators 1999*. Washington, D.C.: The World Bank.
World Bank. 2000. *World Development Indicators 2000*. Washington, D.C.: The World Bank.
World Meteorological Organization (WMO). 1997. *Comprehensive Assessment of the Freshwater Resources of the World*. Geneva: WMO.
World Resources Institute in collaboration with the United Nations Environment Programme and the United Nations Development Programme and the World Bank. 1990. *World Resources 1990-91*. New York: Oxford University Press.
Young, A. 1994. Land Degradation in South Asia: Its Severity, Causes, and Effects upon the People. Final report prepared for submission to the Economic and Social Council for the United Nations (ECOSOC). Rome: FAO, UNDP, and UNEP.

Coastal

Alexander, C. E. 1998. Classified Shellfish Growing Waters. State of the Coast Report. NOAA, Silver Spring, MD. Online at: http://state-of-coast.noaa.gov/bulletins/html/sgw_04/sgw.html.
Alverson, D. L., M. H. Freeberg, S. A. Murawski and J. G. Pope. 1994. A Global Assessment of Fisheries Bycatch and Discards. FAO Fisheries Technical Paper 339. Rome: FAO.
BAP Planning Team. 1993. Biodiversity Action Plan for Vietnam. Hanoi: BAP Planning Team. December 1993.
Berg, H., M. C. Öhman, S. Troëng and O. Lindén. 1998. Environmental economics of coral reef destruction in Sri Lanka. *Ambio* 27(8):627-634.
Bright, C. 1998. *Life Out of Bounds: Bioinvasion in a Borderless World*. New York: W.W. Norton & Company.
Bryant, D., L. Burke, J. McManus and M. Spalding. 1998. *Reefs at Risk: A Map-Based Indicator of Threats to the World's Coral Reefs*. Washington, D.C.: World Resources Institute.
Burke, L., Y. Kura, K. Kassem, M. Spalding and C. Revenga. 2000. Pilot Analysis of Global Ecosystems: Coastal Ecosystems Technical Report. Washington, D.C.: World Resources Institute.
Caddy, J. F., J. Csirke, S. M. Garcia and R. J. R. Grainger. 1998. How pervasive is "Fishing down marine food webs"? *Science* 282:1383. Online at: http://www.sciencemag.org/cgi/content/full/282/5393/1383a.
Caribbean Tourism Organization (CTO). 1997. *Caribbean Tourism Statistical Report*. St. Michael, Barbados: CTO.
Center for International Earth Science Information Network (CIESIN), Columbia University, International Food Policy Research Institute (IFPRI) and World Resources Institute (WRI).

References

2000. Gridded Population of the World, Version 2 alpha. Palisades, NY: CIESIN, Columbia University. Online at: http://sedac.ciesin.org/plue/gpw.

Davidson, I. and M. Gauthier. 1993. Wetland Conservation in Central America. Report No. 93-3. Ottawa: North American Wetlands Conservation Council (Canada).

Diaz, R., Virginia Institute of Marine Science, College of William and Mary. 1999. Personal Communication. E-mail.

Diaz, R. and R. Rosenberg. 1995. Marine benthic hypoxia: A review of its ecological effects and the behavioural responses of benthic macrofauna. *Oceanography and Marine Biology: an Annual Review* 33:245–303.

Etkin, D. S. 1998. International oil spill statistics: 1997. Arlington, MA: Cutter Information Corp.

European Environment Agency (EEA). 1998. *Europe's Environment: The Second Assessment*. Oxford, UK: Elsevier Science Ltd.

European Environment Agency (EEA). 1999. *Environment in the European Union at the Turn of the Century*. Environmental Assessment Report No. 2. Copenhagen: EEA.

Food and Agriculture Organization of the United Nations (FAO). 1995. Review of the State of World Fishery Resources: Marine Fisheries. FAO Fisheries Circular No. 884. Rome: FAO.

Food and Agriculture Organization of the United Nations (FAO). 1997. *State of the World Fisheries and Aquaculture, 1996*. Rome: FAO.

Food and Agriculture Organization of the United Nations (FAO). 1999a. *The State of World Fisheries and Aquaculture, 1998*. Rome: FAO.

Food and Agriculture Organization of the United Nations (FAO). 1999b. Fisheries and food security. Online at: http://www.fao.org/focus/e/fisheries/intro.htm.

Food and Agriculture Organization of the United Nations (FAO). 1999c. FISHSTAT PLUS. Version 2.19 by Yury Shatz. Rome: FAO.

Food and Agriculture Organization of the United Nations (FAO). 1999d. Projection of World Fishery Production in 2010. Online at: http://www.fao.org/fi/highligh/2010.asp.

Garcia, S. M. and R. J. R. Grainger. 1996. Fisheries Management and Sustainability: A New Perspective of an Old Problem. Paper prepared for the Second World Fisheries Congress, Brisbane, Australia. Rome: FAO. July, 1996.

Garcia, S. M. and I. De Leiva Moreno. 2000. Trends in world fisheries and their resources: 1974–1999. In *The State of Fisheries and Aquaculture 2000*. Rome: FAO.

Grainger, R. J. R. and S. M. Garcia. 1996. Chronicles of Marine Fishery Landings (1950–1994): Trend Analysis and Fisheries Potential. FAO Fisheries Technical Paper 359. Rome: FAO.

Green, E. P. and A. W. Bruckner. In Press. The significance of coral disease epizootiology for coral reef conservation. *Biological Conservation*.

Harvell, C. D., K. Kim, J. M. Burkholder, R. R. Colwell, P. R. Epstein, D. J. Grimes, E. E. Hofmann, E. K. Lipp, A. D. M. E. Osterhaus, R. M. Overstreet, et al. 1999. Emerging marine diseases: Climate links and anthropogenic factors. *Science* 285:1505–1510.

Health Ecological and Economic Dimensions of Global Change (HEED). 1998. Marine Ecological Disturbance Database.

Heywood, V. H. and W. T. Watson, U.N. Environment Programme. (UNEP). 1995. *Global Biodiversity Assessment*. Cambridge, UK: Cambridge University Press.

Hoegh-Guldberg, O. 1999. Climate Change, Coral Bleaching and the Future of the World's Coral Reefs. Washington, D.C.: Greenpeace.

Intergovernmental Panel on Climate Change (IPCC) (R. Watson, M. C. Z., R. H. Moss eds.). 1996. *Climate Change 1995: Impacts, Adaptations and Mitigation of Climate Change: Scientific-Technical Analyses*. New York: Cambridge University Press.

International Union for Conservation of Nature and Natural Resources (IUCN). 1996. *1996 IUCN Red List of Threatened Animals*. Gland, Switzerland: IUCN.

International Tanker Owners Pollution Federation Limited (ITOPF). 1999. Past Spills. Online at: www.itopf.com/stats.html.

Island Resources Foundation. 1996. Tourism and Coastal Resources Degradation in the Wider Caribbean. Online at: http://www.org/irtourdg.html.

Japanese Ministry of Construction (JMC). 1998. General Information on Present Situation and Issues of Infrastructure. Fiscal Year 1998 version. Online at: http://www.moc.go.jp/cgi-bin/ids_binran.pl.

Joint Group of Experts on the Scientific Aspects of Marine Pollution (GESAMP). 1990. The State of the Marine Environment. Reports and Studies No. 39. Nairobi, Kenya: UNEP.

Kelleher, G., C. Bleakley and S. Wells, 1995. *A Global Representative System of Marine Protected Areas, Volume 1*. A joint publication of the World Bank, The Great Barrier Reef Marine Park Authority and the World conservation Union (IUCN). Washington, D.C.: The World Bank.

Kleypas, J. A., R. W. Buddemeier, D. Archer, J.-P. Gattuso, C. Langdon and B. N. Opdyke. 1999. Geochemical consequences of increased atmospheric carbon dioxide on coral reefs. *Science* 284:118–120.

Laureti, E. 1999. *1961–1997 Fish and Fishery Products: World Apparent Consumption Statistics Based on Food Balance Sheets*. FAO Fisheries Circular No. 821 Revision 5. Rome: FAO.

Loveland, T. R., B. C. Reed, J. F. Brown, D. O. Ohlen, Z. Zhu, L. Yang and J. Merchant. 2000. Development of a Global Land Cover Characteristics Database and IGBP DISCover from 1 km AVHRR data. *International Journal of Remote Sensing* 21(6):1303–1330. Online at: http://edcdaac.usgs.gov/glcc/glcc.html.

MacKinnon, J., 1997. *Protected Areas Systems Review of the Indo-Malayan Realm*. Caterbury, UK: The Asian Bureau for Conservation.

McAllister, D. E., J. Baquero, G. Spiller and R. Campbell. 1999. A Global Trawling Ground Survey. Unpublished Report prepared as part of the Pilot Assessment of Global Ecosystems (PAGE).

McGinn, A. P. 1999. Safeguarding the Health of Oceans. Worldwatch Paper 145. Washington, D.C.: Worldwatch Institute.

National Climatic Data Center (NCDC). 2000. Billion Dollar U.S. Weather Disasters 1980–1999. NCDC, Asheville, NC. Online at: http://www.ncdc.noaa.gov/ol/reports/billionz.html. (10 April 2000).

National Oceanic and Atmospheric Administration (NOAA). 1999. Trends in U.S. Coastal Regions, 1970–1998. Draft January 1999. Silver Spring, MD: NOAA.

National Oceanic and Atmospheric Administration, National Environmental Satellite Data and Information Service (NOAA/NESDIS). 2000. Seasurface Temperature Anomalies. Unpublished Data provided by Marguerite Toscano.

National Oceanic and Atmospheric Administration, National Environmental Satellite Data and Information Service (NOAA/NESDIS) and World Conservation Monitoring Centre (WCMC). 1999. Unpublished Data integrated at WCMC from: Hendee, J. 1999. Coral-list listserver coral bleaching archives. Online at: ftp://coral.aoml.noaa.gov/pub/champ/bleach, Wilkinson, C. 1998. Status of the coral reefs of the world, and McClannahan, T. CORDIO data set for the Indian Ocean.

National Research Council (NRC). 1985. Oil in the Sea. Washington, D.C.: National Academy Press.

National Research Council (NRC). 1999. From Monsoons to Microbes: Understanding the Ocean's Role in Human Health. Washington, D.C.: National Academy Press.

Norse, E. A., 1993. *Global Marine Biological Diversity: A Strategy for Building Conservation into Decision Making*. Washington, D.C.: Island Press.

O'Conner, T. 1998. Chemical Contaminants in Oysters and Mussels. State of the Coast Report. NOAA, Silver Spring, MD. Online at: http://state-of-coast.noaa.gov/bulletins/html/ccom_05/ccom.html.

Pauly, D., V. Christensen, J. Dalsgaard, R. Froese and F. Torres Jr. 1998. Fishing down marine food webs. *Science* 279:860–863.

Rabalais, N. and D. Scavia. 1999. Origin, impact and implications of the "Dead Zone" in the Gulf of Mexico. Presented at the US Global Change Program Seminar Series. 19 July.

Reaka-Kudla, M. L. 1997. The global biodiversity of coral reefs: A comparison with rain forests. Pp: 83–108 in *Biodiversity II: Understanding and Protecting Our Biological Resources*. M. L. Reaka-Kudla, D. E. Wilson and E. O. Wilson, eds. Washington, D.C.: Joseph Henry Press.

Salm, R. V. and J. R. Clark. 2000. *Marine and Coastal Protected Areas: A Guide for Planners and Managers*. Third Edition. Gland, Switzerland: IUCN.

Sherman, K. 1993. Large marine ecosystems as global units for marine resource management: An ecological perspective. Pp: 3–14 in *Large Marine Ecosystems: Stress, Mitigation and Sustainability*. K. Sherman, L. M. Alexander and B. D. Gold, eds. Washington, D.C.: AAAS Press.

Spalding, M. D., F. Blasco and C. D. Field, 1997. *World Mangrove Atlas*. Okinawa, Japan: The International Society for Mangrove Ecosystems.

Spalding, M. D. and A. M. Grenfell. 1997. New estimates of global and regional coral reef areas. *Coral Reefs* 16:225–230.

Thorne-Miller, B. and J. G. Catena. 1991. *The Living Ocean: Understanding and Protecting Marine Biodiversity*. Washington, D.C.: Island Press.

Travis, J. 1993. Invader threatens Black, Azov Seas. *Science* 262(26 November):1366–1367.

U.N. Environment Programme and Caribbean Environment Programme (UNEP/CEP). 1994. Coastal Tourism in the Wider Caribbean Region: Impacts and Best Management Practices. Technical Report No. 38. Kingston, Jamaica: UNEP/CEP.

Walting, L. and E. A. Norse. 1998. Disturbance of the seabed by mobile fishing gear: A comparison to forest clearcutting. *Conservation Biology* 12(6):1180–1197.

Wells, M. 1997. *Economic Perspectives on Nature Tourism, Conservation and Development*. Washington D.C.: The World Bank.

Williams, M. 1996. The Transition in the Contribution of Living Aquatic Resources to Food Security. Food, Agriculture and the Environment Discussion Paper 13. Washington, D.C.: IFPRI.

World Bank. 1989. Philippines: Environment and Natural Resource Management Study. Washington, D.C.: The World Bank.

World Conservation Monitoring Centre (WCMC). 1999. Unpublished data. Cambridge, UK: WCMC. May/August 1999.

World Conservation Monitoring Centre (WCMC). 2000. Protected Area Database. Cambridge, UK: WCMC.

World Conservation Monitoring Centre (WCMC). In preparation. *Global Biodiversity Assessment 2000*. Cambridge, UK: WCMC.

World Travel and Tourism Council/WETA (WTTC/WETA). 1998. Satellite Accounting Research-Caribbean Economic Impact. London: WTTC.

World Travel and Tourism Council (WTTC). 1999. Travel and Tourism's Economic Impact: Regional/National Statistics. Online at: http://wttc.org/economic_research/sat_accounting_research.html. (June).

Forests

Barber. 2000. *Trial By Fire: Forest Fires and Forestry Policy in Indonesia's Era of Crisis and Reform*. Washington, D.C.: World Resources Institute.

Brown, C. 1999. Global Forest Products Outlook Study: Thematic Study on Plantations. Working paper No. GFPOS/WP/03. Rome: FAO.

Bryant, D., D. Nielsen and L. Tangley. 1997. *The Last Frontier Forests*. Washington D.C.: World Resources Institute.

CARPE CD-ROM. 1998. Digital Chart of the World (DCW) CD-ROM 1993 (modified).

Cochrane, M. A., A. Alencar, M. D. Schulze, C. M. Souza Jr., D. C. Nepstad, P. Lefebvre and E. A. Davidson. 1999. Positive feedbacks in the fire dynamic of closed canopy tropical forests. *Science* 284:1832–1835.

Couzin, J. 1999. Landscape changes make regional climate run hot and cold. *Science* 283:317–319.

Davis, S. D., V. H. Heywood and A. C. Hamilton. 1994. *Centres of Plant Diversity: A Guide and Strategy for their Conservation*. Vol. 1. Gland, Switzerland: World Wide Fund for Nature.

DeFries, R. S., M. C. Hanson, J. R. G. Townshend, A. C. Janetos and T. R. Loveland. 2000. A new global 1-km data set of percentage tree cover derived from remote sensing. *Global Change Biology* 6:247–254.

Denniston, D. 1995. High Priorities: Conserving Mountain Ecosystems and Cultures. Worldwatch Paper No. 123. Washington, D.C.: Worldwatch Institute. February.

Dombeck, M. 1999. The United States Forest Service: The World's Largest Water Company. Sioux Falls, SD: Paper presented to the Outdoor Writers Association of America Conference.

Economy and Environment Programme for Southeast Asia (EEPSA) and the World Wildlife Fund (WWF). 1998. Interim Results of a Study on the Economic Value of Haze Damages in SE Asia. Unpublished Report. EEPSA and WWF.

Elvidge, C. D. et al. 1999. DMSP-OLS Estimation of tropical forest area impacted by ground fires in Roriama, Brazil. Submitted to *International Journal of Remote Sensing* 23 July 1999.

Food and Agriculture Organization (FAO). 1993. Forest Resources Assessment 1990-Tropical Countries. Forestry Paper No. 112. Rome: FAO.

Food and Agriculture Organization (FAO). 1997a. *State of the World's Forests 1997*. Rome: FAO.

Food and Agriculture Organization (FAO). 1997b. Wood Energy Today for Tomorrow, Regional Studies: The Role of Wood Energy in Europe and OECD. Forestry Department Working Paper FOPW/97/1. Rome: FAO.

Food and Agriculture Organization (FAO). 1997c. Wood Energy Today for Tomorrow, Regional Studies: The Role of Wood Energy in Asia. Forestry Department Working Paper FOPW/97/2. Rome: FAO.

Food and Agriculture Organization (FAO). 1998. Global Fibre Supply Model. Online at: http://www.fao.org/forestry/FOP/FOPW/GFSM/gfsmint-e.stm.

Food and Agriculture Organization (FAO). 1999. *State of the World's Forests 1999*. Rome: FAO.

Food and Agriculture Organization (FAO). 2000. Statistical Databases. Roundwood, Sawnwood, Wood-based Panels. Online at: http://apps.fao.org. (20 January).

Gaston, G., S. Brown, M. Lorenzini and K. D. Singh. 1998. State and change in carbon pools in the forests of tropical Africa. *Global Change Biology* 4:97–114.

Garnier, J.-Y. 1997. Statistics' role in policy development: The case of the Ivory Coast. Pp: 49–56 in *Biomass Energy: Key Issues and Priority Needs. Conference Proceedings*. Paris, 3–5 February 1997. Paris: IEA/OECD.

Haynes, R. W., D. M. Adams and J. R. Mills. 1995. The 1993 RPA Timber Assessment Update. U.S. Department of Agriculture Forest Service General Technical Report RM-259. Fort Collins, CO.

Heywood, V. H. and W. T. Watson, U.N. Environment Programme (UNEP). 1995. *Global Biodiversity Assessment*. Cambridge: Cambridge University Press.

Holmes, D. 2000. Deforestation in Indonesia: A Review of the Situation in Sumatra, Kalimantan, and Sulawesi. Draft report in preparation for the World Bank, based on mapping carried out by the

References

Indonesian Ministry of Forestry and Estate Crops. Washington, D.C.: The World Bank.
Houghton, R. A. 1999. The annual net flux of carbon to the atmosphere from changes in land use 1850–1990. *Tellus* 50B(298–313).
Houghton, R. A. and J. L. Hackler. 1999. Emissions of carbon from forestry and land-use change in tropical Asia. *Global Change Biology* 5(481–492).
Intergovernmental Panel on Climate Change (IPCC) (R. Watson, I. N., B. Bolin, N. Ravindranath, D. Verardo, and D. Dokken, eds.),. 2000. *Land Use, Land-Use Change, and Forestry*. Cambridge, UK: Cambridge University Press.
International Energy Agency (IEA). 1996. *Energy Statistics and Balances of Non-OECD Countries 1994–1995*. Paris: IEA.
Johnson, N. and D. Ditz. 1997. Challenges to sustainability in the U.S. forest sector. Pp: 191–280 in *Frontiers of Sustainability: Environmentally Sound Agriculture, Forestry, Transportation, and Power Production*. R. Dower, D. Ditz, P. Faethet al, eds. Washington, D.C.: World Resources Institute.
Kasischke, E. S., K. Bergen, R. Fennimore, F. Sotelo, G. Stephens, A. Janetos and H. H. Shugart. 1999. Satellite imagery gives a clear picture of Russia's boreal forest fires. *EOS-Transactions of the American Geophysical Union* 80:141–147
Levine, J. S., T. Bobbe, N. Ray, A. Singh and R. G. Witt. 1999. Wildland Fires and the Environment: A Global Synthesis. UNEP /DEIAEW/TR.99-1. Nairobi: UNEP.
Loveland, T. R., B. C. Reed, J. F. Brown, D. O. Ohlen, Z. Zhu, L. Yang and J. Merchant. 2000. Development of a Global Land Cover Characteristics Database and IGBP DISCover from 1 km AVHRR data. *International Journal of Remote Sensing* 21(6):1303–1330. Online at: http://edcdaac.usgs.gov/glcc/glcc.html.
Matthews, E. 1983. Global vegetation and land use: New high-resolution data bases for climatic studies. *Journal of Climate and Applied Meteorology* 22:474–487.
Matthews, E., R. Payne, M. Rohweder and S. Murray. 2000. Pilot Analysis of Global Ecosystems: Forest Ecosystems. Washington, D.C.: World Resources Institute.
Nilsson, S. 1996. Do We Have Enough Forests? IUFRO Occasional Paper No. 5. Laxenburg, Austria: International Institute for Applied Systems Analysis.
Oldfield, S., C. Lusty and A. MacKiven, 1998. *The World List of Threatened Trees*. World Conservation Press.
Olson, D. M. and E. Dinerstein. 1998. The Global 200: A representation approach to conserving the earth's most biologically valuable ecoregions. *Conservation Biology* 12(3):502–515.
Reid, W. V. and K. R. Miller. 1989. *Keeping Options Alive: The Scientific Basis for Conserving Biodiversity*. Washington, D.C.: World Resources Institute.
Revenga, C., S. Murray, J. Abramovitz and A. Hammond. 1998. *Watersheds of the World: Ecological Value and Vulnerability*. Washington D.C.: World Resources Institute and Worldwatch Institute.
Ricketts, T., E. Dinerstein, D. Olson, C. Loucks, W. Eichbaum, K. Kavanagh, P. Hedao, P. Hurley, K. Carney, R. Abell, et al. 1997. A Conservation Assessment of the Terrestrial Ecosystems of North America. Vol. 1: The United States and Canada. Washington, D.C.: World Wildlife Fund.
Solberg, B., D. J. Brooks, H. Pajuoja, T. Peck and P. Wardle. 1996. An overview of factors affecting the long-term trends of non-industrial and industrial wood supply and demand. Pp: 45–74 in *Long-term Trends and Prospects in World Supply and Demand for Wood and Implications for Sustainable Forest Management* (European Forest Institute Research Report No. 6). B. Solberg, ed. European Forest Institute.
Stattersfield, A. J., J. J. Crosby, A. J. Long and D. C. Wege. 1998. *Endemic Bird Areas of the World: Priorities for Biodiversity Conservation*. Cambridge, UK: BirdLife International.
Tucker, C. J. and J. R. G. Townshend. 2000. Strategies for monitoring tropical deforestation using satellite data. *International Journal of Remote Sensing* 21(6):1461–1472.
Wege, D. C. and A. J. Long. 1995. *Key Areas for Threatened Birds in the Neotropics*. Washington, D.C.: Smithsonian Institution Press.

Freshwater

Abell, R. A., D. M. Olson, E. Dinerstein, P. T. Hurley, J. T. Diggs, W. Eichbaum, S. Walters, W. Wettengel, T. Allnutt, C. Loucks, et al. 2000. *Freshwater Ecoregions of North America: A Conservation Assessment*. Washington, D.C.: Island Press.
Abramovitz, J. N. 1996. Imperiled Waters, Impoverished Future: The Decline of Freshwater Ecosystems. World Watch Paper 128. Washington, D.C.: Worldwatch Institute.
Bacalbasa-Dobrovici, N. 1989. The Danube River and its fisheries. In *Proceedings of the International Large River Symposium*. D. P. Dodge, ed. Canadian Special Publication of Fisheries and Aquatic Science 106. Ottawa, Canada: Department of Fisheries and Oceans.
Beveridge, M. C. M., L. G. Ross and L. A. Kelly. 1994. Aquaculture and biodiversity. *Ambio* 23(8):497–502.
Bos, R. 1997. The human health impact of aquatic weeds. In *Proceedings of the International Water Hyacinth Consortium*. E. S. Delfosse and N. R. Spencer, eds. Washington, D.C.: The World Bank.
Bräutigam, A. 1999. The Freshwater biodiversity crisis. *World Conservation* 30(2):4–5.
Brunner, J., Y. Kura and K. Thompson. 2000. Water Scarcity, Water Resources Management, and Hydrological Monitoring (Draft). Unpublished Report. Washington, D.C.: World Resources Institute.
Carlson, C. A. and R. T. Muth. 1989. The Colorado River: Lifeline of the American Southwest. In *Proceedings of the International Large River Symposium*. D. P. Dodge, ed. Canadian Special Publication of Fisheries and Aquatic Science 106. Ottawa, Canada: Department of Fisheries and Oceans.
Center for International Earth Science Information Network (CIESIN), Columbia University, International Food Policy Research Institute (IFPRI) and World Resources Institute (WRI). 2000. Gridded Population of the World, Version 2 alpha. Palisades, NY: CIESIN, Columbia University. Online at: http://sedac.ciesin.org/plue/gpw.
Dahl, T. E. 1990. Wetlands Losses in the United States 1780s to 1980s. Washington, D.C.: U.S. Department of the Interior, Fish and Wildlife Service.
Declining Amphibian Populations Task Force (DAPTF). 1999. What are Amphibian Declines and their Causes? Online at: http://www2.open.ac.uk/Ecology/J_Baker/DAPTF.What_are_ADs.html.
Dynesius, M. and C. Nilsson. 1994. Fragmentation and flow regulation of river systems in the northern third of the world. *Science* 266:753–762.
European Environment Agency (EEA). 1994. *European Rivers and Lakes: Assessment of their Environmental State*. Copenhagen: European Environment Agency.
European Environment Agency (EEA). 1999. *Environment in the European Union at the Turn of the Century*. Environmental Assessment Report No. 2. Copenhagen: European Environment Agency.
Evans, M. I. 1994. *Important Bird Areas in the Middle East*. BirdLife Conservation Series No. 2. Cambridge, UK: BirdLife International.
Fekete, B. M., C. J. Vörösmarty and W. Grabs. 1999. Global, Composite Runoff Fields Based on Observed River Discharge and Simu-

lated Water Balances. University of New Hampshire Data Set. Durham, NH: Complex Systems Research Center.

Finlayson, C. M. and N. C. Davidson. 1999. Global Review of Wetland Resources and Priorities for Wetland Inventory. Summary Report. Australia: Wetlands International and the Environmental Research Institute of the Supervising Scientists.

Food and Agriculture Organization of the United Nations (FAO). 1996. Fishery Country Profile: The Republic of Malawi. Online at: http://www.fao.org/fi/fcp/malawie.asp.

Food and Agriculture Organization of the United Nations (FAO). 1998. Aquaculture Quantities Dataset 1984-1997. Fishery Statistics Databases, downloadable with Fishstat-Plus software, Version 2.19 by Yury Shatz. Online at: http://www.fao.org/WAICENT/FAOINFO/FISHERY/statist/FISOFT/FISHPLUS.HTM.

Food and Agriculture Organization of the United Nations (FAO). 1999a. *The State of World Fisheries and Aquaculture, 1998*. Rome: FAO.

Food and Agriculture Organization of the United Nations (FAO). 1999b. Review of the State of World Fishery Resources: Inland Fisheries. Fisheries Circular No. 942. Rome: FAO.

Fuller, P. L., L. G. Nico and J. D. Williams. 1999. *Nonindigenous Fishes Introduced into Inland Waters of the United States*. American Fisheries Society, Special Publication No. 27. Bethesda, MD: American Fisheries Society.

Garibaldi, L. and D. M. Bartley. 1998. The database on introductions of aquatic species (DIAS). Food and Agriculture Organization of the United Nations (FAO) Aquaculture Newsletter (FAN) no. 20. Online at: http://www.fao.org/waicent/faoinfo/fishery/statist/fishoft/dias/index.htm.

Gopal, B. 1987. *Water Hyacinth*. Aquatic Plant Studies 1. Amsterdam: Elsevier Science.

Hill, G., J. Waage and G. Phiri. 1997. The water hyacinth problem in tropical Africa. In *Proceedings of the International Water Hyacinth Consortium*. E. S. Delfosse and N. R. Spencer, eds. Washington, D.C.: The World Bank.

Hinrichsen, D., R. Robey and U. D. Upadhyay. 1998. Solutions for a Water-Short World. Population Reports, Series M, No. 14. Baltimore, MD: Johns Hopkins University School of Public Health, Population Information Program. September 1998.

Houlahan, J., C. Findlay, B. Schmidt, A. Meyer and S. Kuzmin. 2000. Quantitative evidence for global amphibian population declines. *Nature* 404:752-755.

Hughes, R. M. and R. F. Noss. 1992. Biological diversity and biological integrity: Current concerns for lakes and streams. *Fisheries* May-June.

International Commission on Large Dams (ICOLD). 1998. *World Register of Dams 1998*. Paris: ICOLD.

International Journal on Hydropower and Dams (IJHD). 1998. *1998 World Atlas and Industry Guide*. Surrey, UK: Aqua-Media International.

International Union for Conservation of Nature and Natural Resources (IUCN). 1996. *1996 IUCN Red List of Threatened Animals*. Gland, Switzerland: IUCN.

Kapetsky, Chief Fisheries Officer, Inland Water Resources and Aquaculture Service, Fisheries Resources Division FAO. 1999. Personal Communication. E-mail. 27 August.

Karr, J. R. and E. W. Chu. 1999. *Restoring Life in Running Waters: Better Biological Monitoring*. Washington, D.C.: Island Press.

Kaufman, L. 1992. Catastrophic change in species-rich freshwater ecosystems: The lessons from Lake Victoria. *BioScience* 42(11): 846-858.

Liao, G. Z., K. X. Lu and X. Z. Xiao. 1989. Fisheries resources of the Pearl River and their exploitation. In *Proceedings of the International Large River Symposium*. D. P. Dodge, ed. Canadian Special Publication of Fisheries and Aquatic Science 106. Ottawa, Canada: Department of Fisheries and Oceans.

Lips, K. R. 1998. Decline of tropical montane amphibian fauna. *Conservation Biology* 12(1):106-117.

L'vovich, M. I. and G. F. White. 1990. Use and transformation of terrestrial water systems. Pp: 235-252 in *The Earth as Transformed by Human Action: Global and Regional Changes in the Biosphere Over the Past 300 Years*. B. L. Turner II, W. C. Clark, R. W. Kateset al, eds. Cambridge, UK: Cambridge University Press.

Lyons, J., S. Navarro-Perez, P. A. Cochran, E. Santana and M. Guzman-Arroyo. 1995. Index of biotic integrity based on fish assemblages for the conservation of streams and rivers in West-Central Mexico. *Conservation Biology* 9(3):569-584.

Master, L. L., S. R. Flack and B. A. Stein. 1998. *Rivers of Life: Critical Watersheds for Protecting Freshwater Biodiversity*. Arlington, VA: The Nature Conservancy.

McAllister, D. E., A. L. Hamilton and B. Harvey. 1997. Global freshwater biodiversity: Striving for the integrity of freshwater ecosystems. *Sea Wind-Bulletin of Ocean Voice International* 11(3): 1-140.

Mekong River Commission (MRC). 1997. Greater Mekong Sub-Region: State of the Environment Report. Bangkok: MRC.

Missouri River Coalition. 1995. Comments on the Missouri River Master Water Control Manual Review and Update Draft Environmental Impact Assessment. 1 March 1995.

Myers, N. 1997. The rich diversity of biodiversity issues. Pp: 125-138 in *Biodiversity II: Understanding and Protecting Our Biological Resources*. M. L. Reaka-Kudla, D. E. Wilson and E. O. Wilson, eds. Washington, D.C.: Joseph Henry Press.

Miller, R. R., J. D. Williams and J. E. Williams. 1989. Extinctions of North American fishes during the past century. *Fisheries* 14(6): 22-38.

Moyle, P. B. and R. A. Leidy. 1992. Loss of biodiversity in aquatic ecosystems: Evidence from fish faunas. In *Conservation Biology: The Theory and Practice of Nature Conservation, Preservation and Management*. P. L. Fiedler and S. K. Jain, eds. New York: Chapman and Hall.

Nilsson, C., M. Svedmark, P. Hansson, S. Xiong and K. Berggren. 1999. Fragmentation and flow regulation of Southern Rivers. Unpublished Report Commissioned by PAGE. Landscape Ecology, Umeå University, Sweden.

Oberdorff, T. and R. M. Hughes. 1992. Modification of an index of biotic integrity based on fish assemblages to characterize rivers of the Seine Basin, France. *Hydrobiologia* 228:117-130.

O'Neill, C., Coastal Resources Specialist, New York Sea Grant. 1999. Personal Communication.

O'Neill, C. R. 1996. Economic Impact of Zebra Mussels: The 1995 National Zebra Mussel Information Clearinghouse Study. *New York Sea Grant Extension*.

Pelly, J. 1998. No simple answer to recent amphibian declines. *Environmental Science and Technology* 32(15):352-353.

Postel, S. 1995. Where have all the rivers gone? *World Watch* 8(3):9-19.

Postel, S. and S. Carpenter. 1997. Freshwater ecosystem services. Pp: 195-214 in *Nature's Services: Societal Dependence on Natural Ecosystems*. G. C. Daily, ed. Washington, D.C.: Island Press.

Reaka-Kudla, M. L. 1997. The global biodiversity of coral reefs: A comparison with rain forests. Pp: 83-108 in *Biodiversity II: Understanding and Protecting Our Biological Resources*. M. L. Reaka-Kudla, D. E. Wilson and E. O. Wilson, eds. Washington, D.C.: Joseph Henry Press.

Revenga, C., S. Murray, J. Abramovitz and A. Hammond. 1998. *Watersheds of the World: Ecological Value and Vulnerability*. Washington, D.C.: World Resources Institute and Worldwatch Institute.

Revenga, C., J. Brunner, N. Henninger, K. Kassem and R. Payne. 2000. Pilot Analysis of Global Ecosystems: Freshwater Ecosystems Technical Report. Washington, D.C.: World Resources Institute.

Ricciardi, A. and J. B. Rasmussen. 1999. Extinction rates of North American freshwater fauna. *Conservation Biology* 15(5): 1220-1222.

Ross, S. T. 1991. Mechanisms structuring stream fish assemblages: Are there lessons from introduced species? *Environmental Biology of Fishes* 30:359-368.

References

Shiklomanov, I. A. 1997. Comprehensive Assessment of the Freshwater Resources of the World: Assessment of Water Resources and Water Availability in the World. Stockholm, Sweden: WMO and Stockholm Environment Institute.

Sparks, R. E. 1992. The Illinois River floodplain ecosystem. In *Restoration of Aquatic Ecosystems: Science, Technology and Public Policy*. National Research Council (NRC), ed. Washington, D.C.: National Academy Press.

U.N. Environment Programme (UNEP) and Global Environment Monitoring System (GEMS). 1995. Water Quality of World River Basins. Nairobi: UNEP.

U.N. Environment Programme (UNEP). 1996. *Groundwater: A Threatened Resource*. Nairobi, Kenya: UNEP.

Vörösmarty, C. J., K. P. Sharma, B. M. Fekete, A. H. Copeland, J. Holden, J. Marble and J. A. Lough. 1997. The storage and aging of continental runoff in large reservoir systems of the world. *Ambio* 26(4):210–219.

Watson, R. T., M. C. Zinyowera and R. H. Moss, 1996. *Climate Change 1995, Impacts, Adaptations and Mitigation of Climate Change: Scientific Technical Analyses*. Contribution of Working Group II to the Second Assessment Report of the Intergovernmental Panel on Climate Change. Cambridge, UK: Cambridge University Press.

Welcomme, R. L. 1988. International Introductions of Inland Aquatic Species. Technical Series Paper 294. Rome: Food and Agriculture Organization of the United Nations.

World Meteorological Organization (WMO). 1997. *Comprehensive Assessment of the Freshwater Resources of the World*. Geneva: WMO.

Grasslands

Andreae, M. O. 1991. Biomass burning: Its history, use and distribution and its impact on environmental quality and global climate. Pp: 3–21 in *Global Biomass Burning*. J. S. Levine, ed. London: MIT Press.

Arino and Melinotte. 1998. The 1993 Africa Fire Map. *International Journal of Remote Sensing* 19(11):2019–2023.

Atjay, G. L., P. Ketner and P. Duvigneaud. 1979. Terrestrial primary production and phytomass. Pp: 129–181 in *The Global Carbon Cycle*. B. Bolin, E. T. Degens, S. Kempe and P. Ketner, eds. Chichester, UK: John Wiley & Sons.

Campbell, K. and M. Borner. 1995. Population trends and distribution of Serengeti herbivores: Implications for management. Pp: 117–145 in *P. Arcese*. A. R. E. Sinclair, ed. Chicago: University of Chicago Press.

Christian, J. M. and S. D. Wilson. 1999. Long-term ecosystem impacts of an introduced grass in the Northern Great Plains. *Ecology* 80(7):2397–2404.

de Haan, C., H. Steinfeld and H. Blackburn. 1997. Livestock and the Environment: Finding a Balance. Brussels, Belgium: European Commission Directorate-General for Development, Development Policy Sustainable Development and Natural Resources.

Dinerstein, E., D. M. Olson, D. J. Graham, A. L. Webster, S. A. Primm, M. P. Bookbinder and G. Ledec. 1995. *A Conservation Assessment of the Terrestrial Ecoregions of Latin America and the Caribbean*. Washington, D.C.: World Wildlife Fund and The World Bank.

Ehrlich, D., E. F. Lambin and J. Malingreau. 1997. Biomass burning and broad-scale land-cover changes in Western Africa. *Remote Sens. Environ.* 61:201–209.

Evans, R. 1998. The erosional impacts of grazing animals. *Progress in Physical Geography* 22(2):251–268.

Food and Agriculture Organization of the United Nations Statistical Databases (FAOSTAT). 1999. Online at: http://apps.fao.org.

Frank, D. A., S. J. McNaughton and B. F. Tracy. 1998. The ecology of the Earth's grazing ecosystems. *BioScience* 48(7):513–521.

Frost, P. G. H. 1985. The responses of savanna organisms to fire. Pp: 232–237 in *Ecology and Management of the World's Savannas*. J. C. Tothill and J. J. Mott, eds. Canberra: Australian Academy of Science.

Goldammer, J. P. 1995. Biomass burning and the atmosphere. Paper presented at "Forests and Global Climate Change: Forests and the Global Carbon Cycle."

Honey, M. 1999. *Ecotourism and Sustainable Development: Who Owns Paradise?* Washington, D.C.: Island Press.

International Livestock Research Institute (ILRI). 1998. *Cattle Density Database*. Nairobi, Kenya: ILRI.

Levine, J. S., T. Bobbe, N. Ray, A. Singh and R. G. Witt. 1999. Wildland Fires and the Environment: A Global Synthesis. UNEP /DEIAEW/TR.99-1. Nairobi: UNEP.

Loveland, T. R., B. C. Reed, J. F. Brown, D. O. Ohlen, Z. Zhu, L. Yang and J. Merchant. 2000. Development of a Global Land Cover Characteristics Database and IGBP DISCover from 1 km AVHRR data. *International Journal of Remote Sensing* 21(6):1303–1330. Online at: http://edcdaac.usgs.gov/glcc/glcc.html.

McNaughton, S. J. 1993. Grasses and grazers, science and management. *Ecological Applications* 3:17–20.

Menaut, J. C., L. Abbadie, F. Lavenu, P. Loudjani and A. Podaire. 1991. Biomass burning in West African savannas. Pp: 131–142 in *Global Biomass Burning*. J. S. Levine, ed. London: MIT Press.

Middleton, N. and D. Thomas, 1997. *World Atlas of Desertification (Second Edition)* London: UN Environment Programme (UNEP).

Ojima, D. S., B. O. M. Dirks, E. P. Glenn, C. E. Owensby and J. M. O. Scurlock. 1993. Assessment of C budget for grasslands and drylands of the world. *Water, Air and Soil Pollution* 70:643–657.

Oldeman, L. R., R. T. A. Hakkeling and W. G. Sombroek. 1991. World Map of the Status of Human-Induced Soil Degradation: An Explanatory Note. Global Assessment of Soil Degradation (GLASOD), International Soil Reference Information Centre (ISRIC), and United Nations Environment Program (UNEP).

Olson, J. S., J. A. Watts and L. J. Allison. 1983. Carbon in Live Vegetation of Major World Ecosystems. Report ORNL-5862. Oak Ridge, Tennessee: Oak Ridge National Laboratory.

Planning Assessment for Wildlife Management. 1996. Returns from tourist hunting in Tanzania. Pp: 71–80 in *Tourist Hunting in Tanzania*. N. Leader-Williams, J. A. Kayera and G. L. Overton, eds. Occasional Paper of the IUCN Species Survival Commission No. 14. Gland, Switzerland: The World Conservation Union (IUCN).

Price Waterhouse. 1996. The hunting industry in Zimbabwe. Pp: 81–93 in *Tourist Hunting in Tanzania*. N. Leader-Williams, J. A. Kayera and G. L. Overton, eds. Occasional Paper of the IUCN Species Survival Commission No. 14. Gland, Switzerland: The World Conservation Union (IUCN).

Ricketts, T., E. Dinerstein, D. Olson, C. Loucks, W. Eichbaum, K. Kavanagh, P. Hedao, P. Hurley, K. Carney, R. Abell, et al. 1997. A Conservation Assessment of the Terrestrial Ecosystems of North America. Vol. 1: The United States and Canada. Washington, D.C.: World Wildlife Fund.

Risser, P. G. 1996. A new framework for prairie conservation. Pp: 261–274 in *Prairie Conservation: Preserving North America's Most Endangered Ecosystem*. F. B. Samson and F. L. Knopf, eds. Washington, D.C.: Island Press.

Sala, O. E. and J. M. Paruelo. 1997. Ecosystem services in grasslands. Pp: 237–252 in *Nature's Services: Societal Dependence on Natural Ecosystems*. G. C. Daily, ed. Washington, D.C.: Island Press.

Scholes, R. J. and B. H. Walker. 1993. *An African Savanna*. Cambridge, UK: Cambridge University Press.

Seré, C. and H. Steinfeld. 1996. World Livestock Production Systems: Current Status, Issues and Trends. Rome: FAO.

Sneath, D. 1998. State policy and pasture degradation in Inner Asia. *Science* 281:1147–1148.

United States Congress Office of Technology Assessment (USCOTA). 1993. Harmful Non-Indigenous Species in the United States. OTA-F-565. Washington, D.C.: U.S. Government Printing Office.

Walker, B. H. 1985. Structure and function of savannas: An overview. Pp: 83–92 in *Ecology and Management of the World's Savannas*. J. C. Tothill and J. J. Mott, eds. Canberra: Australian Academy of Science.

White, R., S. Murray and M. Rohweder. 2000. Pilot Analysis of Global Ecosystems: Grassland Ecosystems Technical Report. Washington, D.C.: World Resources Institute.

Whittaker, R. H. and E. Likens. 1975. The biosphere and man. Pp: 305–328 in *Primary Productivity of the Biosphere*. H. Lieth and R. H. Whittaker, eds. Ecological Studies No. 14. Berlin: Springer-Verlag.

Williams, J. R. and P. L. Diebel. 1996. The economic value of the prairie. Pp: 19–35 in *Prairie Conservation: Preserving North America's Most Endangered Ecosystem*. F. B. Samson and F. L. Knopf, eds. Washington, D.C.: Island Press.

Mountain Ecosystems

Carlson, C. 2000. Money pits. *Mineral Policy Center Newsletter* Spring:9–12.

D'Esposito, S. and J. Feiler. 2000. Lessons from the disasters on the Danube: Is modern mining safe? *Mineral Policy Center Newsletter* Spring:1, 4–5, 17.

Federal Research Centre for Forestry and Forest Products (FRCFFP). 1998. Forest Condition in Europe: Results of the 1997 Crown Condition Survey. 1998 Technical Report. Geneva and Brussels: United Nations Economic Commission for Europe (UN/ECE) and European Commission (EC).

Food and Agriculture Organization of the United Nations (FAO). 1993. Forest Resources Assessment 1990-Tropical Countries. Forestry Paper No. 112. Rome: FAO.

Food and Agriculture Organization of the United Nations (FAO). 1995. Agenda 21: Chapter 13–Sustainable Mountain Development. FAO Progress Report. Online at: http://www.fao.org /WAICENT/faoinfo/sustdev/epdirect/EPRE0005.htm. (April).

Grötzbach, E. and C. Stadel. 1997. Mountain peoples and cultures. Pp: 17–38 in *Mountains of the World: A Global Priority*. B. Messerli and J. D. Ives, eds. New York: The Parthenon Publishing Group.

International Potato Center (CIP). 2000. Breeding and Conservation. Online at: http://www.cipotato.org/market/Brochure99 /dynamic2.htm. (17 July 2000).

Ives, J. D., B. Messerli and E. Spiess. 1997. Mountains of the world – A global priority. Pp: 1–15 in *Mountains of the World: A Global Priority*. B. Messerli and J. D. Ives, eds. New York: The Parthenon Publishing Group.

Jeník, J. 1997. The diversity of mountain life. Pp: 199–235 in *Mountains of the World: A Global Priority*. B. Messerli and J. D. Ives, eds. New York: The Parthenon Publishing Group.

Liniger, H., R. Weingartner and M. Grosjean. 1998. *Mountains of the World: Water Towers for the 21st Century*. Berne, Switzerland: Mountain Agenda.

Messerli, B. and J. D. Ives, 1997. *Mountains of the World: Challenges for the 21st Century*. Berne, Switzerland: Mountain Agenda.

Price, M., T. Wachs and E. Byers, 1999. *Mountains of the World: Tourism and Sustainable Mountain Development*. Berne, Switzerland: Mountain Agenda.

Schaaf, T., UNESCO. 1999. Personal Communication. E-mail. 29 March 1999.

Tripp, R. and W. van der Heide. 1996. The erosion of crop genetic diversity: Challenges, strategies and uncertainties. *Natural Resource Perspectives* 7(March).

World Conservation Monitoring Centre (WCMC). 1997. Tropical Montane Cloud Forests: An Urgent Priority for Conservation. WCMC Biodiversity Bulletin No. 2. Cambridge, UK: WCMC.

Wuetrich, B. 1993. Forests in the clouds face stormy future. *Science News* 144(2):23.

Polar Ecosystems

Arctic Monitoring and Assessment Programme (AMAP). 1997. *Arctic Pollution Issues: A State of the Arctic Report*. Oslo: AMAP.

Fergusson, A. and D. I. Wardle. 1998. *Arctic Ozone: The Sensitivity of the Ozone Layer to Chemical Depletion and Climate Change*. Environment Canada.

GLACIER, Rice University. 1998. Introduction: How Big is the Ice? Online at: www.glacier.rice.edu/invitation/1_ice.html.

Hamilton, L., C. M. Duncan and N. Flanders. 1998. Northern Atlantic fishing communities in and era of ecological change. *New Hampshire Sea Grant* (1999):28–30.

International Association of Antarctica Tour Operators (IAATO). 1999. Tourism Statistics. Online at: http://www.iaato.org/tour_ stats.html. (15 May).

Petit, J. R., J. Jouzel, D. Raynaud, N. I. Barkov, J.-M. Barnola, I. Basile, M. Bender, J. Chappellaz, M. Davis, G. Delaygue, et al. 1999. Climate and atmospheric history of the past 420,000 years from the Vostok ice core, Antarctica. *Nature* 399:429–436.

Rothrock, D. A., Y. Yu and G. A. Maykut. 1999. Thinning of the Arctic sea-ice cover. *Geophysical Research Letters* 26(23):3469–3472.

Stauffer, B. 1999. Cornucopia of ice core results. *Nature* 399:412–413.

U.N. Environment Programme (UNEP). 1999. *Global Environment Outlook 2000*. London, UK: Earthscan Publications Ltd.

U.N. Environment Programme (UNEP). 1998. *Environmental Effects of Ozone Depletion: 1998 Assessment*. Nairobi: UNEP.

U.S. Global Change Research Program (USGCRP). 1999. Arctic Sea-Ice: Changes, Causes and Implications. Briefing paper for the 20 April US Global Change Seminar. Washington, D.C.: USGCRP.

Watson, R. T., M. C. Zinyowera and R. H. Moss, 1998. *The Regional Impacts of Climate Change: An Assessment of Vulnerability*. A Special Report of IPCC Working Group II. Cambridge, UK: Cambridge University Press.

Urban Ecosystems

Adams, L. W. 1994. *Urban Wildlife Habitats: A Landscape Perspective*. Minneapolis: University of Minnesota Press.

American Forests. 1999. Regional Ecosystem Analysis Chesapeake Bay Region and the Baltimore-Washington Corridor: Calculating the Value of Nature. Washington, D.C.: American Forests. 22 March 1999.

Bolund, P. and S. Hunhammar. 1999. Ecosystem services in urban areas. *Ecological Economics* 29:293–301.

Bryson, R. and J. Ross. 1972. The climate of the city. Pp: 52–76 in *Urbanization and Environment*. T. Detwyler and M. Marens, eds. Belmont, CA: Duxbury Press.

Chaplowe, S. G. 1998. Havana's popular gardens: Sustainable prospects for urban agriculture. *The Environmentalist* 18(1):47–57.

Douglas, I. 1983. *The Urban Environment*. London, UK: Edward Arnold.

Eurostat, European Environment Agency Task Force, DG XI and PHARE European Commission, U.N. Economic Commission for Europe, Organization for Economic Cooperation and Development, and World Health Organization. 1995. *Europe's Environment: Statistical Compendium for the Dobris Assessment*. Luxembourg: Office for Official Publications of the European Communities.

Folke, C., Å. Jansson, J. Larsson and R. Costanza. 1997. Ecosystem appropriation of cities. *Ambio* 26(3):167–172.

Food and Agriculture Organization of the United Nations (FAO). 1999. Urban and Peri-Urban Agriculture. Report to the FAO Committee on Agriculture (COAG). Online at: http://www.fao.org /unfao/bodies/COAG/COAG15/X0076e.htm.

Goudie, A. 2000. *The Human Impact on the Natural Environment*. Cambridge, MA: MIT Press.

References

Kowarik, I. 1990. Some responses of flora and vegetation to urbanization in Central Europe. Pp: 45–74 in *Urban Ecology: Plants and Plant Communities in Urban Environments*. H. Sukopp and S. Hejný, eds. The Hague: SBP Academic Publishing.

Lyle, J. and R. D. Quinn. 1991. Ecological corridors in urban southern California. In: *Wildlife Conservation in Metropolitan Environments: Proceedings of a National Symposium on Urban Wildlife*. L. W. Adams and D. L. Leedy, eds. Columbia, MD: National Institute for Urban Wildlife.

Margolis, M. 1992. A third world city that works. *World Monitor* March: 42–50.

Miller, R. W. 1983. Multiple use urban forest management in the Federal Republic of Germany. Pp: 21–24 in *Management of Outlying Forests for Metropolitan Populations*. Milwaukee, WI: Man and the Biosphere Seminar.

Miller, R. W. 1997. *Urban Forestry: Planning and Managing Urban Greenspaces*. Second Edition. Upper Saddle River, New Jersey: Prentice Hall.

Mountford, D., U.S. Environmental Protection Agency. 1999. Personal Communication. E-mail. 12 March.

Nowak, D. J. 1994. Air pollution removal by Chicago's urban forest. Pp: 63–81 in *Chicago's Urban Forest Ecosystem: Results of the Chicago Urban Forest Climate Project*. E. G. McPherson, D. J. Nowak and R. A. Rowntree, eds. Gen. Tech. Report NE-186. Radnor, PA: U.S. Department of Agriculture, Forest Service, Northeastern Forest Experiment Station.

Nowak, D. J. and J. F. Dwyer. 1996. Urban Forestry. Pp: 470–472 in *McGraw-Hill Yearbook of Science and Technology*. New York: McGraw-Hill.

Nowak, D. J., R. A. Rowntree, E. G. McPherson, S. M. Sisinni, E. R. Kerkmann and J. C. Stevens. 1996. Measuring and analyzing urban tree cover. *Landscape and Urban Planning* 36:49–57.

Rees, W. E. 1992. Ecological footprints and appropriated carrying capacity: What urban economics leaves out. *Environment and Urbanization* 4(2):121–130.

Sampson, R. N. 1994. Making cities safe for trees. Pp: 157–170 in *The City as a Human Environment*. D. G. LeVine and A. C. Upton, eds. Westport, CT: Praeger Publishers.

Smit, J. and J. Nasr. 1992. Urban Agriculture for sustainable cities: Using wastes and idle land and water bodies as resources. *Environment and Urbanization* 4(2):141–154.

Smith, D. 1999. The case for greener cities. *American Forests Magazine* Autumn 1999:35–37.

Stanners, D. and P. Bordeau, 1995. *Europe's Environment: The Dobris Assessment*. Copenhagen: European Environment Agency.

The Mega Cities Project, The Centre for Community Studies, Action and Development. 1994. Urban Market Gardens: Accra. Urban Environment-Poverty Case Study Series. New York: The Mega Cities Project. July, 1994.

U.N. Centre for Human Settlements (Habitat). 1996. *An Urbanizing World: Global Report on Human Settlements*. Oxford, UK: Oxford University Press.

U.N. Population Division (UNPD). 1996. Urban and Rural Areas 1950–2030: (The 1996 Revision). On Diskette. New York: UNPD.

U.S. Census Bureau. 1995. Urban and Rural Definitions. Online at: http://www.census.gov/population/censusdata/urdef.txt.

U.S. Department of the Interior, U.S. Fish and Wildlife Service, U.S. Department of Commerce and Bureau of the Census. 1997. 1996 National Survey of Fishing, Hunting, and Wildlife-Associated Recreation. Washington, D.C.: U.S. Government Printing Office.

U.S. National Biological Survey. 2000. Washington D.C. Project Birdscape. Online at: http://www.im.nbs.gov/birdscap/birdscap .html.

World Bank. 2000. *World Development Indicators 2000*. Washington, D.C.: The World Bank.

World Resources Institute in collaboration with the United Nations Environment Programme and the United Nations Development Programme and the World Bank. 1998. *World Resources 1998–99*. New York: Oxford University Press.

Chapter 3

Regaining the High Ground: Reviving the Hillsides of Machakos

African Development and Economic Consultants. 1986. Machakos Integrated Development Programme Socio-Economic Survey: Final Report. Nairobi and Machakos: Ministry of Planning and National Development.

Huxley, E. 1960. *A New Earth*. London.

Jaetzold, R. and H. Schmidt. 1983. *Natural Conditions and Farm Management Information, Part IIC: East Kenya (Eastern Coast Provinces)*. Vol. 2 of Farm Management Handbook of Kenya. Nairobi: Ministry of Agriculture.

Kenya Web. 1999. Machakos District: Economic Potential. Online at: http://www.kenyaweb.com/ourland/eastern/machakos/ma_econ p .html. (20 July).

Lindblom, K. G. 1920. *The Akamba of British East Africa*. Uppsala: Appelborgs Boktrycheri Aktieborg.

Mbate, G., Economist with US Agency for International Development, Regional EDSO. 1999. Personal Communication. Interview. 19 February.

Mortimore, M. and M. Tiffen. 1994. Population growth and a sustainable environment. *Environment* 36(8): 10–32.

Mullei, M. 1999. Agricultural Officer, USAID. Personal Communication. Interview. 17 March.

Murton, J. 1999. Population growth and poverty in Machakos District, Kenya. *The Geographical Journal* 165(1).

Mutiso, S., Geography Department, University of Nairobi. 1999. Personal Communication. Interview. 25 February.

Ndambuki, A. M., Machakos District Agricultural Officer for Machakos. 1999. Personal Communication. Interview. 1 March.

Peberdy, J. 1958. Machakos District Gazetteer. Department of Agriculture Mimeo. Machakos District Office.

Tiffen, M. 1995. Population density, economic growth and societies in transition: Boserup reconsidered in a Kenyan case-study. *Development and Change* 26: 31–66.

Tiffen, M. and M. Mortimore. 1992. Environment, population growth and productivity in Kenya: A case study of Machakos District. *Development Policy Review* 10: 359–387.

Tiffen, M., M. Mortimore and F. Gichuki. 1994. *More People, Less Erosion: Environmental Recovery in Kenya*. Chichester, UK: John Wiley & Sons Ltd.

Zaal, F. 1999. Driving forces of sustainable agriculture; Results from a farmer survey in Machakos and Kitui Districts, Kenya. Unpublished report. Amsterdam: AGIDS/University of Amsterdam.

Cuba's Agricultural Revolution: A Return to Oxen and Organics

Bourque, M., Sustainable Agriculture Program Director, Institute for Food and Development Policy. 1999. Personal Communication. Interview. 27 April.

Food and Agriculture Organization of the United Nations (FAO). 1999. The State of Food Insecurity in the World. Online at: http: //www.fao.org/news/1999/img/SOFI99-E.PDF.

Gellerman, B. 1996. Organics in Cuba. Living on Earth (National Public Radio, Cambridge, MA). Online at: http://www.loe.org /html/susag/cuba.html.

Monzote, F. F. No Date. Cuban Agriculture Alternatives: An Overview of Cuba's Experience in Organic Agriculture. Havana: Pastures and Forages Research Institute.

Moskow, A. 1999. Havana's self provision gardens. *Environment and Urbanization* 11(2):127–133.

Mueller, J. P., Sustainable Agriculture Coordinator, North Carolina State University. 1999. Personal Communication. Interview. 10 February.

Murphy, C. 1999. Cultivating Havana: Urban Agriculture and Food Security in the Years of Crisis. Development Report #12. Oakland, CA: Institute for Food and Development Policy.

Rosset, P. 1996. Cuba: Alternative agriculture during crisis. Pp: 64–74 in *New Partnerships for Sustainable Agriculture*. L. A. Thrupp, ed. Washington, D.C.: World Resources Institute.

Rosset, P. 1998. Alternative agriculture works: The case of Cuba. *Monthly Review* 50(3).

Rosset, P. and M. Benjamin. 1993. Two Steps Backward, One Step Forward: Cuba's Nationwide Experiment with Organic Agriculture. San Francisco: Global Exchange.

World Bank. 2000. *World Development Indicators 2000*. Washington, D.C.: The World Bank.

Coastal

Replumbing the Everglades: Large-Scale Wetlands Restoration in South Florida

Armentano, T., Chief of Biological Resources Branch, Everglades National Park. 1998. Personal Communication. Interview. 11 December.

Aumen, N. G., Research Program Director, South Florida Water Management District. 1998. Personal Communication. Interview. 13–14 December.

Birbeck. 1990. Birbeck College, Department of Geography. World Cities Population Database (WCPD). London, UK: University of London.

Davis, S. M. and J. C. Ogden, 1994. *Everglades: The Ecosystem and Its Restoration*. Delray Beach, Florida: St. Lucie Press.

Davis, S. M., Lead Ecologist, South Florida Water Management District. 1998. Personal Communication. Interview. 14 December.

de Golia, J. 1997. *Everglades: The Story Behind the Scenery*. Las Vegas: KC Publications, Inc.

Environmental Systems Research Institute (ESRI). 1993. Digital Chart of the World (DCW). Redlands, CA: ESRI.

Florida Department of Environmental Protection. 1996a. A Digital Spatial Database of Existing and Proposed Conservation Lands for the State. Tallahassee, FL: Florida DEP.

Florida Department of Environmental Protection. 1996b. US Highways for Florida. Tallahassee, FL: Florida DEP.

Governor's Commission for a Sustainable South Florida (GCSSF). 1995. Initial Report of the Governor's Commission for a Sustainable South Florida. Coral Gables, Florida: GCSSF.

Jones, R., Director and Professor, Southeast Environmental Research Center and Department of Biological Sciences Florida International University. 1999. Personal Communication. E-mail. 2 August 1999.

Light, S. S. and J. W. Dineen. 1994. Water control in the Everglades: A historical perspective. Pp: 47–83 in *Everglades: The Ecosystem and Its Restoration*. S. M. Davis and J. C. Ogden, eds. Delray Beach, Florida: St. Lucie Press.

Light, S. S., L. H. Gunderson and C. S. Holling. 1995. The Everglades: Evolution of management in a turbulent ecosystem. Pp: 103–168 in *Barriers and Bridges to the Renewal of Ecosystem and Institutions*. L. H. Gunderson, C. S. Holling and S. S. Light, eds. New York: Columbia University Press.

McClure, R. 1999a. Critics wary of Army Corps role in restoration. Sun-Sentinel (Fort Lauderdale, FL). Online at: http://www.sun-sentinel.com/news. (27 March).

McClure, R. 1999b. Sweet deal purchases big sugars' land for conservation. Sun-Sentinel (Fort Lauderdale, FL). Online at: http://www.sun-sentinel.com/news. (26 March).

McPherson, B. F. and R. Halley. 1996. The South Florida environment: A region under stress. *United States Geological Survey Circular* 1134.

Ogden, J. C. 1994. A comparison of wading bird nesting colony dynamics (1931–1946 and 1974–1989) as an indication of ecosystem conditions in the southern Everglades. Pp: 533–570 in *Everglades: The Ecosystem and Its Restoration*. S. M. Davis and J. C. Ogden, eds. Delray Beach, Florida: St. Lucie Press.

Ogden, J. C. 1999. Status of wading bird recovery-1999. *South Florida Wading Bird Report* 5(1):16–18.

Santaniello, N. 1998. Sierra Club faults Everglades restoration plan. Sun-Sentinel (Fort Lauderdale, FL). Online at: http://www.sun-sentinel.com/news. (26 March).

Santaniello, N. 1999. Glades restoration schedule criticized by environmentalists. Sun-Sentinel (Fort Lauderdale, FL). Online at: http://www.sun-sentinel.com/news. (1 February).

Snyder, G. H. and J. M. Davidson. 1994. Everglades agriculture: Past, present and future. Pp: 85–115 in *Everglades: The Ecosystem and Its Restoration*. S. M. Davis and J. C. Ogden, eds. Delray Beach, FL: St. Lucie Press.

South Florida Ecosystem Restoration Task Force (SFERTF) Working Group. 1998a. *Maintaining the Momentum: South Florida Ecosystem Restoration Task Force Biennial Report (Draft)*. Miami: SFERTF.

South Florida Ecosystem Restoration Task Force (SFERTF) Working Group. 1998b. *Success in the Making: An Integrated Plan for South Florida Ecosystem Restoration and Sustainability*. Miami: SFERTF.

South Florida Water Management District (SFWMD). 1998a. 1997 Everglades Annual Report. West Palm Beach: SFWMD.

South Florida Water Management District (SFWMD). 1998b. Everglades Interim Report (Executive Summary). West Palm Beach: SFWMD.

South Florida Water Management District (SFWMD). 2000a. Everglades Consolidated Report. West Palm Beach: SFWMD.

South Florida Water Management District (SFWMD). 2000b. Everglades Nutrient Removal Project: 5-Year Synopsis. West Palm Beach: SFWMD.

Stevens, W. K. 1999. Everglades restoration plan does too little, experts say. *New York Times* (22 February):A-1.

Tebeau, C. W. 1968. *Man in the Everglades: 2000 Years of Human History in the Everglades National Park*. Miami: University of Miami Press.

U.S. Army Corps of Engineers (USACE). 1998. Central and Southern Florida Project Comprehensive Review Study: Draft Integrated Feasibility Report and Programmatic Environmental Impact Statement. Jacksonville: USACE.

Managing Mankòtè Mangrove

Brown, N. A. 1996. The Caribbean Natural Resources Institute: Working towards participation and collaboration in the Caribbean. Islander Magazine, Issue 2, July. Online at: http://www.islandstudies.org/islander/issue2/canari.htm.

Goeghegan, T. and A. H. Smith. 1998. Conservation and Sustainable Livelihoods: Collaborative Management of the Mankòtè Mangrove, St. Lucia. Caribbean Natural Resources Institute. August.

Smith, A. H., Research Scientist, Caribbean Natural Resources Institute (CANARI). 1999. Personal Communication. E-mail. 15 February.

Smith, A. H., Research Scientist, Caribbean Natural Resources Institute (CANARI). 2000. Personal Communication. E-mail. 13 April.

Smith, A. H. and F. Berkes. 1993. Community-based use of mangrove resources in St. Lucia. *International Journal of Environmental Studies* 43:123–131.

References

Bolinao Rallies Around Its Reefs

Environmental Building News. 1993. Cement and concrete: Environmental considerations. *Environmental Business News* 2(2).

Maragos, J. E., M. P. Crosby and J. W. McManus. 1996. Coral reefs and biodiversity: A critical and threatened relationship. *Oceanography* 9(1):83–99.

McManus, J. W., C. L. Nañola, R. B. Reyes Jr. and K. N. Kesner. 1992. Resource ecology of the Bolinao coral reef system. *ICLARM Stud. Rev.* 22:117.

Ramos, V. O., Department of Environment and Natural Resources. 1996. Personal Communication. Letter to Mr. Andrew E. J. Wang Re: ECC Application for the Proposed Pangasinan Cement Complex in Bolinao, Pangasinan. 6 August.

Surbano, M. A. 1998. Cement makers are top polluters: Study. *Business Daily* March 10.

Talaue-McManus, L., Associate Professor, Marine Science Institute, University of the Philippines. 1999. Personal Communication. Interview. 29 June.

Talaue-McManus, L. and K. P. N. Kesner. 1995. Valuation of a Philippine municipal sea urchin fishery and implications of its collapse. *Philippine Coastal Resources Under Stress. Selected Papers from the Fourth Annual Common Property Conference*, Manila, Philippines. 16–19 June 1993

Talaue-McManus, L., A. C. Yambao, S. G. Salmo III and P. M. Aliño. 1999. Participatory coastal development planning in Bolinao, Northern Philippines: A potent tool for conflict resolution. Pp: 149–157 in *Community-Based Natural Resource Management.* D. Buckles, ed. Ottawa: International Development Research Centre of Canada and the World Bank.

Forests

Up From the Roots: Regenerating Dhani Forest Through Community Action

Gadgil, M. and R. Guha. 1992. *This Fissured Land: An Ecological History of India.* Delhi: Oxford University Press.

Gadgil, M. 1999. Co-Management of Forest Resources: The Indian Experience. Unpublished paper provided to WRI.

Jodha, N. S. 1990. Rural Common Property Resources: Contributions and Crisis. Society for Promotion of Wastelands Development, Foundation Day Lecture. 16 May 1990.

Loveland, T. R., B. C. Reed, J. F. Brown, D. O. Ohlen, Z. Zhu, L. Yang and J. Merchant. 2000. Development of a Global Land Cover Characteristics Database and IGBP DISCover from 1 km AVHRR data. *International Journal of Remote Sensing* 21(6):1303–1330. Online at: http://edcdaac.usgs.gov/glcc/glcc.html.

MacKinnon, J., 1997. *Protected Areas Systems Review of the Indo-Malayan Realm.* Canterbury, UK: The Asian Bureau for Conservation.

Mahapatra, R. 1999. On the Warpath. *Down to Earth* 8(9):32–42.

Nayak, P. and N. Singh, 1999. *Dhani Panch Mauza Jungle Surakhya Samiti: A Case of Community Forest Management in Orissa.* Study Paper. Bhubaneswar, India: Vasundhara.

Pachauri, R. K. and P. V. Sridharan, 1998. *Looking Back to Think Ahead: GREEN India 2047.* New Delhi: Tata Energy Research Institute.

Panagrahi, R. and Y. Rao. 1996. *Conserving Biodiversity: A Decade's Experience of Dhani Panch Mauza Jungle Surakhya Samiti.* Study Paper. Bhubaneswar, India: Vasundhara.

Singh, N., Vasundhara. 2000. Personal Communication. E-mail. January.

Singh, N. and P. Nayak. 1999. Community Forestry in Dhani. Unpublished paper provided to WRI.

Watts, H. 1999. Indian State Faces Ecological Crisis After Cyclone. Reuters News Service. Online at: http://www.planetark.org/daily newstory.cfm?newsid+4293. (3 December).

Freshwater

Working for Water, Working for Human Welfare in South Africa

Basson, M. S., P. H. van Niekerk and J. A. van Rooyen. 1997. *Overview of Water Resources Availability and Utilisation in South Africa.* Pretoria, South Africa: Department of Water Affairs and Forestry.

Botha, M., Conservation Officer, Botanical Society of South Africa. 1999. Personal Communication. E-mail. 17 November.

Department of Water Affairs and Forestry (DWAF). 1994. Water Supply and Sanitation Policy. White paper: Water–An indivisible national asset. Cape Town: Republic of South Africa. November 1994.

de Wit, M. P., D. J. Crookes and B. W. van Wilgen. (Forthcoming). Conflicts of interest in environmental management: Estimating the costs and benefits of black wattle (*Acacia mearnsii*) in South Africa. *Environment and Development.*

Dye, P. and A. Poulter. 1995. Field demonstrations of the effect on streamflow of clearing invasive pine and wattle trees from a riparian zone. *South African Forestry Journal* 173:27–30.

Fynbos Working for Water Allied Industries. 1998. Overview for the Development of Allied Industries. *Job Summit Investor Conferences*, unpublished.

Gelderblom, C. Consultant. 2000. Personal Communication. E-mail. 13 January.

Higgins, S. I., J. K. Turpie, R. Costanza, R. M. Cowling, D. C. le Maitre, C. Marais and G. F. Midgley. 1997. An ecological simulation model of mountain fynbos ecosystems. *Ecological Economics* 22:155–169.

Hilton-Taylor, C. 1996. Red Data List of Southern African Plants. Pretoria: National Botanical Institute.

International Water Management Institute (IWMI). 1999. South Africa's Progressive New Water Law. Online at: http://www.cgiar .org/iwmi. (October).

Koch, E. 1996. A watershed for apartheid. *New Scientist* 150 (2025):12–13.

le Maitre, D. C., B. W. van Wilgen, C. M. Gelderblom, C. Bailey, R. A. Chapman and J. A. Nel. (Forthcoming). Invasive alien trees and water resources in South Africa: Case studies of the costs and benefits of management. *Forest Ecology and Management.*

Marais, C. 1998. An Economic Evaluation of Invasive Alien Plant Control Programmes in the Mountain Catchment areas of the Western Cape Province, South Africa. Ph.D. Dissertation. University of Stellenbosch.

Marais, C., Manager: Programme Development and Planning: National Working for Water Programme. 1999. Personal Communication. E-mail. 23 November.

Marais, C., Manager: Programme Development and Planning: National Working for Water Programme. 2000. Personal Communication. E-mail. 19 January.

May, J. 1998. Poverty and Inequality in South Africa. Indicator South Africa **15**(2). Online at: http://www.und.ac.zo/und/indic /archives/indicator/winter98/Fmay.htm.

Nel, J. L., B. W. van Wilgen and C. M. Gelderblom. 1999. The Contribution of Plantation Forestry to the Problem of Invading Alien Trees in South Africa: A Preliminary Assessment. Unpublished CSIR Report ENV/S-C 93003. Department of Water Affairs and Forestry.

Raddock, G., National Parks Service. 1999. Personal Communication. Interview. March.

Republic of South Africa. 1997. Water Services Act: Act 108 as of 1997.

Republic of South Africa. 1998. National Water Act: Act 36 as of 1998.

Saleth, R. M. and A. Dinar. 1999. Water Challenge and Institutional Response: A Cross-Country Perspective. Draft Mimeo. Washington, D.C.: The World Bank. 5 February 1999.

Scott, D. F. 1999. Managing riparian zone vegetation to sustain streamflow: Results of paired catchment experiments in South Africa. *Can. J. For. Res.* 29:1149–1157.

Shaughnessy, G. A. 1986. A case study of some woody plant introductions to the Cape Town area. Pp: 37–43 in *The Ecology and Management of Biological Invasions in Southern Africa*. I. A. W. Macdonald, F. J. Kruger and A. A. Ferrar, eds. Cape Town: Oxford University Press.

South African Institute for Race Relations. 1998. South Africa Survey 1997/1998. Johannesburg.

Spies, R. E. and J. B. Barriage. 1991. Western Cape System Analysis: Long-Term Urban Water Demand in the Western Cape Metropolitan Region 1990–2020. Unpublished Report. Ninham Shand Consulting Engineers for DWAF and City of Cape Town.

U.N. Environment Programme. 1999. State of the Environment: South Africa. Online at: http://www.ngo.grida.no/soesa/nsoer/issues/social/state.htm.

U.S. Geological Survey (USGS). 1997. Hydro1K Data Set, Africa. Online at: http://edcdaac.usgs.gov/gtopo30/hydro.

Van der Zel, D. W. 1981. Optimum mountain catchment management in Southern Africa. *South African Forestry Journal* 116:75–81.

van Wilgen, B. W. Scientific Advisor to the Working for Water Programme. 1999. Personal Communication. E-mail. 22 October and 28 November.

van Wilgen, B. W. Scientific Advisor to the Working for Water Programme. 2000. Personal Communication. E-mail. 10 April.

van Wilgen, B. W., W. J. Bond and D. M. Richardson. 1992. Ecosystem management. Pp: 345–371 in *The Ecology of Fynbos: Nutrients, Fire, and Diversity*. R. M. Cowling, ed. Cape Town: Oxford University Press.

van Wilgen, B. W., R. M. Cowling and C. J. Burgers. 1996. Valuation of ecosystem services: A case study from South African fynbos ecosystem. *BioScience* 46(3):184–189.

van Wilgen, B. W., P. R. Little, R. A. Chapman, A. H. M. Görgens, T. Willems and C. Marais. 1997. The sustainable development of water resources: History, financial costs, and benefits of alien plant control programmes. *South African Journal of Science* 93:404–411.

van Wilgen, B. W. and E. van Wyk. 1999. Invading alien plants in South Africa: Impacts and solutions. Pp: 566–571 in *The VI International Rangeland Congress*. Townsville, Australia.

Versveld, D. B., D. C. le Maitre and R. A. Chapman. 1998. Alien Invading Plants and Water Resources in South Africa: A Preliminary Assessment. TT 99/98. WRC Report. Pretoria: Water Research Commission.

Versveld, D. B. and B. W. van Wilgen. 1986. Impact of woody aliens on ecosystem properties. Pp: 239–246 in *The Ecology and Management of Biological Invasions in Southern Africa*. I. A. W. Macdonald, F. J. Kruger and A. A. Ferrar, eds. Cape Town: Oxford University Press.

Wells, M. J., R. J. Poynton, A. A. Balsinhas, K. J. Musil, H. Joffe and E. van Hoepen. 1986. The history of introductions of alien plants to South Africa. Pp: 21–35 in *The Ecology and Management of Biological Invasions in Southern Africa*. I. A. W. Macdonald, F. J. Kruger and A. A. Ferrar, eds. Cape Town: Oxford University Press.

Working for Water Programme. N.D. The Environmental Impacts of Invading Alien Plants in South Africa. Pretoria: Working for Water Programme.

Working for Water Programme. 1998. The Working for Water Programme 1997/98 Annual Report.

Working for Water Programme. 1999. The Working for Water Programme 1998/99 Annual Report.

Managing the Mekong River: Will a Regional Approach Work?

Center for International Earth Science Information Network (CIESIN). 1999. Gridded Population of the World: Provisional Release of Updated Database of 1990 and 1995 Estimates. Palisades, NY and Washington D.C.: Columbia University and World Resources Institute. 22 November.

China Environment Series. 1998. Chinese Transboundary Water Issues. China Environment Series 2. Summer 1998. Woodrow Wilson Center. Online at: http://ecsp.si.edu/ecsplib.nsf/6b5e482eec6e8a27852565d1000e1a4c/ebcb218fc7fe824985256677007c6181?OpenDocument.

Elvidge, C. D., K. E. Baugh, V. R. Hobson, E. A. Kihn, H. W. Kroehl, E. R. Davis and D. Cocero. 1997. Satellite inventory of human settlements using nocturnal radiation emissions: A contribution for the global toolchest. *Global Change Biology* 3(5):387–396.

Energy Information Administration (EIA). 1999. International Energy Outlook 1999. Report: # DOE/EIA-0484(99). Online at: http://www.eia.doe.gov/oiaf/ieo99/electricity.html.

Food and Agriculture Organization of the United Nations (FAO). 1999. *State of the World's Forests 1999*. Rome: Food and Agriculture Organization of the United Nations.

Friederich, H. 2000. The biodiversity of the wetlands in the Lower Mekong Basin. *Paper submitted to the World Commission on Dams, Presented at the Commission's East/Southeast Asia Regional Consultation*, Hanoi, Vietnam. 26–27 February

Institute for Development Anthropology. 1998. Environment and Society in the Lower Mekong Basin: A Landscaping Review of the Literature, Volume I. Prepared for the Mekong River Basin Research and Capacity Building Initiative, Oxfam-America.

Mekong River Commission (MRC). 1997. *Mekong River Basin Diagnostic Study: Final Report*. Bangkok: Mekong River Commission.

Nguyen, B. T. 1998. The Mekong Delta in Vietnam: Hydrology and Livelihoods. Prepared for Oxfam-America. Boston: Oxfam-America.

U.N. Population Division (UNPD). 1998. *Annual Populations 1950–2050 (The 1998 Revision on diskette)*. 1. New York: U.N. Population Division.

World Conservation Monitoring Centre (WCMC). 1994. *Biodiversity Source Book*. WCMC Biodiversity Series No. 1. Cambridge, UK: World Conservation Press.

World Bank. 1999. *World Development Indicators 1999*. Washington, D.C.: The World Bank.

Xie, M. 1996. Water resources in Vietnam. *Vietnam Water Resources Sector Review: Selected Working Papers*. A Joint Report by The World Bank, Asian Development Bank, FAO, UNDP, and the NGO Water Resources Group in cooperation with the Institute of Water Resources Planning, Vietnam.

New York City's Watershed Protection Plan

Gratz, J., New York City Watershed Team Leader, EPA Region 2. 1999. Personal Communication. E-mail. 4 January.

Izeman, M., Attorney, Natural Resources Defense Council. 1999. Personal Communication. E-mail. 20 January.

National Research Council (NRC). 1999. *Watershed Management for Potable Water Supply: Assessing New York City's Approach*. Washington, D.C.: National Academy Press.

Revkin, A. 1995. Rates to rise 2 percent at most under plan to protect city reservoirs. *The New York Times* (3 November):1, sec. B.

Revkin, A. 1997. Troubled headwaters: A special report: Billion dollar plan to clean New York City water at its source. *The New York Times* (31 August):sec. 1.

Ryan, G., New York City Department of Environmental Protection, Bureau of Water Supply and Wastewater Collection. 1998. Personal Communication. Interview. 2 December.

State of New York. 1998. The New York City Watershed Agreement: Memorandum of Final Agreement Final Draft. Online at: http://www.state.ny.us/watershed/overview.html.

References

Grasslands

Sustaining the Steppe: The Future of Mongolia's Grasslands

Asian Development Bank/PALD. 1993. Improved Livestock Feed Production, Management and Use in Mongolia: Socio-Economic Potentials and Constraints. Report of a survey undertaken by the Policy Alternatives for Livestock Development in Mongolia (PALD) project for the ADB. Sussex: IDS.

Chang, V. T. 1933. *The Economic Development and Prospects of Inner Mongolia*. Shanghai: Commercial Press; Reprinted by Taipei: Ch'eng Wen Publishing (1971).

Economic and Social Commission for Asia and the Pacific. 1999. Statistical Yearbook for Asia and the Pacific. Bangkok: United Nations.

Food and Agriculture Organization (FAO). 2000. Special Alert No. 303: Concerns Mount over Serious Food Shortages in Mongolia Following the Harshest Winter in 30 Years. Online at: http://www.fao.org/waicent/faoinfo/economic/giews/english/alertes/2000/SA303MON.htm. (10 March).

Gomboev, B. O. 1996. The structure and process of land use in Inner Asia. Pp: 12–57 in *Culture and Environment in Inner Asia, Volume 1: The Pastoral Economy and the Environment*. C. Humphrey and D. Sneath, eds. Cambridge, UK: The White Horse Press.

Government of Mongolia. 1995. National Environmental Action Plan: Towards Mongolia's Environmentally Sound Sustainable Development. Ulaanbaatar. February.

Hasbagan and C. Shan. 1996. The cultural importance of animals in traditional Mongolian plant nomenclature. Pp: 25–29 in *Culture and Environment in Inner Asia, Volume 2: Society and Culture*. C. Humphrey and D. Sneath, eds. Cambridge, UK: The White Horse Press.

Ho, P. 2000. China's rangelands under stress: A comparative study of pasture commons in the Ningzia Hui autonomous region. *Development and Change* 31:385–412.

Humphrey, C. and D. Sneath. 1999. *The End of Nomadism? Society, State and the Environment in Inner Asia*. Durham, NC: Duke University Press.

Inner Mongolian Territorial Resources Compilation Committee. 1987. *Nei Menggu Guotu Ziyuan (Inner Mongolian Territorial Resources Survey)*. Hohhot: Inner Mongolian People's Press.

Li, O., M. Rong and J. R. Simpson. 1993. Changes in the nomadic pattern and its impact on the Inner Mongolian grassland ecosystem. *Nomadic Peoples* 33:66–72.

MacArthur Environmental and Cultural Conservation in Inner Asia Project (MECCIA). 1995. Mongolia and Inner Asia Studies Unit. Cambridge, UK: Cambridge University.

Mearns, R. 1991. Pastoralists, patch ecology and perestroika: Understanding potentials for change in Mongolia. *IDS Bulletin* 22(4):25–33.

Mearns, R. 1996. Community, collective action and common grazing: The case of post-socialist Mongolia. *The Journal of Development Studies* 32:297–339.

Ministry of Agriculture and Industry of Mongolia. 1998. Mongolian Agriculture and Agro-industry. Online at: http://www.agriculture.mn/agroindustry.htm#2.

National Statistical Office of Mongolia. 1999. Mongol Ülsyn Statistikiin Emhtgel, (Mongolian Statistical Yearbook), 1998. Ulaanbaatar.

Neupert, R. 1999. Population, nomadic pastoralism and the environment in the Mongolian plateau. *Population and Environment: A Journal of Interdisciplinary Studies* 20(5):413–441.

Palmer, W. A. 1991. National Report: Mongolia. (Draft). UN Development Programme, Regional Bureau for Asia and the Pacific.

Sheehy, D. 1995. Grazingland Interactions among Large Wild and Domestic Herbivores in Mongolia. Report for the Mongolian Ministry of Nature and Environment Biodiversity Project. October 1995.

Simukov, A. D. 1936. Materialy po kochevomu bytu naseleniya MNR (Materials concerning the nomadic life of the population of Mongolia). *Sovremennaya Mongoliya (Contemporary Mongolia)* 2(15).

Sneath, D. 1993. *Database of results from fieldwork in Sumber sum, Dornogov' aimag, Mongolia*. Unpublished Working Materials. Cambridge, UK: MECCIA.

Sneath, D. 1998. State Policy and Pasture Degradation in Inner Asia. *Science* 281:1147–1148.

Statistical Office of Mongolia. 1993. Mongolyn Ediin Zasag, Niigem 1992 (Mongolian Economy and Society in 1992). Ulaanbaatar: J.L.D Gurval.

Tserendash, S. and B. Erdenebaatar. 1993. Performance and management of natural pasture in Mongolia. *Nomadic Peoples* 33.

U.N. Development Programme (UNDP). 2000. New web-services provide faster news on Mongolia disaster. Online at: http://www.un-mongolia.mn/undp/news/undp-news.htm. (12 April 2000).

Ward, G. 1996. Education systems in Inner Asia: An examination of the interface between social practice and cultural representations. Pp: 30–48 in *Culture and Environment in Inner Asia Volume 2: Society and Culture*. C. Humphrey and D. Sneath, eds. Cambridge, UK: The White Horse Press.

Whitten, T. 1999. Mongolia: Environment and Natural Resources: Opportunities for Investment. (Draft). Environment Unit, East Asia and the Pacific Region, The World Bank. 24 September.

World Bank. 2000. *World Development Indicators 2000*. Washington, D.C.: The World Bank.

World Conservation Monitoring Centre (WCMC). 1992. *Global Biodiversity: Status of the Earth's Living Resources*. London: Chapman and Hall.

Yenhu, T. 1996. A comparative study of the attitudes of the peoples of pastoral areas of Inner Asia towards their environments. Pp: 1–24 in *Culture and Environment in Inner Asia Volume 2: Society and Culture*. C. Humphrey and D. Sneath, eds. Cambridge, UK: The White Horse Press.

Zasagyn Gazar Medeel. 1992. (Mongolian Newspaper). 2(63):1992.

Chapter 4

Bengston, D. N. 1994. Changing forest values and ecosystem management. *Society and Natural Resources* 7(6):515–533.

Conference of the Parties to the Convention on Biological Diversity, Fifth Meeting (COP-5. 2000). Annex III. Nairobi, Kenya. Online at: http://www/biodiv.org/Decisions/COP5/pdf/COP-5-Dec-All-e.pdf.

Environmental Valuation Reference Inventory (EVRI). 2000. Online at: http://www.evri.ec.gc.ca/evri.

European Environment Agency (EEA). 1999. *Environment in the European Union at the Turn of the Century*. Environmental Assessment Report No. 2. Copenhagen: EEA.

McManus, J. W., C. L. Nañola, R. B. Reyes Jr. and K. N. Kesner. 1995. The Bolinao coral reef resource system. Pp: 193–204 in *Philppine Coasal Resources Under Stress*. M. A. Juinio-Meñez and G. F. NewKirk, eds. Selected papers from the Fourth Annual Common Property Conference held in Manila, Philippines. June 16–19.

Stokstad, E. 1999. Scarcity of Rain, Stream Gages Threatens Forecasts. *Science* 285:1199–1200.

U.N. Environment Programme (UNEP). 1999. *Global Environment Outlook 2000*. London, UK: Earthscan Publications Ltd.

Index

Index

Index